STUDENT'S SOLUTIONS MANUAL

GARRET ETGEN

University of Houston

CALCULUS FOR BUSINESS, ECONOMICS, LIFE SCIENCES, AND SOCIAL SCIENCES

THIRTEENTH EDITION

Raymond Barnett

Merritt College

Michael Ziegler

Marquette University

Karl Byleen

Marquette University

PEARSON

Boston Columbus Indianapolis New York San Francisco Upper Saddle River
Amsterdam Cape Town Dubai London Madrid Milan Munich Paris Montreal Toronto
Delhi Mexico City São Paulo Sydney Hong Kong Seoul Singapore Taipei Tokyo

Reproduced by Pearson from electronic files supplied by the author.

Publishing as Pearson, 75 Arlington Street, Boston, MA 02116.

ISBN-13: 978-0-321-93173-3
ISBN-10: 0-321-93173-4

4
www.pearsonhighered.com

Table of Contents

1 FUNCTIONS AND GRAPHS

EXERCISE 1-1

Things to remember:

1. POINT-BY-POINT PLOTTING

   To sketch the graph of an equation in two variables, plot enough points from its solution set in a rectangular coordinate system so that the total graph is apparent and then connect these points with a smooth curve.

2. A FUNCTION is a correspondence between one set of elements, called the DOMAIN, and a second set of elements, called the RANGE, such that to each element in the domain there corresponds one and only one element in the range.

3. EQUATIONS AND FUNCTIONS

   Given an equation in two variables: If there corresponds exactly one value of the dependent variable (output) to each value of the independent variable (input), then the equation specifies a function. If there is more than one output for at least one input, then the equation does not specify a function.

4. VERTICAL LINE TEST FOR A FUNCTION

   An equation specifies a function if each vertical line in the coordinate system passes through at most one point on the graph of the equation. If any vertical line passes through two or more points on the graph of an equation, then the equation does not specify a function.

5. AGREEMENT ON DOMAINS AND RANGES

   If a function is specified by an equation and the domain is not given explicitly, then assume that the domain is the set of all real number replacements of the independent variable (inputs) that produce real values for the dependent variable (outputs). The range is the set of all outputs corresponding to input values.

   In many applied problems, the domain is determined by practical considerations within the problem.

6. FUNCTION NOTATION — THE SYMBOL $f(x)$

   For any element x in the domain of the function f, the symbol $f(x)$ represents the element in the range of f corresponding to x in the domain of f. If x is an input value, then $f(x)$ is the corresponding output value. If x is an element which is not in the domain of f, then f is NOT DEFINED at x and $f(x)$ DOES NOT EXIST.

1. $y = x + 1$:

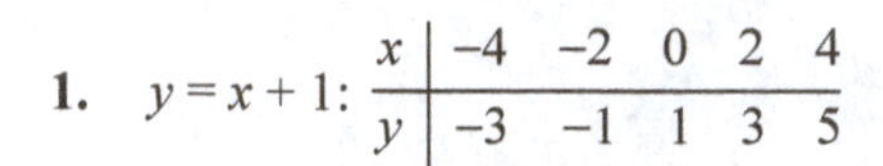

x	−4	−2	0	2	4
y	−3	−1	1	3	5

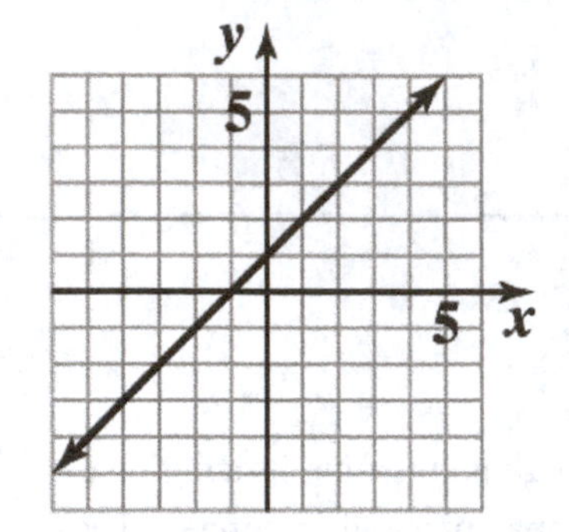

3. $x = y^2$:

x	0	1	4	9	16
y	0	±1	±2	±3	±4

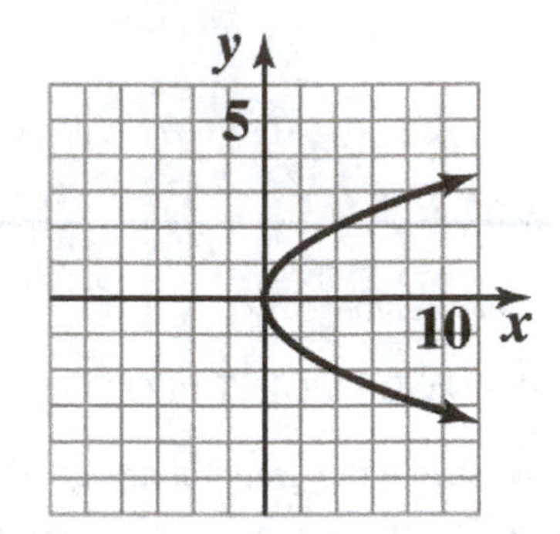

5. $y = x^3$:

x	−2	−1	0	1	2
y	−8	−1	0	1	8

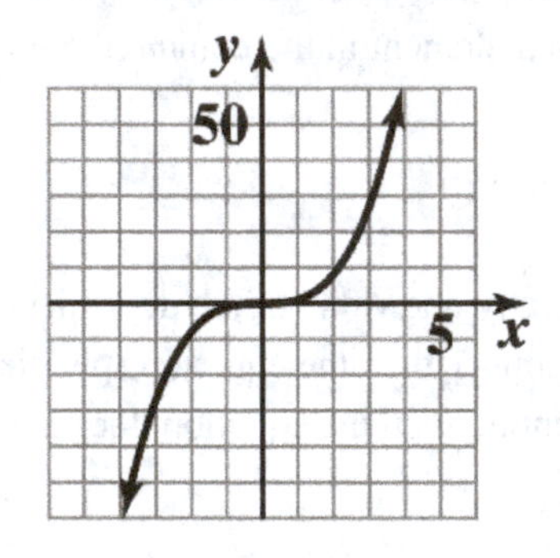

7. $xy = -6$:

x	−6	−3	−1	1	3	6
y	1	2	6	−6	−2	−1

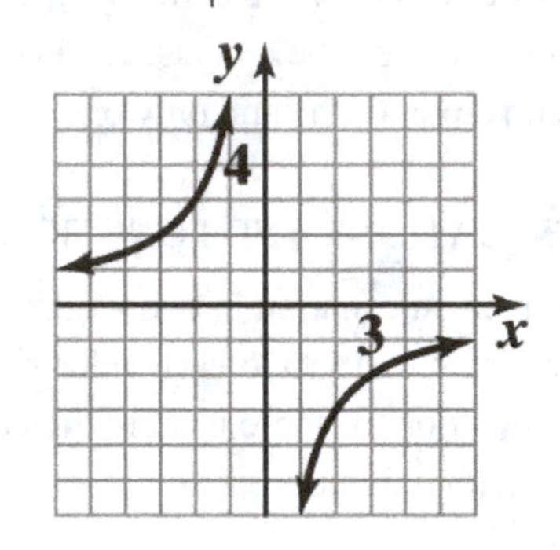

9. The table specifies a function, since for each domain value there corresponds one and only one range value.

11. The table does not specify a function, since more than one range value corresponds to a given domain value. (Range values 5, 6 correspond to domain value 3; range values 6, 7 correspond to domain value 4.)

13. This is a function.

15. The graph specifies a function; each vertical line in the plane intersects the graph in at most one point.

17. The graph does not specify a function. There are vertical lines which intersect the graph in more than one point. For example, the y-axis intersects the graph in three points.

19. The graph specifies a function.

21. $y - 2x = 7$ or $y = 2x + 7$; a linear function.

23. $xy - 4 = 0$ or $y = \frac{4}{x}$; neither a linear nor a constant function.

25. $y = 5x + \frac{1}{2}(7 - 10x) = 5x + \frac{7}{2} - 5x = \frac{7}{2}$; a constant function.

27. $3x + 4y = 5$ or $y = -\frac{3}{4}x + \frac{5}{4}$; a linear function

29. $f(x) = 1 - x$: Since f is a linear function, we only need to plot two points.

x	$f(x)$
–2	3
2	–1

31. $f(x) = x^2 - 1$:

x	–3	–2	–1	0	1	2	3
$f(x)$	8	3	0	–1	0	3	8

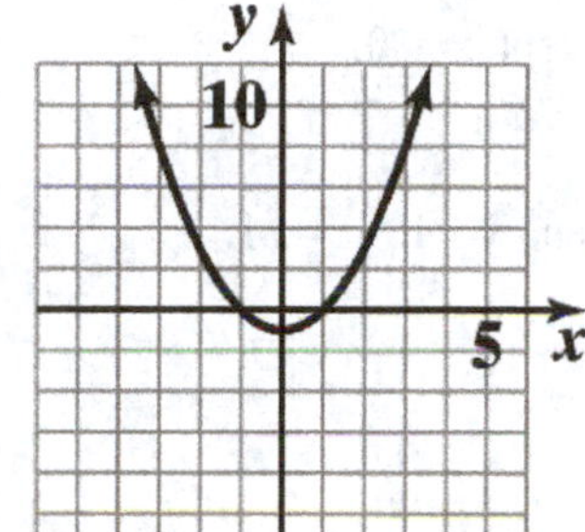

33. $f(x) = 4 - x^3$:

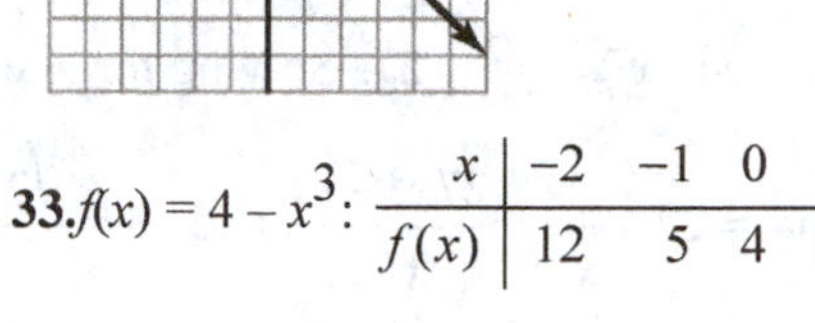

x	–2	–1	0	1	2
$f(x)$	12	5	4	3	–4

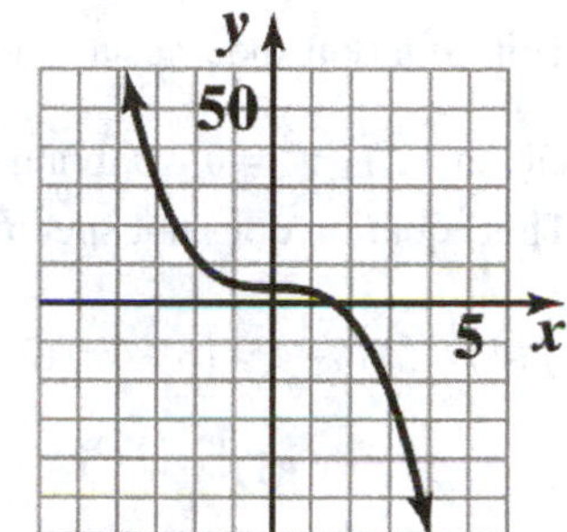

35. $f(x) = \frac{8}{x}$

x	–8	–4	–2	–1	1	2	4	8
$f(x)$	–1	–2	–4	–8	8	4	2	1

37. The graph of f is:

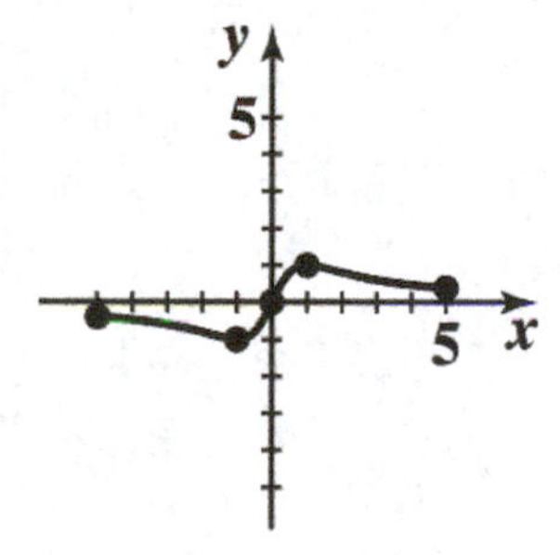

39. $y = f(-5) = 0$

41. $y = f(5) = 4$

43. $f(x) = 0$ at $x = -5, 0, 4$

45. $f(x) = -4$ at $x = -6$

47. domain: all real numbers or $(-\infty, \infty)$

49. domain: all real numbers except -4

51. domain: $x \le 7$

53. Given $2x + 5y = 10$. Solving for y, we have: $5y = 10 - 2x$ and $y = 2 - \frac{2}{5}x$.

This equation specifies a function. The domain is R, the set of real numbers.

55. Given $y(x+y)=4$. Solving for *y*, we have:

$y^2+xy=4$ or $y^2+xy-4=0$. So $y=\dfrac{-x\pm\sqrt{x^2+16}}{2}$

This equation does not specify y as a function x. For example, when $x=0$, $y=\pm 2$, when $x=3$, $y=-4,1$.

57. Given $x^{-3}+y^3=27$. Solving for *y*, we have:

$y^3=27-\dfrac{1}{x^3}=\dfrac{27x^3-1}{x^3}$ and $y=\dfrac{\sqrt[3]{27x^3-1}}{x}$.

This equation specifies a function. The domain is all real numbers except $x=0$.

59. Given $x^3-y^2=0$. Solving for *y*, we have: $y=\pm\sqrt{x^3}$.
This equation does not specify y as a function x. For example, when $x=1$, $y=\pm 1$.

61. $f(4)=(4)^2-4=16-4=12$

63. $f(x+1)=(x+1)^2-4=x^2+2x+1-4=x^2+2x-3$

65. $f(-6x)=(-6x)^2-4=36x^2-4$

67. $f(x^3)=(x^3)^2-4=x^6-4$

69. $f(2)+f(h)=[(2)^2-4]+[(h)^2-4]=[4-4]+[h^2-4]=h^2-4$

71. $f(2+h)=(2+h)^2-4=4+4h+h^2-4=4h+h^2$

73. $f(2+h)-f(2)=[(2+h)^2-4]-[(2)^2-4]=[4+4h+h^2-4]-0=4h+h^2$

75. $f(x)=4x-3$

(A) $f(x+h)=4(x+h)-3=4x+4h-3$

(B) $f(x+h)-f(x)=4x+4h-3-(4x-3)=4h$

(C) $\dfrac{f(x+h)-f(x)}{h}=\dfrac{4h}{h}=4$

77. $f(x)=4x^2-7x+6$

(A) $f(x+h)=4(x+h)^2-7(x+h)+6=4(x^2+2xh+h^2)-7x-7h+6$
$=4x^2+8xh+4h^2-7x-7h+6$

(B) $f(x+h)-f(x)=4x^2+8xh+4h^2-7x-7h+6-(4x^2-7x+6)=8xh+4h^2-7h$

(C) $\dfrac{f(x+h)-f(x)}{h}=\dfrac{8xh+4h^2-7h}{h}=\dfrac{h(8x+4h-7)}{h}=8x+4h-7$

79. $f(x)=x(20-x)=20x-x^2$

(A) $f(x+h)=20(x+h)-(x+h)^2=20x+20h-x^2-2xh-h^2$

(B) $f(x+h)-f(x)=20x+20h-x^2-2xh-h^2-(20x-x^2)=20h-2xh-h^2$

(C) $\dfrac{f(x+h)-f(x)}{h} = \dfrac{20h-2xh-h^2}{h} = \dfrac{h(20-2x-h)}{h} = 20 - 2x - h$

81. Given $A = \ell\, w = 25$.

Thus, $\ell = \dfrac{25}{w}$. Now $P = 2\,\ell + 2w$

$$= 2\left(\frac{25}{w}\right) + 2w = \frac{50}{w} + 2w.$$

The domain is $w > 0$.

83. Given $P = 2\,\ell + 2w = 100$ or $\ell + w = 50$ and $w = 50 - \ell$.

Now $A = \ell\, w = \ell\,(50 - \ell)$ and $A = 50\,\ell - \ell^2$.

The domain is $0 \le \ell \le 50$. [Note: $\ell \le 50$ since $\ell > 50$ implies $w < 0$.]

85.

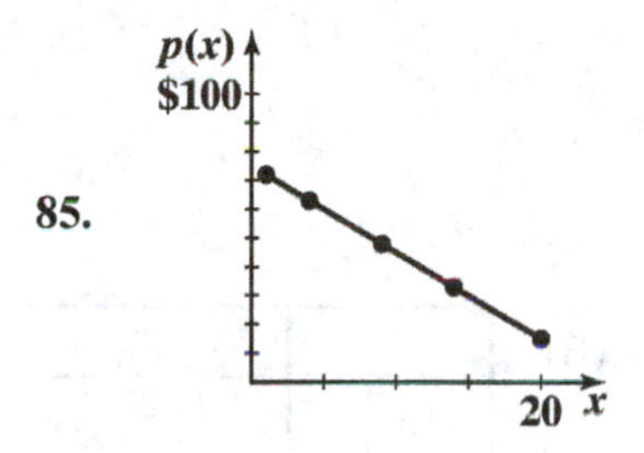

(B) $p(x) = 75 - 3x$
$p(7) = 75 - 3(7) = 75 - 21 = 54$;
$p(11) = 75 - 3(11) = 75 - 33 = 42$
Estimated price per chip for a demand of 7 million chips: \$54;
for a demand of 11 million chips: \$42

87. (A) $R(x) = x \cdot p(x) = x(75 - 3x) = 75x - 3x^2,\ 1 \le x \le 20$

(B)

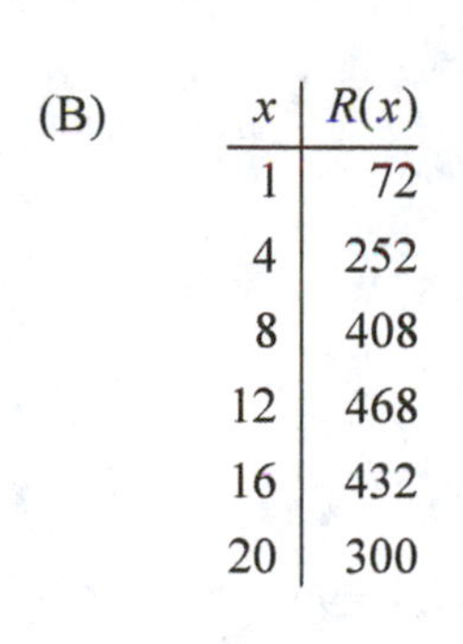

x	$R(x)$
1	72
4	252
8	408
12	468
16	432
20	300

(C)

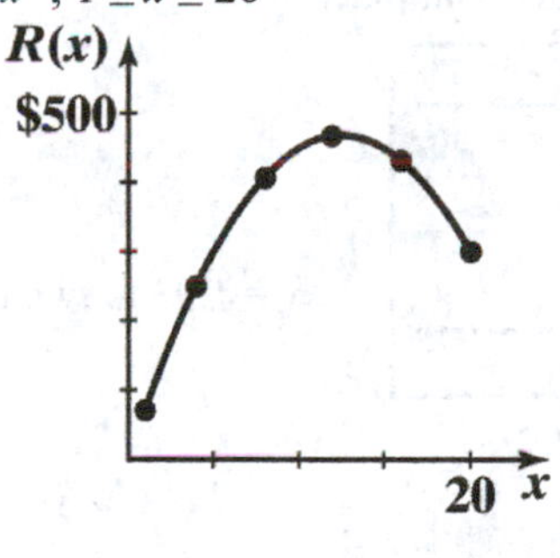

89. (A) Profit: $P(x) = R(x) - C(x) = 75x - 3x^2 - [125 + 16x] = 59x - 3x^2 - 125,\ 1 \le x \le 20$

(B)

x	$P(x)$
1	−69
4	63
8	155
12	151
16	51
20	−145

(C)

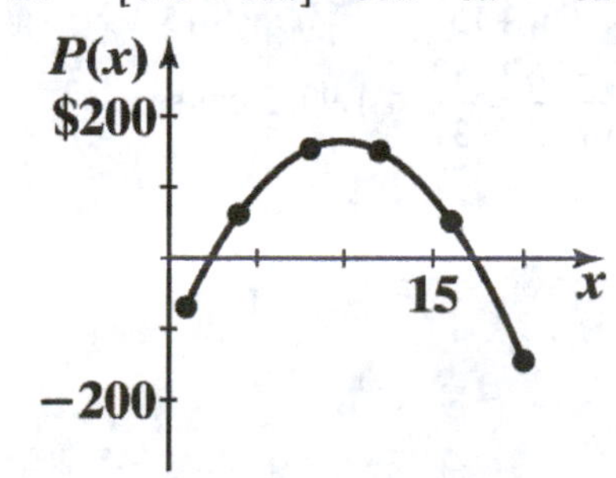

91.

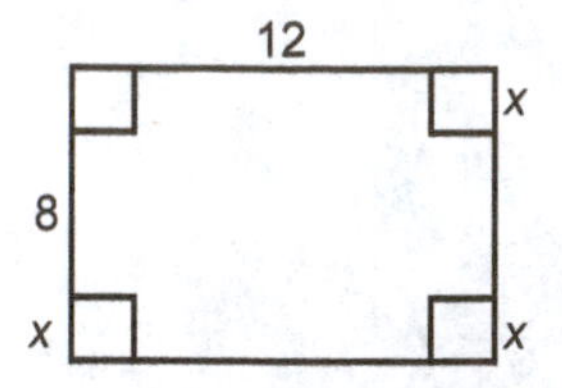

(A) $V = \text{(length)(width)(height)}$

$V(x) = (12 - 2x)(8 - 2x)x$

$= x(8 - 2x)(12 - 2x)$

(B) Domain: $0 \le x \le 4$

(C) $V(1) = (12 - 2)(8 - 2)(1)$

$= (10)(6)(1) = 60$

$V(2) = (12 - 4)(8 - 4)(2)$

$= (8)(4)(2) = 64$

$V(3) = (12 - 6)(8 - 6)(3)$

$= (6)(2)(3) = 36$

Thus,

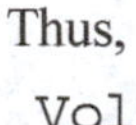

Volume

x	$V(x)$
1	60
2	64
3	36

(D)

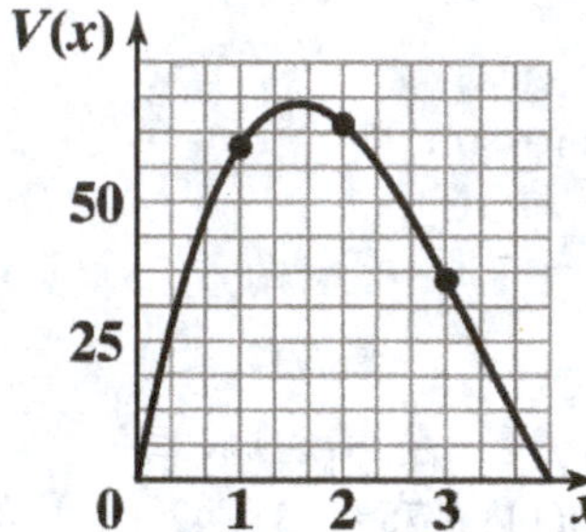

93. (A) The graph indicates that there is a value of x near 2, and slightly less than 2, such that $V(x) = 65$. The table is shown at the right.

Thus, $x = 1.9$ to one decimal place.

(B)

x	y_1
1.7	67.252
1.8	66.528
1.9	65.436
2.0	64.000

(C)

X	Y1
1.9	65.436
1.91	65.307
1.92	65.176
1.93	65.04
1.94	64.902
1.95	64.76
1.96	64.614

X=1.93

$x = 1.93$ to two decimal places.

95. Given $(w + a)(v + b) = c$. Let $a = 15, b = 1,$ and $c = 90$. Then: $(w + 15)(v + 1) = 90$
Solving for v, we have

$$v + 1 = \frac{90}{w+15} \quad \text{and} \quad v = \frac{90}{w+15} - 1 = \frac{90-(w+15)}{w+15} = \frac{75-w}{w+15}.$$

If $w = 16$, then $v = \dfrac{75-16}{16+15} = \dfrac{59}{31} \approx 1.9032$ cm/sec.

EXERCISE 1-2

Things to remember:

1. LIBRARY OF ELEMENTARY FUNCTIONS

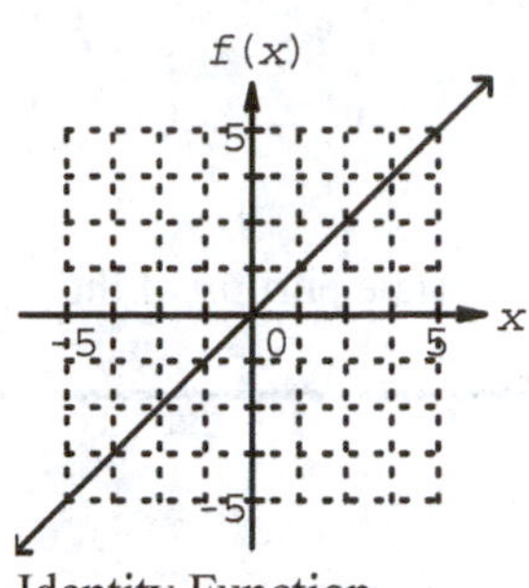

Identity Function

$f(x) = x$

Domain: All real numbers

Range: All real numbers

(a)

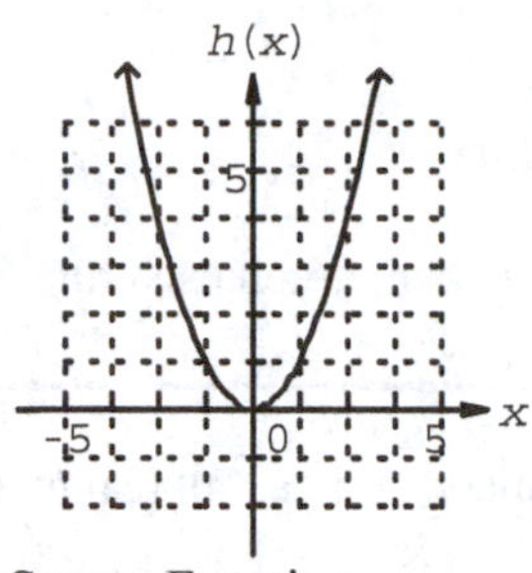

Square Function

$h(x) = x^2$

Domain: All real numbers

Range: $[0, \infty)$

(b)

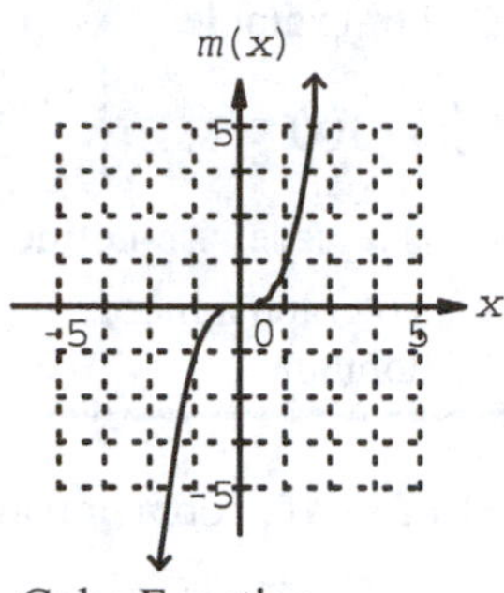

Cube Function

$m(x) = x^3$

Domain: All real numbers

Range: All real numbers

(c)

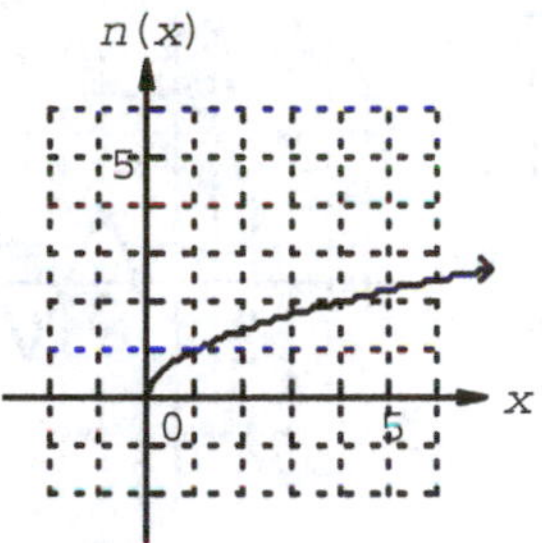

Square-Root Function

$n(x) = \sqrt{x}$

Domain: $[0, \infty)$

Range: $[0, \infty)$

(d)

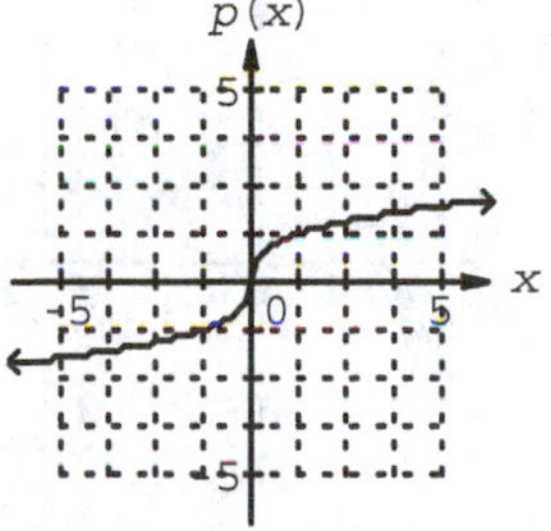

Cube-Root Function

$p(x) = \sqrt[3]{x}$

Domain: All real numbers

Range: All real numbers

(e)

Absolute Value Function

$g(x) = |x|$

Domain: All real numbers

Range: $[0, \infty)$

(f)

NOTE: Letters used to designate the above functions may vary from context to context.

2. GRAPH TRANSFORMATIONS SUMMARY

Vertical Translation:

$y = f(x) + k$ $\begin{cases} k > 0 & \text{Shift graph of } y = f(x) \text{ up } k \text{ units} \\ k < 0 & \text{Shift graph of } y = f(x) \text{ down } |k| \text{ units} \end{cases}$

Horizontal Translation:

$y = f(x + h)$ $\begin{cases} h > 0 & \text{Shift graph of } y = f(x) \text{ left } h \text{ units} \\ h < 0 & \text{Shift graph of } y = f(x) \text{ right } |h| \text{ units} \end{cases}$

Reflection:

$y = -f(x)$ Reflect the graph of $y = f(x)$ in the x axis

Vertical Stretch and Shrink:

$y = Af(x)$ $\begin{cases} A > 1 & \text{Stretch graph of } y = f(x) \text{ vertically by multiplying each ordinate value by } A \\ 0 < A < 1 & \text{Shrink graph of } y = f(x) \text{ vertically by multiplying each ordinate value by } A \end{cases}$

3. PIECEWISE-DEFINED FUNCTIONS

Functions whose definitions involve more than one rule are called PIECEWISE-DEFINED FUNCTIONS.

For example,

$$f(x) = |x| = \begin{cases} -x & \text{if} \quad x < 0 \\ x & \text{if} \quad x \geq 0 \end{cases}$$

is a piecewise-defined function.

To graph a piecewise-defined function, graph each rule over the appropriate portion of the domain.

---

1. $f(x) = 5x - 10$; domain: all real numbers; range: all real numbers

3. $f(x) = 15 - \sqrt{x}$; domain: $[0, \infty)$; range: $(-\infty, 15]$

5. $f(x) = 2|x| + 7$; domain: all real numbers; range: $[7, \infty)$

7. $f(x) = -\sqrt[3]{x} + 100$; domain: all real numbers; range: all real numbers

9.

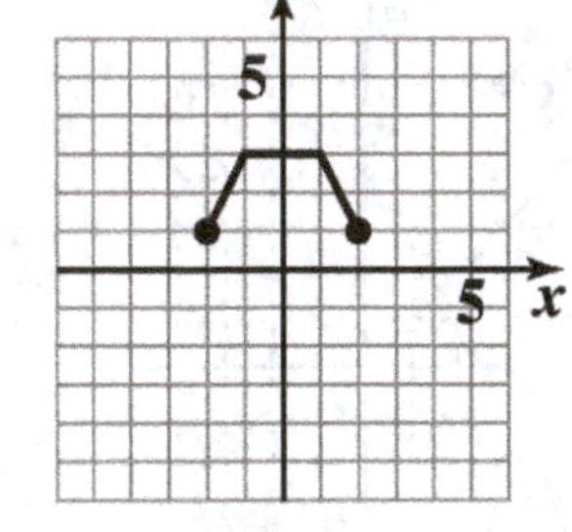

11.

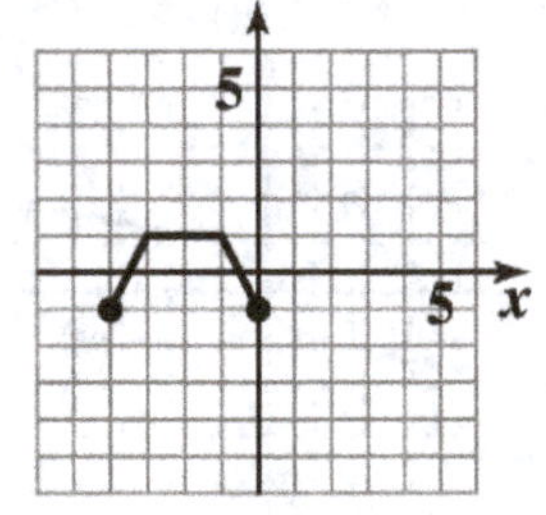

13.

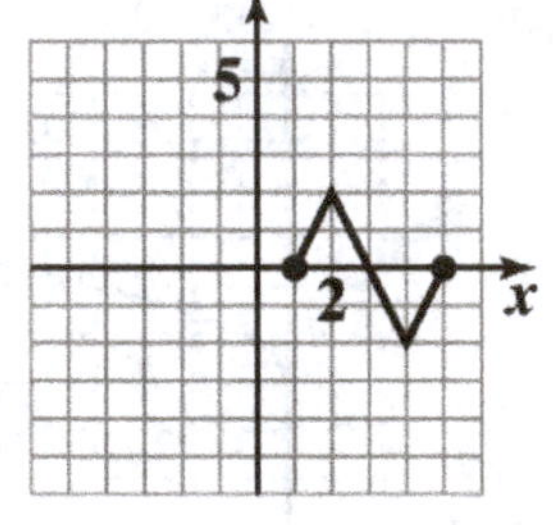

15.

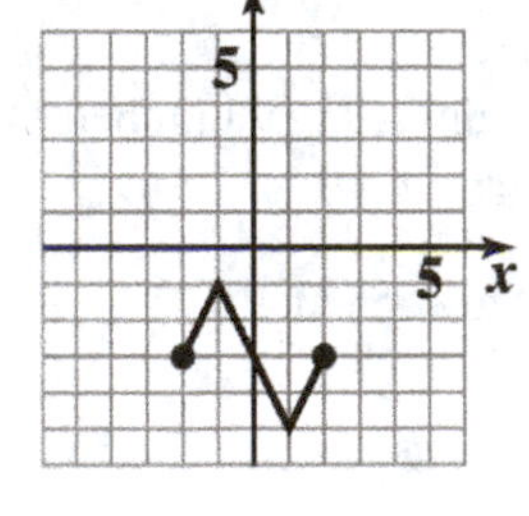

17.

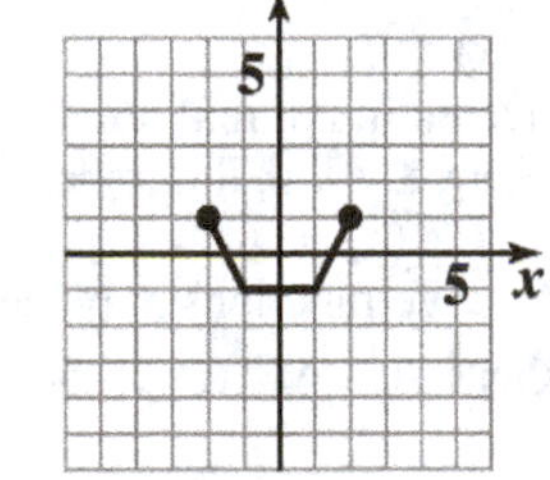

19.

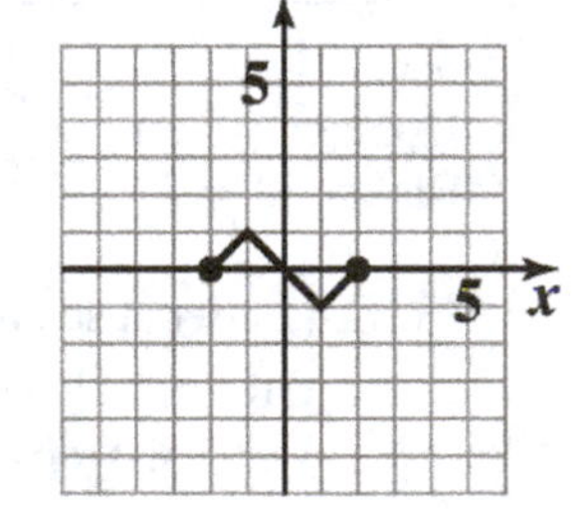

21.

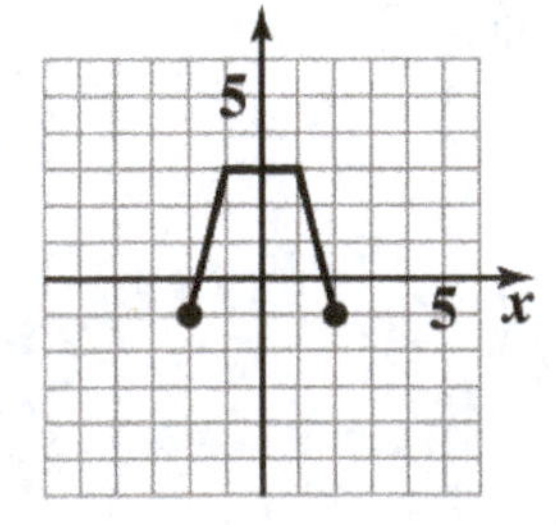

23.

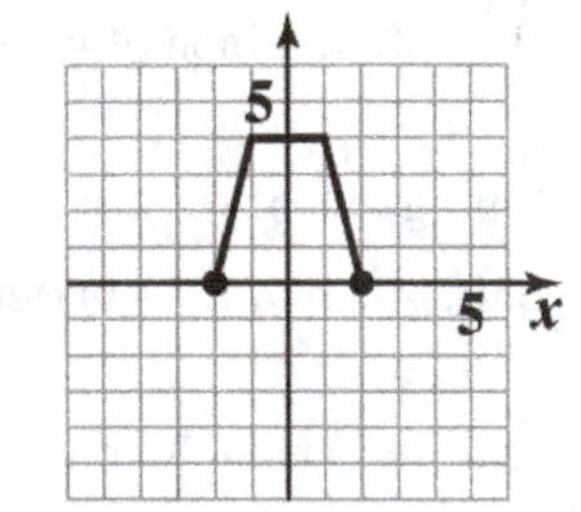

25. The graph of $g(x) = -|x + 3|$ is the graph of $y = |x|$ reflected in the x axis and shifted 3 units to the left.

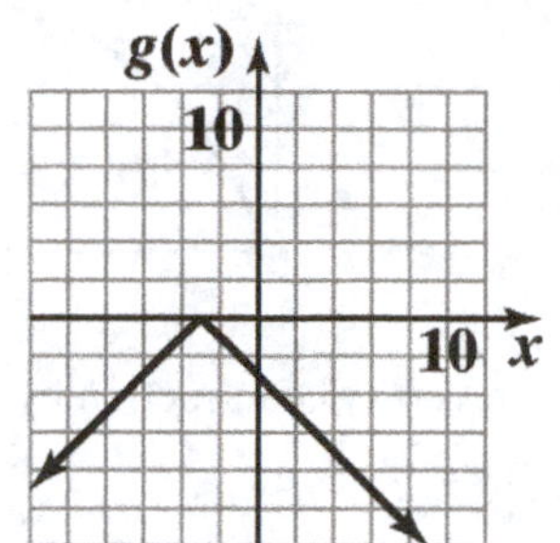

27. The graph of $f(x) = (x - 4)^2 - 3$ is the graph of $y = x^2$ shifted 4 units to the right and 3 units down.

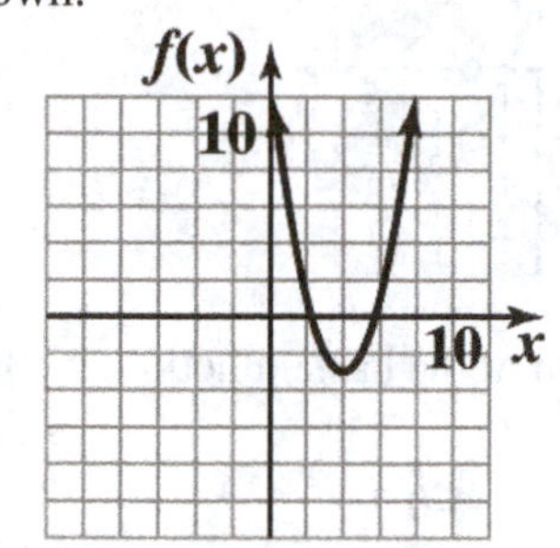

29. The graph of $f(x) = 7 - \sqrt{x}$ is the graph of $y = \sqrt{x}$ reflected in the x axis and shifted 7 units up.

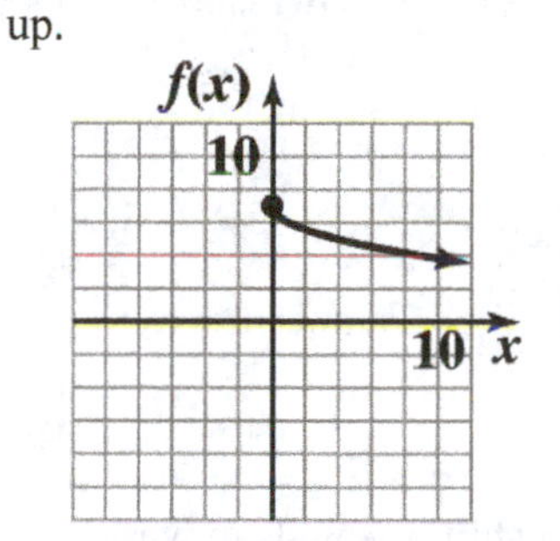

31. The graph of $h(x) = -3|x|$ is the graph of $y = |x|$ reflected in the x axis and vertically expanded by a factor of 3.

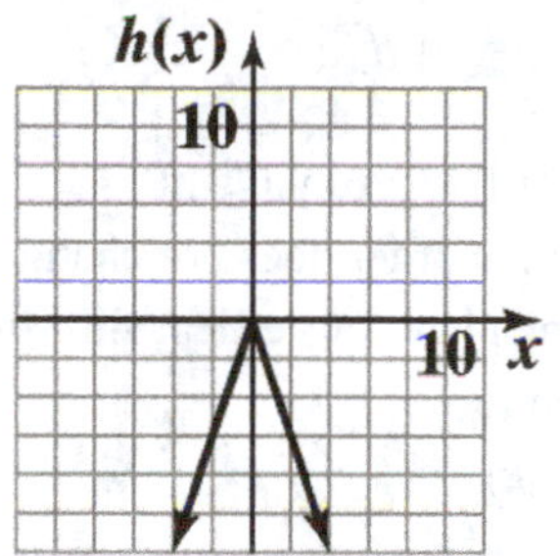

33. The graph of the basic function $y = x^2$ is shifted 2 units to the left and 3 units down.
Equation: $y = (x + 2)^2 - 3$.

35. The graph of the basic function $y = x^2$ is reflected in the x axis, shifted 3 units to the right and 2 units up.
Equation: $y = 2 - (x - 3)^2$.

37. The graph of the basic function $y = \sqrt{x}$ is reflected in the x axis and shifted 4 units up.
Equation: $y = 4 - \sqrt{x}$.

39. The graph of the basic function $y = x^3$ is shifted 2 units to the left and 1 unit down.
Equation: $y = (x + 2)^3 - 1$.

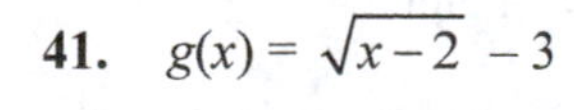

41. $g(x) = \sqrt{x-2} - 3$

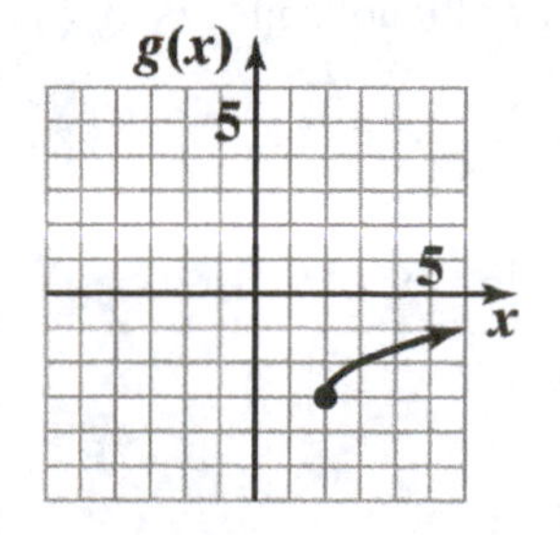

43. $g(x) = -|x + 3|$

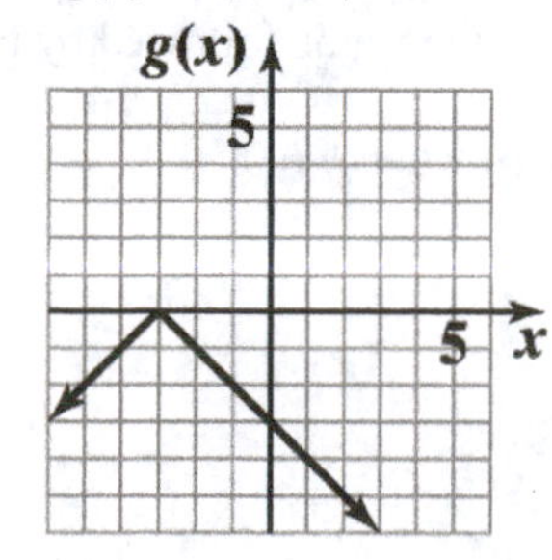

45. $g(x) = -(x - 2)^3 - 1$

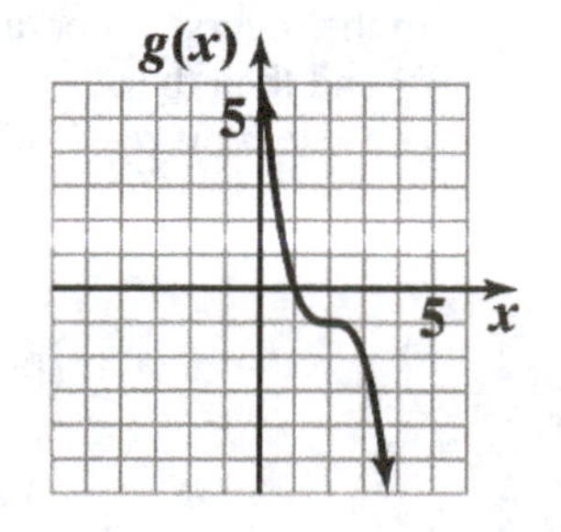

47.

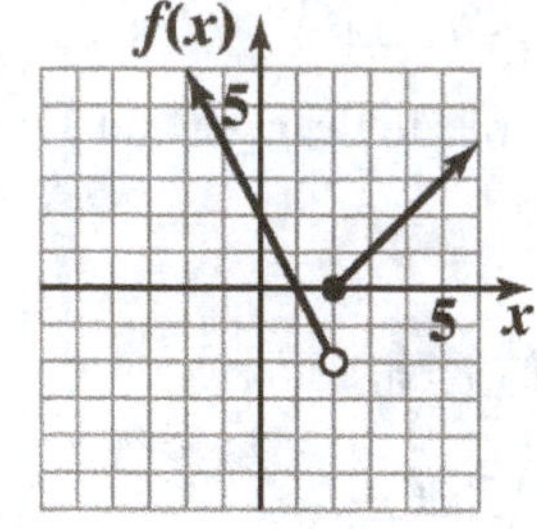

49.

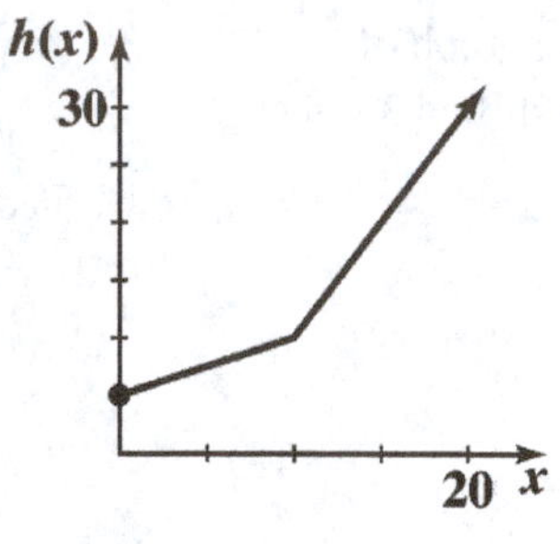

51.

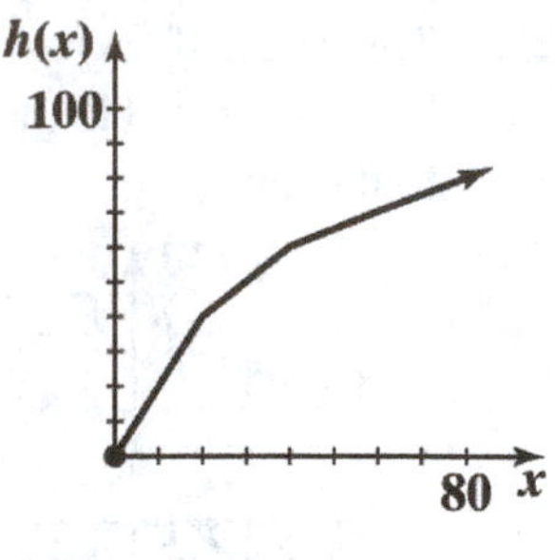

53. The graph of the basic function: $y = |x|$ is reflected in the x axis and has a vertical contraction by the factor 0.5.
Equation: $y = -0.5|x|$.

55. The graph of the basic function $y = x^2$ is reflected in the x axis and is vertically expanded by the factor 2.
Equation: $y = -2x^2$.

57. The graph of the basic function $y = \sqrt[3]{x}$ is reflected in the x axis and is vertically expanded by the factor 3.
Equation: $y = -3\sqrt[3]{x}$.

59. Vertical shift, horizontal shift.
Reversing the order does not change the result. Consider a point
(a, b) in the plane. A vertical shift of k units followed by a horizontal shift of h units moves (a, b) to $(a, b + k)$ and then to
$(a + h, b + k)$.

In the reverse order, a horizontal shift of h units followed by a vertical shift of k units moves (a, b) to $(a + h, b)$ and then to
$(a + h, b + k)$. The results are the same.

61. Vertical shift, reflection in the x axis.
Reversing the order can change the result. For example, let (a, b) be a point in the plane with $b > 0$. A vertical shift of k units, $k \neq 0$, followed by a reflection in the x axis moves (a, b) to $(a, b + k)$ and then to $(a, -[b + k]) = (a, -b - k)$.
In the reverse order, a reflection in the x axis followed by the vertical shift of k units moves (a, b) to $(a, -b)$ and then to
$(a, -b + k)$; $(a, -b - k) \neq (a, -b + k)$ when $k \neq 0$.

63. Horizontal shift, reflection in y axis.
Reversing the order can change the result. For example, let (a, b) be a point in the plane with $a > 0$. A horizontal shift of h units followed by a reflection in the y axis moves (a, b) to the point
$(a + h, b)$ and then to $(-[a + h], b) = (-a - h, b)$.
In the reverse order, a reflection in the y axis followed by the horizontal sift of h units moves (a, b) to $(-a, b)$ and then the
$(-a + h, b)$, $(-a - h, b) \neq (-a + h, b)$ when $h \neq 0$.

65. (A) The graph of the basic function $y = \sqrt{x}$ is reflected in the x axis, vertically expanded by a factor of 4, and shifted up 115 units.

(B)

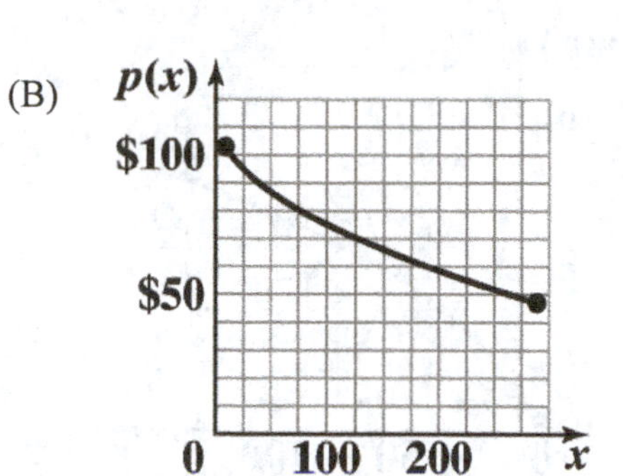

67. (A) The graph of the basic function $y = x^3$ is vertically contracted by a factor of 0.00048 and shifted right 500 units and up 60,000 units.

(B)

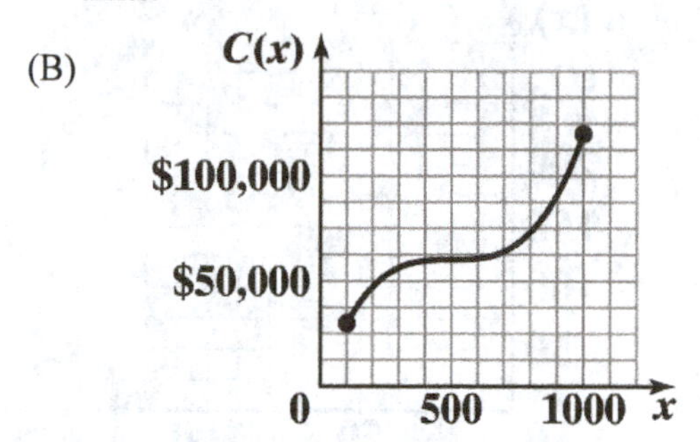

69. (A) $S(x) = \begin{cases} 8.50 + 0.0650x & \text{if } 0 \le x \le 700 \\ 8.50 + 0.0650(700) + 0.09(x - 700) & \text{if } x > 700 \end{cases}$

$= \begin{cases} 8.50 + 0.0650x & \text{if } 0 \le x \le 700 \\ -9 + 0.09x & \text{if } x > 700 \end{cases}$

(B)

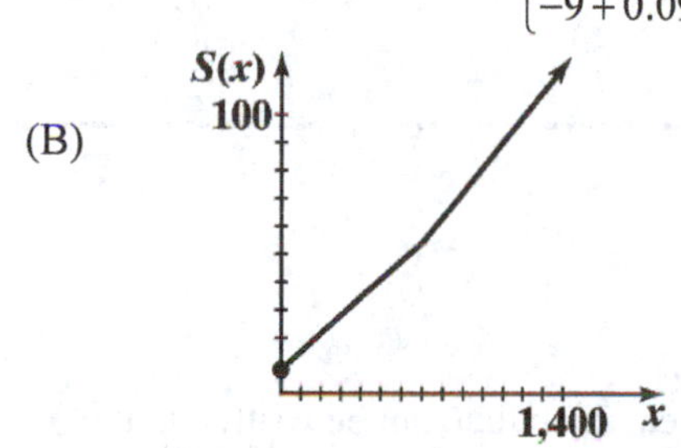

71. (A) If $0 \le x \le 30{,}000$, $T(x) = 0.035x$, and $T(30{,}000) = 1{,}050$.

If $30{,}000 < x \le 60{,}000$, $T(x) = 1{,}050 + 0.0625(x - 30{,}000) = 0.0625x - 825$, and $T(60{,}000) = 2{,}925$.

If $x > 60{,}000$, $T(x) = 2{,}925 + 0.0645(x - 60{,}000) = 0.0645x - 945$.

Thus, $T(x) = \begin{cases} 0.035x & \text{if } 0 \le x \le 30{,}000 \\ 0.0625x - 825 & \text{if } 30{,}000 < x \le 60{,}000 \\ 0.0645x - 945 & \text{if } x > 60{,}000 \end{cases}$

(B)

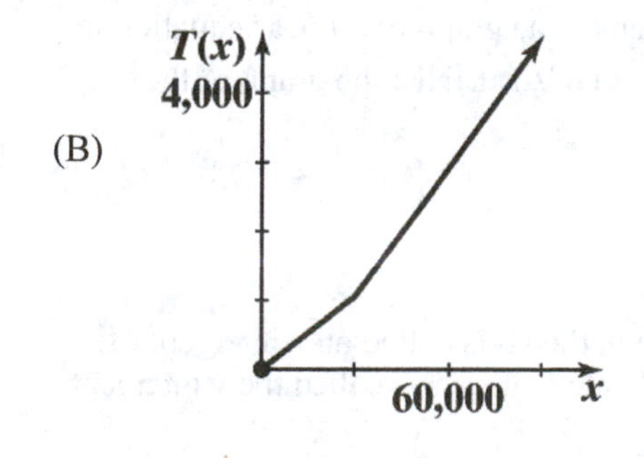

(C) $T(40{,}000) = 0.0625(40{,}000) - 825 = 1{,}675$; $1,675

$T(70{,}000) = 0.0645(70{,}000) - 945 = 3{,}570$; $3,570

73. (A) The graph of the basic function $y = x$ is vertically expanded by a factor of 5.5 and shifted down 220 units.

(B)

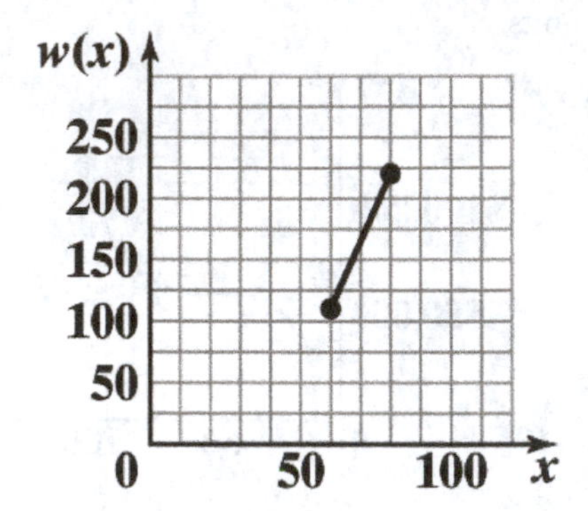

75. (A) The graph of the basic function $y = \sqrt{x}$ is vertically expanded by a factor of 7.08.

(B)

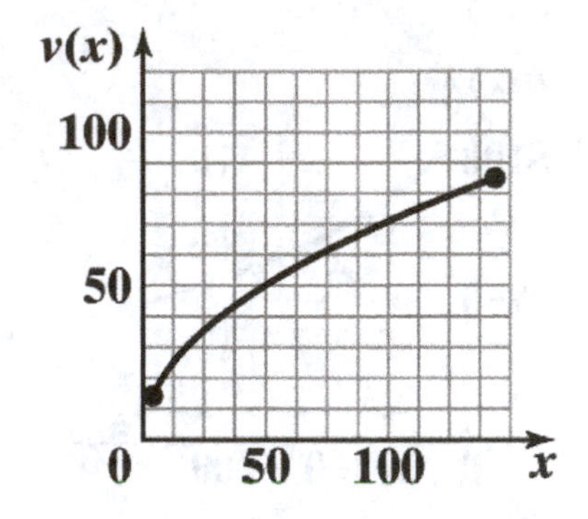

EXERCISE 1-3

Things to remember:

1. LINEAR EQUATIONS IN TWO VARIABLES

A LINEAR EQUATION IN TWO VARIABLES is an equation that can be written in the STANDARD FORM

$$Ax + By = C$$

where A, B, and C are constants and A and B are not both zero.

2. GRAPH OF A LINEAR EQUATION IN TWO VARIABLES

The graph of any equation of the form

$$Ax + By = C \quad (A \text{ and } B \text{ not both } 0)$$

is a line, and every line in a Cartesian coordinate system is the graph of a linear equation in two variables. The graph of the equation $x = a$ is a VERTICAL LINE; the graph of the equation $y = b$ is a HORIZONTAL LINE.

3. INTERCEPTS

If a line crosses the x-axis at a point with x coordinate a, then a is called an x intercept of the line; if it crosses the y-axis at a point with y coordinate b, then b is called the y intercept.

4. SLOPE OF A LINE

If a line passes through two distinct points $P_1(x_1, y_1)$ and $P_2(x_2, y_2)$, then its slope is given by the formula

$$m = \frac{y_2 - y_1}{x_2 - x_1} = \frac{\text{vertical change (rise)}}{\text{horizontal change (run)}}$$

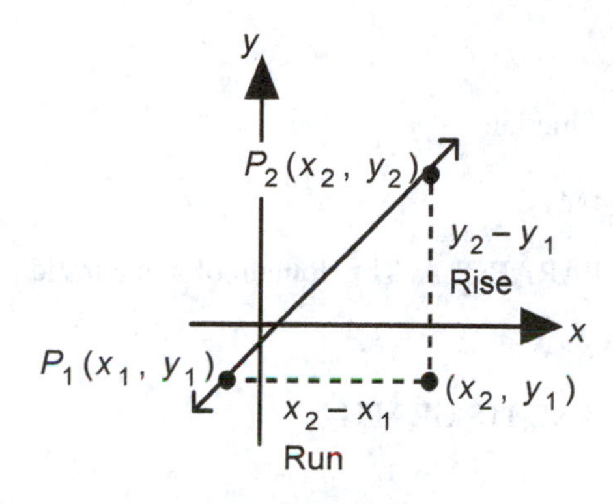

5. GEOMETRIC INTERPRETATION OF SLOPE

Line	Slope	Example
Rising as x moves from left to right	Positive	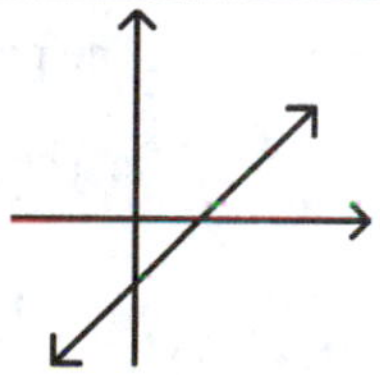
Falling as x moves from left to right	Negative	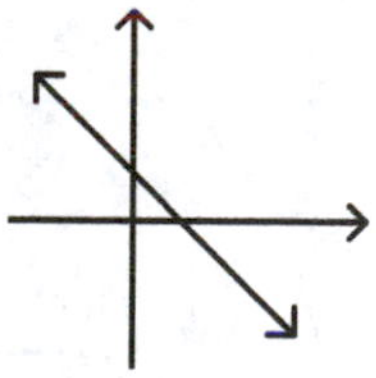
Horizontal	0	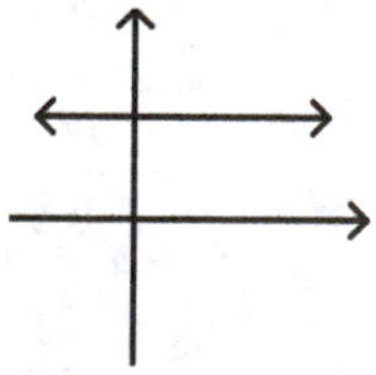
Vertical	Not defined	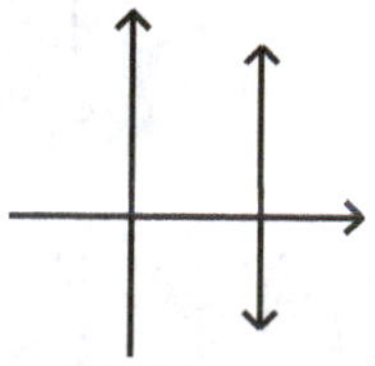

6. EQUATIONS OF A LINE; SPECIAL FORMS

Standard form	$Ax + By = C$	A and B not both 0
Slope-intercept form	$y = mx + b$	Slope: m; y-intercept: b
Point-slope form	$y - y_1 = m(x - x_1)$	Slope: m; point: (x_1, y_1)
Horizontal line	$y = b$	Slope: 0
Vertical line	$x = a$	Slope: Undefined

7. QUADRATIC FUNCTION

If a, b, and c are real numbers with $a \neq 0$, then the function

$$f(x) = ax^2 + bx + c \qquad \text{STANDARD FORM}$$

is a QUADRATIC FUNCTION and its graph is a PARABOLA. The domain of a quadratic function is the set of all real numbers.

8. PROPERTIES OF A QUADRATIC FUNCTION AND ITS GRAPH

Given a quadratic function

$$f(x) = ax^2 + bx + c, \quad a \neq 0,$$

and the VERTEX FORM obtained by completing the square

$$f(x) = a(x - h)^2 + k.$$

The general properties of f are as follows:

a. The graph of f is a parabola:

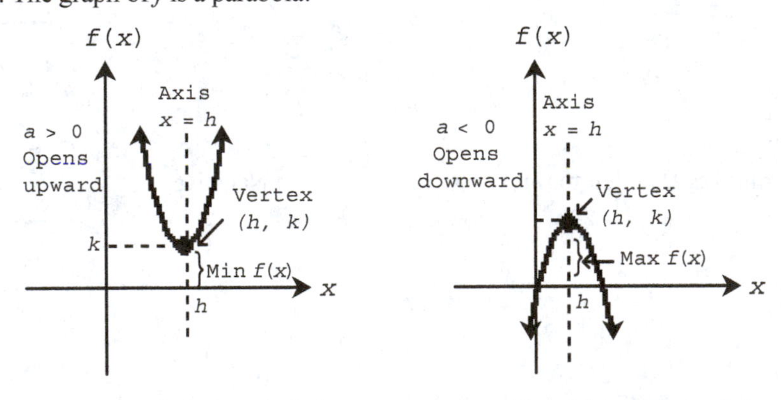

b. Vertex: (h, k) [parabola increases on one side of the vertex and decreases on the other]

c. Axis (of symmetry): $x = h$ (parallel to y axis)

d. $f(h) = k$ is the minimum if $a > 0$ and the maximum if $a < 0$

e. Domain: All real numbers
Range: $(-\infty, k]$ if $a < 0$ or $[k, \infty)$ if $a > 0$

f. The graph of f is the graph of $g(x) = ax^2$ translated horizontally h units and vertically k units.

1. $y = 2x - 3$

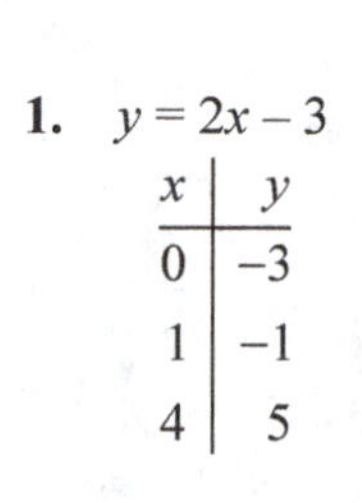

x	y
0	−3
1	−1
4	5

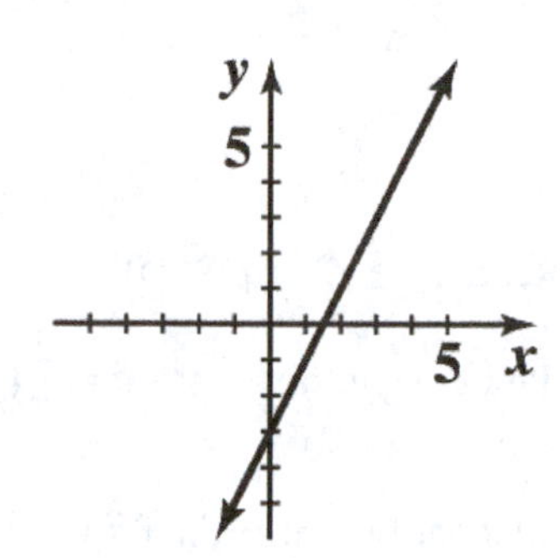

3. $2x + 3y = 12$

x	y
0	4
6	0
9	−2

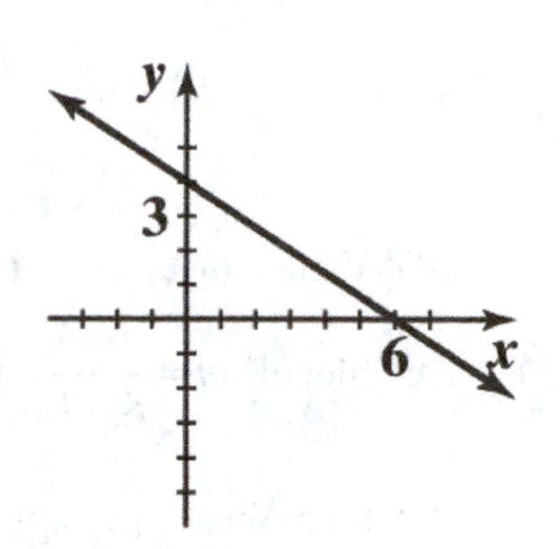

5. $y = 5x - 7$; slope: $m = 5$; y intercept: $b = -7$

7. $y = -\frac{5}{2}x - 9$; slope: $m = -\frac{5}{2}$; y intercept: $b = -9$

9. $m = 2$, $b = 1$; using 7, equation: $y = 2x + 1$

11. $m = -\frac{1}{3}$, $b = 6$; using 7, equation: $y = -\frac{1}{3}x + 6$

13. x intercept: −1 [or (−1, 0)]
y intercept: −2 [or (0, −2)]
slope: $m = \frac{-2-0}{0-(-1)} = \frac{-2}{1} = -2$
equation: $y = -2x - 2$

15. x intercept: −3 [or (−3, 0)]
y intercept: 1 [or (0, 1)]
slope: $m = \frac{1-0}{0-(-3)} = \frac{1}{3}$
equation: $y = \frac{1}{3}x + 1$

17. (A) m (B) g (C) f (D) n

19. (A) x-intercepts: 1, 3; y-intercept: −3 (B) Vertex: (2, 1)
(C) Maximum: 1 (D) Range: $y \le 1$ or $(-\infty, 1]$

21. (A) x-intercepts: −3, −1; $y$intercept: 3 (B) Vertex: (−2, −1)
(C) Minimum: −1 (D) Range: $y \ge -1$ or $[-1, \infty)$

23. $f(x) = -(x-3)^2 + 2$

(A) y-intercept: $f(0) = -(0-3)^2 + 2 = -7$
x-intercepts: $f(x) = 0$

$$-(x-3)^2 + 2 = 0$$
$$(x-3)^2 = 2$$
$$x - 3 = \pm\sqrt{2}$$
$$x = 3 \pm \sqrt{2}$$

(B) Vertex: (3, 2) (C) Maximum: 2 (D) Range: $y \le 2$ or $(-\infty, 2]$

25. $m(x) = (x+1)^2 - 2$

(A) y-intercept: $m(0) = (0+1)^2 - 2 = 1 - 2 = -1$
x-intercepts: $m(x) = 0$

$$(x+1)^2 - 2 = 0$$

$(x+1)^2 = 2$

$x+1 = \pm\sqrt{2}$

$x = -1 \pm \sqrt{2}$

(B) Vertex: $(-1, -2)$ (C) Minimum: -2 (D) Range: $y \geq -2$ or $[-2, \infty)$

27. (A) Slope: $m = \dfrac{7-5}{5-2} = \dfrac{2}{3}$. (B) Point-slope form: $y - 5 = \dfrac{2}{3}(x-2)$.

(C) Slope-intercept form: $y = \dfrac{2}{3}x + \dfrac{11}{3}$. (D) Standard form: $-2x + 3y = 11$.

29. (A) Slope: $m = \dfrac{-6-(-1)}{2-(-2)} = -\dfrac{5}{4}$. (B) Point-slope form: $y + 1 = -\dfrac{5}{4}(x+2)$.

(C) Slope-intercept form: $y = -\dfrac{5}{4}x - \dfrac{7}{2}$. (D) Standard form: $5x + 4y = -14$.

31. (A) Slope: $m = \dfrac{-3-3}{0}$ not defined. (B) Point-slope form: none.

(C) Slope-intercept form: none. (D) Standard form: $x = 5$

33. (A) Slope: $m = \dfrac{5-5}{3-(-2)} = 0$ (B) Point-slope form: $y - 5 = 0$

(C) Slope-intercept form: $y = 5$. (D) Standard form: $y = 5$.

35. $f(x) = x^2 - 8x + 12 = (x^2 - 8x) + 12$

$= (x^2 - 8x + 16) + 12 - 16$

$= (x-4)^2 - 4$ (vertex form)

(A) y-intercept: $f(0) = 0^2 - 8(0) + 12 = 12$

x-intercepts: $f(x) = 0$

$(x-4)^2 - 4 = 0$

$(x-4)^2 = 4$

$x - 4 = \pm 2$

$x = 2, 6$

(B) Vertex: $(4, -4)$ (C) Minimum: -4 (D) Range: $y \geq -4$ or $[-4, \infty)$

37. $r(x) = -4x^2 + 16x - 15 = -4(x^2 - 4x) - 15$

$= -4(x^2 - 4x + 4) - 15 + 16$

$= -4(x-2)^2 + 1$ (vertex form)

(A) y-intercept: $r(0) = -4(0)^2 + 16(0) - 15 = -15$

x-intercepts: $r(x) = 0$

$-4(x-2)^2 + 1 = 0$

$(x-2)^2 = \dfrac{1}{4}$

$$x - 2 = \pm\frac{1}{2}$$
$$x = \frac{3}{2}, \frac{5}{2}$$

(B) Vertex: (2, 1) (C) Maximum: 1 (D) Range: $y \le 1$ or $(-\infty, 1]$

39. $u(x) = 0.5x^2 - 2x + 5 = 0.5(x^2 - 4x) + 5$
$= 0.5(x^2 - 4x + 4) + 3$
$= 0.5(x - 2)^2 + 3$ (vertex form)

(A) y-intercept: $u(0) = 0.5(0)^2 - 2(0) + 5 = 5$
x-intercepts: $u(x) = 0$
$0.5(x - 2)^2 + 3 = 0$
$(x - 2)^2 = -6$; no solutions.
There are no x-intercepts.

(B) Vertex: (2, 3) (C) Minimum: 3 (D) Range: $y \ge 3$ or $[3, \infty)$

41. $2x + 7 < 0$, $2x < -7$, $x < -\frac{7}{2}$; solution set: $\left(-\infty, -\frac{7}{2}\right)$

43. $-5x + 3 \ge 6$, $-5x \ge 3$, $x \le -\frac{3}{5}$; solution set: $\left(-\infty, -\frac{3}{5}\right]$

45.

$x^2 - 10x > 0$
$x(x - 10) > 0$

Therefore, either $x > 0$ and $x - 10 > 0$ or $x < 0$ and $x - 10 < 0$. The first case implies $x > 10$, and the second case implies $x < 0$. Solution set: $(-\infty, 0) \cup (10, \infty)$.

47.

$x^2 + 20x + 50 \le 0$
$x^2 + 20 + 100 - 50 \le 0$
$(x + 10)^2 \le 50$
$-5\sqrt{2} \le x + 10 \le 5\sqrt{2}$
$-10 - 5\sqrt{2} \le x \le -10 + 5\sqrt{2}$

Solution set: $\left[-10 - 5\sqrt{2}, -10 + 5\sqrt{2}\right]$

49.

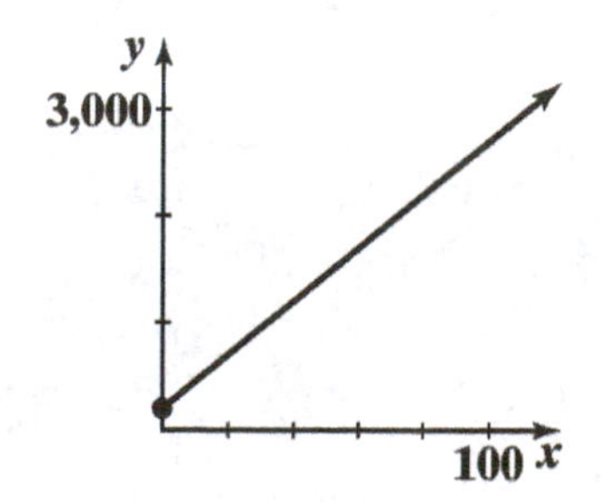

51. (A)

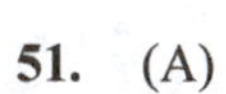

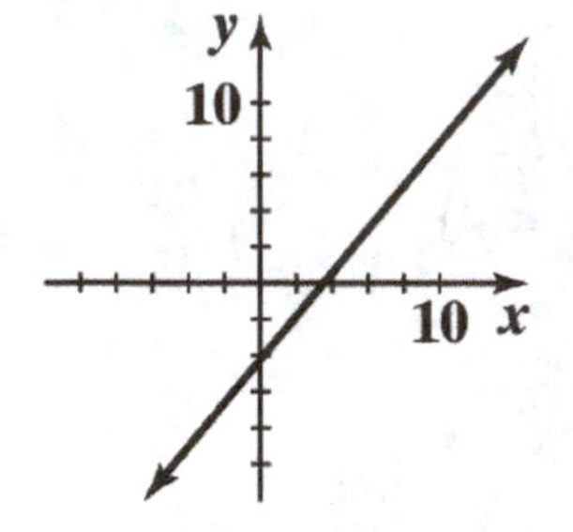

(B) x- intercept — set $y = 0$:

$1.2x - 4.2 = 0$

$x = 3.5$

y- intercept — set $x = 0$:

$y = -4.2$

(C)

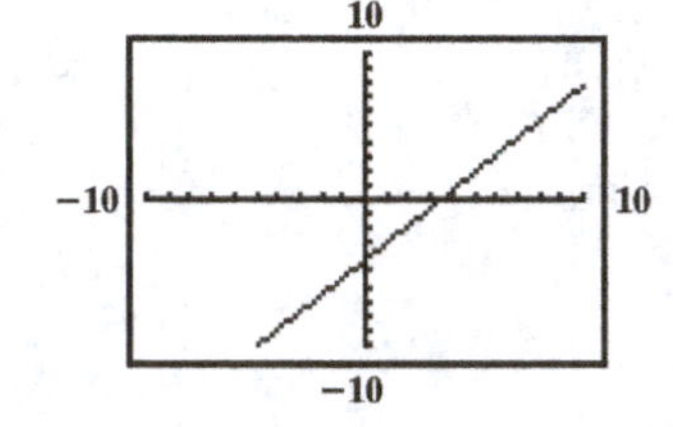

(D) x-intercept: 3.5; y-intercept: −4.2

53. $f(x) = 0.3x^2 - x - 8$

(A) $f(x) = 4$: $0.3x^2 - x - 8 = 4$

$0.3x^2 - x - 12 = 0$

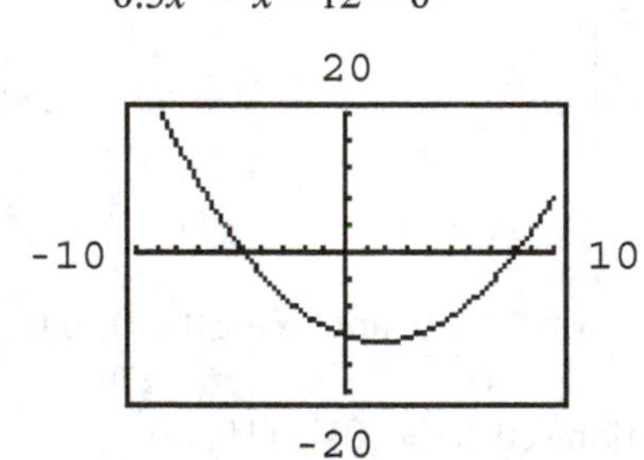

$x = -4.87, 8.21$

(B) $f(x) = -1$: $0.3x^2 - x - 8 = -1$

$0.3x^2 - x - 7 = 0$

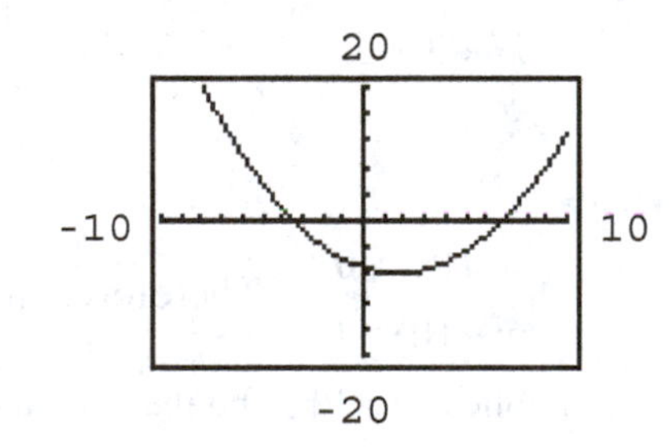

$x = -3.44, 6.78$

(C) $f(x) = -9$: $0.3x^2 - x - 8 = -9$

$0.3x^2 - x + 1 = 0$

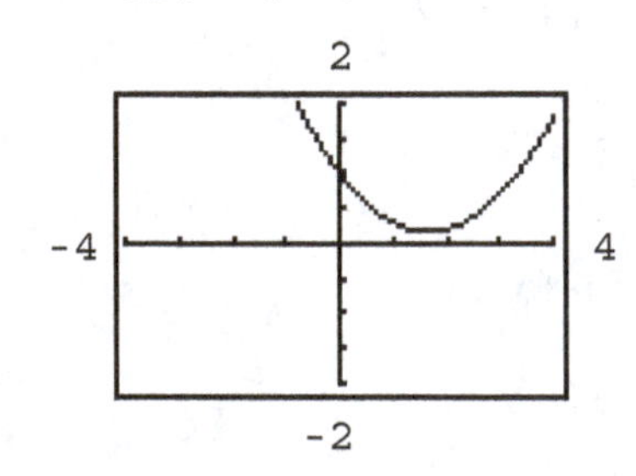

No solutions.

55.

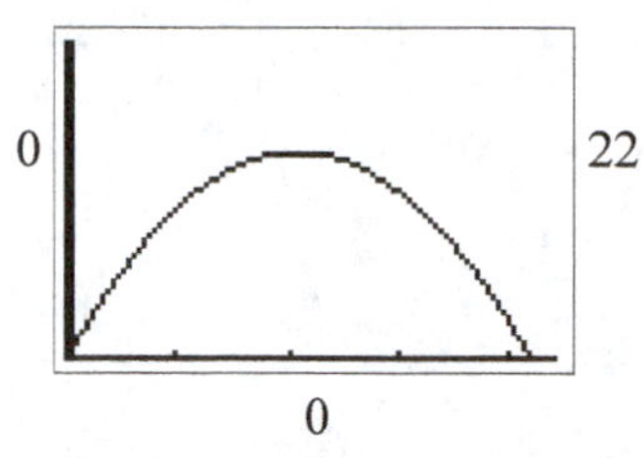

maximum value $f(10.41667) = 651.0417$

57. $Ax_1 + By_1 = C$ and $Ax_2 + By_2 = C$. Subtracting the first equation from the second gives

$Ax_2 + By_2 - (Ax_1 + By_1) = 0$

$A(x_2 - x_1) + B(y_2 - y_1) = 0$ and

$\frac{y_2 - y_1}{x_2 - x_1} = -\frac{A}{B}$

59. (A) $P = md + b$

At $d = 0$, $P = 14.7$. Therefore, $P = md + 14.7$.

At $d = 33$, $P = 29.4$. Therefore,

slope $m = \frac{29.4 - 14.7}{33 - 0} = \frac{14.7}{33} = 0.4\overline{45}$

and $P = 0.4\overline{45}\,d + 14.7$.

(B) The rate of change of pressure with respect to depth is $0.4\overline{45}$ lbs/in^2 per foot.

(C) At $d = 50$, $P = 0.4\overline{45}(50) + 14.7 \approx 37$ lbs/in^2.

(D) 4 atmospheres = 14.7(4) = 58.8 lbs/in^2.

Solve $58.8 = 0.4\overline{45}\,d + 14.7$ for d:

$0.4\overline{45}\,d = 58.8 - 14.7 = 44.1$

$d \approx 99$ feet

61. (A) $a = mt + b$

At time $t = 0$, $a = 2880$. Therefore, $b = 2880$.

At time $t = 120$, $a = 0$. The rate of change of altitude with respect to time is:

$$m = \frac{0 - 2880}{120 - 0} = -24$$

Therefore, $a = -24t + 2880$.

(B) The rate of descent is –24 ft/sec.

(C) The speed at landing is 24 ft/sec.

63. (A) Since daily cost and production are linearly related,

$$C = mx + b$$

From the given information, the points (80, 7647) and (100, 9147) satisfy this equation. Therefore:

$$\text{slope } m = \frac{9147 - 7647}{100 - 80} = \frac{1500}{20} = 75$$

Using the point-slope form with $(x_1, C_1) = (80, 7647)$:

$$C - 7467 = 75(x - 80)$$
$$C = 75x - 6000 + 7647$$
$$C = 75x + 1647$$

(B)

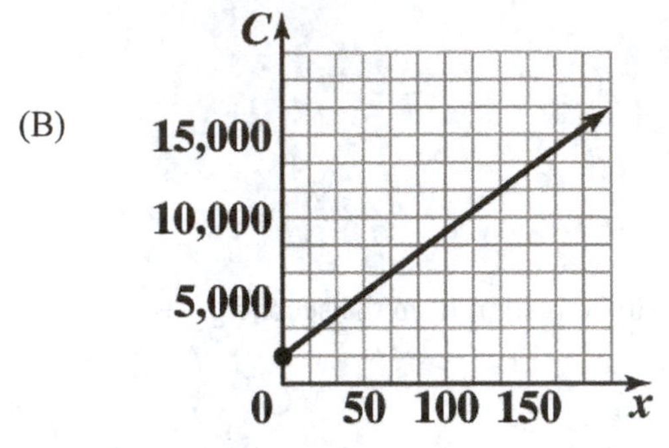

(C) The y-intercept, \$1,647, is the fixed cost and the slope, \$75, is the cost per club.

65. (A) $V = mt + 157{,}000$. At $t = 10$,

$$V = 10m + 157{,}000 = 82{,}000$$
$$10m = 82{,}000 - 157{,}000 = -75{,}000$$
$$m = -7{,}500$$
$$V = -7{,}500t + 157{,}000$$

(B) At $t = 6$, $V = -7{,}500(6) + 157{,}000 = 112{,}000$

The value of the tractor after 6 years is \$112,000.

(C) Solve $-7{,}500t + 157{,}000 < 70{,}000$ for t:

$$-7{,}500t < 70{,}000 - 157{,}000 = -87{,}000$$
$$t > 11.6$$

The value of the tractor will fall below \$70,000 in the 12th year.

(D)

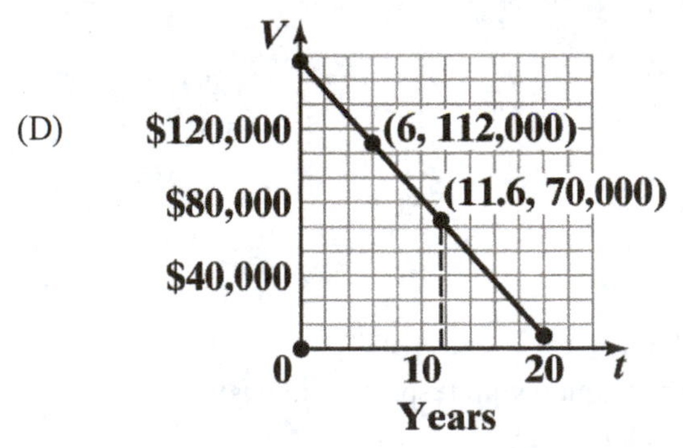

67. (A) $T = -3.6A + b$

At $A = 0$, $T = 70$. Therefore, $T = -3.6A + 70$.

(B) Solve $34 = -3.6A + 70$ for A:

$$-3.6A = 34 - 70 = -36$$
$$A = 10$$

The altitude of the aircraft is 10,000 feet.

69. (A) $p = mx + b$

At $x = 7{,}500$, $p = 2.28$; at $x = 7{,}900$, $p = 2.37$.

Therefore, slope $m = \dfrac{2.37 - 2.28}{7{,}900 - 7{,}500} = \dfrac{0.09}{400} = 0.000225$

Using the point-slope form with $(x_1, p_1) = (7{,}500, 2.28)$:

$p - 2.28 = 0.000225(x - 7{,}500) = 0.000225x + 0.5925$

$p = 0.000225x + 0.5925$ Price-supply equation

(B) $p = mx + b$

At $x = 7{,}900$, $p = 2.28$; at $x = 7{,}800$, $p = 2.37$.

Therefore, slope $m = \dfrac{2.37 - 2.28}{7{,}800 - 7{,}900} = -\dfrac{0.09}{100} = -0.0009$

Using the point-slope form with $(x_1, p_1) = (7{,}900, 2.37)$:

$p - 2.28 = -0.0009(x - 7{,}900) = -0.0009x + 9.39$

$p = -0.0009x + 9.39$ Price-demand equation

(C) To find the equilibrium point, solve

$0.000225x + 0.5925 = -0.0009x + 9.39$

$0.001125x = 9.39 - 0.5925 = 8.7975$

$x = 7{,}820$

At $x = 7{,}820$, $p = 0.000225(7{,}820) + 0.5925 = 2.352$

The equilibrium point is (7,820, 2.352).

(D)

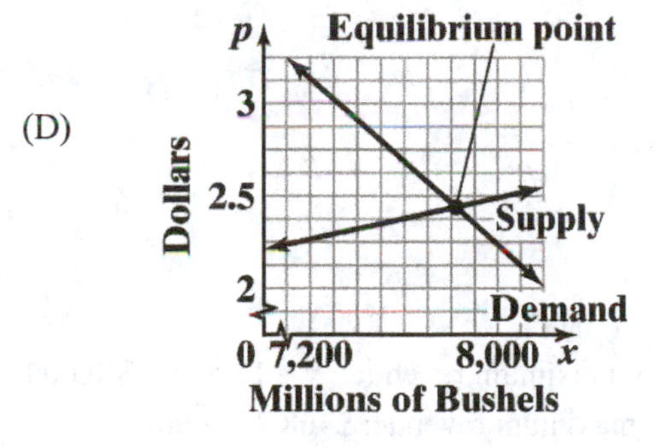

71. $y = 0.75x$

(A)

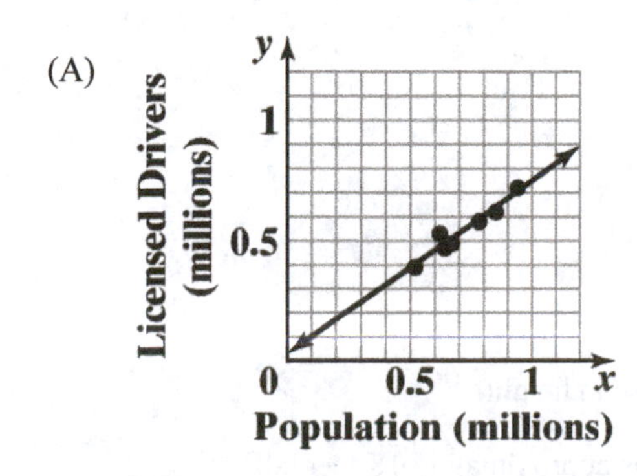

(B) At $x = 1.6$, $y = 0.75(1.6) = 1.2$. There were approximately 1,200,000 licensed drivers in Idaho in 2010.

(C) Solve $0.75 = 0.75x$ for x: $x = 1$

The population of Rhode Island in 2010 was approximately 1,000,000.

73. Mathematical model: $f(x) = -0.518x^2 + 33.3x - 481$

(A)

x	28	30	32	34	36
Mileage	45	52	55	51	47
$f(x)$	45.3	51.8	54.2	52.4	46.5

(B)

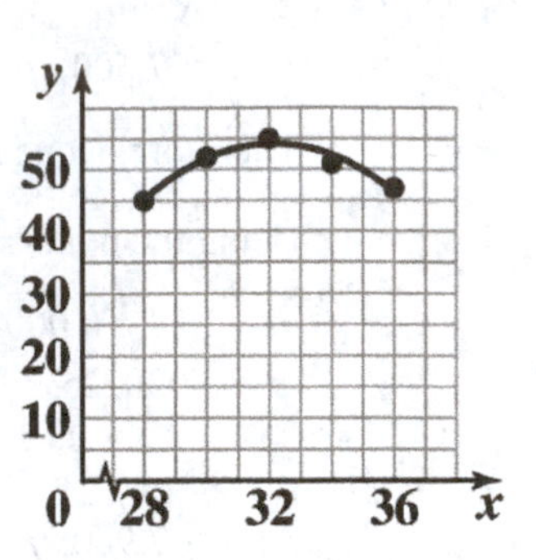

(C) $x = 31: f(31) = -0.518(31)^2 + 33.3(31) - 481 = 53.502$
$f(31) \approx 53.50$ thousand miles
$x = 35: f(35) = -0.518(35)^2 + 33(35) - 481 \approx 49.95$ thousand miles

(D) The maximum mileage is achieved at 32 lb/in^2 pressure. Increasing the pressure or decreasing the pressure reduces the mileage.

75. Quadratic regression using the data in Problem 73:
$f(x) = -0.518x^2 + 33.3x - 481$

77. $p(x) = 75 - 3x$; $R(x) = xp(x)$; $1 \le x \le 20$

$$R(x) = x(75 - 3x) = 75x - 3x^2$$
$$= -3(x^2 - 25x)$$
$$= -3\left(x^2 - 25x + \frac{625}{4}\right) + \frac{1875}{4}$$
$$= -3\left(x - \frac{25}{2}\right)^2 + \frac{1875}{4}$$
$$= -3(x - 12.5)^2 + 468.75$$

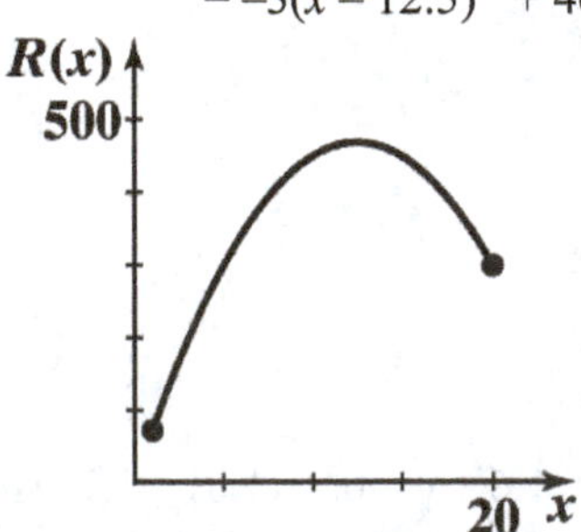

(B) Output for maximum revenue: $x = 12.5$ (12,500,000 chips); maximum revenue: \$468,750,000

(C) Wholesale price per chip at maximum revenue:
$p(12.5) = 75 - 3(12.5) = 37.5$ or \$37.50

79. $y = 1.37x - 2.58$

(A) The rate of change of height with respect to Dbh is 1.37 ft/in.

(B) A 1 in. increase in Dbh produces a 1.37 foot increase in height.

(C) At $x = 15$, $y = 1.37(15) - 2.58 = 17.97$. The spruce is approximately 18 feet tall.

(D) Solve $25 = 1.37x - 2.58$ for x:
$1.37x = 25 + 2.58 = 27.8$

$x \approx 20.3$

The Dbh is approximately 20 inches.

81. $y = 1.70x + 30.90$

(A) The average monthly price is increasing at the rate of \$1.70 per year.

(B) At $x = 24$, $y = 1.70(24) + 30.90 = 71.70$. The average monthly price in 2024 will be \$71.70.

83. Linear regression $y = a + bx$

Men: the regression formula is:

$y = -0.087x + 49.207$

Women: the regression formula is:

$y = -0.088x + 54.884$

Yes; the women's times are decreasing at a faster rate than the men's times.

85. Quadratic regression model

```
QuadReg
 y=ax²+bx+c
 a=1.4E-6
 b=-.00266
 c=5.4
```

$y \approx 0.0000014x^2 - 0.00266 + 5.4 \approx 10.6$ mph.

EXERCISE 1-4

Things to remember:

1. POLYNOMIAL FUNCTION

   A POLYNOMIAL FUNCTION is a function that can be written in the form

   $$f(x) = a_n x^n + a_{n-1} x^{n-1} + \ldots + a_1 x + a_0$$

   for n a nonnegative integer, called the DEGREE of the polynomial. The coefficients $a_0, a_1, \ldots, a_n$ are real numbers with $a_n \neq 0$, a_n is called the LEADING COEFFICIENT of f. The DOMAIN of a polynomial function is the set of all real numbers. The graph of a polynomial function is continuous, with no holes or breaks.

2. A RATIONAL FUNCTION is any function that can be written in the form

   $$f(x) = \frac{n(x)}{d(x)} \quad d(x) \neq 0$$

   where $n(x)$ and $d(x)$ are polynomials. The DOMAIN is the set of all real numbers such that $d(x) \neq 0$.

3. PROCEDURE: VERTICAL AND HORIZONTAL ASYMPTOTES OF RATIONAL FUNCTIONS

   Consider the rational function

$$f(x) = \frac{n(x)}{d(x)}$$

where $n(x)$ and $d(x)$ are polynomials.

VERTICAL ASYMPTOTES:

Case 1. Suppose $n(x)$ and $d(x)$ have no real zeros in common. If c is a number such that $d(c) =$ 0,
then the line $x = c$ is a vertical asymptote of the graph of f.

Case 2. If $n(x)$ and $d(x)$ have one or more real zeros in common, cancel the common linear factors and
apply Case 1 to the reduced function. (The reduced function has the same asymptotes as f.)

HORIZONTAL ASYMPTOTES:

Case 1. If degree $n(x) <$ degree $d(x)$, then $y = 0$ is the horizontal asymptote.

Case 2. If degree $n(x) =$ degree $d(x)$, then $y = a/b$ is the horizontal asymptote, where a is the leading
coefficient of $n(x)$ and b is the leading coefficient of $d(x)$.

Case 3. If degree $n(x) >$ degree $d(x)$, then there is no horizontal asymptote.

---

1. $f(x) = 50 - 5x$

(A) degree 1

(B) x-intercept: $f(x) = 0$
$50 - 5x = 0$
$x = 10$

(C) y-intercept: $f(0) = 50$

3. $f(x) = x^4(x-1)$

(A) degree: 5

(B) x-intercepts: $f(x) = 0$
$x^4(x-1) = 0$
$x = 0, 1$

(C) y-intercept: $f(0) = 0$

5. $f(x) = x^2 + 3x + 2 = (x+2)(x+1)$

(A) degree: 2

(B) x-intercepts: $f(x) = 0$
$x = -2, -1$

(C) y-intercept: $f(0) = 2$

7. $f(x) = (x^2 - 1)(x^2 - 9)$

(A) degree: 4

(B) x-intercepts: $f(x) = 0$
$x = -1, 1, -3, 3$

(C) y-intercept: $f(0) = 9$

9. $f(x) = (2x+3)^4(x-5)^5$

(A) degree: 9

(B) x-intercepts: $f(x) = 0$
$x = -\frac{3}{2}, 5$

(C) y-intercept: $f(0) = 3^4(-5)^5 = -253{,}125$

11. (A) 4 (B) negative

13. (A) 5 (B) negative

15. (A) 1 (B) negative

17. (A) 6 (B) positive

19. 10

21. 1; polynomials of odd degree cross the x-axis at least once.

23. $f(x) = \dfrac{x+2}{x-2}$

(A) *Intercepts:*

x-intercepts: $f(x) = 0$ only if $x + 2 = 0$ or $x = -2$.
The x intercept is -2.

y-intercept: $f(0) = \dfrac{0+2}{0-2} = -1$
The y intercept is -1.

(B) *Domain:* The denominator is 0 at $x = 2$. Thus, the domain is the set of all real numbers except 2.

(C) *Asymptotes:*

Vertical asymptotes: $f(x) = \dfrac{x+2}{x-2}$ The denominator is 0 at $x = 2$. Therefore, the line $x = 2$ is a vertical asymptote.

Horizontal asymptotes: $f(x) = \dfrac{x+2}{x-2} = \dfrac{1+\frac{2}{x}}{1-\frac{2}{x}}$

As x increases or decreases without bound, the numerator tends to 1 and the denominator tends to 1. Therefore, the line $y = 1$ is a horizontal asymptote.

(D)

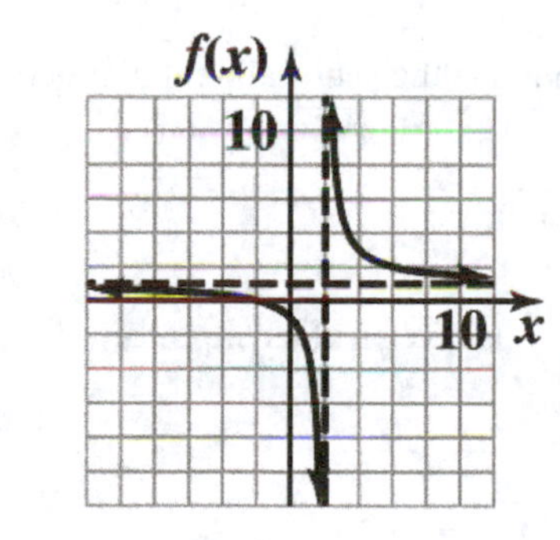

(E)

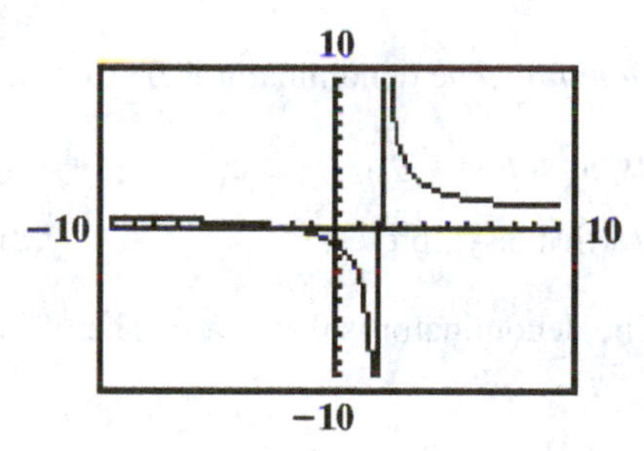

25. $f(x) = \dfrac{3x}{x+2}$

(A) *Intercepts:*

x-intercepts: $f(x) = 0$ only if $3x = 0$ or $x = 0$.
The x-intercept is 0.

y-intercept: $f(0) = \dfrac{3 \cdot 0}{0+2} = 0$
The y-intercept is 0.

(B) *Domain:* The denominator is 0 at $x = -2$. Thus, the domain is the set of all real numbers except -2.

(C) *Asymptotes:*

Vertical asymptotes: $f(x) = \dfrac{3x}{x+2}$
The denominator is 0 at $x = -2$; $x = -2$ is a vertical asymptote.

Horizontal asymptotes: $f(x) = \frac{3x}{x+2} = \frac{3}{1+\frac{2}{x}}$

As x increases or decreases without bound, the numerator is 3 and the denominator tends to 1. Therefore, the line $y = 3$ is a horizontal asymptote.

(D)

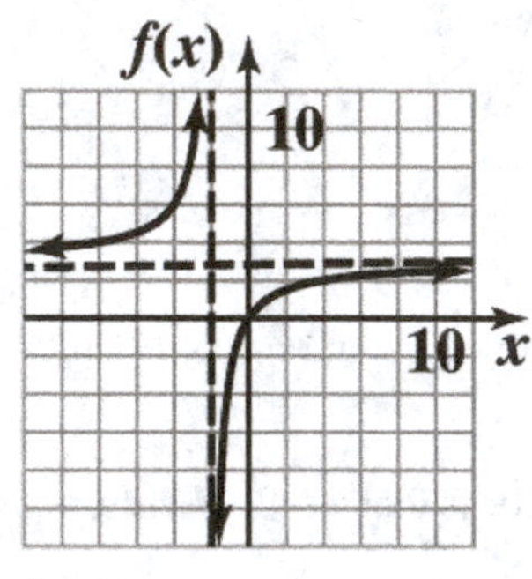

(E)

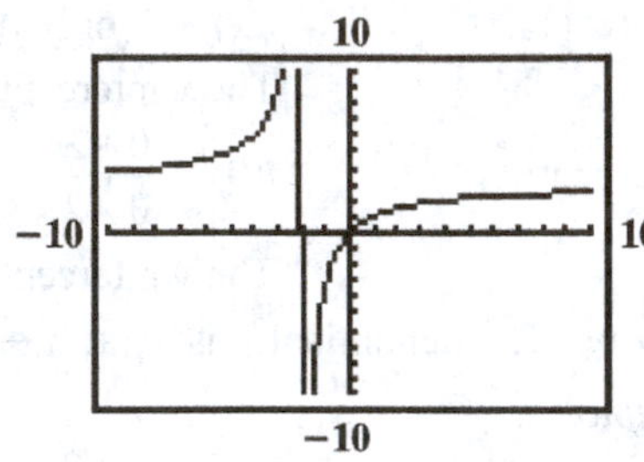

27. $f(x) = \frac{4 - 2x}{x - 4}$

(A) *Intercepts:*

x-intercepts: $f(x) = 0$ only if $4 - 2x = 0$ or $x = 2$.
The *x*-intercept is 2.

y-intercept: $f(0) = \frac{4 - 2 \cdot 0}{0 - 4} = -1$
The *y*-intercept is -1.

(B) *Domain:* The denominator is 0 at $x = 4$. Thus, the domain is the set of all real numbers except 4.

(C) *Asymptotes:*

Vertical asymptotes: $f(x) = \frac{4-2x}{x-4}$

The denominator is 0 at $x = 4$. Therefore, the line $x = 4$ is a vertical asymptote.

Horizontal asymptotes: $f(x) = \frac{4-2x}{x-4} = \frac{\frac{4}{x}-2}{1-\frac{4}{x}}$

As x increases or decreases without bound, the numerator tends to -2 and the denominator tends to 1. Therefore, the line $y = -2$ is a horizontal asymptote.

(D)

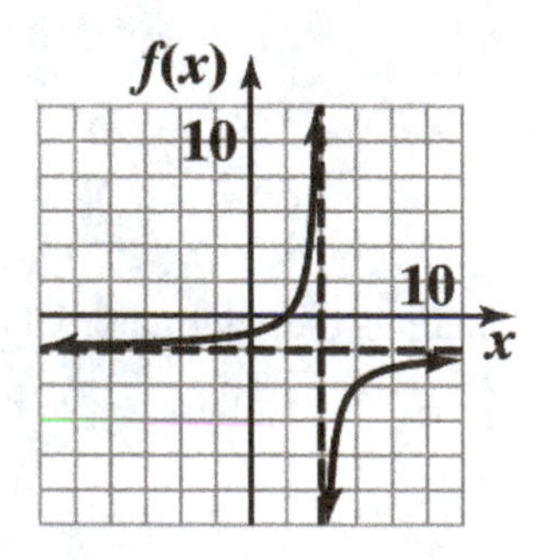

(E)

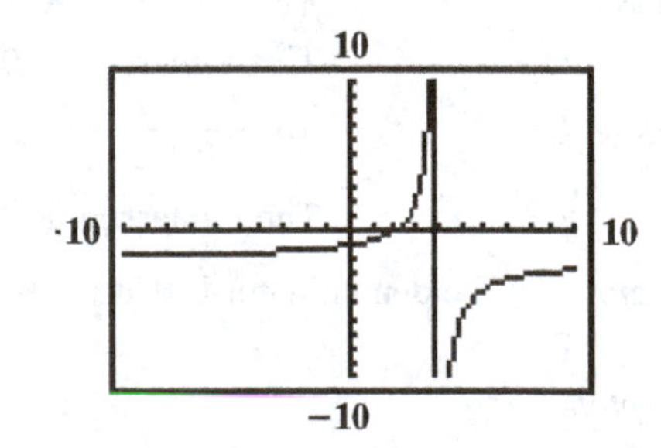

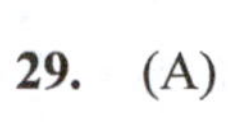

29. (A)

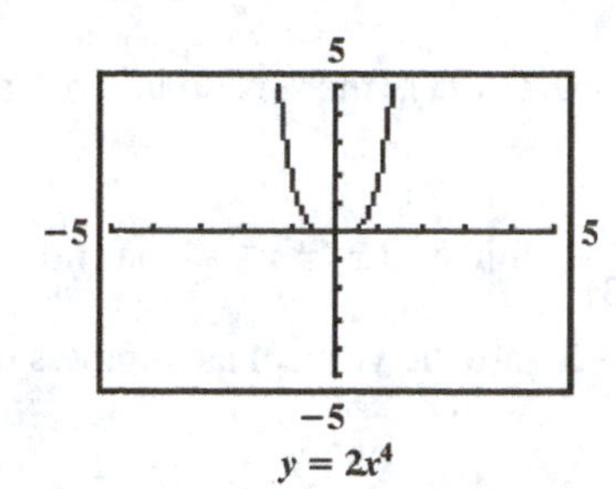

$y = 2x^4$

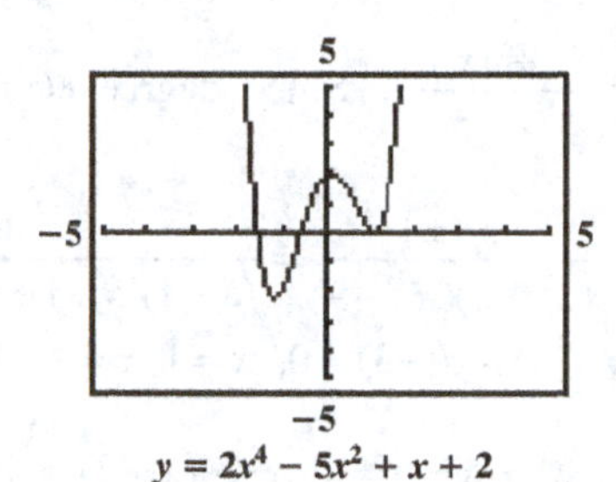

$y = 2x^4 - 5x^2 + x + 2$

(B)

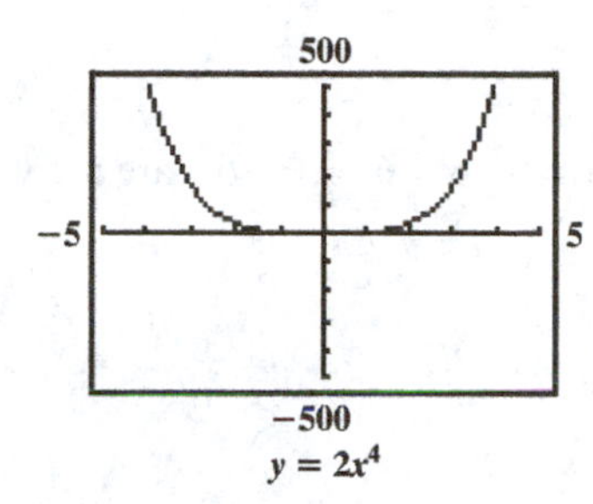

$y = 2x^4$

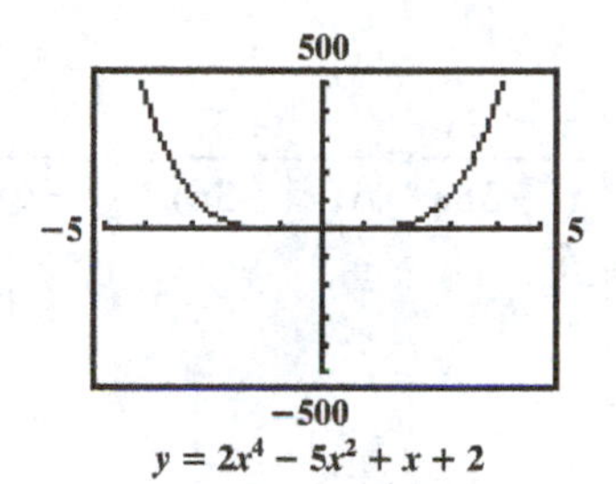

$y = 2x^4 - 5x^2 + x + 2$

31. (A)

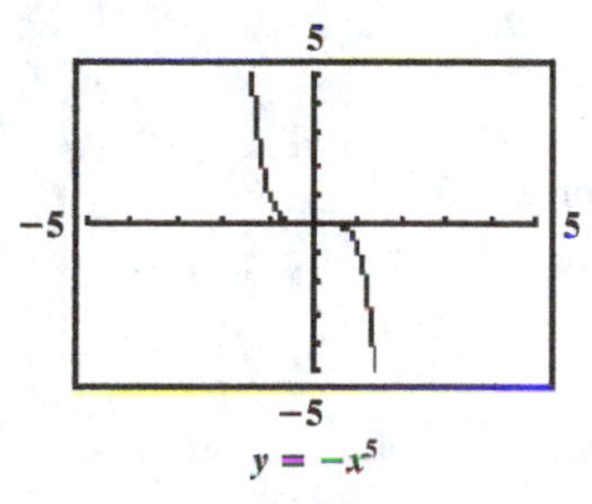

$y = -x^5$

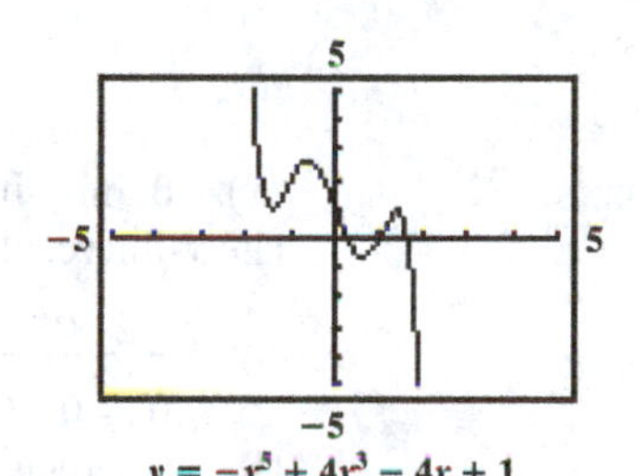

$y = -x^5 + 4x^3 - 4x + 1$

(B)

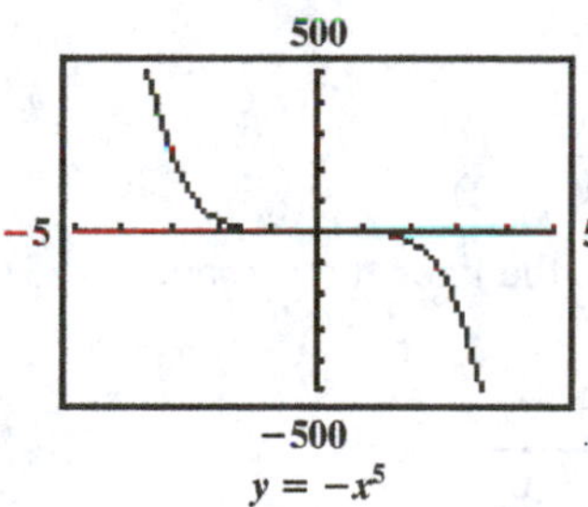

$y = -x^5$

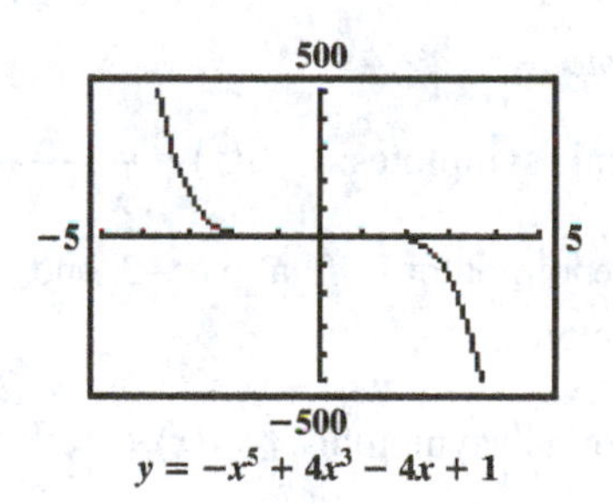

$y = -x^5 + 4x^3 - 4x + 1$

33. $f(x) = \frac{n(x)}{d(x)} = \frac{5x^3 + 2x - 3}{6x^3 - 7x + 1}$. Since degree $n(x) = 3 =$ degree $d(x)$, $y = \frac{5}{6}$ is the horizontal asymptote.

35. $f(x) = \frac{n(x)}{d(x)} = \frac{1 - 5x + x^2}{2 + 3x + 4x^2}$. Since degree $n(x) = 2 =$ degree $d(x)$, $y = \frac{1}{4}$ is the horizontal asymptote.

37. $f(x) = \frac{n(x)}{d(x)} = \frac{x^4 + 2x^2 + 1}{1 - x^5}$. Since degree $n(x) = 4 < 5 =$ degree $d(x)$, $y = 0$ is the horizontal asymptote.

39. $f(x)=\frac{n(x)}{d(x)}=\frac{x^2+6x+1}{x-5}$. Since degree $n(x)=2>1=$ degree $d(x)$, there is no horizontal asymptote.

41. $f(x)=\frac{n(x)}{d(x)}=\frac{x^2+1}{(x^2-1)(x^2-9)}=\frac{x^2+1}{(x-1)(x+1)(x-3)(x+3)}$. Since $n(x)=x^2+1$ has no real zeros and $d(1)=d(-1)=d(3)=d(-3)=0$, $x=1$, $x=-1$, $x=3$, $x=-3$ are the vertical asymptotes of the graph of *f*.

43. $f(x)=\frac{n(x)}{d(x)}=\frac{x^2-x-6}{x^2-3x-10}=\frac{(x-3)(x+2)}{(x-5)(x+2)}=\frac{x-3}{x-5}$, $x\neq-2$. $x=5$ is a vertical asymptote of the graph of *f*.

45. $f(x)=\frac{n(x)}{d(x)}=\frac{x^2+3x}{x^3-36x}=\frac{x(x+3)}{x(x^2-36)}=\frac{x+3}{(x-6)(x+6)}$, $x\neq0$. $x=6$, $x=-6$ are the vertical asymptotes of the graph of *f*.

47. $f(x)=\frac{2x^2}{x^2-x-6}$

(A) *Intercepts:*

x-intercepts: $f(x)=0$ only if $2x^2=0$ or $x=0$. The *x*-intercept is 0.

y-intercept: $f(0)=\frac{2\cdot0^2}{0^2-0-6}=0$ The *y*-intercept is 0.

(B) *Asymptotes:*

Vertical asymptotes: $f(x)=\frac{2x^2}{x^2-x-6}=\frac{2x^2}{(x-3)(x+2)}$

The denominator is 0 at $x=-2$ and $x=3$. Thus, the lines $x=-2$ and $x=3$ are vertical asymptotes.

Horizontal asymptotes: $f(x)=\frac{2x^2}{x^2-x-6}=\frac{2}{1-\frac{1}{x}-\frac{6}{x^2}}$

As *x* increases or decreases without bound, the numerator is 2 and the denominator tends to 1. Therefore, the line $y=2$ is a horizontal asymptote.

(C)

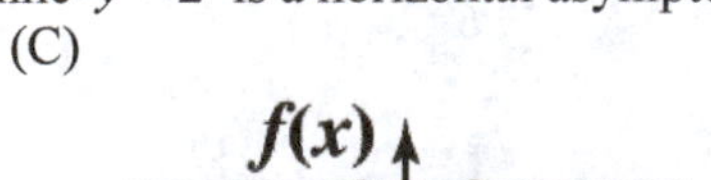

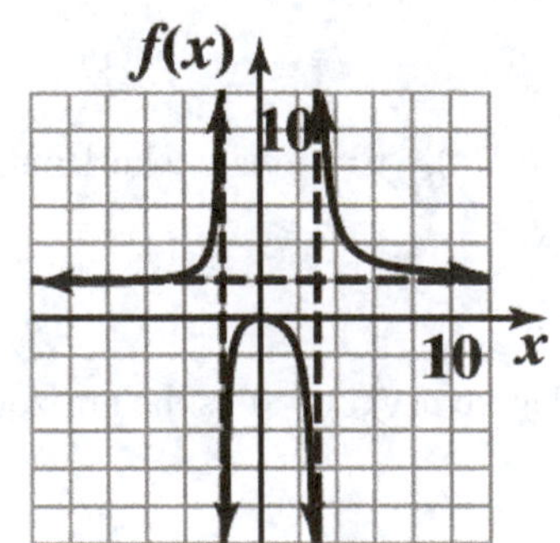

(D)

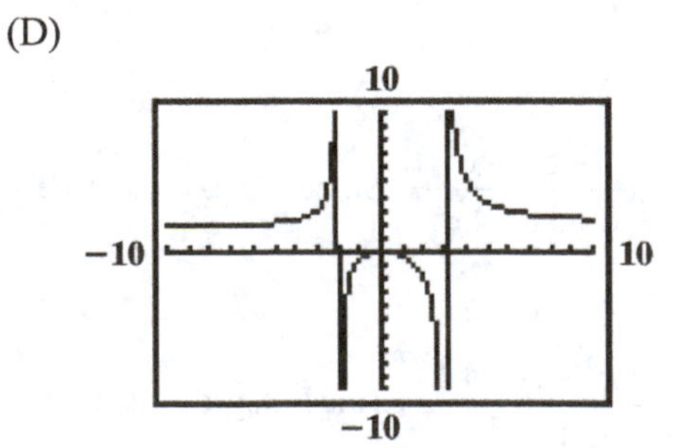

49. $f(x) = \frac{6 - 2x^2}{x^2 - 9}$

(A) *Intercepts:*

x-intercepts: $f(x) = 0$ only if $6 - 2x^2 = 0$

$$2x^2 = 6$$
$$x^2 = 3$$
$$x = \pm\sqrt{3}$$

The x-intercepts are $\pm\sqrt{3}$.

y-intercept: $f(0) = \frac{6 - 2 \cdot 0^2}{0^2 - 9} = -\frac{2}{3}$

The y-intercept is $-\frac{2}{3}$.

(B) *Asymptotes:*

Vertical asymptotes: $f(x) = \frac{6 - 2x^2}{x^2 - 9} = \frac{6 - 2x^2}{(x-3)(x+3)}$

The denominator is 0 at $x = -3$ and $x = 3$. Thus, the lines $x = -3$ and $x = 3$ are vertical asymptotes.

Horizontal asymptotes: $f(x) = \frac{6 - 2x^2}{x^2 - 9} = \frac{\frac{6}{x^2} - 2}{1 - \frac{9}{x^2}}$

As x increases or decreases without bound, the numerator tends to -2 and the denominator tends to 1. Therefore, the line $y = -2$ is a horizontal asymptote.

(C)

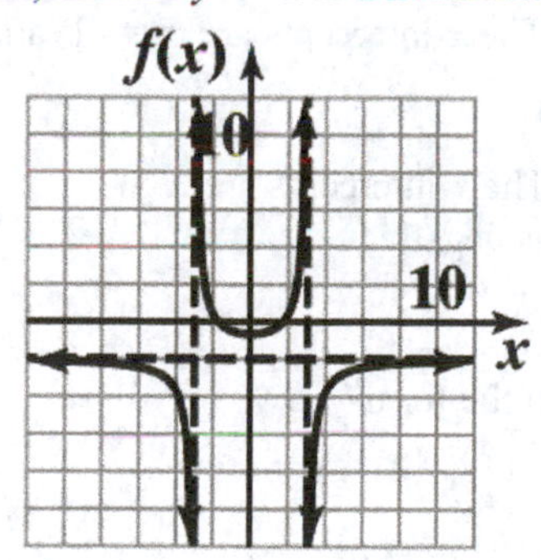

(D)

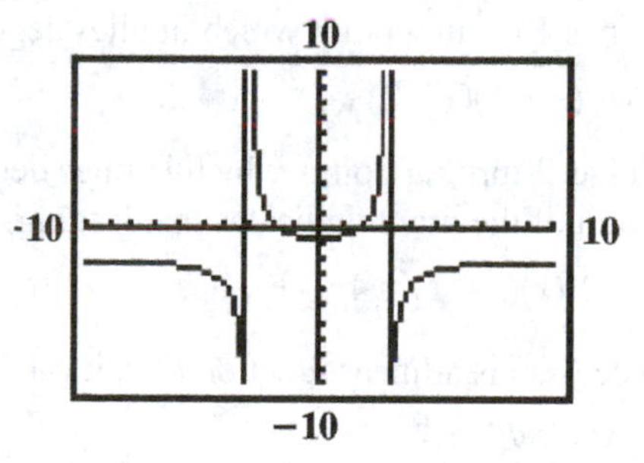

51. $f(x) = \frac{-4x + 24}{x^2 + x - 6}$

(A) *Intercepts:*

x-intercepts: $f(x) = 0$ only if $-4x + 24 = 0$ or $x = 6$.

The x-intercept is 6.

y-intercept: $f(0) = \frac{-4(0) + 24}{0^2 + 0 - 6} = -4$

The y-intercept is -4.

(B) *Asymptotes:*

Vertical asymptotes: $f(x) = \dfrac{-4x+24}{x^2+x-6} = \dfrac{-4x+24}{(x+3)(x-2)}$

The denominator is 0 at $x = -3$ and $x = 2$. Thus, the lines $x = -3$ and $x = 2$ are vertical asymptotes.

Horizontal asymptotes: $f(x) = \dfrac{-4x+24}{x^2+x-6} = \dfrac{-\frac{4}{x}+\frac{24}{x^2}}{1+\frac{1}{x}+\frac{6}{x^2}}$

As x increases or decreases without bound, the numerator tends to 0 and the denominator tends to 1. Therefore, the line $y = 0$ (the x axis) is a horizontal asymptote.

(C)

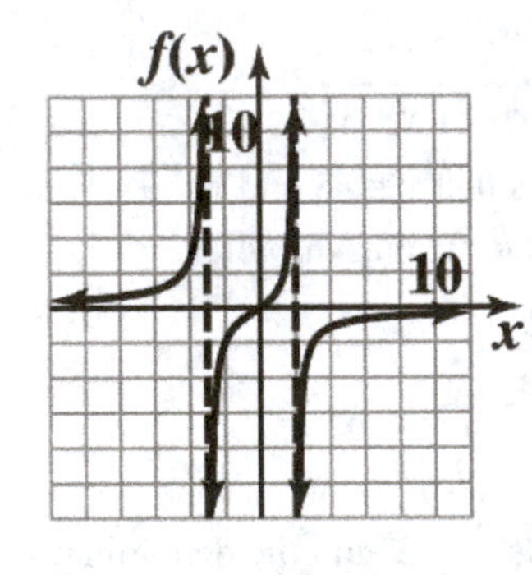

(D)

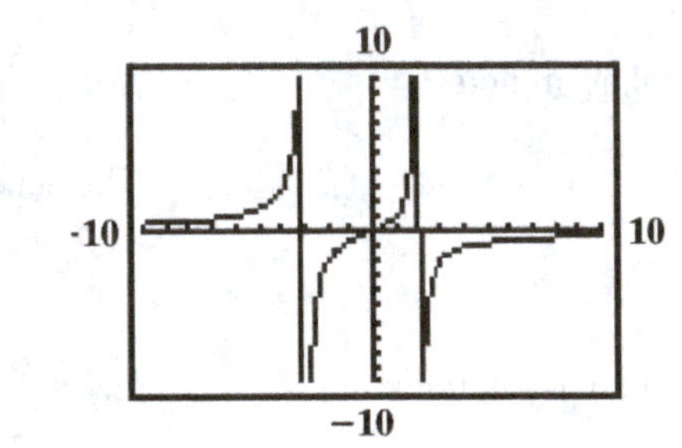

53. The graph has 1 turning point which implies degree $n = 2$. The x-intercepts are $x = -1$ and $x = 2$. Thus, $f(x) = (x+1)(x-2) = x^2 - x - 2$.

55. The graph has 2 turning points which implies degree $n = 3$. The x-intercepts are $x = -2$, $x = 0$, and $x = 2$. The direction of the graph indicates that leading coefficient is negative $f(x) = -(x+2)(x)(x-2) = 4x - x^3$.

57. (A) Since $C(x)$ is a linear function of x, it can be written in the form

$C(x) = mx + b$

Since the fixed costs are \$200, $b = 200$.

Also, $C(20) = 3800$, so

$3800 = m(20) + 200$

$20m = 3600$

$m = 180$

Therefore, $C(x) = 180x + 200$

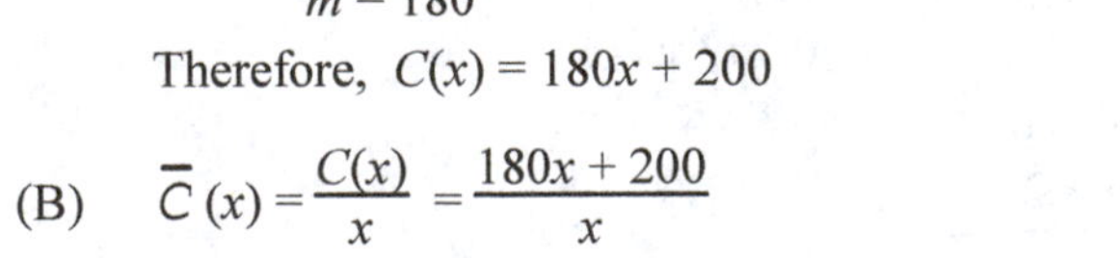

(B) $\bar{C}(x) = \dfrac{C(x)}{x} = \dfrac{180x+200}{x}$

(C)

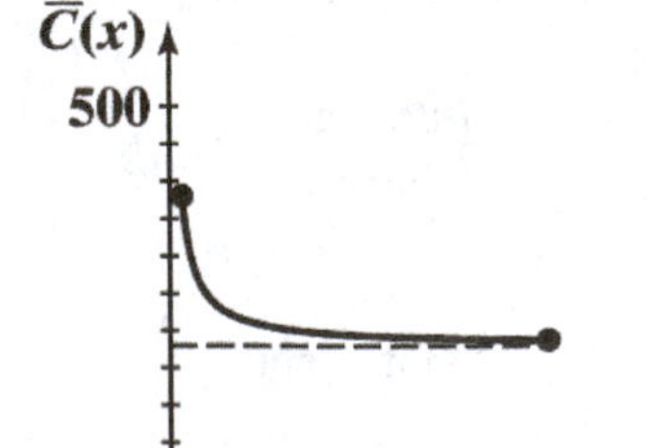

(D) $\bar{C}(x) = \dfrac{180x+200}{x} = \dfrac{180+\frac{200}{x}}{1}$

As x increases, the numerator tends to 180 and the denominator is 1. Therefore, $\bar{C}(x)$ tends to 180 or \$180 per board.

59. (A) $\bar{C}(n) = \dfrac{2500 + 175n + 25n^2}{n}$

(B)

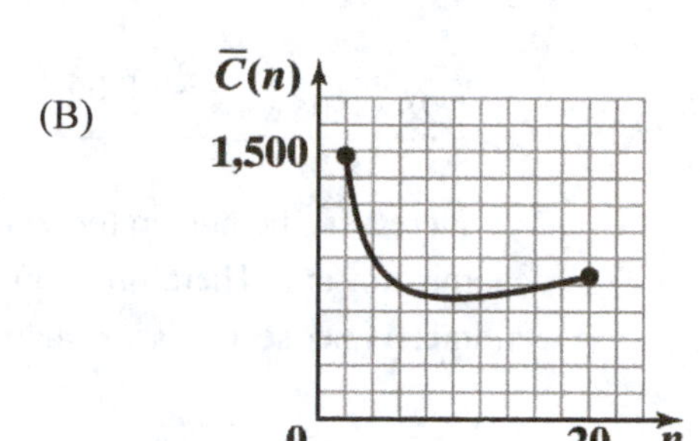

(C) Using the graph, we calculate

$$\bar{C}(8) = \frac{2500 + 175(8) + 25(8)^2}{8} = 687.50$$

$$\bar{C}(9) = \frac{2500 + 175(9) + 25(9)^2}{9} = 677.78$$

$$\bar{C}(10) = \frac{2500 + 175(10) + 25(10)^2}{10} = 675.00$$

$$\bar{C}(11) = \frac{2500 + 175(11) + 25(11)^2}{11} = 677.27$$

$$\bar{C}(12) = \frac{2500 + 175(12) + 25(12)^2}{12} = 683.33$$

Thus, it appears that the average cost per year is a minimum at $n = 10$ years; at 10 years, the average minimum cost is $675.00 per year.

(D) 10 years; $675.00 per year

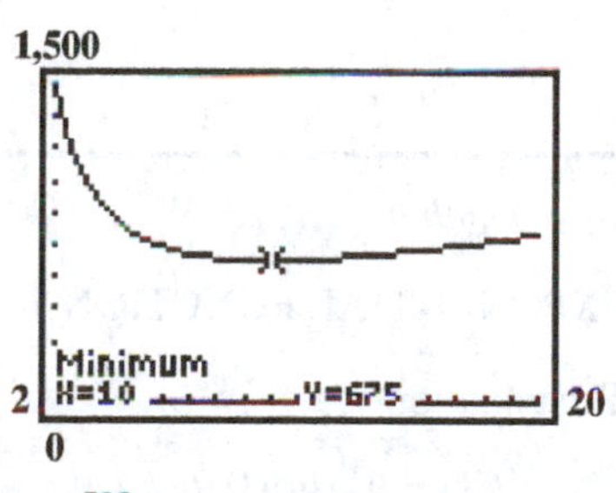

61. (A) $\bar{C}(x) = \dfrac{0.00048(x - 500)^3 + 60,000}{x}$

(B)

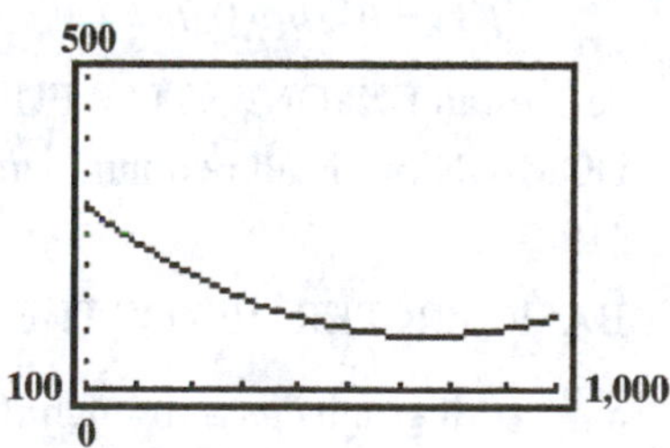

(C) The caseload which yields the minimum average cost per case is 750 cases per month. At 750 cases per month, the average cost per case is $90.

63. (A) Cubic regression model

```
CubicReg
 y=ax³+bx²+cx+d
 a=⁻2.666667E⁻4
 b=.0096666667
 c=⁻.2011904762
 d=17.84761905
```

(B) Per capita consumption of ice cream in 2025: $y(45) \approx 4.1$ lbs.

65. (A) $v(x) = \dfrac{26 + 0.06x}{x} = \dfrac{\frac{26}{x} + 0.06}{1}$

As x increases, the numerator tends to 0.06 and the denominator is 1. Therefore, $v(x)$ approaches 0.06 centimeters per second as x increases.

(B)

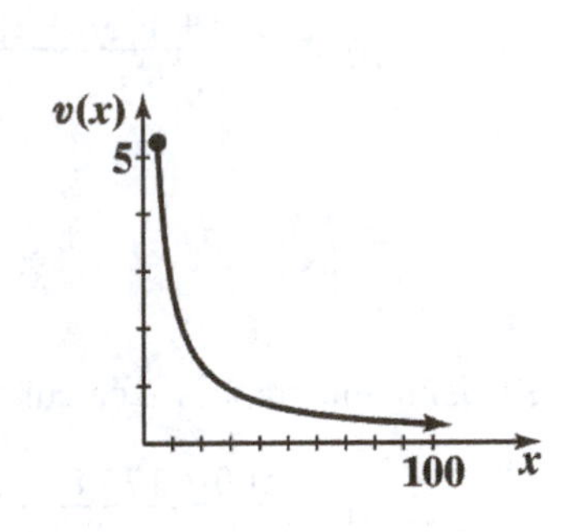

67. (A) Cubic regression model

```
CubicReg
 y=ax³+bx²+cx+d
 a=8.7037037E-5
 b=-.0108492063
 c=.2907407407
 d=8.546031746
```

(B) The marriage rate for 2025 will be 5.5.

EXERCISE 1-5

Things to remember:

1. EXPONENTIAL FUNCTION

The equation

$f(x) = b^x, b > 0, b \neq 1$

defines an EXPONENTIAL FUNCTION for each different constant b, called the BASE. The DOMAIN of f is all real numbers, and the RANGE of f is the set of positive real numbers.

2. BASIC PROPERTIES OF THE GRAPH OF $f(x) = b^x, b > 0, b \neq 1$

a. All graphs pass through (0,1); $b^0 = 1$ for any base b.

b. All graphs are continuous curves; there are no holes or jumps.

c. The x-axis is a horizontal asymptote.

d. If $b > 1$, then b^x increases as x increases.

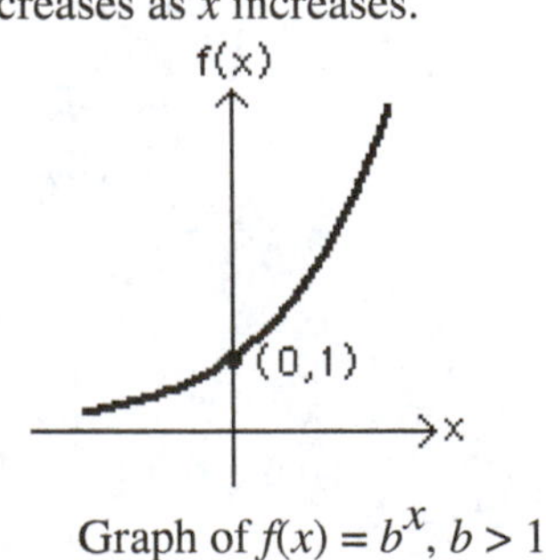

Graph of $f(x) = b^x, b > 1$

e. If $0 < b < 1$, then b^x decreases as x increases.

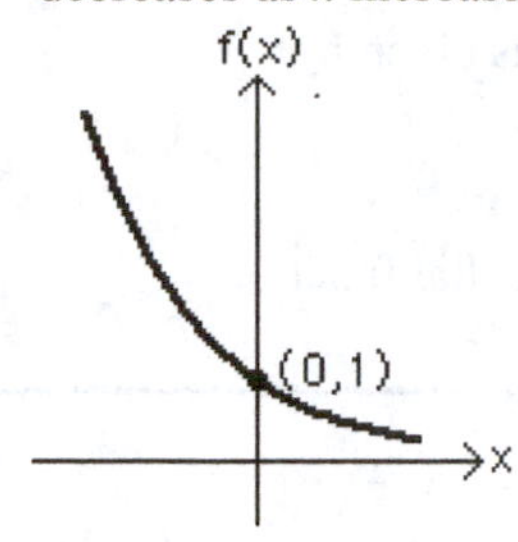

Graph of $f(x) = b^x$, $0 < b < 1$

3. PROPERTIES OF EXPONENTIAL FUNCTIONS

For $a, b > 0$, $a \neq 1$, $b \neq 1$, and x, y real numbers:

a. EXPONENT LAWS

(i) $a^x a^y = a^{x+y}$

(ii) $\dfrac{a^x}{a^y} = a^{x-y}$

(iii) $(a^x)^y = a^{xy}$

(iv) $(ab)^x = a^x b^x$

(v) $\left(\dfrac{a}{b}\right)^x = \dfrac{a^x}{b^x}$

b. $a^x = a^y$ if and only if $x = y$.

c. For $x \neq 0$, $a^x = b^x$ if and only if $a = b$.

4. EXPONENTIAL FUNCTION WITH BASE $e = 2.71828\ldots$

Exponential functions with base e and base $1/e$ are respectively defined by $y = e^x$ and $y = e^{-x}$.

Domain: $(-\infty, \infty)$

Range: $(0, \infty)$

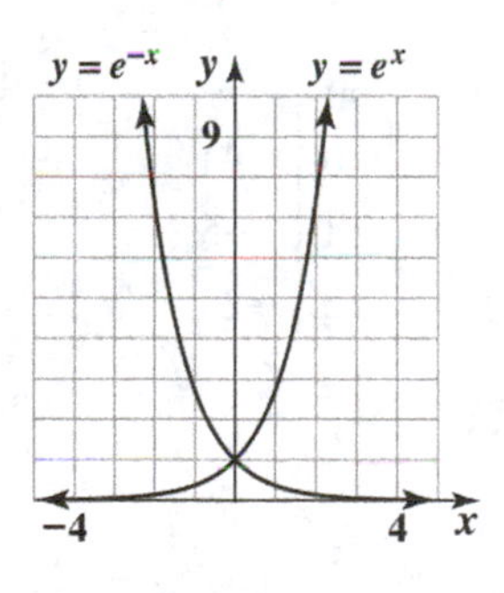

5. Functions of the form $y = ce^{kt}$, where c and k are constants and the independent variable t represents time, are often used to model population growth and radioactive decay. Since $y(0) = c$, c represents the initial population or initial amount. The constant k represents the growth or decay rate; $k > 0$ in the case of population growth, $k < 0$ in the case of radioactive decay.

6. COMPOUND INTEREST

If a principal P (present value) is invested at an annual rate r (expressed as a decimal) compounded m times per year, then the amount A (future value) in the account at the end of t years is given by:

$$A = P\left(1 + \frac{r}{m}\right)^{mt}.$$

If a principal P is invested at an annual rate r (expressed as a decimal) compounded continuously, then the amount in the account at the end of t years is given by

$$A = Pe^{rt}$$

where $e \approx 2.71828$ is the base of the exponential function.

---

1. (A) k (B) g (C) h (D) f

3. $y = 5^x$, $-2 \le x \le 2$

x	y
-2	$\frac{1}{25}$
-1	$\frac{1}{5}$
0	1
1	5
2	25

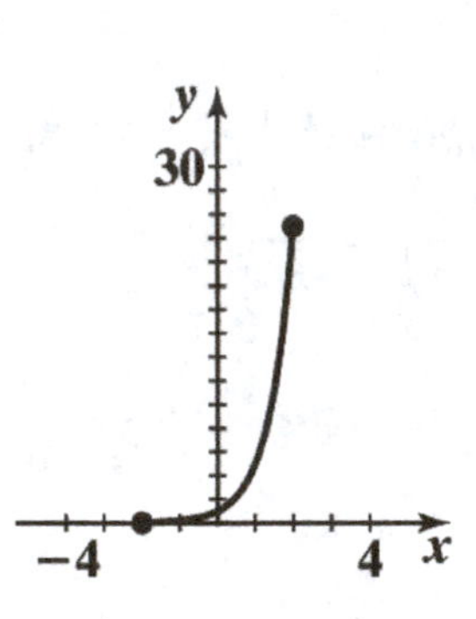

5. $y = \left(\frac{1}{5}\right)^x = 5^{-x}$, $-2 \le x \le 2$

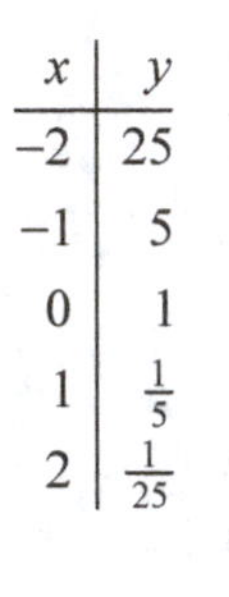

x	y
-2	25
-1	5
0	1
1	$\frac{1}{5}$
2	$\frac{1}{25}$

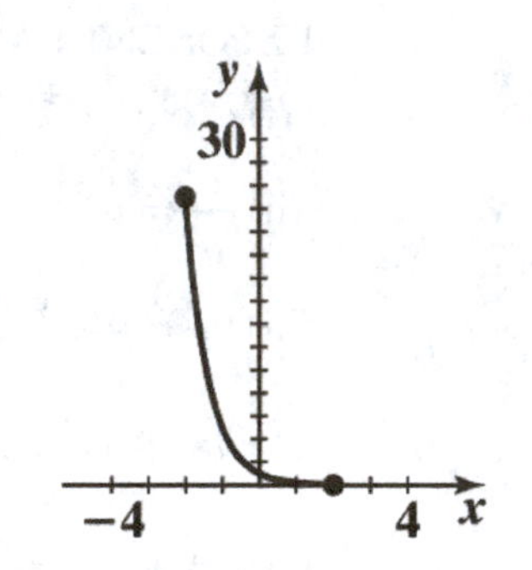

7. $f(x) = -5^x$, $-2 \le x \le 2$

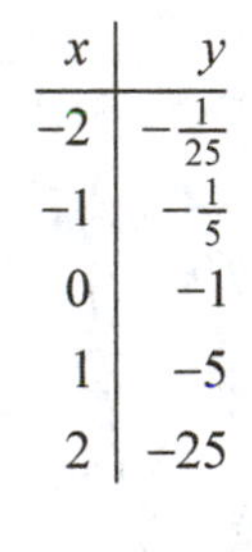

x	y
-2	$-\frac{1}{25}$
-1	$-\frac{1}{5}$
0	-1
1	-5
2	-25

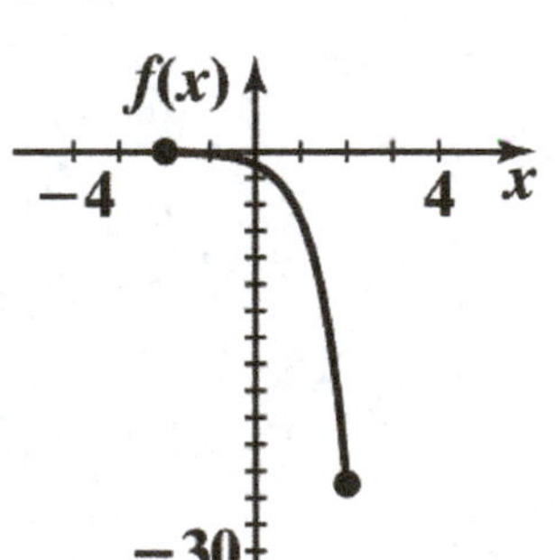

9. $y = -e^{-x}$, $-3 \le x \le 3$

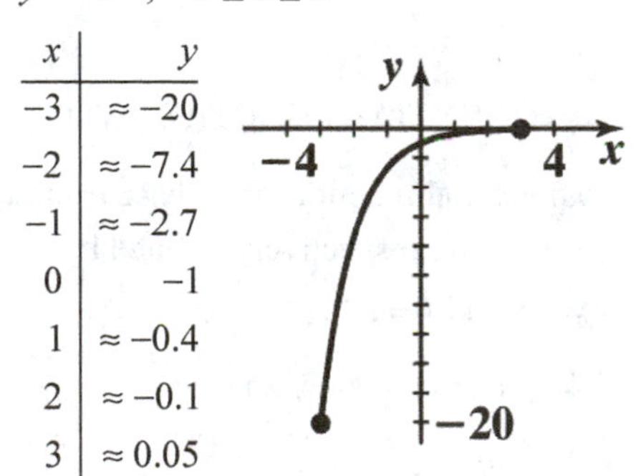

x	y
-3	≈ -20
-2	≈ -7.4
-1	≈ -2.7
0	-1
1	≈ -0.4
2	≈ -0.1
3	≈ 0.05

11. $g(x) = -f(x)$; the graph of g is the graph of f reflected in the x axis.

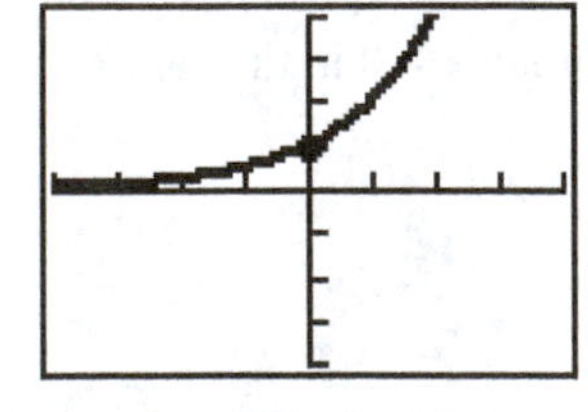

13. $g(x) = f(x + 1)$; the graph of g is the graph of f shifted one unit to the left.

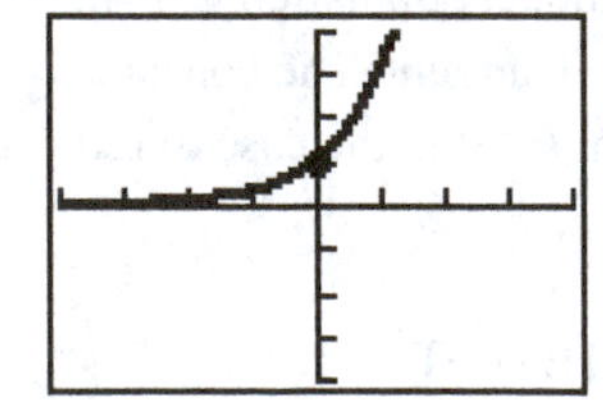

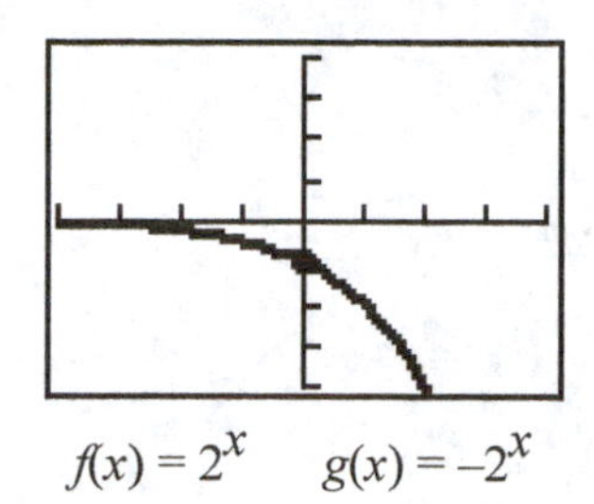

$f(x) = 2^x \quad g(x) = -2^x$

$f(x) = 3^x \quad g(x) = 3^{x+1}$

15. $g(x) = f(x) + 1$; the graph of g is the graph of f shifted one unit up.

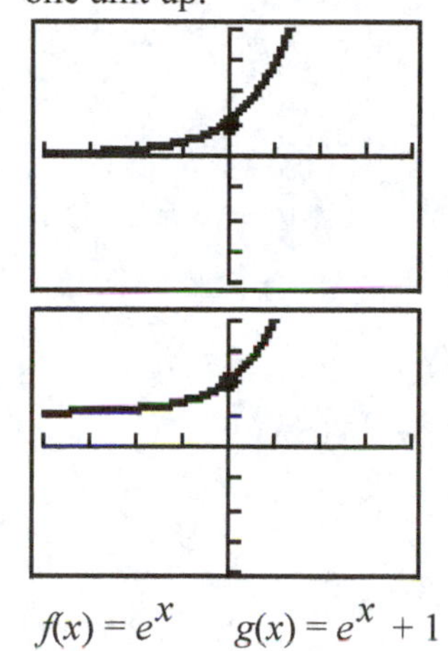

$f(x) = e^x \quad g(x) = e^x + 1$

17. $g(x) = 2f(x + 2)$; the graph of g is the graph of f vertically expanded by a factor of 2 and shifted to the left 2 units.

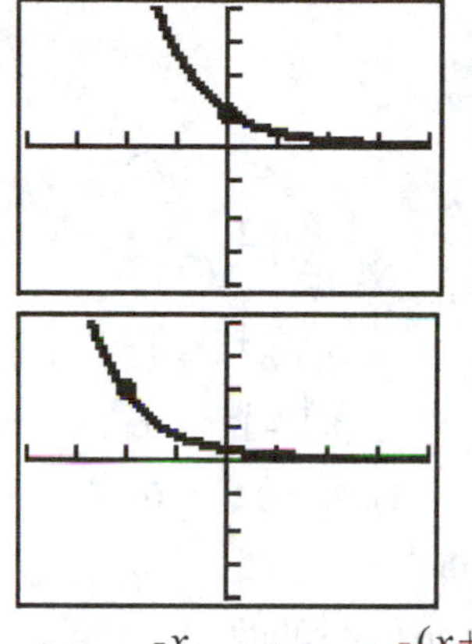

$f(x) = e^{-x} \quad g(x) = 2e^{-(x+2)}$

19. (A) $y = f(x) - 1$

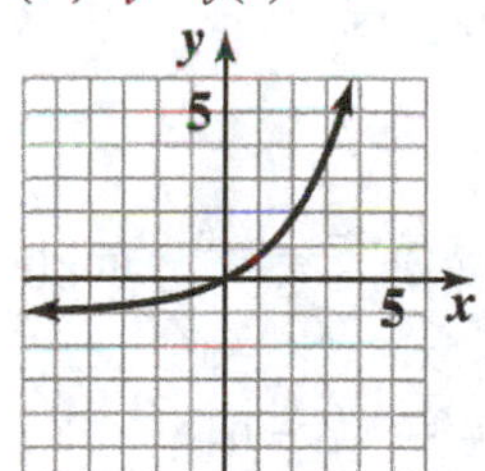

(B) $y = f(x + 2)$

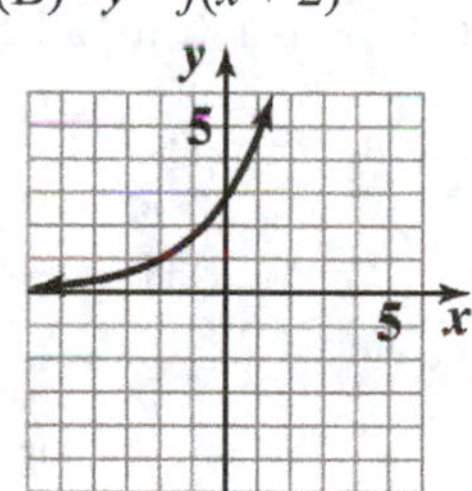

(C) $y = 3f(x) - 2$

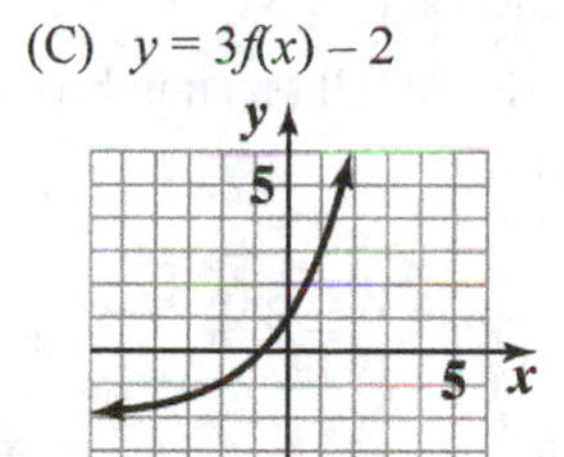

(D) $y = 2 - f(x - 3)$

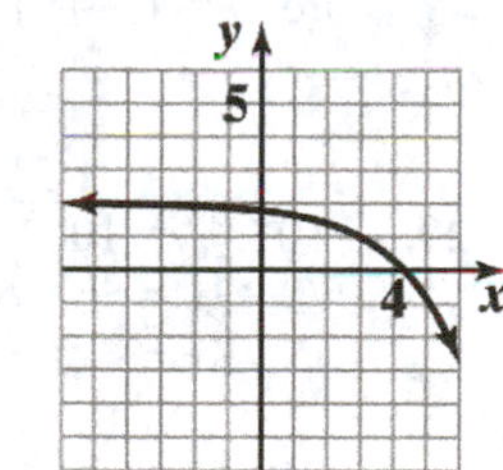

21. $f(t) = 2^{t/10}, \ -30 \le t \le 30$

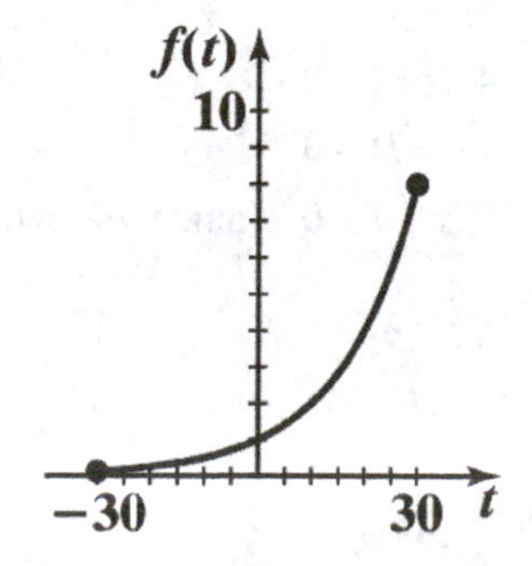

23. $y = -3 + e^{1+x}, \ -4 \le x \le 2$

x	y
−4	≈ −3
−2	≈ −2.6
−1	−2
0	≈ −0.3
1	≈ 4.4
2	≈ 17.1

25. $y = e^{|x|}$, $-3 \le x \le 3$

x	y
-3	≈ 20.1
-1	≈ 2.7
0	1
1	≈ 2.7
3	≈ 20.1

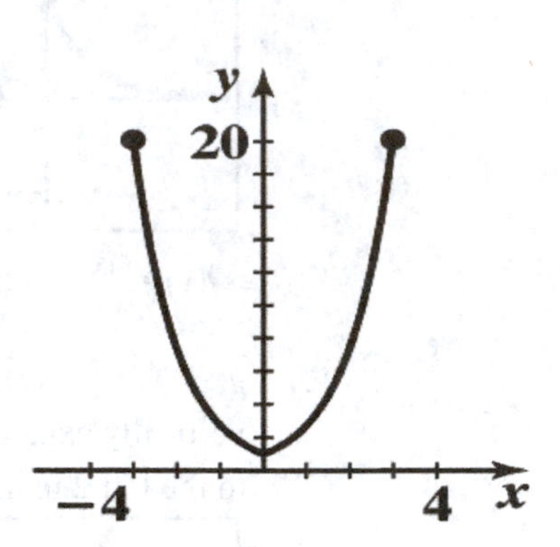

27. Solve

$$a^2 = a^{-2}$$
$$a^2 = \frac{1}{a^2}$$
$$a^4 = 1$$
$$a^4 - 1 = 0$$
$$(a^2 - 1)(a^2 + 1) = 0$$

$a^2 - 1 = 0$ implies $a = 1, -1$

$a^2 + 1 = 0$ has no real solutions

The exponential function property: $a^x = a^y$ if and only if $x = y$ assumes $a > 0$ and $a \neq 1$. Our solutions are $a = 1, -1$; $1^x = 1^y$ for all real numbers x, y, $(-1)^x = (-1)^y$ for all even integers.

29. $10^{2-3x} = 10^{5x-6}$ implies (see 3b)

$$2 - 3x = 5x - 6$$
$$-8x = -8$$
$$x = 1$$

31. $4^{5x-x^2} = 4^{-6}$ implies

$$5x - x^2 = -6$$

or $-x^2 + 5x + 6 = 0$

$$x^2 - 5x - 6 = 0$$
$$(x - 6)(x + 1) = 0$$
$$x = 6, -1$$

33. $5^3 = (x + 2)^3$ implies (by property 3c)

$$5 = x + 2$$

Thus, $x = 3$.

35. $xe^{-x} + 7e^{-x} = 0$

$$e^{-x}(x+7) = 0$$
$$x + 7 = 0 \quad (\text{since } e^{-x} \neq 0)$$
$$x = -7$$

37. $2x^2e^x - 8e^x = 0$

$$2e^x(x^2 - 4) = 0$$
$$x^2 - 4 = 0 \quad (\text{since } e^x \neq 0)$$
$$x = -2, 2$$

39. $e^{4x} - e = 0$

$$e^{4x} = e^1$$
$$4x = 1$$
$$x = \tfrac{1}{4}$$

41. $e^{3x-1}+e>0$ for all x. Therefore, $e^{3x-1}+e=0$ has no solutions.

43. $h(x)=x2^x,\ -5\le x\le 0$

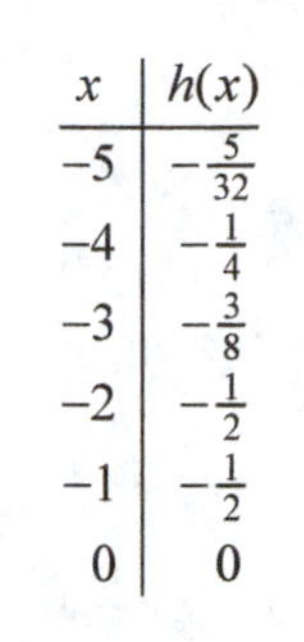

x	$h(x)$
-5	$-\frac{5}{32}$
-4	$-\frac{1}{4}$
-3	$-\frac{3}{8}$
-2	$-\frac{1}{2}$
-1	$-\frac{1}{2}$
0	0

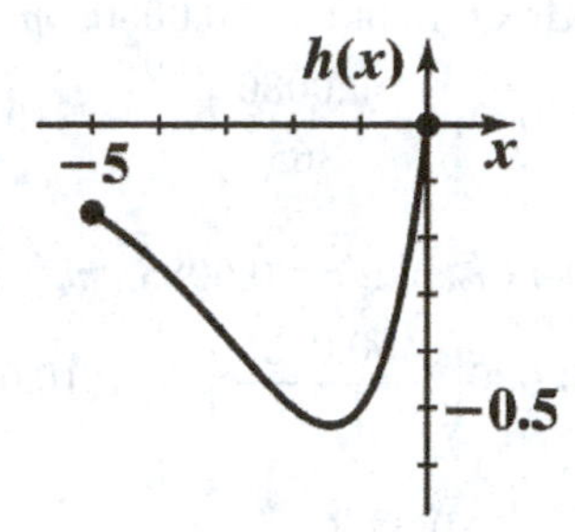

45. $N=\dfrac{100}{1+e^{-t}},\ 0\le t\le 5$

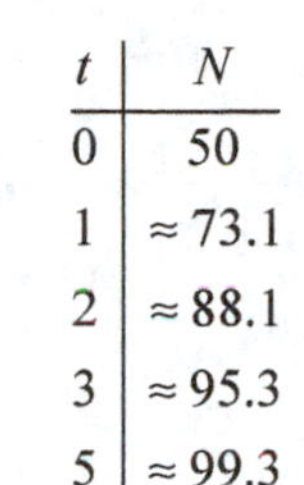

t	N
0	50
1	≈ 73.1
2	≈ 88.1
3	≈ 95.3
5	≈ 99.3

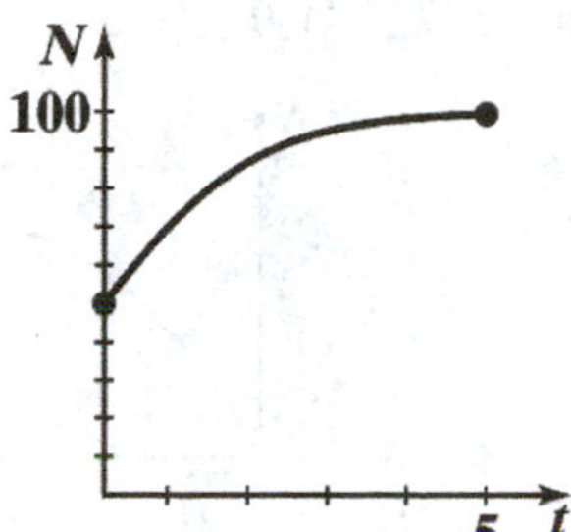

47. Use $A=Pe^{rt}$, $P=10{,}000$, $r=0.0395$, and $t=12$.

$A=10{,}000e^{0.0395(12)}\approx 10{,}000e^{0.474}\approx \$16{,}064.07$.

49. $A=P\left(1+\dfrac{r}{m}\right)^{mt}$, we have:

(A) $P=2{,}500,\ r=0.07,\ m=4,\ t=\dfrac{3}{4}$

$$A=2{,}500\left(1+\frac{0.07}{4}\right)^{4\cdot 3/4}=2{,}500(1+0.0175)^3=2{,}633.56$$

Thus, $A=\$2{,}633.56$.

(B) $$A=2{,}500\left(1+\frac{0.07}{4}\right)^{4\cdot 15}=2{,}500(1+0.0175)^{60}=7{,}079.54$$

Thus, $A=\$7{,}079.54$.

51. $A=P\left(1+\dfrac{r}{m}\right)^{mt}$. With $A=15{,}000$, $r=0.0675$, and $m=52$, we have:

$$15{,}000=P\left(1+\frac{0.0675}{52}\right)^{52(5)}=P(1.0411).$$

Therefore, $P=\dfrac{15{,}000}{1.4011}\approx \$10{,}706$.

53. $A=P\left(1+\dfrac{r}{m}\right)^{mt}$. $P=10{,}000$, $t=1$.

(A) Stonebridge Bank: $r=0.0095$, $m=12$

$$A = 10{,}000\left(1+\frac{0.0095}{12}\right)^{12(1)} = \$10{,}095.41$$

(B) Deep Green Bank: $r = 0.0080,\ m = 365$

$$A = 10{,}000\left(1+\frac{0.0080}{365}\right)^{365(1)} = \$10{,}080.32$$

(C) Provident Bank: $r = 0.0085,\ m = 4$

$$A = 10{,}000\left(1+\frac{0.0085}{4}\right)^{4(1)} = \$10{,}085.27$$

55. Given $N = 2(1 - e^{-0.037t}),\ 0 \le t \le 50$

t	N
0	0
10	≈ 0.62
30	≈ 1.34
50	≈ 1.69

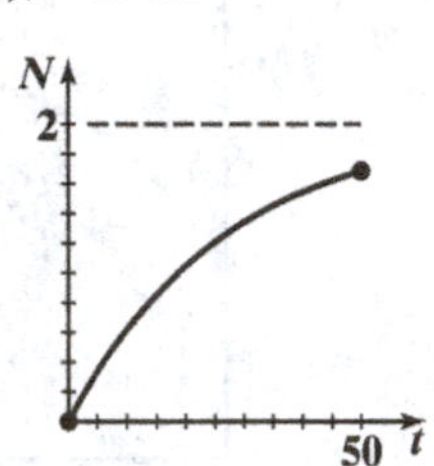

N approaches 2 as t increases without bound.

57. (A) Exponential regression model

```
ExpReg
 y=a*b^x
 a=732.7877195
 b=1.084347949
```

$y = ab^x$; 2022 is year 32;
$y(32) \approx \$9{,}781{,}000$

(B) According to the model, $y(10) \approx \$1{,}647{,}000$.

59. Given $I = I_0 e^{-0.23d}$

(A) $I = I_0 e^{-0.23(10)} = I_0 e^{-2.3} \approx I_0(0.10)$

Thus, about 10% of the surface light will reach a depth of 10 feet.

(B) $I = I_0 e^{-0.23(20)} = I_0 e^{-4.6} \approx I_0(0.010)$

Thus, about 1% of the surface light will reach a depth of 20 feet.

61. (A) Model: $P(t) = 7.1.8e^{0.011t}$

(B) In the year 2025, $t = 12$; $P(12) = 7.1e^{0.011(12)} = 7.1e^{0.132} \approx 8{,}100{,}000{,}000$ (nearest 100 million)
In the year 2035, $t = 22$; $P(22) = 7.1e^{0.011(22)} = 7.1e^{0.242} \approx 9{,}000{,}000{,}000$ (nearest 100 million).

63. (A) Exponential regression model

```
ExpReg
 y=a*b^x
 a=5.729914723
 b=1.375060902
```

2022: $t = 28,\ y(28) \approx 42{,}772{,}000{,}000$

From Problem 61, the world population in 2022 is projected to be 7,8000,000. The model implies that the number of internet hosts will be almost 6 times larger than the estimated world population in 2022.

EXERCISE 1-6

Things to remember:

1. ONE-TO-ONE FUNCTIONS

   A function *f* is said to be ONE-TO-ONE if each range value corresponds to exactly one domain value.

2. INVERSE OF A FUNCTION

   If *f* is a one-to-one function, then the INVERSE of *f* is the function formed by interchanging the independent and dependent variables for *f*. Thus, if (*a*, *b*) is a point on the graph of *f*, then (*b*, *a*) is a point on the graph of the inverse of *f*.

   Note: If *f* is not one-to-one, then *f* DOES NOT HAVE AN INVERSE.

3. LOGARITHMIC FUNCTIONS

   The inverse of an exponential function is called a LOGARITHMIC FUNCTION. For $b > 0$ and $b \neq 1$,

Logarithmic form		Exponential form
$y = \log_b x$	is equivalent to	$x = b^y$

   The LOG TO THE BASE *b* OF *x* is the exponent to which *b* must be raised to obtain *x*. [Remember: A logarithm is an exponent.] The DOMAIN of the logarithmic function is the range of the corresponding exponential function, and the RANGE of the logarithmic function is the domain of the corresponding exponential function. Typical graphs of an exponential function and its inverse, a logarithmic function, for $b > 1$, are shown in the figure below:

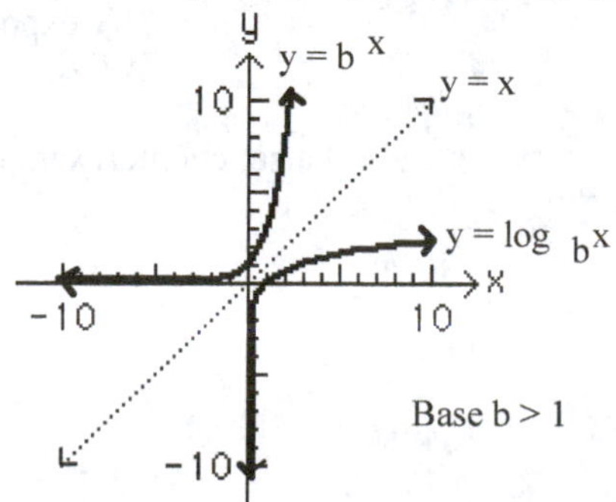

4. PROPERTIES OF LOGARITHMIC FUNCTIONS

If b, M, and N are positive real numbers, $b \neq 1$, and p and x are real numbers, then:

a. $\log_b 1 = 0$

b. $\log_b b = 1$

c. $\log_b b^x = x$

d. $b^{\log_b x} = x, x > 0$

e. $\log_b MN = \log_b M + \log_b N$

f. $\log_b \frac{M}{N} = \log_b M - \log_b N$

g. $\log_b M^p = p \log_b M$

h. $\log_b M = \log_b N$ if and only if $M = N$

5. LOGARITHMIC NOTATION; LOGARITHMIC-EXPONENTIAL RELATIONSHIPS

Common logarithm: $\log x$ means $\log_{10} x$
Natural logarithm: $\ln x$ means $\log_e x$

$\log x = y$ is equivalent to $x = 10^y$
$\ln x = y$ is equivalent to $x = e^y$

---

1. $27 = 3^3$ (using 3)

3. $1 = 10^0$

5. $8 = 4^{3/2}$

7. $\log_7 49 = 2$

9. $\log_4 8 = \frac{3}{2}$

11. $\log_b A = u$

13. $\log_{10} 100 = \log_{10} 10^2 = 2$

15. $\log_2 16 = \log_2 2^4 = 4$

17. $\log_5 \frac{1}{25} = \log_5 5^{-2} = -2$

19. $\ln \frac{1}{e^4} = \ln e^{-4} = -4$
(using 2a)

21. $\log_{10} 1{,}000 = \log_{10} 10^3 = 3$

23. $\log_b L^5 = 5 \log_b L$

25. $3^{p\log_3 q} = 3^{\log_3 q^p} = q^p$ (using 4g and 4d)

27. $\log_3 x = 2$
$x = 3^2$
$x = 9$

29. $\log_7 49 = y$
$\log_7 7^2 = y$
$2 = y$
Thus, $y = 2$.

31. $\log_b 10^{-4} = -4$
$10^{-4} = b^{-4}$
This equality implies $b = 10$ (since the exponents are the same).

33. $\log_4 x = \frac{1}{2}$; $x = 4^{1/2}$; $x = 2$

35. False; counterexample: $y = x^2$.

37. True; if g is the inverse of f, then f is the inverse of g so g must be one-to-one.

39. True; if $y = 2x$, then $x = 2y$ implies $y = \frac{x}{2}$.

41. False; $f(x) = \ln x$ is one-to-one; domain of $f = (0, \infty)$, range of $f = (-\infty, \infty)$.

43. $\log_b x = \frac{2}{3}\log_b 8 + \frac{1}{2}\log_b 9 - \log_b 6 = \log_b 8^{2/3} + \log_b 9^{1/2} - \log_b 6$

$= \log_b 4 + \log_b 3 - \log_b 6 = \log_b \frac{4\cdot 3}{6}$

$\log_b x = \log_b 2$

$x = 2$

45. $\log_b x = \frac{3}{2}\log_b 4 - \frac{2}{3}\log_b 8 + 2\log_b 2 = \log_b 4^{3/2} - \log_b 8^{2/3} + \log_b 2^2$

$= \log_b 8 - \log_b 4 + \log_b 4 = \log_b 8$

$\log_b x = \log_b 8$

$x = 8$

47. $\log_b x + \log_b(x-4) = \log_b 21$

$\log_b x(x-4) = \log_b 21$

Therefore, $x(x-4) = 21$

$x^2 - 4x - 21 = 0$

$(x-7)(x+3) = 0$

Thus, $x = 7$.

[Note: $x = -3$ is not a solution since $\log_b(-3)$ is not defined.]

49. $\log_{10}(x-1) - \log_{10}(x+1) = 1$

$\log_{10}\left(\frac{x-1}{x+1}\right) = 1$

Therefore, $\frac{x-1}{x+1} = 10^1 = 10$

$x - 1 = 10(x+1)$

$x - 1 = 10x + 10$

$-9x = 11$

$x = -\frac{11}{9}$

There is *no solution*, since

$\log_{10}\left(-\frac{11}{9}-1\right) = \log_{10}\left(-\frac{20}{9}\right)$

is not defined. Similarly,

$\log_{10}\left(-\frac{11}{9}+1\right) = \log_{10}\left(-\frac{2}{9}\right)$

is not defined.

51. $y = \log_2(x-2)$

$x - 2 = 2^y$

$x = 2^y + 2$

x	y
$\frac{9}{4}$	-2
$\frac{5}{2}$	-1
3	0
4	1
6	2
18	4

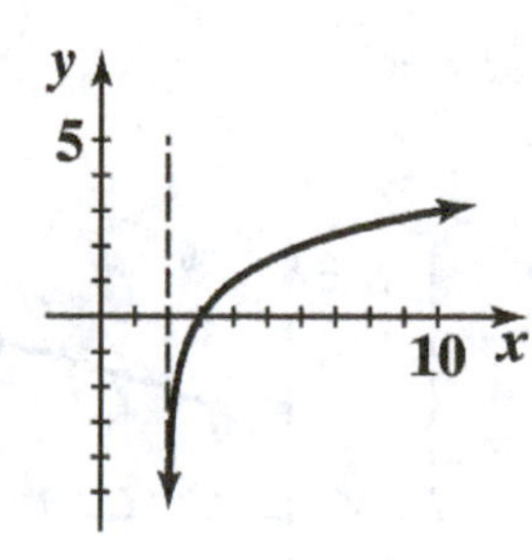

53. The graph of $y = \log_2(x-2)$ is the graph of $y = \log_2 x$ shifted to the right 2 units.

55. Since logarithmic functions are defined only for positive "inputs", we must have $x + 1 > 0$ or $x > -1$; domain: $(-1, \infty)$. The range of $y = 1 + \ln(x+1)$ is the set of all real numbers.

57. (A) 3.54743

(B) −2.16032

59. (A) $\log x = 1.1285$

$x = 13.4431$

(C) 5.62629

(D) –3.19704

(B) $\log x = -2.0497$
$x = 0.0089$

(C) $\ln x = 2.7763$
$x = 16.0595$

(D) $\ln x = -1.8879$
$x = 0.1514$

61. $10^x = 12$ (Take common logarithms of both sides)
$\log 10^x = \log 12 \approx 1.0792$
$x \approx 1.0792$ ($\log 10^x = x \log 10 = x$; $\log 10 = 1$)

63. $e^x = 4.304$ (Take natural logarithms of both sides)
$\ln e^x = \ln 4.304 \approx 1.4595$
$x \approx 1.4595$ ($\ln e^x = x \ln e = x$; $\ln e = 1$)

65. $1.005^{12t} = 3$ (Take either common or natural logarithms of both sides; here we'll use natural logarithms.)
$\ln 1.005^{12t} = \ln 3$

$$12t = \frac{\ln 3}{\ln 1.005} \approx 220.2713$$

$t = 18.3559$

67. $y = \ln x, x > 0$

x	y
0.5	≈ -0.69
1	0
2	≈ 0.69
4	≈ 1.39
5	≈ 1.61

increasing $(0, \infty)$

69. $y = |\ln x|, x > 0$

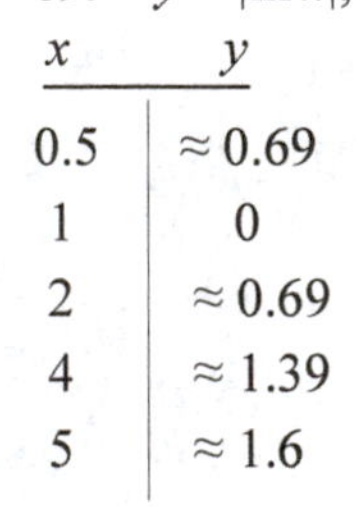

x	y
0.5	≈ 0.69
1	0
2	≈ 0.69
4	≈ 1.39
5	≈ 1.6

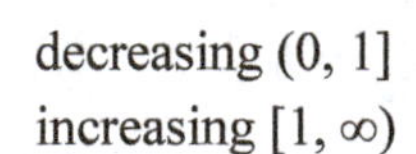

decreasing $(0, 1]$
increasing $[1, \infty)$

71. $y = 2 \ln(x + 2),\ x > -2$

x	y
–1.5	≈ -1.39
–1	0
0	≈ 1.39
1	≈ 2.2
5	≈ 3.89
10	≈ 4.97

increasing $(-2, \infty)$

73. $y = 4 \ln x - 3,\ x > 0$

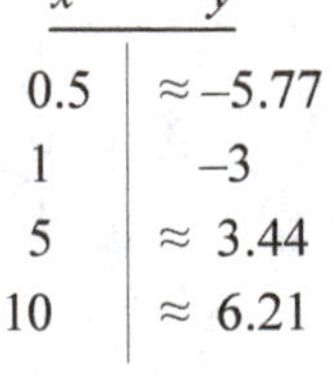

x	y
0.5	≈ -5.77
1	–3
5	≈ 3.44
10	≈ 6.21

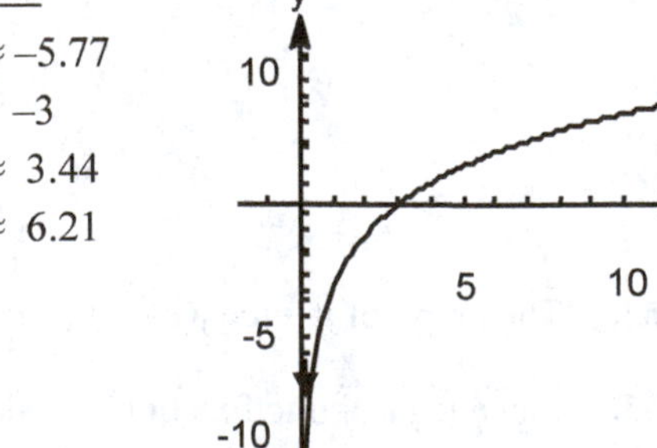

increasing $(0, \infty)$

75. For any number b, $b > 0$, $b \neq 1$, $\log_b 1 = y$ is equivalent to $b^y = 1$ which implies $y = 0$. Thus, $\log_b 1 = 0$ for any permissible base b.

77. 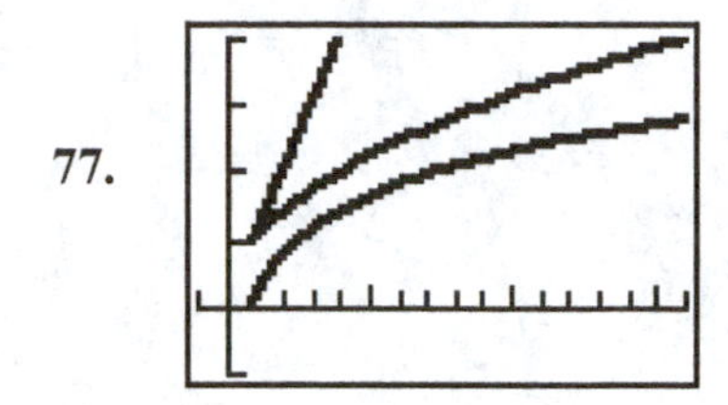

A function f is "larger than" a function g on an interval $[a, b]$ if $f(x) > g(x)$ for $a \le x \le b$.
$r(x) > q(x) > p(x)$ for $1 < x \le 16$, that is $x > \sqrt{x} > \ln x$ for $1 < x \le 16$

79. From the compound interest formula $A = P(1 + r)^t$, we have:
$2P = P(1 + 0.2136)^t$ or $(1.2136)^t = 2$
Take the natural log of both sides of this equation:
$\ln(1.2136)^t = \ln 2$ [Note: the common log could have been used instead of the natural log.]

$t \ln(1.2136) = \ln 2$

$$t = \frac{\ln 2}{\ln 1.2136} \approx \frac{0.69135}{0.19359} = 3.58 \approx 4 \text{ years}$$

81. $A = P\left(1+\frac{r}{m}\right)^{mt}$, $r = 0.06$, $P = 1000$, $A = 1800$.

Quarterly compounding: $m = 4$

$$1800 = 1000\left(1+\frac{0.06}{4}\right)^{4t} = 1000(1.015)^{4t}$$

$$(1.015)^{4t} = \frac{1800}{1000} = 1.8$$

$$4t \ln(1.015) = \ln(1.8)$$

$$t = \frac{\ln(1.8)}{4\ln(1.015)} \approx 9.87$$

\$1000 at 6% compounded quarterly will grow to \$1800 in 9.87 years.
Daily compounding: $m = 365$

$$1800 = 1000\left(1+\frac{0.06}{365}\right)^{365t} = 1000(1.0001644)^{365t}$$

$$(1.0001644)^{365t} = \frac{1800}{1000} = 1.8$$

$$365t \ln(1.0001644) = \ln 1.8$$

$$t = \frac{\ln 1.8}{365\ln(1.0001644)} \approx 9.80$$

\$1000 at 6% compounded daily will grow to
\$1800 in 9.80 years.

83. $A = Pe^{rt}$; $r = 0.0475$, $P = 35{,}000$, $A = 50{,}000$

$$50{,}000 = 35{,}000e^{0.0475t}$$

$$e^{0.0475t} = \frac{50{,}000}{35{,}000} = \frac{10}{7}$$

$$0.0475t = \ln\left(\frac{10}{7}\right)$$

$$t = \frac{\ln(10/7)}{0.0475} \approx 7.51$$

85. (A) Logarithmic regression model,

Table 1:

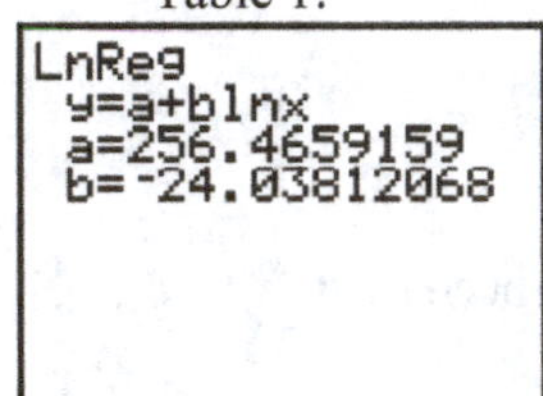
LnReg
y=a+blnx
a=256.4659159
b=-24.03812068

To estimate the demand at a price level of \$50, we solve

$a + b\ln x = 50$

for x. The result is $x \approx 5{,}373$ screwdrivers per month.

(B) Logarithmic regression model,

Table 2:

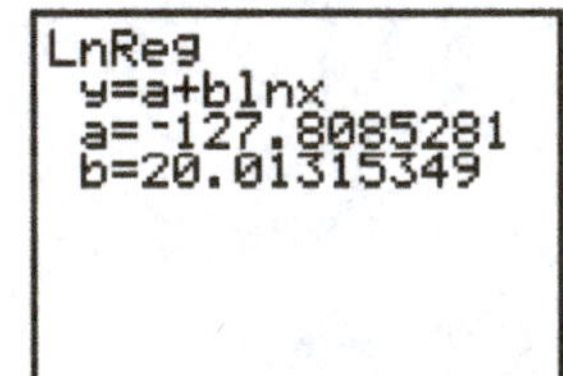
LnReg
y=a+blnx
a=-127.8085281
b=20.01315349

To estimate the supply at a price level of \$50, we solve

$a + b\ln x = 50$

for x. The result is $x \approx 7{,}220$ screwdrivers per month.

(C) The condition is not stable, the price is likely to decrease since the demand at a price level of \$50 is much lower than the supply at this level.

87. $I = I_0 10^{N/10}$

Take the common log of both sides of this equation. Then:

$$\log I = \log(I_0 10^{N/10}) = \log I_0 + \log 10^{N/10}$$

$$= \log I_0 + \frac{N}{10}\log \ 10 = \log I_0 + \frac{N}{10} \text{ (since } \log 10 = 1)$$

So, $\frac{N}{10} = \log I - \log I_0 = \log\left(\frac{I}{I_0}\right)$ and $N = 10\log\left(\frac{I}{I_0}\right)$.

89. Logarithmic regression model

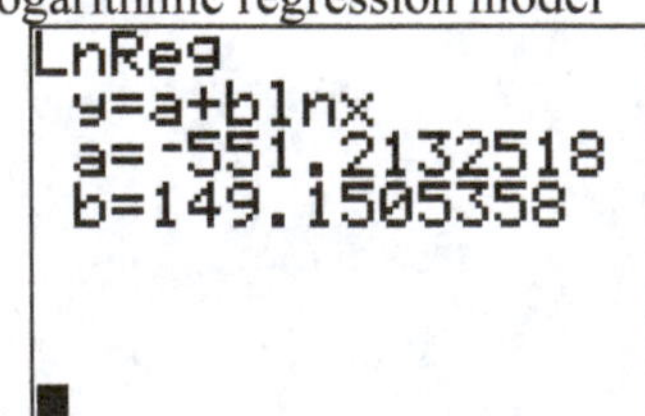
LnReg
y=a+blnx
a=-551.2132518
b=149.1505358

The yield in 2024: $t = 124$, $y(124) \approx 168$ bushels/acre.

91. Assuming that the current population is 7.1 billion and that the growth rate is 1.1% compounded continuously, the population after t years will be

$$P(t) = 7.1e^{0.011t}$$

Given that there are $1.68 \times 10^{14} = 168{,}000$ billion square yards of land, solve

$$7.1e^{0.011t} = 168{,}000$$

for t:

$$e^{0.011t} = \frac{168{,}000}{7.1} \approx 23{,}662$$

$$0.011t = \ln 23{,}662$$

$$t = \frac{\ln 23{,}662}{0.011} \approx 916$$

It will take approximately 916 years.

CHAPTER 1 REVIEW

1.

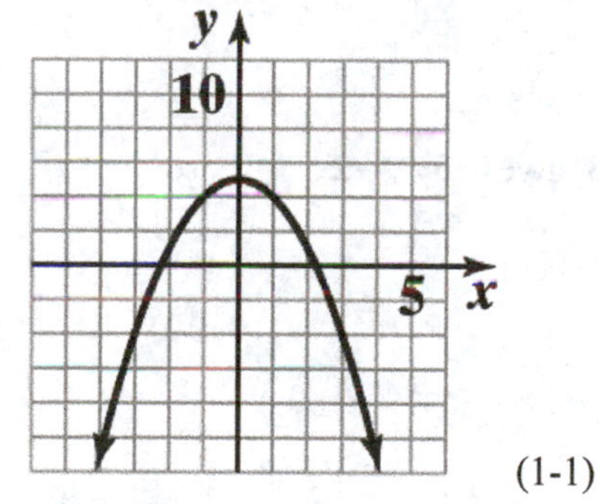

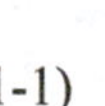

(1-1)

2. $x^2 = y^2$:

x	−3	−2	−1	0	1	2	3
y	±3	±2	±1	0	±1	±2	±3

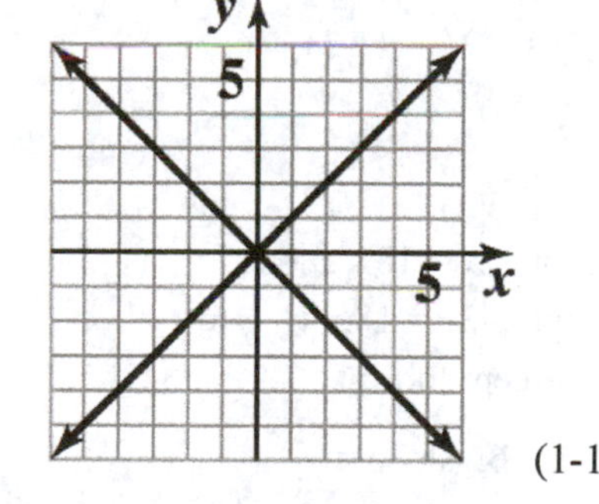

(1-1)

3. $y^2 = 4x^2$:

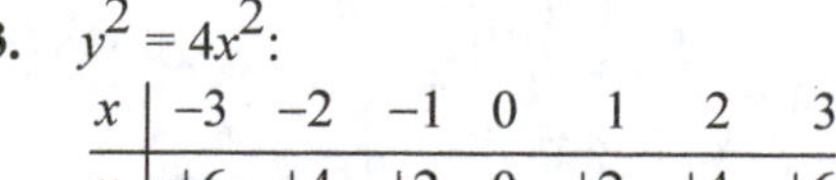

x	−3	−2	−1	0	1	2	3
y	±6	±4	±2	0	±2	±4	±6

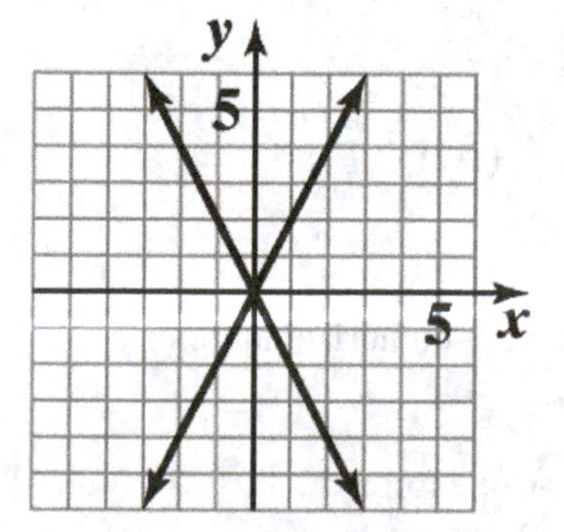

(1-1)

4. (A) Not a function; fails vertical line test

(B) A function

(C) A function

(D) Not a function; fails vertical line test (1-1)

5. $f(x) = 2x - 1$, $g(x) = x^2 - 2x$

(A) $f(-2) + g(-1) = 2(-2) - 1 + (-1)^2 - 2(-1) = -2$

(B) $f(0) \cdot g(4) = (2 \cdot 0 - 1)(4^2 - 2 \cdot 4) = -8$

(C) $\dfrac{g(2)}{f(3)} = \dfrac{2^2 - 2 \cdot 2}{2 \cdot 3 - 1} = 0$

(D) $\dfrac{f(3)}{g(2)}$ not defined because $g(2) = 0$ (1-1)

6. $3x + 2y = 9$

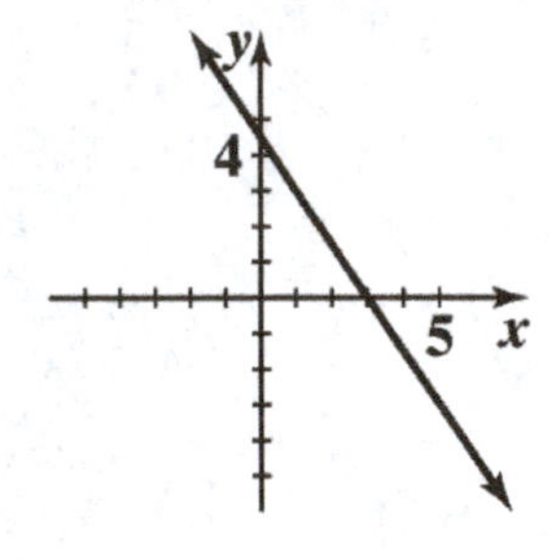

(1-2)

7. The line passes through (6, 0) and (0, 4)

slope $m = \dfrac{4-0}{0-6} = -\dfrac{2}{3}$

From the slope-intercept form: $y = -\dfrac{2}{3}x + 4$; multiplying by 3 gives: $3y = -2x + 12$, so

$2x + 3y = 12$ (1-2)

8. x-intercept: $2x = 18$, $x = 9$;
y-intercept: $-3y = 18$, $y = -6$;
slope-intercept form:

$y = \dfrac{2}{3}x - 6$; slope $= \dfrac{2}{3}$

Graph:

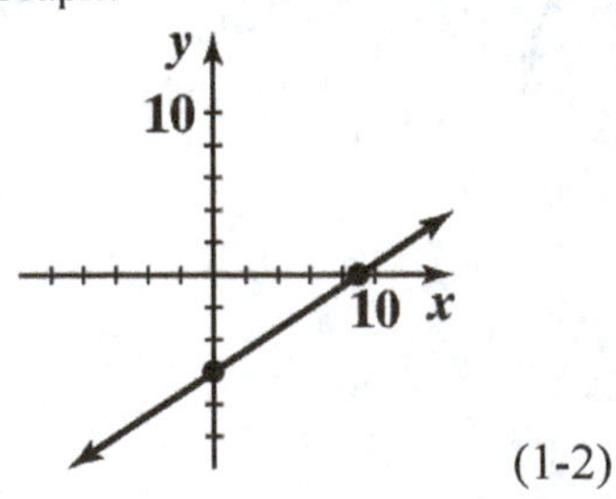

(1-2)

9. $y = -\dfrac{2}{3}x + 6$ (1-2)

10. Vertical line: $x = -6$; horizontal line: $y = 5$ (1-2)

11. Use the point-slope form:

(A) $y - 2 = -\dfrac{2}{3}[x - (-3)]$

(B) $y - 3 = 0(x - 3)$

$y - 2 = -\frac{2}{3}(x + 3)$ $\qquad$ $y = 3$

$y = -\frac{2}{3}x$ $\qquad$ (1-2)

12. (A) Slope: $\frac{-1-5}{1-(-3)} = -\frac{3}{2}$

$y - 5 = -\frac{3}{2}(x + 3)$

$3x + 2y = 1$

(B) Slope: $\frac{5-5}{4-(-1)} = 0$

$y - 5 = 0(x - 1)$

$y = 5$

(C) Slope: $\frac{-2-7}{-2-(-2)}$ not defined since $-2-(-2) = 0$

$x = -2$ $\qquad$ (1-2)

13. $u = e^v$

$v = \ln u$ (1-6)

14. $x = 10^y$

$y = \log x$ (1-6)

15. $\ln M = N$

$M = e^N$ (1-6)

16. $\log u = v$

$u = 10^v$ (1-6)

17. $\log_3 x = 2$

$x = 3^2 = 9$ (1-6)

18. $\log_x 36 = 2$

$x^2 = 36$

$x = 6$

(1-6)

19 $\log_2 16 = x$

$2^x = 16$

$x = 4$

(1-6)

20. $10^x = 143.7$

$x = \log\ 143.7$

$x \approx 2.157$

(1-6)

21. $e^x = 503{,}000$

$x = \ln 503{,}000 \approx 13.128$ (1-6)

22. $\log x = 3.105$

$x = 10^{3.105} \approx 1273.503$ (1-6)

23. $\ln x = -1.147$

$x = e^{-1.147} \approx 0.318$ (1-6)

24. (A) $y = 4$ (B) $x = 0$ (C) $y = 1$ (D) $x = -1$ or 1
(E) $y = -2$ (F) $x = -5$ or 5 $\qquad$ (1-1)

25. (A)

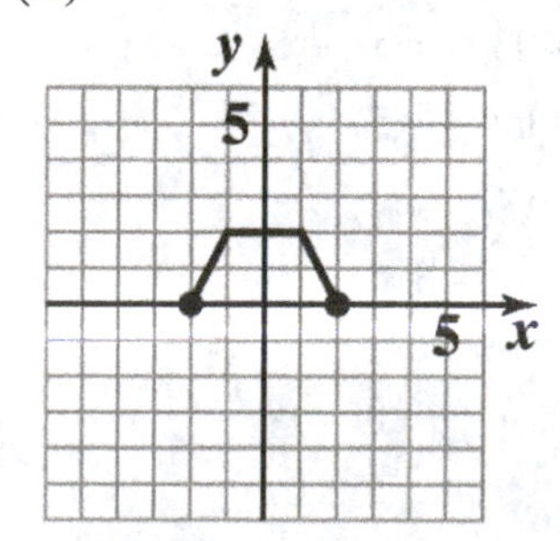

(B)

y
5
5
x

(C)

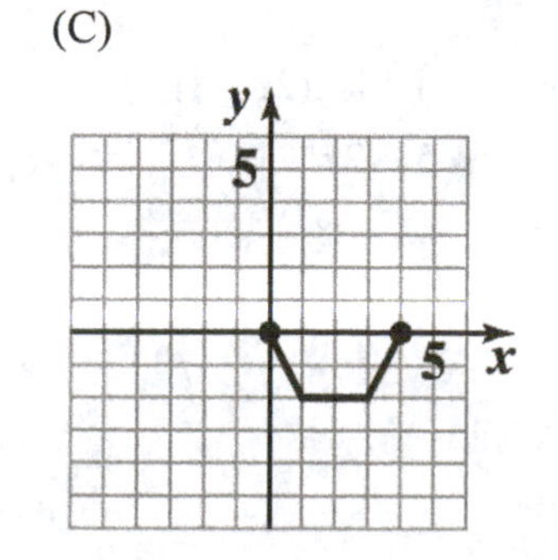

(D)

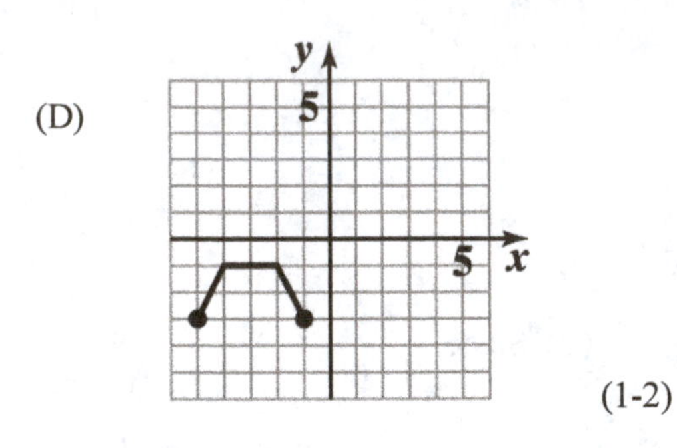

(1-2)

26. $f(x) = -x^2 + 4x = -(x^2 - 4x)$

$= -(x^2 - 4x + 4) + 4$

$= -(x - 2)^2 + 4$ (vertex form)

The graph of $f(x)$ is the graph of $y = x^2$ reflected in the x axis, then shifted right 2 units and up 4 units. (1-3)

27. (A) g (B) m (C) n (D) f (1-2, 1-3)

28. $y = f(x) = (x + 2)^2 - 4$

(A) x intercepts: $(x + 2)^2 - 4 = 0$; y intercept: 0

$(x + 2)^2 = 4$

$x + 2 = -2$ or 2

$x = -4, 0$

(B) Vertex: $(-2, -4)$ (C) Minimum: -4 (D) Range: $y \geq -4$ or $[-4, \infty)$ (1-3)

29. $y = 4 - x + 3x^2 = 3x^2 - x + 4$; quadratic function. (1-3)

30. $y = \frac{1 + 5x}{6} = \frac{5}{6}x + \frac{1}{6}$; linear function. (1-1, 1-3)

31. $y = \frac{7 - 4x}{2x} = \frac{7}{2x} - 2$; none of these. (1-1), (1-3)

32. $y = 8x + 2(10 - 4x) = 8x + 20 - 8x = 20$; constant function (1-1)

33. $\log(x + 5) = \log(2x - 3)$

$x + 5 = 2x - 3$

$-x = -8$

$x = 8$ (1-6)

34. $2 \ln(x - 1) = \ln(x^2 - 5)$

$\ln(x - 1)^2 = \ln(x^2 - 5)$

$(x - 1)^2 = x^2 - 5$

$x^2 - 2x + 1 = x^2 - 5$

$-2x = -6$

$x = 3$ (1-6)

35. $2x^2e^x = 3xe^x$

$2x^2 = 3x$

$2x^2 - 3x = 0$

$x(2x - 3) = 0$

$x = 0,\ 3/2$ (1-5)

36. $\log_{1/3} 9 = x$

$\left(\frac{1}{3}\right)^x = 9$

$\frac{1}{3^x} = 9$

$3^x = \frac{1}{9}$

$x = -2$ (1-6)

37. $35 = 7(3^x)$

$3^x = 5$

$\ln 3^x = \ln\ 5$

$x \ln 3 = \ln\ 5$

$x = \frac{\ln 5}{\ln 3} \approx 1.4650$ (1-6)

38. $0.01 = e^{-0.05x}$

$\ln(0.01) = \ln(e^{-0.05x}) = -0.05x$

Thus, $x = \frac{\ln(0.01)}{-0.05} \approx 92.1034$

(1-6)

39. $8{,}000 = 4{,}000(1.08)^x$

$(1.08)^x = 2$

$\ln(1.08)^x = \ln\ 2$

$x \ln 1.08 = \ln\ 2$

$x = \frac{\ln 2}{\ln 1.08} \approx 9.0065$ (1-6)

40. $5^{2x-3} = 7.08$

$\ln(5^{2x-3}) = \ln 7.08$

$(2x - 3) \ln 5 = \ln 7.08$

$2x \ln 5 - 3 \ln 5 = \ln 7.08$

$x = \frac{\ln 7.08 + 3\ln 5}{2\ln 5}$

$x \approx 2.1081$ (1-6)

41. (A) $x^2 - x - 6 = 0$ at $x = -2, 3$

Domain: all real numbers except $x = -2, 3$

(B) $5 - x > 0$ for $x < 5$

Domain: $x < 5$ or $(-\infty, 5)$ (1-1)

42. $f(x) = 4x^2 + 4x - 3 = 4(x^2 + x) - 3$

$= 4\left(x^2 + x + \frac{1}{4}\right) - 3 - 1$

$= 4\left(x + \frac{1}{2}\right)^2 - 4$ (vertex form)

Intercepts:

y intercept: $f(0) = 4(0)^2 + 4(0) - 3 = -3$

x intercepts: $f(x) = 0$

$4\left(x + \frac{1}{2}\right)^2 - 4 = 0$

$\left(x + \frac{1}{2}\right)^2 = 1$

$x + \frac{1}{2} = \pm 1$

$$x = -\frac{1}{2} \pm 1 = -\frac{3}{2}, \frac{1}{2}$$

Vertex: $\left(-\frac{1}{2}, -4\right)$; minimum: -4; range: $y \geq -4$ or $[-4, \infty)$ (1-3)

43. $f(x) = e^x - 1$, $g(x) = \ln(x + 2)$

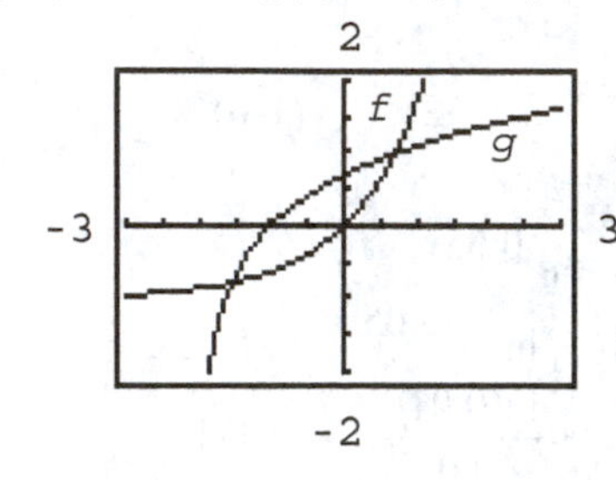

Points of intersection:
$(-1.54, -0.79)$, $(0.69, 0.99)$

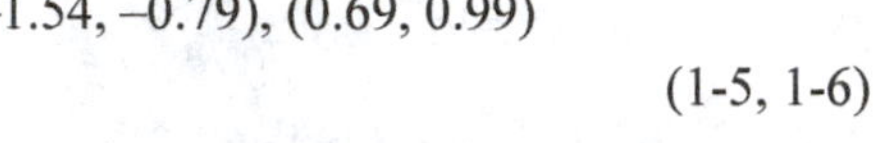

(1-5, 1-6)

44. $f(x) = \dfrac{50}{x^2 + 1}$:

x	-3	-2	-1	0	1	2	3
$f(x)$	5	10	25	50	25	10	5

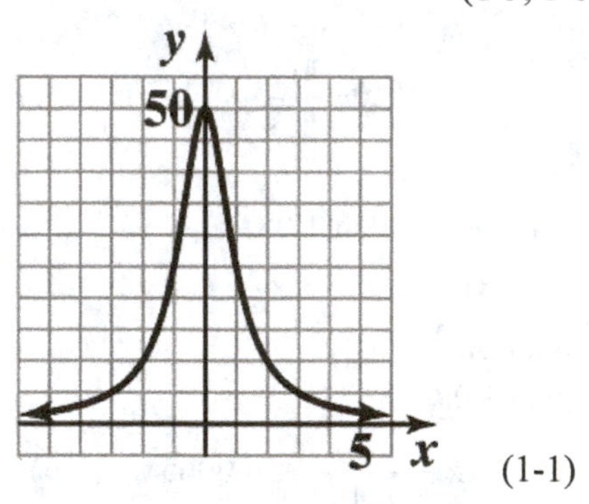

(1-1)

45. $f(x) = \dfrac{-66}{2 + x^2}$:

x	-3	-2	-1	0	1	2	3
$f(x)$	-6	-11	-22	-66	-22	-11	-6

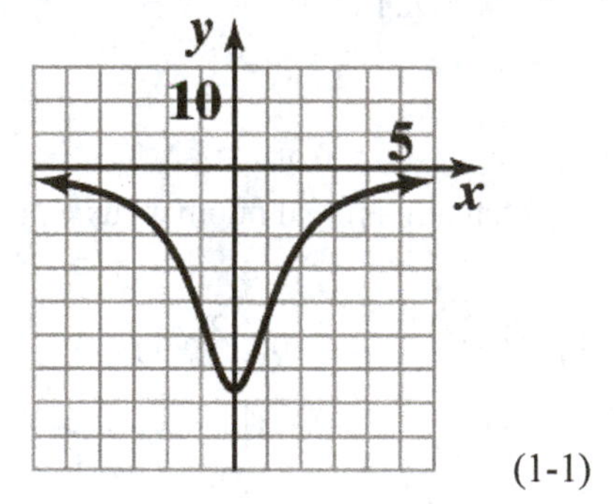

(1-1)

For Problems 46 – 50, $f(x) = 5x + 1$.

46. $f(f(0)) = f(5(0) + 1) = f(1) = 5(1) + 1 = 6$ (1-1)

47. $f(f(-1)) = f(5(-1) + 1) = f(-4) = 5(-4) + 1 = -19$ (1-1)

48. $f(2x - 1) = 5(2x - 1) + 1 = 10x - 4$ (1-1)

49. $f(4 - x) = 5(4 - x) + 1 = 20 - 5x + 1 = 21 - 5x$ (1-1)

50. $f(x) = 3 - 2x$

(A) $f(2) = 3 - 2(2) = 3 - 4 = -1$

(B) $f(2 + h) = 3 - 2(2 + h) = 3 - 4 - 2h = -1 - 2h$

(C) $f(2 + h) - f(2) = -1 - 2h - (-1) = -2h$

(D) $\frac{f(2+h)-f(2)}{h} = -\frac{2h}{h} = -2$ (1-1)

51. The graph of m is the graph of $y = |x|$ reflected in the x axis and shifted 4 units to the right. (2–2)

52. The graph of g is the graph of $y = x^3$ vertically contracted by a factor of 0.3 and shifted up 3 units. (1-2)

53. $f(x) = \frac{n(x)}{d(x)} = \frac{5x+4}{x^2-3x+1}$. Since degree $n(x) = 1 < 2 =$ degree $d(x)$, $y = 0$ is the horizontal asymptote. (1-4)

54. $f(x) = \frac{n(x)}{d(x)} = \frac{3x^2+2x-1}{4x^2-5x+3}$. Since degree $n(x) = 2 =$ degree $d(x)$, $y = \frac{3}{4}$ is the horizontal asymptote. (1-4)

55. $f(x) = \frac{n(x)}{d(x)} = \frac{x^2+4}{100x+1}$. Since degree $n(x) = 2 > 1 =$ degree $d(x)$, there is no horizontal asymptote (1-4)

56. $f(x) = \frac{n(x)}{d(x)} = \frac{x^2+100}{x^2-100} = \frac{x^2+100}{(x-10)(x+10)}$. Since $n(x) = x^2+100$ has no real zeros and $d(10) = d(-10) = 0$, $x = 10$ and $x = -10$ are the vertical asymptotes of the graph of f. (1-4)

57. $f(x) = \frac{n(x)}{d(x)} = \frac{x^2+3x}{x^2+2x} = \frac{x(x+3)}{x(x+2)} = \frac{x+3}{x+2}$, $x \neq 0$. $x = -2$ is a vertical asymptote of the graph of f. (1-4)

58. True; $p(x) = \frac{p(x)}{1}$ is a rational function for every polynomial p. (1-4)

59. False; $f(x) = \frac{1}{x} = x^{-1}$ is not a polynomial function. (1-4)

60. False; $f(x) = \frac{1}{x^2+1}$ has no vertical asymptotes. (1-4)

61. True; $f(x) = \frac{x}{x-1}$ has vertical asymptote $x = 1$ and horizontal asymptote $y = 1$. (1-4)

62.

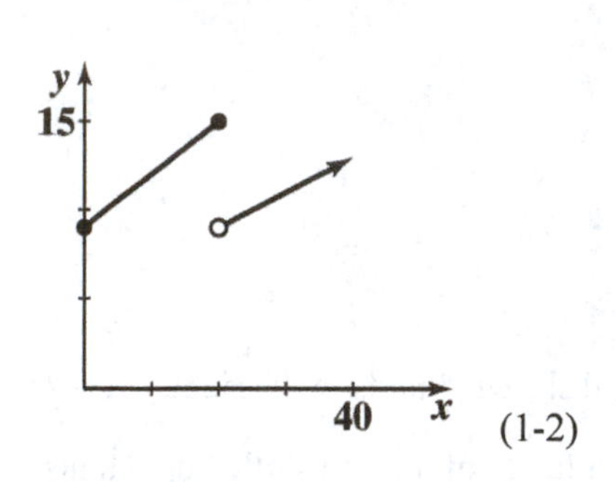

(1-2)

63.

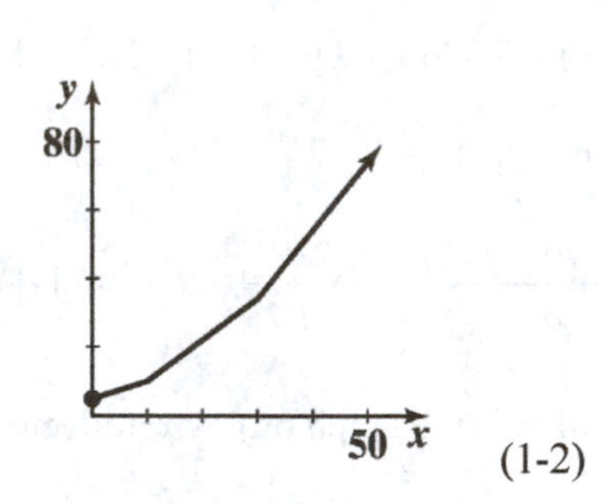

(1-2)

64. $y = -(x-4)^2 + 3$ (1-2, 1-3)

65. $f(x) = -0.4x^2 + 3.2x + 1.2 = -0.4(x^2 - 8x + 16) + 7.6$

$= -0.4(x-4)^2 + 7.6$

(A) y intercept: 1.2

x intercepts: $-0.4(x-4)^2 + 7.6 = 0$

$(x-4)^2 = 19$

$x = 4 + \sqrt{19} \approx 8.359, \quad 4 - \sqrt{19} \approx -0.359$

(B) Vertex: (4.0, 7.6) (C) Maximum: 7.6 (D) Range: $y \leq 7.6$ or $(-\infty, 7.6]$ (1-3)

66.

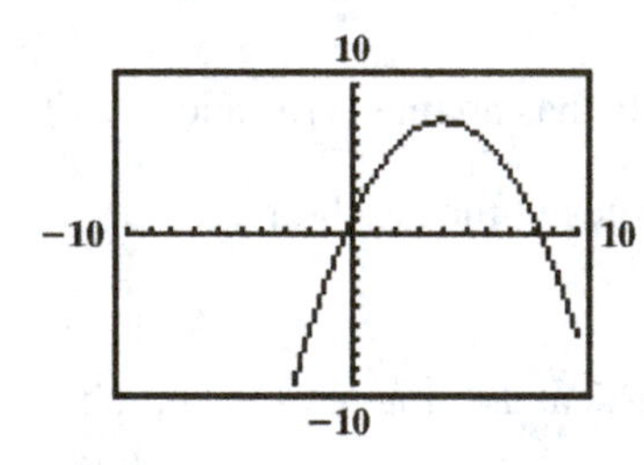

(A) y intercept: 1.2

x intercepts: −0.359, 8.359

(B) Vertex: (4.0, 7.6)

(C) Maximum: 7.6

(D) Range: $y \leq 7.6$ or $(-\infty, 7.6]$ (1-3)

67. $\log 10^{\pi} = \pi \log 10 = \pi$ (see logarithm properties 4.b & g, Section 1-5)

$10^{\log\sqrt{2}} = y$ is equivalent to $\log y = \log\sqrt{2}$

which implies $y = \sqrt{2}$

Similarly, $\ln e^{\pi} = \pi \ln e = \pi$ (Section 1-5, 4.b & g) and $e^{\ln\sqrt{2}} = y$ implies $\ln y = \ln\sqrt{2}$ and $y = \sqrt{2}$. (1-6)

68. $\log x - \log 3 = \log 4 - \log(x+4)$

$$\log\frac{x}{3} = \log\frac{4}{x+4}$$

$$\frac{x}{3} = \frac{4}{x+4}$$

$$x(x+4) = 12$$

$$x^2 + 4x - 12 = 0$$

$$(x+6)(x-2) = 0$$

$$x = -6, 2$$

Since log(−6) is not defined, −6 is not a solution. Therefore, the solution is $x = 2$. (1-6)

69. $\ln(2x-2)-\ln(x-1)=\ln x$

$$\ln\left(\frac{2x-2}{x-1}\right)=\ln x$$

$$\ln\left[\frac{2(x-1)}{x-1}\right]=\ln x$$

$$\ln 2=\ln x$$

$$x=2 \qquad (1\text{-}6)$$

70. $\ln(x+3)-\ln x=2\ln 2$

$$\ln\left(\frac{x+3}{x}\right)=\ln(2^2)$$

$$\frac{x+3}{x}=4$$

$$x+3=4x$$

$$3x=3$$

$$x=1 \qquad (1\text{-}6)$$

71. $\log 3x^2=2+\log 9x$

$$\log 3x^2-\log 9x=2$$

$$\log\left(\frac{3x^2}{9x}\right)=2$$

$$\log\left(\frac{x}{3}\right)=2$$

$$\frac{x}{3}=10^2=100$$

$$x=300 \qquad (1\text{-}6)$$

72. $\ln y=-5t+\ln c$

$$\ln y-\ln c=-5t$$

$$\ln\frac{y}{c}=-5t$$

$$\frac{y}{c}=e^{-5t}$$

$$y=ce^{-5t} \qquad (1\text{-}6)$$

73. Let x be *any* positive real number and suppose $\log_1 x=y$. Then $1^y=x$. But, $1^y=1$, so $x=1$, i.e., $x=1$ for all positive real numbers x. This is clearly impossible. (1-6)

74. The graph of $y=\sqrt[3]{x}$ is vertically expanded by a factor of 2, reflected in the x axis, shifted 1 unit to the left and 1 unit down.
Equation: $y=-2\sqrt[3]{x+1}-1$ (1-2)

75. $G(x)=0.3x^2+1.2x-6.9=0.3(x^2+4x+4)-8.1$
$=0.3(x+2)^2-8.1$

(A) y intercept: -6.9

x intercepts: $0.3(x+2)^2-8.1=0$

$$(x+2)^2=27$$

$$x=-2+\sqrt{27}\approx 3.196,\quad -2-\sqrt{27}\approx -7.196$$

(B) Vertex: $(-2,-8.1)$ (C) Minimum: -8.1 (D) Range: $y\geq -8.1$ or $[-8.1,\infty)$ (1-3)

76.

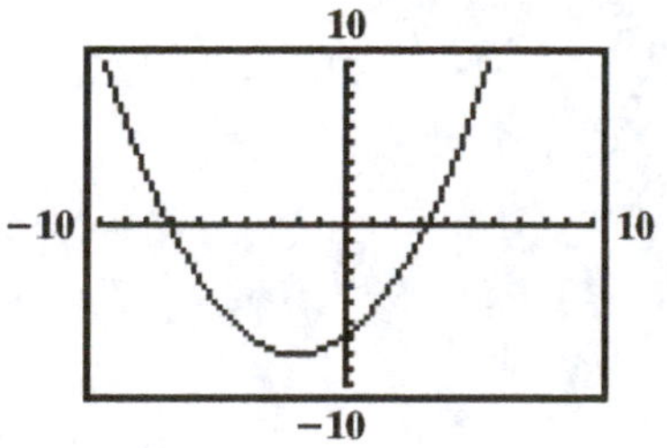

(A) y intercept: -6.9
x intercept: -7.196, 3.196

(B) Vertex: $(-2,-8.1)$

(C) Minimum: -8.1

(D) Range: $y\geq -8.1$ or $[-8.1,\infty)$ (1-3)

77. (A) $S(x)=3$ if $0\leq x\leq 20$;

$S(x)=3+0.057(x-20)$
$=0.057x+1.86$ if $20<x\leq 200$;

$S(200)=13.26$

$S(x)=13.26+0.0346(x-200)$

$= 0.0346x + 6.34$ if $200 < x \le 1000$;

$S(1000) = 40.94$

$S(x) = 40.94 + 0.0217(x - 1000)$

$= 0.0217x + 19.24$ if $x > 1000$

$$\text{Therefore, } S(x) = \begin{cases} 3 & \text{if } 0 \le x \le 20 \\ 0.057x + 1.86 & \text{if } 20 < x \le 200 \\ 0.0346x + 6.34 & \text{if } 200 < x \le 1000 \\ 0.0217x + 19.24 & \text{if } x > 1000 \end{cases}$$

(B)

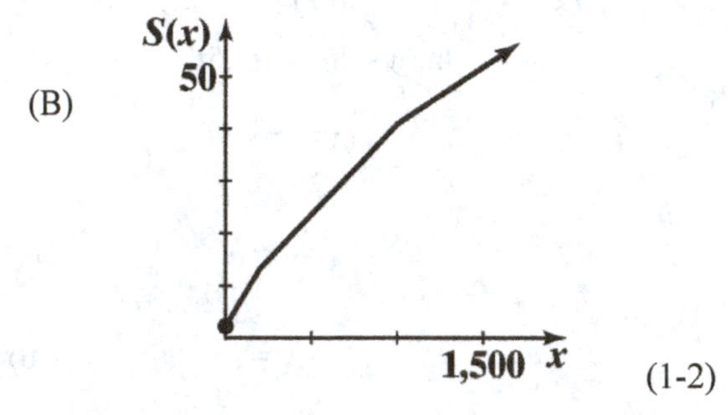

(1-2)

78. $A = P\left(1+\frac{r}{m}\right)^{mt}$; $P = 5{,}000$, $r = 0.0125$, $m = 4$, $t = 5$.

$$A = 5000\left(1+\frac{0.0125}{4}\right)^{4(5)} = 5000\left(1+\frac{0.0125}{4}\right)^{20} \approx 5321.95$$

After 5 years, the CD will be worth \$5,321.95 (1-5)

79. $A = A = P\left(1+\frac{r}{m}\right)^{mt}$; $P = 5{,}000$, $r = 0.0105$, $m = 365$, $t = 5$

$$A = 5000\ A = 5000\left(1+\frac{0.0105}{365}\right)^{365(5)} = 5000\left(1+\frac{0.0105}{365}\right)^{1825} \approx 5269.51$$

After 5 years, the CD will be worth \$5,269.51. (1-5)

80. $A = P\left(1+\frac{r}{m}\right)^{mt}$; $r = 0.0659$, $m = 12$

Solve $P\left(1+\frac{0.0659}{12}\right)^{12t} = 3P$ or $(1.005492)^{12t} = 3$

for t:

$12t \ln(1.005492) = \ln 3$,

$$t = \frac{\ln 3}{12\ln(1.005492)} \approx 16.7 \text{ year.}$$ (1-5)

81. $A = Pe^{rt}$; $r = 0.0739$. Solve $2P = Pe^{0.0739t}$ for t.

$2P = Pe^{0.0739t}$

$e^{0.0739t} = 2$

$0.0739t = \ln 2$

$t = \frac{\ln 2}{0.0739} \approx 9.38$ years.

(1-5)

82. Let x = person's age in years.

(A) Minimum heart rate: $m = (220 - x)(0.6) = 132 - 0.6x$

(B) Maximum heart rate: $M = (220 - x)(0.85) = 187 - 0.85x$

(C) At $x = 20$, $m = 132 - 0.6(20) = 120$

$M = 187 - 0.85(20) = 170$

range – between 120 and 170 beats per minute.

(D) At $x = 50$, $m = 132 - 0.6(50) = 102$

$M = 187 - 0.85(50) = 144.5$

range – between 102 and 144.5 beats per minute. (1-3)

83. $V = mt + b$

(A) At $t = 0$, $V = 224{,}000$; at $t = 8$, $V = 100{,}000$

$$\text{slope } m = \frac{100{,}000 - 224{,}000}{8 - 0} = -\frac{124{,}000}{8} = -15{,}500$$

$V = -15{,}500t + 224{,}000$

(B) At $t = 12$, $V = -15{,}500(12) + 224{,}000 = 38{,}000$.

The bulldozer will be worth $38,000 after 12 years. (1-2)

84. $r = -0.198t + 14.2$

(A)

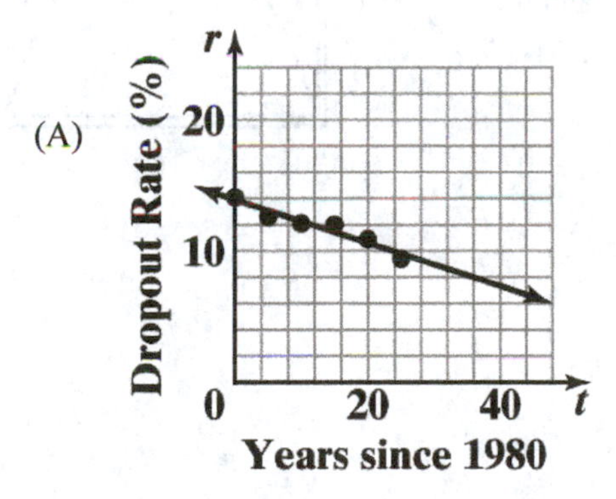

(C) Solve $5 = -0.198t + 14.2$ for t:

$-0.198t = -9.2$

$t = 46.\overline{46}$

The dropout rate will be less than 5% in 2027.

(B) The dropout rate is decreasing at the rate of 0.198 percentage points per year. (1-3)

85. $y = 4.75x + 171$

(A) The CPI is increasing at the rate of $4.75 units per year.

(B) At $x = 24$, $y = 4.75(24) + 171 = 285$; the CPI in 2024 will be 285.00. (1-3)

86. (A) The area enclosed by the pens is given by

$A = (2y)x$

Now, $3x + 4y = 840$

so $y = 210 - \frac{3}{4}x$

Thus $A(x) = 2\left(210 - \frac{3}{4}x\right)x$

$= 420x - \frac{3}{2}x^2$

(B) Clearly x and y must be nonnegative; the fact that $y \geq 0$ implies

$210 - \frac{3}{4}x \geq 0$

and $210 \geq \frac{3}{4}x$

$840 \geq 3x$

$280 \geq x$

Thus, domain A: $0 \leq x \leq 280$

(C)

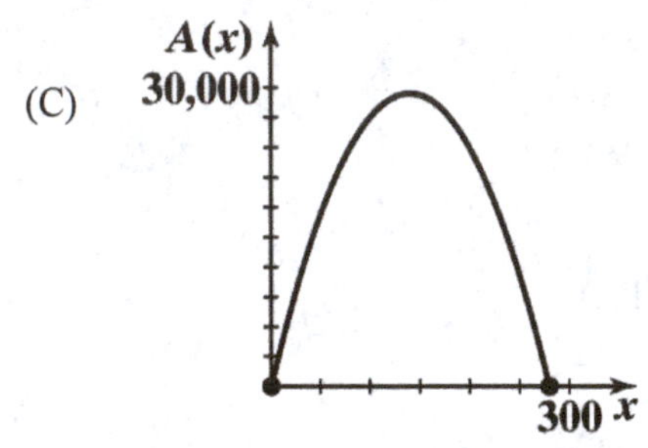

(D) Graph $A(x) = 420x - \frac{3}{2}x^2$ and $y = 25{,}000$ together.

There are two values of x that will produce storage areas with a combined area of 25,000 square feet, one near $x = 90$ and the other near $x = 190$.

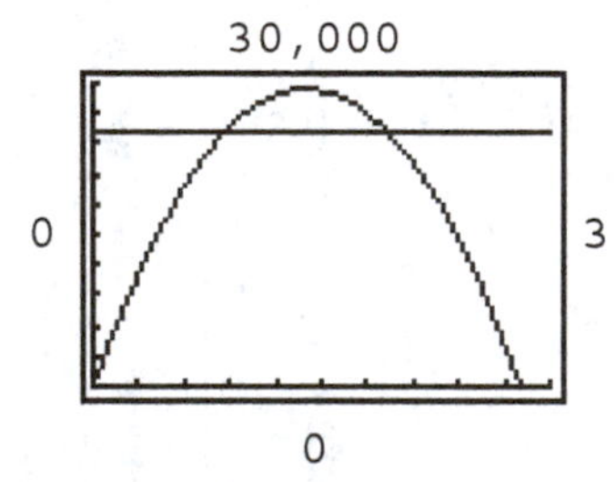

(E) $x = 86, x = 194$

(F) $A(x) = 420x - \frac{3}{2}x^2 = -\frac{3}{2}(x^2 - 280x)$

Completing the square, we have

$A(x) = -\frac{3}{2}(x^2 - 280x + 19{,}600 - 19{,}600)$

$= -\frac{3}{2}[(x - 140)^2 - 19{,}600]$

$= -\frac{3}{2}(x - 140)^2 + 29{,}400$

The dimensions that will produce the maximum combined area are: $x = 140$ ft, $y = 105$ ft. The maximum area is 29,400 sq. ft. (1-3)

87. (A) Quadratic regression model,

To estimate the demand at price level of \$180, we solve the equation

$ax^2 + bx + c = 180$

QuadReg
$y=ax^2+bx+c$
a=5.9477212E-6
b=-.1024018814
c=422.3467853

for x. The result is $x \approx 2{,}833$ sets.

(B) Linear regression model,
Table 2:

LinReg
$y=ax+b$
a=.0387421907
b=-7.364689544

To estimate the supply at a price level of \$180, we solve the equation

$$ax + b = 180$$

for x. The result is $x \approx 4{,}836$ sets.

(C) The condition is not stable; the price is likely to decrease since the supply at the price level of \$180 exceeds the demand at this level.

(D) Equilibrium price: \$131.59
Equilibrium quantity: 3,587 cookware sets. (1-3)

88. (A) Cubic Regression

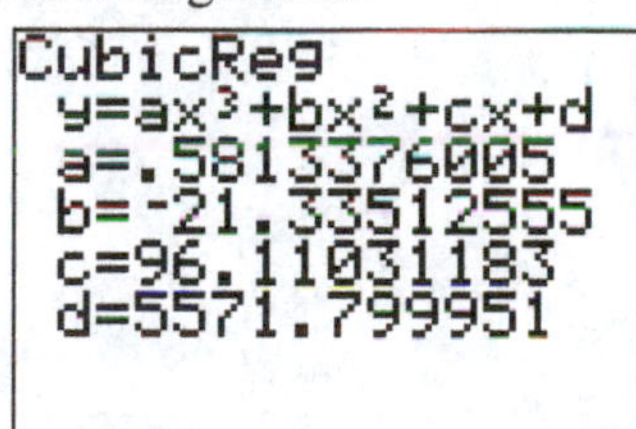

(B) $y(35) = 7725$ (?) (1-5)

89. (A) $N(0) = 1$

$$N\left(\frac{1}{2}\right) = 2$$

$$N(1) = 4 = 2^2$$

$$N\left(\frac{3}{2}\right) = 8 = 2^3$$

$$N(2) = 16 = 2^4$$

$$\vdots$$

Thus, we conclude that
$N(t) = 2^{2t}$ or $N = 4^t$.

(B) We need to solve:

$$2^{2t} = 10^9$$

$$\log 2^{2t} = \log 10^9 = 9$$

$$2t \log 2 = 9$$

$$t = \frac{9}{2\log 2} \approx 14.95$$

Thus, the mouse will die in 15 days.

(1-6)

90. Given $I = I_0 e^{-kd}$. When $d = 73.6$, $I = \frac{1}{2} I_0$. Thus, we have:

$$\frac{1}{2} I_0 = I_0 e^{-k(73.6)}$$

$$e^{-k(73.6)} = \frac{1}{2}$$

$$-k(73.6) = \ln \frac{1}{2}$$

$$k = \frac{\ln(0.5)}{-73.6} \approx 0.00942$$

Thus, $k \approx 0.00942$.

To find the depth at which 1% of the surface light remains, we set $I = 0.01 I_0$ and solve

$$0.01 I_0 = I_0 e^{-0.00942d} \text{ for } d:$$

$$0.01 = e^{-0.00942d}$$

$$-0.00942d = \ln \ 0.01$$

$$d = \frac{\ln 0.01}{-0.00942} \approx 488.87$$

Thus, 1% of the surface light remains at approximately 489 feet. (1-6)

91. (A) Logarithmic regression model:

```
LnReg
 y=a+blnx
 a=42400.65695
 b=-8207.259234
```

Year 2023 corresponds to $x = 83$; $y(83) \approx 6{,}134{,}000$ cows.

(B) ln (0) is not defined. (1-6)

92. Using the continuous compounding model, we have:

$$2P_0 = P_0 e^{0.03t}$$

$$2 = e^{0.03t}$$

$$0.03t = \ln \ 2$$

$$t = \frac{\ln 2}{0.03} \approx 23.1$$

Thus, the model predicts that the population will double in approximately 23.1 years. (1-5)

93. (A) Exponential regression model

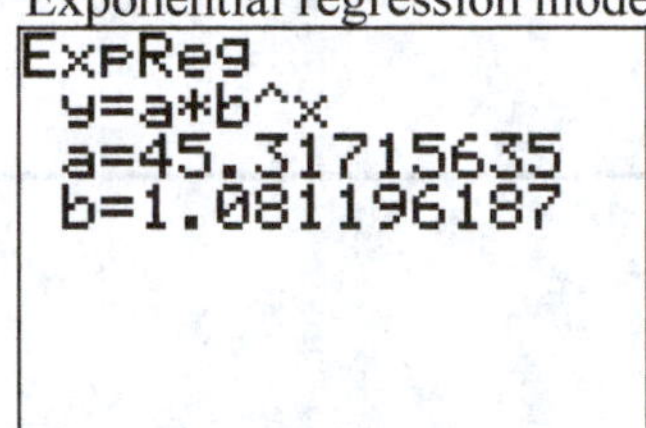

Year 2022 corresponds to $x = 42$; $y(42) \approx \$1{,}203$ billion.

(B) To find when the expenditures will reach two trillion, solve $ab^x = 2{,}000$ for x. The result is $x \approx 48.51$ years; that is, in 2028. (1-5)

2 LIMITS AND THE DERIVATIVE

EXERCISE 2-1

Things to remember:

1. LIMIT

We write

$$\lim_{x \to c} f(x) = L \text{ or } f(x) \to L \text{ as } x \to c$$

if the functional value $f(x)$ is close to the single real number L whenever x is close to but not equal to c (on either side of c).

[Note: The existence of a limit at c has nothing to do with the value of the function at c. In fact, c may not even be in the domain of f. However, the function must be defined on both sides of c.]

2. ONE-SIDED LIMITS

We write $\lim_{x \to c^-} f(x) = K$ [$x \to c^-$ is read "x approaches c from the left" and means $x \to c$ and $x < c$] and call K the LIMIT FROM THE LEFT or LEFT-HAND LIMIT if $f(x)$ is close to K whenever x is close to c, but to the left of c on the real number line.

We write $\lim_{x \to c^+} f(x) = L$ [$x \to c^+$ is read "x approaches c from the right" and means $x \to c$ and $x > c$] and call L the LIMIT FROM THE RIGHT or RIGHT-HAND LIMIT if $f(x)$ is close to L whenever x is close to c, but to the right of c on the real number line.

3. EXISTENCE OF A LIMIT

In order for a limit to exist, the limit from the left and the limit from the right must both exist, and must be equal. That is,

$$\lim_{x \to c} f(x) = L \quad \text{if and only if} \quad \lim_{x \to c^-} f(x) = \lim_{x \to c^+} f(x) = L.$$

4. PROPERTIES OF LIMITS

(a) $\lim_{x \to c} k = k$ for any constant k

(b) $\lim_{x \to c} x = c$

Let f and g be two functions and assume that

$$\lim_{x \to c} f(x) = L \qquad \lim_{x \to c} g(x) = M$$

where L and M are real numbers (both limits exist). Then:

(c) $\lim_{x \to c} [f(x) + g(x)] = \lim_{x \to c} f(x) + \lim_{x \to c} g(x) = L + M.$

(d) $\lim_{x \to c} [f(x) - g(x)] = \lim_{x \to c} f(x) - \lim_{x \to c} g(x) = L - M.$

(e) $\lim_{x \to c} kf(x) = k \lim_{x \to c} f(x) = kL$ for any constant k.

(f) $\lim_{x \to c} [f(x)g(x)] = \left(\lim_{x \to c} f(x)\right)\left(\lim_{x \to c} g(x)\right) = LM.$

(g) $\lim_{x\to c} \dfrac{f(x)}{g(x)} = \dfrac{L}{M}$ if $M \neq 0$; $\lim_{x\to c} \dfrac{f(x)}{g(x)}$ does not exist if $L \neq 0$ and $M = 0$; $\lim_{x\to c} \dfrac{f(x)}{g(x)}$ is a 0/0 INDETERMINATE FORM if $L = M = 0$.

(h) $\lim_{x\to c} \sqrt[n]{f(x)} = \sqrt[n]{\lim_{x\to c} f(x)} = \sqrt[n]{L}$ ($L \geq 0$ for n even).

5. LIMITS OF POLYNOMIAL AND RATIONAL FUNCTIONS

(a) $\lim_{x\to c} f(x) = f(c)$ for f any polynomial function

(b) $\lim_{x\to c} r(x) = r(c)$ for r any rational function with nonzero denominator at $x = c$.

6. DIFFERENCE QUOTIENT

Let the function f be defined in an open interval containing the number a. The expression

$$\frac{f(a+h)-f(a)}{h}$$

is called the DIFFERENCE QUOTIENT. One of the most important limits in calculus is the limit of the difference quotient:

$$\lim_{h\to 0} \frac{f(a+h)-f(a)}{h}$$

---

1. $x^2 - 81 = (x-9)(x+9)$

3. $x^2 - 4x - 21 = (x-7)(x+3)$

5. $x^3 - 7x^2 + 12x = x(x^2 - 7x + 12) = x(x-3)(x-4)$

7. $6x^2 - x - 1 = (2x-1)(3x+1)$

9. $f(-0.5) = 2$

11. $f(1.75) = 1.25$

13. (A) $\lim_{x\to 0^-} f(x) = 2$ (B) $\lim_{x\to 0^+} f(x) = 2$ (C) $\lim_{x\to 0} f(x) = 2$ (D) $f(0) = 2$

15. (A) $\lim_{x\to 2^-} f(x) = 1$ (B) $\lim_{x\to 2^+} f(x) = 2$ (C) $\lim_{x\to 2} f(x)$ does not exist

(D) $f(2) = 2$ (E) No, because $\lim_{x\to 2^-} f(x) = 1 \neq \lim_{x\to 2^+} f(x) = 2$

17. $g(1.9) = 2$

19. $g(3.5) = 0.5$

21. (A) $\lim_{x\to 1^-} g(x) = 1$ (B) $\lim_{x\to 1^+} g(x) = 2$ (C) $\lim_{x\to 1} g(x) =$ does not exist

(D) $g(1)$ does not exist (E) No, because $\lim_{x\to 1^-} g(x) = 1 \neq \lim_{x\to 1^+} g(x) = 2$

23. (A) $\lim_{x\to 3^-} g(x) = 1$ (B) $\lim_{x\to 3^+} g(x) = 1$ (C) $\lim_{x\to 3} g(x) = 1$ (D) $g(3) = 3$

(E) Yes, define $g(3) = 1$.

25. (A) $\lim_{x\to -3^+} f(x) = -2$ (B) $\lim_{x\to -3^-} f(x) = -2$ (C) $\lim_{x\to -3} f(x) = -2$

(D) $f(-3) = 1$ (E) Yes, set $f(-3) = -2$.

27. (A) $\lim_{x\to 0^+} f(x) = 2$ (B) $\lim_{x\to 0^-} f(x) = 2$ (C) $\lim_{x\to 0} f(x) = 2$

(D) $f(0)$ does not exist. (E) Yes, define $f(0) = 2$.

29. $\lim_{x\to 3} 4x = 4\cdot 3 = 12$ (use 5)

31. $\lim_{x\to -4} (x+5) = -4+5 = 1$ (use 5)

33. $\lim_{x\to 2} x(x-4) = 2(2-4) = 2(-2) = -4$ (use 4f and 5)

35. $\lim_{x\to -3} \frac{x}{x+5} = \frac{-3}{-3+5} = -\frac{3}{2} = -1.5$ (use 4g and 5)

37. $\lim_{x\to 1} \sqrt{5x+4} = \sqrt{5+4} = \sqrt{9} = 3$ (use 4h and 5)

39. $\lim_{x\to 1} -3f(x) = -3\lim_{x\to 1} f(x) = -3(-5) = 15$

41. $\lim_{x\to 1} [2f(x) + g(x)] = 2\lim_{x\to 1} f(x) + \lim_{x\to 1} g(x) = 2(-5) + 4 = -6$

43. $\lim_{x\to 1} \frac{2-f(x)}{x+g(x)} = \frac{\lim_{x\to 1}[2-f(x)]}{\lim_{x\to 1}[x+g(x)]} = \frac{2-\lim_{x\to 1} f(x)}{1+\lim_{x\to 1} g(x)} = \frac{2-(-5)}{1+4} = \frac{7}{5}$

45. $\lim_{x\to 1} \sqrt{g(x)-f(x)} = \sqrt{\lim_{x\to 1}[g(x)-f(x)]} = \sqrt{\lim_{x\to 1} g(x) - \lim_{x\to 1} f(x)} = \sqrt{4-(-5)} = \sqrt{9} = 3$

Note: Answers for Problems 47 and 49 may vary.

47.

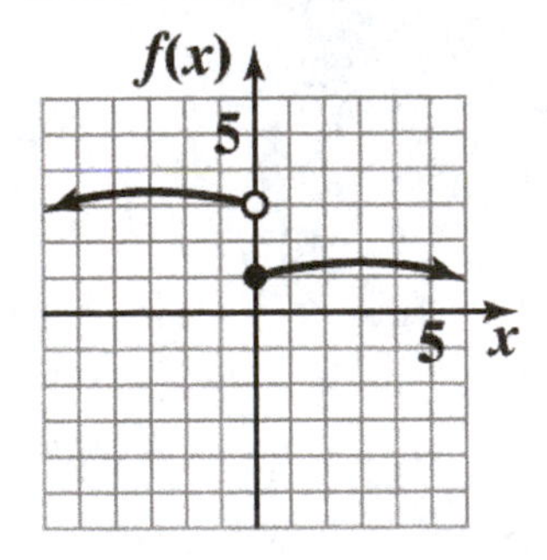

49.

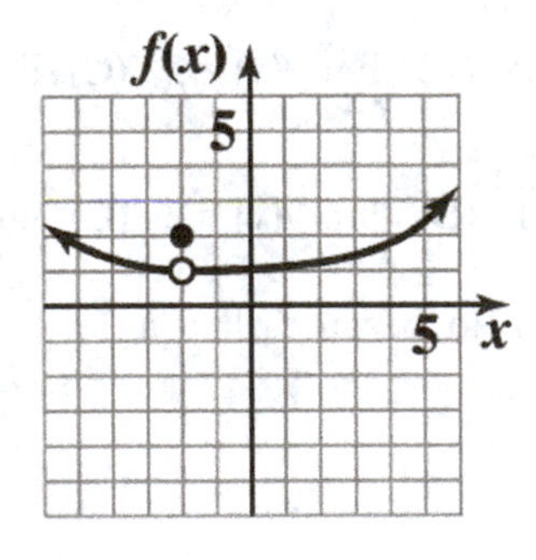

51. $f(x) = \begin{cases} 1-x^2 & \text{if } x \le 0 \\ 1+x^2 & \text{if } x > 0 \end{cases}$

(A) $\lim_{x\to 0^+} f(x) = \lim_{x\to 0^+} (1+x^2) = 1$

(B) $\lim_{x\to 0^-} f(x) = \lim_{x\to 0^-} (1-x^2) = 1$

(C) $\lim_{x\to 0} f(x) = 1$

(D) $f(0) = 1$

53. $f(x) = \begin{cases} x^2 & \text{if } x < 1 \\ 2x & \text{if } x > 1 \end{cases}$

(A) $\lim_{x\to 1^+} f(x) = \lim_{x\to 1^+} 2x = 2$

(B) $\lim_{x\to 1^-} f(x) = \lim_{x\to 1^-} x^2 = 1$

(C) $\lim_{x\to 1} f(x)$ does not exist

(D) $f(1)$ does not exist

55. $f(x) = \begin{cases} \dfrac{x^2-9}{x+3} & \text{if } x < 0 \\ \dfrac{x^2-9}{x-3} & \text{if } x > 0 \end{cases}$

(A) $\lim_{x\to -3} f(x) = \lim_{x\to -3} \dfrac{x^2-9}{x+3} = \lim_{x\to -3} \dfrac{(x-3)(x+3)}{x+3} = \lim_{x\to -3} (x-3) = -6$

(B) $\lim_{x\to 0^-} f(x) = \lim_{x\to 0^-} \dfrac{x^2-9}{x+3} = \dfrac{\lim_{x\to 0^-}(x^2-9)}{\lim_{x\to 0^-}(x+3)} = \dfrac{-9}{3} = -3$

$\lim_{x\to 0^+} f(x) = \lim_{x\to 0^+} \dfrac{x^2-9}{x-3} = \dfrac{\lim_{x\to 0^+}(x^2-9)}{\lim_{x\to 0^+}(x-3)} = \dfrac{-9}{-3} = 3$

$\lim_{x\to 0} f(x)$ does not exist

(C) $\lim_{x\to 3} f(x) = \lim_{x\to 3} \dfrac{x^2-9}{x-3} = \lim_{x\to 3} \dfrac{(x-3)(x+3)}{x-3} = \lim_{x\to 3} (x+3) = 6$

57. $f(x) = \dfrac{|x-1|}{x-1}$

(A) For $x > 1$, $|x-1| = x-1$.

Thus, $\lim_{x\to 1^+} \dfrac{|x-1|}{x+1} = \lim_{x\to 1^+} \dfrac{x-1}{x-1} = \lim_{x\to 1^+} 1 = 1.$

(B) For $x < 1$, $|x-1| = -(x-1)$.

Thus, $\lim_{x\to 1^-} \dfrac{|x-1|}{x-1} = \lim_{x\to 1^-} \dfrac{-(x-1)}{x-1} = \lim_{x\to 1^-} -1 = -1$

(C) $\lim_{x\to 1} f(x)$ does not exist

(D) $f(1)$ does not exist

59. $f(x) = \dfrac{x-2}{x^2-2x} = \dfrac{x-2}{x(x-2)} = \dfrac{1}{x}, x \neq 2;$ $f(2)$ does not exist.

(A) $\lim_{x\to 0} f(x) = \lim_{x\to 0} \dfrac{1}{x}$ does not exist

(B) $\lim_{x\to 2} f(x) = \lim_{x\to 2} \dfrac{1}{x} = \dfrac{1}{2}$

(C) $\lim_{x\to 4} f(x) = \lim_{x\to 4} \dfrac{1}{x} = \dfrac{1}{4}$

61. $f(x) = \frac{x^2 - x - 6}{x+2} = \frac{(x-3)(x+2)}{x+2} = x - 3, x \neq -2$; $f(-2)$ does not exist

(A) $\lim_{x\to -2} f(x) = \lim_{x\to -2} (x-3) = -5$

(B) $\lim_{x\to 0} f(x) = \lim_{x\to 0} (x-3) = -3$

(C) $\lim_{x\to 3} f(x) = \lim_{x\to 3} (x-3) = 0$

63. $f(x) = \frac{(x+2)^2}{x^2-4} = \frac{(x+2)^2}{(x-2)(x+2)} = \frac{x+2}{x-2}, x \neq -2$; $f(-2)$ does not exist

(A) $\lim_{x\to -2} f(x) = \lim_{x\to -2} \frac{x+2}{x-2} = \frac{0}{-4} = 0$

(B) $\lim_{x\to 0} f(x) = \lim_{x\to 0} \frac{x+2}{x-2} = \frac{2}{-2} = -1$

(C) $\lim_{x\to 2} f(x) = \lim_{x\to 2} \frac{x+2}{x-2}$ does not exist

65. $f(x) = \frac{2x^2 - 3x - 2}{x^2 + x - 6} = \frac{(2x+1)(x-2)}{(x+3)(x-2)} = \frac{2x+1}{x+3}, x \neq 2$; $f(2)$ does not exist

(A) $\lim_{x\to 2} f(x) = \lim_{x\to 2} \frac{2x+1}{x+3} = \frac{5}{5} = 1$

(B) $\lim_{x\to 0} f(x) = \lim_{x\to 0} \frac{2x+1}{x+3} = \frac{1}{3}$

(C) $\lim_{x\to 1} f(x) = \lim_{x\to 1} \frac{2x+1}{x+3} = \frac{3}{4}$

67. False. Set $f(x) = x^2 - 1$, $g(x) = x - 1$. Then $\lim_{x\to 1} \frac{x^2-1}{x-1} = \lim_{x\to 1} x + 1 = 2$.

69. True. $\lim_{x\to 0} f(x) = f(0)$ for any polynomial function f.

71. False. Set $f(x) = \frac{x^2}{x}$. Then $f(0)$ does not exist, but $\lim_{x\to 0} \frac{x^2}{x} = \lim_{x\to 0} x = 0$.

73. $\lim_{x\to 7} \frac{(x-7)^2}{x^2 - 4x - 21}$ has the form $\frac{0}{0}$; $\frac{(x-7)^2}{x^2-4x-21} = \frac{(x-7)^2}{(x-7)(x+3)} = \frac{x-7}{x+3}$.

Therefore, $\lim_{x\to 7} \frac{(x-7)^2}{x^2-4x-21} = \lim_{x\to 7} \frac{x-7}{x+3} = 0$.

75. $\lim_{x\to 4} \frac{x^2+4}{(x+4)^2}$ does not have the form $\frac{0}{0}$; $\lim_{x\to 4} \frac{x^2+4}{(x+4)^2} = \frac{16+4}{8^2} = \frac{20}{64} = \frac{5}{16}$.

77. $\lim_{x\to -6} \frac{x^2+36}{x+6}$ does not have the form $\frac{0}{0}$; $\lim_{x\to -6} \frac{x^2+36}{x+6}$ has the form $\frac{72}{0}$, the limit does not exist.

79. $\lim_{x\to 8}\frac{x-8}{x^2-64}$ has the form $\frac{x-8}{x^2-64}=\frac{x-8}{(x-8)(x+8)}=\frac{1}{x+8}$.

Therefore, $\lim_{x\to 8}\frac{x-8}{x^2-64}=\lim_{x\to 8}\frac{1}{x+8}=\frac{1}{16}$.

81. $f(x) = 3x + 1$

$$\lim_{h\to 0}\frac{f(2+h)-f(2)}{h}=\lim_{h\to 0}\frac{3(2+h)+1-(3\cdot 2+1)}{h}=\lim_{h\to 0}\frac{6+3h+1-7}{h}=\lim_{h\to 0}\frac{3h}{h}=\lim_{h\to 0}3=3$$

83. $f(x) = x^2 + 1$

$$\lim_{h\to 0}\frac{f(2+h)-f(2)}{h}=\lim_{h\to 0}\frac{(2+h)^2+1-(2^2+1)}{h}=\lim_{h\to 0}\frac{4+4h+h^2+1-5}{h}=\lim_{h\to 0}\frac{4h+h^2}{h}=\lim_{h\to 0}(4+h)=4$$

85. $f(x)=-7x+9$

$$\lim_{h\to 0}\frac{f(2+h)-f(2)}{h}=\lim_{h\to 0}\frac{-7(2+h)+9-[-7(2)+9]}{h}=\frac{-7h}{h}=-7.$$

87. $f(x+1)=|x+1|$. For $x\geq -1$, $f(x)=x+1$. Therefore,

$$\lim_{h\to 0}\frac{f(2+h)-f(2)}{h}=\lim_{h\to 0}\frac{(2+h)+1-[2+1]}{h}=\frac{h}{h}=1.$$

89. (A) $\lim_{x\to 1^-}f(x)=\lim_{x\to 1^-}(1+x)=2$

$\lim_{x\to 1^+}f(x)=\lim_{x\to 1^+}(4-x)=3$

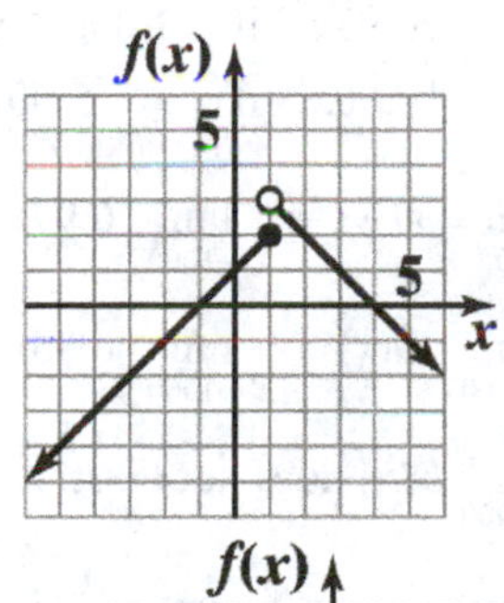

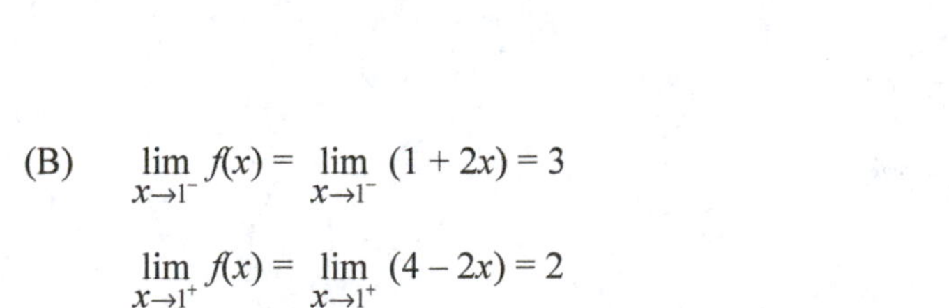

(B) $\lim_{x\to 1^-}f(x)=\lim_{x\to 1^-}(1+2x)=3$

$\lim_{x\to 1^+}f(x)=\lim_{x\to 1^+}(4-2x)=2$

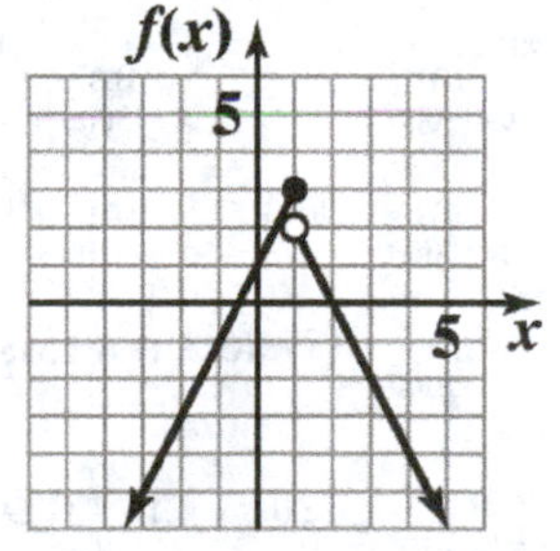

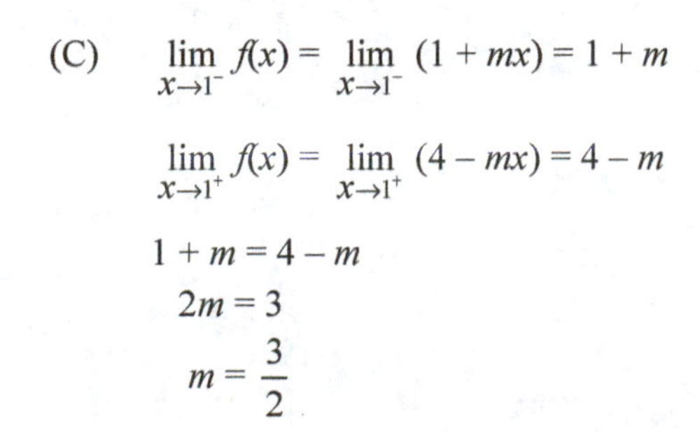

(C) $\lim_{x\to 1^-}f(x)=\lim_{x\to 1^-}(1+mx)=1+m$

$\lim_{x\to 1^+}f(x)=\lim_{x\to 1^+}(4-mx)=4-m$

$1+m=4-m$

$2m=3$

$m=\frac{3}{2}$

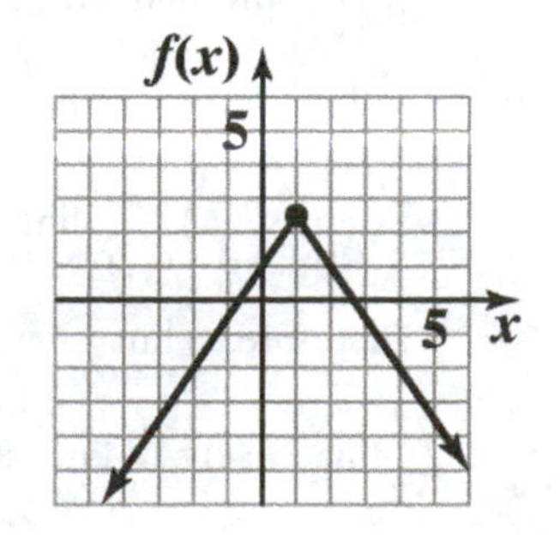

(D) The graph in (A) is broken at $x = 1$; it jumps up from $(1, 2)$ to $(1, 3)$.

The graph in (B) is also broken at $x = 1$; it jumps down from $(1, 3)$ to $(1, 2)$.

The graph in (C) is not broken; the two pieces meet at $\left(1, \frac{5}{2}\right)$.

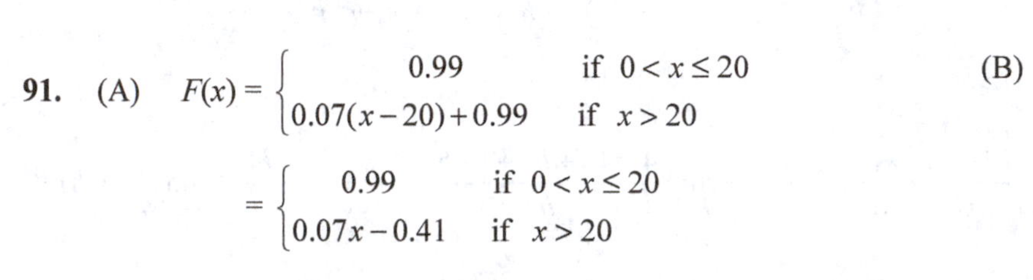

91. (A) $F(x) = \begin{cases} 0.99 & \text{if } 0 < x \le 20 \\ 0.07(x-20)+0.99 & \text{if } x > 20 \end{cases}$

$= \begin{cases} 0.99 & \text{if } 0 < x \le 20 \\ 0.07x - 0.41 & \text{if } x > 20 \end{cases}$

(B)

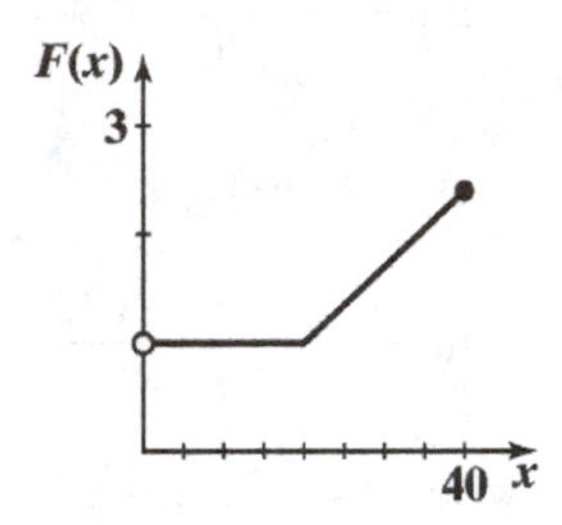

(C) $\lim_{x \to 20^-} F(x) = 0.99 = \lim_{x \to 20^+} F(x)$. Therefore, $\lim_{x \to 20} F(x) = 0.99$

93. At $x = 20$ minutes, the first service charge is \$0.99 and the second service charge is \$2.70. Also, after 20 minutes, the first service charge is \$0.07 per minute versus \$0.09 per minute for the second service. The second service is much more expensive than the first (unless the call is 10 minutes or less).

95. (A) $D(x) = \begin{cases} x & \text{if} & 0 \le x < 300 \\ 0.97x & \text{if} & 300 \le x < 1{,}000 \\ 0.95x & \text{if} & 1{,}000 \le x < 3{,}000 \\ 0.93x & \text{if} & 3{,}000 \le x < 5{,}000 \\ 0.90x & \text{if} & 5{,}000 \le x \end{cases}$

(B) $\lim_{x \to 1000^-} D(x) = \lim_{x \to 1000^-} 0.97x = 970,$

$\lim_{x \to 1000^+} D(x) = \lim_{x \to 1000^+} 0.95x = 950,$

$\lim_{x \to 1000} D(x)$ does not exist;

$\lim_{x \to 3000^-} D(x) = \lim_{x \to 3000^-} 0.95x = 2850,$

$\lim_{x \to 3000^+} D(x) = \lim_{x \to 3000^+} 0.93x = 2790,$

$\lim_{x \to 3000} D(x)$ does not exist.

97. (A) $F(x) = \begin{cases} 20x & \text{if} & 0 \le x \le 4{,}000 \\ 80{,}000 & \text{if} & x > 4{,}000 \end{cases}$

(B) $\lim_{x \to 4000^-} F(x) = \lim_{x \to 4000^-} 20x = 80{,}000,$

$\lim_{x \to 4000^+} F(x) = \lim_{x \to 4000^+} 80{,}000 = 80{,}000.$

Therefore, $\lim_{x \to 4000} F(x) = 80{,}000.$

$\lim_{x \to 8000} F(x) = \lim_{x \to 8000} 80{,}000 = 80{,}000.$

99. $\lim_{x\to5^-} f(x) = \lim_{x\to5^-} 0 = 0,$

$\lim_{x\to5^+} f(x) = \lim_{x\to5^+} (0.8 - 0.08x) = 0.4.$

Therefore, $\lim_{x\to5} f(x)$ does not exist.

$\lim_{x\to5^-} g(x) = \lim_{x\to5^-} 0 = 0,$

$\lim_{x\to5^+} g(x) = \lim_{x\to5^+} (0.8x - 0.04x^2 - 3) = 0.$

Therefore, $\lim_{x\to5} g(x) = 0.$

$\lim_{x\to10^-} f(x) = \lim_{x\to10^-} (0.8 - 0.08x) = 0,$

$\lim_{x\to10^+} f(x) = \lim_{x\to10^+} 0 = 0.$

Therefore, $\lim_{x\to10} f(x) = 0.$

$\lim_{x\to10^-} g(x) = \lim_{x\to10^-} (0.8x - 0.04x^2 - 3) = 1,$

$\lim_{x\to10^+} g(x) = \lim_{x\to0^+} 1 = 1.$

Therefore, $\lim_{x\to10} g(x) = 1.$

EXERCISE 2-2

Things to remember:

1. VERTICAL ASYMPTOTES

   The vertical line $x = a$ is a VERTICAL ASYMPTOTE for the graph of $y = f(x)$ if

   $$f(x) \to \infty \text{ or } f(x) \to -\infty \text{ as } x \to a^+ \text{ or } x \to a^-$$

   That is, if $f(x)$ either increases or decreases without bound as x approaches a from the right or from the left.

2. LOCATING VERTICAL ASYMPTOTES

   A polynomial function has no vertical asymptotes.

   If $f(x) = n(x)/d(x)$ is a rational function, $d(c) = 0$ and $n(c) \neq 0$, then the line $x = c$ is a vertical asymptote of the graph of f.

3. HORIZONTAL ASYMPTOTES

   The horizontal line $y = b$ is a HORIZONTAL ASYMPTOTE for the graph of $y = f(x)$ if

   $$\lim_{x\to-\infty} f(x) = b \text{ or } \lim_{x\to\infty} f(x) = b.$$

4. LIMITS AT INFINITY FOR POWER FUNCTIONS

   If p is a positive real number and k is a nonzero constant, then

(a) $\lim_{x\to-\infty} \frac{k}{x^p} = 0$ (b) $\lim_{x\to\infty} \frac{k}{x^p} = 0$

(c) $\lim_{x\to-\infty} kx^p = \pm\infty$ (d) $\lim_{x\to\infty} kx^p = \pm\infty$

provided that x^p is defined for negative values of x. The limits in (c) and (d) will be either $-\infty$ or ∞, depending on k and p.

5. LIMITS AT INFINITY FOR POLYNOMIAL FUNCTIONS

If

$$p(x) = a_n x^n + a_{n-1}x^{n-1} + \ldots + a_1 x + a_0,\ a_n \neq 0,\ n \geq 1,$$

then

$$\lim_{x\to\infty} p(x) = \lim_{x\to\infty} a_n x^n = \pm\infty$$

and

$$\lim_{x\to-\infty} p(x) = \lim_{x\to-\infty} a_n x^n = \pm\infty$$

Each limit will be either $-\infty$ or ∞, depending on a_n and n.

6. LIMITS AT INFINITY AND HORIZONTAL ASYMPTOTES FOR RATIONAL FUNCTIONS

If $f(x) = \dfrac{a_m x^m + a_{m-1}x^{m-1} + \ldots + a_1 x + a_0}{b_n x^n + b_{n-1}x^{n-1} + \ldots + b_1 x + b_0}$, $a_m \neq 0,\ b_n \neq 0$

then $\lim_{x\to\infty} f(x) = \lim_{x\to\infty} \dfrac{a_m x^m}{b_n x^n}$ and $\lim_{x\to-\infty} f(x) = \lim_{x\to-\infty} \dfrac{a_m x^m}{b_n x^n}$

(a) If $m < n$, then $\lim_{x\to\infty} f(x) = \lim_{x\to-\infty} f(x) = 0$ and the line $y = 0$ (the x-axis) is a horizontal asymptote for $f(x)$.

(b) If $m = n$, then $\lim_{x\to\infty} f(x) = \lim_{x\to-\infty} f(x) = \dfrac{a_m}{b_n}$ and the line $y = \dfrac{a_m}{b_n}$ is a horizontal asymptote for $f(x)$.

(c) If $m > n$, then each limit will be ∞ or $-\infty$, depending on m, n, a_m, and b_n, and $f(x)$ does not have a horizontal asymptote

---

1. $y = 4$

3. $x = -6$

5. $y - 9 = 2(x + 2)$ (point-slope form); $2x - y = -13$.

7. Slope: $m = \dfrac{7-0}{0-9} = \dfrac{-7}{9}$; $y - 0 = \dfrac{-7}{9}(x - 9)$ (point-slope form); $7x + 9y = 63$.

9. $\lim_{x\to\infty} f(x) = -2$

11. $\lim_{x\to-2^+} f(x) = -\infty$

13. $\lim_{x\to-2} f(x)$ does not exist

15. $\lim_{x\to2^-} f(x) = 0$

17. $f(x) = \dfrac{x}{x-5}$

(A) $\lim_{x\to5^-} f(x) = -\infty$ (B) $\lim_{x\to5^+} f(x) = \infty$

(C) $\lim_{x\to5} f(x)$ does not exist.

19. $f(x) = \dfrac{2x-4}{(x-4)^2}$

(A) $\lim_{x\to4^-} f(x) = \infty$ (B) $\lim_{x\to4^+} f(x) = \infty$ (C) $\lim_{x\to4} f(x) = \infty$

21. $f(x) = \dfrac{x^2+x-2}{(x-1)} = \dfrac{(x-1)(x+2)}{x-1} = x+2$, provided $x \neq 1$

(A) $\lim_{x\to1^-} f(x) = \lim_{x\to1^-}(x+2) = 3$

(B) $\lim_{x\to1^+} f(x) = \lim_{x\to1^+}(x+2) = 3$

(C) $\lim_{x\to1} f(x) = \lim_{x\to1}(x+2) = 3$

23. $f(x) = \dfrac{x^2-3x+2}{x+2}$

(A) $\lim_{x\to-2^-} f(x) = -\infty$ (B) $\lim_{x\to-2^+} f(x) = \infty$

(C) $\lim_{x\to-2} f(x)$ does not exist.

25. $p(x) = 15+3x^2-5x^3 = -5x^3+3x^2+15$

(A) Leading term: $-5x^3$ (B) $\lim_{x\to\infty} p(x) = \lim_{x\to\infty}(-5x^3) = -\infty$ (C) $\lim_{x\to-\infty} p(x) = \lim_{x\to-\infty}(-5x^3) = \infty$

27. $p(x) = 9x^2-6x^4+7x = -6x^4+9x^2+7x$

(A) Leading term: $-6x^4$ (B) $\lim_{x\to\infty} p(x) = \lim_{x\to\infty}(-6x^4) = -\infty$ (C) $\lim_{x\to-\infty} p(x) = \lim_{x\to-\infty}(-6x^4) = -\infty$

29. $p(x) = x^2+7x+12$

(A) Leading term: x^2 (B) $\lim_{x\to\infty} p(x) = \lim_{x\to\infty}(x^2) = \infty$ (C) $\lim_{x\to-\infty} p(x) = \lim_{x\to-\infty}(x^2) = \infty$

31. $p(x) = x^4+2x^5-11x = 2x^5+x^4-11x$

(A) Leading term: $2x^5$ (B) $\lim_{x\to\infty} p(x) = \lim_{x\to\infty}(2x^5) = \infty$ (C) $\lim_{x\to-\infty} p(x) = \lim_{x\to-\infty}(2x^5) = -\infty$

33. $f(x) = \dfrac{1}{x+3}$; f is discontinuous at $x = -3$.

$\lim_{x\to-3^-} f(x) = -\infty$, $\lim_{x\to-3^+} f(x) = \infty$; $x = -3$ is a vertical asymptote.

35. $h(x) = \dfrac{x^2+4}{x^2-4} = \dfrac{x^2+4}{(x-2)(x+2)}$; h is discontinuous at $x = -2, x = 2$.

At $x = -2$:
$\lim\limits_{x\to -2^-} h(x) = \infty$, $\lim\limits_{x\to -2^+} h(x) = -\infty$; $x = -2$ is a vertical asymptote.

At $x = 2$:
$\lim\limits_{x\to 2^-} h(x) = -\infty$, $\lim\limits_{x\to 2^+} h(x) = \infty$; $x = 2$ is a vertical asymptote.

37. $F(x) = \dfrac{x^2-4}{x^2+4}$. Since $x^2 + 4 \neq 0$ for all x, F is continuous for all x; there are no vertical asymptotes.

39. $H(x) = \dfrac{x^2-2x-3}{x^2-4x+3} = \dfrac{(x-3)(x+1)}{(x-3)(x-1)}$; H is discontinuous at $x = 1, x = 3$.

At $x = 1$:
$\lim\limits_{x\to 1^-} H(x) = -\infty$, $\lim\limits_{x\to 1^+} H(x) = \infty$; $x = 1$ is a vertical asymptote.

At $x = 3$:
Since $\dfrac{(x-3)(x+1)}{(x-3)(x-1)} = \dfrac{x+1}{x-1}$ provided $x \neq 3$,

$\lim\limits_{x\to 3} H(x) = \lim\limits_{x\to 3}\left(\dfrac{x+1}{x-1}\right) = \dfrac{4}{2} = 2$; H does not have a vertical asymptote at $x = 3$.

41. $T(x) = \dfrac{8x-16}{x^4-8x^3+16x^2} = \dfrac{8(x-2)}{x^2(x^2-8x-16)} = \dfrac{8(x-2)}{x^2(x-4)^2}$

T is discontinuous at $x = 0, x = 4$.

At $x = 0$:
$\lim\limits_{x\to 0^-} T(x) = -\infty$, $\lim\limits_{x\to 0^+} T(x) = -\infty$, $x = 0$ is a vertical asymptote.

At $x = 4$:
$\lim\limits_{x\to 4^-} T(x) = \infty$, $\lim\limits_{x\to 4^+} T(x) = \infty$; $x = 4$ is a vertical asymptote.

43. $f(x) = \dfrac{4x+7}{5x-9}$

(A) $f(10) = \dfrac{4(10)+7}{5(10)-9} = \dfrac{47}{41} \approx 1.146$

(B) $f(100) = \dfrac{4(100)+7}{5(100)-9} = \dfrac{407}{491} \approx 0.829$

(C) $\lim\limits_{x\to\infty} \dfrac{4x+7}{5x-9} = \dfrac{4}{5} = 0.8$

45. $f(x) = \dfrac{5x^2+11}{7x-2}$

(A) $f(20) = \dfrac{5(20)^2+11}{7(20)-2} = \dfrac{2011}{138} \approx 14.572$

(B) $f(50) = \dfrac{5(50)^2+11}{7(50)-2} = \dfrac{12,511}{348} \approx 35.951$

(C) $\lim_{x\to\infty} \frac{5x^2+11}{7x-2} = \lim_{x\to\infty} \frac{5x^2}{7x} = \infty$ since $m = 2 > n = 1$

47. $f(x) = \frac{7x^4 - 14x^2}{6x^5 + 3}$

(A) $f(-6) = \frac{7(-6)^4 - 14(-6)^2}{6(-6)^5 + 3} = \frac{9{,}072 - 504}{-46{,}656 + 3} = -\frac{8{,}568}{46{,}653} \approx -0.184$

(B) $f(-12) = \frac{7(-12)^4 - 14(-12)^2}{6(-12)^5 + 3} = \frac{145{,}152 - 2{,}016}{-1{,}492{,}992 + 3} = -\frac{143{,}136}{1{,}492{,}989} \approx -0.096$

(C) $\lim_{x\to\infty} \frac{7x^4 - 14x^2}{6x^5 + 3} = \lim_{x\to\infty} \frac{7x^4}{6x^5} = 0$ since $m = 4 < n = 5$

49. $f(x) = \frac{10 - 7x^3}{4 + x^3}$

(A) $f(-10) = \frac{10 - 7(-10)^3}{4 + (-10)^3} = -\frac{7{,}010}{996} \approx -7.038$

(B) $f(-20) = \frac{10 - 7(-20)^3}{4 + (-20)^3} = -\frac{56{,}010}{7{,}996} \approx -7.005$

(C) $\lim_{x\to-\infty} \frac{10 - 7x^3}{4 + x^3} = \lim_{x\to-\infty} \frac{-7x^3}{x^3} = -7$

51. $f(x) = \frac{2x}{x+2}$; f is discontinuous at $x = -2$

$\lim_{x\to-2^-} f(x) = \infty$, $\lim_{x\to-2^+} f(x) = -\infty$; $x = -2$ is a vertical asymptote.

$\lim_{x\to\infty} \frac{2x}{x+2} = \lim_{x\to\infty} \frac{2x}{x} = 2$; $y = 2$ is a horizontal asymptote.

53. $f(x) = \frac{x^2+1}{x^2-1} = \frac{x^2+1}{(x-1)(x+1)}$; f is discontinuous at $x = -1, x = 1$.

At $x = -1$:
$\lim_{x\to-1^-} f(x) = \infty$, $\lim_{x\to-1^+} f(x) = -\infty$; $x = -1$ is a vertical asymptote.

At $x = 1$:
$\lim_{x\to1^-} f(x) = -\infty$, $\lim_{x\to1^+} f(x) = \infty$; $x = 1$ is a vertical asymptote.

$\lim_{x\to\infty} \frac{x^2+1}{x^2-1} = \lim_{x\to\infty} \frac{x^2}{x^2} = 1$; $y = 1$ is a horizontal asymptote.

55. $f(x)=\dfrac{x^3}{x^2+6}$. Since $x^2+6\neq 0$ for all x, f is continuous for all x; there are no vertical asymptotes.

$\lim\limits_{x\to\infty}\dfrac{x^3}{x^2+6}=\lim\limits_{x\to\infty}\dfrac{x^3}{x^2}=\lim\limits_{x\to\infty}x=\infty$; there are no horizontal asymptotes.

57. $f(x)=\dfrac{x}{x^2+4}$. Since $x^2+4\neq 0$ for all x, f is continuous for all x; there are no vertical asymptotes.

$\lim\limits_{x\to\infty}\dfrac{x}{x^2+4}=\lim\limits_{x\to\infty}\dfrac{x}{x^2}=\lim\limits_{x\to\infty}\dfrac{1}{x}=0$; $y=0$ is a horizontal asymptote.

59. $f(x)=\dfrac{x^2}{x-3}$; f is discontinuous at $x=3$.

At $x=3$:
$\lim\limits_{x\to 3^-}f(x)=-\infty$, $\lim\limits_{x\to 3^+}f(x)=\infty$; $x=3$ is a vertical asymptote.

$\lim\limits_{x\to\infty}\dfrac{x^2}{x-3}=\lim\limits_{x\to\infty}\dfrac{x^2}{x}=\lim\limits_{x\to\infty}x=\infty$; there are no horizontal asymptotes.

61. $f(x)=\dfrac{2x^2+3x-2}{x^2-x-2}=\dfrac{(x+2)(2x-1)}{(x-2)(x+1)}$; f is discontinuous at $x=-1$, $x=2$.

At $x=-1$:
$\lim\limits_{x\to -1^-}f(x)=-\infty$, $\lim\limits_{x\to -1^+}f(x)=\infty$; $x=-1$ is a vertical asymptote.

At $x=2$:
$\lim\limits_{x\to 2^-}f(x)=-\infty$, $\lim\limits_{x\to 2^+}f(x)=\infty$; $x=2$ is a vertical asymptote.

$\lim\limits_{x\to\infty}\dfrac{2x^2+3x-2}{x^2-x-2}=\lim\limits_{x\to\infty}\dfrac{2x^2}{x^2}=2$; $y=2$ is a horizontal asymptote.

63. $f(x)=\dfrac{2x^2-5x+2}{x^2-x-2}=\dfrac{(x-2)(2x-1)}{(x-2)(x+1)}$; f is discontinuous at $x=-1$, $x=2$.

At $x=-1$:
$\lim\limits_{x\to -1^-}f(x)=\infty$, $\lim\limits_{x\to -1^+}f(x)=-\infty$; $x=-1$ is a vertical asymptote.

At $x=2$:

$\lim\limits_{x\to 2}f(x)=\lim\limits_{x\to 2}\left(\dfrac{2x-1}{x+1}\right)=1$

$\lim\limits_{x\to\infty}\dfrac{2x^2-5x+2}{x^2-x-2}=\lim\limits_{x\to\infty}\dfrac{2x^2}{x^2}=2$; $y=2$ is a horizontal asymptote.

65. $f(x)=\dfrac{x+3}{x^2-5}$; $\lim\limits_{x\to\infty}f(x)=\lim\limits_{x\to\infty}\dfrac{x+3}{x^2-5}=\lim\limits_{x\to\infty}\dfrac{x}{x^2}=\lim\limits_{x\to\infty}\dfrac{1}{x}=0$

67. $f(x)=\dfrac{x^2-5}{x+3}$; $\lim\limits_{x\to\infty}f(x)=\lim\limits_{x\to\infty}\dfrac{x^2-5}{x+3}=\lim\limits_{x\to\infty}\dfrac{x^2}{x}=\lim\limits_{x\to 0}x=\infty$

69. $f(x)=\dfrac{5-2x^2}{1+8x^2}$; $\lim\limits_{x\to-\infty} f(x)=\lim\limits_{x\to-\infty}\dfrac{5-2x^2}{1+8x^2}=\lim\limits_{x\to-\infty}\dfrac{-2x^2}{8x^2}=-\dfrac{1}{4}$

71. $f(x)=\dfrac{x^2+4x}{3x+2}$; $\lim\limits_{x\to-\infty} f(x)=\lim\limits_{x\to-\infty}\dfrac{x^2+4x}{3x+2}=\lim\limits_{x\to-\infty}\dfrac{x^2}{3x}=\lim\limits_{x\to-\infty}\dfrac{x}{3}=-\infty$

73. $f(x)=x^3-3x+1$;

$\lim\limits_{x\to\infty} f(x)=\lim\limits_{x\to\infty}(x^3-3x+1)=\lim\limits_{x\to\infty}(x^3)=\infty$; $\lim\limits_{x\to-\infty} f(x)=\lim\limits_{x\to-\infty}(x^3-3x+1)=\lim\limits_{x\to-\infty}(x^3)=-\infty$

75. $f(x)=\dfrac{2+5x}{1-x}$;

$\lim\limits_{x\to\infty} f(x)=\lim\limits_{x\to\infty}\dfrac{2+5x}{1-x}=\lim\limits_{x\to\infty}\dfrac{5x}{-x}=-5$; $\lim\limits_{x\to-\infty} f(x)=\lim\limits_{x\to-\infty}\dfrac{2+5x}{1-x}=\lim\limits_{x\to-\infty}\dfrac{5x}{-x}=-5$

77. False. $f(x)=\dfrac{1}{x^2+1}$ has no vertical asymptotes.

79. False. $f(x)=\dfrac{x^2+1}{x+1}$ has no horizontal asymptote.

81. True. Since the domain of a polynomial function is all real numbers, a polynomial has no vertical asymptotes. Also, a polynomial function of degree $n\ge 1$ has no horizontal asymptotes.

83. If $n\ge 1$ and $a_n>0$, then

$\lim\limits_{n\to\infty}(a_nx^n+a_{n-1}x^{n-1}+\ldots+a_0)=\infty$

If $n\ge 1$ and $a_n<0$, then

$\lim\limits_{n\to\infty}(a_nx^n+a_{n-1}x^{n-1}+\ldots+a_0)=-\infty$

85. (A) Since $C(x)$ is a linear function of x, it can be written in the form

$$C(x)=mx+b$$

Since the fixed costs are \$200, $b=200$.

Also, $C(20)=3800$, so

$3800=m(20)+200$

$20m=3600$

$m=180$

Therefore, $C(x)=180x+200$

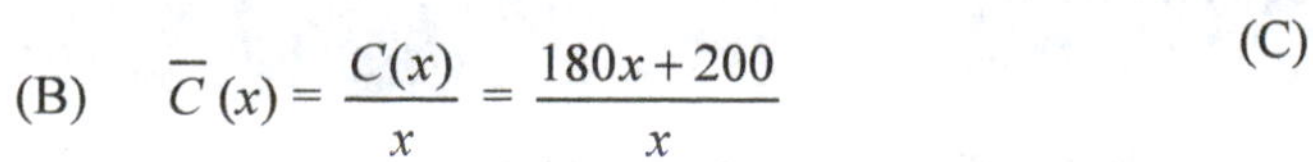

(B) $\overline{C}(x)=\dfrac{C(x)}{x}=\dfrac{180x+200}{x}$

(C)

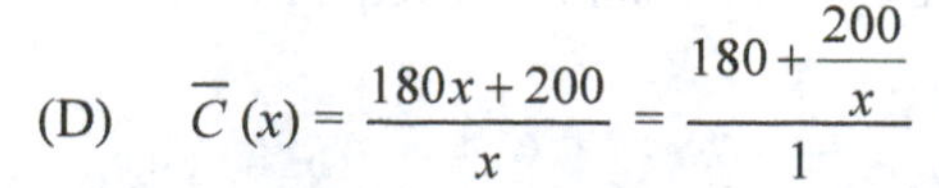

(D) $\overline{C}(x)=\dfrac{180x+200}{x}=\dfrac{180+\frac{200}{x}}{1}$

As x increases, the numerator tends to 180 and the denominator is 1. Therefore, $\overline{C}(x)$ tends to 180 or \$180 per board.

87. (A) $C_e(x) = 950 + 56x$; $\overline{C}_e(x) = \frac{C_e(x)}{x} = \frac{950+56x}{x} = \frac{950}{x} + 56$

(B) $C_c(x) = 900 + 66x$; $\overline{C}_c(x) = \frac{C_c(x)}{x} = \frac{900+66x}{x} = \frac{900}{x} + 66$

(C) Set $C_c(x) = C_e(x)$ and solve for x:

$$900 + 66x = 950 + 56x$$
$$10x = 50$$
$$x = 5$$

The total costs for the two models are equal at $x = 5$ years.

(D) Set $\overline{C}_c(x) = \overline{C}_e(x)$ and solve for x.

$$\frac{900}{x} + 66 = \frac{950}{x} + 56$$
$$\frac{900-950}{x} = -10$$
$$-\frac{50}{x} = -10$$
$$-10x = -50$$
$$x = 5$$

The average costs for the two models are equal at $x = 5$ years.

(E) $\lim_{x\to\infty} \overline{C}_e(x) = \lim_{x\to\infty}\left(\frac{950}{x}+56\right) = 56$

$\lim_{x\to\infty} \overline{C}_c(x) = \lim_{x\to\infty}\left(\frac{900}{x}+66\right) = 66$

For large x, the energy efficient model is approximately \$10 per year cheaper to operate than the conventional model.

89. $C(t) = \frac{5t^2(t+50)}{t^3+100} = \frac{5t^3+250t^2}{t^3+100}$

$\lim_{t\to\infty} C(t) = \lim_{t\to\infty} \frac{5t^3}{t^3} = 5$; the long-term drug concentration is 5 mg/ml.

91. $P(x) = \frac{2x}{1-x}, 0 \le x < 1$

(A) $P(0.9) = \frac{2(0.9)}{1-0.9} = \frac{1.8}{0.1} = 18$; \$18 million

(B) $P(0.95) = \frac{2(0.95)}{1-0.95} = \frac{1.9}{0.05} = 38$; \$38 million

(C) $\lim_{x\to1^-} P(x) = \lim_{x\to1^-} \frac{2x}{1-x} = \infty$; removal of 100% of the contamination would require an infinite amount of money; impossible.

93. $V(s) = \dfrac{V_{\max} s}{K_M + s}$

(A) $\lim_{s\to\infty} V(s) = \lim_{s\to\infty} \dfrac{V_{\max} s}{K_M + s} = \lim_{s\to\infty} \dfrac{V_{\max} s}{s} = V_{\max}$

(B) $V(K_M) = \dfrac{V_{\max} \cdot K_M}{K_M + K_M} = \dfrac{V_{\max} K_M}{2K_M} = \dfrac{V_{\max}}{2}$

(C)

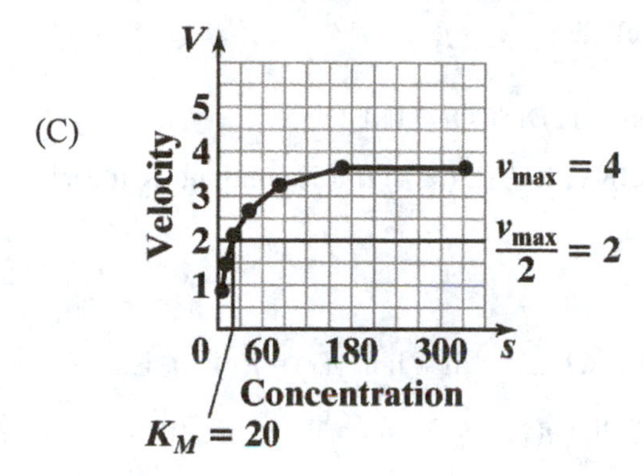

(D) $V(s) = \dfrac{4s}{20+s}$

(E) $V(15) = \dfrac{4(15)}{20+15} = \dfrac{60}{35} = \dfrac{12}{7}$

Set $V = 3$ and solve for s:

$3 = \dfrac{4s}{20+s}$

$60 + 3s = 4s$

$s = 60$

Thus, $s = 60$ when $V = 3$.

95. (A) $C_{\max} = 18, M = 150$

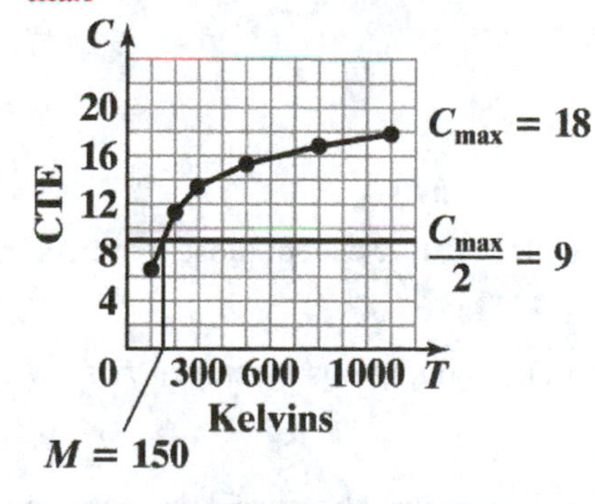

(B) $C(T) = \dfrac{18T}{150+T}$

(C) $C(600) = \dfrac{18(600)}{150+600} = 14.4$

To find T when $C(T) = 12$, solve $\dfrac{18T}{150+T} = 12$ for T:

$18T = 1800 + 12T$

$6T = 1800$

$T = 300$

Thus, $C(T) = 12$ at $T = 300\text{K}$

EXERCISE 2-3

Things to remember:

1. CONTINUITY

A function f is CONTINUOUS AT THE POINT $x = c$ if:

(a) $\lim_{x \to c} f(x)$ exists; (b) $f(c)$ exists; (c) $\lim_{x \to c} f(x) = f(c)$

If one or more of the three conditions fails, then f is DISCONTINUOUS at $x = c$.

A function is CONTINUOUS ON THE OPEN INTERVAL (a, b) if it is continuous at each point on the interval.

2. ONE-SIDED CONTINUITY

A function f is CONTINUOUS ON THE LEFT AT $x = c$ if $\lim_{x \to c^-} f(x) = f(c)$; f is CONTINUOUS ON THE RIGHT AT $x = c$ if $\lim_{x \to c^+} f(x) = f(c)$.

The function f is continuous on the closed interval $[a, b]$ if it is continuous on the open interval (a, b), and is continuous on the right at a and continuous on the left at b.

3. CONTINUITY PROPERTIES OF SOME SPECIFIC FUNCTIONS

(a) A constant function, $f(x) = k$, is continuous for all x.

(b) For n a positive integer, $f(x) = x^n$ is continuous for all x.

(c) A polynomial function

$P(x) = a_n x^n + a_{n-1} x^{n-1} + \dots + a_1 x + a_0$

is continuous for all x.

(d) A rational function

$$R(x) = \frac{P(x)}{Q(x)},$$

P and Q polynomial functions, is continuous for all x except those numbers $x = c$ such that $Q(c) = 0$.

(e) For n an odd positive integer, $n > 1$, $\sqrt[n]{f(x)}$ is continuous wherever f is continuous.

(f) For n an even positive integer, $\sqrt[n]{f(x)}$ is continuous wherever f is continuous and non-negative.

4. SIGN PROPERTIES ON AN INTERVAL (a, b)

If f is continuous or (a, b) and $f(x) \neq 0$ for all x in (a, b), then either $f(x) > 0$ for all x in (a, b) or $f(x) < 0$ for all x in (a, b).

5. CONSTRUCTING SIGN CHARTS

Given a function f:

Step 1. Find all partition numbers. That is:

(A) Find all numbers where f is discontinuous. (Rational functions are discontinuous for values of x that make a denominator 0.)

(B) Find all numbers where $f(x) = 0$. (For a rational function, this occurs where the numerator is 0 and the denominator is not 0.)

Step 2. Plot the numbers found in step 1 on a real number line, dividing the number line into intervals.

Step 3. Select a test number in each open interval determined in step 2, and evaluate $f(x)$ at each test number to determine whether $f(x)$ is positive (+) or negative (–) in each interval.

Step 4. Construct a sign chart using the real number line in step 2. This will show the sign of $f(x)$ on each open interval.

[*Note*: From the sign chart, it is easy to find the solution for the inequality $f(x) < 0$ or $f(x) > 0$.]

---

1. $[-3,5]$ **3** $(-10,100)$ **5.** $(-\infty,-5)\cup(5,\infty)$ **7.** $(-\infty,-1]\cup(2,\infty)$

Note: Answers to problems 9, 11, 13 may vary.

9. f is continuous at $x = 1$, since $\lim_{x\to1} f(x) = f(1) = 2$

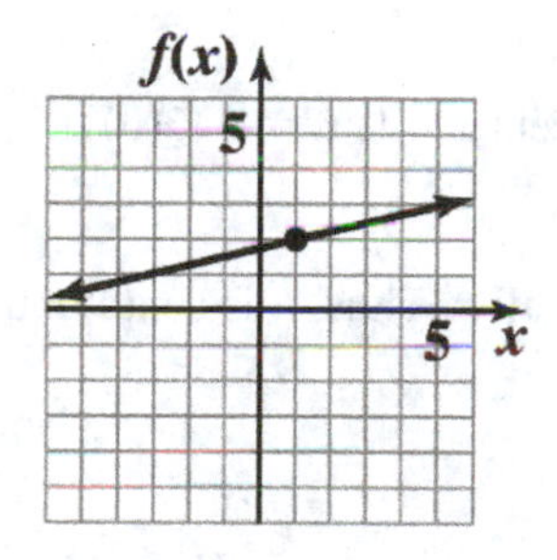

11. f is discontinuous at $x = 1$, since $\lim_{x\to1} f(x) \neq f(1)$

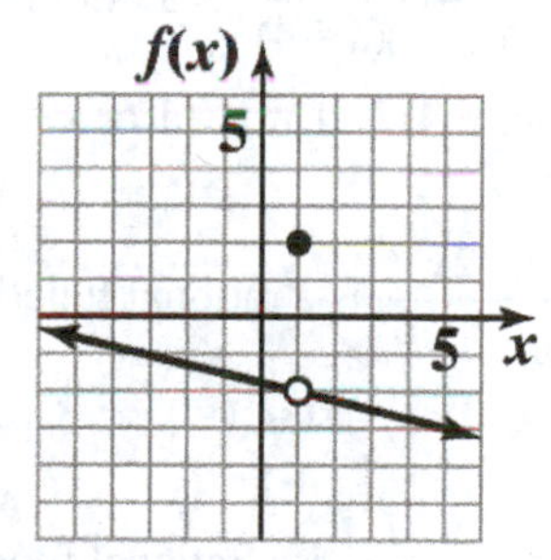

13. $\lim_{x\to1^-} f(x) = 2$, $\lim_{x\to1^+} f(x) = -2$

implies $\lim_{x\to1} f(x)$ does not exist;

f is discontinuous at $x = 1$, since $\lim_{x\to1} f(x)$ does not exist

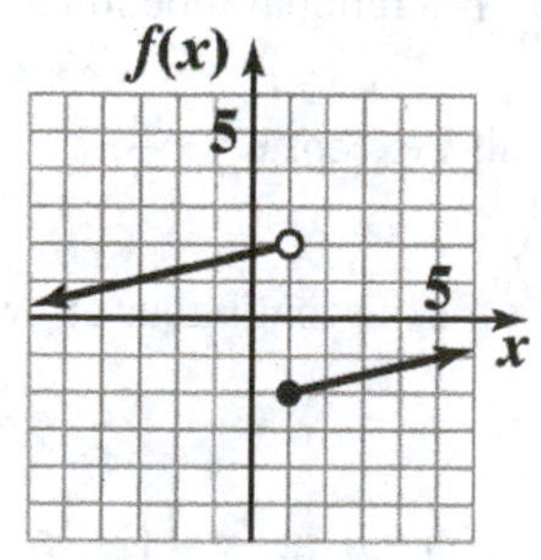

15. $f(0.9) \approx 1.9$ **17.** $f(-1.9) \approx 0.9$

19. (A) $\lim_{x\to1^-} f(x) = 2$ (B) $\lim_{x\to1^+} f(x) = 1$ (C) $\lim_{x\to1} f(x)$ does not exist (D) $f(1) = 1$

(E) No, because $\lim_{x\to1} f(x)$ does not exist.

21. (A) $\lim_{x\to -2^-} f(x) = 1$ (B) $\lim_{x\to -2^+} f(x) = 1$ (C) $\lim_{x\to -2} f(x) = 1$ (D) $f(-2) = 3$

(E) No, because $\lim_{x\to -2} f(x) \neq f(-2)$.

23. $g(-3.1) \approx 0.9$ **25.** $g(1.9) \approx 2.05$

27. (A) $\lim_{x\to -3^-} g(x) = 1$ (B) $\lim_{x\to -3^+} g(x) = 1$ (C) $\lim_{x\to -3} g(x) = 1$ (D) $g(-3) = 3$

(E) No, because $\lim_{x\to -3} g(x) \neq g(-3)$.

29. (A) $\lim_{x\to 2^-} g(x) = 2$ (B) $\lim_{x\to 2^+} g(x) = -1$ (C) $\lim_{x\to 2} g(x)$ does not exist (D) $g(2) = 2$

(E) No, because $\lim_{x\to 2} g(x)$ does not exist.

31. $f(x) = 3x - 4$ is a polynomial function. Therefore, f is continuous for all x [3(c)].

33. $g(x) = \dfrac{3x}{x+2}$ is a rational function and the denominator $x + 2$ is 0 at $x = -2$. Thus, g is continuous for all x except $x = -2$ [3(d)].

35. $m(x) = \dfrac{x+1}{(x-1)(x+4)}$ is a rational function and the denominator $(x - 1)(x + 4)$ is 0 at $x = 1$ or $x = -4$. Thus, m is continuous for all x except $x = 1, x = -4$ [3(d)].

37. $F(x) = \dfrac{2x}{x^2+9}$ is a rational function and the denominator $x^2 + 9 \neq 0$ for all x. Thus, F is continuous for all x.

39. $M(x) = \dfrac{x-1}{4x^2-9}$ is a rational function and the denominator $4x^2 - 9 = 0$ at $x = \dfrac{3}{2}, -\dfrac{3}{2}$. Thus, M is continuous for all x except $x = \pm\dfrac{3}{2}$.

41. $f(x) = \dfrac{3x+8}{x-4}$; f is discontinuous at $x = 4$; $f(x) = 0$ at $x = \dfrac{-8}{3}$. Partition numbers $4, \dfrac{-8}{3}$.

43. $f(x) = \dfrac{1-x^2}{1+x^2}$; $f(x) = 0$ at $x = 1, -1$; $1 + x^2 \neq 0$ for all x. Partition numbers $1, -1$.

45. $f(x) = \dfrac{x^2+4x-45}{x^2+6x} = \dfrac{(x+9)(x-5)}{x(x+6)}$; f is discontinuous at $x = 0, -6$; $f(x) = 0$ at $x = -9, 5$. Partition numbers $-9, -6, 0, 5$.

47. $x^2 - x - 12 < 0$

Let $f(x) = x^2 - x - 12 = (x-4)(x+3)$. Then f is continuous for all x and $f(-3) = f(4) = 0$. Thus, $x = -3$ and $x = 4$ are partition numbers.

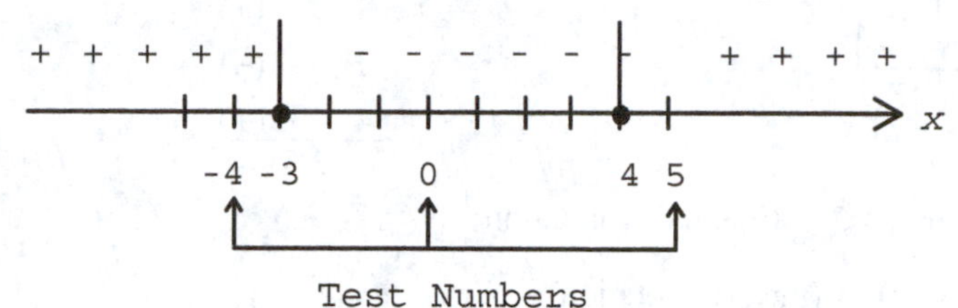

Test Numbers

x	$f(x)$
−4	8(+)
0	−12 (−)
5	8(+)

Thus, $x^2 - x - 12 < 0$ for:

$-3 < x < 4$ (inequality notation)

$(-3, 4)$ (interval notation)

49. $x^2 + 21 > 10x$ or $x^2 - 10x + 21 > 0$

Let $f(x) = x^2 - 10x + 21 = (x-7)(x-3)$. Then f is continuous for all x and $f(3) = f(7) = 0$. Thus, $x = 3$ and $x = 7$ are partition numbers.

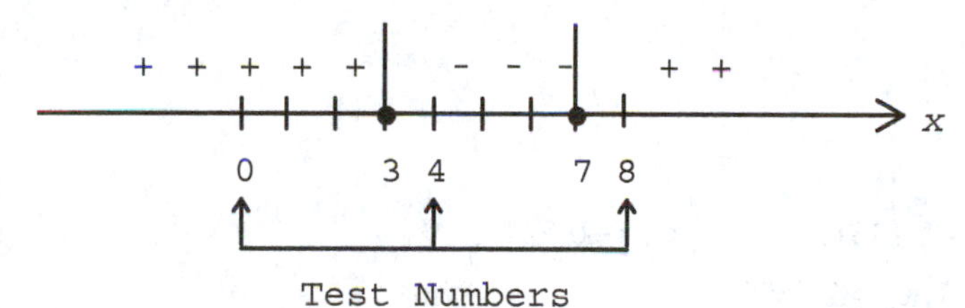

Test Numbers

x	$f(x)$
0	21(+)
4	−3 (−)
8	5(+)

Thus, $x^2 - 10x + 21 > 0$ for:

$x < 3$ or $x > 7$ (inequality notation)

$(-\infty, 3) \cup (7, \infty)$ (interval notation)

51. $x^3 < 4x$ or $x^3 - 4x < 0$

Let $f(x) = x^3 - 4x = x(x^2 - 4) = x(x-2)(x+2)$. Then f is continuous for all x and $f(-2) = f(0) = f(2) = 0$. Thus, $x = -2$, $x = 0$ and $x = 2$ are partition numbers.

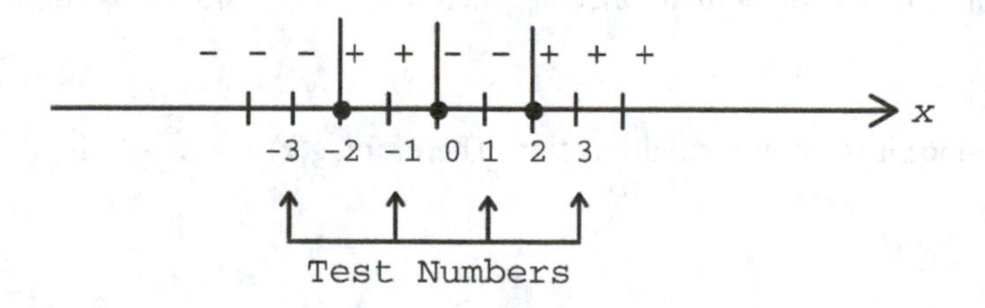

Test Numbers

x	$f(x)$
−3	−15(−)
−1	3 (+)
1	−3(−)
3	15(+)

Thus, $x^3 < 4x$ for:

$-\infty < x < -2$ or $0 < x < 2$ (inequality notation)

$(-\infty, -2) \cup (0, 2)$ (interval notation)

53. $\dfrac{x^2 + 5x}{x - 3} > 0$

Let $f(x) = \dfrac{x^2 + 5x}{x - 3} = \dfrac{x(x+5)}{x-3}$. Then f is discontinuous at $x = 3$ and $f(0) = f(-5) = 0$. Thus, $x = -5$, $x = 0$, and $x = 3$ are partition numbers.

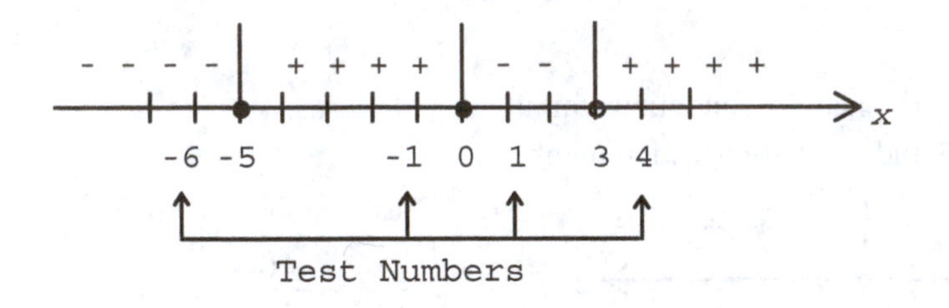

Test Numbers

x	$f(x)$
–6	$-\frac{2}{3}(-)$
–1	1(+)
1	–3(–)
4	36(+)

Thus, $\frac{x^2+5x}{x-3} > 0$ for: $-5 < x < 0$ or $x > 3$ (inequality notation)

$(-5, 0) \cup (3, \infty)$ (interval notation)

55. (A) $f(x) > 0$ on $(-4, -2) \cup (0, 2) \cup (4, \infty)$
(B) $f(x) < 0$ on $(-\infty, -4) \cup (-2, 0) \cup (2, 4)$

57. $f(x) = x^4 - 6x^2 + 3x + 5$
Partition numbers: $x_1 \approx -2.5308$, $x_2 \approx -0.7198$

(A) $f(x) > 0$ on $(-\infty, -2.5308) \cup (-0.7198, \infty)$

(B) $f(x) < 0$ on $(-2.5308, -0.7198)$

59. $f(x) = \frac{3+6x-x^3}{x^2-1}$
Partition numbers: $x_1 \approx -2.1451$, $x_2 = -1$, $x_3 \approx -0.5240$,
$x_4 = 1$, $x_5 \approx 2.6691$

(A) $f(x) > 0$ on $(-\infty, -2.1451) \cup (-1, -0.5240) \cup (1, 2.6691)$

(B) $f(x) < 0$ on $(-2.1451, -1) \cup (-0.5240, 1) \cup (2.6691, \infty)$

61. $f(x) = x - 6$ is continuous for all x since it is a polynomial function. Therefore, $g(x) = \sqrt{x-6}$ is continuous for all x such that $x - 6 \geq 0$, that is, for all x in $[6, \infty)$ [see 3(f)].

63. $f(x) = 5 - x$ is continuous for all x since it is a polynomial function. Therefore, $F(x) = \sqrt[3]{5-x}$ is continuous for all x, that is, for all x in $(-\infty, \infty)$.

65. $f(x) = x^2 - 9$ is continuous for all x since it is a polynomial function. Therefore, $g(x) = \sqrt{x^2-9}$ is continuous for all x such that
$x^2 - 9 = (x-3)(x+3) \geq 0$.

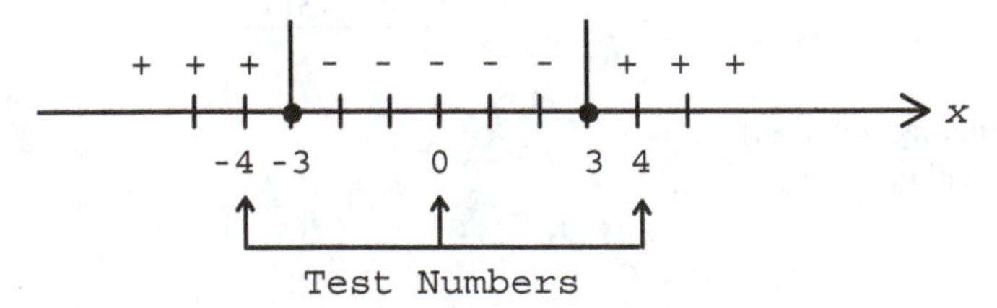

Test Numbers

x	$f(x)$
–4	7
0	–9
4	7

$\sqrt{x^2-9}$ is continuous on $(-\infty, -3] \cup [3, \infty)$.

67. $f(x) = x^2 + 1$ is continuous for all x since it is a polynomial function. Also $x^2 + 1 \geq 1 > 0$ for all x. Therefore, $\sqrt{x^2+1}$ is continuous for all x, that is, for all x in $(-\infty, \infty)$.

69. The graph of f is shown at the right. This function is discontinuous at $x = 1$. [$\lim_{x\to 1} f(x)$ does not exist.]

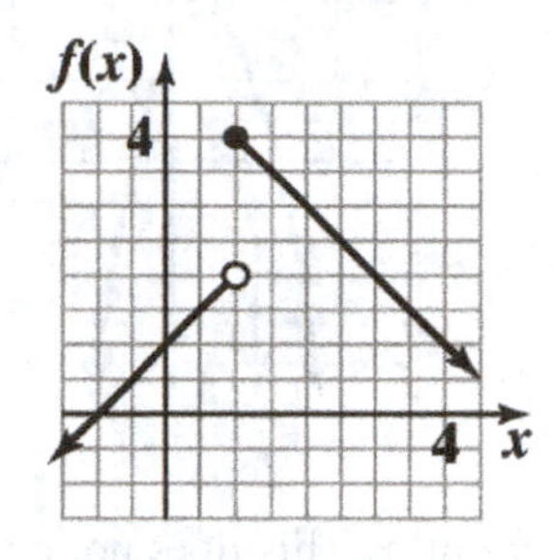

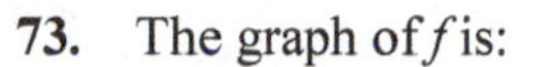

71. The graph of f is:

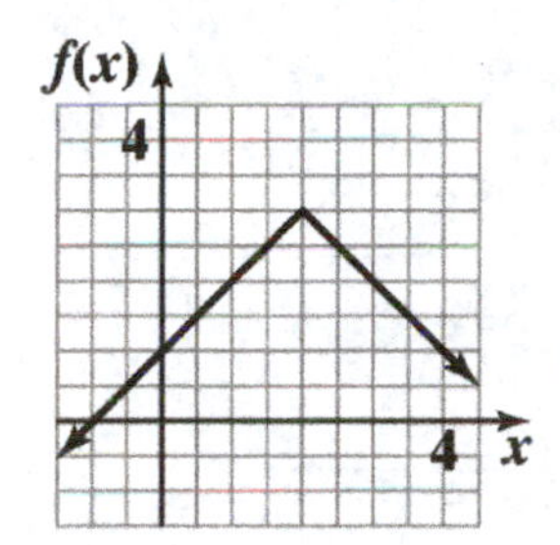

This function is continuous for all x.

$\left[\lim_{x\to 2} f(x) = f(2) = 3.\right]$

73. The graph of f is:

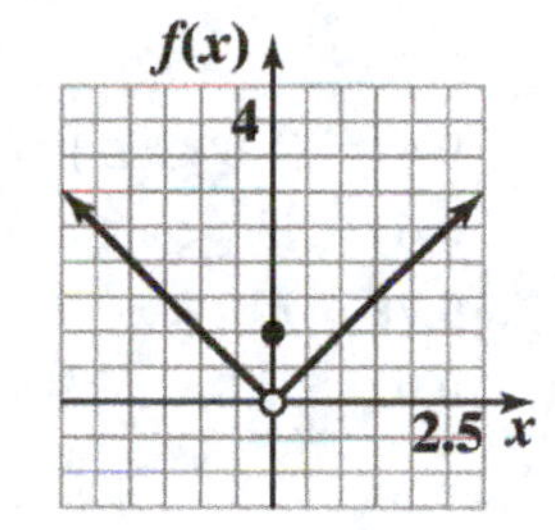

This function is discontinuous at $x = 0$.

$\left[\lim_{x\to 0} f(x) = 0 \neq f(0) = 1.\right]$

75. (A) Since $\lim_{x\to 0^+} f(x) = f(0) = 0$, f is continuous from the right at $x = 0$.

(B) Since $\lim_{x\to 0^-} f(x) = -1 \neq f(0) = 0$, f is not continuous from the left at $x = 0$.

(C) f is continuous on the open interval $(0, 1)$.

(D) f is *not* continuous on the closed interval $[0, 1]$ since $\lim_{x\to 1^-} f(x) = 0 \neq f(1) = 1$, i.e., f is not continuous from the left at $x = 1$.

(E) f is continuous on the half-closed interval $[0, 1)$.

77. True. Theorem 1(c) in the text and 3c in *Things to Remember*.

79. False. Set $f(x) = \dfrac{1}{x-1}$. Then f is continuous at $x = 0$ and $x = 2$, but f is not continuous at $x = 1$.

81. True. If f has no partition numbers, then f has no points of discontinuity.

83. 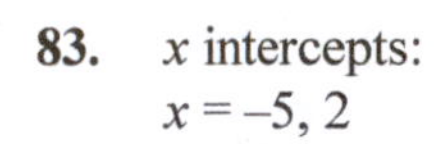
x intercepts: $x = -5, 2$

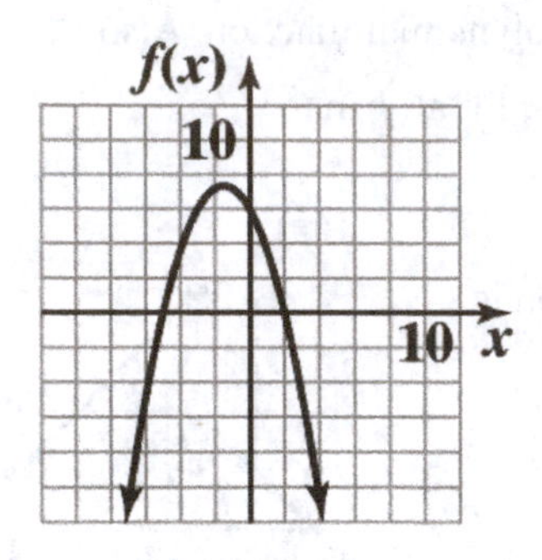

85. x intercepts: $x = -6, -1, 4$

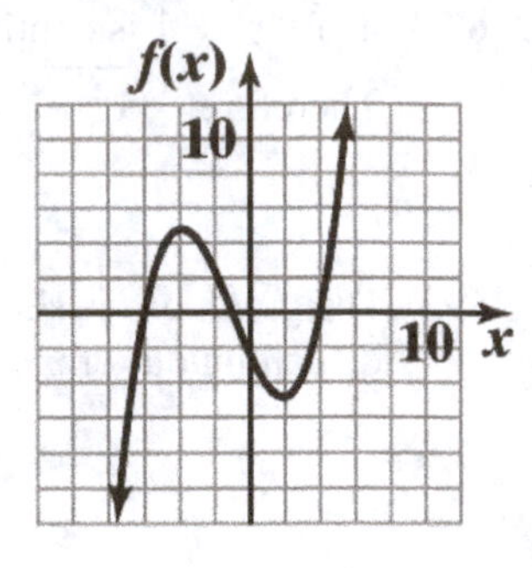

87. $f(x) = \dfrac{2}{1-x} \neq 0$ for all x. This does not contradict Theorem 2 because f is not continuous on $(-1, 3)$; f is discontinuous at $x = 1$.

89. (A)

$$P(x) = \begin{cases} 0.44 & \text{if } 0 < x \leq 1 \\ 0.61 & \text{if } 1 < x \leq 2 \\ 0.78 & \text{if } 2 < x \leq 3 \\ 0.95 & \text{if } 3 < x < 3.5 \end{cases}$$

(B)

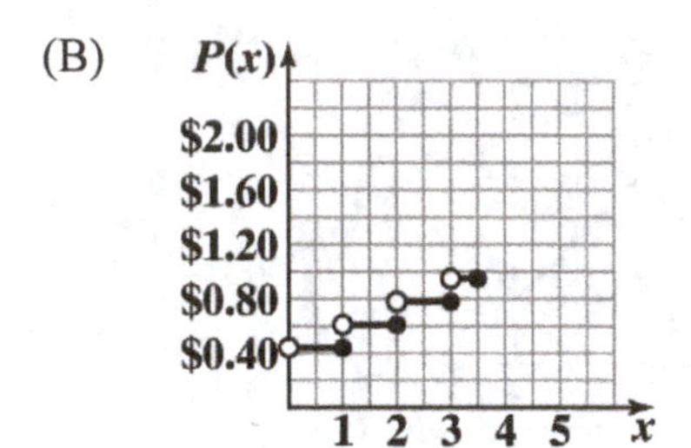

(C) P is continuous at $x = 2.5$ since P is a continuous function on $(2, 3]$; $P(x) = 0.88$ for $2 < x \leq 3$.
P is not continuous at $x = 3$ since $\lim_{x\to 3^-} P(x) = 0.78 = P(3) \neq \lim_{x\to 3^+} P(x) = 0.95$.

91. Q is defined for all real numbers whereas P is defined only for $x \in (0, 3.5]$.

93. (A) $S(x) = 5.00 + 0.63x$ if $0 \leq x \leq 50$;
$S(50) = 36.50$;
$S(x) = 36.50 + 0.45(x - 50)$
$= 14 + 0.45x$ if $x > 50$

Therefore, $S(x) = \begin{cases} 5.00 + 0.63x & \text{if } 0 \leq x \leq 50 \\ 14.00 + 0.45x & \text{if } x > 50 \end{cases}$

(B)

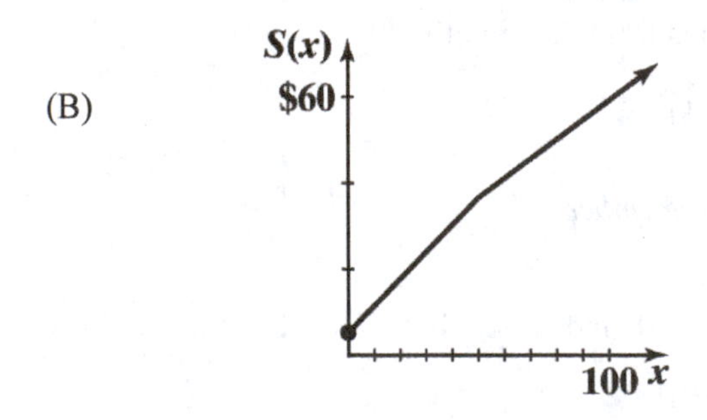

(C) $S(x)$ is continuous at $x = 50$;

$$\lim_{x\to 50^-} S(x) = \lim_{x\to 50^+} S(x)$$
$$= \lim_{x\to 50} S(x) = S(50) = 36.5.$$

95. (A) $E(s) = \begin{cases} 1000, 0 \le s \le 10,000 \\ 1000 + 0.05(s - 10,000), 10,000 < s < 20,000 \\ 1500 + 0.05(s - 10,000), s \ge 20,000 \end{cases}$

The graph of E is:

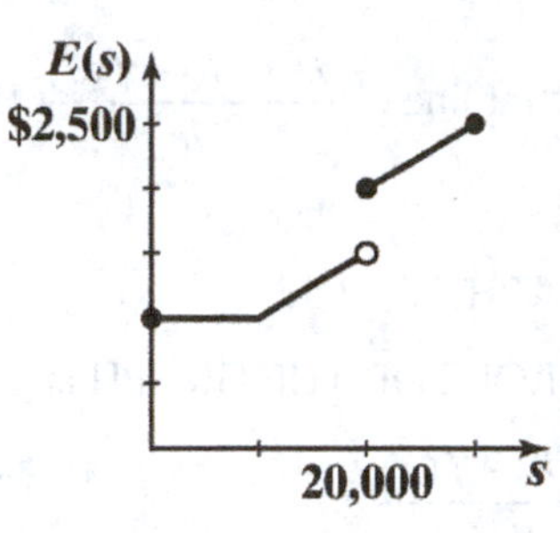

(B) From the graph, $\lim_{s \to 10,000} E(s) = \$1000$ and $E(10,000) = \$1000$.

(C) From the graph, $\lim_{s \to 20,000} E(s)$ does not exist. $E(20,000) = \$2000$.

(D) E is continuous at 10,000; E is not continuous at 20,000.

97. (A) From the graph, N is discontinuous at $t = t_2$, $t = t_3$, $t = t_4$, $t = t_6$, and $t = t_7$.

(B) From the graph, $\lim_{t \to t_5} N(t) = 7$ and $N(t_5) = 7$.

(C) From the graph, $\lim_{t \to t_3} N(t)$ does not exist; $N(t_3) = 4$.

EXERCISE 2-4

Things to remember:

1. AVERAGE RATE OF CHANGE

For $y = f(x)$, the AVERAGE RATE OF CHANGE FROM $x = a$ TO $x = a + h$ is

$$\frac{f(a+h) - f(a)}{(a+h) - a} = \frac{f(a+h) - f(a)}{h} \quad h \neq 0$$

The expression $\frac{f(a+h) - f(a)}{h}$ is called the DIFFERENCE QUOTIENT.

2. INSTANTANEOUS RATE OF CHANGE

For $y = f(x)$, the INSTANTANEOUS RATE OF CHANGE AT $x = a$ is

$$\lim_{h \to 0} \frac{f(a+h) - f(a)}{h}$$

if the limit exists.

3. SECANT LINE

A line through two points on the graph of a function is called a SECANT LINE. If $(a, f(a))$ and $((a + h), f(a + h))$ are two points on the graph of $y = f(x)$, then

$$\text{Slope of secant line} = \frac{f(a+h)-f(a)}{h} \quad \text{[Difference quotient]}$$

4. SLOPE OF A GRAPH

For $y = f(x)$, the SLOPE OF THE GRAPH at the point $(a, f(a))$ is given by

$$\lim_{h\to 0} \frac{f(a+h)-f(a)}{h}$$

provided the limit exists. The slope of the graph is also the SLOPE OF THE TANGENT LINE at the point $(a, f(a))$.

5. THE DERIVATIVE

For $y = f(x)$, we define THE DERIVATIVE OF f AT x, denoted by $f'(x)$, to be

$$f'(x) = \lim_{h\to 0} \frac{f(x+h)-f(x)}{h} \quad \text{if the limit exists.}$$

If $f'(x)$, exists for each x in the open interval (a, b), then f is said to be DIFFERENTIABLE OVER (a, b).

6. INTERPRETATIONS OF THE DERIVATIVE

The derivative of a function f is a new function f'. The domain of f' is a subset of the domain of f. Interpretations of the derivative are:

a. Slope of the tangent line. For each x in the domain of f', $f'(x)$ is the slope of the line tangent to the graph of f at the point $(x, f(x))$.

b. Instantaneous rate of change. For each x in the domain of f', $f'(x)$ is the instantaneous rate of change of $y = f(x)$ with respect to x.

c. Velocity. If $f(x)$ is the position of a moving object at time x, then $v = f'(x)$, is the velocity of the object at that time.

7. THE FOUR STEP PROCESS FOR FINDING THE DERIVATIVE OF A FUNCTION f.

Step 1. Find $f(x + h)$.

Step 2. Find $f(x + h) - f(x)$.

Step 3. Find $\dfrac{f(x+h)-f(x)}{h}$.

Step 4. Find $\displaystyle\lim_{h\to 0} \frac{f(x+h)-f(x)}{h}$.

---

1. Slope $m = \dfrac{16-7}{6-2} = \dfrac{9}{4}$, 2.25

3. Slope $m = \dfrac{68-14}{0-10} = \dfrac{-54}{10} = \dfrac{-27}{5}$; -5.4

5. $\frac{1}{\sqrt{3}} = \frac{1}{\sqrt{3}} \cdot \frac{\sqrt{3}}{\sqrt{3}} = \frac{\sqrt{3}}{3}$

7. $\frac{5}{3+\sqrt{7}} = \frac{5}{3+\sqrt{7}} \cdot \frac{3-\sqrt{7}}{3-\sqrt{7}} = \frac{15-5\sqrt{7}}{2} = \frac{15}{2} - \frac{5}{2}\sqrt{7}$

9. (A) $\frac{f(2)-f(1)}{2-1} = \frac{1-4}{1} = -3$ is the slope of the secant line through $(1, f(1))$ and $(2, f(2))$.

(B) $\frac{f(1+h)-f(1)}{h} = \frac{5-(1+h)^2-4}{h} = \frac{5-[1+2h+h^2]-4}{h} = \frac{-2h-h^2}{h} = -2-h$;

slope of the secant line through $(1, f(1))$ and $(1+h, f(1+h))$

(C) $\lim_{h\to 0} \frac{f(1+h)-f(1)}{h} = \lim_{h\to 0} (-2-h) = -2$;

slope of the tangent line at $(1, f(1))$

11. $f(x) = 3x^2$

(A) Slope of secant line through $(1, f(1))$ and $(4, f(4))$:

$$\frac{f(4)-f(1)}{4-1} = \frac{3(4)^2-3(1)^2}{4-1} = \frac{48-3}{3} = \frac{45}{3} = 15.$$

(B) Slope of secant line through $(1, f(1))$ and $(1+h, f(1+h))$:

$$\frac{3(1+h)^2-3(1)^2}{1+h-1} = \frac{3(1+2h+h^2)-3}{h} = \frac{6h+3h^2}{h} = 6+3h$$

(C) Slope of the graph at $(1, f(1))$: $\lim_{h\to 0} \frac{f(1+h)-f(1)}{h} = \lim_{h\to 0}(6+3h) = 6.$

13. (A) Distance traveled for $0 \le t \le 2$: 80 km; average velocity: $v = \frac{80}{2} = 40$ km/h..

(B) $\frac{f(2)-f(0)}{2-0} = \frac{80}{2} = 40.$

(C) Slope at $x = 2$: m = 45. Equation of tangent line at $(2, f(2))$: $y-80 = 45(x-2)$.

15. $f(x) = \frac{1}{1+x^2}$; $f(1) = \frac{1}{2}$. Equation of tangent line: $y - \frac{1}{2} = -\frac{1}{2}(x-1)$.

17. $f(x) = x^4$; $f(-2) = 16$. Equation of tangent line: $y-16 = -32(x+2)$ or $y = -32x-48$.

19. $f(x) = -5$

Step 1. Find $f(x+h)$:
$f(x+h) = -5$

Step 2. Find $f(x+h) - f(x)$:
$f(x+h) - f(x) = -5 - (-5) = 0$

Step 3. Find $\frac{f(x+h)-f(x)}{h}$:

$$\frac{f(x+h)-f(x)}{h} = \frac{0}{h} = 0$$

Step 4. Find $\lim_{h\to 0}\frac{f(x+h)-f(x)}{h}$:

$$\lim_{h\to 0}\frac{f(x+h)-f(x)}{h} = \lim_{h\to 0} 0 = 0$$

Thus, $f'(x) = 0$.

$f'(1) = 0,\ f'(2) = 0,\ f'(3) = 0$

21. $f(x) = 3x - 7$

Step 1. Find $f(x+h)$:

$f(x+h) = 3(x+h) - 7 = 3x + 3h - 7$

Step 2. Find $f(x+h) - f(x)$:

$f(x+h) - f(x) = 3x + 3h - 7 - (3x - 7) = 3h$

Step 3. Find $\frac{f(x+h)-f(x)}{h}$:

$$\frac{f(x+h)-f(x)}{h} = \frac{3h}{h} = 3$$

Step 4. Find $\lim_{h\to 0}\frac{f(x+h)-f(x)}{h}$:

$$\lim_{h\to 0}\frac{f(x+h)-f(x)}{h} = \lim_{h\to 0} 3 = 3$$

Thus, $f'(x) = 3$.

$f'(1) = 3,\ f'(2) = 3,\ f'(3) = 3$

23. $f(x) = 2 - 3x^2$

Step 1. Find $f(x+h)$:

$f(x+h) = 2 - 3(x+h)^2 = 2 - 3(x^2 + 2xh + h^2) = 2 - 3x^2 - 6xh - 3h^2$

Step 2. Find $f(x+h) - f(x)$:

$f(x+h) - f(x) = 2 - 3x^2 - 6xh - 3h^2 - (2 - 3x^2) = -6xh - 3h^2$

Step 3. Find $\frac{f(x+h)-f(x)}{h}$:

$$\frac{f(x+h)-f(x)}{h} = \frac{-6xh-3h^2}{h} = -6x - 3h$$

Step 4. Find $\lim_{h\to 0}\frac{f(x+h)-f(x)}{h}$:

$$\lim_{h\to 0}\frac{f(x+h)-f(x)}{h} = \lim_{h\to 0}(-6x - 3h) = -6x$$

Thus, $f'(x) = -6x$.
$f'(1) = -6$, $f'(2) = -12$, $f'(3) = -18$

25. $f(x) = x^2 + 6x - 10$

Step 1. Find $f(x + h)$:
$f(x + h) = (x + h)^2 + 6(x + h) - 10 = x^2 + 2xh + h^2 + 6x + 6h - 10$

Step 2. Find $f(x + h) - f(x)$:
$f(x + h) - f(x) = x^2 + 2xh + h^2 + 6x + 6h - 10 - (x^2 + 6x - 10) = 2xh + h^2 + 6h$

Step 3. Find $\frac{f(x+h)-f(x)}{h}$:
$$\frac{f(x+h)-f(x)}{h} = \frac{2xh+h^2+6h}{h} = 2x + h + 6$$

Step 4. Find $\lim_{h\to 0} \frac{f(x+h)-f(x)}{h}$:
$$\lim_{h\to 0} \frac{f(x+h)-f(x)}{h} = \lim_{h\to 0}(2x + h + 6) = 2x + 6$$
Thus, $f'(x) = 2x + 6$.
$f'(1) = 8$, $f'(2) = 10$, $f'(3) = 12$

27. $f(x) = 2x^2 - 7x + 3$

Step 1. Find $f(x + h)$:
$$f(x + h) = 2(x + h)^2 - 7(x + h) + 3 = 2(x^2 + 2xh + h^2) - 7x - 7h + 3$$
$$= 2x^2 + 4xh + 2h^2 - 7x - 7h + 3$$

Step 2. Find $f(x + h) - f(x)$:
$$f(x + h) - f(x) = 2x^2 + 4xh + 2h^2 - 7x - 7h + 3 - (2x^2 - 7x + 3)$$
$$= 4xh + 2h^2 - 7h$$

Step 3. Find $\frac{f(x+h)-f(x)}{h}$:
$$\frac{f(x+h)-f(x)}{h} = \frac{4xh+2h^2-7h}{h} = 4x + 2h - 7$$

Step 4. Find $\lim_{h\to 0} \frac{f(x+h)-f(x)}{h}$:
$$\lim_{h\to 0} \frac{f(x+h)-f(x)}{h} = \lim_{h\to 0}(4x + 2h - 7) = 4x - 7$$
Thus, $f'(x) = 4x - 7$.
$f'(1) = -3$, $f'(2) = 1$, $f'(3) = 5$

29. $f(x) = -x^2 + 4x - 9$

Step 1. Find $f(x + h)$:

$$f(x+h) = -(x+h)^2 + 4(x+h) - 9 = -(x^2 + 2xh + h^2) + 4x + 4h - 9$$
$$= -x^2 - 2xh - h^2 + 4x + 4h - 9$$

Step 2. Find $f(x + h) - f(x)$:

$$f(x+h) - f(x) = -x^2 - 2xh - h^2 + 4x + 4h - 9 - (-x^2 + 4x - 9)$$
$$= -2xh - h^2 + 4h$$

Step 3. Find $\frac{f(x+h)-f(x)}{h}$:

$$\frac{f(x+h)-f(x)}{h} = \frac{-2xh - h^2 + 4h}{h} = -2x - h + 4$$

Step 4. Find $\lim_{h\to 0} \frac{f(x+h)-f(x)}{h}$:

$$\lim_{h\to 0} \frac{f(x+h)-f(x)}{h} = \lim_{h\to 0} (-2x - h + 4) = -2x + 4$$

Thus, $f'(x) = -2x + 4$.

$f'(1) = 2,\ \ f'(2) = 0,\ \ f'(3) = -2$

31. $f(x) = 2x^3 + 1$

Step 1. Find $f(x + h)$:

$$f(x+h) = 2(x+h)^3 + 1 = 2(x^3 + 3x^2h + 3xh^2 + h^3) + 1$$
$$= 2x^3 + 6x^2h + 6xh^2 + 2h^3 + 1$$

Step 2. Find $f(x + h) - f(x)$:

$$f(x+h) - f(x) = 2x^3 + 6x^2h + 6xh^2 + 2h^3 + 1 - (2x^3 + 1)$$
$$= 6x^2h + 6xh^2 + 2h^3$$

Step 3. Find $\frac{f(x+h)-f(x)}{h}$:

$$\frac{f(x+h)-f(x)}{h} = \frac{6x^2h + 6xh^2 + 2h^3}{h} = 6x^2 + 6xh + 2h^2$$

Step 4. Find $\lim_{h\to 0} \frac{f(x+h)-f(x)}{h}$:

$$\lim_{h\to 0} \frac{f(x+h)-f(x)}{h} = \lim_{h\to 0} (6x^2 + 6xh + 2h^2) = 6x^2$$

Thus, $f'(x) = 6x^2$.

$f'(1) = 6,\ \ f'(2) = 24,\ \ f'(3) = 54$

33. $f(x) = 4 + \frac{4}{x}$

Step 1. Find $f(x+h)$:

$$f(x+h) = 4 + \frac{4}{x+h}$$

Step 2. Find $f(x+h) - f(x)$:

$$f(x+h) - f(x) = 4 + \frac{4}{x+h} - \left(4 + \frac{4}{x}\right) = \frac{4}{x+h} - \frac{4}{x}$$

$$= \frac{4x - 4(x+h)}{x(x+h)} = -\frac{4h}{x(x+h)}$$

Step 3. Find $\frac{f(x+h) - f(x)}{h}$:

$$\frac{f(x+h) - f(x)}{h} = \frac{-\frac{4h}{x(x+h)}}{h} = -\frac{4}{x(x+h)}$$

Step 4. Find $\lim_{h \to 0} \frac{f(x+h) - f(x)}{h}$:

$$\lim_{h \to 0} \frac{f(x+h) - f(x)}{h} = \lim_{h \to 0} -\frac{4}{x(x+h)} = -\frac{4}{x^2}$$

Thus, $f'(x) = -\frac{4}{x^2}$.

$f'(1) = -4, \quad f'(2) = -1, \quad f'(3) = -\frac{4}{9}$

35. $f(x) = 5 + 3\sqrt{x}$

Step 1. Find $f(x+h)$:

$$f(x+h) = 5 + 3\sqrt{x+h}$$

Step 2. Find $f(x+h) - f(x)$:

$$f(x+h) - f(x) = 5 + 3\sqrt{x+h} - (5 + 3\sqrt{x}) = 3(\sqrt{x+h} - \sqrt{x})$$

Step 3. Find $\frac{f(x+h) - f(x)}{h}$:

$$\frac{f(x+h) - f(x)}{h} = \frac{3(\sqrt{x+h} - \sqrt{x})}{h} = \frac{3(\sqrt{x+h} - \sqrt{x})}{h} \cdot \frac{(\sqrt{x+h} + \sqrt{x})}{(\sqrt{x+h} + \sqrt{x})}$$

$$= \frac{3(x+h-x)}{h(\sqrt{x+h} + \sqrt{x})} = \frac{3h}{h(\sqrt{x+h} + \sqrt{x})} = \frac{3}{\sqrt{x+h} + \sqrt{x}}$$

Step 4. Find $\lim_{h \to 0} \frac{f(x+h) - f(x)}{h}$:

$$\lim_{h \to 0} \frac{f(x+h) - f(x)}{h} = \lim_{h \to 0} \frac{3}{\sqrt{x+h} + \sqrt{x}} = \frac{3}{2\sqrt{x}}$$

Thus, $f'(x) = \dfrac{3}{2\sqrt{x}}$.

$f'(1) = \dfrac{3}{2}$, $f'(2) = \dfrac{3}{2\sqrt{2}} = \dfrac{3\sqrt{2}}{4}$, $f'(3) = \dfrac{3}{2\sqrt{3}} = \dfrac{\sqrt{3}}{2}$

37. $f(x) = 10\sqrt{x+5}$

Step 1. Find $f(x+h)$:
$f(x+h) = 10\sqrt{x+h+5}$

Step 2. Find $f(x+h) - f(x)$:
$f(x+h) - f(x) = 10\sqrt{x+h+5} - 10\sqrt{x+5} = 10\left(\sqrt{x+h+5} - \sqrt{x+5}\right)$

Step 3. Find $\dfrac{f(x+h)-f(x)}{h}$:

$$\frac{f(x+h)-f(x)}{h} = \frac{10\left(\sqrt{x+h+5}-\sqrt{x+5}\right)}{h}$$
$$= \frac{10\left(\sqrt{x+h+5}-\sqrt{x+5}\right)}{h} \cdot \frac{\left(\sqrt{x+h+5}+\sqrt{x+5}\right)}{\left(\sqrt{x+h+5}+\sqrt{x+5}\right)}$$
$$= \frac{10[x+h+5-(x+5)]}{h\left(\sqrt{x+h+5}+\sqrt{x+5}\right)} = \frac{10h}{h\left(\sqrt{x+h+5}+\sqrt{x+5}\right)}$$
$$= \frac{10}{\sqrt{x+h+5}+\sqrt{x+5}}$$

Step 4. Find $\lim_{h\to 0}\dfrac{f(x+h)-f(x)}{h}$:

$$\lim_{h\to 0}\frac{f(x+h)-f(x)}{h} = \lim_{h\to 0}\frac{10}{\sqrt{x+h+5}+\sqrt{x+5}} = \frac{10}{2\sqrt{x+5}} = \frac{5}{\sqrt{x+5}}$$

Thus, $f'(x) = \dfrac{5}{\sqrt{x+5}}$.

$f'(1) = \dfrac{5}{\sqrt{6}} = \dfrac{5\sqrt{6}}{6}$, $f'(2) = \dfrac{5}{\sqrt{7}} = \dfrac{5\sqrt{7}}{7}$, $f'(3) = \dfrac{5}{\sqrt{8}} = \dfrac{5}{2\sqrt{2}} = \dfrac{5\sqrt{2}}{4}$

39. $f(x) = \dfrac{1}{x-4}$.

Step 1. $f(x+h) = \dfrac{1}{x-4+h}$

Step 2. $f(x+h) - f(x) = \dfrac{1}{x-4+h} - \dfrac{1}{x-4} = \dfrac{x-4-(x-4+h)}{(x-4+h)(x-4)} = \dfrac{-h}{(x-4+h)(x-4)}$

Step3. $\dfrac{f(x+h)-f(x)}{h} = \dfrac{-h}{h(x-4+h)(x-4)} = \dfrac{-1}{(x-4+h)(x-4)}$

Step 4. $f'(x) = \lim_{h\to 0}\frac{f(x+h)-f(x)}{h} = \lim_{h\to 0}\frac{-1}{(x-4+h)(x-4)} = \frac{-1}{(x-4)^2}$.

$f'(1) = \frac{-1}{9}$, $f'(2) = \frac{-1}{4}$, $f'(3) = -1$

41. $f(x) = \frac{x}{x+1}$

Step1. $f(x+h) = \frac{x+h}{x+1+h}$

Step 2. $f(x+h)-f(x) = \frac{x+h}{x+1+h} - \frac{x}{x+1} = \frac{(x+h)(x+1)-x(x+1+h)}{(x+1+h)(x+1)} = \frac{h}{(x+1+h)(x+1)}$

Step 3. $\frac{f(x+h)-f(x)}{h} = \frac{h}{h(x+1+h)(x+1)} = \frac{1}{(x+1+h)(x+1)}$

Step 4. $f'(x) = \lim_{h\to 0}\frac{f(x+h)-f(x)}{h} = \lim_{h\to 0}\frac{1}{(x+1+h)(x+1)} = \frac{1}{(x+1)^2}$.

$f'(1) = \frac{1}{4}$, $f'(2) = \frac{1}{9}$, $f'(3) = \frac{1}{16}$

43. $y = f(x) = x^2 + x$

(A) $f(1) = 1^2 + 1 = 2, f(3) = 3^2 + 3 = 12$

Slope of secant line: $\frac{f(3)-f(1)}{3-1} = \frac{12-2}{2} = 5$

(B) $f(1) = 2, f(1+h) = (1+h)^2 + (1+h) = 1 + 2h + h^2 + 1 + h = 2 + 3h + h^2$

Slope of secant line: $\frac{f(1+h)-f(1)}{h} = \frac{2+3h+h^2-2}{h} = 3 + h$

(C) Slope of tangent line at $(1, f(1))$:

$\lim_{h\to 0}\frac{f(1+h)-f(1)}{h} = \lim_{h\to 0}(3+h) = 3$

(D) Equation of tangent line at $(1, f(1))$:
$y - f(1) = f'(1)(x-1)$ or $y - 2 = 3(x-1)$ and $y = 3x - 1$.

45. $f(x) = x^2 + x$

(A) Average velocity: $\frac{f(3)-f(1)}{3-1} = \frac{3^2+3-(1^2+1)}{2} = \frac{12-2}{2} = 5$ meters/sec.

(B) Average velocity: $\frac{f(1+h)-f(1)}{h} = \frac{(1+h)^2+(1+h)-(1^2+1)}{h} = \frac{1+2h+h^2+1+h-2}{h}$

$= \frac{3h+h^2}{h} = 3 + h$ meters/sec.

(C) Instantaneous velocity: $\lim_{h\to 0}\frac{f(1+h)-f(1)}{h} = \lim_{h\to 0}(3+h) = 3$ m/sec.

47. $F'(x)$ does exist at $x = a$.

49. $F'(x)$ does not exist at $x = c$; the graph has a vertical tangent line at $(c, F(c))$.

51. $F'(x)$ does exist at $x = e$; $F'(e) = 0$.

53. $F'(x)$ does exist at $x = g$.

55. $f(x) = x^2 - 4x$

(A) Step 1. Find $f(x + h)$:

$f(x + h) = (x + h)^2 - 4(x + h) = x^2 + 2xh + h^2 - 4x - 4h$

Step 2. Find $f(x + h) - f(x)$:

$f(x + h) - f(x) = x^2 + 2xh + h^2 - 4x - 4h - (x^2 - 4x) = 2xh + h^2 - 4h$

Step 3. Find $\frac{f(x+h)-f(x)}{h}$:

$$\frac{f(x+h)-f(x)}{h} = \frac{2xh+h^2-4h}{h} = 2x + h - 4$$

Step 4. Find $\lim_{h\to 0} \frac{f(x+h)-f(x)}{h}$:

$$\lim_{h\to 0} \frac{f(x+h)-f(x)}{h} = \lim_{h\to 0} (2x + h - 4) = 2x - 4$$

Thus, $f'(x) = 2x - 4$.

(B) $f'(0) = -4$, $f'(2) = 0$, $f'(4) = 4$

(C) Since f is a quadratic function, the graph of f is a parabola.

y intercept: $y = 0$
x intercepts: $x = 0$, $x = 4$
Vertex: $(2, -4)$

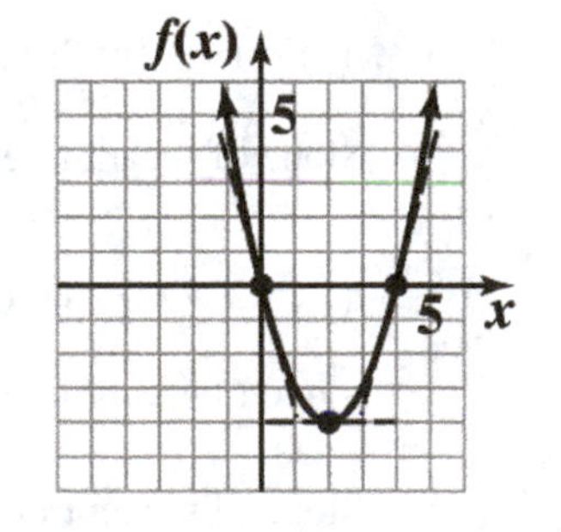

57. To find $v = f'(x)$, use the four-step process on the position function $f(x) = 4x^2 - 2x$.

Step 1. Find $f(x + h)$:

$f(x + h) = 4(x + h)^2 - 2(x + h) = 4(x^2 + 2xh + h^2) - 2x - 2h$
$= 4x^2 + 8xh + 4h^2 - 2x - 2h$

Step 2. Find $f(x + h) - f(x)$:

$f(x + h) - f(x) = 4x^2 + 8xh + 4h^2 - 2x - 2h - (4x^2 - 2x) = 8xh + 4h^2 - 2h$

Step 3. Find $\frac{f(x+h)-f(x)}{h}$:

$$\frac{f(x+h)-f(x)}{h} = \frac{8xh+4h^2-2h}{h} = 8x + 4h - 2$$

Step 4. Find $\lim_{h\to 0} \dfrac{f(x+h)-f(x)}{h}$:

$$\lim_{h\to 0} \frac{f(x+h)-f(x)}{h} = \lim_{h\to 0} (8x + 4h - 2) = 8x - 2$$

Thus, the velocity, $v(x) = f'(x) = 8x - 2$

$f'(1) = 8 \cdot 1 - 2 = 6$ ft/sec, $f'(3) = 8 \cdot 3 - 2 = 22$ ft/sec, $f'(5) = 8 \cdot 5 - 2 = 38$ ft/sec

59. (A) The graphs of g and h are vertical translations of the graph of f. All three functions should have the same derivative.

(B) $m(x) = x^2 + C$

Step 1. Find $m(x + h)$:

$$m(x + h) = (x + h)^2 + C$$

Step 2. Find $m(x + h) - m(x)$:

$$m(x + h) - m(x) = (x + h)^2 + C - (x^2 + C) = x^2 + 2xh + h^2 + C - x^2 - C$$
$$= 2xh + h^2$$

Step 3. Find $\dfrac{m(x+h)-m(x)}{h}$:

$$\frac{m(x+h)-m(x)}{h} = \frac{2xh+h^2}{h} = 2x + h$$

Step 4. $\lim_{h\to 0} \dfrac{m(x+h)-m(x)}{h}$:

$$\lim_{h\to 0} \frac{m(x+h)-m(x)}{h} = \lim_{h\to 0} (2x + h) = 2x$$

Thus, $m'(x) = 2x$.

61. True. The graph of a constant function is a horizontal line. The slope of a horizontal line is 0.

63. False. The function $f(x) = |x|$ is continuous on (-1,1) but it is not differentiable at $x = 0$.

65. False. Set $f(x) = x^3$ on [0,1]. The average rate of change of f:

$$\frac{f(1)-f(0)}{1} = 1;$$

$f'(x) = 3x^2$. The instantaneous rate of change of f at $x = 1/2$ is $f'(1/2) = 3/4 < 1$.

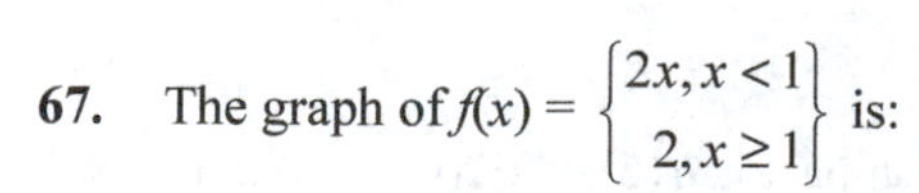

67. The graph of $f(x) = \begin{Bmatrix} 2x, x < 1 \\ 2, x \geq 1 \end{Bmatrix}$ is:

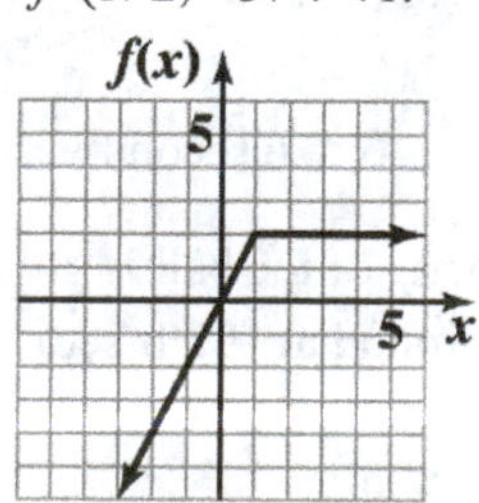

f is not differentiable at $x = 1$ because the graph of f has a sharp corner at this point.

69. $f(x) = \begin{cases} x^2+1 & \text{if } x < 0 \\ 1 & \text{if } x \geq 0 \end{cases}$

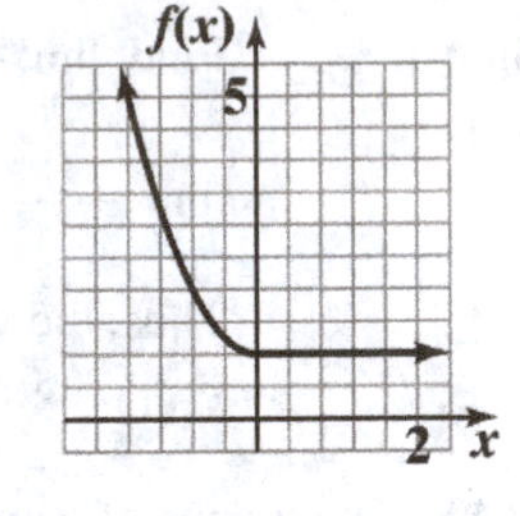

It is clear that $f'(x) = \begin{cases} 2x & \text{if } x < 0 \\ 0 & \text{if } x > 0 \end{cases}$

Thus, the only question is $f'(0)$. Since

$$\lim_{x\to 0^-} f'(x) = \lim_{x\to 0^-} 2x = 0 \text{ and } \lim_{x\to 0^+} f'(x) = \lim_{x\to 0^+} 0 = 0$$

f is differentiable at 0 as well; f is differentiable for all real numbers.

71. $f(x) = |x|$

$$\lim_{h\to 0} \frac{f(0+h)-f(0)}{h} = \lim_{h\to 0} \frac{|0+h|-|0|}{h} = \lim_{h\to 0} \frac{|h|}{h}$$

The limit does not exist. Thus, f is not differentiable at $x = 0$.

73. $f(x) = \sqrt[3]{x} = x^{1/3}$

$$\lim_{h\to 0} \frac{f(0+h)-f(0)}{h} = \lim_{h\to 0} \frac{(0+h)^{1/3}-0^{1/3}}{h} = \lim_{h\to 0} \frac{h^{1/3}}{h} = \lim_{h\to 0} \frac{1}{h^{2/3}}$$

The limit does not exist. Thus, f is not differentiable at $x = 0$.

75. $f(x) = \sqrt{1-x^2}$

$$\frac{f(0+h)-f(0)}{h} = \frac{\sqrt{1-h^2}-1}{h} = \frac{\sqrt{1-h^2}-1}{h} \cdot \frac{\sqrt{1-h^2}+1}{\sqrt{1-h^2}+1} = \frac{1-h^2-1}{h\left(\sqrt{1-h^2}+1\right)} = \frac{-h}{\sqrt{1-h^2}+1}$$

$$\lim_{h\to 0} \frac{f(0+h)-f(0)}{h} = \lim_{h\to 0} \frac{-h}{\sqrt{1-h^2}+1} = 0$$

f is differentiable at 0; $f'(0) = 0$.

77. The height of the ball at x seconds is $h(x) = 576 - 16x^2$. To find when the ball hits the ground, we solve:

$576 - 16x^2 = 0$

$16x^2 = 576$

$x^2 = 36$

$x = 6$ seconds

The velocity of the ball is given by $h'(x) = -32x$. The velocity at impact is $h'(6) = -32(6) = -192$; the ball hits the ground at 192 ft/sec.

79. $R(x) = 60x - 0.025x^2 \quad 0 \le x \le 2{,}400.$

(A) Average rate of change:

$$\frac{R(1{,}050) - R(1{,}000)}{1{,}050 - 1{,}000}$$
$$= \frac{60(1{,}050) - 0.025(1{,}050)^2 - [60(1{,}000) - 0.025(1{,}000)^2]}{50}$$
$$= \frac{35{,}437.50 - 35{,}000}{50} = \$8.75 \text{ per car seat}$$

(B) Step 1. Find $R(x + h)$:

$$R(x+h) = 60(x+h) - 0.025(x+h)^2$$
$$= 60x + 60h - 0.025(x^2 + 2xh + h^2)$$
$$= 60x + 60h - 0.025x^2 - 0.050xh - 0.025h^2$$

Step 2. Find $R(x + h) - R(x)$:

$$R(x+h) - R(x) = 60x + 60h - 0.025x^2 - 0.050xh - 0.025h^2 - (60x - 0.025x^2)$$
$$= 60h - 0.050xh - 0.025h^2$$

Step 3. Find $\frac{R(x+h) - R(x)}{h}$:

$$\frac{R(x+h) - R(x)}{h} = \frac{60h - 0.050xh - 0.025h^2}{h} = 60 - 0.050x - 0.025h$$

Step 4. Find $\lim_{h \to 0} \frac{R(x+h) - R(x)}{h}$:

$$\lim_{h \to 0} \frac{R(x+h) - R(x)}{h} = \lim_{h \to 0} (60 - 0.050x - 0.025h) = 60 - 0.050x$$

Thus, $R'(x) = 60 - 0.050x$.

(C) $R(1{,}000) = 60(1{,}000) - 0.025(1{,}000)^2 = \$35{,}000$;
$R'(1{,}000) = 60 - 0.05(1{,}000) = \$10$;
at a production level of 1,000 car seats, the revenue is \$35,000 and is increasing at the rate of \$10 per

car seat.

81. (A) $S(t) = 2\sqrt{t+10}$

Step 1. Find $S(t + h)$:

$$S(t+h) = 2\sqrt{t+h+10}$$

Step 2. Find $S(t + h) - S(t)$:

$$S(t+h) - S(t) = 2\sqrt{t+h+10} - 2\sqrt{t+10} = 2(\sqrt{t+h+10} - \sqrt{t+10})$$

Step 3. Find $\frac{S(t+h) - S(t)}{h}$:

$$\frac{S(t+h) - S(t)}{h} = \frac{2\left(\sqrt{t+h+10} - \sqrt{t+10}\right)}{h}$$

$$= \frac{2\left(\sqrt{t+h+10}-\sqrt{t+10}\right)}{h} \cdot \frac{\left(\sqrt{t+h+10}+\sqrt{t+10}\right)}{\left(\sqrt{t+h+10}+\sqrt{t+10}\right)}$$

$$= \frac{2[t+h+10-(t+10)]}{h\left(\sqrt{t+h+10}+\sqrt{t+10}\right)} = \frac{2h}{h\left(\sqrt{t+h+10}+\sqrt{t+10}\right)}$$

$$= \frac{2}{\sqrt{t+h+10}+\sqrt{t+10}}$$

Step 4. Find $\lim\limits_{h\to 0} \frac{S(t+h)-S(t)}{h}$:

$$\lim_{h\to 0} \frac{S(t+h)-S(t)}{h} = \lim_{h\to 0} \frac{2}{\sqrt{t+h+10}+\sqrt{t+10}} = \frac{1}{\sqrt{t+10}}$$

Thus, $S'(t) = \frac{1}{\sqrt{t+10}}$.

(B) $S(15) = 2\sqrt{15+10} = 2\sqrt{25} = 10;$

$S'(15) = \frac{1}{\sqrt{15+10}} = \frac{1}{\sqrt{25}} = \frac{1}{5} = 0.2$

After 15 months, the total sales are \$10 million and are INCREASING at the rate of \$0.2 million or \$200,000 per month.

(C) The estimated total sales are \$10.2 million after 16 months and \$10.4 million after 17 months.

83. $p(t) = 138t^2 + 1{,}072t + 14{,}917$

(A) Step 1. Find $p(t+h)$:

$$p(t+h) = 138(t+h)^2 + 1{,}072(t+h) + 14{,}917$$

Step 2. Find $p(t+h) - p(t)$:

$$p(t+h) - p(t) = 138(t+h)^2 + 1{,}072(t+h) + 14{,}917 - (138t^2 + 1{,}072t + 14{,}917)$$
$$= 276th + 138h^2 + 1{,}072h$$

Step 3. Find $\frac{p(t+h)-p(t)}{h}$:

$$\frac{p(t+h)-p(t)}{h} = \frac{276th + 138h^2 + 1{,}072h}{h} = 276t + 138h + 1{,}072$$

Step 4. Find $\lim\limits_{h\to 0} \frac{p(t+h)-p(t)}{h}$:

$$\lim_{h\to 0} \frac{p(t+h)-p(t)}{h} = \lim_{h\to 0} (276t + 138h + 1{,}072) = 276t + 1{,}072$$

Thus, $p'(t) = 276t + 1{,}072$

(B) The year 2020 corresponds to $t = 10$.

$p(10) = 138(10)^2 + 1{,}072(10) + 14{,}917 = 39{,}437$ metric tons;

$p'(t) = 276t + 1{,}072$,

$p'(10) = 276(10) + 1{,}072 = 3{,}832$.

In 2020 the US will produce 39,437 metric tons of tungsten and this quantity is increasing at the rate of 3,832 metric tons/year.

85. (A) Quadratic regression model

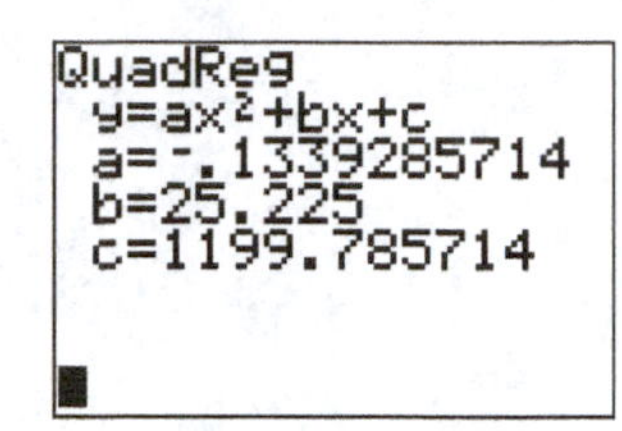

(B)

$R(x) \approx -0.1339205714\,x^2 + 25.225\,x + 1199.785714$;
$R(20) \approx 1{,}650.7$;
$R'(x) \approx -0.267857142\,x + 25.225$;
$R'(20) \approx 19.9$

In 2020, 1,650.7 billion residential kilowatts will be sold and the amount sold is increasing at the rate of 19.9 billion kilowatts per year.

87. (A) $P(t) = 80 + 12t - t^2$

Step 1. Find $P(t+h)$:

$P(t+h) = 80 + 12(t+h) - (t+h)^2 = 80 + 12t + 12h - (t+h)^2$

Step 2. Find $P(t+h) - P(t)$:

$$P(t+h) - P(t) = 80 + 12t + 12h - (t+h)^2 - (80 + 12t - t^2)$$
$$= 12h - 2th - h^2$$

Step 3. Find $\dfrac{P(t+h)-P(t)}{h}$:

$$\frac{P(t+h)-P(t)}{h} = \frac{12h - 2th - h^2}{h} = 12 - 2t - h$$

Step 4. Find $\lim_{h\to 0} \dfrac{P(t+h)-P(t)}{h}$:

$$\lim_{h\to 0} \frac{P(t+h)-P(t)}{h} = \lim_{h\to 0} (12 - 2t - h) = 12 - 2t$$

Thus, $P'(t) = 12 - 2t$.

(B) $P(3) = 80 + 12(3) - (3)^2 = 107$; $P'(3) = 12 - 2(3) = 6$

After 3 hours, the ozone level is 107 ppb and is INCREASING at the rate of 6 ppb per hour.

EXERCISE 2-5

Things to remember:

1. DERIVATIVE NOTATION

Given $y = f(x)$, then

$$f'(x),\ \ y',\ \ \frac{dy}{dx}$$

all represent the derivative of f at x.

2. CONSTANT FUNCTION RULE

If $f(x) = C$, C a constant, then $f'(x) = 0$. Also

$$y' = 0 \text{ and } \frac{dy}{dx} = 0.$$

3. POWER RULE

If $f(x) = x^n$, n any real number, then

$$f'(x) = nx^{n-1}.$$

Also, $y' = nx^{n-1}$ and $\frac{dy}{dx} = nx^{n-1}$

4. CONSTANT MULTIPLE PROPERTY

If $y = f(x) = ku(x)$, where k is a constant, then

$$f'(x) = ku'(x).$$

Also,

$$y' = ku' \text{ and } \frac{dy}{dx} = k\frac{du}{dx}.$$

5. SUM AND DIFFERENCE PROPERTY

If $y = f(x) = u(x) \pm v(x)$, then

$$f'(x) = u'(x) \pm v'(x).$$

Also,

$$y' = u' \pm v' \text{ and } \frac{dy}{dx} = \frac{du}{dx} \pm \frac{dv}{dx}$$

[Note: This rule generalizes to the sum and difference of any given number of functions.]

---

1. $\sqrt{x} = x^{1/2}$ **3.** $\frac{1}{x^5} = x^{-5}$ **5.** $(x^4)^3 = x^{12}$ **7.** $\frac{1}{\sqrt[4]{x}} = \frac{1}{x^{1/4}} = x^{-1/4}$

9. $f(x) = 7; f'(x) = 0$ (using 2) **11.** $y = x^9;\ \frac{dy}{dx} = 9x^8$ (using 3)

13. $\frac{d}{dx}x^3 = 3x^2$ (using 3)

15. $y = x^{-4}$; $y' = -4x^{-5}$ (using 3)

17. $g(x) = x^{8/3}$; $g'(x) = \frac{8}{3}x^{5/3}$ (using 3)

19. $y = \frac{1}{x^{10}}$; $\frac{dy}{dx} = -10x^{-11} = \frac{-10}{x^{11}}$

21. $f(x) = 5x^2$; $f'(x) = 5(2x) = 10x$ (using 4)

23. $y = 0.4x^7$; $y' = 0.4(7x^6) = 2.8x^6$

25. $\frac{d}{dx}\left(\frac{x^3}{18}\right) = \frac{1}{18}(3x^2) = \frac{1}{6}x^2$

27. $h(x) = 4f(x)$; $h'(2) = 4 \cdot f'(2) = 4(3) = 12$

29. $h(x) = f(x) + g(x)$; $h'(2) = f'(2) + g'(2) = 3 + (-1) = 2$

31. $h(x) = 2f(x) - 3g(x) + 7$; $h'(2) = 2f'(2) - 3g'(2) = 2(3) - 3(-1) = 9$

33. $\frac{d}{dx}(2x - 5) = \frac{d}{dx}(2x) - \frac{d}{dx}(5) = 2$

35. $f(t) = 2t^2 - 3t + 1$; $f'(t) = (2t^2)' - (3t)' + (1)' = 4t - 3$

37. $y = 5x^{-2} + 9x^{-1}$; $y' = -10x^{-3} - 9x^{-2}$

39. $\frac{d}{du}(5u^{0.3} - 4u^{2.2}) = \frac{d}{du}(5u^{0.3}) - \frac{d}{du}(4u^{2.2}) = 1.5u^{-0.7} - 8.8u^{1.2}$

41. $h(t) = 2.1 + 0.5t - 1.1t^3$; $h'(t) = 0.5 - (1.1)3t^2 = 0.5 - 3.3t^2$

43. $y = \frac{2}{5x^4} = \frac{2}{5}x^{-4}$; $y' = \frac{2}{5}(-4x^{-5}) = -\frac{8}{5}x^{-5} = \frac{-8}{5x^5}$

45. $\frac{d}{dx}\left(\frac{3x^2}{2} - \frac{7}{5x^2}\right) = \frac{d}{dx}\left(\frac{3}{2}x^2\right) - \frac{d}{dx}\left(\frac{7}{5}x^{-2}\right) = 3x + \frac{14}{5}x^{-3} = 3x + \frac{14}{5x^3}$

47. $G(w) = \frac{5}{9w^4} + 5\sqrt[3]{w} = \frac{5}{9}w^{-4} + 5w^{1/3}$;

$G'(w) = -\frac{20}{9}w^{-5} + \frac{5}{3}w^{-2/3} = \frac{-20}{9w^5} + \frac{5}{3w^{2/3}}$

49. $\frac{d}{du}(3u^{2/3} - 5u^{1/3}) = \frac{d}{du}(3u^{2/3}) - \frac{d}{du}(5u^{1/3}) = 2u^{-1/3} - \frac{5}{3}u^{-2/3} = \frac{2}{u^{1/3}} - \frac{5}{3u^{2/3}}$

51. $h(t) = \frac{3}{t^{3/5}} - \frac{6}{t^{1/2}} = 3t^{-3/5} - 6t^{-1/2}$;

$$h'(t) = 3\left(-\frac{3}{5}t^{-8/5}\right) - 6\left(-\frac{1}{2}t^{-3/2}\right) = -\frac{9}{5}t^{-8/5} + 3t^{-3/2} = \frac{-9}{5t^{8/5}} + \frac{3}{t^{3/2}}$$

53. $y = \frac{1}{\sqrt[3]{x}} = \frac{1}{x^{1/3}} = x^{-1/3}; y' = -\frac{1}{3}x^{-4/3} = \frac{-1}{3x^{4/3}}$

55. $\frac{d}{dx}\left(\frac{1.2}{\sqrt{x}} - 3.2x^{-2} + x\right) = \frac{d}{dx}(1.2x^{-1/2} - 3.2x^{-2} + x) = \frac{d}{dx}(1.2x^{-1/2}) - \frac{d}{dx}(3.2x^{-2}) + \frac{d}{dx}(x)$

$$= -0.6x^{-3/2} + 6.4x^{-3} + 1 = \frac{-0.6}{x^{3/2}} + \frac{6.4}{x^3} + 1$$

57. $f(x) = 6x - x^2$

(A) $f'(x) = 6 - 2x$

(B) Slope of the graph of f at $x = 2$: $f'(2) = 6 - 2(2) = 2$
Slope of the graph of f at $x = 4$: $f'(4) = 6 - 2(4) = -2$

(C) Tangent line at $x = 2$: $y - y_1 = m(x - x_1)$

$x_1 = 2$, $y_1 = f(2) = 6(2) - 2^2 = 8$

$m = f'(2) = 2$

Thus, $y - 8 = 2(x - 2)$ or $y = 2x + 4$.

Tangent line at $x = 4$: $y - y_1 = m(x - x_1)$

$x_1 = 4$, $y_1 = f(4) = 6(4) - 4^2 = 8$

$m = f'(4) = -2$

Thus, $y - 8 = -2(x - 4)$ or $y = -2x + 16$

(D) The tangent line is horizontal at the values $x = c$ such that $f'(c) = 0$. Thus, we must solve the following:

$$f'(x) = 6 - 2x = 0$$
$$2x = 6$$
$$x = 3$$

59. $f(x) = 3x^4 - 6x^2 - 7$

(A) $f'(x) = 12x^3 - 12x$

(B) Slope of the graph at $x = 2$: $f'(2) = 12(2)^3 - 12(2) = 72$

Slope of the graph of $x = 4$: $f'(4) = 12(4)^3 - 12(4) = \frac{72}{0}$, 720

(C) Tangent line at $x = 2$: $y - y_1 = m(x - x_1)$, where $x_1 = 2$,

$y_1 = f(2) = 3(2)^4 - 6(2)^2 - 7 = 17$, $m = 72$.

$y - 17 = 72(x - 2)$ or $y = 72x - 127$

Tangent line at $x = 4$: $y - y_1 = m(x - x_1)$, where $x_1 = 4$,

$y_1 = f(4) = 3(4)^4 - 6(4)^2 - 7 = 665$, $m = 720$.

$y - 665 = 720(x - 4)$ or $y = 720x - 2215$

(D) Solve $f'(x) = 0$ for x:

$$12x^3 - 12x = 0$$
$$12x(x^2 - 1) = 0$$
$$12x(x - 1)(x + 1) = 0$$
$$x = -1, \quad x = 0, \quad x = 1$$

61. $f(x) = 176x - 16x^2$

(A) $v = f'(x) = 176 - 32x$

(B) $v\big|_{x=0} = f'(0) = 176$ ft/sec.

$v\big|_{x=3} = f'(3) = 176 - 32(3) = 80$ ft/sec.

(C) Solve $v = f'(x) = 0$ for x:

$$176 - 32x = 0$$
$$32x = 176$$
$$x = 5.5 \text{ sec.}$$

63. $f(x) = x^3 - 9x^2 + 15x$

(A) $v = f'(x) = 3x^2 - 18x + 15$

(B) $v\big|_{x=0} = f'(0) = 15$ feet/sec.

$v\big|_{x=3} = f'(3) = 3(3)^2 - 18(3) + 15 = -12$ feet/sec.

(C) Solve $v = f'(x) = 0$ for x:

$$3x^2 - 18x + 15 = 0$$
$$3(x^2 - 6x + 5) = 0$$
$$3(x - 5)(x - 1) = 0$$
$$x = 1, \quad x = 5$$

65. $f(x) = x^2 - 3x - 4\sqrt{x} = x^2 - 3x - 4x^{1/2}$

$f'(x) = 2x - 3 - 2x^{-1/2}$

The graph of f has a horizontal tangent line at the value(s) of x where $f'(x) = 0$. Thus, we need to solve the equation

$$2x - 3 - 2x^{-1/2} = 0$$

By graphing the function $y = 2x - 3 - 2x^{-1/2}$, we see that there is one zero. To four decimal places, it is $x = 2.1777$.

67. $f(x) = 3\sqrt[3]{x^4} - 1.5x^2 - 3x = 3x^{4/3} - 1.5x^2 - 3x$

$f'(x) = 4x^{1/3} - 3x - 3$

The graph of f has a horizontal tangent line at the value(s) of x where $f'(x) = 0$. Thus, we need to solve the equation

$$4x^{1/3} - 3x - 3 = 0$$

Graphing the function $y = 4x^{1/3} - 3x - 3$, we see that there is one zero. To four decimal places, it is $x = -2.9018$.

69. $f(x) = 0.05x^4 - 0.1x^3 - 1.5x^2 - 1.6x + 3$

$f'(x) = 0.2x^3 + 0.3x^2 - 3x - 1.6$

The graph of f has a horizontal tangent line at the value(s) of x where $f'(x) = 0$. Thus, we need to solve the equation

$$0.2x^3 + 0.3x^2 - 3x - 1.6 = 0$$

By graphing the function $y = 0.2x^3 + 0.3x^2 - 3x - 1.6$, we see that there are three zeros. To four decimal places, they are

$$x_1 = -4.4607, \quad x_2 = -0.5159, \quad x_3 = 3.4765$$

71. $f(x) = 0.2x^4 - 3.12x^3 + 16.25x^2 - 28.25x + 7.5$

$f'(x) = 0.8x^3 - 9.36x^2 + 32.5x - 28.25$

The graph of f has a horizontal tangent line at the value(s) of x where $f'(x) = 0$. Thus, we need to solve the equation

$$0.8x^3 - 9.36x^2 + 32.5x - 28.25 = 0$$

Graphing the function $y = 0.8x^3 - 9.36x^2 + 32.5x - 28.25$, we see that there is one zero. To four decimal places, it is $x = 1.3050$.

73. $f(x) = ax^2 + bx + c; f'(x) = 2ax + b.$

The derivative is 0 at the vertex of the parabola:

$$2ax + b = 0$$
$$x = -\frac{b}{2a}$$

75. (A) $f(x) = x^3 + x$ (B) $f(x) = x^3$ (C) $f(x) = x^3 - x$

77. $f(x) = (2x - 1)^2 = 4x^2 - 4x + 1$; $f'(x) = 8x - 4$

79. $\frac{d}{dx}\left(\frac{10x+20}{x}\right) = \frac{d}{dx}\left(10+\frac{20}{x}\right) = \frac{d}{dx}(10) + \frac{d}{dx}(20x^{-1}) = -20x^{-2} = -\frac{20}{x^2}$

81. $y = \frac{3x-4}{12x^2} = \frac{3x}{12x^2} - \frac{4}{12x^2} = \frac{1}{4}x^{-1} - \frac{1}{3}x^{-2}$

$\frac{dy}{dx} = -\frac{1}{4}x^{-2} + \frac{2}{3}x^{-3} = -\frac{1}{4x^2} + \frac{2}{3x^3}$

83. False. Set $f(x) = x^2$, $g(x) = x^3$. Then $f(x)\cdot g(x) = x^5$ and $[f(x)\cdot g(x)]' = 5x^4$;

$f'(x) = 2x$, $g'(x) = 3x^2$ and $f'(x)\cdot g'(x) = 6x^3 \neq [f(x)\cdot g(x)]'$.

85. True. Theorem 1, which is also 2 in *Things to Remember.*

87. $f(x) = u(x) + v(x)$

Step1. $f(x+h) = u(x+h) + v(x+h)$

Step 2. $f(x+h) - f(x) = u(x+h) + v(x+h) - [u(x) + v(x)] = u(x+h) - u(x) + [v(x+h) - v(x)]$

Step 3. $\dfrac{f(x+h)-f(x)}{h}=\dfrac{u(x+h)-u(x)+[v(x+h)-v(x)]}{h}=\dfrac{u(x+h)-u(x)}{h}+\dfrac{v(x+h)-v(x)}{h}$

Step 4. $\lim\limits_{h\to 0}\dfrac{f(x+h)-f(x)}{h}=\lim\limits_{h\to 0}\left[\dfrac{u(x+h)-u(x)}{h}+\dfrac{v(x+h)-v(x)}{h}\right]$

$=\lim\limits_{h\to 0}\dfrac{u(x+h)-u(x)}{h}+\lim\limits_{h\to 0}\dfrac{v(x+h)-v(x)}{h}=u'(x)+v'(x)$

89. (A) $S(t)=0.03t^3+0.5t^2+2t+3$

$S'(t)=0.09t^2+t+2$

(B) $S(5)=0.03(5)^3+0.5(5)^2+2(5)+3=29.25$

$S'(5)=0.09(5)^2+5+2=9.25$

After 5 months, sales are \$29.25 million and are increasing at the rate of \$9.25 million per month.

(C) $S(10)=0.03(10)^3+0.5(10)^2+2(10)+3=103$

$S'(10)=0.09(10)^2+10+2=21$

After 10 months, sales are \$103 million and are increasing at the rate of \$21 million per month.

91. (A) $N(x)=1{,}000-\dfrac{3{,}780}{x}=1{,}000-3{,}780x^{-1}$

$N'(x)=3{,}780x^{-2}=\dfrac{3{,}780}{x^2}$

(B) $N'(10)=\dfrac{3{,}780}{(10)^2}=37.8$

At the \$10,000 level of advertising, sales are INCREASING at the rate of 37.8 boats per \$1000 spent on advertising.

$N'(20)=\dfrac{3{,}780}{(20)^2}=9.45$

At the \$20,000 level of advertising, sales are INCREASING at the rate of 9.45 boats per \$1000 spent on advertising.

93. (A) Cubic regression model

```
CubicReg
 y=ax³+bx²+cx+d
 a=-8.083333E-4
 b=.0624285714
 c=-1.081309524
 d=40.57571429
```

(B) $M(x)\approx -0.0008083x^3+0.0624x^2-1.081x+40.576$

$M'(x)\approx -0.0024249x^2+0.1248x-1.081$

In 2020, 41.5% of male high school graduates will enroll in college and the percentage is decreasing at the rate of 0.9% per year.

95. $y=590x^{-1/2},\ 30\le x\le 75$

First, find $\dfrac{dy}{dx}=\dfrac{d}{dx}590x^{-1/2}=-295x^{-3/2}=\dfrac{-295}{x^{3/2}}$, the instantaneous rate of change of pulse when a person is x inches tall.

(A) The instantaneous rate of change of pulse rate at $x = 36$ is:
$\frac{-295}{(36)^{3/2}} = \frac{-295}{216} = -1.37$ (1.37 decrease in pulse rate)

(B) The instantaneous rate of change of pulse rate at $x = 64$ is:
$\frac{-295}{(64)^{3/2}} = \frac{-295}{512} = -0.58$ (0.58 decrease in pulse rate)

97. $y = 50\sqrt{x}$, $0 \le x \le 9$

First, find $y' = (50\sqrt{x})' = (50x^{1/2})' = 25x^{-1/2}$

$= \frac{25}{\sqrt{x}}$, the rate of learning at the end of x hours.

(A) Rate of learning at the end of 1 hour: $\frac{25}{\sqrt{1}} = 25$ items/hr

(B) Rate of learning at the end of 9 hours: $\frac{25}{\sqrt{9}} = \frac{25}{3} = 8.33$ items/hr

EXERCISE 2-6

Things to remember:

1. INCREMENTS

For $y = f(x)$, $\Delta x = x_2 - x_1$, $\Delta y = y_2 - y_1$, so $x_2 = x_1 + \Delta x$, and

$$\begin{aligned}\Delta y &= y_2 - y_1 \\ &= f(x_2) - f(x_1) \\ &= f(x_1 + \Delta x) - f(x_1)\end{aligned}$$

Δy represents the change in y corresponding to a Δx change in x. Δx can be either positive or negative.

[*Note*: Δy depends on the function f, the input x, and the increment Δx.]

2. DIFFERENTIALS

If $y = f(x)$ defines a differentiable function, then the **differential** dy, **or** df, is defined as the product of $f'(x)$ and dx, where $dx = \Delta x$. Symbolically,

$$dy = f'(x)dx \quad \text{or} \quad df = f'(x)dx$$

where

$$dx = \Delta x$$

[*Note*: The differential dy (or df) is actually a function involving two independent variables, x and dx; a change in either one or both will affect dy (or df).]

1. $f(x) = 0.1x + 3;\ f(0) = 3,\ \ f(0.1) = 0.1(0.1) + 3 = 3.01$

3. $f(x) = 0.1x + 3;\ f(-2) = 0.1(-2) + 3 = 2.8,\ \ f(-2.1) = 0.1(-2.1) + 3 = 2.79$

5. $g(x) = x^2;\ g(0) = 0^2 = 0,\ \ g(0.1) = (0.1)^2 = 0.01$

7. $g(x) = x^2;\quad g(10) = (10)^2 = 100,\quad g(10.1) = (10.1)^2 = (10 + 0.1)^2 = 102.01$

9. $\Delta x = x_2 - x_1 = 4 - 1 = 3$

$\Delta y = f(x_2) - f(x_1) = 3 \cdot 4^2 - 3 \cdot 1^2 = 48 - 3 = 45$

$\frac{\Delta y}{\Delta x} = \frac{45}{3} = 15$

11. $\frac{f(x_1 + \Delta x) - f(x_1)}{\Delta x} = \frac{f(1+2) - f(1)}{2} = \frac{3 \cdot 3^2 - 3 \cdot 1^2}{2} = \frac{24}{2} = 12$

13. $\Delta y = f(x_2) - f(x_1) = f(3) - f(1) = 3 \cdot 3^2 - 3 \cdot 1^2 = 27 - 3 = 24$

$\Delta x = x_2 - x_1 = 3 - 1 = 2$

$\frac{\Delta y}{\Delta x} = \frac{24}{2} = 12$

15. $y = 30 + 12x^2 - x^3$

$dy = (30 + 12x^2 - x^3)'dx = (24x - 3x^2)dx$

17. $y = x^2\left(1 - \frac{x}{9}\right) = x^2 - \frac{x^3}{9}$

$dy = \left[x^2\left(1 - \frac{x}{9}\right)\right]' dx = \left[x^2 - \frac{x^3}{9}\right]' dx = \left(2x - \frac{1}{3}x^2\right)dx$

19. $y = \frac{590}{\sqrt{x}} = 590x^{-1/2}$

$dy = 590\left(-\frac{1}{2}\right)x^{-3/2}dx = \frac{-295}{x^{3/2}}dx$

21. (A) $\frac{f(2 + \Delta x) - f(2)}{\Delta x} = \frac{3(2 + \Delta x)^2 - 3 \cdot 2^2}{\Delta x}$

$= \frac{3(2^2 + 4\Delta x + \Delta x^2) - 12}{\Delta x}$

$= \frac{\Delta x(12 + 3\Delta x)}{\Delta x}$

$= 12 + 3\Delta x,\ \Delta x \neq 0$

(B) As Δx tends to zero, then, clearly, $12 + 3\Delta x$ tends to 12. Note the values in the following table:

Δx	$12 + 3\Delta x$
1	15
0.1	12.3
0.01	12.03
0.001	12.003

23. $y = (2x + 1)^2 = 4x^2 + 4x + 1$

$dy = (8x + 4)\,dx$

25. $y = \dfrac{x^2+9}{x} = x + \dfrac{9}{x} = x + 9x^{-1}$

$dy = \left(1 - 9x^{-2}\right)dx = \left(1 - \dfrac{9}{x^2}\right)dx$

27. $y = f(x) = x^2 - 3x + 2$

$\Delta y = f(5 + 0.2) - f(5)$ (using 1)

$= f(5.2) - f(5)$

$= (5.2)^2 - 3(5.2) + 2 - (5^2 - 3\cdot 5 + 2) = 1.44$

$dy = (x^2 - 3x)'\big|_{x=5}\,\Delta x = (2x-3)\big|_{x=5}\,\Delta x = 7(0.2) = 1.4$

29. $y = f(x) = 75\left(1 - \dfrac{2}{x}\right)$

$\Delta y = f[5 + (-0.5)] - f(5) = f(4.5) - f(5) = 75\left(1 - \dfrac{2}{4.5}\right) - 75\left(1 - \dfrac{2}{5}\right) = 41.67 - 45 = -3.33$

$dy = \left[75\left(1 - \dfrac{2}{x}\right)\right]'_{x=5}(-0.5) = \dfrac{150}{x^2}\bigg]_{x=5}(-0.5) = 6\left(-\dfrac{1}{2}\right) = -3$

31. A cube with sides of length x has volume $V(x) = x^3$. If we increase the length of each side by an amount $\Delta x = dx$, then the approximate change in volume is:

$dV = 3x^2dx$

Therefore, letting $x = 10$ and $dx = 0.4$ [$= 2(0.2)$], since each face has a 0.2 inch coating], we have

$dV = 3(10)^2(0.4) = 120$ cubic inches,

which is the approximate volume of the fiberglass shell.

33. $f(x) = x^2 + 2x + 3;\ f'(x) = 2x + 2;\ x = -0.5;\ \Delta x = dx$

(A) $\Delta y = f(-0.5 + \Delta x) - f(-0.5)$

$= (-0.5 + \Delta x)^2 + 2(-0.5 + \Delta x) + 3 - [(-0.5)^2 + 2(-0.5) + 3]$

$= (-0.5 + \Delta x)^2 + 2(-0.5 + \Delta x) + 0.75 = \Delta x + (\Delta x)^2$

$dy = f'(-0.5)dx = 1 \cdot dx = dx = \Delta x$

(B)

(C) $\Delta y(0.1) = (-0.5 + 0.1)^2 + 2(-0.5 + 0.1) + 0.75 = 0.11$

$dy(0.1) = 0.1$

$\Delta y(0.2) = (-0.5 + 0.2)^2 + 2(-0.5 + 0.2) + 0.75 = 0.24$

$dy(0.2) = 0.2$

$\Delta y(0.3) = (-0.5 + 0.3)^2 + 2(-0.5 + 0.3) + 0.75 = 0.39$

$dy(0.3) = 0.3$

Δx	Δy	dy
.1	[illegible]	.1
.2	.24	.2
.3	.39	.3

35. $f(x) = x^3 - 2x^2$; $f'(x) = 3x^2 - 4x$; $x = 1$; $\Delta x = dx$

(A) $\Delta y = f(1 + \Delta x) - f(1) = (1 + \Delta x)^3 - 2(1 + \Delta x)^2 - [1^3 - 2(1)^2]$

$= (1 + \Delta x)^3 - 2(1 + \Delta x)^2 + 1$

$= -\Delta x + (\Delta x)^2 + (\Delta x)^3$

$dy = f'(1)dx = (-1)dx = -dx$

(B)

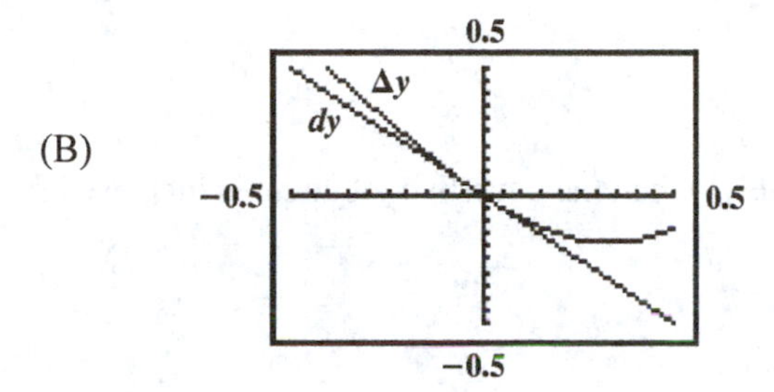

(C) $\Delta y(0.05) = -0.05 + (0.05)^2 + (0.05)^3 = -0.047375$

$dy(0.05) = -0.05$

$\Delta y(0.10) = -0.10 + (0.10)^2 + (0.10)^3 = -0.089$

$dy(0.10) = -0.10$

$\Delta y(0.15) = -0.15 + (0.15)^2 + (0.15)^3 = -0.124125$

$dy(0.15) = -0.15$ 0.3

37. True

$f(x) = mx + b$; $f'(x) = m$; $x = 3$; $\Delta x = dx$

$\Delta y = f(3 + \Delta x) - f(3) = m(3 + \Delta x) + b - (3m + b) = m\,\Delta x$

$dy = (3)dx = m\,dx$

Thus, $\Delta y = dy$

39. False.

At $x = 2$, $dy = f'(2)dx$. $Dy = 0$ for all dx implies only that $f'(2) = 0$.

Example. Let $f(x) = x^2 - 4x$

41. $y = (1-2x)\sqrt[3]{x^2} = (1-2x)x^{2/3} = x^{2/3} - 2x^{5/3}$

$$Dy = \left(\frac{2}{3}x^{-1/3} - \frac{10}{3}x^{2/3}\right)dx$$

43. $y = f(x) = 52\sqrt{x} = 52x^{1/2}$; $x = 4$, $\Delta x = dx = 0.3$

$\Delta y = f(x + \Delta x) - f(x)$

Let $x = 4$, $\Delta x = 0.3$. Then:

$\Delta y = 52(4 + 0.3)^{1/2} - 52(4)^{1/2}$

$dy = f'(x)dx = \dfrac{26}{x^{1/2}}dx$

Let $x = 4$, $dx = 0.3$. Then:

$dy = \dfrac{26}{4^{1/2}}(0.3)$

$= 52(4.3)^{1/2} - 104$ $\qquad\qquad$ $= \frac{26}{2}(0.3) = 3.9$

$\approx 107.83 - 104$

≈ 3.83

45. Given $N(x) = 60x - x^2$, $5 \le x \le 30$. Then $N'(x) = 60 - 2x$. The approximate change in the sales dN corresponding to a change $\Delta x = dx$ in the amount x (in thousands of dollars) spent on advertising is:

$$dN = N'(x)dx$$

Thus, letting $x = 10$ and $dx = 1$, we get:

$$dN = N'(10)\cdot 1 = (60 - 2\cdot 10) = 40$$

There will be a 40-unit increase in sales (approximately) when the advertising budget is increased from \$10,000 to \$11,000.

Similarly, letting $x = 20$ and $dx = 1$, we get:

$$dN = N'(20)\cdot 1 = (60 - 2\cdot 20) = 20\text{-unit increase.}$$

47. $\bar{C}(x) = \frac{400}{x} + 5 + \frac{1}{2}x,\ x \ge 1.$

If we increase production per hour by an amount $\Delta x = dx$, then the approximate change in average cost is:

$$d\bar{C} = \bar{C}'(x)\,dx = \left(-\frac{400}{x^2} + \frac{1}{2}\right)dx$$

Thus, letting $x = 20$ and $dx = 5$, we have

$$d\bar{C} = \left(-\frac{400}{(20)^2} + \frac{1}{2}\right)5 = \left(-1 + \frac{1}{2}\right)5 = -2.50,$$

that is, the average cost per racket will *decrease* \$2.50.

Letting $x = 40$ and $dx = 5$, we have

$$d\bar{C} = \left(-\frac{400}{(40)^2} + \frac{1}{2}\right)5 = \left(-\frac{1}{4} + \frac{1}{2}\right)5 = 1.25,$$

that is, the average cost per racket will *increase* \$1.25.

49. $y = \frac{590}{\sqrt{x}}$, $30 \le x \le 75$. Thus, $y = 590x^{-1/2}$ and $y' = -295x^{-3/2} = \frac{-295}{x^{3/2}}$.

The approximate change in pulse rate for a height change from 36 to 37 inches is given by:

$$dy = \left.\frac{-295}{x^{3/2}}\right]_{x=36} = -1.37 \text{ per minute.}$$ [Note: $\Delta x = dx = 1$.]

Similarly, the approximate change in pulse rate for a height change from 64 to 65 inches is given by:

$$dy = \left.\frac{-295}{x^{3/2}}\right]_{x=64} = -0.58 \text{ per minute.}$$

51. Area: $A(r) = \pi r^2 \approx 3.14r^2$; $A'(r) = 6.28r$.

An approximate increase in cross-sectional area when the radius of an artery is increased from 2 mm to 2.1 mm is given by:

$dA = A'(r)\Big|_{r=2} \times 0.1$ [Note: $\Delta r = 0.1$.]

$= \left(6.28r\Big|_{r=2}\right) \times 0.1 = 12.56 \times 0.1 = 1.256 \text{ mm}^2 \approx 1.26 \text{ mm}^2$

53. $N(t) = 75\left(1 - \frac{2}{t}\right)$, $3 \le t \le 20$; $N'(t) = \frac{150}{t^2}$.

The approximate improvement from 5 to 5.5 weeks of practice is given by:

$dN = N'(t)\Big|_{t=5} \times 0.5$ [Note: $\Delta t = 0.5$.]

$= \frac{150}{t^2}\Big|_{t=5} \times 0.5 = 6 \times 0.5 = 3$ words per minute

55. $N(t) = 30 + 12t^2 - t^3$, $0 \le t \le 8$; $N'(t) = 24t - 3t^2$.

(A) The approximate change in votes when time changes from 1 to 1.1 years is

$= N'(t)\Big|_{t=1} \times 0.1$ [Note: $\Delta t = 0.1$.]

$= \left(24t - 3t^2\Big|_{t=1}\right) \times 0.1 = 21 \times 0.1 = 2.1$ thousand or 2100 increase.

(B) The approximate change in votes when time changes from 4 to 4.1 years is

$= N'(t)\Big|_{t=4} \times 0.1$ [Note: $\Delta t = 0.1$.]

$= \left(24t - 3t^2\Big|_{t=4}\right) \times 0.1 = 48 \times 0.1 = 4.8$ thousand or 4800 increase.

(C) The approximate change in votes when time changes from 7 to 7.1 years is

$= N'(t)\Big|_{t=7} \times 0.1$ [Note: $\Delta t = 0.1$.]

$= \left(24t - 3t^2\Big|_{t=7}\right) \times 0.1 = 2.1$ thousand or 2100 increase.

EXERCISE 2-7

Things to remember:

1. MARGINAL COST, REVENUE, AND PROFIT

   If x is the number of units of a product produced in some time interval, then:

   Total Cost = $C(x)$
   Marginal Cost = $C'(x)$
   Total Revenue = $R(x)$
   Marginal Revenue = $R'(x)$

Total Profit = $P(x) = R(x) - C(x)$

Marginal Profit = $P'(x) = R'(x) - C'(x)$

= (Marginal Revenue) – (Marginal Cost)

Marginal cost (or revenue or profit) is the instantaneous rate of change of cost (or revenue or profit) relative to production at a given production level.

2. MARGINAL COST AND EXACT COST

If $C(x)$ is the cost of producing x items, then the marginal cost function approximates the exact cost of producing the
$(x + 1)$st item:

Marginal Cost		Exact Cost
$C'(x)$	$\approx$	$C(x+1) - C(x)$

Similar interpretations can be made for total revenue and total profit functions.

3. BREAK-EVEN POINTS

The BREAK-EVEN POINTS are the points where total revenue equals total cost.

4. MARGINAL AVERAGE COST, REVENUE, AND PROFIT

If x is the number of units of a product produced in some time interval, then:

Average Cost = $\overline{C}(x) = \frac{C(x)}{x}$ Cost per unit

Marginal Average Cost = $\overline{C}'(x)$

Average Revenue = $\overline{R}(x) = \frac{R(x)}{x}$ Revenue per unit

Marginal Average Revenue = $\overline{R}'(x)$

Average Profit = $\overline{P}(x) = \frac{P(x)}{x}$ Profit per unit

Marginal Average Profit = $\overline{P}'(x)$

---

1. $C(99) = 10,000 + 150(99) - 0.2(99)^2 = 22,889.80$, \$22,889.80

3. $C(100) = 10,000 + 150(100) - 0.2(100)^2 = 23,000$; $C(100) - C(99) = 23,000 - 22889.80 = 110.20$, \$110.20

5. $C(200) = 10,000 + 150(200) - 0.2(200)^2 = 32,000$, \$32,000

7. Average cost of producing 100 bicycles: $\frac{C(100)}{100} = \frac{23,000}{100} = 230$, \$230

9. $C(x) = 175 + 0.8x$; $C'(x) = 0.8$

11. $C(x) = 210 + 4.6x - 0.01x^2$; $C'(x) = 4.6 - 0.02x$

13. $R(x) = 4x - 0.01x^2$; $R'(x) = 4 - 0.02x$

15. $R(x) = x(12 - 0.04x) = 12x - 0.04x^2$; $R'(x) = 12 - 0.08x$

17. $P(x) = R(x) - C(x) = 4x - 0.01x^2 - [175 + 0.8x] = 3.2x - 0.01x^2 - 175;$

$P'(x) = 3.2 - 0.02x$

19. $P(x) = R(x) - C(x) = x(12 - 0.04x) - [210 + 4.6x - 0.01x^2] = 7.4x - 0.03x^2 - 210;$

$P'(x) = 7.4 - 0.06x$

21. $C(x) = 145 + 1.1x;\quad \overline{C}(x) = \dfrac{145 + 1.1x}{x} = 1.1 + \dfrac{145}{x}$

23. $\overline{C}(x) = 1.1 + \dfrac{145}{x};\quad \overline{C}'(x) = -\dfrac{145}{x^2}$

25. $P(x) = R(x) - C(x) = 5x - 0.02x^2 - [145 + 1.1x] = 3.9x - 0.02x^2 - 145$

27. $\overline{P}(x) = \dfrac{P(x)}{x} = 3.9 - 0.02x - \dfrac{145}{x}$

29. True. $C(x) = ax + b,\quad C'(x) = a,$ constant.

31. False. $P(x) = R(x) - C(x);\quad P'(x) = R'(x) - C'(x) =$ marginal revenue − marginal cost.

33. $C(x) = 2000 + 50x - 0.5x^2$

(A) The exact cost of producing the 21st food processor is:

$$C(21) - C(20) = 2000 + 50(21) - \frac{(21)^2}{2} - \left[2000 + 50(20) - \frac{(20)^2}{2}\right]$$

$$= 2829.50 - 2800$$
$$= 29.50 \text{ or } \$29.50$$

(B) $C'(x) = 50 - x$

$C'(20) = 50 - 20 = 30$ or \$30

35. $C(x) = 60{,}000 + 300x$

(A) $\overline{C}(x) = \dfrac{60{,}000 + 300x}{x} = \dfrac{60{,}000}{x} + 300 = 60{,}000x^{-1} + 300$

$\overline{C}(500) = \dfrac{60{,}000 + 300(500)}{500} = \dfrac{210{,}000}{500} = 420$ or \$420

(B) $\overline{C}'(x) = -60{,}000x^{-2} = \dfrac{-60{,}000}{x^2}$

$\overline{C}'(500) = \dfrac{-60{,}000}{(500)^2} = -0.24$ or $-\$0.24$

Interpretation: At a production level of 500 frames, average cost is decreasing at the rate of 24¢ per frame.

(C) The average cost per frame if 501 frames are produced is approximately \$420 – \$0.24 = \$419.76.

37. $P(x) = 30x - 0.3x^2 - 250,\ 0 \le x \le 100$

(A) The exact profit from the sale of the 26th skateboard is:

$P(26) - P(25) = 30(26) - 0.3(26)^2 - 250 - [30(25) - 0.3(25)^2 - 250]$

$= 327.20 - 312.50 = \$14.70$

(B) Marginal profit: $P'(x) = 30 - 0.6x;\quad P'(25) = \$15$

39. $P(x) = 5x - \frac{x^2}{200} - 450,\ 0 \le x \le 1000$; $P'(x) = 5 - \frac{x}{100}$

(A) $P'(450) = 5 - \frac{450}{100} = 0.5$ or \$0.50

Interpretation: At a production level of 450 DVD's, profit is increasing at the rate of 50¢ per DVD.

(B) $P'(750) = 5 - \frac{750}{100} = -2.5$ or $-\$2.50$

Interpretation: At a production level of 750 DVD's, profit is decreasing at the rate of \$2.50 per DVD.

41. $P(x) = 30x - 0.03x^2 - 750,\ 0 \le x \le 1000$

Average profit: $\overline{P}(x) = \frac{P(x)}{x} = 30 - 0.03x - \frac{750}{x} = 30 - 0.03x - 750x^{-1}$

(A) At $x = 50$, $\overline{P}(50) = 30 - (0.03)50 - \frac{750}{50} = 13.50$ or \$13.50.

(B) $\overline{P}'(x) = -0.03 + 750x^{-2} = -0.03 + \frac{750}{x^2}$

$\overline{P}'(50) = -0.03 + \frac{750}{(50)^2} = -0.03 + 0.3 = 0.27$ or \$0.27; at a production level of 50 mowers, the average profit per mower is INCREASING at the rate of \$0.27 per mower.

(C) The average profit per mower if 51 mowers are produced is approximately \$13.50 + \$0.27 = \$13.77.

43. $x = 4{,}000 - 40p$

(A) Solving the given equation for p, we get $40p = 4{,}000 - x$

and $p = 100 - \frac{1}{40}x$ or $p = 100 - 0.025x$

Since $p \ge 0$, the domain is: $0 \le x \le 4{,}000$

(B) $R(x) = xp = 100x - 0.025x^2,\ 0 \le x \le 4{,}000$

(C) $R'(x) = 100 - 0.05x$; $R'(1{,}600) = 100 - 80 = 20$

At a production level of 1,600 pairs of running shoes, revenue is INCREASING at the rate of \$20 per pair.

(D) $R'(2{,}500) = 100 - 125 = -25$

At a production level of 2,500 pairs of running shoes, revenue is DECREASING at the rate of \$25 per pair.

45. Price-demand equation: $x = 6{,}000 - 30p$

Cost function: $C(x) = 72{,}000 + 60x$

(A) Solving the price-demand equation for p, we get

$p = 200 - \frac{1}{30}x$; domain: $0 \le x \le 6{,}000$

(B) Marginal cost: $C'(x) = 60$

(C) Revenue function: $R(x) = 200x - \frac{1}{30}x^2$; domain: $0 \le x \le 6{,}000$

(D) Marginal revenue: $R'(x) = 200 - \frac{1}{15}x$

(E) $R'(1{,}500) = 100$; at a production level of 1,500 saws, revenue is INCREASING at the rate of \$100 per saw.

$R'(4{,}500) = -100$; at a production level of 4,500 saws, revenue is DECREASING at the rate of \$100 per saw.

(F)

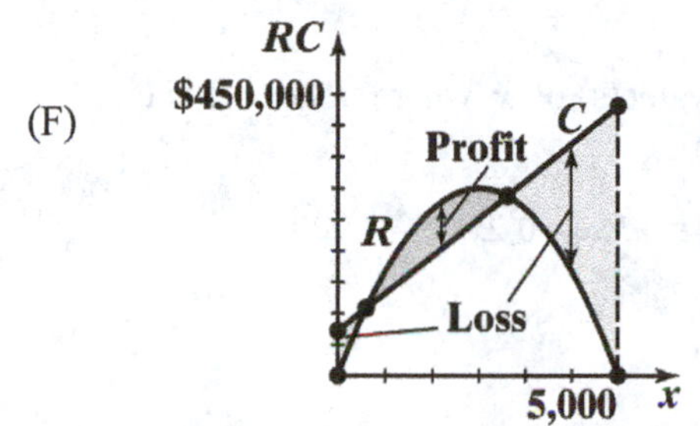

(G) Profit function: $P(x) = R(x) - C(x) = 200x - \frac{1}{30}x^2 - [72{,}000 + 60x] = 140x - \frac{1}{30}x^2 - 72{,}000$

(H) Marginal profit: $P'(x) = 140 - \frac{1}{15}x$

(I) $P'(1{,}500) = 140 - 100 = 40$; at a production level of 1,500 saws, profit is INCREASING at the rate of \$40 per saw.

$P'(3{,}000) = 140 - 200 = -60$; at a production level of 3,000 saws, profit is DECREASING at the rate of \$60 per saw.

47. (A) Assume $p = mx + b$. We are given

$$16 = m \cdot 200 + b$$

and

$$14 = m \cdot 300 + b$$

Subtracting the second equation from the first, we get

$$-100m = 2 \text{ so } m = -\frac{1}{50} = -0.02$$

Substituting this value into either equation yields $b = 20$. Therefore,

$P = 20 - 0.02x$; domain: $0 \le x \le 1{,}000$

(B) Revenue function: $R(x) = xp = 20x - 0.02x^2$, domain: $0 \le x \le 1{,}000$.

(C) $C(x) = mx + b$. From the finance department's estimates, $m = 4$ and $b = 1{,}400$. Thus, $C(x) = 4x + 1{,}400$.

(D)

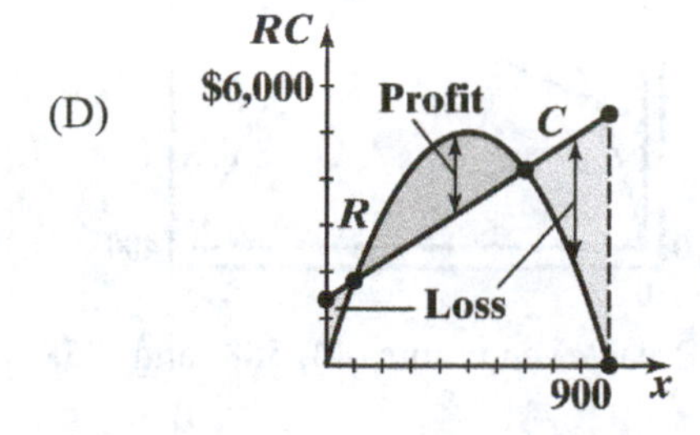

(E) Profit function: $P(x) = R(x) - C(x) = 20x - 0.02x^2 - [4x + 1{,}400]$
$= 16x - 0.02x^2 - 1{,}400$

(F) Marginal profit: $P'(x) = 16 - 0.04x$
$P'(250) = 16 - 10 = 6$; at a production level of 250 toasters, profit is INCREASING at the rate of \$6 per toaster.
$P'(475) = 16 - 19 = -3$; at a production level of 475 toasters, profit is DECREASING at the rate of \$3 per toaster.

49. Total cost: $C(x) = 24x + 21{,}900$
Total revenue: $R(x) = 200x - 0.2x^2$, $0 \le x \le 1{,}000$

(A) $R'(x) = 200 - 0.4x$
The graph of R has a horizontal tangent line at the value(s) of x where $R'(x) = 0$, i.e.,
$$200 - 0.4x = 0 \quad \text{or} \quad x = 500$$

(B) $P(x) = R(x) - C(x) = 200x - 0.2x^2 - (24x + 21{,}900) = 176x - 0.2x^2 - 21{,}900$

(C) $P'(x) = 176 - 0.4x$. Setting $P'(x) = 0$, we have
$$176 - 0.4x = 0 \quad \text{or} \quad x = 440$$

(D) The graphs of C, R and P are shown below.

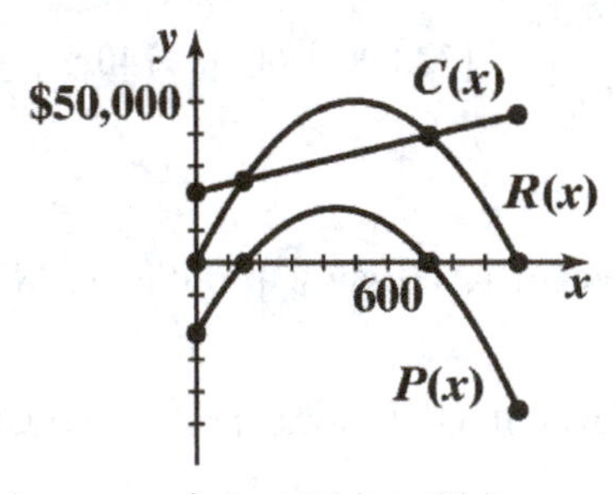

Break-even points: $R(x) = C(x)$
$$200x - 0.2x^2 = 24x + 21{,}900$$
$$0.2x^2 - 176x + 21{,}900 = 0$$
$$x = \frac{176 \pm \sqrt{(-176)^2 - (4)(0.2)(21{,}900)}}{2(0.2)} \quad \text{(quadratic formula)}$$
$$= \frac{176 \pm \sqrt{30{,}976 - 17{,}520}}{0.4} = \frac{176 \pm \sqrt{13{,}456}}{0.4} = \frac{176 \pm 116}{0.4} = 730,\ 150$$

Thus, the break-even points are: (730, 39,420) and (150, 25,500).

x-intercepts for P: $-0.2x^2 + 176x - 21{,}900 = 0$ or $0.2x^2 - 176x + 21{,}900 = 0$
which is the same as the equation above. Thus, $x = 150$ and $x = 730$.

51. Demand equation: $p = 20 - \sqrt{x} = 20 - x^{1/2}$
Cost equation: $C(x) = 500 + 2x$

(A) Revenue $R(x) = xp = x(20 - x^{1/2})$
or $R(x) = 20x - x^{3/2}$

(B) The graphs for R and C for $0 \le x \le 400$ are shown at the right.

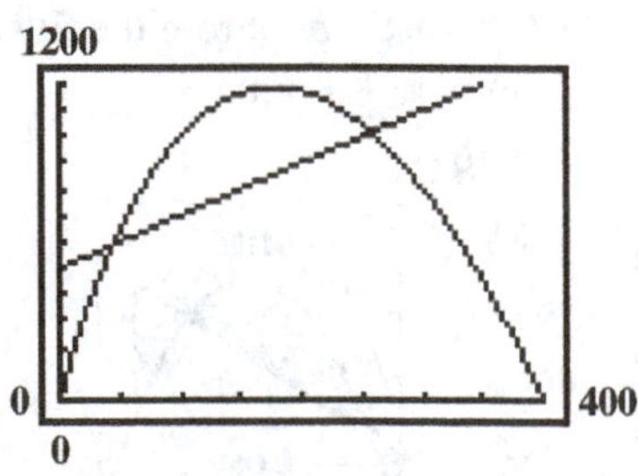

Break-even points (44, 588) and (258, 1,016).

53. (A)

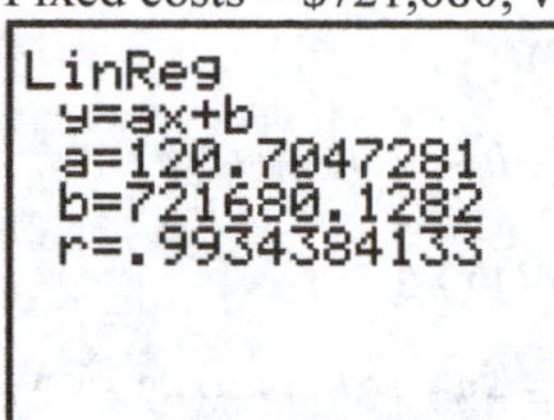

(B) Fixed costs ≈ \$721,680; variable costs ≈ \$121 per projector

(C) Let $y = p(x)$ be the quadratic regression equation found in part (A) and let $y = C(x)$ be the linear regression equation found in part (B). Then revenue $R(x) = xp(x)$, and the break-even points are the points where $R(x) = C(x)$.

break-even points: (713, 807,703), (5,423, 1,376,227)

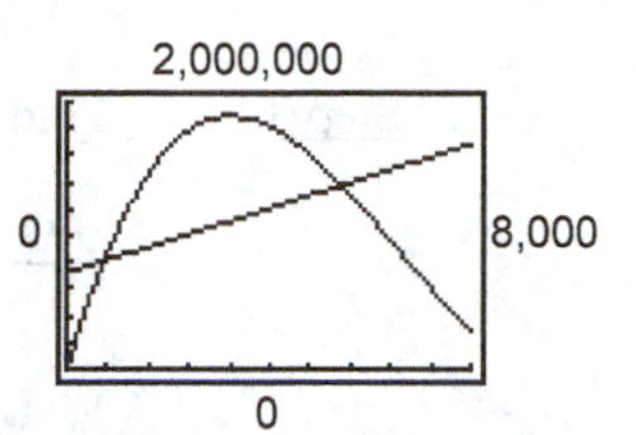

(D) The company will make a profit when $713 \le x \le 5{,}423$. From part (A), $p(713) \approx 1{,}133$ and $p(5{,}423) \approx 254$. Thus, the company will make a profit for the price range $\$254 \le p \le \$1{,}133$.

CHAPTER 2 REVIEW

1. $f(x) = 2x^2 + 5$

(A) $f(3) - f(1) = 2(3)^2 + 5 - [2(1)^2 + 5] = 16$

(B) Average rate of change: $\dfrac{f(3) - f(1)}{3-1} = \dfrac{16}{2} = 8$

(C) Slope of secant line: $\dfrac{f(3) - f(1)}{3-1} = \dfrac{16}{2} = 8$

(D) Instantaneous rate of change at $x = 1$:

Step 1. $\dfrac{f(1+h) - f(1)}{h} = \dfrac{2(1+h)^2 + 5 - [2(1)^2 + 5]}{h} = \dfrac{2(1+2h+h^2)+5-7}{h} = \dfrac{4h+2h^2}{h} = 4 + 2h$

Step 2. $\lim\limits_{h\to 0} \dfrac{f(1+h) - f(1)}{h} = \lim\limits_{h\to 0}(4+2h) = 4$

(E) Slope of the tangent line at $x = 1$: 4

(F) $f'(1) = 4$ (2-2)

2. $f(x) = -3x + 2$

Step 1. Find $f(x+h)$

$$f(x+h) = -3(x+h)+2 = -3x-3h+2$$

Step 2. Find $f(x+h)-f(x)$

$$f(x+h)-f(x) = -3x-3h+2-(-3x+2) = -3x-3h+2+3x-2 = -3h$$

Step 3. Find $\frac{f(x+h)-f(x)}{h}$

$$\frac{f(x+h)-f(x)}{h} = \frac{-3h}{h} = -3$$

Step 4. Find $\lim_{h\to 0}\frac{f(x+h)-f(x)}{h}$.

$$\lim_{h\to 0}\frac{f(x+h)-f(x)}{h} = \lim_{h\to 0}(-3) = -3$$ (2-2)

3. (A) $\lim_{x\to 1}(5f(x)+3g(x)) = 5\lim_{x\to 1}f(x) + 3\lim_{x\to 1}g(x) = 5\cdot 2 + 3\cdot 4 = 22$

(B) $\lim_{x\to 1}[f(x)g(x)] = [\lim_{x\to 1}f(x)][\lim_{x\to 1}g(x)] = 2\cdot 4 = 8$

(C) $\lim_{x\to 1}\frac{g(x)}{f(x)} = \frac{\lim_{x\to 1}g(x)}{\lim_{x\to 1}f(x)} = \frac{4}{2} = 2$

(D) $\lim_{x\to 1}[5+2x-3g(x)] = \lim_{x\to 1}5 + \lim_{x\to 1}2x - 3\lim_{x\to 1}g(x) = 5+2-3(4) = -5$ (2-1)

4. $f(1.5) \approx 1.5$ (2-1)

5. $f(2.5) \approx 3.5$ (2-1)

6. $f(2.75) \approx 3.75$ (2-1)

7. $f(3.25) \approx 3.75$ (2-1)

8. (A) $\lim_{x\to 1^-}f(x) = 1$ (B) $\lim_{x\to 1^+}f(x) = 1$ (C) $\lim_{x\to 1}f(x) = 1$ (D) $f(1) = 1$ (2-1)

9. (A) $\lim_{x\to 2^-}f(x) = 2$ (B) $\lim_{x\to 2^+}f(x) = 3$ (C) $\lim_{x\to 2}f(x)$ does not exist (D) $f(2) = 3$ (2-1)

10. (A) $\lim_{x\to 3^-}f(x) = 4$ (B) $\lim_{x\to 3^+}f(x) = 4$ (C) $\lim_{x\to 3}f(x) = 4$ (D) $f(3)$ does not exist (2-1)

11. (A) From the graph, $\lim_{x\to 1}f(x)$ does not exist since

$\lim_{x\to 1^-}f(x) = 2 \neq \lim_{x\to 1^+}f(x) = 3$.

(B) $f(1) = 3$

(C) f is NOT continuous at $x = 1$, since $\lim_{x\to 1}f(x)$ does not exist. (2-3)

12. (A) $\lim_{x\to 2}f(x) = 2$ (B) $f(2)$ is not defined

(C) f is NOT continuous at $x = 2$ since $f(2)$ is not defined. (2-3)

13. (A) $\lim_{x\to 3} f(x) = 1$ (B) $f(3) = 1$

(C) f is continuous at $x = 3$ since $\lim_{x\to 3} f(x) = f(3)$. (2-3)

14. $\lim_{x\to\infty} f(x) = 5$ (2-2)

15. $\lim_{x\to-\infty} f(x) = 5$ (2-2)

16. $\lim_{x\to 2^+} f(x) = \infty$ (2-2)

17. $\lim_{x\to 2^-} f(x) = -\infty$ (2-2)

18. $\lim_{x\to 0^-} f(x) = 0$ (2-1)

19. $\lim_{x\to 0^+} f(x) = 0$ (2-1)

20. $\lim_{x\to 0} f(x) = 0$ (2-1)

21. $x = 2$ is a vertical asymptote (2-3)

22. $y = 5$ is a horizontal asymptote (2-2)

23. f is discontinuous at $x = 2$ (2-3)

24. $f(x) = 5x^2$

Step 1. Find $f(x + h)$:

$f(x + h) = 5(x + h)^2 = 5(x^2 + 2xh + h^2) = 5x^2 + 10xh + 5h^2$

Step 2. Find $f(x + h) - f(x)$:

$f(x + h) - f(x) = 5x^2 + 10xh + 5h^2 - 5x^2 = 10xh + 5h^2$

Step 3. Find $\dfrac{f(x+h)-f(x)}{h}$:

$$\frac{f(x+h)-f(x)}{h} = \frac{10xh+5h^2}{h} = 10x + 5h$$

Step 4. Find $\lim_{h\to 0} \dfrac{f(x+h)-f(x)}{h}$

$$\lim_{h\to 0} \frac{f(x+h)-f(x)}{h} = \lim_{h\to 0} (10x + 5h) = 10x$$

Thus, $f'(x) = 10x$. (2-4)

25. (A) $h'(x) = (3f(x))' = 3f'(x)$; $h'(5) = 3f'(5) = 3(-1) = -3$

(B) $h'(x) = (-2g(x))' = -2g'(x)$; $h'(5) = -2g'(5) = -2(-3) = 6$

(C) $h'(x) = 2f'(x)$; $h'(5) = 2(-1) = -2$

(D) $h'(x) = -g'(x)$; $h'(5) = -(-3) = 3$

(E) $h'(x) = 2f'(x) + 3g'(x)$; $h'(5) = 2(-1) + 3(-3) = -11$ (2-5)

26. $f(x) = \frac{1}{3}x^3 - 5x^2 + 1$; $f'(x) = x^2 - 10x$ (2-5)

27. $f(x) = 2x^{1/2} - 3x$; $f'(x) = 2\cdot\frac{1}{2}x^{-1/2} - 3 = \frac{1}{x^{1/2}} - 3$ (2-5)

28. $f(x) = 5$

$f'(x) = 0$

(2-5)

29. $f(x) = \frac{3}{2x} + \frac{5x^3}{4} = \frac{3}{2}x^{-1} + \frac{5}{4}x^3;$

$f'(x) = -\frac{3}{2}x^{-2} + \frac{15}{4}x^2 = -\frac{3}{2x^2} + \frac{15}{4}x^2$ (2-5)

30. $f(x) = \frac{0.5}{x^4} + 0.25x^4 = 0.5x^{-4} + 0.25x^4$

$f'(x) = 0.5(-4)x^{-5} + 0.25(4x^3) = -2x^{-5} + x^3 = -\frac{2}{x^5} + x^3$ (2-5)

31. $f(x) = (3x^3 - 2)(x + 1) = 3x^4 + 3x^3 - 2x - 2$

$f'(x) = 12x^3 + 9x^2 - 2$ (2-5)

For Problems 32 – 35, $f(x) = x^2 + x$.

32. $\Delta x = x_2 - x_1 = 3 - 1 = 2$, $\Delta y = f(x_2) - f(x_1) = 12 - 2 = 10$,

$\frac{\Delta y}{\Delta x} = \frac{10}{2} = 5.$ (2-6)

33. $\frac{f(x_1 + \Delta x) - f(x_1)}{\Delta x} = \frac{f(1+2) - f(1)}{2} = \frac{f(3) - f(1)}{2} = \frac{12-2}{2} = 5$ (2-6)

34. $dy = f'(x)dx = (2x + 1)dx$. For $x_1 = 1, x_2 = 3$,

$dx = \Delta x = 3 - 1 = 2$, $dy = (2\cdot 1 + 1)\cdot 2 = 3\cdot 2 = 6$ (2-6)

35. $\Delta y = f(x + \Delta x) - f(x)$; at $x = 1$, $\Delta x = 0.2$,

$\Delta y = f(1.2) - f(1) = 0.64$

$dy = f'(x)dx$ where $f'(x) = 2x + 1$; at $x = 1$

$dy = 3(0.2) = 0.6$ (2-6)

36. From the graph:

(A) $\lim_{x\to 2^-} f(x) = 4$

(B) $\lim_{x\to 2^+} f(x) = 6$

(C) $\lim_{x\to 2} f(x)$ does not exist since $\lim_{x\to 2^-} f(x) \neq \lim_{x\to 2^+} f(x)$

(D) $f(2) = 6$

(E) No, since $\lim_{x\to 2} f(x)$ does not exist. (2-3)

37. From the graph:

(A) $\lim_{x\to 5^-} f(x) = 3$ (B) $\lim_{x\to 5^+} f(x) = 3$ (C) $\lim_{x\to 5} f(x) = 3$ (D) $f(5) = 3$

(E) Yes, since $\lim_{x\to 5} f(x) = f(5) = 3$. (2-3)

38. (A) $f(x) < 0$ on $(8, \infty)$ (B) $f(x) \geq 0$ on $[0, 8]$ (2-3)

39. $x^2 - x < 12$ or $x^2 - x - 12 < 0$

Let $f(x) = x^2 - x - 12 = (x+3)(x-4)$. Then f is continuous for all x and $f(-3) = f(4) = 0$. Thus, $x = -3$ and $x = 4$ are partition numbers.

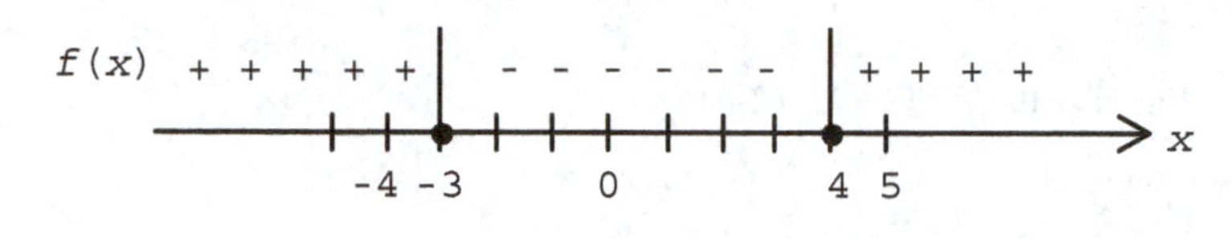

Test Numbers

x	$f(x)$
-4	$8\ (+)$
0	$-12\ (-)$
5	$8\ (+)$

Thus, $x^2 - x < 12$ for: $-3 < x < 4$ or $(-3, 4)$. (2-3)

40. $\dfrac{x-5}{x^2+3x} > 0$ or $\dfrac{x-5}{x(x+3)} > 0$

Let $f(x) = \dfrac{x-5}{x(x+3)}$. Then f is discontinuous at $x = 0$ and $x = -3$, and $f(5) = 0$. Thus, $x = -3$, $x = 0$, and $x = 5$ are partition numbers.

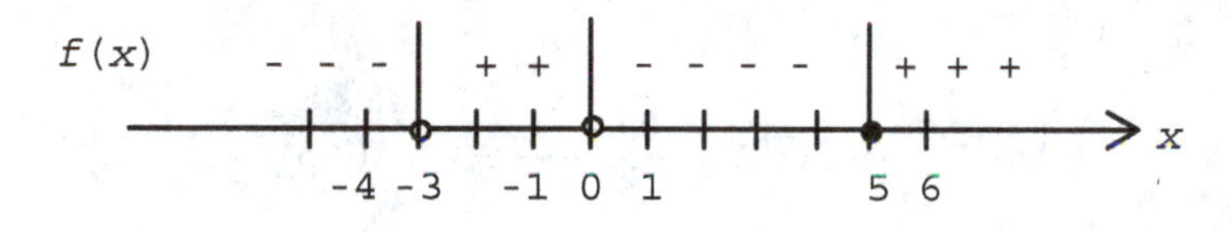

Test Numbers

x	$f(x)$
-4	$-\frac{9}{4}(-)$
-1	$3(+)$
1	$-1(-)$
6	$\frac{1}{54}(+)$

Thus, $\dfrac{x-5}{x^2+3x} > 0$ for $-3 < x < 0$ or $x > 5$, or $(-3, 0) \cup (5, \infty)$. (2-3)

41. $x^3 + x^2 - 4x - 2 > 0$

Let $f(x) = x^3 + x^2 - 4x - 2$. Then f is continuous for all x and $f(x) = 0$ at $x = -2.3429$, -0.4707 and 1.8136.

f(x) - - - 0 + + + 0 - - - - 0 + + +

x

-2.34 -0.47 0 1.81

Thus, $x^3 + x^2 - 4x - 2 > 0$ for $-2.3429 < x < -0.4707$ or $1.8136 < x < \infty$, or $(-2.3429, -0.4707) \cup (1.8136, \infty)$. (2-3)

42. $f(x) = 0.5x^2 - 5$

(A) $\dfrac{f(4) - f(2)}{4-2} = \dfrac{0.5(4)^2 - 5 - [0.5(2)^2 - 5]}{2} = \dfrac{8-2}{2} = 3$

(B) $\frac{f(2+h)-f(2)}{h} = \frac{0.5(2+h)^2 - 5 - [0.5(2)^2 - 5]}{h} = \frac{0.5(4+4h+h^2) - 5 + 3}{h}$

$= \frac{2h+0.5h^2}{h} = \frac{h(2+0.5h)}{h} = 2 + 0.5h$

(C) $\lim_{h\to 0} \frac{f(2+h)-f(2)}{h} = \lim_{h\to 0} (2 + 0.5h) = 2$ (2-4)

43. $y = \frac{1}{3}x^{-3} - 5x^{-2} + 1;$

$\frac{dy}{dx} = \frac{1}{3}(-3)x^{-4} - 5(-2)x^{-3} = -x^{-4} + 10x^{-3}$ (2-5)

44. $y = \frac{3\sqrt{x}}{2} + \frac{5}{3\sqrt{x}} = \frac{3}{2}x^{1/2} + \frac{5}{3}x^{-1/2};$

$y' = \frac{3}{2}\left(\frac{1}{2}x^{-1/2}\right) + \frac{5}{3}\left(-\frac{1}{2}x^{-3/2}\right) = \frac{3}{4x^{1/2}} - \frac{5}{6x^{3/2}} = \frac{3}{4\sqrt{x}} - \frac{5}{6\sqrt{x^3}}$ (2-5)

45. $g(x) = 1.8\sqrt[3]{x} + \frac{0.9}{\sqrt[3]{x}} = 1.8x^{1/3} + 0.9x^{-1/3}$

$g'(x) = 1.8\left(\frac{1}{3}x^{-2/3}\right) + 0.9\left(-\frac{1}{3}x^{-4/3}\right) = 0.6x^{-2/3} - 0.3x^{-4/3} = \frac{0.6}{x^{2/3}} - \frac{0.3}{x^{4/3}}$ (2-5)

46. $y = \frac{2x^3 - 3}{5x^3} = \frac{2}{5} - \frac{3}{5}x^{-3};\ y' = -\frac{3}{5}(-3x^{-4}) = \frac{9}{5x^4}$ (2-5)

47. $f(x) = x^2 + 4$
$f'(x) = 2x$

(A) The slope of the graph at $x = 1$ is $m = f'(1) = 2$.

(B) $f(1) = 1^2 + 4 = 5$
The tangent line at (1, 5), where the slope $m = 2$, is:
$(y - 5) = 2(x - 1)$ [Note: $(y - y_1) = m(x - x_1)$.]

$y = 5 + 2x - 2$
$y = 2x + 3$ (2-4, 2-5)

48. $f(x) = 10x - x^2$
$f'(x) = 10 - 2x$
The tangent line is horizontal at the values of x such that
$f'(x) = 0$:
$10 - 2x = 0$
$x = 5$ (2-4)

49. $f(x)=x^3+3x^2-45x-135$

$f'(x) = 3x^2 + 6x - 45$

Set $f'(x) = 0$:

$3x^2 + 6x - 45 = 0$

$x^2 + 2x - 15 = 0$

$(x-3)(x+5) = 0$

$x = 3, \quad x = -5$ (2-5)

50. $f(x) = x^4 - 2x^3 - 5x^2 + 7x$

$f'(x) = 4x^3 - 6x^2 - 10x + 7$

Set $f'(x) = 4x^3 - 6x^2 - 10x + 7 = 0$ and solve for x using a root-approximation routine on a graphing utility:

$f'(x) = 0$ at $x = -1.34, \quad x = 0.58, \quad x = 2.26$ (2-5)

51. $f(x) = x^5 - 10x^3 - 5x + 10$

$f'(x) = 5x^4 - 30x^2 - 5 = 5(x^4 - 6x^2 - 1)$

Let $f'(x) = 5(x^4 - 6x^2 - 1) = 0$ and solve for x using a root-approximation routine on a graphing utility; that is, solve $x^4 - 6x^2 - 1 = 0$ for x.

$f'(x) = 0$ at $x = \pm 2.4824$ (2-5)

52. $y = f(x) = 8x^2 - 4x + 1$

(A) Instantaneous velocity function; $v(x) = f'(x) = 16x - 4$.

(B) $v(3) = 16(3) - 4 = 44$ ft/sec. (2-5)

53. $y = f(x) = -5x^2 + 16x + 3$

(A) Instantaneous velocity function: $v(x) = f'(x) = -10x + 16$.

(B) $v(x) = 0$ when $-10x + 16 = 0$

$10x = 16$

$x = 1.6$ sec (2-5)

54. (A) $f(x) = x^3$, $g(x) = (x-4)^3$, $h(x) = (x+3)^3$

The graph of g is the graph of f shifted 4 units to the right; the graph of h is the graph of f shifted 3 units to the left.

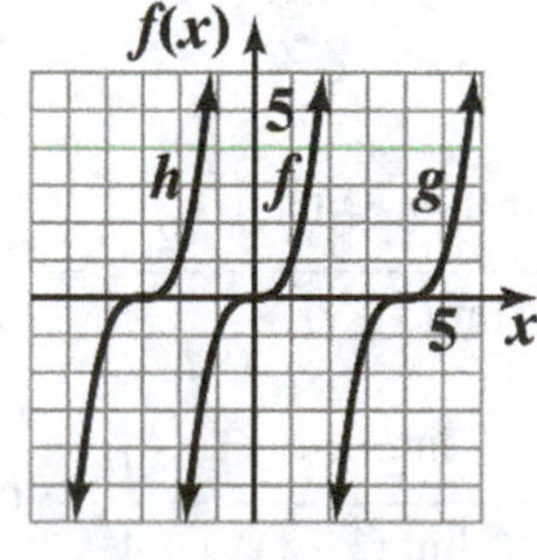

(B) $f'(x) = 3x^2$, $g'(x) = 3(x-4)^2$, $h'(x) = 3(x+3)^2$

The graph of g' is the graph of f' shifted 4 units to the right; the graph of h' is the graph of f' shifted 3 units to the left.

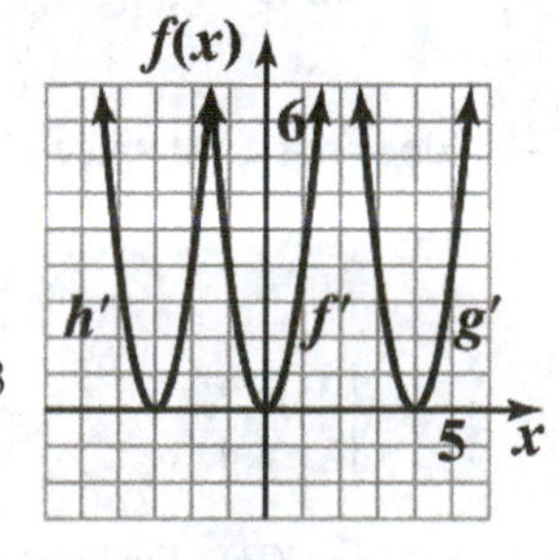

(1-2, 2-5, 2-6)

55. $f(x) = x^2 - 4$ is a polynomial function; f is continuous on $(-\infty, \infty)$. (2-3)

56. $f(x) = \frac{x+1}{x-2}$ is a rational function and the denominator $x - 2$ is 0 at $x = 2$. Thus f is continuous for all x such that $x \neq 2$, i.e., on $(-\infty, 2) \cup (2, \infty)$. (2-3)

57. $f(x) = \frac{x+4}{x^2+3x-4}$ is a rational function and the denominator $x^2 + 3x - 4 = (x + 4)(x - 1)$ is 0 at $x = -4$ and $x = 1$. Thus, f is continuous for all x except $x = -4$ and $x = 1$, i.e., on $(-\infty, -4) \cup (-4, 1) \cup (1, \infty)$. (2-2)

58. $f(x) = \sqrt[3]{4-x^2}$; $g(x) = 4 - x^2$ is continuous for all x since it is a polynomial function. Therefore, $f(x) = \sqrt[3]{g(x)}$ is continuous for all x, i.e., on $(-\infty, \infty)$. (2-3)

59. $f(x) = \sqrt{4-x^2}$; $g(x) = 4 - x^2$ is continuous for all x and $g(x)$ is nonnegative for $-2 \leq x \leq 2$. Therefore, $f(x) = \sqrt{g(x)}$ is continuous for $-2 \leq x \leq 2$, i.e., on $[-2, 2]$. (2-3)

60. $f(x) = \frac{2x}{x^2-3x} = \frac{2x}{x(x-3)} = \frac{2}{x-3}, x \neq 0$

(A) $\lim_{x\to1} f(x) = \lim_{x\to1} \frac{2}{x-3} = \frac{\lim_{x\to1} 2}{\lim_{x\to1}(x-3)} = \frac{2}{-2} = -1$

(B) $\lim_{x\to3} f(x) = \lim_{x\to3} \frac{2}{x-3}$ does not exist since $\lim_{x\to3} 2 = 2$ and $\lim_{x\to3} (x-3) = 0$

(C) $\lim_{x\to0} f(x) = \lim_{x\to0} \frac{2}{x-3} = -\frac{2}{3}$ (2-1)

61. $f(x) = \frac{x+1}{(3-x)^2}$

(A) $\lim_{x\to1} \frac{x+1}{(3-x)^2} = \frac{\lim_{x\to1}(x+1)}{\lim_{x\to1}(3-x)^2} = \frac{2}{2^2} = \frac{1}{2}$

(B) $\lim_{x\to-1} \frac{x+1}{(3-x)^2} = \frac{\lim_{x\to-1}(x+1)}{\lim_{x\to-1}(3-x)^2} = \frac{0}{4^2} = 0$

(C) $\lim_{x\to3} \frac{x+1}{(3-x)^2}$ does not exist since $\lim_{x\to3} (x+1) = 4$ and $\lim_{x\to3} (3-x)^2 = 0$ (2-1)

62. $f(x) = \frac{|x-4|}{x-4} = \begin{cases} -1 & \text{if } x < 4 \\ 1 & \text{if } x > 4 \end{cases}$

(A) $\lim_{x\to4^-} f(x) = -1$ (B) $\lim_{x\to4^+} f(x) = 1$ (C) $\lim_{x\to4} f(x)$ does not exist. (2-1)

63. $f(x) = \frac{x-3}{9-x^2} = \frac{x-3}{(3+x)(3-x)} = \frac{-(3-x)}{(3+x)(3-x)} = \frac{-1}{3+x}, x \neq 3$

(A) $\lim_{x\to 3} f(x) = \lim_{x\to 3} \frac{-1}{3+x} = -\frac{1}{6}$

(B) $\lim_{x\to -3} f(x) = \lim_{x\to -3} \frac{-1}{3+x}$ does not exist

(C) $\lim_{x\to 0} f(x) = \lim_{x\to 0} \frac{-1}{3+x} = -\frac{1}{3}$ (2-1)

64. $f(x) = \frac{x^2-x-2}{x^2-7x+10} = \frac{(x-2)(x+1)}{(x-2)(x-5)} = \frac{x+1}{x-5}, x \neq 2$

(A) $\lim_{x\to -1} f(x) = \lim_{x\to -1} \frac{x+1}{x-5} = 0$

(B) $\lim_{x\to 2} f(x) = \lim_{x\to 2} \frac{x+1}{x-5} = \frac{3}{-3} = -1$

(C) $\lim_{x\to 5} f(x) = \lim_{x\to 5} \frac{x+1}{x-5}$ does not exist (2-1)

65. $f(x) = \frac{2x}{3x-6} = \frac{2x}{3(x-2)}$

(A) $\lim_{x\to \infty} \frac{2x}{3x-6} = \lim_{x\to \infty} \frac{2x}{3x} = \frac{2}{3}$

(B) $\lim_{x\to -\infty} \frac{2x}{3x-6} = \lim_{x\to -\infty} \frac{2x}{3x} = \frac{2}{3}$

(C) $\lim_{x\to 2^-} \frac{2x}{3x-6} = \lim_{x\to 2^-} \frac{2x}{3(x-2)} = -\infty$

$\lim_{x\to 2^+} \frac{2x}{3(x-2)} = \infty$; $\lim_{x\to 2} \frac{2x}{3x-6}$ does not exist. (2-2)

66. $f(x) = \frac{2x^3}{3(x-2)^2} = \frac{2x^3}{3x^2-12x+12}$

(A) $\lim_{x\to \infty} \frac{2x^3}{3x^2-12x+12} = \lim_{x\to \infty} \frac{2x^3}{3x^2} = \lim_{x\to \infty} \frac{2x}{3} = \infty$

(B) $\lim_{x\to -\infty} \frac{2x^3}{3x^2-12x+12} = \lim_{x\to -\infty} \frac{2x^3}{3x^2} = \lim_{x\to -\infty} \frac{2x}{3} = -\infty$

(C) $\lim_{x\to 2^-} \frac{2x^3}{3(x-2)^2} = \lim_{x\to 2^+} \frac{2x^3}{3(x-2)^2} = \infty$; $\lim_{x\to 2} \frac{2x^3}{3(x-2)^2} = \infty$ (2-2)

67. $f(x) = \dfrac{2x}{3(x-2)^3}$

(A) $\lim_{x\to\infty} \dfrac{2x}{3(x-2)^3} = \lim_{x\to\infty} \dfrac{2x}{3x^3} = \lim_{x\to\infty} \dfrac{2}{3x^2} = 0$

(B) $\lim_{x\to-\infty} \dfrac{2x}{3(x-2)^3} = \lim_{x\to-\infty} \dfrac{2x}{3x^3} = \lim_{x\to-\infty} \dfrac{2}{3x^2} = 0$

(C) $\lim_{x\to 2^-} \dfrac{2x}{3(x-2)^3} = -\infty$, $\lim_{x\to 2^+} \dfrac{2x}{3(x-2)^3} = \infty$; $\lim_{x\to 2} \dfrac{2x}{3(x-2)^3}$ does not exist. (2-2)

68. $f(x) = x^2 + 4$

$$\lim_{h\to 0} \frac{f(2+h)-f(2)}{h} = \lim_{h\to 0} \frac{[(2+h)^2+4]-[2^2+4]}{h} = \lim_{h\to 0} \frac{4+4h+h^2+4-8}{h} = \lim_{h\to 0} \frac{4h+h^2}{h}$$
$$= \lim_{h\to 0}(4+h) = 4 \qquad \text{(2-1)}$$

69. Let $f(x) = \dfrac{1}{x+2}$

$$\lim_{h\to 0} \frac{f(x+h)-f(x)}{h} = \lim_{h\to 0} \frac{\frac{1}{(x+h)+2} - \frac{1}{x+2}}{h} = \lim_{h\to 0} \frac{x+2-(x+h+2)}{h(x+h+2)(x+2)} = \lim_{h\to 0} \frac{-h}{h(x+h+2)(x+2)}$$
$$= \lim_{h\to 0} \frac{-1}{(x+h+2)(x+2)} = \frac{-1}{(x+2)^2} \qquad \text{(2-1)}$$

70. $f(x) = x^2 - x$

Step 1. Find $f(x+h)$.

$f(x+h) = (x+h)^2 - (x+h) = x^2 + 2xh + h^2 - x - h$

Step 2. Find $f(x+h) - f(x)$

$f(x+h) - f(x) = x^2 + 2xh + h^2 - x - h - (x^2 - x) = x^2 + 2xh + h^2 - x - h - (x^2 - x)$
$= x^2 + 2xh + h^2 - x - h - x^2 + x = 2xh + h^2 - h$

Step 3. Find $\dfrac{f(x+h)-f(x)}{h}$.

$\dfrac{f(x+h)-f(x)}{h} = \dfrac{2xh+h^2-h}{h} = 2x + h - 1$

Step 4. Find $\lim_{h\to 0} \dfrac{f(x+h)-f(x)}{h}$.

$\lim_{h\to 0} \dfrac{f(x+h)-f(x)}{h} = \lim_{h\to 0} (2x + h - 1) = 2x - 1$

Thus, $f'(x) = 2x - 1$. (2-4)

71. $f(x) = \sqrt{x} - 3$

Step 1. Find $f(x+h)$.

$f(x+h) = \sqrt{x+h} - 3$

Step 2. Find $f(x+h)-f(x)$

$$f(x+h)-f(x)=\sqrt{x+h}-3-(\sqrt{x}-3)=\sqrt{x+h}-3-\sqrt{x}+3=\sqrt{x+h}-\sqrt{x}$$

Step 3. Find $\frac{f(x+h)-f(x)}{h}$.

$$\frac{f(x+h)-f(x)}{h}=\frac{\sqrt{x+h}-\sqrt{x}}{h}=\frac{\sqrt{x+h}-\sqrt{x}}{h}\cdot\frac{\sqrt{x+h}+\sqrt{x}}{\sqrt{x+h}+\sqrt{x}}=\frac{1}{\sqrt{x+h}+\sqrt{x}}$$

Step 4. Find $\lim_{h\to 0}\frac{f(x+h)-f(x)}{h}$.

$$\lim_{h\to 0}\frac{f(x+h)-f(x)}{h}=\lim_{h\to 0}\frac{1}{\sqrt{x+h}+\sqrt{x}}=\frac{1}{2\sqrt{x}}$$ (2-4)

72. Yes, $f'(-1)=0$. (2-4)

73. No. f is not differentiable at $x=0$ since it is not continuous at $x=0$. (2-4)

74. No. f has a vertical tangent at $x=1$. (2-4)

75. No. f is not differentiable at $x=2$; the curve has a "corner" at this point. (2-4)

76. Yes. f is differentiable at $x=3$. In fact, $f'(3)=0$. (2-4)

77. Yes. f is differentiable at $x=4$. (2-4)

78. $f(x)=\frac{5x}{x-7}$; f is discontinuous at $x=7$

$$\lim_{x\to 7^-}\frac{5x}{x-7}=-\infty,\quad \lim_{x\to 7^+}\frac{5x}{x-7}=\infty;$$ $x=7$ is a vertical asymptote

$$\lim_{x\to\infty}f(x)=\lim_{x\to\infty}\frac{5x}{x-7}=\lim_{x\to\infty}\frac{5x}{x}=5;$$ $y=5$ is a horizontal asymptote. (2-2)

(2-3)

79. $f(x)=\frac{-2x+5}{(x-4)^2}$; f is discontinuous at $x=4$.

$$\lim_{x\to 4^-}\frac{-2x+5}{(x-4)^2}=-\infty,\quad \lim_{x\to 4^+}\frac{-2x+5}{(x-4)^2}=-\infty;$$ $x=4$ is a vertical asymptote.

$$\lim_{x\to\infty}\frac{-2x+5}{(x-4)^2}=\lim_{x\to\infty}\frac{-2x}{x^2}=\lim_{x\to\infty}\frac{-2}{x}=0;$$ $y=0$ is a horizontal asymptote. (2-2)

80. $f(x) = \frac{x^2+9}{x-3}$; f is discontinuous at $x = 3$.

$\lim_{x\to 3^-} \frac{x^2+9}{x-3} = -\infty$, $\lim_{x\to 3^+} \frac{x^2+9}{x-3} = \infty$; $x = 3$ is a vertical asymptote.

$\lim_{x\to\infty} \frac{x^2+9}{x-3} = \lim_{x\to\infty} \frac{x^2}{x} = \lim_{x\to\infty} x = \infty$; no horizontal asymptotes. (2-2)

81. $f(x) = \frac{x^2-9}{x^2+x-2} = \frac{x^2-9}{(x+2)(x-1)}$; f is discontinuous at $x = -2$, $x = 1$.

At $x = -2$:

$\lim_{x\to -2^-} \frac{x^2-9}{(x+2)(x-1)} = -\infty$, $\lim_{x\to -2^+} \frac{x^2-9}{(x+2)(x-1)} = \infty$; $x = -2$ is a vertical asymptote.

At $x = 1$

$\lim_{x\to 1^-} \frac{x^2-9}{(x+2)(x-1)} = \infty$, $\lim_{x\to 1^+} \frac{x^2-9}{(x+2)(x-1)} = -\infty$; $x = 1$ is a vertical asymptote.

$\lim_{x\to\infty} \frac{x^2-9}{x^2+x-2} = \lim_{x\to\infty} \frac{x^2}{x^2} = \lim_{x\to\infty} 1 = 1$; $y = 1$ is a horizontal asymptote. (2-2)

82. $f(x) = \frac{x^3-1}{x^3-x^2-x+1} = \frac{(x-1)(x^2+x+1)}{(x-1)(x^2-1)} = \frac{(x-1)(x^2+x+1)}{(x-1)^2(x+1)} = \frac{x^2+x+1}{(x-1)(x+1)}, x \neq 1.$

f is discontinuous at $x = 1$, $x = -1$.

At $x = 1$:

$\lim_{x\to 1^-} f(x) = \lim_{x\to 1^-} \frac{x^2+x+1}{(x-1)(x+1)} = -\infty$, $\lim_{x\to 1^+} f(x) = \infty$; $x = 1$ is a vertical asymptote.

At $x = -1$:

$\lim_{x\to -1^-} \frac{x^2+x+1}{(x-1)(x+1)} = \infty$, $\lim_{x\to -1^+} \frac{x^2+x+1}{(x-1)(x+1)} = -\infty$; $x = -1$ is a vertical asymptote.

$\lim_{x\to\infty} \frac{x^3-1}{x^3-x^2-x+1} = \lim_{x\to\infty} \frac{x^3}{x^3} = \lim_{x\to\infty} 1 = 1$; $y = 1$ is a horizontal asymptote. (2-2)

83. $f(x) = x^{1/5}$; $f'(x) = \frac{1}{5}x^{-4/5} = \frac{1}{5x^{4/5}}$

The domain of f' is all real numbers except $x = 0$. At $x = 0$, the graph of f is smooth, but the tangent line to the graph at (0, 0) is vertical. (2-4)

84. $f(x) = \begin{cases} x^2 - m & \text{if } x \le 1 \\ -x^2 + m & \text{if } x > 1 \end{cases}$

(A)

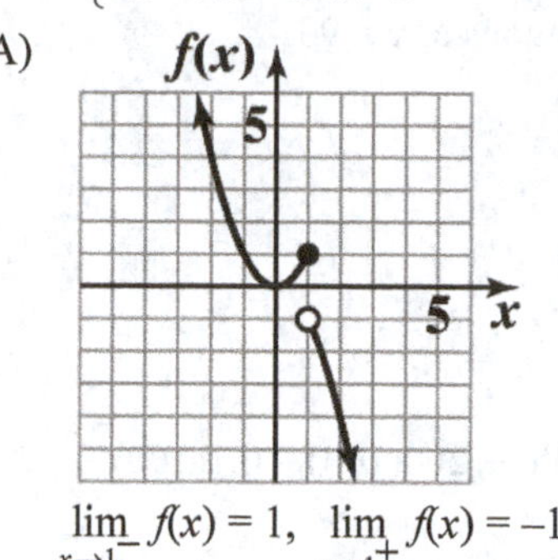

$\lim_{x\to1^-} f(x) = 1, \quad \lim_{x\to1^+} f(x) = -1$

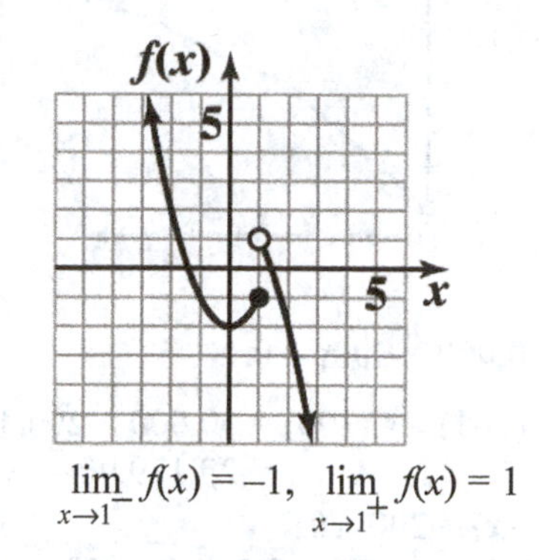

$\lim_{x\to1^-} f(x) = -1, \quad \lim_{x\to1^+} f(x) = 1$

(C) $\lim_{x\to1^-} f(x) = 1 - m, \quad \lim_{x\to1^+} f(x) = -1 + m$

We want $1 - m = -1 + m$ which implies $m = 1$.

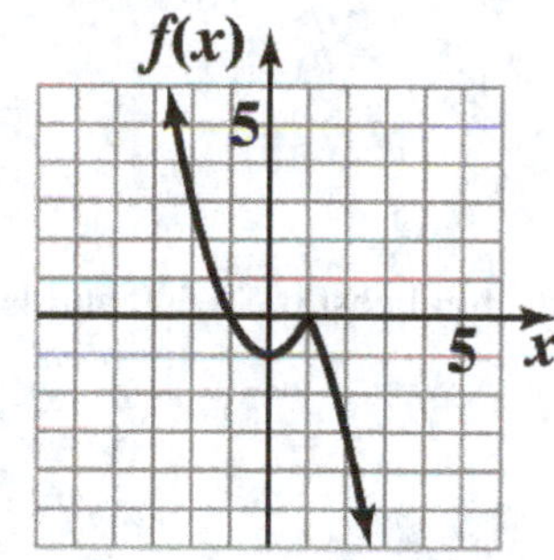

(D) The graphs in (A) and (B) have jumps at $x = 1$; the graph in (C) does not. (2-2)

85. $f(x) = 1 - |x - 1|, 0 \le x \le 2$

(A) $\lim_{h\to0^-} \frac{f(1+h)-f(1)}{h} = \lim_{h\to0^-} \frac{1-|1+h-1|-1}{h} = \lim_{h\to0^-} \frac{-|h|}{h} = \lim_{h\to0^-} \frac{h}{h} = 1$ ($|h| = -h$ if $h < 0$)

(B) $\lim_{h\to0^+} \frac{f(1+h)-f(1)}{h} = \lim_{h\to0^+} \frac{1-|1+h-1|-1}{h} = \lim_{h\to0^+} \frac{-|h|}{h} = \lim_{h\to0^+} \frac{-h}{h} = -1$ ($|h| = h$ if $h > 0$)

(C) $\lim_{h\to0} \frac{f(1+h)-f(1)}{h}$ does not exist, since the left limit and the right limit are not equal.

(D) $f'(1)$ does not exist. (2-4)

86. (A) $S(x) = 7.47 + 0.4000x$ for $0 \le x \le 90$; $S(90) = 43.47$;
$S(x) = 43.47 + 0.2076(x - 90) = 24.786 + 0.2076x, x > 90$
Therefore,

$$S(x) = \begin{cases} 7.47 + 0.4000x & \text{if } 0 \le x \le 90 \\ 24.786 + 0.2076x & \text{if } x > 90 \end{cases}$$

(B)

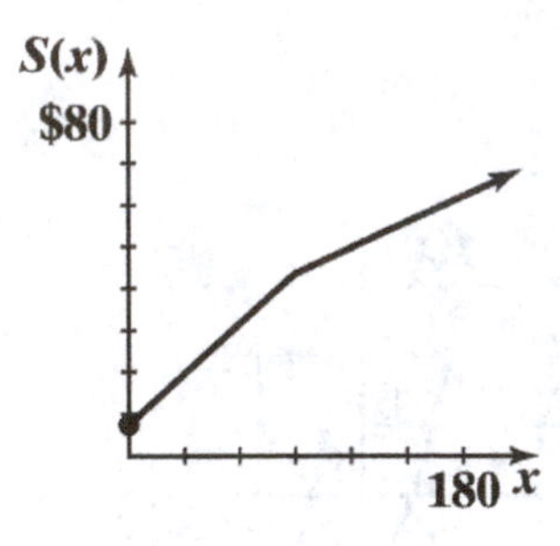

(C) $\lim_{x\to 90^-} S(x) = \lim_{x\to 90^+} S(x) = 43.47 = S(90)$;
$S(x)$ is continuous at $x = 90$. (2-2)

87. $C(x) = 10{,}000 + 200x - 0.1x^2$

(A) $C(101) - C(100) = 10{,}000 + 200(101) - 0.1(101)^2 - [10{,}000 + 200(100) - 0.1(100)^2]$
$= 29{,}179.90 - 29{,}000 = \$179.90$

(B) $C'(x) = 200 - 0.2x$
$C'(100) = 200 - 0.2(100) = 200 - 20 = \$180$ (2-7)

88. $C(x) = 5{,}000 + 40x + 0.05x^2$

(A) Cost of producing 100 bicycles:
$C(100) = 5{,}000 + 40(100) + 0.05(100)^2 = 9000 + 500 = 9500$
Marginal cost:
$C'(x) = 40 + 0.1x$
$C'(100) = 40 + 0.1(100) = 40 + 10 = 50$
Interpretation: At a production level of 100 bicycles, the total cost is \$9,500 and is increasing at the rate of \$50 per additional bicycle.

(B) Average cost: $\overline{C}(x) = \dfrac{C(x)}{x} = \dfrac{5000}{x} + 40 + 0.05x$

$$\overline{C}(100) = \frac{5000}{100} + 40 + 0.05(100) = 50 + 40 + 5 = 95$$

Marginal average cost: $\overline{C}'(x) = -\dfrac{5000}{x^2} + 0.05$ and

$$\overline{C}'(100) = -\frac{5000}{(100)^2} + 0.05 = -0.5 + 0.05 = -0.45$$

Interpretation: At a production level of 100 bicycles, the average cost is \$95 and the average cost is decreasing at a rate of \$0.45 per additional bicycle. (2-7)

89. The approximate cost of producing the 201st printer is greater than that of producing the 601st printer (the slope of the tangent line at $x = 200$ is greater than the slope of the tangent line at $x = 600$). Since the marginal costs are decreasing, the manufacturing process is becoming more efficient. (2-7)

90. $p = 25 - 0.01x$, $C(x) = 2x + 9{,}000$

(A) Marginal cost: $C'(x) = 2$

Average cost: $\overline{C}(x) = \dfrac{C(x)}{x} = 2 + \dfrac{9{,}000}{x}$

Marginal average cost: $\overline{C}' = -\dfrac{9{,}000}{x^2}$

(B) Revenue: $R(x) = xp = 25x - 0.01x^2$
Marginal revenue: $R'(x) = 25 - 0.02x$

Average revenue: $\overline{R}(x) = \dfrac{R(x)}{x} = 25 - 0.01x$

Marginal average revenue: $\overline{R}'(x) = -0.01$

(C) Profit: $P(x) = R(x) - C(x) = 25x - 0.01x^2 - (2x + 9{,}000) = 23x - 0.01x^2 - 9{,}000$
Marginal profit: $P'(x) = 23 - 0.02x$

Average profit: $\overline{P}(x) = \frac{P(x)}{x} = 23 - 0.01x - \frac{9{,}000}{x}$

Marginal average profit: $\overline{P}'(x) = -0.01 + \frac{9{,}000}{x^2}$

(D) Break-even points:

$$\begin{aligned} R(x) &= C(x) \\ 25x - 0.01x^2 &= 2x + 9{,}000 \\ 0.01x^2 - 23x + 9{,}000 &= 0 \\ x^2 - 2{,}300x + 900{,}000 &= 0 \\ (x - 500)(x - 1{,}800) &= 0 \end{aligned}$$

Thus, the break-even points are at $x = 500$, $x = 1{,}800$;
break-even points: (500, 10,000), (1,800, 12,600).

(E) $P'(1{,}000) = 23 - 0.02(1000) = 3$; profit is increasing at the rate of \$3 per umbrella.

$P'(1{,}150) = 23 - 0.02(1{,}150) = 0$; profit is flat.

$P'(1{,}400) = 23 - 0.02(1{,}400) = -5$; profit is decreasing at the rate of \$5 per umbrella.

(F)

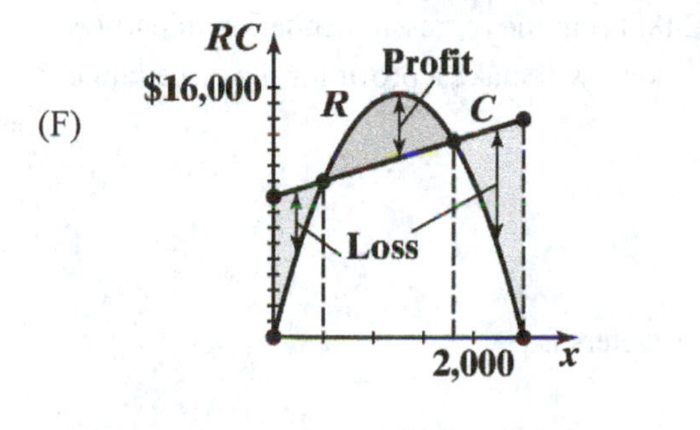

(2-7)

91. $N(t) = \frac{40t - 80}{t} = 40 - \frac{80}{t}, t \geq 2$

(A) Average rate of change from $t = 2$ to $t = 5$:

$$\frac{N(5) - N(2)}{5 - 2} = \frac{\frac{40(5) - 80}{5} - \frac{40(2) - 80}{2}}{3} = \frac{120}{15} = 8 \text{ components per day.}$$

(B) $N(t) = 40 - \frac{80}{t} = 40 - 80t^{-1}$; $N'(t) = 80t^{-2} = \frac{80}{t^2}$.

$N'(2) = \frac{80}{4} = 20$ components per day. (2-5)

92. $N(t) = 2t + \frac{1}{3}t^{3/2}$, $N'(t) = 2 + \frac{1}{2}t^{1/2} = \frac{4 + \sqrt{t}}{2}$

$N(9) = 18 + \frac{1}{3}(9)^{3/2} = 27$, $N'(9) = \frac{4 + \sqrt{9}}{2} = \frac{7}{2} = 3.5$

After 9 months, 27,000 pools have been sold and the total sales are increasing at the rate of 3,500 pools per month. (2-5)

93. (A)

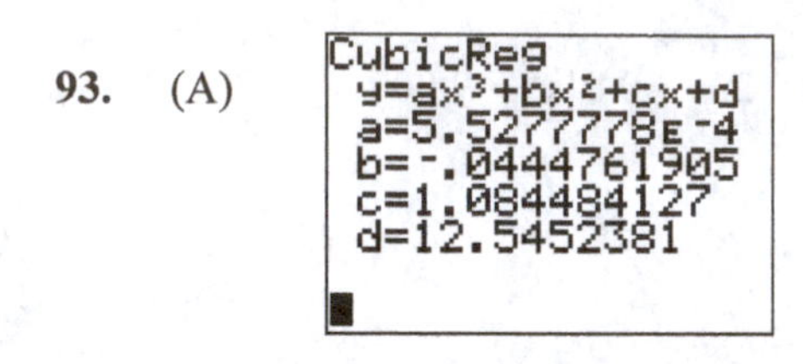

(B) $N(x) \approx 0.0005528x^3 - 0.044x^2 + 1.084x + 12.545$

$N'(x) \approx 0.0016584x^2 - 0.088x + 1.084$

$N(60) \approx 36.9$, $N'(60) \approx 1.7$. In 2020, natural gas consumption will be 36.9 trillion cubic feet and will be INCREASING at the rate of 1.7 trillion cubic feet per year. (2-4)

94. (A)

```
LinReg
 y=ax+b
 a=-.0384180791
 b=13.59887006
 r=-.9897782666
```

(B) Fixed costs: $484.21; variable cost per kringle: $2.11.

```
LinReg
 y=ax+b
 a=2.107344633
 b=484.2090395
 r=.9939318704
```

(C) Let $p(x)$ be the linear regression equation found in part (A) and let $C(x)$ be the linear regression equation found in part (B). Then revenue $R(x) = xp(x)$ and the break-even points are the points where $R(x) = C(x)$.

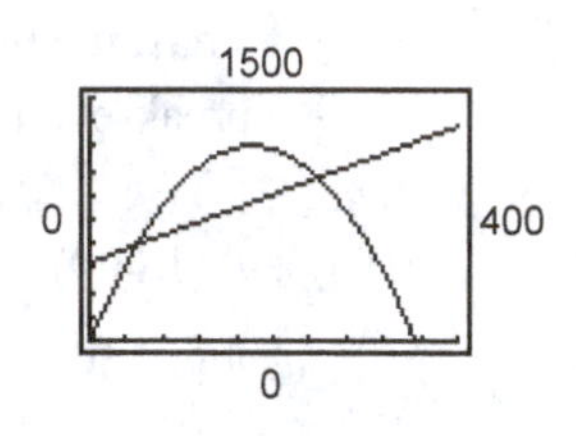

Using an intersection routine on a graphing utility, the break-even points are: (51, 591.15) and (248, 1,007.62).

(D) The bakery will make a profit when $51 < x < 248$. From the regression equation in part (A), $p(51) = 11.64$ and $p(248) = 4.07$. Thus, the bakery will make a profit for the price range $\$4.07 < p < \$11.64$. (2-7)

95. $C(x) = \dfrac{500}{x^2} = 500x^{-2}, x \geq 1.$

The instantaneous rate of change of concentration at x meters is:

$C'(x) = 500(-2)x^{-3} = \dfrac{-1000}{x^3}$

The rate of change of concentration at 10 meters is:

$C'(10) = \dfrac{-1000}{10^3} = -1$ parts per million per meter

The rate of change of concentration at 100 meters is:

$C'(100) = \dfrac{-1000}{(100)^3} = \dfrac{-1000}{100{,}000{,}000} = -\dfrac{1}{1000} = -0.001$ parts per million per meter. (2-5)

96. $F(t) = 0.16t^2 - 1.6t + 102$, $F'(t) = 0.32t - 1.6$

$F(4) = 98.16$, $F'(4) = -0.32$.

After 4 hours the patient's temperature is 98.16°F and is decreasing at the rate of 0.32°F per hour. (2-5)

97. $N(t) = 20\sqrt{t} = 20t^{1/2}$

The rate of learning is $N'(t) = 20\left(\dfrac{1}{2}\right)t^{-1/2} = 10t^{-1/2} = \dfrac{10}{\sqrt{t}}$.

(A) The rate of learning after one hour is $N'(1) = \dfrac{10}{\sqrt{1}} = 10$ items per hour.

(B) The rate of learning after four hours is $N'(4) = \frac{10}{\sqrt{4}} = \frac{10}{2} = 5$ items per hour. (2-5)

98. (A)

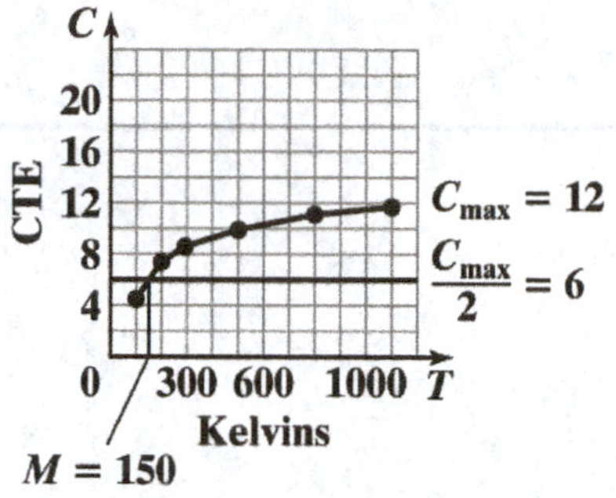

(B) $C(T) = \frac{12T}{150+T}$

(C) $C(600) = \frac{12(600)}{150+600} = 9.6$

To find T when $C = 10$, solve $\frac{12T}{150+T} = 10$ for T.

$$\frac{12T}{150+T} = 10$$
$$12T = 1500 + 10T$$
$$2T = 1500$$
$$T = 750$$

$T = 750$ when $C = 10$. (2-3)

3 ADDITIONAL DERIVATIVE TOPICS

EXERCISE 3-1

Things to remember:

1. THE NUMBER e

The irrational number e is defined by

$$e = \lim_{n\to\infty}\left(1+\frac{1}{n}\right)^n$$

or alternatively,

$$e = \lim_{s\to 0}(1+s)^{1/s}$$

Both limits are equal to $e = 2.718\ 281\ 828\ 459\ \ldots$

2. CONTINUOUS COMPOUND INTEREST FORMULA

$A = Pe^{rt}$

where P = Principal
r = Annual nominal interest rate compounded continuously
t = Time in years
A = Amount at time t

---

1. $A = 1{,}200e^{0.04(5)} = 1{,}200e^{0.2} \approx 1{,}465.68$

3. $9{,}827.30 = Pe^{0.025(3)};\quad P = \dfrac{9{,}827.30}{e^{0.075}} \approx 9{,}117.21$

5. $6{,}000 = 5{,}000e^{0.0325t},\quad e^{0.0325t} = \dfrac{6{,}000}{5{,}000},\quad 0.0325t = \ln\left(\dfrac{6{,}000}{5{,}000}\right),\quad t = \dfrac{\ln 1.2}{0.0325} \approx 5.61$

7. $956 = 900e^{1.5r},\quad e^{1.5r} = \dfrac{956}{900},\quad 1.5r = \ln\left(\dfrac{956}{900}\right),\quad r \approx 0.04$

9. $A = \$1000e^{0.1t}$
When $t = 2$, $A = \$1000e^{(0.1)2} = \$1000e^{0.2} = \$1221.40$.
When $t = 5$, $A = \$1000e^{(0.1)5} = \$1000e^{0.5} = \$1648.72$.
When $t = 8$, $A = \$1000e^{(0.1)8} = \$1000e^{0.8} = \$2225.54$

11.

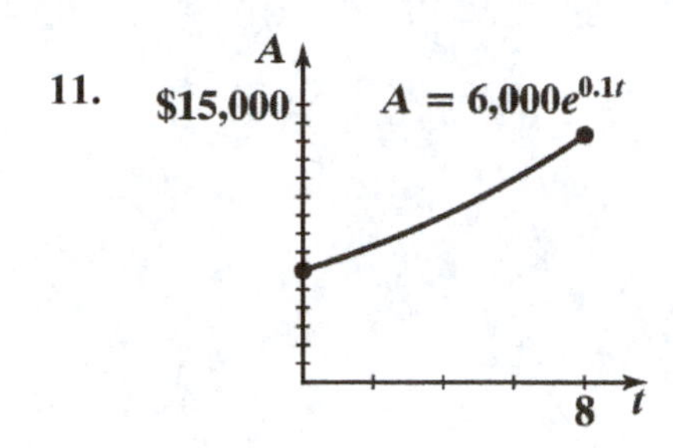

13. $2 = e^{0.06t}$

Take the natural log of both sides of this equation

$\ln(e^{0.06t}) = \ln 2$

$0.06t \ln e = \ln 2$

$0.06t = \ln 2 \quad (\ln e = 1)$

$t = \dfrac{\ln 2}{0.06} \approx 11.55$

15. $3 = e^{0.1t}$

$\ln(e^{0.1t}) = \ln 3$

$0.1t = \ln 3$

$t = \dfrac{\ln 3}{0.1} \approx 10.99$

17. $2 = e^{5r}$

$\ln(e^{5r}) = \ln 2$

$5r = \ln 2$

$r = \dfrac{\ln 2}{5} \approx 0.14$

19.

n	$\left(1+\frac{1}{n}\right)^n$
10	2.59374
100	2.70481
1000	2.71692
10,000	2.71815
100,000	2.71827
1,000,000	2.71828
10,000,000	2.71828
↓	↓
∞	$e = 2.7182818\ldots$

21.

n	4	16	64	256	1024	4096
$(1+n)^{1/n}$	1.495349	1.193722	1.067399	1.021913	1.006793	1.002033

$\lim_{n\to\infty} (1+n)^{1/n} = 1$

23. The graphs of $y_1 = \left(1+\dfrac{1}{n}\right)^n$, $y_2 = 2.718281828 \approx e$, and

$y_3 = \left(1+\dfrac{1}{n}\right)^{n+1}$ for $0 \le n \le 20$ are given at the right.

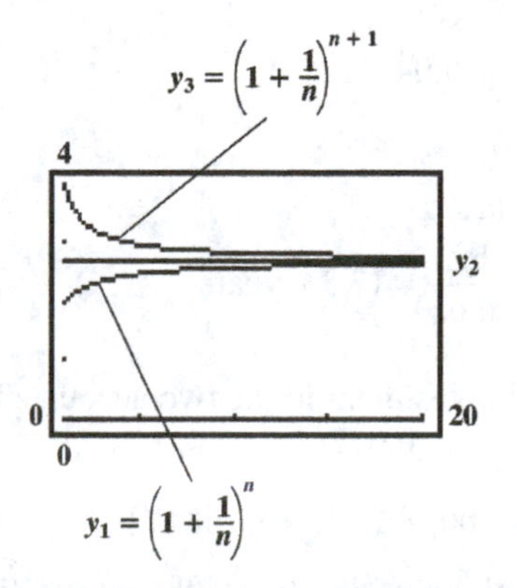

25. (A) $A = Pe^{rt}$; $P = \$10,000$, $r = 2.15\% = 0.0215$, $t = 10$:

$A = 10,000e^{(0.0215)10} = 10,000e^{0.215} = \$12,398.62$

(B) $A = \$18{,}000,\ P = \$10{,}000,\ r = 0.0215$:

$$18{,}000 = 10{,}000e^{0.0215t}$$
$$e^{0.0215t} = 1.8$$
$$0.0215t = \ln(1.8)$$
$$t = \frac{\ln(1.8)}{0.0215} \approx 27.34 \text{ years}$$

27. $A = Pe^{rt}$; $A = \$20{,}000,\ r = 0.052,\ t = 10$:

$$20{,}000 = Pe^{(0.052)10} = Pe^{0.52}$$
$$P = \frac{20{,}000}{e^{0.52}} = 20{,}000e^{-0.52} \approx \$11{,}890.41$$

29. $30{,}000 = 20{,}000e^{5r}$

$$e^{5r} = 1.5$$
$$5r = \ln(1.5)$$
$$r = \frac{\ln 1.5}{5} \approx 0.0811 \text{ or } 8.11\%$$

31. 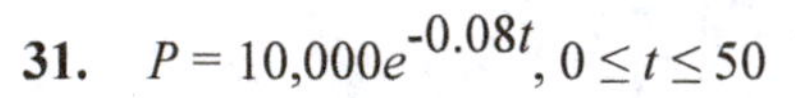 $P = 10{,}000e^{-0.08t},\ 0 \le t \le 50$

(A) 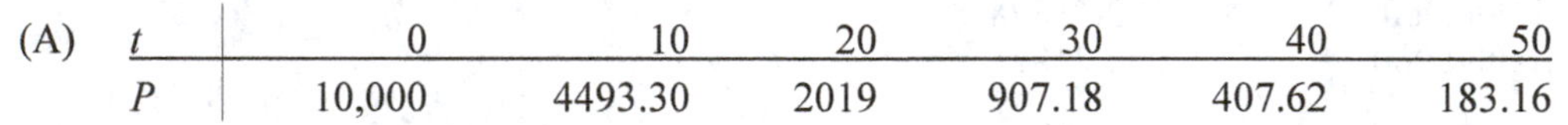

t	0	10	20	30	40	50
P	10,000	4493.30	2019	907.18	407.62	183.16

The graph of P is shown at the right.

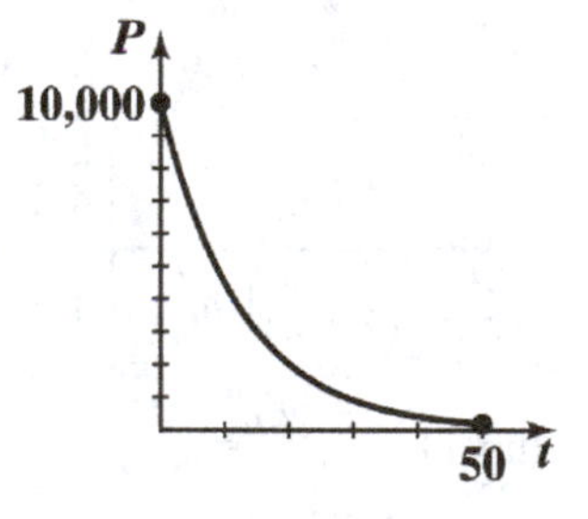

(B) 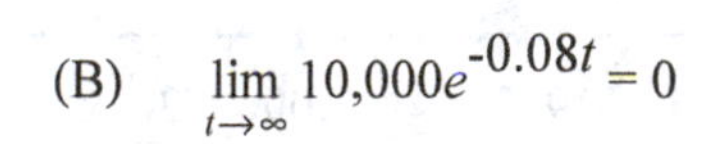 $\lim_{t \to \infty} 10{,}000e^{-0.08t} = 0$

33.
$$2P = Pe^{0.04t}$$
$$e^{0.04t} = 2$$
$$0.04t = \ln 2$$
$$t = \frac{\ln 2}{0.04} \approx 17.33 \text{ years}$$

35. 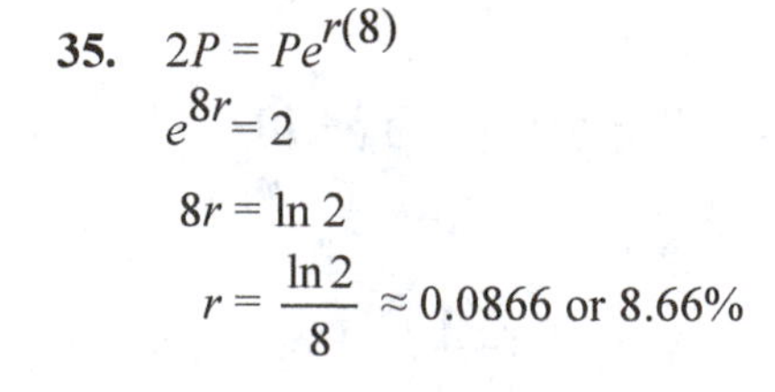
$$2P = Pe^{r(8)}$$
$$e^{8r} = 2$$
$$8r = \ln 2$$
$$r = \frac{\ln 2}{8} \approx 0.0866 \text{ or } 8.66\%$$

37. The total investment in the two accounts is given by

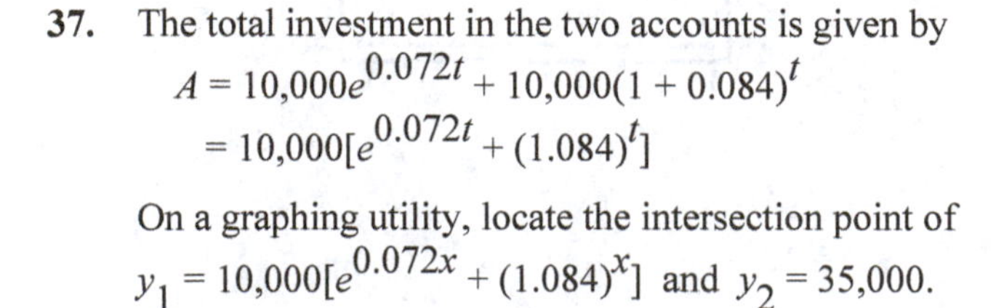
$$A = 10{,}000e^{0.072t} + 10{,}000(1 + 0.084)^t$$
$$= 10{,}000[e^{0.072t} + (1.084)^t]$$

On a graphing utility, locate the intersection point of $y_1 = 10{,}000[e^{0.072x} + (1.084)^x]$ and $y_2 = 35{,}000$.

The result is: $x = t \approx 7.3$ years.

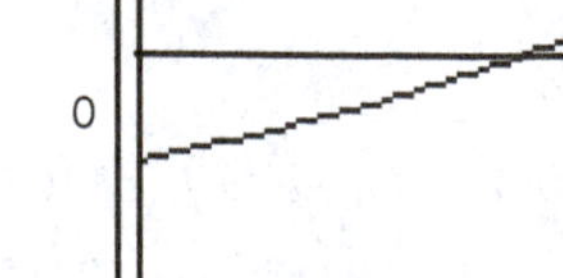

39. (A) $A = Pe^{rt}$; set $A = 2P$

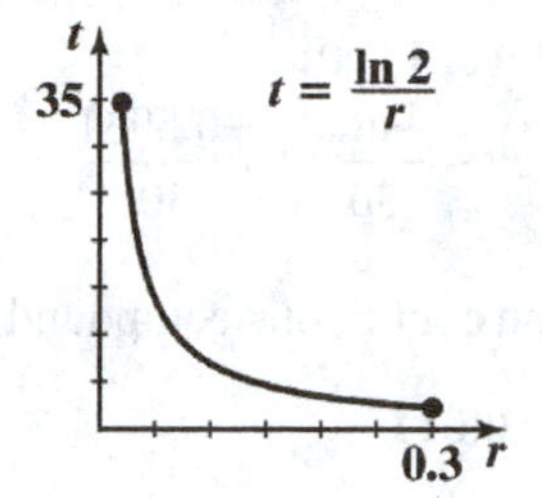

(B) $2P = Pe^{rt}$

$e^{rt} = 2$

$rt = \ln 2$

$t = \frac{\ln 2}{r}$

In theory, r could be any positive number. However, the restrictions on r are reasonable in the sense that most investments would be expected to earn between 2% and 30%.

(C) $r = 5\%;\ t = \frac{\ln 2}{0.05} \approx 13.86$ years

$r = 10\%;\ t = \frac{\ln 2}{0.10} \approx 6.93$ years

$r = 15\%;\ t = \frac{\ln 2}{0.15} \approx 4.62$ years

$r = 20\%;\ t = \frac{\ln 2}{0.20} \approx 3.47$ years

$r = 25\%;\ t = \frac{\ln 2}{0.25} \approx 2.77$ years

$r = 30\%;\ t = \frac{\ln 2}{0.30} \approx 2.31$ years

41.

$$Q = Q_0 e^{-0.0004332t}$$

$$\frac{1}{2}Q_0 = Q_0 e^{-0.0004332t}$$

$$e^{-0.0004332t} = \frac{1}{2}$$

$$\ln(e^{-0.0004332t}) = \ln\left(\frac{1}{2}\right) = \ln 1 - \ln 2$$

$$-0.0004332t = -\ln 2 \quad (\ln 1 = 0)$$

$$t = \frac{\ln 2}{0.0004332} \approx \frac{0.6931}{0.0004332} \approx 1599.95$$

Thus, the half-life of radium is approximately 1600 years.

43.

$$Q = Q_0 e^{rt} \quad (r < 0)$$

$$\frac{1}{2}Q_0 = Q_0 e^{r(30)}$$

$$e^{30r} = \frac{1}{2}$$

$$\ln(e^{30r}) = \ln\frac{1}{2} = \ln 1 - \ln 2$$

$30r = -\ln 2$ $(\ln 1 = 0)$

$$r = \frac{-\ln 2}{30} \approx \frac{-0.6931}{30} \approx -0.0231$$

Thus, the continuous compound rate of decay of the cesium isotope is approximately –0.0231.

45. $2P_0 = P_0 e^{0.013t}$

$e^{0.013t} = 2$

$0.013t = \ln 2$

$$t = \frac{\ln 2}{0.013} \approx 53.3$$

It will take approximately 53.3 years.

47. $2P_0 = P_0 e^{r(50)}$

$e^{50r} = 2$

$50r = \ln 2$

$$r = \frac{\ln 2}{50} \approx 0.0139 \text{ or } 1.39\%$$

EXERCISE 3-2

Things to remember:

1. DERIVATIVES OF EXPONENTIAL FUNCTIONS

(a) $\frac{d}{dx} e^x = e^x$ (b) $\frac{d}{dx} b^x = b^x \ln b$

2. LOGARITHMIC FUNCTIONS

The inverse of an exponential function is called a LOGARITHMIC FUNCTION. For $b > 0$, $b \neq 1$,

Logarithmic form		Exponential form
$y = \log_b x$	is equivalent to	$x = b^y$
Domain: $(0, \infty)$		Domain: $(-\infty, \infty)$
Range: $(-\infty, \infty)$		Range: $(0, \infty)$

The graphs of $y = \log_b x$ and $y = b^x$ are symmetric with respect to the line $y = x$.

The two most commonly used logarithmic functions are:

$\log x = \log_{10} x$ Common logarithm (base 10)

$\ln x = \log_e x$ Natural logarithm (base e)

3. DERIVATIVES OF LOGARITHMIC FUNCTIONS

(a) $\frac{d}{dx} \ln x = \frac{1}{x}$ (b) $\frac{d}{dx} \log_b x = \frac{1}{\ln b} \cdot \frac{1}{x}$

1. $y = \log_2 128 = \log_2 2^7 = 7$

3. $\log_3 x = 4$, $x = 3^4 = 81$

5. $\log_b 64 = 2$, $b^2 = 64$, $b = 8$

7. $y = \ln\sqrt{e} = \ln e^{1/2} = 1/2$

9. $f(x) = 5e^x + 3x + 1$

$f'(x) = 5e^x + 3$

11. $f(x) = -2\ln x + x^2 - 4$

$f'(x) = -2\left(\frac{1}{x}\right) + 2x = \frac{-2}{x} + 2x$

13. $f(x) = x^3 - 6e^x$

$f'(x) = 3x^2 - 6e^x$

15. $f(x) = e^x + x - \ln x$

$f'(x) = e^x + 1 - \frac{1}{x}$

17. $f(x) = \ln x^3 = 3\ln x$

$f'(x) = 3\left(\frac{1}{x}\right) = \frac{3}{x}$

19. $f(x) = 5x - \ln x^5 = 5x - 5\ln x$

$f'(x) = 5 - 5\left(\frac{1}{x}\right) = 5 - \frac{5}{x}$

21. $f(x) = \ln x^2 + 4e^x = 2\ln x + 4e^x$

$f'(x) = \frac{2}{x} + 4e^x$

23. $f(x) = e^x + x^e$

$f'(x) = e^x + ex^{e-1}$

25. $f(x) = x\,x^e = x^{e+1}, \quad f'(x) = (e+1)x^e$

27. $f(x) = 3 + \ln x;\ f(1) = 3$

$f'(x) = \frac{1}{x};\ f'(1) = 1$

Equation of the tangent line: $y - 3 = 1(x - 1)$ or $y = x + 2$

29. $f(x) = 3e^x;\ f(0) = 3$

$f'(x) = 3e^x;\ f'(0) = 3$

Tangent line: $y - 3 = 3(x - 0)$ or $y = 3x + 3$

31. $f(x) = \ln x^3 = 3\ln x;\ f(e) = 3\ln e = 3$

$f'(x) = \frac{3}{x};\ f'(e) = \frac{3}{e}$

Tangent line: $y - 3 = \frac{3}{e}(x - e)$ or $y = \frac{3}{e}x$

33. $f(x) = 2 + e^x;\ f(1) = 2 + e$

$f'(x) = e^x;\ f'(1) = e$

Tangent line: $y - (2 + e) = e(x - 1)$ or $y = ex + 2$

35. An equation for the tangent line to the graph of $f(x) = e^x$ at the point $(3, f(3)) = (3, e^3)$ is:

$$y - e^3 = e^3(x - 3)$$

or

$$y = xe^3 - 2e^3 = e^3(x - 2)$$

Clearly, $y = 0$ when $x = 2$, that is the tangent line passes through the point (2, 0). In general, an equation for the tangent line to the graph of $f(x) = e^x$ at the point $(c, f(c)) = (c, e^c)$ is:

$$y - e^c = e^c(x - c)$$

or

$$y = e^c(x - [c - 1])$$

Thus, the tangent line at the point (c, e^c) passes through $(c - 1, 0)$; the tangent line at the point $(4, e^4)$ passes through (3, 0).

37. An equation for the tangent line to the graph of $g(x) = \ln x$ at the point $(3, g(3)) = (3, \ln 3)$ is:

$$y - \ln 3 = m(x - 3) \text{ where } m = g'(3)$$

$g'(x) = \frac{d}{dx} \ln x = \frac{1}{x}$; $g'(3) = \frac{1}{3}$. Thus,

$y - \ln 3 = \frac{1}{3}(x - 3)$

For $x = 0$, $y = \ln 3 - 1$, so this tangent line does not pass through the origin. In fact, for any real number c, the tangent line to $g(x) = \ln x$ at the point $(c, \ln c)$ has equation $y - \ln c = \frac{1}{c}(x - c)$, and thus the only tangent line which passes through the origin is the tangent line at $(e, 1)$.

39. $f(x) = 10x + \ln 10x = 10x + \ln 10 + \ln x$

$f'(x) = 10 + \frac{1}{x}$

41. $f(x) = \ln\left(\frac{4}{x^3}\right) = \ln 4 - \ln x^3 = \ln 4 - 3\ln x$; $f'(x) = -\frac{3}{x}$

43. $y = \log_2 x$; $\frac{dy}{dx} = \frac{1}{\ln 2} \cdot \frac{1}{x} = \frac{1}{x \ln 2}$

45. $y = 3^x$; $\frac{dy}{dx} = 3^x \ln 3$

47. $y = 2x - \log x = 2x - \log_{10} x$; $\frac{dy}{dx} = 2 - \frac{1}{x \ln 10}$

49. $y = 10 + x + 10^x$; $y' = 1 + 10^x \ln 10$

51. $y = 3\ln x + 2\log_3 x$; $y' = \frac{3}{x} + \frac{2}{x \ln 3}$

53. $y = 2^x + e^2$; $y' = 2^x \ln 2$ (e^2 is a constant; $\frac{d}{dx}(e^2) = 0$)

55. On a graphing utility, graph $y_1 = e^x$ and $y_2 = x^4$. Rounded off to two decimal places, the points of intersection are: (–0.82, 0.44), (1.43, 4.18), (8.61, 5503.66).

57. On a graphing utility, graph $y_1 = (\ln x)^2$ and $y_2(x) = x$. The curves intersect at (0.49, 0.49) (two decimal places).

59. On a graphing utility, graph $y_1 = \ln x$ and $y_2 = x^{1/5}$. There is a point of intersection at (3.65, 1.30) (two decimal places). Using the hint that $\ln x < x^{1/5}$ for large x, we find a second point of intersection at (332,105.11, 12.71) (two decimal places).

61. Assume $c \neq 0$. Then $\frac{e^{ch} - 1}{h} = c\frac{e^{ch} - 1}{ch}$.

Let $t = ch$. Then $t \to 0$ if and only if $h \to 0$. Therefore,

$$\lim_{h \to 0} \frac{e^{ec} - 1}{h} = \lim_{h \to 0} c\frac{e^{ec} - 1}{ch} = c \lim_{h \to 0} \frac{e^{ec} - 1}{ch} = c \lim_{t \to 0} \frac{e^t - 1}{t} = c \cdot 1 = c$$

63. $S(t) = 300{,}000(0.9)^t, \quad t \ge 0$

$S'(t) = 300{,}000(0.9)^t \ln(0.9) = -31{,}608.15(0.9)^t$

The rate of depreciation after 1 year is:

$S'(1) = -31{,}608.15(0.9) \approx -28{,}447.34$; rate of depreciation $28,447.34.

The rate of depreciation after 5 years is:

$S'(5) = -31{,}608.15(0.9)^5 \approx -18{,}664.30$; rate of depreciation $18,664.30.

The rate of depreciation after 10 years is:

$S'(10) = -31{,}608.15(0.9)^{10} \approx -11{,}021.08$; rate of depreciation $11,021.08

65. $A(t) = 5{,}000 \cdot 4^t$; $A'(t) = 5{,}000 \cdot 4^t (\ln 4)$

$A'(1) = 5{,}000 \cdot 4(\ln 4) \approx 27{,}726$ – the rate of change of the bacteria population at the end of the first hour.

$A'(5) = 5{,}000 \cdot 4^5(\ln 4) \approx 7{,}097{,}827$ – the rate of change of the bacteria population at the end of the fifth hour.

67. $P(x) = 17.5(1 + \ln x), \quad 10 \le x \le 100, \quad P'(x) = \dfrac{17.5}{x}$

$$P'(40) = \frac{17.5}{40} \approx 0.44$$

$$P'(90) = \frac{17.5}{90} \approx 0.19$$

Thus, at the 40 pound weight level, blood pressure would increase at the rate of 0.44 mm of mercury per pound of weight gain; at the 90 pound weight level, blood pressure would increase at the rate of 0.19 mm of mercury per pound of weight gain.

69. $R = k \ln(S/S_0) = k[\ln S - \ln S_0]$; $\dfrac{dR}{dS} = \dfrac{k}{S}$

71. $P(t) = 10{,}000e^{0.075t}, \quad P'(t) = 750e^{0.075t}$

(A) $P'(1) = 750e^{0.075(1)} \approx 808.41$; $808.41 per year.

(B) First solve $10{,}000e^{0.075t} = 12{,}500$ for t:

$$e^{0.075t} = \frac{12{,}500}{10{,}000} = 1.25$$

$$0.075t = \ln 1.25$$

$$t = \frac{\ln 1.25}{0.075} \approx 2.975247351$$

$P'(2.98) = 750e^{0.075(2.975247351)} \approx 937.50$; $937.50 per year.

EXERCISE 3-3

Things to remember:

1. PRODUCT RULE

   If

   $$y = f(x) = F(x)S(x)$$

   and if $F'(x)$ and $S'(x)$ exist, then

   $$f'(x) = F(x)\, S'(x) + S(x)\, F'(x).$$

   Also,

   $$y' = FS' + SF';$$

   $$\frac{dy}{dx} = F\frac{dS}{dx} + S\frac{dF}{dx}$$

2. QUOTIENT RULE

   If

   $$y = f(x) = \frac{T(x)}{B(x)}$$

   and if $T'(x)$ and $B'(x)$ exist, then

   $$f'(x) = \frac{B(x)T'(x) - T(x)B'(x)}{[B(x)]^2}.$$

   Also,

   $$y' = \frac{BT' - TB'}{B^2};$$

   $$\frac{dy}{dx} = \frac{B\frac{dT}{dx} - T\frac{dB}{dx}}{B^2}.$$

---

1. $f(x) = 5x^3 - 4x^3 \ln x = x^3(5 - 4\ln x)$, $F(x) = x^3$, $S(x) = 5 - 4\ln x$

3. $f(x) = x^3e^x + 2x^3 + 3e^x + 6 = x^3(e^x + 2) + 3(e^x + 2) = (x^3 + 3)(e^x + 2)$

$F(x) = x^3 + 3$, $S(x) = e^x + 2$

5. $f(x) = 9x^2e^{-5x} = \dfrac{9x^2}{e^{5x}}$, $T(x) = 9x^2$, $B(x) = e^{5x}$

7. $f(x) = \dfrac{3}{x^2} + \dfrac{e^x}{x^4} = \dfrac{3x^2 + e^x}{x^4}$; $T(x) = 3x^2 + e^x$, $B(x) = x^4$

9. $f(x) = 2x^3(x^2 - 2)$

$f'(x) = 2x^3(x^2 - 2)' + (x^2 - 2)(2x^3)'$ [using 1 with $F(x) = 2x^3$, $S(x) = x^2 - 2$]

$= 2x^3(2x) + (x^2 - 2)(6x^2) = 4x^4 + 6x^4 - 12x^2 = 10x^4 - 12x^2$

11. $f(x) = (x - 3)(2x - 1)$

$f'(x) = (x - 3)(2x - 1)' + (2x - 1)(x - 3)'$ (using 1)

$= (x - 3)(2) + (2x - 1)(1) = 2x - 6 + 2x - 1 = 4x - 7$

13. $f(x) = \dfrac{x}{x - 3}$

$f'(x) = \dfrac{(x-3)(x)' - x(x-3)'}{(x-3)^2}$ [using 2 with $T(x) = x$, $B(x) = x - 3$]

$= \dfrac{(x - 3)(1) - x(1)}{(x - 3)^2} = \dfrac{-3}{(x - 3)^2}$

15. $f(x) = \dfrac{2x + 3}{x - 2}$

$f'(x) = \dfrac{(x-2)(2x+3)' - (2x+3)(x-2)'}{(x-2)^2}$ (using 2)

$= \dfrac{(x-2)(2) - (2x+3)(1)}{(x-2)^2} = \dfrac{2x - 4 - 2x - 3}{(x-2)^2} = \dfrac{-7}{(x-2)^2}$

17. $f(x) = 3xe^x$

$f'(x) = 3x(e^x)' + e^x(3x)'$ (using 1)

$= 3xe^x + e^x(3) = 3(x + 1)e^x$

19. $f(x) = x^3 \ln x$

$f'(x) = x^3(\ln x)' + \ln x(x^3)'$ (using 1)

$= x^3 \cdot \dfrac{1}{x} + \ln x(3x^2) = x^2(1 + 3 \ln x)$

21. $f(x) = (x^2 + 1)(2x - 3)$

$f'(x) = (x^2 + 1)(2x - 3)' + (2x - 3)(x^2 + 1)'$ (using 1)

$= (x^2 + 1)(2) + (2x - 3)(2x) = 2x^2 + 2 + 4x^2 - 6x = 6x^2 - 6x + 2$

23. $f(x) = (0.4x + 2)(0.5x - 5)$

$f'(x) = (0.4x + 2)(0.5x - 5)' + (0.5x - 5)(0.4x + 2)'$

$= (0.4x + 2)(0.5) + (0.5x - 5)(0.4) = 0.2x + 1 + 0.2x - 2 = 0.4x - 1$

25. $f(x) = \dfrac{x^2 + 1}{2x - 3}$

$f'(x) = \dfrac{(2x-3)(x^2+1)' - (x^2+1)(2x-3)'}{(2x-3)^2}$ (using 2)

$= \dfrac{(2x-3)(2x) - (x^2+1)(2)}{(2x-3)^2} = \dfrac{4x^2 - 6x - 2x^2 - 2}{(2x-3)^2} = \dfrac{2x^2 - 6x - 2}{(2x-3)^2}$

27. $f(x) = (x^2 + 2)(x^2 - 3)$

$f'(x) = (x^2 + 2)(x^2 - 3)' + (x^2 - 3)(x^2 + 2)'$

$= (x^2 + 2)(2x) + (x^2 - 3)(2x) = 2x^3 + 4x + 2x^3 - 6x = 4x^3 - 2x$

29. $f(x) = \dfrac{x^2+2}{x^2-3}$

$f'(x) = \dfrac{(x^2-3)(x^2+2)' - (x^2+2)(x^2-3)'}{(x^2-3)^2}$

$= \dfrac{(x^2-3)(2x) - (x^2+2)(2x)}{(x^2-3)^2} = \dfrac{2x^3-6x-2x^3-4x}{(x^2-3)^2} = \dfrac{-10x}{(x^2-3)^2}$

31. $f(x) = \dfrac{e^x}{x^2+1}$

$f'(x) = \dfrac{(x^2+1)(e^x)' - e^x(x^2+1)'}{(x^2+1)^2} = \dfrac{(x^2+1)e^x - e^x(2x)}{(x^2+1)^2} = \dfrac{e^x(x^2-2x+1)}{(x^2+1)^2} = \dfrac{e^x(x-1)^2}{(x^2+1)^2}$

33. $f(x) = \dfrac{\ln x}{x+1}$

$f'(x) = \dfrac{(x+1)(\ln x)' - \ln x(x+1)'}{(x+1)^2} = \dfrac{(x+1)\frac{1}{x} - \ln x(1)}{(x+1)^2} = \dfrac{x+1-x\ln x}{x(x+1)^2}$

35. $h(x) = xf(x)$; $h'(x) = xf'(x) + f(x)$

37. $h(x) = x^3f(x)$; $h'(x) = x^3f'(x) + f(x)(3x^2) = x^3f'(x) + 3x^2f(x)$

39. $h(x) = \dfrac{f(x)}{x^2}$; $h'(x) = \dfrac{x^2f'(x) - f(x)(2x)}{(x^2)^2} = \dfrac{x^2f'(x) - 2xf(x)}{x^4} = \dfrac{xf'(x) - 2f(x)}{x^3}$

or $h(x) = x^{-2}f(x)$; $h'(x) = x^{-2}f'(x) + f(x)(-2x^{-3}) = \dfrac{xf'(x) - 2f(x)}{x^3}$

41. $h(x) = \dfrac{x}{f(x)}$; $h'(x) = \dfrac{f(x) - xf'(x)}{[f(x)]^2}$

43. $h(x) = e^xf(x)$

$h'(x) = e^xf'(x) + f(x)e^x = e^x[f'(x) + f(x)]$

45. $h(x) = \dfrac{\ln x}{f(x)}$

$h'(x) = \dfrac{f(x)\frac{1}{x} - \ln x(f'(x))}{[f(x)]^2} = \dfrac{f(x) - (x\ln x)f'(x)}{x[f(x)]^2}$

47. $f(x) = (2x+1)(x^2-3x)$

$f'(x) = (2x+1)(x^2-3x)' + (x^2-3x)(2x+1)'$

$= (2x+1)(2x-3) + (x^2-3x)(2) = 6x^2 - 10x - 3$

49. $y = (2.5t - t^2)(4t + 1.4)$

$$\frac{dy}{dt} = (2.5t - t^2)\frac{d}{dt}(4t + 1.4) + (4t + 1.4)\frac{d}{dt}(2.5t - t^2)$$
$$= (2.5t - t^2)(4) + (4t + 1.4)(2.5 - 2t) = 10t - 4t^2 + 10t - 2.8t + 3.5 - 8t^2 = -12t^2 + 17.2t + 3.5$$

51. $y = \dfrac{5x-3}{x^2+2x}$

$$y' = \frac{(x^2+2x)(5x-3)' - (5x-3)(x^2+2x)'}{(x^2+2x)^2}$$
$$= \frac{(x^2+2x)(5) - (5x-3)(2x+2)}{(x^2+2x)^2} = \frac{-5x^2+6x+6}{(x^2+2x)^2}$$

53. $$\frac{d}{dw}\left[\frac{w^2-3w+1}{w^2-1}\right] = \frac{(w^2-1)\frac{d}{dw}(w^2-3w+1) - (w^2-3w+1)\frac{d}{dw}(w^2-1)}{(w^2-1)^2}$$
$$= \frac{(w^2-1)(2w-3) - (w^2-3w+1)(2w)}{(w^2-1)^2} = \frac{3w^2-4w+3}{(w^2-1)^2}$$

55. $y = (1 + x - x^2)\,e^x$
$y' = (1 + x - x^2)e^x + e^x(1 - 2x)$
$= e^x(1 + x - x^2 + 1 - 2x) = (2 - x - x^2)e^x$

57. $f(x) = \dfrac{1}{x}$

(A) $f'(x) = \dfrac{x\frac{d}{dx}(1) - 1\frac{d}{dx}(x)}{x^2} = \dfrac{-1}{x^2}$

(B) $f(x) = \dfrac{1}{x} = x^{-1}$, $f'(x) = -x^{-2} = \dfrac{-1}{x^2}$ (power rule)

59. $f(x) = \dfrac{-3}{x^4}$

(A) $f'(x) = \dfrac{x^4\frac{d}{dx}(-3) - (-3)\frac{d}{dx}(x^4)}{(x^4)^2} = \dfrac{12x^3}{x^8} = \dfrac{12}{x^5}$

(B) $f(x) = \dfrac{-3}{x^4} = -3x^{-4}$, $f'(x) = 12x^{-5} = \dfrac{12}{x^5}$ (power rule)

61. $f(x) = (1 + 3x)(5 - 2x)$

First find $f'(x)$:

$f'(x) = (1 + 3x)(5 - 2x)' + (5 - 2x)(1 + 3x)'$
$= (1 + 3x)(-2) + (5 - 2x)(3) = -2 - 6x + 15 - 6x = 13 - 12x$

An equation for the tangent line at $x = 2$ is:

$y - y_1 = m(x - x_1)$
where $x_1 = 2$, $y_1 = f(x_1) = f(2) = 7$, and $m = f'(x_1) = f'(2) = -11$.

Thus, we have:

$y - 7 = -11(x - 2)$ or $y = -11x + 29$

63. $f(x) = \dfrac{x-8}{3x-4}$

First find $f'(x)$:

$$f'(x) = \frac{(3x-4)(x-8)' - (x-8)(3x-4)'}{(3x-4)^2}$$
$$= \frac{(3x-4)(1) - (x-8)(3)}{(3x-4)^2} = \frac{20}{(3x-4)^2}$$

An equation for the tangent line at $x = 2$ is: $y - y_1 = m(x - x_1)$
where $x_1 = 2$, $y_1 = f(x_1) = f(2) = -3$, and $m = f'(x_1) = f'(2) = 5$.
Thus, we have: $y - (-3) = 5(x - 2)$ or $y = 5x - 13$

65. $f(x) = \dfrac{x}{2^x}$; $f'(x) = \dfrac{2^x(1) - x \cdot 2^x \cdot \ln 2}{[2^x]^2} = \dfrac{2^x(1 - x\ln 2)}{[2^x]^2} = \dfrac{1 - x\ln 2}{2^x}$

$f(2) = \dfrac{2}{4} = \dfrac{1}{2}$; $f'(2) = \dfrac{1 - 2\ln 2}{4}$

Tangent line: $y - \dfrac{1}{2} = \dfrac{1 - 2\ln 2}{4}(x - 2)$ or $y = \dfrac{1 - 2\ln 2}{4}x + \ln 2$.

67. $f(x) = (2x - 15)(x^2 + 18)$

$f'(x) = (2x - 15)(x^2 + 18)' + (x^2 + 18)(2x - 15)'$
$= (2x - 15)(2x) + (x^2 + 18)(2) = 6x^2 - 30x + 36$

To find the values of x where $f'(x) = 0$, set: $f'(x) = 6x^2 - 30x + 36 = 0$

or $x^2 - 5x + 6 = 0$
$(x - 2)(x - 3) = 0$

Thus, $x = 2$, $x = 3$.

69. $f(x) = \dfrac{x}{x^2 + 1}$

$$f'(x) = \frac{(x^2+1)(x)' - x(x^2+1)'}{(x^2+1)^2} = \frac{(x^2+1)(1) - x(2x)}{(x^2+1)^2} = \frac{1 - x^2}{(x^2+1)^2}$$

Now, set $f'(x) = \dfrac{1 - x^2}{(x^2+1)^2} = 0$

or $1 - x^2 = 0$
$(1 - x)(1 + x) = 0$

Thus, $x = 1$, $x = -1$.

71. $f(x) = x^3(x^4 - 1)$

First, we use the product rule:

$f'(x) = x^3(x^4 - 1)' + (x^4 - 1)(x^3)'$
$= x^3(4x^3) + (x^4 - 1)(3x^2) = 7x^6 - 3x^2$

Next, simplifying $f(x)$, we have $f(x) = x^7 - x^3$.

Thus, $f'(x) = 7x^6 - 3x^2$.

73. $f(x) = \dfrac{x^3 + 9}{x^3}$

First, we use the quotient rule:

$$f'(x) = \frac{x^3(x^3+9)' - (x^3+9)(x^3)'}{(x^3)^2} = \frac{x^3(3x^2) - (x^3+9)(3x^2)}{x^6} = \frac{-27x^2}{x^6} = \frac{-27}{x^4}$$

Next, simplifying $f(x)$, we have $f(x) = \dfrac{x^3+9}{x^3} = 1 + \dfrac{9}{x^3} = 1 + 9x^{-3}$

Thus, $f'(x) = -27x^{-4} = -\dfrac{27}{x^4}$.

75. $f(w) = (w + 1)2^w$

$f'(w) = (w + 1)2^w(\ln 2) + 2^w(1) = [(w + 1)\ln 2 + 1]2^w = 2^w(w \ln 2 + \ln 2 + 1)$

77. $y = 9x^{1/3}(x^3 + 5)$

$$\frac{dy}{dx} = 9x^{1/3}\frac{d}{dx}(x^3 + 5) + (x^3 + 5)\frac{d}{dx}(9x^{1/3})$$

$$= 9x^{1/3}(3x^2) + (x^3 + 5)\left(9 \cdot \frac{1}{3}x^{-2/3}\right) = 27x^{7/3} + (x^3 + 5)(3x^{-2/3}) = 27x^{7/3} + \frac{3x^3 + 15}{x^{2/3}} = \frac{30x^3 + 15}{x^{2/3}}$$

79. $y = \dfrac{\log_2 x}{1 + x^2}$

$$y' = \frac{(1+x^2)\cdot\dfrac{1}{x\ln 2} - \log_2 x(2x)}{(1+x^2)^2}$$

$$= \frac{1 + x^2 - 2x^2 \ln 2 \log_2 x}{x(1+x^2)^2 \ln 2} = \frac{1 + x^2 - 2x^2 \ln x}{x(1+x^2)^2 \ln 2}$$

81. $f(x) = \dfrac{6\sqrt[3]{x}}{x^2 - 3} = \dfrac{6x^{1/3}}{x^2 - 3}$

$$f'(x) = \frac{(x^2 - 3)(6x^{1/3})' - 6x^{1/3}(x^2 - 3)'}{(x^2 - 3)^2}$$

$$= \frac{(x^2 - 3)\left(6 \cdot \frac{1}{3}x^{-2/3}\right) - 6x^{1/3}(2x)}{(x^2 - 3)^2} = \frac{(x^2 - 3)(2x^{-2/3}) - 12x^{4/3}}{(x^2 - 3)^2}$$

$$= \frac{\dfrac{2(x^2 - 3)}{x^{2/3}} - 12x^{4/3}}{(x^2 - 3)^2} = \frac{2x^2 - 6 - 12x^2}{(x^2 - 3)^2 x^{2/3}} = \frac{-10x^2 - 6}{(x^2 - 3)^2 x^{2/3}}$$

83. $g(t) = \dfrac{0.2t}{3t^2 - 1}$; $g'(t) = \dfrac{(3t^2 - 1)(0.2) - (0.2t)(6t)}{(3t^2 - 1)^2} = \dfrac{-0.6t^2 - 0.2}{(3t^2 - 1)^2}$

85. $\frac{d}{dx}[4x \log x^5] = 4x\frac{d}{dx}[\log x^5] + \log x^5 \frac{d}{dx}[4x] = 4x\frac{d}{dx}[5 \log x] + 4 \log x^5$

$$= 4x \cdot \left(\frac{5}{x \ln 10}\right) + 4 \log x^5 = \frac{20}{\ln 10} + 20 \log x = \frac{20(1 + \ln x)}{\ln 10}$$

87. $\frac{d}{dx}\frac{x^3 - 2x^2}{\sqrt[3]{x^2}} = \frac{d}{dx}\frac{x^3 - 2x^2}{x^{2/3}} = \frac{x^{2/3}\frac{d}{dx}(x^3 - 2x^2) - (x^3 - 2x^2)\frac{d}{dx}(x^{2/3})}{(x^{2/3})^2}$

$$= \frac{x^{2/3}(3x^2 - 4x) - (x^3 - 2x^2)\left(\frac{2}{3}x^{-1/3}\right)}{x^{4/3}} = x^{-2/3}(3x^2 - 4x) - \frac{2}{3}x^{-5/3}(x^3 - 2x^2)$$

$$= 3x^{4/3} - 4x^{1/3} - \frac{2}{3}x^{4/3} + \frac{4}{3}x^{1/3} = -\frac{8}{3}x^{1/3} + \frac{7}{3}x^{4/3}$$

89. $f(x) = \frac{(2x^2 - 1)(x^2 + 3)}{x^2 + 1}$

$$f'(x) = \frac{(x^2 + 1)[(2x^2 - 1)(x^2 + 3)]' - (2x^2 - 1)(x^2 + 3)(x^2 + 1)'}{(x^2 + 1)^2}$$

$$= \frac{(x^2 + 1)[(2x^2 - 1)(x^2 + 3)' + (x^2 + 3)(2x^2 - 1)'] - (2x^2 - 1)(x^2 + 3)(2x)}{(x^2 + 1)^2}$$

$$= \frac{(x^2 + 1)[(2x^2 - 1)(2x) + (x^2 + 3)(4x)] - (2x^2 - 1)(x^2 + 3)(2x)}{(x^2 + 1)^2}$$

$$= \frac{(x^2 + 1)[4x^3 - 2x + 4x^3 + 12x] - [2x^4 + 5x^2 - 3](2x)}{(x^2 + 1)^2} = \frac{(x^2 + 1)(8x^3 + 10x) - 4x^5 - 10x^3 + 6x}{(x^2 + 1)^2}$$

$$= \frac{8x^5 + 10x^3 + 8x^3 + 10x - 4x^5 - 10x^3 + 6x}{(x^2 + 1)^2} = \frac{4x^5 + 8x^3 + 16x}{(x^2 + 1)^2}$$

91. $y = \frac{t \ln t}{e^t}$

$$y' = \frac{e^t\left[t\left(\frac{1}{t}\right) + \ln t\right] - t \ln t(e^t)}{[e^t]^2} = \frac{e^t(1 + \ln t - t \ln t)}{[e^t]^2} = \frac{1 + \ln t - t \ln t}{e^t}$$

93. $S(t) = \frac{90t^2}{t^2 + 50}$

(A) $S'(t) = \frac{(t^2 + 50)(180t) - 90t^2(2t)}{(t^2 + 50)^2} = \frac{9000t}{(t^2 + 50)^2}$

(B) $S(10) = \frac{90(10)^2}{(10)^2 + 50} = \frac{9000}{150} = 60;$

$$S'(10) = \frac{9000(10)}{[(10)^2 + 50]^2} = \frac{90,000}{22,500} = 4$$

After 10 months, the total sales are 60,000 DVD's and the sales are INCREASING at the rate of 4,000 DVD's per month.

(C) The total sales after 11 months will be approximately 64,000 DVD's.

95. $x = \frac{4,000}{0.1p+1}$, $10 \le p \le 70$

(A) $\frac{dx}{dp} = \frac{(0.1p+1)(0) - 4,000(0.1)}{(0.1p+1)^2} = \frac{-400}{(0.1p+1)^2}$

(B) $x(40) = \frac{4,000}{0.1(40)+1} = \frac{4,000}{5} = 800;$

$\frac{dx}{dp} = \frac{-400}{[0.1(40)+1]^2} = \frac{-400}{25} = -16$

At a price level of \$40, the demand is 800 DVD players and the demand is DECREASING at the rate of 16 CD players per dollar.

(C) At a price of \$41, the demand will be approximately 784 CD players.

97. $C(t) = \frac{0.14t}{t^2+1}$

(A) $C'(t) = \frac{(t^2+1)(0.14t)' - (0.14t)(t^2+1)'}{(t^2+1)^2}$

$= \frac{(t^2+1)(0.14) - (0.14t)(2t)}{(t^2+1)^2} = \frac{0.14 - 0.14t^2}{(t^2+1)^2} = \frac{0.14(1-t^2)}{(t^2+1)^2}$

(B) $C'(0.5) = \frac{0.14(1-[0.5]^2)}{([0.5]^2+1)^2} = \frac{0.14(1-0.25)}{(1.25)^2} = 0.0672$

Interpretation: At $t = 0.5$ hours, the concentration is increasing at the rate of 0.0672 mg/cm^3 per hour.

$C'(3) = \frac{0.14(1-3^2)}{(3^2+1)^2} = \frac{0.14(-8)}{100} = -0.0112$

Interpretation: At $t = 3$ hours, the concentration is decreasing at the rate of 0.0112 mg/cm^3 per hour.

EXERCISE 3-4

Things to remember:

1. COMPOSITE FUNCTIONS

A function m is a COMPOSITE of functions f and g if

$m(x) = f[g(x)]$

The domain of m is the set of all numbers x such that x is in the domain of g and $g(x)$ is in the domain of f.

2. GENERAL POWER RULE

If $u(x)$ is a differentiable function, n is any real number, and

$$y = f(x) = [u(x)]^n$$

then

$$f'(x) = n[u(x)]^{n-1} u'(x)$$

This rule is often written more compactly as

$$y' = nu^{n-1}u' \quad \text{or} \quad \frac{d}{dx}(u^n) = nu^{n-1}\frac{du}{dx}, \quad u = u(x)$$

3. THE CHAIN RULE: GENERAL FORM

If $y = f(u)$ and $u = g(x)$, define the composite function

$$y = m(x) = f[g(x)],$$

then

$$\frac{dy}{dx} = \frac{dy}{du}\frac{du}{dx} \text{ provided that } \frac{dy}{du} \text{ and } \frac{du}{dx} \text{ exist.}$$

Or, equivalently,

$$m'(x) = f'[g(x)]g'(x) \text{ provided that } f'[g(x)] \text{ and } g'(x) \text{ exist.}$$

4. GENERAL DERIVATIVE RULES

(a) $\frac{d}{dx}[f(x)]^n = n[f(x)]^{n-1}f'(x)$

(b) $\frac{d}{dx}\ln[f(x)] = \frac{1}{f(x)}f'(x)$

(c) $\frac{d}{dx}e^{f(x)} = e^{f(x)}f'(x)$

---

1. $y = f(u) = 3u + 5, \quad u = g(x) = x^3; \quad y = f(u) = f[g(x)] = 3x^3 + 5$

3. $y = f(u) = 2u + \ln u, \quad u = g(x) = x^2e^x; \quad y = f(u) = f[g(x)] = 2x^2e^x + \ln(x^2e^x)$

5. $y = \ln(x^3 - 6x + 10)$. Let $y = E(u) = \ln u, \quad u = I(x) = x^3 - 6x + 10$.
Then $y = E(u) = E[I(x)] = \ln(x^3 - 6x + 10)$

7. $y = \sqrt{x^2 + 4}$. Let $y = E(u) = \sqrt{u}, \quad u = I(x) = x^2 + 4$. Then $y = E(u) = E[I(x)] = \sqrt{x^2 + 4}$

9. $3; \frac{d}{dx}(3x + 4)^4 = 4(3x + 4)^3(3) = 12(3x + 4)^3$

11. $-4x; \frac{d}{dx}(4 - 2x^2)^3 = 3(4 - 2x^2)^2(-4x) = -12x(4 - 2x^2)^2$

13. $2x; \frac{d}{dx}(e^{x^2+1}) = e^{x^2+1}\frac{d}{dx}(x^2 + 1) = e^{x^2+1}(2x) = 2xe^{x^2+1}$

15. $4x^3$; $\frac{d}{dx}[\ln(x^4+1)] = \frac{1}{x^4+1}\frac{d}{dx}(x^4+1) = \frac{1}{x^4+1}(4x^3) = \frac{4x^3}{x^4+1}$

17. $f(x) = (5-2x)^4$
$f'(x) = 4(5-2x)^3(5-2x)' = 4(5-2x)^3(-2) = -8(5-2x)^3$

19. $f(x) = (4+0.2x)^5$
$f'(x) = 5(4+0.2x)^4(4+0.2x)' = 5(4+0.2x)^4(0.2) = (4+0.2x)^4$

21. $f(x) = (3x^2+5)^5$
$f'(x) = 5(3x^2+5)^4(3x^2+5)' = 5(3x^2+5)^4(6x) = 30x(3x^2+5)^4$

23. $f(x) = 5e^x$
$f'(x) = 5e^x + e^x(0) = 5e^x$

25. $f(x) = e^{5x}$
$f'(x) = e^{5x}(5x)' = e^{5x}(5) = 5e^{5x}$

27. $f(x) = 3e^{-6x}$
$f'(x) = 3e^{-6x}(-6x)' = 3e^{-6x}(-6) = -18e^{-6x}$

29. $f(x) = (2x-5)^{1/2}$
$f'(x) = \frac{1}{2}(2x-5)^{-1/2}(2x-5)' = \frac{1}{2}(2x-5)^{-1/2}(2) = \frac{1}{(2x-5)^{1/2}}$

31. $f(x) = (x^4+1)^{-2}$
$f'(x) = -2(x^4+1)^{-3}(x^4+1)' = -2(x^4+1)^{-3}(4x^3) = -8x^3(x^4+1)^{-3} = \frac{-8x^3}{(x^4+1)^3}$

33. $f(x) = 4 - 2\ln x$
$f'(x) = -\frac{2}{x}$

35. $f(x) = 3\ln(1+x^2)$
$f'(x) = 3\cdot\frac{1}{1+x^2}\cdot(1+x^2)' = \frac{3}{1+x^2}(2x) = \frac{6x}{1+x^2}$

37. $f(x) = (1+\ln x)^3$
$f'(x) = 3(1+\ln x)^2(1+\ln x)' = 3(1+\ln x)^2\cdot\frac{1}{x} = \frac{3}{x}(1+\ln x)^2$

39. $f(x) = (2x-1)^3$
$f'(x) = 3(2x-1)^2(2) = 6(2x-1)^2$

Tangent line at $x = 1$: $y - y_1 = m(x - x_1)$ where $x_1 = 1$, $y_1 = f(1) = (2(1)-1)^3 = 1$, $m = f'(1) = 6[2(1)-1]^2 = 6$. Thus, $y - 1 = 6(x-1)$ or $y = 6x - 5$.

The tangent line is horizontal at the value(s) of x such that $f'(x) = 0$:

$$6(2x-1)^2 = 0$$
$$2x - 1 = 0$$
$$x = \frac{1}{2}$$

41. $f(x) = (4x-3)^{1/2}$

$$f'(x) = \frac{1}{2}(4x-3)^{-1/2}(4) = \frac{2}{(4x-3)^{1/2}}$$

Tangent line at $x = 3$: $y - y_1 = m(x - x_1)$ where $x_1 = 3, y_1 = f(3) = (4\cdot 3 - 3)^{1/2} = 3$,

$f'(3) = \frac{2}{(4\cdot 3-3)^{1/2}} = \frac{2}{3}$. Thus, $y - 3 = \frac{2}{3}(x-3)$ or $y = \frac{2}{3}x + 1$.

The tangent line is horizontal at the value(s) of x such that

$f'(x) = 0$. Since $\frac{2}{(4x-3)^{1/2}} \neq 0$ for all x $\left(x \neq \frac{3}{4}\right)$, there are no values of x where the tangent line is horizontal.

43. $f(x) = 5e^{x^2-4x+1}$

$$f'(x) = 5e^{x^2-4x+1}(2x-4) = 10(x-2)e^{x^2-4x+1}$$

Tangent line at $x = 0$: $y - y_1 = m(x - x_1)$ where $x_1 = 0$, $y_1 = f(0) = 5e$, $f'(0) = -20e$.

Thus, $y - 5e = -20ex$ or $y = -20ex + 5e$.

The tangent line is horizontal at the value(s) of x such that $f'(x) = 0$:

$$10(x-2)e^{x^2-4x+1} = 0$$
$$x - 2 = 0$$
$$x = 2$$

45. $y = 3(x^2-2)^4$

$$\frac{dy}{dx} = 3\cdot 4(x^2-2)^3(2x) = 24x(x^2-2)^3$$

47. $\frac{d}{dt}[2(t^2+3t)^{-3}] = 2(-3)(t^2+3t)^{-4}(2t+3) = \frac{-6(2t+3)}{(t^2+3t)^4}$

49. $h(w) = \sqrt{w^2+8} = (w^2+8)^{1/2}$;

$$h'(w) = \frac{1}{2}(w^2+8)^{-1/2}(2w) = \frac{w}{(w^2+8)^{1/2}} = \frac{w}{\sqrt{w^2+8}}.$$

51. $g(x) = 4xe^{3x}$

$$g'(x) = 4x\cdot e^{3x}(3) + e^{3x}\cdot 4 = 12xe^{3x} + 4e^{3x} = 4(3x+1)e^{x}$$

53. $\frac{d}{dx}\left[\frac{\ln(1+x)}{x^3}\right] = \frac{x^3\cdot\frac{1}{1+x}(1) - \ln(1+x)3x^2}{(x^3)^2} = \frac{x^2\left[\frac{x}{1+x} - 3\ln(1+x)\right]}{x^6} = \frac{x-3(1+x)\ln(1+x)}{x^4(1+x)}$

55. $F(t) = (e^{t^2+1})^3 = e^{3t^2+3}$

$F'(t) = e^{3t^2+3}(6t) = 6te^{3(t^2+1)}$

57. $y = \ln(x^2 + 3)^{3/2} = \frac{3}{2}\ln(x^2 + 3)$

$y' = \frac{3}{2} \cdot \frac{1}{x^2+3}(2x) = \frac{3x}{x^2+3}$

59. $\frac{d}{dw}\left[\frac{1}{(w^3+4)^5}\right] = \frac{d}{dw}[(w^3+4)^{-5}] = -5(w^3+4)^{-6}(3w^2) = \frac{-15w^2}{(w^3+4)^6}$.

61. $f(x) = x(4-x)^3$

$f'(x) = x[(4-x)^3]' + (4-x)^3(x)'$

$= x(3)(4-x)^2(-1) + (4-x)^3(1) = (4-x)^3 - 3x(4-x)^2 = (4-x)^2[4-x-3x] = 4(4-x)^2(1-x)$

An equation for the tangent line to the graph of f at $x = 2$ is:

$y - y_1 = m(x - x_1)$ where $x_1 = 2$, $y_1 = f(x_1) = f(2) = 16$, and $m = f'(x_1) = f'(2) = -16$.

Thus, $y - 16 = -16(x - 2)$ or $y = -16x + 48$.

63. $f(x) = \frac{x}{(2x-5)^3}$

$f'(x) = \frac{(2x-5)^3(1) - x(3)(2x-5)^2(2)}{[(2x-5)^3]^2}$

$= \frac{(2x-5)^3 - 6x(2x-5)^2}{(2x-5)^6} = \frac{(2x-5)-6x}{(2x-5)^4} = \frac{-4x-5}{(2x-5)^4}$

An equation for the tangent line to the graph of f at $x = 3$ is:

$y - y_1 = m(x - x_1)$ where $x_1 = 3$, $y_1 = f(x_1) = f(3) = 3$, and $m = f'(x_1) = f'(3) = -17$.

Thus, $y - 3 = -17(x - 3)$ or $y = -17x + 54$.

65. $f(x) = \sqrt{\ln x} = (\ln x)^{1/2}$

$f'(x) = \frac{1}{2}(\ln x)^{-1/2} \cdot \frac{1}{x} = \frac{1}{2x\sqrt{\ln x}}$

Tangent line at $x = e$:

$f(e) = \sqrt{\ln e} = \sqrt{1} = 1$, $f'(e) = \frac{1}{2e\sqrt{\ln e}} = \frac{1}{2e}$

$y - 1 = \frac{1}{2e}(x - e)$ or $y = \frac{1}{2e}x + \frac{1}{2}$

67. $f(x) = x^2(x-5)^3$

$f'(x) = x^2[(x-5)^3]' + (x-5)^3(x^2)' = x^2(3)(x-5)^2(1) + (x-5)^3(2x)$

$= 3x^2(x-5)^2 + 2x(x-5)^3 = x(x-5)^2[3x + 2(x-5)] = x(x-5)^2[5x-10] = 5x(x-5)^2(x-2)$

The tangent line to the graph of f is horizontal at the values of x such that $f'(x) = 0$. Thus, we set $5x(x-5)^2(x-2) = 0$ which implies $x = 0$, $x = 2$, $x = 5$.

69. $f(x) = \dfrac{x}{(2x+5)^2}$

$$f'(x) = \frac{(2x+5)^2(x)' - x[(2x+5)^2]'}{[(2x+5)^2]^2} = \frac{(2x+5)^2(1) - x(2)(2x+5)(2)}{(2x+5)^4} = \frac{2x+5-4x}{(2x+5)^3} = \frac{5-2x}{(2x+5)^3}$$

The tangent line to the graph of f is horizontal at the values of x such that $f'(x) = 0$. Thus, we set

$\dfrac{5-2x}{(2x+5)^3} = 0$ which implies $5 - 2x = 0$ and $x = \dfrac{5}{2}$.

71. $f(x) = \sqrt{x^2 - 8x + 20} = (x^2 - 8x + 20)^{1/2}$

$f'(x) = \dfrac{1}{2}(x^2 - 8x + 20)^{-1/2}(2x - 8) = \dfrac{x-4}{(x^2 - 8x + 20)^{1/2}}$

The tangent line to the graph of f is horizontal at the values of x such that $f'(x) = 0$. Thus, we set

$\dfrac{x-4}{(x^2 - 8x + 20)^{1/2}} = 0$ which implies $x - 4 = 0$ and $x = 4$.

73. $f'(x) = \dfrac{1}{5(x^2+3)^4}[20(x^2 + 3)^3](2x) = \dfrac{8x}{x^2+3}$;

$g'(x) = 4 \cdot \dfrac{1}{x^2+3}(2x) = \dfrac{8x}{x^2+3}$

For another way to see this, recall the properties of logarithms discussed in Section 2-3:

$f(x) = \ln[5(x^2 + 3)^4] = \ln 5 + \ln(x^2 + 3)^4 = \ln 5 + 4\ln(x^2 + 3) = \ln 5 + g(x)$

Now $\dfrac{d}{dx}f(x) = \dfrac{d}{dx}\ln 5 + \dfrac{d}{dx}g(x) = 0 + \dfrac{d}{dx}g(x) = \dfrac{d}{dx}g(x)$

Conclusion: f and g differ by a constant; $f'(x)$ and $g'(x)$ ARE the same function.

75. $f(u) = \ln u$, domain of $f: (0,\infty)$; $g(x) = 4 - x^2$, domain of $g: (-\infty,\infty)$

$m(x) = f[g(x)] = \ln(4 - x^2)$, domain of $m: (-2,2)$.

77. $f(u) = \dfrac{1}{u^2-1}$, domain of f: all real numbers except $x = \pm 1$; $g(x) = \ln x$, domain of $g: (0,\infty)$

$m(x) = f[g(x)] = \dfrac{1}{(\ln x)^2 - 1}$, domain of m: all real numbers except $x = e, e^{-1}$.

79. $\dfrac{d}{dx}[3x(x^2 + 1)^3] = 3x\dfrac{d}{dx}(x^2 + 1)^3 + (x^2 + 1)^3\dfrac{d}{dx}3x = 3x\cdot 3(x^2 + 1)^2(2x) + (x^2 + 1)^3(3)$

$= 18x^2(x^2 + 1)^2 + 3(x^2 + 1)^3 = (x^2 + 1)^2[18x^2 + 3(x^2 + 1)]$

$= (x^2 + 1)^2(21x^2 + 3) = 3(x^2 + 1)^2(7x^2 + 1)$

81. $\frac{d}{dx}\frac{(x^3-7)^4}{2x^3} = \frac{2x^3\frac{d}{dx}(x^3-7)^4-(x^3-7)^4\frac{d}{dx}2x^3}{(2x^3)^2} = \frac{2x^3\cdot 4(x^3-7)^3(3x^2)-(x^3-7)^4 6x^2}{4x^6}$

$= \frac{3(x^3-7)^3x^2[8x^3-2(x^3-7)]}{4x^6} = \frac{3(x^3-7)^3(6x^3+14)}{4x^4} = \frac{3(x^3-7)^3(3x^3+7)}{2x^4}$

83. $\frac{d}{dx}\log_2(3x^2-1) = \frac{1}{\ln 2}\cdot\frac{1}{3x^2-1}\cdot 6x = \frac{1}{\ln 2}\cdot\frac{6x}{3x^2-1}$

85. $\frac{d}{dx}10^{x^2+x} = 10^{x^2+x}(\ln 10)(2x+1) = (2x+1)10^{x^2+x}\ln 10$

87. $\frac{d}{dx}\log_3(4x^3+5x+7) = \frac{1}{\ln 3}\cdot\frac{1}{4x^3+5x+7}(12x^2+5) = \frac{12x^2+5}{\ln 3(4x^3+5x+7)}$

89. $\frac{d}{dx}2^{x^3-x^2+4x+1} = 2^{x^3-x^2+4x+1}\ln 2(3x^2-2x+4) = \ln 2(3x^2-2x+4)2^{x^3-x^2+4x+1}$

91. $C(x) = 10+\sqrt{2x+16} = 10+(2x+16)^{1/2},\ 0 \le x \le 50$

(A) $C'(x) = \frac{1}{2}(2x+16)^{-1/2}(2) = \frac{1}{(2x+16)^{1/2}}$

(B) $C'(24) = \frac{1}{[2(24)+16]^{1/2}} = \frac{1}{(64)^{1/2}} = \frac{1}{8}$ or \$12.50; at a production level of 24 cell phones, total costs are INCREASING at the rate of \$12.50 per phone; also, the cost of producing the 25th phone is approximately \$12.50.

$C'(42) = \frac{1}{[2(42)+16]^{1/2}} = \frac{1}{(100)^{1/2}} = \frac{1}{10}$ or \$10.00; at a production level of 42 cell phones, total costs are INCREASING at the rate of \$10.00 per phone; also the cost of producing the 43rd phone is approximately \$10.00.

93. $x = 80\sqrt{p+25}-400 = 80(p+25)^{1/2}-400,\ 20 \le p \le 100$

(A) $\frac{dx}{dp} = 80\left(\frac{1}{2}\right)(p+25)^{-1/2}(1) = \frac{40}{(p+25)^{1/2}}$

(B) At $p = 75$, $x = 80\sqrt{75+25}-400 = 400$ and

$\frac{dx}{dp} = \frac{40}{(75+25)^{1/2}} = \frac{40}{(100)^{1/2}} = 4.$

At a price of \$75, the supply is 400 bicycle helments, and the supply is INCREASING at a rate of 4 helmet per dollar.

95. $C(t) = 4.35e^{-t},\quad 0 \le t \le 5$

(A) $C'(t) = -4.35e^{-t}$

$C'(1) = -4.35e^{-1} \approx -1.60$

$C'(4) = -4.35e^{-4} \approx -0.08$

Thus, after one hour, the concentration is decreasing at the rate of 1.60 mg/ml per hour; after four hours, the concentration is decreasing at the rate of 0.08 mg/ml per hour.

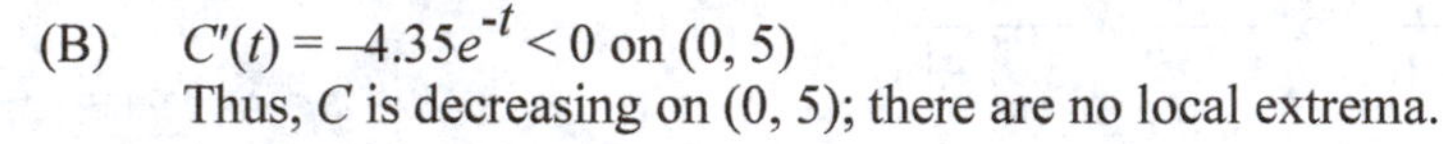

(B) $C'(t) = -4.35e^{-t} < 0$ on (0, 5)

Thus, C is decreasing on (0, 5); there are no local extrema.

The graph of C is shown at the right.

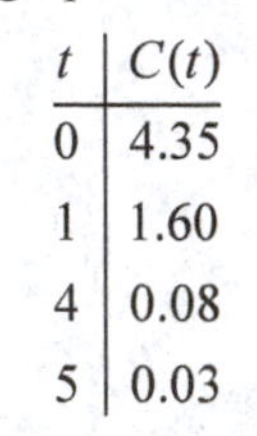

t	$C(t)$
0	4.35
1	1.60
4	0.08
5	0.03

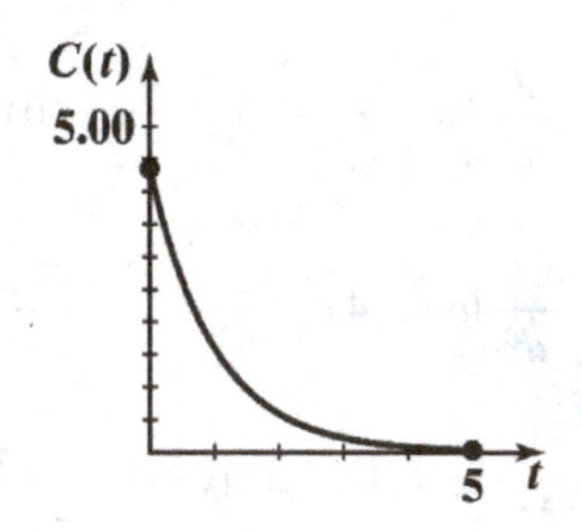

97. $P(x) = 40 + 25\ln(x + 1) \quad 0 \le x \le 65$

$$P'(x) = 25\left(\frac{1}{x+1}\right)(1) = \frac{25}{x+1}$$

$$P'(10) = \frac{25}{11} \approx 2.27$$

$$P'(30) = \frac{25}{31} \approx 0.81$$

$$P'(60) = \frac{25}{61} \approx 0.41$$

Thus, the rate of change of pressure at the end of 10 years is 2.27 millimeters of mercury per year; at the end of 30 years the rate of change is 0.81 millimeters of mercury per year; at the end of 60 years the rate of change is 0.41 millimeters of mercury per year.

EXERCISE 3-5

Things to remember:

1. Let $y = y(x)$. Then

   (a) $\frac{d}{dx}y^n = ny^{n-1}y'$ (General Power Rule)

   (b) $\frac{d}{dx}\ln y = \frac{1}{y}\cdot y' = \frac{y'}{y}$

   (c) $\frac{d}{dx}e^y = e^y \cdot y' = y'e^y$

---

1. $3x + 2y - 20 = 0; \quad 2y = 20 - 3x, \quad y = 10 - \frac{3}{2}x$

3. $\frac{x^2}{9}+\frac{y^2}{16}=1;\ \frac{y^2}{16}=1-\frac{x^2}{9}=\frac{1}{9}(9-x^2),\ y^2=\frac{16}{9}\left(9-x^2\right),\ y=\pm\frac{4}{3}\sqrt{9-x^2}$

5. $x^2+xy+y^2=1;\ y^2+xy+x^2-1=0,$

$y=\frac{-x\pm\sqrt{x^2-4(x^2-1)}}{2}=\frac{-x\pm\sqrt{4-3x^2}}{2}$ (quadratic formula)

7. $5x+3y=e^y$, impossible, cannot be solved for y.

9. $3x+5y+9=0$

(A) Implicit differentiation:

$$\frac{d}{dx}(3x)+\frac{d}{dx}(5y)+\frac{d}{dx}(9)=\frac{d}{dx}(0)$$

$$3+5y'+0=0$$

$$y'=-\frac{3}{5}$$

(B) Solve for y:

$$5y=-9-3x$$

$$y=-\frac{9}{5}-\frac{3}{5}x$$

$$y'=-\frac{3}{5}$$

11. $3x^2-4y-18=0$

(A) Implicit differentiation:

$$\frac{d}{dx}(3x^2)-\frac{d}{dx}(4y)-\frac{d}{dx}(18)=\frac{d}{dx}(0)$$

$$6x-4y'-0=0$$

$$y'=\frac{6}{4}x=\frac{3}{2}x$$

(B) Solve for y:

$$-4y=18-3x^2$$

$$y=\frac{3}{4}x^2-\frac{9}{2}$$

$$y'=\frac{6}{4}x=\frac{3}{2}x$$

13. $y-5x^2+3=0;\ (1,2)$

Using implicit differentiation:

$$\frac{d}{dx}(y)-\frac{d}{dx}(5x^2)+\frac{d}{dx}(3)=\frac{d}{dx}(0)$$

$$y'-10x=0$$

$$y'=10x$$

$$y'\Big|_{(1,2)}=10(1)=10$$

15. $x^2-y^3-3=0;\ (2,1)$

$$\frac{d}{dx}(x^2)-\frac{d}{dx}(y^3)-\frac{d}{dx}(3)=\frac{d}{dx}(0)$$

$$2x-3y^2y'=0$$

$$3y^2y'=2x$$

$$y'=\frac{2x}{3y^2}$$

$$y'\Big|_{(2,1)}=\frac{4}{3}$$

17. $y^2 + 2y + 3x = 0; (-1, 1)$

$$\frac{d}{dx}(y^2) + \frac{d}{dx}(2y) + \frac{d}{dx}(3x) = \frac{d}{dx}(0)$$

$$2yy' + 2y' + 3 = 0$$

$$2y'(y + 1) = -3$$

$$y' = -\frac{3}{2(y+1)}; \quad y'\Big|_{(-1,1)} = \frac{-3}{2(2)} = -\frac{3}{4}$$

19. $xy - 6 = 0$

$$\frac{d}{dx}xy - \frac{d}{dx}6 = \frac{d}{dx}(0)$$

$$xy' + y - 0 = 0$$

$$xy' = -y$$

$$y' = -\frac{y}{x}; \quad y' \text{ at } (2, 3) = -\frac{3}{2}$$

21. $2xy + y + 2 = 0$

$$2\frac{d}{dx}xy + \frac{d}{dx}y + \frac{d}{dx}2 = \frac{d}{dx}(0)$$

$$2xy' + 2y + y' + 0 = 0$$

$$y'(2x + 1) = -2y$$

$$y' = \frac{-2y}{2x+1}$$

$$y' \text{ at } (-1, 2) = \frac{-2(2)}{2(-1)+1} = 4$$

23. $x^2y - 3x^2 - 4 = 0$

$$\frac{d}{dx}x^2y - \frac{d}{dx}3x^2 - \frac{d}{dx}4 = \frac{d}{dx}(0)$$

$$x^2y' + y\frac{d}{dx}(x^2) - 6x - 0 = 0$$

$$x^2y' + y2x - 6x = 0$$

$$x^2y' = 6x - 2yx$$

$$y' = \frac{6x - 2yx}{x^2} \text{ or } \frac{6-2y}{x}$$

$$y'\Big|_{(2,4)} = \frac{6-2\cdot 4}{2^2} = \frac{6-8}{4} = -1$$

25. $e^y = x^2 + y^2$

$$\frac{d}{dx}e^y = \frac{d}{dx}x^2 + \frac{d}{dx}y^2$$

$$e^y y' = 2x + 2yy'$$

$$y'(e^y - 2y) = 2x$$

$$y' = \frac{2x}{e^y - 2y}$$

$$y'\bigg|_{(1,0)} = \frac{2\cdot 1}{e^0 - 2\cdot 0} = \frac{2}{1} = 2$$

27. $x^3 - y = \ln y$

$$\frac{d}{dx}x^3 - \frac{d}{dx}y = \frac{d}{dx}\ln y$$

$$3x^2 - y' = \frac{y'}{y}$$

$$3x^2 = \left(1 + \frac{1}{y}\right)y'$$

$$3x^2 = \frac{y+1}{y}y'$$

$$y' = \frac{3x^2 y}{y+1}$$

$$y'\bigg|_{(1,1)} = \frac{3\cdot 1^2\cdot 1}{1+1} = \frac{3}{2}$$

29. $x \ln y + 2y = 2x^3$

$$\frac{d}{dx}[x \ln y] + \frac{d}{dx}2y = \frac{d}{dx}2x^3$$

$$\ln y\cdot \frac{d}{dx}x + x\frac{d}{dx}\ln y + 2y' = 6x^2$$

$$\ln y\cdot 1 + x\cdot \frac{y'}{y} + 2y' = 6x^2$$

$$y'\left(\frac{x}{y} + 2\right) = 6x^2 - \ln y$$

$$y' = \frac{6x^2 y - y\ln y}{x + 2y}$$

$$y'\bigg|_{(1,1)} = \frac{6\cdot 1^2\cdot 1 - 1\cdot \ln 1}{1 + 2\cdot 1} = \frac{6}{3} = 2$$

31. $x^2 - t^2x + t^3 + 11 = 0$

$$\frac{d}{dt}x^2 - \frac{d}{dt}(t^2x) + \frac{d}{dt}t^3 + \frac{d}{dt}11 = \frac{d}{dt}0$$

$$2xx' - [t^2x' + x(2t)] + 3t^2 + 0 = 0$$

$$2xx' - t^2x' - 2tx + 3t^2 = 0$$

$$x'(2x - t^2) = 2tx - 3t^2$$

$$x' = \frac{2tx - 3t^2}{2x - t^2}$$

$$x'\bigg|_{(-2,1)} = \frac{2(-2)(1) - 3(-2)^2}{2(1) - (-2)^2} = \frac{-4-12}{2-4} = \frac{-16}{-2} = 8$$

33. $(x-1)^2+(y-1)^2=1.$

Differentiating implicitly, we have:

$$\begin{aligned}\frac{d}{dx}(x-1)^2+\frac{d}{dx}(y-1)^2 &= \frac{d}{dx}(1)\\ 2(x-1)+2(y-1)y' &= 0\\ y' &= -\frac{(x-1)}{(y-1)}\end{aligned}$$

To find the points on the graph where $x = 1.6$, we solve the given equation for y:

$$\begin{aligned}(y-1)^2 &= 1-(x-1)^2\\ y-1 &= \pm\sqrt{1-(x-1)^2}\\ y &= 1\pm\sqrt{1-(x-1)^2}\end{aligned}$$

Now, when $x = 1.6$, $y = 1+\sqrt{1-0.36} = 1+\sqrt{0.64} = 1.8$ and $y = 1-\sqrt{0.64} = 0.2$. Thus, the points are (1.6, 1.8) and (1.6, 0.2). These values can be verified on the graph.

$$\left.y'\right|_{(1.6,1.8)} = -\frac{(1.6-1)}{(1.8-1)} = -\frac{0.6}{0.8} = -\frac{3}{4}$$

$$\left.y'\right|_{(1.6,0.2)} = -\frac{(1.6-1)}{(0.2-1)} = -\frac{0.6}{(-0.8)} = \frac{3}{4}$$

35. $xy-x-4=0$

When $x=2$, $2y-2-4=0$, so $y=3$. Thus, we want to find the equation of the tangent line at (2, 3).

First, find y'.

$$\begin{aligned}\frac{d}{dx}xy-\frac{d}{dx}x-\frac{d}{dx}4 &= \frac{d}{dx}0\\ xy'+y-1-0 &= 0\\ xy' &= 1-y\\ y' &= \frac{1-y}{x}\\ \left.y'\right|_{(2,3)} &= \frac{1-3}{2}=-1\end{aligned}$$

Thus, the slope of the tangent line at (2, 3) is $m=-1$. The equation of the line through (2, 3) with slope $m=-1$ is:

$(y-3)=-1(x-2)$ or $y=-x+5$

37. $y^2 - xy - 6 = 0$

When $x = 1$,

$$y^2 - y - 6 = 0$$
$$(y-3)(y+2) = 0$$
$$y = 3 \text{ or } -2.$$

Thus, we want to find the equations of the tangent lines at (1, 3) and (1, –2). First, find y'.

$$\frac{d}{dx}y^2 - \frac{d}{dx}xy - \frac{d}{dx}6 = \frac{d}{dx}0$$
$$2yy' - xy' - y - 0 = 0$$
$$y'(2y - x) = y$$
$$y' = \frac{y}{2y - x}$$
$$y'\Big|_{(1,3)} = \frac{3}{2(3)-1} = \frac{3}{5} \quad \text{[Slope at (1, 3)]}$$

The equation of the tangent line at (1, 3) with $m = \frac{3}{5}$ is:

$$(y-3) = \frac{3}{5}(x-1)$$
$$y - 3 = \frac{3}{5}x - \frac{3}{5}$$
$$y = \frac{3}{5}x + \frac{12}{5}$$
$$y'\Big|_{(1,-2)} = \frac{-2}{2(-2)-1} = \frac{2}{5} \quad \text{[Slope at (1, –2)]}$$

Thus, the equation of the tangent line at (1, –2) with $m = \frac{2}{5}$ is:

$$(y+2) = \frac{2}{5}(x-1)$$
$$y + 2 = \frac{2}{5}x - \frac{2}{5}$$
$$y = \frac{2}{5}x - \frac{12}{5}$$

39. $xe^y = 1$

Implicit differentiation: $x \cdot \frac{d}{dx}e^y + e^y\frac{d}{dx}x = \frac{d}{dx}1$

$$xe^y y' + e^y = 0$$
$$y' = -\frac{e^y}{xe^y} = -\frac{1}{x}$$

Solve for y: $e^y = \frac{1}{x}$

$$y = \ln\left(\frac{1}{x}\right) = -\ln x \quad \text{(see Section 2-3)}$$
$$y' = -\frac{1}{x}$$

In this case, solving for y first and then differentiating is a little easier than differentiating implicitly.

41. $(1+y)^3 + y = x + 7$

$$\frac{d}{dx}(1+y)^3 + \frac{d}{dx}y = \frac{d}{dx}x + \frac{d}{dx}7$$

$$3(1+y)^2y' + y' = 1$$

$$y'[3(1+y)^2 + 1] = 1$$

$$y' = \frac{1}{3(1+y)^2+1}$$

$$y'\Big|_{(2,1)} = \frac{1}{3(1+1)^2+1} = \frac{1}{13}$$

43. $(x-2y)^3 = 2y^2 - 3$

$$\frac{d}{dx}(x-2y)^3 = \frac{d}{dx}(2y^2) - \frac{d}{dx}(3)$$

$$3(x-2y)^2(1-2y') = 4yy' - 0$$ [Note: The chain rule is applied to the left-hand side.]

$$3(x-2y)^2 - 6(x-2y)^2y' = 4yy'$$

$$-6(x-2y)^2y' - 4yy' = -3(x-2y)^2$$

$$-y'[6(x-2y)^2 + 4y] = -3(x-2y)^2$$

$$y' = \frac{3(x-2y)^2}{6(x-2y)^2+4y}$$

$$y'\Big|_{(1,1)} = \frac{3(1-2\cdot1)^2}{6(1-2)^2+4} = \frac{3}{10}$$

45. $\sqrt{7+y^2} - x^3 + 4 = 0$ or $(7+y^2)^{1/2} - x^3 + 4 = 0$

$$\frac{d}{dx}(7+y^2)^{1/2} - \frac{d}{dx}x^3 + \frac{d}{dx}4 = \frac{d}{dx}0$$

$$\frac{1}{2}(7+y^2)^{-1/2}\frac{d}{dx}(7+y^2) - 3x^2 + 0 = 0$$

$$\frac{1}{2}(7+y^2)^{-1/2}2yy' - 3x^2 = 0$$

$$\frac{yy'}{(7+y^2)^{1/2}} = 3x^2$$

$$y' = \frac{3x^2(7+y^2)^{1/2}}{y}$$

$$y'\Big|_{(2,3)} = \frac{3\cdot2^2(7+3^2)^{1/2}}{3} = \frac{12(16)^{1/2}}{3} = 16$$

47. $\ln(xy) = y^2 - 1$

$$\frac{d}{dx}[\ln(xy)] = \frac{d}{dx}y^2 - \frac{d}{dx}1$$

$$\frac{1}{xy} \cdot \frac{d}{dx}(xy) = 2yy'$$

$$\frac{1}{xy}(xy' + y) = 2yy'$$

$$\frac{1}{y} \cdot y' - 2yy' + \frac{1}{x} = 0$$

$$xy' - 2xy^2y' + y = 0$$

$$y'(x - 2xy^2) = -y$$

$$y' = \frac{-y}{x - 2xy^2} = \frac{y}{2xy^2 - x}$$

$$y'\Big|_{(1,1)} = \frac{1}{2 \cdot 1 \cdot 1^2 - 1} = 1$$

49. First find point(s) on the graph of the equation with abscissa $x = 1$:
Setting $x = 1$, we have

$y^3 - y - 1 = 2$ or $y^3 - y - 3 = 0$

Graphing this equation on a graphing utility, we get $y \approx 1.67$.

Now, differentiate implicitly to find the slope of the tangent line at the point (1, 1.67):

$3y^2y' -$

$$3y^2y' - xy' - y - 3x^2 = 0$$

$$(3y^2 - x)y' = 3x^2 + y$$

$$y' = \frac{3x^2 + y}{3y^2 - x};$$

$$y'\Big|_{(1,1.67)} = \frac{3 + 1.67}{3(1.67)^2 - 1} = \frac{4.67}{7.37} \approx 0.63$$

Tangent line: $y - 1.67 = 0.63(x - 1)$ or $y = 0.63x + 1.04$

51. $x = p^2 - 2p + 1000$

$$\frac{d(x)}{dx} = \frac{d(p^2)}{dx} - \frac{d(2p)}{dx} + \frac{d(1000)}{dx}$$

$$1 = 2p\frac{dp}{dx} - 2\frac{dp}{dx} + 0$$

$$1 = (2p - 2)\frac{dp}{dx}$$

Thus, $\frac{dp}{dx} = p' = \frac{1}{2p - 2}$.

53. $x = \sqrt{10{,}000 - p^2} = (10{,}000 - p^2)^{1/2}$

$$\frac{d}{dx}x = \frac{d}{dx}(10{,}000 - p^2)^{1/2}$$

$$1 = \frac{1}{2}(10{,}000 - p^2)^{-1/2}\frac{d}{dx}[10{,}000 - p^2]$$

$$1 = \frac{1}{2(10{,}000 - p^2)^{1/2}} \cdot (-2pp')$$

$$1 = \frac{-pp'}{\sqrt{10{,}000 - p^2}}$$

$$p' = \frac{-\sqrt{10{,}000 - p^2}}{p}$$

55. $(L + m)(V + n) = k$

$$(L + m)(1) + (V + n)\frac{dL}{dV} = 0$$

$$\frac{dL}{dV} = -\frac{(L + m)}{V + n}$$

57. $v = \sqrt{kT} = \sqrt{k}\,T^{1/2}$

$$\frac{d}{dv}(v) = \frac{d}{dv}\left(\sqrt{k}\,T^{1/2}\right)$$

$$1 = \frac{1}{2}\sqrt{k}\,T^{-1/2} \cdot \frac{dT}{dv} = \frac{\sqrt{k}}{2\sqrt{T}} \cdot \frac{dT}{dv}$$

$$\frac{dT}{dv} = \frac{2\sqrt{T}}{\sqrt{k}}$$

59. $v = \sqrt{kT} = \sqrt{k}\,T^{1/2}$

$$\frac{dv}{dT} = \frac{1}{2}\sqrt{k}\,T^{-1/2} = \frac{\sqrt{k}}{2\sqrt{T}}$$

$\frac{dT}{dv}$ is the reciprocal of $\frac{dv}{dT}$; that is $\frac{dT}{dv} = \frac{1}{dv / dT}$

EXERCISE 3-6

Things to remember:

1. SUGGESTIONS FOR SOLVING RELATED RATE PROBLEMS

   Step 1. Sketch a figure.

   Step 2. Identify all relevant variables, including those whose rates are given and those whose rates are to be found.

   Step 3. Express all given rates and rates to be found as derivatives.

   Step 4. Find an equation connecting the variables in Step 2.

Step 5. Implicitly differentiate the equation found in Step 4, using the chain rule where appropriate,
and substitute in all given values.

Step 6. Solve for the derivative that will give the unknown rate.

---

1. $A = \pi r^2$; $\pi r^2 = 300$, $r^2 = \dfrac{300}{\pi}$, $r = \sqrt{\dfrac{300}{\pi}} \approx 9.77$ ft, diameter ≈ 19.5 ft.

3. $a^2 + b^2 = c^2$, $a = 20$, $c = 50$; $b = \sqrt{(50)^2 - (20)^2} = \sqrt{2100} \approx 46$ m.

5. Let h be the height of the streetlight. The distance from the base of the streetlight to the tip of the shadow is 40 ft + 96 in = 48 ft. By similar triangles:
$\dfrac{h}{48} = \dfrac{69}{96}$; $h = 48 \cdot \dfrac{69}{96} = \dfrac{69}{2} = 34.5$ ft.

7. Sphere: $V = \dfrac{4}{3}\pi r^3 = \dfrac{4}{3}\pi(12)^3 = 2304\pi$. Cylinder: $V = 2304\pi(2) = 4608\pi$

$V = \pi r^2 h = \pi(12)^2 h = 4608\pi$; $h = \dfrac{4608\pi}{144\pi} = 32$ ft.

9. $y = x^2 + 2$

Differentiating with respect to t:

$\dfrac{dy}{dt} = 2x\dfrac{dx}{dt}$; $\dfrac{dy}{dt} = 2(5)(3) = 30$ when $x = 5$, $\dfrac{dx}{dt} = 3$

11. $x^2 + y^2 = 1$

Differentiating with respect to t:

$$2x\frac{dx}{dt} + 2y\frac{dy}{dt} = 0$$

$$2x\frac{dx}{dt} = -2y\frac{dy}{dt}$$

$$\frac{dx}{dt} = -\frac{y}{x}\frac{dy}{dt};\quad \frac{dx}{dt} = -\frac{0.8}{(-0.6)}(-4) = -\frac{16}{3},$$

when $x = -0.6$, $y = 0.8$, $\dfrac{dy}{dt} = -4$

13. $x^2 + 3xy + y^2 = 11$

Differentiating with respect to t:

$$2x\frac{dx}{dt} + 3x\frac{dy}{dt} + 3y\frac{dx}{dt} + 2y\frac{dy}{dt} = 0$$

$$(3x + 2y)\frac{dy}{dt} = -(2x + 3y)\frac{dx}{dt}$$

$$\frac{dy}{dt} = -\frac{(2x+3y)}{3x+2y}\frac{dx}{dt};\quad \frac{dy}{dt} = -\frac{(2\cdot 1 + 3\cdot 2)}{(3\cdot 1 + 2\cdot 2)}2 = -\frac{16}{7}$$

when $x = 1$, $y = 2$, $\dfrac{dx}{dt} = 2$

15. $xy = 36$

Differentiate with respect to t:

$$\frac{d(xy)}{dt} = \frac{d(36)}{dt}$$

$$x\frac{dy}{dt} + y\frac{dx}{dt} = 0$$

Given: $\frac{dx}{dt} = 4$ when $x = 4$ and $y = 9$. Therefore,

$$4\frac{dy}{dt} + 9(4) = 0$$

$$4\frac{dy}{dt} = -36 \text{ and } \frac{dy}{dt} = -9.$$

The y coordinate is decreasing at 9 units per second.

17.

z = rope
y = 4
x

From the triangle,

$$x^2 + y^2 = z^2$$

or $x^2 + 16 = z^2$, since $y = 4$.

Differentiate with respect to t:

$$2x\frac{dx}{dt} = 2z\frac{dz}{dt}$$

or $x\frac{dx}{dt} = z\frac{dz}{dt}$

Given: $\frac{dz}{dt} = -3$. Also, when $x = 30$, $900 + 16 = z^2$ or $z = \sqrt{916}$.

Therefore,

$$30\frac{dx}{dt} = \sqrt{916}(-3) \text{ and } \frac{dx}{dt} = \frac{-3\sqrt{916}}{30} = \frac{-\sqrt{916}}{10} \approx \frac{-30.27}{10} \approx -3.03 \text{ feet/second.}$$

[Note: The negative sign indicates that the distance between the boat and the dock is decreasing.]

19. Area: $A = \pi R^2$

$$\frac{dA}{dt} = \frac{d\pi R^2}{dt} = \pi \cdot 2R\frac{dR}{dt}$$

Given: $\frac{dR}{dt} = 2$ ft/sec

$$\frac{dA}{dt} = 2\pi R \cdot 2 = 4\pi R$$

$$\left.\frac{dA}{dt}\right|_{R=10\text{ft}} = 4\pi(10) = 40\pi \text{ ft}^2/\text{sec} \approx 126 \text{ ft}^2/\text{sec}$$

21. $V = \frac{4}{3}\pi R^3$

$$\frac{dV}{dt} = \frac{4}{3}\pi 3R^2\frac{dR}{dt} = 4\pi R^2\frac{dR}{dt}$$

Given: $\frac{dR}{dt} = 3$ cm/min

$$\frac{dV}{dt} = 4\pi R^2 3 = 12\pi R^2$$

$$\left.\frac{dV}{dt}\right|_{R=10\text{cm}} = 12\pi(10)^2 = 1200\pi \approx 3770 \text{ cm}^3/\text{min}$$

23. $\frac{P}{T} = k$ (1)

$P = kT$

Differentiate with respect to t:

$$\frac{dP}{dt} = k\frac{dT}{dt}$$

Given: $\frac{dT}{dt} = 3$ degrees per hour, $T = 250°$, $P = 500$ pounds per square inch.

From (1), for $T = 250$ and $P = 500$,

$$k = \frac{500}{250} = 2.$$

Thus, we have

$$\frac{dP}{dt} = 2\frac{dT}{dt}$$

$$\frac{dP}{dt} = 2(3) = 6$$

Pressure increases at 6 pounds per square inch per hour.

25. By the Pythagorean theorem,

$$x^2 + y^2 = 10^2$$

or $x^2 + y^2 = 100$ (1)

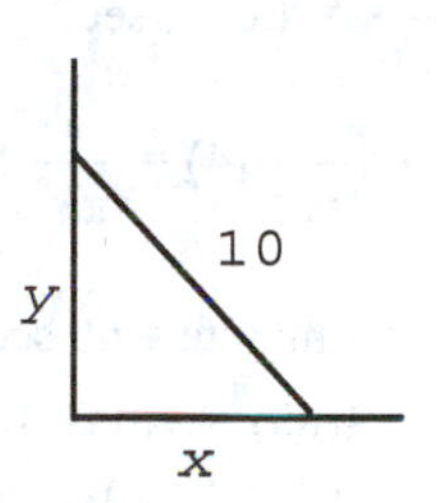

Differentiate with respect to t:

$$2x\frac{dx}{dt} + 2y\frac{dy}{dt} = 0$$

Therefore, $\frac{dy}{dt} = -\frac{x}{y}\frac{dx}{dt}$. Given: $\frac{dx}{dt} = 3$. Thus, $\frac{dy}{dt} = \frac{-3x}{y}$.

From (1), $y^2 = 100 - x^2$ and, when $x = 6$,

$$y^2 = 100 - 6^2$$
$$= 100 - 36 = 64.$$

Thus, $y = 8$ when $x = 6$, and

$$\left.\frac{dy}{dt}\right|_{(6,8)} = \frac{-3(6)}{8} = \frac{-18}{8} = \frac{-9}{4} \text{ ft/sec.}$$

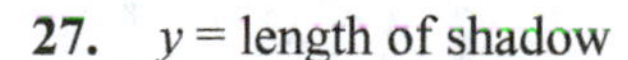

27. y = length of shadow
x = distance of man from light
z = distance of tip of shadow from light

We want to compute $\frac{dz}{dt}$. Triangles ABE and CDE are similar triangles; thus, the ratios of corresponding sides are equal.

Therefore, $\frac{z}{20} = \frac{y}{5} = \frac{z-x}{5}$ [Note: $y = z - x$.]

or

$$\frac{z}{20} = \frac{z-x}{5}$$
$$z = 4(z - x)$$
$$z = 4z - 4x$$
$$4x = 3z$$

Differentiate with respect to t:

$$4\frac{dx}{dt} = 3\frac{dz}{dt}$$

$$\frac{dz}{dt} = \frac{4}{3}\frac{dx}{dt}$$

Given: $\frac{dx}{dt} = 5$. Thus, $\frac{dz}{dt} = \frac{4}{3}(5) = \frac{20}{3}$ ft/sec.

29. $V = \frac{4}{3}\pi r^3 \quad (1)$

Differentiate with respect to t:

$$\frac{dV}{dt} = 4\pi r^2\frac{dr}{dt} \quad \text{and} \quad \frac{dr}{dt} = \frac{1}{4\pi r^2}\cdot\frac{dV}{dt}$$

Since $\frac{dV}{dt} = 4$ cu ft/sec,

$$\frac{dr}{dt} = \frac{1}{4\pi r^2}(4) = \frac{1}{\pi r^2} \text{ ft/sec} \quad (2)$$

At $t = 1$ minute = 60 seconds,

$V = 4(60) = 240$ cu ft and, from (1),

$$r^3 = \frac{3V}{4\pi} = \frac{3(240)}{4\pi} = \frac{180}{\pi};\ r = \left(\frac{180}{\pi}\right)^{1/3} \approx 3.855.$$

From (2)

$$\frac{dr}{dt} = \frac{1}{\pi(3.855)^2} \approx 0.0214 \text{ ft/sec}$$

At $t = 2$ minutes = 120 seconds,

$V = 4(120) = 480$ cu ft and

$$r^3 = \frac{3V}{4\pi} = \frac{3(480)}{4\pi} = \frac{360}{\pi};\ r = \left(\frac{360}{\pi}\right)^{1/3} \approx 4.857$$

From (2),

$$\frac{dr}{dt} = \frac{1}{\pi(4.857)^2} \approx 0.0135 \text{ ft/sec}$$

To find the time at which $\frac{dr}{dt} = 100$ ft/sec, solve

$$\frac{1}{\pi r^2} = 100$$

$$r^2 = \frac{1}{100\pi} \quad \text{and} \quad r = \frac{1}{\sqrt{100\pi}} = \frac{1}{10\sqrt{\pi}}$$

Now, when $r = \frac{1}{10\sqrt{\pi}}$,

$$V = \frac{4}{3}\pi\left(\frac{1}{10\sqrt{\pi}}\right)^3 = \frac{4}{3} \cdot \frac{1}{1000\sqrt{\pi}} = \frac{1}{750\sqrt{\pi}}$$

Since the volume at time t is $4t$, we have

$$4t = \frac{1}{750\sqrt{\pi}} \text{ and } t = \frac{1}{3000\sqrt{\pi}} \approx 0.00019 \text{ secs.}$$

31. $y = e^x + x + 1; \frac{dx}{dt} = 3.$

Differentiate with respect to t:

$$\frac{dy}{dt} = e^x\frac{dx}{dt} + \frac{dx}{dt} = e^x(3) + 3 = 3(e^x + 1)$$

To find where the point crosses the x axis, use a graphing utility to solve

$e^x + x + 1 = 0$

The result is $x \approx -1.278$.

Now, at $x = -1.278$,

$$\frac{dy}{dt} = 3(e^{-1.278} + 1) \approx 3.835 \text{ units/sec.}$$

33. $C = 90{,}000 + 30x$ (1)

$R = 300x - \frac{x^2}{30}$ (2)

$P = R - C$ (3)

(A) Differentiating (1) with respect to t:

$$\frac{dC}{dt} = \frac{d(90{,}000)}{dt} + \frac{d(30x)}{dt}$$

$$\frac{dC}{dt} = 30\frac{dx}{dt}$$

Thus, $\frac{dC}{dt} = 30(500) \quad \left(\frac{dx}{dt} = 500\right)$

$= \$15{,}000$ per week.

Costs are increasing at the rate of \$15,000 per week at this production level.

(B) Differentiating (2) with respect to t:

$$\frac{dR}{dt} = \frac{d(300x)}{dt} - \frac{d\frac{x^2}{30}}{dt}$$

$$= 300\frac{dx}{dt} - \frac{2x}{30}\frac{dx}{dt}$$

$$= \left(300 - \frac{x}{15}\right)\frac{dx}{dt}$$

Thus, $\frac{dR}{dt} = \left(300 - \frac{6000}{15}\right)(500) \quad \left(x = 6000, \frac{dx}{dt} = 500\right)$

$= (-100)500 = -50{,}000.$

Revenue is decreasing at the rate of \$50,000 per week at this production level.

(C) Differentiating (3) with respect to t:

$$\frac{dP}{dt} = \frac{dR}{dt} - \frac{dC}{dt}$$

Thus, from parts (A) and (B), we have:

$$\frac{dP}{dt} = -50{,}000 - 15{,}000 = -\$65{,}000$$

Profits are decreasing at the rate of \$65,000 per week at this production level.

35. $s = 60{,}000 - 40{,}000e^{-0.0005x}$

Differentiating implicitly with respect to t, we have

$$\frac{ds}{dt} = -40{,}000(-0.0005)e^{-0.0005x}\frac{dx}{dt} \text{ and } \frac{ds}{dt} = 20e^{-0.0005x}\frac{dx}{dt}$$

Now, for $x = 2000$ and $\frac{dx}{dt} = 300$, we have

$$\frac{ds}{dt} = 20(300)e^{-0.0005(2000)}$$

$$= 6000e^{-1} = 2{,}207$$

Thus, sales are increasing at the rate of \$2,207 per week.

37. Price p and demand x are related by the equation

$$2x^2 + 5xp + 50p^2 = 80{,}000 \qquad (1)$$

Differentiating implicitly with respect to t, we have

$$4x\frac{dx}{dt} + 5x\frac{dp}{dt} + 5p\frac{dx}{dt} + 100p\frac{dp}{dt} = 0 \qquad (2)$$

(A) From (2), $\frac{dx}{dt} = \frac{-(5x+100p)\frac{dp}{dt}}{4x+5p}$

Setting $p = 30$ in (1), we get

$$2x^2 + 150x + 45{,}000 = 80{,}000$$

or $x^2 + 75x - 17{,}500 = 0$

Thus, $x = \frac{-75 \pm \sqrt{(75)^2 + 70{,}000}}{2} = \frac{-75 \pm 275}{2} = 100, \ -175$

Since $x \geq 0$, $x = 100$.

Now, for $x = 100$, $p = 30$ and $\frac{dp}{dt} = 2$, we have

$$\frac{dx}{dt} = \frac{-[5(100)+100(30)]\cdot 2}{4(100)+5(30)} = -\frac{7000}{550} \text{ and } \frac{dx}{dt} = -12.73$$

The demand is decreasing at the rate of 12.73 units/month.

(B) From (2), $\frac{dp}{dt} = \frac{-(4x+5p)\frac{dx}{dt}}{(5x+100p)}$

Setting $x = 150$ in (1), we get

$45{,}000 + 750p + 50p^2 = 80{,}000$

or $p^2 + 15p - 700 = 0$

and $p = \frac{-15 \pm \sqrt{225+2800}}{2} = \frac{-15 \pm 55}{2} = -35, 20$

Since $p \geq 0$, $p = 20$.

Now, for $x = 150$, $p = 20$ and $\frac{dx}{dt} = -6$, we have

$$\frac{dp}{dt} = -\frac{[4(150)+5(20)](-6)}{5(150)+100(20)} = \frac{4200}{2750} \approx 1.53$$

Thus, the price is increasing at the rate of $1.53 per month.

39. Volume $V = \pi R^2 h$, where h = thickness of the circular oil slick.

Since $h = 0.1 = \frac{1}{10}$, we have:

$$V = \frac{\pi}{10}R^2$$

Differentiating with respect to t:

$$\frac{dV}{dt} = \frac{d\left(\frac{\pi}{10}R^2\right)}{dt} = \frac{\pi}{10}2R\frac{dR}{dt} = \frac{\pi}{5}R\frac{dR}{dt}$$

Given: $\frac{dR}{dt} = 0.32$ when $R = 500$. Therefore,

$$\frac{dV}{dt} = \frac{\pi}{5}(500)(0.32) = 100\pi(0.32) \approx 100.53 \text{ cubic feet per minute.}$$

EXERCISE 3-7

Things to remember:

1. RELATIVE AND PERCENTAGE RATES OF CHANGE

   The RELATIVE RATE OF CHANGE of a function $f(x)$ is $\frac{f'(x)}{f(x)}$.

   The PERCENTAGE RATE OF CHANGE is $100 \times \frac{f'(x)}{f(x)}$.

2. ELASTICITY OF DEMAND

   If price and demand are related by $x = f(p)$, then the ELASTICITY OF DEMAND is given by

   $$E(p) = -\frac{pf'(p)}{f(p)}$$

3. INTERPRETATION OF ELASTICITY OF DEMAND

$E(p)$	Demand	Interpretation
$0 < E(p) < 1$	Inelastic	Demand is not sensitive to changes in price. A change in price produces a smaller change in demand.
$E(p) > 1$	Elastic	Demand is sensitive to changes in price. A change in price produces a larger change in demand.
$E(p) = 1$	Unit	A change in price produces the same change in demand.

4. REVENUE AND ELASTICITY OF DEMAND

If $R(p) = pf(p)$ is the revenue function, then $R'(p)$ and $[1 - E(p)]$ always have the same sign.

Demand is inelastic $[E(p) < 1, R'(p) > 0]$:

A price increase will increase revenue.

A price decrease will decrease revenue.

Demand is elastic $[E(p) > 1, R'(p) < 0]$:

A price increase will decrease revenue.

A price decrease will increase revenue.

---

1. $p = 42 - 0.4x,\ 0 \le x \le 105;\quad x = f(p) = \dfrac{42 - p}{0.4} = 105 - 2.5p,\ 0 \le p \le 42.$

3. $p = 50 - 0.5x^2,\ 0 \le x \le 10;\quad x^2 = \dfrac{50 - p}{0.5} = 100 - 2p,\quad x = f(p) = \sqrt{100 - 2p},\ 0 \le p \le 50.$

5. $p = 25e^{-x/20},\ 0 \le x \le 20;\ e^{-x/20} = \dfrac{p}{25},\ -\dfrac{x}{20} = \ln\left(\dfrac{p}{25}\right) = \ln p - \ln 25,$

$x = f(p) = 20(\ln 25 - \ln p),\quad \dfrac{25}{e} \approx 9.2 \le p \le 25.$

7. $p = 80 - 10\ln x,\ 1 \le x \le 30;\ \ln x = \dfrac{80 - p}{10} = 8 - 0.1p,\quad x = e^{8 - 0.1p},$

$x = f(p) = e^{8 - 0.1p},\quad 80 - 10\ln 30 \approx 46 \le p \le 80.$

9. $f(x) = 35x - 0.4x^2$

$f'(x) = 35 - 0.8x$

Relative rate of change of f: $\dfrac{f'(x)}{f(x)} = \dfrac{35 - 0.8x}{35x - 0.4x^2}$

11. $f(x) = 7 + 4e^{-x}$

$f'(x) = -4e^{-x}$

Relative rate of change of f: $\dfrac{f'(x)}{f(x)} = \dfrac{-4e^{-x}}{7 + 4e^{-x}}$

13. $f(x) = 12 + 5\ln x$

$f'(x) = \dfrac{5}{x}$

Relative rate of change of f: $\dfrac{f'(x)}{f(x)} = \dfrac{5/x}{12 + 5\ln x} = \dfrac{5}{x(12 + 5\ln x)}$

15. $f(x) = 45, \quad f'(x) = 0$

$\dfrac{f'(x)}{f(x)} = 0$; relative rate of change of f at $x = 100$: 0

17. $f(x) = 420 - 5x, \quad f'(x) = -5$

$\dfrac{f'(x)}{f(x)} = \dfrac{-5}{420 - 5x}$; relative rate of change of f at $x = 25$: $\dfrac{-5}{420 - 5(25)} = \dfrac{-5}{295} \approx -0.017$

19. $f(x) = 420 - 5x, \quad f'(x) = -5$

$\dfrac{f'(x)}{f(x)} = \dfrac{-5}{420 - 5x}$; relative rate of change of f at $x = 55$: $\dfrac{-5}{420 - 5(55)} = \dfrac{-5}{145} \approx -0.034$

21. $f(x) = 4x^2 - \ln x, \quad f'(x) = 8x - \dfrac{1}{x}$

$\dfrac{f'(x)}{f(x)} = \dfrac{8x - \frac{1}{x}}{4x^2 - \ln x}$, relative rate of change of f at $x = 2$: $\dfrac{16 - \frac{1}{2}}{16 - \ln 2} \approx 1.013$

23. $f(x) = 4x^2 - \ln x, \quad f'(x) = 8x - \dfrac{1}{x}$

$\dfrac{f'(x)}{f(x)} = \dfrac{8x - \frac{1}{x}}{4x^2 - \ln x}$, relative rate of change of f at $x = 5$: $\dfrac{40 - \frac{1}{5}}{100 - \ln 5} \approx 0.405$

25. $f(x) = 225 + 65x, \quad f'(x) = 65$

$\dfrac{f'(x)}{f(x)} = \dfrac{65}{225 + 65x}$; percentage rate of change at $x = 5$: $100 \cdot \dfrac{65}{225 + 65(5)} = 100 \cdot \dfrac{65}{550} \approx 11.8$; 11.8%

27. $f(x) = 225 + 65x, \quad f'(x) = 65$

$\dfrac{f'(x)}{f(x)} = \dfrac{65}{225 + 65x}$; percentage rate of change at $x = 15$: $100 \cdot \dfrac{65}{225 + 65(15)} = 100 \cdot \dfrac{65}{1200} \approx 5.4$; 5.4%

29. $f(x) = 5{,}100 - 3x^2, \quad f'(x) = -6x$

$\dfrac{f'(x)}{f(x)} = \dfrac{-6x}{5{,}100 - 3x^2}$; percentage rate of change at $x = 35$:

$100 \cdot \dfrac{-6(35)}{5{,}100 - 3(35)^2} = 100 \cdot \dfrac{-210}{1425} \approx -14.7$; -14.7%

31. $f(x) = 5{,}100 - 3x^2, \quad f'(x) = -6x$

$\dfrac{f'(x)}{f(x)} = \dfrac{-6x}{5{,}100 - 3x^2}$; percentage rate of change at $x = 41$:

$100 \cdot \dfrac{-6(41)}{5{,}100 - 3(41)^2} = 100 \cdot \dfrac{-246}{57} \approx -431.6$; -431.6%

33. $x = f(p) = 25{,}000 - 450p, \quad f'(p) = -450$

$E(p) = \dfrac{-p f'(p)}{f(p)} = \dfrac{450p}{25{,}000 - 450p} = \dfrac{9p}{5{,}000 - 9p}$

35. $x = f(p) = 4,800 - 4p^2, \quad f'(p) = -8p$

$$E(p) = \frac{-p\,f'(p)}{f(p)} = \frac{8p^2}{4,800 - 4p^2} = \frac{2p^2}{1,200 - p^2}$$

37. $x = f(p) = 98 - 0.6e^p, \quad f'(p) = -0.6e^p$

$$E(p) = \frac{-p\,f'(p)}{f(p)} = \frac{0.6p\,e^p}{98 - 0.6e^p}$$

39. $A(t) = 500e^{0.07t}, \quad A'(t) = 35e^{0.07t}, \quad \frac{A'(t)}{A(t)} = \frac{35e^{0.07t}}{500e^{0.07t}} = \frac{35}{500} = 0.07$

41. $A(t) = 3,500e^{0.15t}, \quad A'(t) = 525e^{0.15t}, \quad \frac{A'(t)}{A(t)} = \frac{525e^{0.15t}}{3,500e^{0.15t}} = \frac{525}{3,500} = 0.15$

43. $f(x) = xe^x, \quad f'(x) = xe^x + e^x; \quad \frac{f'(x)}{f(x)} = \frac{xe^x + e^x}{xe^x} = \frac{x+1}{x}$

45. $f(x) = \ln x, \quad f'(x) = \frac{1}{x}; \quad \frac{f'(x)}{f(x)} = \frac{1/x}{\ln x} = \frac{1}{x\ln x}$

47. $x = f(p) = 12,000 - 10p^2$
$f'(p) = -20p$

Elasticity of demand: $E(p) = \frac{-pf'(p)}{f(p)} = \frac{20p^2}{12,000 - 10p^2}$

(A) At $p = 10$: $E(10) = \frac{2000}{12,000 - 1000} = \frac{2000}{11,000} = \frac{2}{11}$

Demand is inelastic.

(B) At $p = 20$: $E(20) = \frac{8000}{12,000 - 4000} = \frac{8000}{8000} = 1$; unit elasticity.

(C) At $p = 30$: $E(30) = \frac{18,000}{12,000 - 9,000} = \frac{18,000}{3,000} = 6$

Demand is elastic.

49. $x = f(p) = 950 - 2p - 0.1p^2$
$f'(p) = -2 - 0.2p$

Elasticity of demand: $E(p) = \frac{-pf'(p)}{f(p)} = \frac{2p + 0.2p^2}{950 - 2p - 0.1p^2}$

(A) At $p = 30$: $E(30) = \frac{60 + 180}{950 - 60 - 90} = \frac{240}{800} = \frac{3}{10}$

Demand is inelastic.

(B) At $p = 50$: $E(50) = \frac{100 + 500}{950 - 100 - 250} = \frac{600}{600} = 1$; unit elasticity.

(C) At $p = 70$: $E(70) = \frac{140 + 980}{950 - 140 - 490} = \frac{1120}{320} = 3.5$

Demand is elastic.

51. $p + 0.005x = 30$

(A) $x = \dfrac{30 - p}{0.005} = 6000 - 200p,\ 0 \le p \le 30$

(B) $f(p) = 6000 - 200p$
$f'(p) = -200$

Elasticity of demand: $E(p) = \dfrac{-pf'(p)}{f(p)} = \dfrac{200p}{6000 - 200p} = \dfrac{p}{30 - p}$

(C) At $p = 10$: $E(10) = \dfrac{10}{30 - 10} = \dfrac{1}{2} = 0.5$

If the price increases by 10%, the demand will decrease by approximately 0.5(10%) = 5%.

(D) At $p = 25$: $E(25) = \dfrac{25}{30 - 25} = 5$

If the price increases by 10%, the demand will decrease by approximately 5(10%) = 50%.

(E) At $p = 15$: $E(15) = \dfrac{15}{30 - 15} = 1$

If the price increases by 10%, the demand will decrease by approximately 10%.

53. $0.02x + p = 60$

(A) $x = \dfrac{60 - p}{0.02} = 3000 - 50p,\ 0 \le p \le 60$

(B) $R(p) = p(3000 - 50p) = 3000p - 50p^2$

(C) $f(p) = 3000 - 50p$
$f'(p) = -50$

Elasticity of demand: $E(p) = \dfrac{-pf'(p)}{f(p)} = \dfrac{50p}{3000 - 50p} = \dfrac{p}{60 - p}$

(D) Elastic: $E(p) = \dfrac{p}{60 - p} > 1$

$p > 60 - p$
$p > 30, \quad 30 < p < 60$

Inelastic: $E(p) = \dfrac{p}{60 - p} < 1$

$p < 60 - p$
$p < 30, \quad 0 < p < 30$

(E) $R'(p) = f(p)\,[1 - E(p)]$
$R'(p) > 0$ if $E(p) < 1$; $R'(p) < 0$ if $E(p) > 1$
Therefore, revenue is increasing for $0 < p < 30$ and decreasing for $30 < p < 60$.

(F) If $p = \$10$ and the price is decreased, revenue will also decrease.

(G) If $p = \$40$ and the price is decreased, revenue will increase.

55. $x = f(p) = 210 - 30p,\ 0 < p < 7;\ f'(p) = -30$

$$E(p) = \frac{-pf'(p)}{f(p)} = \frac{30p}{210-30p} = \frac{p}{7-p}$$

Elastic:

$$E(p) = \frac{p}{7-p} > 1$$

$$p > 7 - p$$

$$2p > 7$$

$$p > 3.5$$

Demand is elastic for $3.5 < p < 7$; demand is inelastic for $0 < p < 3.5$.

57. $x = f(p) = 3{,}125 - 5p^2,\ 0 < p < 25;\ f'(p) = -10p$

$$E(p) = \frac{-pf'(p)}{f(p)} = \frac{10p^2}{3{,}125-5p^2} = \frac{2p^2}{625-p^2}$$

Elastic:

$$E(p) = \frac{2p^2}{625-p^2} > 1$$

$$2p^2 > 625 - p^2$$

$$3p^2 > 625$$

$$p^2 > \frac{625}{3}$$

$$p > \frac{25}{\sqrt{3}} = \frac{25\sqrt{3}}{3}$$

Demand is elastic for $\frac{25\sqrt{3}}{3} < p < 25$; demand is inelastic for $0 < p < \frac{25\sqrt{3}}{3}$.

59. $x = f(p) = \sqrt{144-2p},\ 0 \le p \le 72$

$$f'(p) = \frac{1}{2}(144-2p)^{-1/2}(-2) = \frac{-1}{\sqrt{144-2p}}$$

Elasticity of demand: $E(p) = \frac{p}{144-2p}$

Elastic: $E(p) = \frac{p}{144-2p} > 1$

$$p > 144 - 2p$$

$$3p > 144$$

$$p > 48, \quad 48 < p < 72$$

Inelastic: $E(p) = \frac{p}{144-2p} < 1$

$$p < 144 - 2p$$

$$3p < 144$$

$$p < 48, \quad 0 < p < 48$$

61. $x = f(p) = \sqrt{2{,}500 - 2p^2} \quad 0 \le p \le 25\sqrt{2}$

$$f'(p) = \frac{1}{2}(2{,}500 - 2p^2)^{-1/2}(-4p) = \frac{-2p}{(2{,}500 - 2p^2)^{1/2}}$$

Elasticity of demand: $E(p) = \dfrac{2p^2}{2{,}500 - 2p^2} = \dfrac{p^2}{1{,}250 - p^2}$

Elastic: $E(p) = \dfrac{p^2}{1{,}250 - p^2} > 1$

$$p^2 > 1{,}250 - p^2$$
$$2p^2 > 1{,}250$$
$$p^2 > 625$$
$$p > 25, \quad 25 < p < 25\sqrt{2}$$

Inelastic: $E(p) = \dfrac{p^2}{1{,}250 - p^2} < 1$

$$p^2 < 1{,}250 - p^2$$
$$2p^2 < 1{,}250$$
$$p^2 < 625$$
$$p < 25, \quad 0 < p < 25$$

63. $x = f(p) = 20(10 - p) \quad 0 \le p \le 10$

$R(p) = pf(p) = 20p(10 - p) = 200p - 20p^2$

$R'(p) = 200 - 40p$

Critical value: $R'(p) = 200 - 40p = 0;\ p = 5$

Sign chart for $R'(p)$:

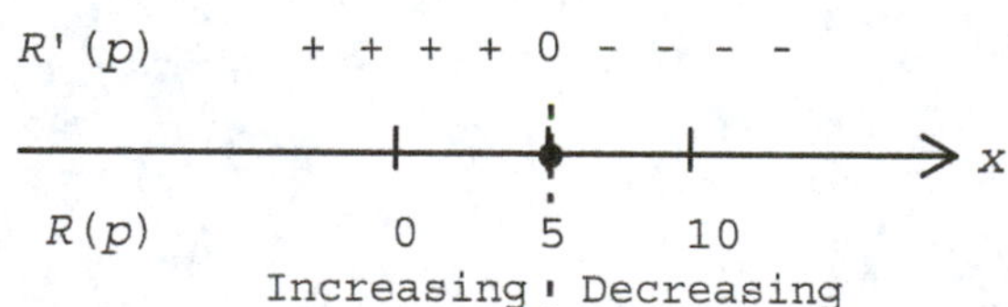

Test Numbers

p	$R'(p)$
0	200(+)
10	−200(−)

Inelastic
$R(p)$
500
Elastic
10 p

65. $x = f(p) = 40(p - 15)^2 \quad 0 \le p \le 15$

$R(p) = pf(p) = 40p(p - 15)^2$

$$\begin{aligned} R'(p) &= 40(p - 15)^2 + 40p(2)(p - 15) \\ &= 40(p - 15)[p - 15 + 2p] \\ &= 40(p - 15)(3p - 15) \\ &= 120(p - 15)(p - 5) \end{aligned}$$

Critical values [in (0, 15)]: $p = 5$

Sign chart for $R'(p)$:

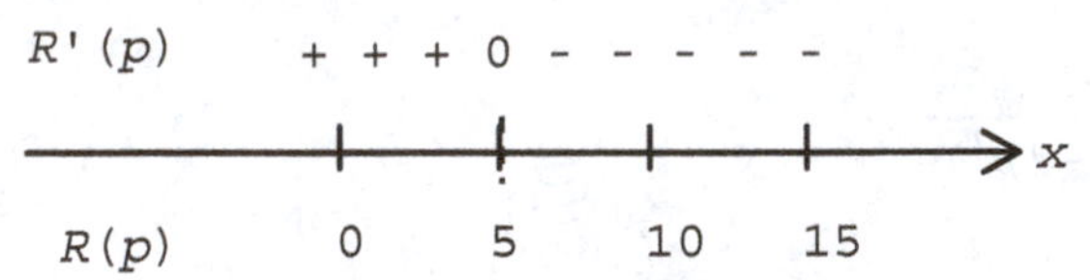

Test Numbers

p	$R'(p)$
0	(+)
10	(−)

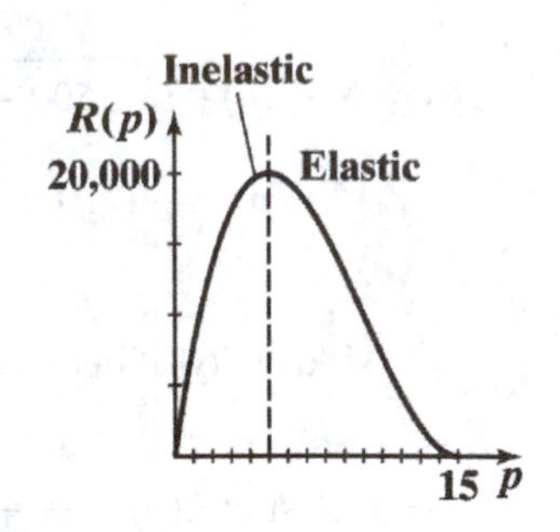

67. $x = f(p) = 30 - 10\sqrt{p} \quad 0 \le p \le 9$

$R(p) = pf(p) = 30p - 10p\sqrt{p}$

$R'(p) = 30 - 10\sqrt{p} - 10p \cdot \frac{1}{2}p^{-1/2}$

$= 30 - 10\sqrt{p} - \frac{5p}{\sqrt{p}} = 30 - 15\sqrt{p}$

Critical values: $R'(p) = 30 - 15\sqrt{p} = 0$

$\sqrt{p} = 2; \quad p = 4$

Sign chart for $R'(p)$:

```
R'(p)        + + + 0 - - -
        ---------+------+-----+----> x
R(p)             0      4     9
             Increasing | Decreasing
Demand:      Inelastic  | Elastic
```

Test Numbers

p	$R'(p)$
0	30(+)
5	(−)

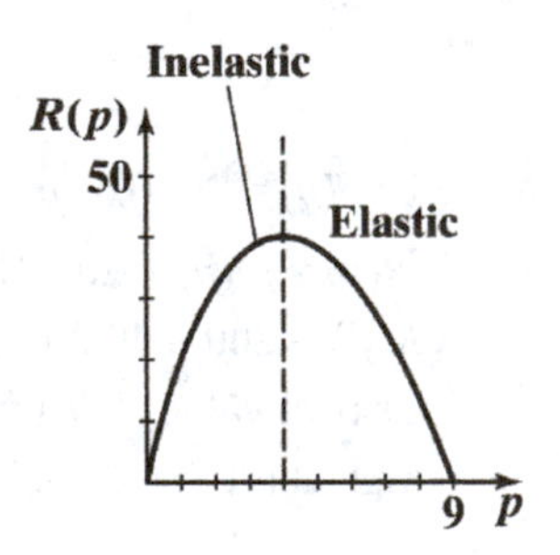

69. $p = g(x) = 50 - 0.1x$

$g'(x) = -0.1$

$E(x) = -\frac{g(x)}{xg'(x)} = -\frac{50 - 0.1x}{-0.1x} = \frac{500}{x} - 1$

$E(200) = \frac{500}{200} - 1 = \frac{3}{2}$

71. $p = g(x) = 50 - 2\sqrt{x}$

$g'(x) = -\frac{1}{\sqrt{x}}$

$E(x) = -\frac{g(x)}{xg'(x)} = -\frac{50 - 2\sqrt{x}}{x\left(-\frac{1}{\sqrt{x}}\right)} = \frac{50}{\sqrt{x}} - 2$

$E(400) = \frac{50}{20} - 2 = \frac{1}{2}$

73. $p = g(x) = 180 - 0.3x,\ \ 0 < x < 600;\ \ g'(x) = -0.3$

$$E(x) = \frac{-g(x)}{xg'(x)} = \frac{-(180-0.3x)}{-0.3x} = \frac{180-0.3x}{0.3x} = \frac{600-x}{x}$$

Elastic:

$$E(x) = \frac{600-x}{x} > 1$$

$$600 - x > x$$

$$2x < 600$$

$$x < 300$$

Demand is elastic for $0 < x < 300$; demand is inelastic for $300 < x < 600$.

75. $p = g(x) = 90 - 0.1x^2,\ \ 0 < x < 30;\ \ g'(x) = -0.2x$

$$E(x) = \frac{-g(x)}{xg'(x)} = \frac{-(90-0.1x^2)}{-0.2x^2} = \frac{90-0.1x^2}{0.2x^2} = \frac{900-x^2}{2x^2}$$

Elastic:

$$E(x) = \frac{900-x^2}{2x^2} > 1$$

$$900 - x^2 > 2x^2$$

$$3x^2 < 900$$

$$x^2 < 300$$

$$x < 10\sqrt{3}$$

Demand is elastic for $0 < x < 10\sqrt{3}$; demand is inelastic for $10\sqrt{3} < x < 30$.

77. $x = f(p) = Ap^{-k}$, A, k positive constants

$f'(p) = -Akp^{-k-1}$

$$E(p) = \frac{-pf'(p)}{f(p)} = \frac{Akp^{-k}}{Ap^{-k}} = k$$

79. The company's daily cost is increasing by $2.50(30) = \$75$ per day.

81. $x + 400p = 3{,}000$

$x = f(p) = 3{,}000 - 400p$

$f'(p) = -400$

Elasticity of demand: $E(p) = \dfrac{400p}{3{,}000 - 400p} = \dfrac{2p}{15 - 2p}$

$E(3) = \dfrac{6}{9} = \dfrac{2}{3} < 1$

The demand is inelastic; a price increase will increase revenue.

83. $x + 1{,}000p = 2{,}500$

$x = f(p) = 2{,}500 - 1{,}000p$

$f'(p) = -1{,}000$

Elasticity of demand: $E(p) = \dfrac{1{,}000p}{2{,}500 - 1{,}000p} = \dfrac{2p}{50 - 2p}$

$E(0.99) = \dfrac{1.98}{5-1.98} \approx 0.66 < 1$

The demand is inelastic; a price decrease will decrease revenue.

85. From Problem 83, $R(p) = pf(p) = 3{,}000p - 400p^2$

$R'(p) = 3{,}000 - 800p$

Critical values: $R'(p) = 3{,}000 - 800p = 0$

$$800p = 3000$$
$$p = 3.75$$

$R''(p) = -800$

Since $p = 3.75$ is the only critical value and $R''(3.75) = -800 < 0$, the maximum revenue occurs when the price $p = \$3.75$.

87. $f(t) = 0.31t + 18.5,\ 0 \le t \le 50$

$f'(t) = 0.31$

Percentage rate of change:

$$100\frac{f'(t)}{f(t)} = \frac{31}{0.31t + 18.5}$$

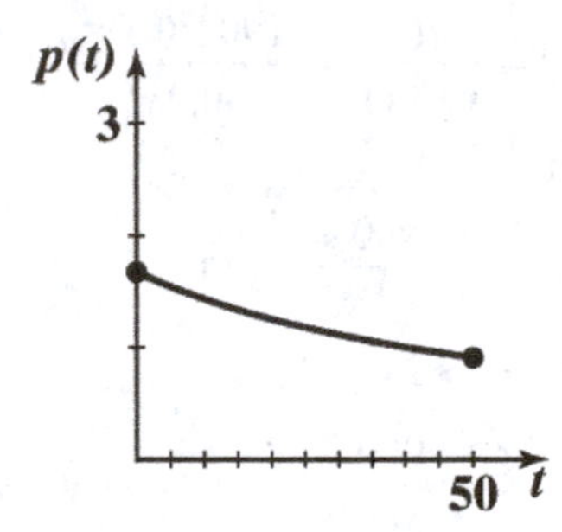

89. $r(t) = 3.3 - 0.7\ln t, \quad r'(t) = -\dfrac{0.7}{t}$

Relative rate of change of $r(t)$: $\dfrac{r'(t)}{r(t)} = \dfrac{\frac{-0.7}{t}}{3.3 - 0.7\ln t} = \dfrac{-0.7}{3.3t - 0.7t\ln t} = C(t)$.

Relative rate of change in 2020: $C(30) = \dfrac{-0.7}{3.3(30) - 0.7(30)\ln(30)} \approx -0.025$

The relative rate of change for robberies annually per 1,000 population is approximately -0.025.

CHAPTER 3 REVIEW

1. $A(t) = 2000e^{0.09t}$

$A(5) = 2000e^{0.09(5)} = 2000e^{0.45} \approx 3136.62$ or $3136.62

$A(10) = 2000e^{0.09(10)} = 2000e^{0.9} \approx 4919.21$ or $4919.21

$A(20) = 2000e^{0.09(20)} = 2000e^{1.8} \approx 12{,}099.29$ or $12,099.29 (3-1)

2. $f(x) = (6x+5)^{3/2}$. Let $y = E(u) = u^{3/2}$, $u = I(x) = 6x + 5$.

Then $y = E(u) = E[I(x)] = (6x+5)^{3/2}$. (3-4)

3. $f(x) = \ln(x^2 + 4)$. Let $y = E(u) = \ln u$, $u = I(x) = x^2 + 4$.

Then $y = E(u) = E[I(x)] = \ln(x^2 + 4)$. (3-4)

4. $f(x) = e^{0.02x}$. Let $y = E(u) = e^u$, $u = I(x) = 0.02x$. Then $y = E(u) = E[I(x)] = e^{0.02x}$. (3-4)

5. $\frac{d}{dx}(2\ln x + 3e^x) = 2\frac{d}{dx}\ln x + 3\frac{d}{dx}e^x = \frac{2}{x} + 3e^x$ (3-2)

6. $\frac{d}{dx}e^{2x-3} = e^{2x-3}\frac{d}{dx}(2x-3)$ (by the chain rule)

$= 2e^{2x-3}$ (3-4)

7. $y = \ln(2x+7)$

$y' = \frac{1}{2x+7}(2)$ (by the chain rule)

$= \frac{2}{2x+7}$ (3-4)

8. $f(x) = \ln(3+e^x)$

$f'(x) = \frac{1}{3+e^x}\cdot e^x$ (by the chain rule)

$= \frac{e^x}{3+e^x}$ (3-4)

9. $\frac{d}{dx}2y^2 - \frac{d}{dx}3x^3 - \frac{d}{dx}5 = \frac{d}{dx}(0)$

$4yy' - 9x^2 - 0 = 0$

$y' = \frac{9x^2}{4y}$

$\left.\frac{dy}{dx}\right|_{(1,2)} = \frac{9\cdot 1^2}{4\cdot 2} = \frac{9}{8}$ (3-5)

10. $y = 3x^2 - 5$

$\frac{dy}{dt} = \frac{d(3x^2)}{dt} - \frac{d(5)}{dt}$

$\frac{dy}{dt} = 6x\frac{dx}{dt}$

$x = 12;\ \frac{dx}{dt} = 3$

$\frac{dy}{dt} = 6\cdot 12\cdot 3 = 216$ (3-6)

11. $25p + x = 1{,}000$

(A) $x = 1{,}000 - 25p$

(B) $x = f(p) = 1{,}000 - 25p$

$f'(p) = -25$

$E(p) = -\frac{pf'(p)}{f(p)} = \frac{25p}{1{,}000-25p} = \frac{p}{40-p}$

(C) $E(15) = \frac{15}{40-15} = \frac{15}{25} = \frac{3}{5} = 0.6$

Demand is inelastic and insensitive to small changes in price.

(D) Revenue: $R(p) = pf(p) = 1{,}000p - 25p^2$

(E) From (B), $E(25) = \frac{25}{40-25} = \frac{25}{15} = \frac{5}{3} = 1.6$

Demand is elastic; a price cut will increase revenue. (3-7)

12. $y = 100e^{-0.1x}$

$y' = 100(-0.1)e^{-0.1x}$; $y'(0) = 100(-0.1) = -10$ (3-2)

13.

n	1000	100,000	10,000,000	100,000,000
$\left(1+\frac{2}{n}\right)^n$	7.374312	7.388908	7.389055	7.389056

$\lim_{n\to\infty}\left(1+\frac{2}{n}\right)^n \approx 7.38906$ (5 decimal places);

$\lim_{n\to\infty}\left(1+\frac{2}{n}\right)^n = e^2$ (3-1)

14. $\frac{d}{dz}[(\ln z)^7 + \ln z^7] = \frac{d}{dz}[\ln z]^7 + \frac{d}{dz}7\ln z = 7[\ln z]^6\frac{d}{dz}\ln z + 7\frac{d}{dz}\ln z = 7[\ln z]^6\frac{1}{z} + \frac{7}{z}$

$= \frac{7(\ln z)^6 + 7}{z} = \frac{7[(\ln z)^6 + 1]}{z}$ (3-4)

15. $\frac{d}{dx}x^6\ln x = x^6\frac{d}{dx}\ln x + (\ln x)\frac{d}{dx}x^6 = x^6\left(\frac{1}{x}\right) + (\ln x)6x^5 = x^5(1 + 6\ln x)$ (3-3)

16. $\frac{d}{dx}\left(\frac{e^x}{x^6}\right) = \frac{x^6\frac{d}{dx}e^x - e^x\frac{d}{dx}x^6}{(x^6)^2} = \frac{x^6e^x - 6x^5e^x}{x^{12}} = \frac{xe^x - 6e^x}{x^7} = \frac{e^x(x-6)}{x^7}$ (3-3)

17. $y = \ln(2x^3 - 3x)$

$y' = \frac{1}{2x^3 - 3x}(6x^2 - 3) = \frac{6x^2 - 3}{2x^3 - 3x}$ (3-4)

18. $f(x) = e^{x^3 - x^2}$

$f'(x) = e^{x^3 - x^2}(3x^2 - 2x)$

$= (3x^2 - 2x)e^{x^3 - x^2}$ (3-4)

19. $y = e^{-2x}\ln 5x$

$\frac{dy}{dx} = e^{-2x}\left(\frac{1}{5x}\right)(5) + (\ln 5x)(e^{-2x})(-2) = e^{-2x}\left(\frac{1}{x} - 2\ln 5x\right) = \frac{1 - 2x\ln 5x}{xe^{2x}}$ (3-4)

20. $f(x) = 1 + e^{-x}$

$f'(x) = e^{-x}(-1) = -e^{-x}$

An equation for the tangent line to the graph of f at $x = 0$ is:

$y - y_1 = m(x - x_1)$,

where $x_1 = 0$, $y_1 = f(0) = 1 + e^0 = 2$, and $m = f'(0) = -e^0 = -1$.

Thus, $y - 2 = -1(x - 0)$ or $y = -x + 2$.

An equation for the tangent line to the graph of f at $x = -1$ is:

$y - y_1 = m(x - x_1)$,

where $x_1 = -1$, $y_1 = f(-1) = 1 + e$, and $m = f'(-1) = -e$. Thus,

$y - (1 + e) = -e[x - (-1)]$ or $y - 1 - e = -ex - e$ and $y = -ex + 1$. (3-4)

21. $x^2 - 3xy + 4y^2 = 23$

Differentiate implicitly:

$$2x - 3(xy' + y \cdot 1) + 8yy' = 0$$
$$2x - 3xy' - 3y + 8yy' = 0$$
$$8yy' - 3xy' = 3y - 2x$$
$$(8y - 3x)y' = 3y - 2x$$
$$y' = \frac{3y - 2x}{8y - 3x}$$
$$y'\Big|_{(-1,2)} = \frac{3 \cdot 2 - 2(-1)}{8 \cdot 2 - 3(-1)} = \frac{8}{19}$$

[Slope at (–1, 2)] (3-5)

22. $x^3 - 2t^2x + 8 = 0$

Differentiate implicitly:

$$3x^2x' - (2t^2x' + x \cdot 4t) + 0 = 0$$
$$3x^2x' - 2t^2x' - 4xt = 0$$
$$(3x^2 - 2t^2)x' = 4xt$$
$$x' = \frac{4xt}{3x^2 - 2t^2}$$
$$x'\Big|_{(-2,2)} = \frac{4 \cdot 2 \cdot (-2)}{3(2^2) - 2(-2)^2} = \frac{-16}{12 - 8} = \frac{-16}{4} = -4$$

(3-5)

23. $x - y^2 = e^y$

Differentiate implicitly:

$$1 - 2yy' = e^y y'$$
$$1 = e^y y' + 2yy'$$
$$1 = y'(e^y + 2y)$$
$$y' = \frac{1}{e^y + 2y}$$
$$y'\Big|_{(1,0)} = \frac{1}{e^0 + 2 \cdot 0} = 1$$

(3-5)

24. $\ln y = x^2 - y^2$

Differentiate implicitly:

$$\frac{y'}{y} = 2x - 2yy'$$
$$y'\left(\frac{1}{y} + 2y\right) = 2x$$
$$y'\left(\frac{1 + 2y^2}{y}\right) = 2x$$
$$y' = \frac{2xy}{1 + 2y^2}$$
$$y'\Big|_{(1,1)} = \frac{2 \cdot 1 \cdot 1}{1 + 2(1)^2} = \frac{2}{3}$$

(3-5)

25. $A(t) = 400e^{0.049t}$, $A'(t) = 19.6e^{0.049t}$; logarithmic derivative: $\frac{A'(t)}{A(t)} = \frac{19.6e^{0.049t}}{400e^{0.049t}} = \frac{19.6}{400} = 0.049$

(3-7)

26. $f(p) = 100 - 3p$, $f'(p) = -3$; logarithmic derivative: $\frac{f'(p)}{f(p)} = \frac{-3}{100 - 3p}$. (3-7)

27. $f(x)=1+x^2,\ \ f'(x)=2x;$ logarithmic derivative: $\dfrac{f'(x)}{f(x)}=\dfrac{2x}{1+x^2}.$ (3-7)

28. $y^2-4x^2=12$

Differentiate with respect to t:

$$2y\frac{dy}{dt}-8x\frac{dx}{dt}=0$$

Given: $\dfrac{dx}{dt}=-2$ when $x=1$ and $y=4$. Therefore,

$$2\cdot 4\frac{dy}{dt}-8\cdot 1\cdot(-2)=0$$

$$8\frac{dy}{dt}+16=0$$

$$\frac{dy}{dt}=-2.$$

The y coordinate is decreasing at 2 units per second. (3-6)

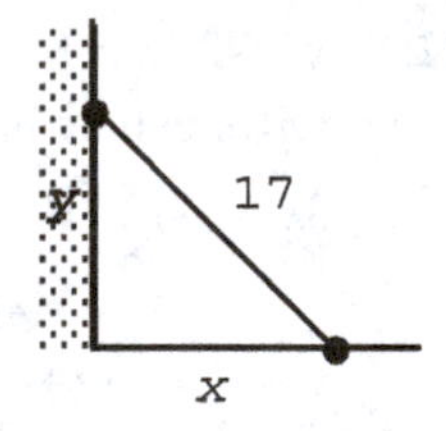

29. 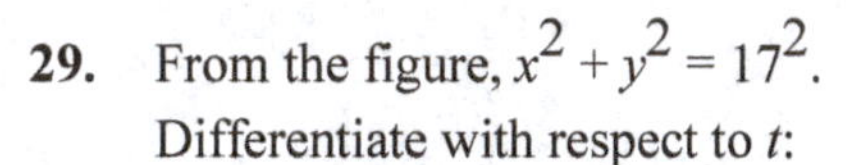 From the figure, $x^2+y^2=17^2$.

Differentiate with respect to t:

$$2x\frac{dx}{dt}+2y\frac{dy}{dt}=0 \quad \text{or} \quad x\frac{dx}{dt}+y\frac{dy}{dt}=0$$

We are given $\dfrac{dx}{dt}=-0.5$ feet per second. Therefore,

$$x(-0.5)+y\frac{dy}{dt}=0 \quad \text{or} \quad \frac{dy}{dt}=\frac{0.5x}{y}=\frac{x}{2y}$$

Now, when $x=8$, we have: $8^2+y^2=17^2$

$$y^2=289-64=225$$

$$y=15$$

Therefore, $\left.\dfrac{dy}{dt}\right|_{(8,15)}=\dfrac{8}{2(15)}=\dfrac{4}{15}\approx 0.27$ ft/sec. (3-5)

30. 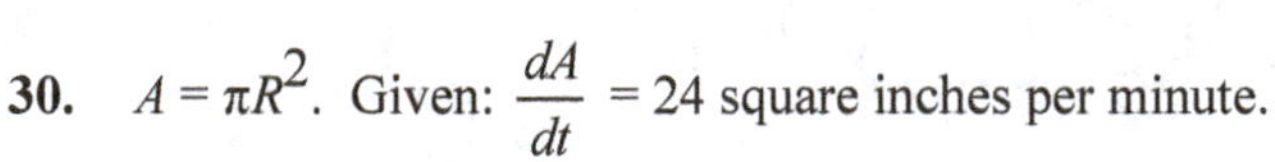 $A=\pi R^2$. Given: $\dfrac{dA}{dt}=24$ square inches per minute.

Differentiate with respect to t:

$$\frac{dA}{dt}=2\pi R\frac{dR}{dt}$$

$$24=2\pi R\frac{dR}{dt}$$

Therefore, $\dfrac{dR}{dt}=\dfrac{24}{2\pi R}=\dfrac{12}{\pi R}.$

$$\left.\frac{dR}{dt}\right|_{R=12}=\frac{12}{\pi\cdot 12}=\frac{1}{\pi}\approx 0.318 \text{ inches per minute}$$ (3-6)

31. $x=f(p)=20(p-15)^2 \quad 0\le p\le 15$

$f'(p)=40(p-15)$

$$E(p) = -\frac{pf'(p)}{f(p)} = \frac{-40p(p-15)}{20(p-15)^2} = \frac{-2p}{p-15}$$

Elastic: $E(p) = \dfrac{-2p}{p-15} > 1$

$-2p < p - 15$ ($p - 15 < 0$ reverses inequality)
$-3p < -15$
$p > 5;\quad 5 < p < 15$

Inelastic: $E(p) = \dfrac{-2p}{p-15} < 1$

$-2p > p - 15$ ($p - 15 < 0$ reverses inequality)
$-3p > -15$
$p < 5;\quad 0 < p < 5$ (3-7)

32. $x = f(p) = 5(20 - p)\ \ 0 \le p \le 20$
$R(p) = pf(p) = 5p(20 - p) = 100p - 5p^2$
$R'(p) = 100 - 10p = 10(10 - p)$
Critical values: $p = 10$
Sign chart for $R'(p)$:

```
R'(p)         + + + 0 - - -
         -------|-----|-----|-----> x
R(p)            0    10    20
            Increasing | Decreasing
Demand:     Inelastic  | Elastic
```

Test Numbers

p	$R'(p)$
5	50(+)
15	−50(−)

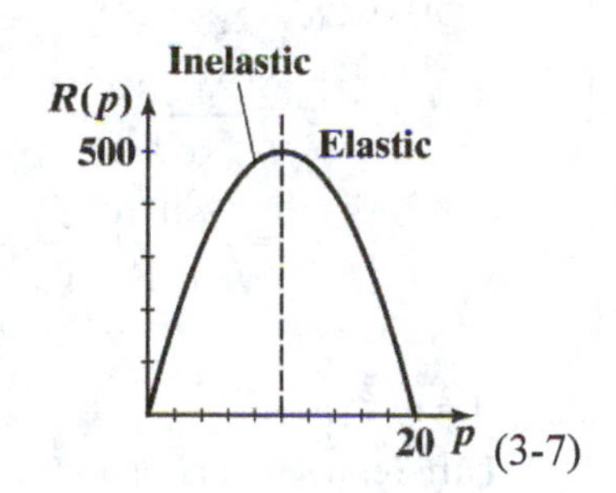

(3-7)

33. $y = w^3,\ w = \ln u,\ u = 4 - e^x$

(A) $y = [\ln(4 - e^x)]^3$

(B) $\dfrac{dy}{dx} = \dfrac{dy}{dw}\cdot\dfrac{dw}{du}\cdot\dfrac{du}{dx}$

$$= 3w^2 \cdot \frac{1}{u} \cdot (-e^x) = 3[\ln(4 - e^x)]^2\left(\frac{1}{4-e^x}\right)(-e^x) = \frac{-3e^x[\ln(4-e^x)]^2}{4-e^x}$$ (3-4)

34. $y = 5^{x^2-1}$

$y' = 5^{x^2-1}(\ln 5)(2x) = 2x5^{x^2-1}(\ln 5)$ (3-4)

35. $\dfrac{d}{dx}\log_5(x^2 - x) = \dfrac{1}{x^2-x}\cdot\dfrac{1}{\ln 5}\cdot\dfrac{d}{dx}(x^2 - x) = \dfrac{1}{\ln 5}\cdot\dfrac{2x-1}{x^2-x}$ (3-4)

36. $$\frac{d}{dx}\sqrt{\ln(x^2+x)} = \frac{d}{dx}[\ln(x^2+x)]^{1/2} = \frac{1}{2}[\ln(x^2+x)]^{-1/2}\frac{d}{dx}\ln(x^2+x)$$
$$= \frac{1}{2}[\ln(x^2+x)]^{-1/2}\frac{1}{x^2+x}\frac{d}{dx}(x^2+x)$$
$$= \frac{1}{2}[\ln(x^2+x)]^{-1/2}\cdot\frac{2x+1}{x^2+x} = \frac{2x+1}{2(x^2+x)[\ln(x^2+x)]^{1/2}} \quad (3\text{-}4)$$

37. $e^{xy} = x^2 + y + 1$

Differentiate implicitly:

$$\frac{d}{dx}e^{xy} = \frac{d}{dx}x^2 + \frac{d}{dx}y + \frac{d}{dx}1$$
$$e^{xy}(xy' + y) = 2x + y'$$
$$xe^{xy}y' - y' = 2x - ye^{xy}$$
$$y' = \frac{2x - ye^{xy}}{xe^{xy} - 1}$$
$$y'\Big|_{(0,0)} = \frac{2\cdot 0 - 0\cdot e^0}{0\cdot e^0 - 1} = 0 \quad (3\text{-}5)$$

38. $A = \pi r^2,\ r \ge 0$

Differentiate with respect to t:

$$\frac{dA}{dt} = 2\pi r\frac{dr}{dt} = 6\pi r \text{ since } \frac{dr}{dt} = 3$$

The area increases at the rate $6\pi r$. This is smallest when $r = 0$; there is no largest value. (3-6)

39. $y = x^3$

Differentiate with respect to t:

$$\frac{dy}{dt} = 3x^2\frac{dx}{dt}$$

Solving for $\frac{dx}{dt}$, we get

$$\frac{dx}{dt} = \frac{1}{3x^2}\cdot\frac{dy}{dt} = \frac{5}{3x^2} \text{ since } \frac{dy}{dt} = 5$$

To find where $\frac{dx}{dt} > \frac{dy}{dt}$, solve the inequality

$$\frac{5}{3x^2} > 5$$
$$\frac{1}{3x^2} > 1$$
$$3x^2 < 1$$
$$-\frac{1}{\sqrt{3}} < x < \frac{1}{\sqrt{3}} \text{ or } \frac{-\sqrt{3}}{3} < x < \frac{\sqrt{3}}{3} \quad (3\text{-}6)$$

40. (A) The compound interest formula is: $A = P(1 + r)^t$. Thus, the time for P to double when $r = 0.05$ and interest is compounded annually can be found by solving

$$2P = P(1 + 0.05)^t \text{ or } 2 = (1.05)^t \text{ for } t.$$

$$\ln(1.05)^t = \ln 2$$

$$t \ln(1.05) = \ln 2$$

$$t = \frac{\ln 2}{\ln(1.05)} \approx 14.2 \text{ or } 15 \text{ years}$$

(B) The continuous compound interest formula is: $A = Pe^{rt}$. Proceeding as above, we have

$$2P = Pe^{0.05t} \text{ or } e^{0.05t} = 2.$$

Therefore, $0.05t = \ln 2$ and $t = \frac{\ln 2}{.05} \approx 13.9$ years (3-1)

41. $A(t) = 100e^{0.1t}$

$A'(t) = 100(0.1)e^{0.1t} = 10e^{0.1t}$

$A'(1) = 11.05$ or \$11.05 per year

$A'(10) = 27.18$ or \$27.18 per year (3-1)

42. $P(t) = 12{,}000e^{0.0395t}$. First solve $12{,}000e^{0.0395t} = 25{,}000$ for t:

$$e^{0.0395t} = \frac{25{,}000}{12{,}000} = \frac{25}{12}$$

$$0.0395t = \ln(25/12)$$

$$t = \frac{\ln(25/12)}{0.0395} \approx 18.58$$

$P'(18.58) = 474e^{0.0395(18.58)} \approx 987.50$; \$987.50 per year. (3-2)

43. $R(x) = xp(x) = 1000xe^{-0.02x}$

$R'(x) = 1000[xD_x e^{-0.02x} + e^{-0.02x}D_x x]$

$= 1000[x(-0.02)e^{-0.02x} + e^{-0.02x}] = (1000 - 20x)e^{-0.02x}$ (3-4)

44. $x = \sqrt{5000 - 2p^3} = (5000 - 2p^3)^{1/2}$

Differentiate implicitly with respect to x:

$$1 = \frac{1}{2}(5000 - 2p^3)^{-1/2}(-6p^2)\frac{dp}{dx}$$

$$1 = \frac{-3p^2}{(5000 - 2p^3)^{1/2}}\frac{dp}{dx}$$

$$\frac{dp}{dx} = \frac{-(5000 - 2p^3)^{1/2}}{3p^2} \quad (3\text{-}5)$$

45. Given: $R(x) = 750x - \frac{x^2}{30}$ and $\frac{dx}{dt} = 3$ when $x = 40$.

Differentiate with respect to t:

$$\frac{dR}{dt} = 750\frac{dx}{dt} - \frac{1}{30}(2x)\frac{dx}{dt} = 750\frac{dx}{dt} - \frac{x}{15}\cdot\frac{dx}{dt}$$

Thus, $\left.\frac{dR}{dt}\right|_{x=40 \text{ and } \frac{dx}{dt}=3} = 750(3) - \frac{40}{15}\cdot 3 = \$2,242$ (3-6)

46. $p = 38.2 - 0.002x$

$x = f(p) = \frac{38.2}{0.002} - \frac{1}{0.002}p = 19,100 - 500p$

$f'(p) = -500$

Elasticity of demand: $E(p) = \frac{-pf'(p)}{f(p)} = \frac{500p}{19,100 - 500p} = \frac{5p}{191 - 5p}$

$E(21) = \frac{105}{191 - 105} = \frac{105}{86} > 1$

Demand is elastic, a (small) price decrease will increase revenue. (3-7)

47. $f(t) = 1,700t + 20,500$

$f'(t) = 1,700$

Relative rate of change: $\frac{f'(t)}{f(t)} = \frac{1,700}{1,700t + 20,500}$

Relative rate of change at $t = 35$: $\frac{1,700}{1,700(35) + 20,500} \approx 0.02125$ (3-7)

48. $C(t) = 5e^{-0.3t}$

$C'(t) = 5e^{-0.3t}(-0.3) = -1.5e^{-0.3t}$

After one hour, the rate of change of concentration is

$C'(1) = -1.5e^{-0.3(1)} = -1.5e^{-0.3} \approx -1.111$ mg/ml per hour.

After five hours, the rate of change of concentration is

$C'(5) = -1.5e^{-0.3(5)} = -1.5e^{-1.5} \approx -0.335$ mg/ml per hour. (3-4)

49. Given: $A = \pi R^2$ and $\frac{dA}{dt} = -45$ mm^2 per day (negative because the area is decreasing).

Differentiate with respect to t:

$$\frac{dA}{dt} = \pi 2R\frac{dR}{dt}$$

$$-45 = 2\pi R\frac{dR}{dt}$$

$$\frac{dR}{dt} = -\frac{45}{2\pi R}$$

$$\left.\frac{dR}{dt}\right|_{R=15} = \frac{-45}{2\pi\cdot 15} = \frac{-3}{2\pi} \approx -0.477 \text{ mm per day}$$ (3-6)

50. $N(t) = 10(1 - e^{-0.4t})$

(A) $N'(t) = -10e^{-0.4t}(-0.4) = 4e^{-0.4t}$

$N'(1) = 4e^{-0.4(1)} = 4e^{-0.4} \approx 2.68.$

Thus, learning is increasing at the rate of 2.68 units per day after 1 day.

$N'(5) = 4e^{-0.4(5)} = 4e^{-2} = 0.54$

Thus, learning is increasing at the rate of 0.54 units per day after 5 days.

(B) We solve $N'(t) = 0.25 = 4e^{-0.4t}$ for t:

$$e^{-0.4t} = \frac{0.25}{4} = 0.0625$$

$$-0.4t = \ln(0.0625)$$

$$t = \frac{\ln(0.0625)}{-0.4} \approx 6.93$$

The rate of learning is less than 0.25 after 7 days. (3-4)

51. Given: $T = 2\left(1 + \frac{1}{x^{3/2}}\right) = 2 + 2x^{-3/2}$, and $\frac{dx}{dt} = 3$ when $x = 9$.

Differentiate with respect to t:

$$\frac{dT}{dt} = 0 + 2\left(-\frac{3}{2}x^{-5/2}\right)\frac{dx}{dt} = -3x^{-5/2}\frac{dx}{dt}$$

$$\left.\frac{dT}{dt}\right|_{x=9 \text{ and } \frac{dx}{dt}=3} = -3(9)^{-5/2}(3) = -3 \cdot 3^{-5} \cdot 3 = -3^{-3} = \frac{-1}{27} \approx -0.037 \text{ minutes per operation}$$

per hour. (3-6)

4 GRAPHING AND OPTIMIZATION

EXERCISE 4-1

Things to remember:

1. INCREASING AND DECREASING FUNCTIONS

   For the interval (a, b):

$f'(x)$	$f(x)$	Graph of f	Examples
+	Increases ↗	Rises ↗	
-	Decreases ↘	Falls ↘	

2. CRITICAL VALUES

   The values of x in the domain of f where $f'(x) = 0$ or where $f'(x)$ does not exist are called the CRITICAL VALUES of f.

   The critical values of f are always in the domain of f and are also partition numbers for f', but f' may have partition numbers that are not critical values.

   If f is a polynomial, then both the partition numbers for f' and the critical values of f are the solutions of $f'(x) = 0$.

3. LOCAL EXTREMA

   Given a function f. The value $f(c)$ is a LOCAL MAXIMUM of f if there is an interval (m, n) containing c such that $f(x) \leq f(c)$ for all x in (m, n). The value $f(e)$ is a LOCAL MINIMUM of f if there is an interval (p, q) containing e such that $f(x) \geq f(e)$ for all x in (p, q). Local maxima and local minima are called LOCAL EXTREMA.

   A point on the graph where a local extremum occurs is also called a TURNING POINT.

4. EXISTENCE OF LOCAL EXTREMA

   If f is continuous on the interval (a, b), c is a number in (a, b) and $f(c)$ is a local extremum, then either $f'(c) = 0$ or $f'(c)$ does not exist (is not defined).

5. FIRST DERIVATIVE TEST FOR LOCAL EXTREMA

   Let c be a critical value of f [$f(c)$ is defined and either $f'(c) = 0$ or $f'(c)$ is not defined.]

   Construct a sign chart for $f'(x)$ close to and on either side of c.

Sign Chart	$f(c)$
$f'(x)$ – – – ¦ + + +, from m to n across c on the x-axis $f(x)$ Decreasing ¦ Increasing	$f(c)$ is a local minimum. If $f'(x)$ changes from negative to positive at c, then $f(c)$ is a local minimum.
$f'(x)$ + + + ¦ – – –, from m to n across c on the x-axis $f(x)$ Increasing ¦ Decreasing	$f(c)$ is a local maximum. If $f'(x)$ changes from positive to negative at c, then $f(c)$ is a local maximum.
$f'(x)$ – – – ¦ – – –, from m to n across c on the x-axis $f(x)$ Decreasing ¦ Decreasing	$f(c)$ is not a local extremum. If $f'(x)$ does not change sign at c, then $f(c)$ is neither a local maximum nor a local minimum.
$f'(x)$ + + + ¦ + + +, from m to n across c on the x-axis $f(x)$ Increasing ¦ Increasing	$f(c)$ is not a local extremum. If $f'(x)$ does not change sign at c, then $f(c)$ is neither a local maximum nor a local minimum.

6. INTERCEPTS AND LOCAL EXTREMA FOR POLYNOMIAL FUNCTIONS

If $f(x) = a_n x^n + a_{n-1}x^{n-1} + \dots + a_1 x + a_0$, $a_n \neq 0$ is an nth degree polynomial then f has at most n x-intercepts and at most $n - 1$ local extrema.

---

1. $g(x) = |x|$ on $(-\infty, 0)$: decreasing

3. $f(x) = x$ on $(-\infty, \infty)$: increasing

5. $p(x) = \sqrt[3]{x}$ on $(-\infty, 0)$: increasing

7. $r(x) = 4 - \sqrt{x}$ on $(0, \infty)$: decreasing

9. (a, b), (d, f), (g, h)

11. (b, c), (c, d), (f, g)

13. $x = c, d, f$

15. $x = b, f$

17. f has a local maximum at $x = a$, and a local minimum at $x = c$; f does not have a local extremum at $x = b$ or at $x = d$.

19. $f(3) = 5$ is a local maximum; (e)

21. No local extrema; (d)

23. $f(3) = 5$ is a local maximum; (f)

25. No local extrema; (c)

27. $f(x) = x^3 - 12x + 8$

(A) $f'(x) = 3x^2 - 12$

(B) Critical values:
$$f'(x) = 0$$
$$3x^2 - 12 = 0$$
$$x^2 = 4$$
$$x = -2,\ 2$$

(C) Partition numbers: $x = -2,\ 2$

29. $f(x)=\frac{6}{x+2}=6(x+2)^{-1}$

(A) $f'(x)=(-1)6(x+2)^{-2}=\frac{-6}{(x+2)^2}$

(B) Critical values: There are no critical values ($x=-2$ is not a critical value since -2 is not in the domain of f.)

(C) Partition numbers: $x=-2$

31. $f(x)=|x|=\begin{cases} x, & x\ge 0 \\ -x, & x<0 \end{cases}$

(A) $f'(x)=\begin{cases} 1, & x>0 \\ -1, & x<0 \end{cases}$

(B) Critical values: $x=0$ ($f'(0)$ does not exist)

(C) Partition numbers: $x=0$.

33. $f(x)=2x^2-4x$; domain of f: $(-\infty,\infty)$
$f'(x)=4x-4$; f' is continuous for all x.
$f'(x)=4x-4=0$
$x=1$
Thus, $x=1$ is a partition number for f', and since 1 is in the domain of f, $x=1$ is a critical value of f.
Sign chart for f':

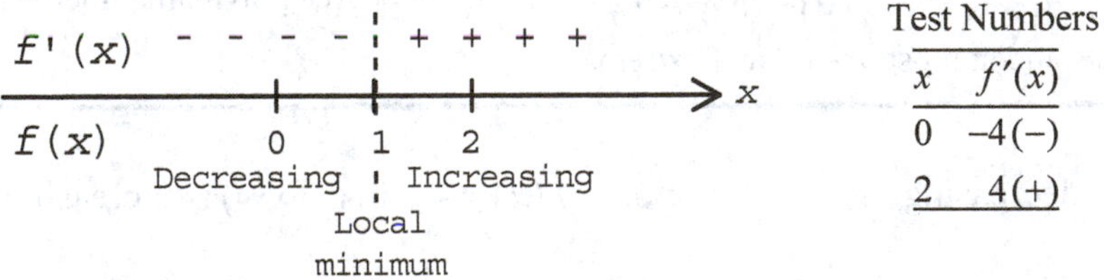

Test Numbers

x	$f'(x)$
0	$-4(-)$
2	$4(+)$

Therefore, f is decreasing on $(-\infty, 1)$; f is increasing on $(1,\infty)$;
$f(1)=-2$ is a local minimum.

35. $f(x)=-2x^2-16x-25$; domain of f: $(-\infty,\infty)$
$f'(x)=-4x-16$; f' is continuous for all x and
$f'(x)=-4x-16=0$
$x=-4$
Thus, $x=-4$ is a partition number for f', and since -4 is in the domain of f, $x=-4$ is a critical value for f.
Sign chart for f':

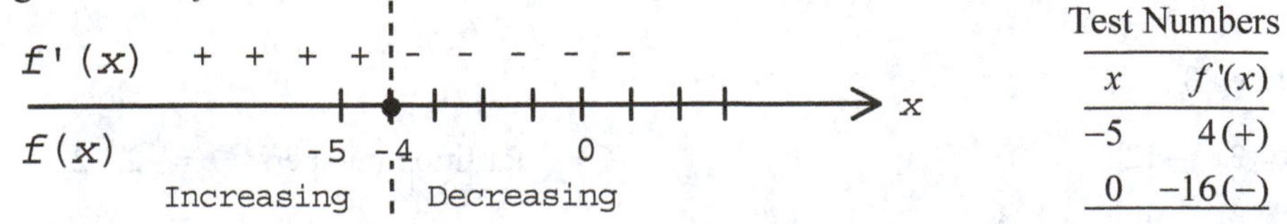

Test Numbers

x	$f'(x)$
-5	$4(+)$
0	$-16(-)$

Therefore, f is increasing on $(-\infty,-4)$; f is decreasing on $(-4,\infty)$; f has a local maximum at $x=-4$.

37. $f(x)=x^3+4x-5$; domain of f: $(-\infty,\infty)$
$f'(x)=3x^2+4\ge 4>0$ for all x; f is increasing on $(-\infty,\infty)$; f has no local extrema.

39. $f(x) = 2x^3 - 3x^2 - 36x$; domain of f: $(-\infty, \infty)$

$f'(x) = 6x^2 - 6x - 36$; f is continuous for all x and

$$f'(x) = 6(x^2 - x - 6) = 0$$
$$6(x-3)(x+2) = 0$$
$$x = -2, 3$$

The partition numbers for f are $x = -2$ and $x = 3$. Since -2 and 3 are in the domain of f, $x = -2, x = 3$ are critical values for f.

Sign chart for f':

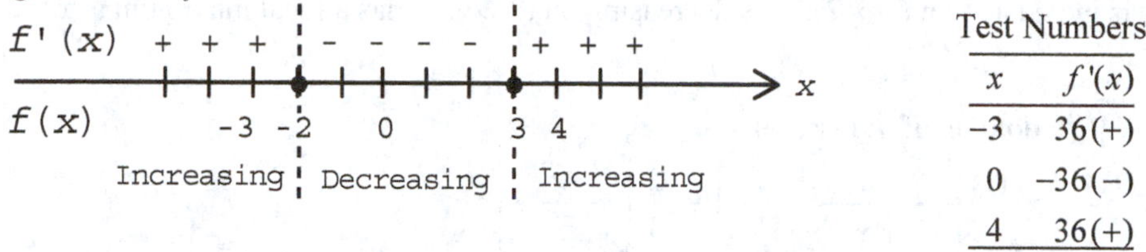

Test Numbers

x	$f'(x)$
-3	$36(+)$
0	$-36(-)$
4	$36(+)$

Therefore, f is increasing on $(-\infty, -2)$ and $(3, \infty)$; f is decreasing on $(-2, 3)$; f has a local maximum at $x = -2$ and a local minimum at $x = 3$.

41. $f(x) = 3x^4 - 4x^3 + 5$; domain of f: $(-\infty, \infty)$

$f'(x) = 12x^3 - 12x^2$; f' is continuous for all x.

$$f'(x) = 12x^3 - 12x^2 = 0$$
$$12x^2(x-1) = 0$$
$$x = 0, 1$$

The partition numbers for f' are $x = 0$ and $x = 1$. Since 0 and 1 are in the domain of f, $x = 0, x = 1$ are critical values of f.

Sign chart for f':

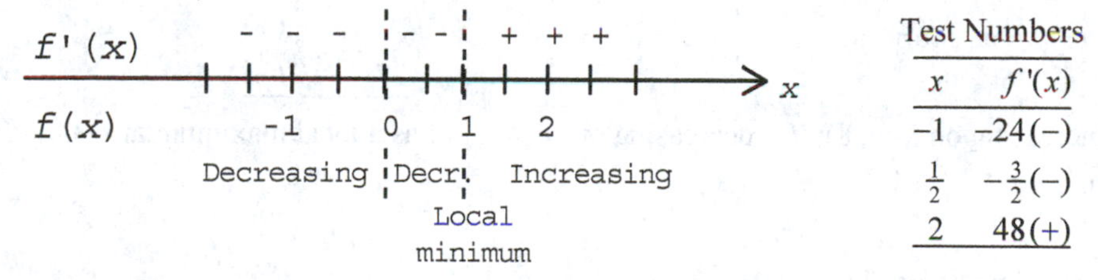

Test Numbers

x	$f'(x)$
-1	$-24(-)$
$\frac{1}{2}$	$-\frac{3}{2}(-)$
2	$48(+)$

Therefore, f is decreasing on $(-\infty, 1)$; f is increasing on $(1, \infty)$; $f(1) = 4$ is a local minimum.

43. $f(x) = (x-1)e^{-x}$; domain of f: $(-\infty, \infty)$

$f'(x) = (x-1)e^{-x}(-1) + e^{-x} = 2e^{-x} - xe^{-x} = (2-x)e^{-x}$

f' is continuous for all x and

$$f'(x) = (2-x)e^{-x} = 0$$
$$2 - x = 0$$
$$x = 2$$

$x = 2$ is a partition number for f'. Since 2 is in the domain of f, $x = 2$ is a critical value of f.

Sign chart for f':

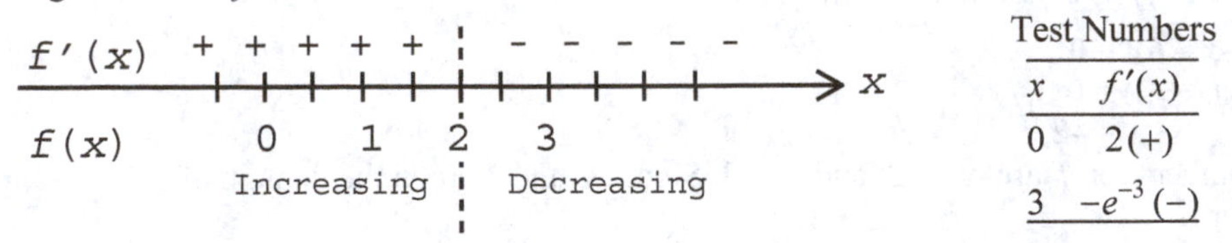

Test Numbers

x	$f'(x)$
0	2(+)
3	$-e^{-3}$ (–)

Therefore, f is increasing on $(-\infty, 2)$; f is decreasing on $(2, \infty)$; f has a local maximum at $x = 2$.

45. $f(x) = 4x^{1/3} - x^{2/3}$; domain of f: $(-\infty, \infty)$

$$f'(x) = \frac{4}{3}x^{-2/3} - \frac{2}{3}x^{-1/3} = \frac{2}{3}\left[\frac{2}{x^{2/3}} - \frac{1}{x^{1/3}}\right] = \frac{2}{3}\left[\frac{2 - x^{1/3}}{x^{2/3}}\right]$$

f' is continuous for all x except $x = 0$.

$$f'(x) = \frac{2}{3}\left[\frac{2 - x^{1/3}}{x^{2/3}}\right] = 0$$

$$2 - x^{1/3} = 0$$

$$x = 8$$

$x = 0$ and $x = 8$ are partition numbers for f'. Since 0 and 8 are in the domain of f, $x = 0$, $x = 8$ are critical values of f ($f'(0)$ does not exist, $f'(8) = 0$).

Sign chart for f':

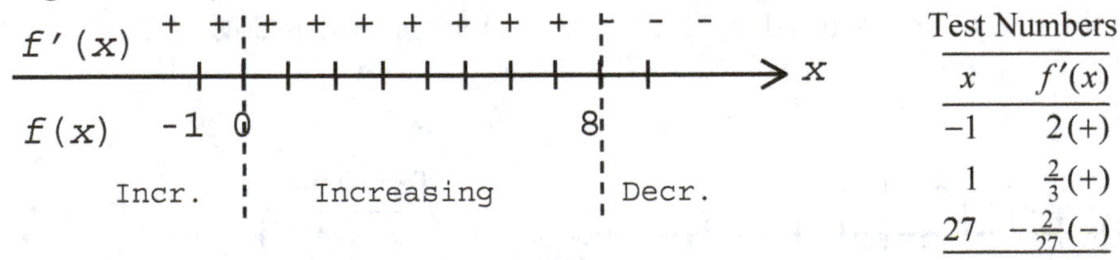

Test Numbers

x	$f'(x)$
–1	2(+)
1	$\frac{2}{3}$(+)
27	$-\frac{2}{27}$(–)

Therefore, f is increasing on $(-\infty, 8)$; f is decreasing on $(8, \infty)$; f has a local maximum at $x = 8$; $f(0)$ is not a local extremum.

47. $f(x) = x^4 - 4x^3 + 9x$; domain of f: $(-\infty, \infty)$

$f'(x) = 4x^3 - 12x^2 + 9$; f' is continuous for all x. Using a root-approximation routine, $f'(x) = 0$ at $x = -0.77$, $x = 1.08$, and $x = 2.69$; critical values.

Sign chart for f':

```
f'(x)  - - | + + + | - - - | + + +
-----------•-------+-------•--------•-------> x
f(x)     -0.77     0     1.08     2.69
     Decr. | Incr.       | Decr. | Incr.
```

f is decreasing on $(-\infty, -0.77)$ and $(1.08, 2.69)$; increasing on $(-0.77, 1.08)$ and $(2.69, \infty)$; f has a local minima at $x = -0.77$ and $x = 2.69$, f has a local maximum at $x = 1.08$.

49. $f(x) = x \ln x - (x - 2)^3$; domain of f: $(0, \infty)$

$f'(x) = 1 + \ln x - 3(x - 2)^2$; f' is continuous on $(0, \infty)$.

Using a root-approximation routine, $f'(x) = 0$ at $x = 1.34$ and $x = 2.82$; critical values.

Sign chart for f':

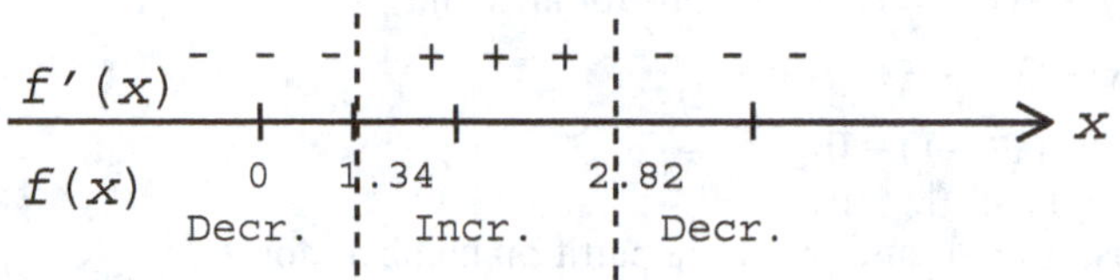

f is decreasing on (0, 1.34) and (2.82, ∞); f is increasing on (1.34, 2.82); f has a local minimum at $x = 1.34$; f has a local maximum at $x = 2.82$.

51. $f(x) = e^x - 2x^2$; domain of f: (–∞, ∞)

$f'(x) = e^x - 4x$; f' is continuous for all x.

Using a root-approximation routine, $f'(x) = 0$ at $x = 0.36$ and $x = 2.15$; critical values.

Sign chart for f':

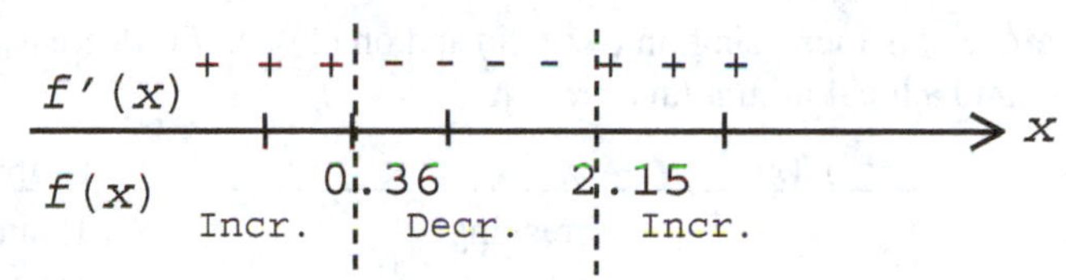

f is increasing on (–∞, 0.36) and (2.15, ∞); f is decreasing on (0.36, 2.15); f has a local maximum at $x = 0.36$; f has a local minimum at $x = 2.15$.

53. $f(x) = 4 + 8x - x^2$

$f'(x) = 8 - 2x$

f' is continuous for all x and

$f'(x) = 8 - 2x = 0$

$x = 4$

Thus, $x = 4$ is a partition number for f.

The sign chart for f' is:

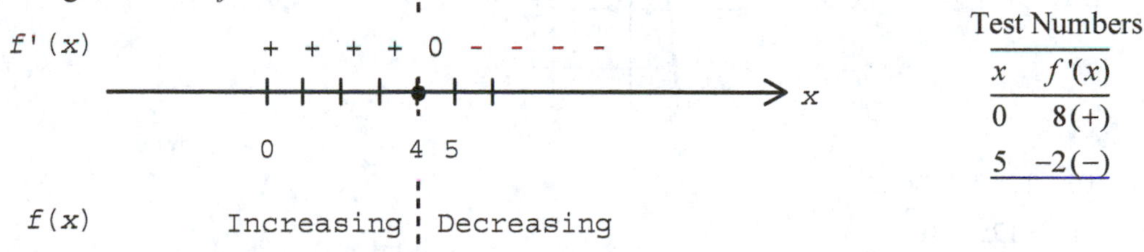

Test Numbers

x	$f'(x)$
0	8(+)
5	–2(–)

Therefore, f is increasing on (–∞, 4) and decreasing on (4, ∞); f has a local maximum at $x = 4$.

x	$f'(x)$	f	GRAPH OF f
(–∞, 4)	+	Increasing	Rising
$x = 4$	0	Local maximum	Horizontal tangent
(4, ∞)	–	Decreasing	Falling

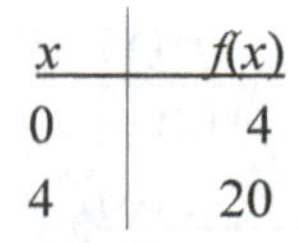

x	$f(x)$
0	4
4	20

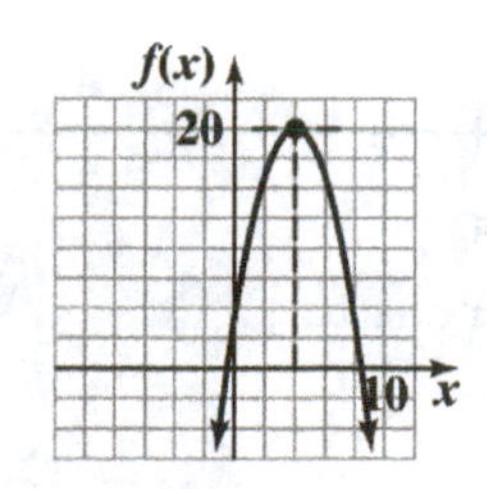

55. $f(x) = x^3 - 3x + 1$

$f'(x) = 3x^2 - 3$ is continuous for all x and

$f'(x) = 3x^2 - 3 = 0$

$3(x^2 - 1) = 0$

$3(x + 1)(x - 1) = 0$

Thus, $x = -1$ and $x = 1$ are partition numbers for f'.

The sign chart for f' is:

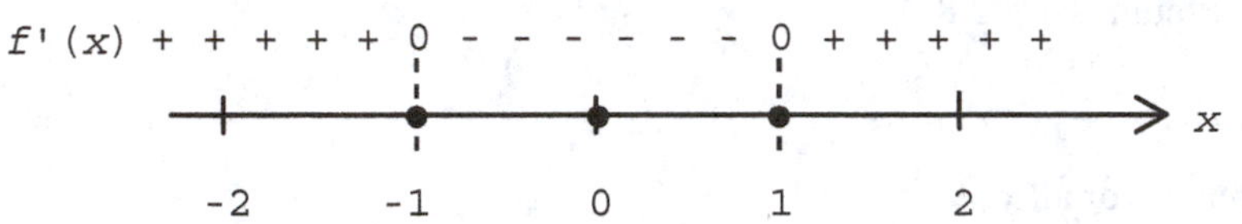

Test Numbers

x	$f(x)$
−2	9 (+)
0	−3 (-)
2	9 (+)

Therefore, f is increasing on $(-\infty, -1)$ and on $(1, \infty)$, f is decreasing on $(-1, 1)$; f has a local maximum at $x = -1$ and a local minimum at $x = 1$.

x	$f'(x)$	f	Graph of f
$(-\infty, -1)$	+	Increasing	Rising
$x = -1$	0	Local maximum	Horizontal tangent
$(-1, 1)$	-	Decreasing	Falling
$x = 1$	0	Local minimum	Horizontal tangent
$(1, \infty)$	+	Increasing	Rising

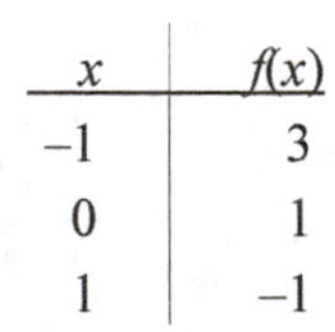

x	$f(x)$
−1	3
0	1
1	−1

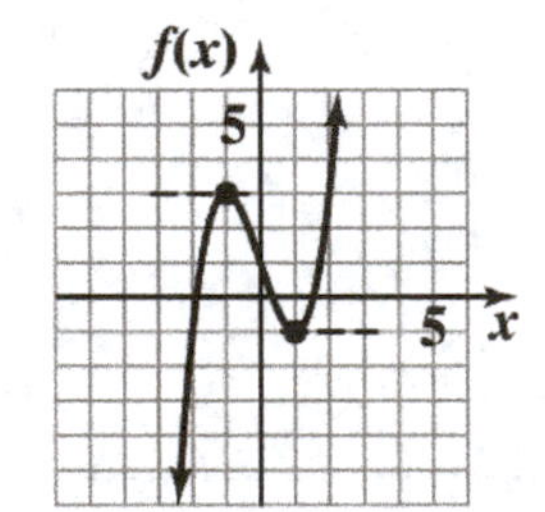

57. $f(x) = 10 - 12x + 6x^2 - x^3$

$f'(x) = -12 + 12x - 3x^2$

f' is continuous for all x and

$f'(x) = -12 + 12x - 3x^2 = 0$

$-3(x^2 - 4x + 4) = 0$

$-3(x - 2)^2 = 0$

Thus, $x = 2$ is a partition number for f'.

The sign chart for f' is:

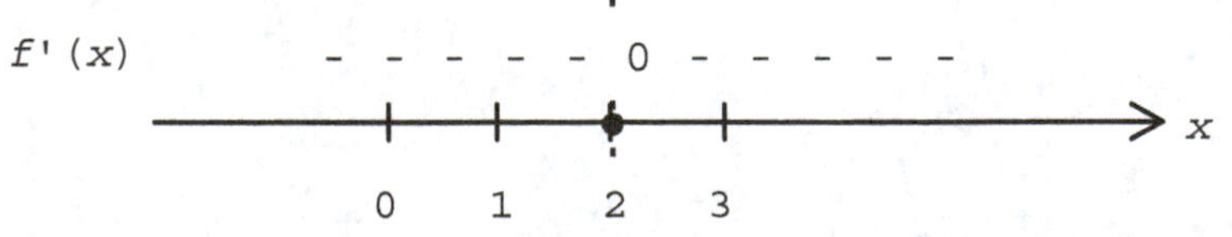

Test Numbers

x	$f'(x)$
0	−12(−)
3	−3(−)

Therefore, f is decreasing for all x, i.e., on $(-\infty, \infty)$, and there is a horizontal tangent line at $x = 2$.

x	$f'(x)$	f	GRAPH of f
$(-\infty, 2)$	-	Decreasing	Falling
$x = 2$	0		Horizontal tangent
$x > 2$	-	Decreasing	Falling

x	$f(x)$
0	10
2	2

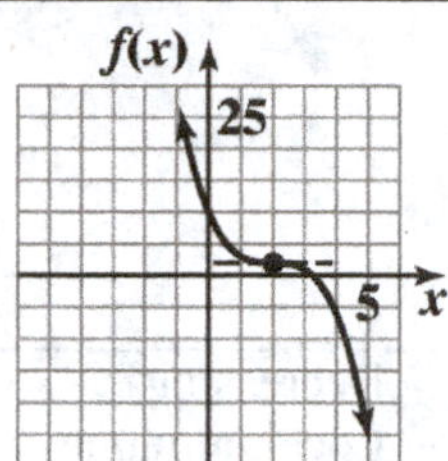

59. $f(x) = x^4 - 18x^2$

$f'(x) = 4x^3 - 36x$

f' is continuous for all x and

$f'(x) = 4x^3 - 36x = 0$

$4x(x^2 - 9) = 0$

$4x(x - 3)(x + 3) = 0$

Thus, $x = -3$, $x = 0$, and $x = 3$ are partition numbers for f'.

Sign chart for f':

```
f'(x)   - - - 0 + + + 0 - - - 0 + + +
        ------+-------+-------+------> x
f(x)         -3       0       3
       Decr.  | Incr. | Decr. | Incr.
```

Test Numbers

x	$f'(x)$
-4	$-112(-)$
-1	$32(+)$
1	$-32(-)$
4	$112(+)$

Therefore, f is increasing on $(-3, 0)$ and on $(3, \infty)$; f is decreasing on $(-\infty, -3)$ and on $(0, 3)$; f has a local maximum at $x = 0$ and local minima at $x = -3$ and $x = 3$.

x	$f'(x)$	f	GRAPH of f
$(-\infty, -3)$	-	Decreasing	Falling
$x = -3$	0	Local minimum	Horizontal tangent
$(-3, 0)$	+	Increasing	Rising
$x = 0$	0	Local maximum	Horizontal tangent
$(0, 3)$	-	Decreasing	Falling
$x = 3$	0	Local minimum	Horizontal tangent
$(3, \infty)$	+	Increasing	Rising

x	$f(x)$
0	0
-3	-81
3	-81

61.

x	$f'(x)$	$f(x)$	GRAPH of f
$(-\infty, -1)$	-	Increasing	Rising
$x = -1$	0	Neither local maximum nor local minimum	Horizontal tangent
$(-1, 1)$	+	Increasing	Rising
$x = 1$	0	Local maximum	Horizontal tangent
$(1, \infty)$	-	Decreasing	Falling

Using this information together with the points (–2, –1), (–1, 1), (0, 2), (1, 3), (2, 1) on the graph, we have

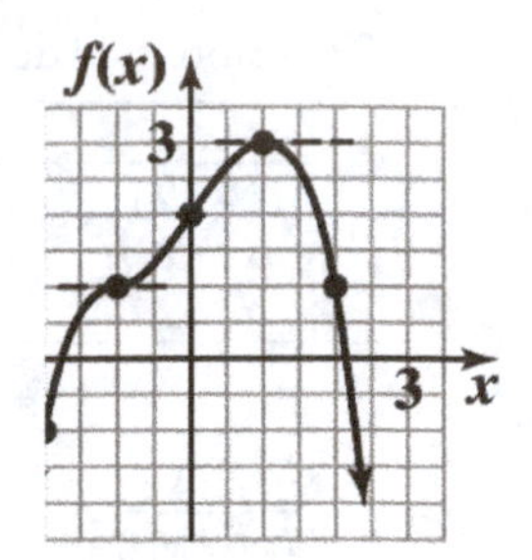

63.

x	$f'(x)$	$f(x)$	GRAPH of $f(x)$
$(-\infty, -1)$	-	Decreasing	Falling
$x = -1$	0	Local minimum	Horizontal tangent
$(-1, 0)$	+	Increasing	Rising
$x = 0$	Not defined	Local maximum	Vertical tangent line
$(0, 2)$	-	Decreasing	Falling
$x = 2$	0	Neither local maximum nor local minimum	Horizontal tangent
$(2, \infty)$	-	Decreasing	Falling

Using this information together with the points (–2, 2), (–1, 1), (0, 2), (2, 1), (4, 0) on the graph, we have

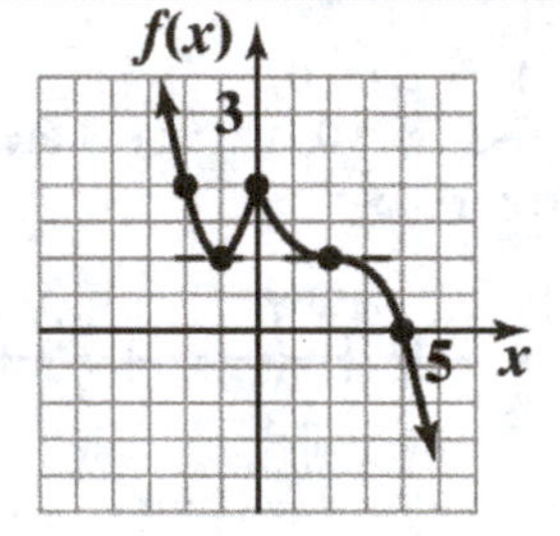

65.

```
f'(x)   + + + + 0 - - - - 0 - - - - 0 + + + + +
        ---------------------------------------> x
f(x)            -2        0         2
  Increasing     ¦                  ¦Increasing
                 Decreasing Decreasing
```

x	-2	0	2
f(x)	4	0	-4

67.

```
f'(x)   + + + + 0 - - - ND - - - 0 + + + + +
        -----------------------------------> x
f(x)            -1       0       1
  Increasing¦ Decr. ¦ Decr. ¦ Increasing
```

x	-1	0	1
f(x)	2	0	2

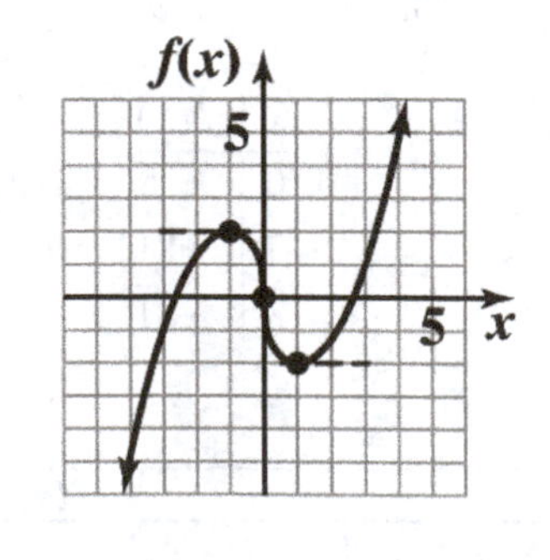

69. $f_1' = g_4$ **71.** $f_3' = g_6$ **73.** $f_5' = g_2$

75. Increasing on $(-1, 2)$ $[f'(x) > 0]$; decreasing on $(-\infty, -1)$ and on $(2, \infty)$ $[f'(x) < 0]$; local minimum at $x = -1$; local maximum at $x = 2$.

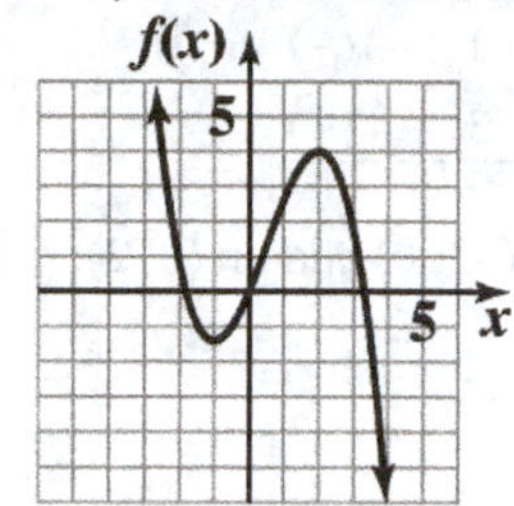

77. Increasing on $(-1, 2)$ and on $(2, \infty)$ $[f'(x) > 0]$; decreasing on $(-\infty, -1)$ $[f'(x) < 0]$; local minimum at $x = -1$.

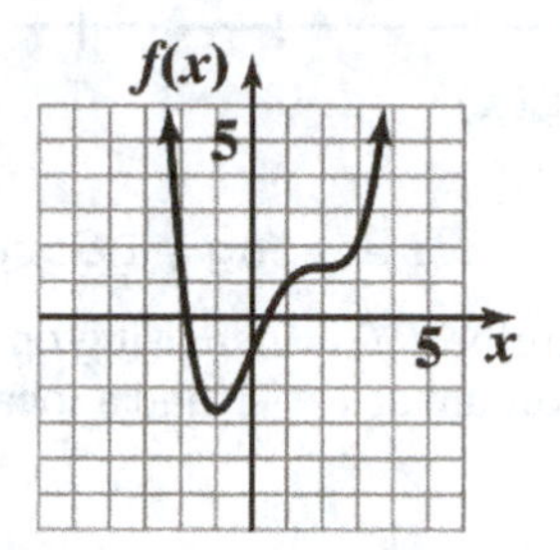

79. Increasing on $(-2, 0)$ and $(3, \infty)$ $[f'(x) > 0]$; decreasing on $(-\infty, -2)$ and $(0, 3)$ $[f'(x) < 0]$; local minima at $x = -2$ and $x = 3$, local maximum at $x = 0$.

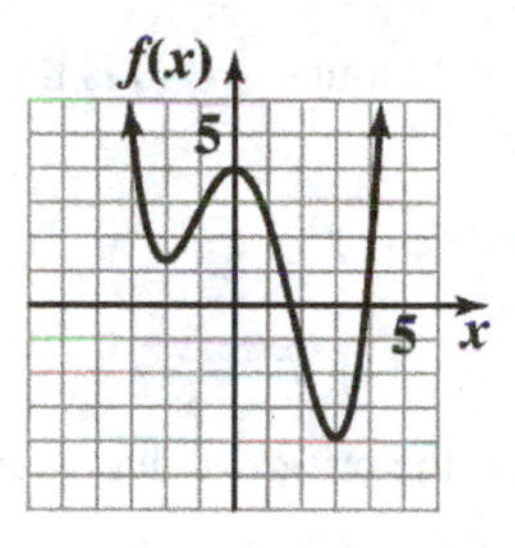

81. $f'(x) > 0$ on $(-\infty, -1)$ and on $(3, \infty)$; $f'(x) < 0$ on $(-1, 3)$; $f'(x) = 0$ at $x = -1$ and $x = 3$.

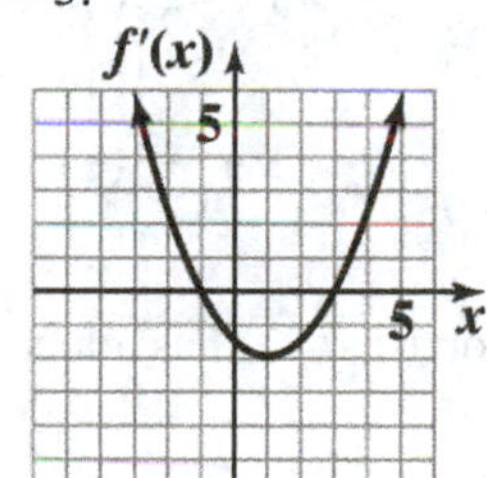

83. $f'(x) > 0$ on $(-2, 1)$ and on $(3, \infty)$; $f'(x) < 0$ on $(-\infty, -2)$ and on $(1, 3)$: $f'(x) = 0$ at $x = -2$, $x = 1$, and $x = 3$.

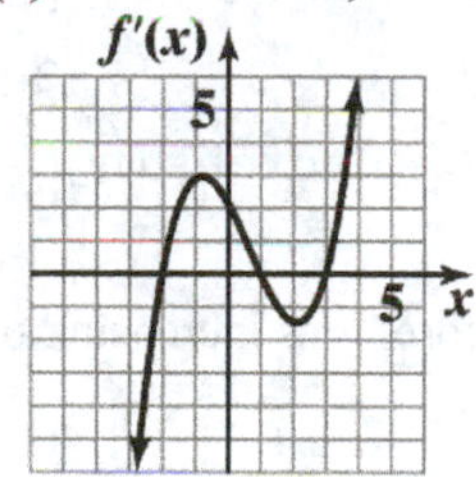

85. $f(x) = x + \frac{4}{x}$ [Note: f is not defined at $x = 0$.]

$$f'(x) = 1 - \frac{4}{x^2}$$

Critical values: $x = 0$ is *not* a critical value of f since 0 is not in the domain of f, but $x = 0$ is a partition number for f'.

$$f'(x) = 1 - \frac{4}{x^2} = 0$$
$$x^2 - 4 = 0$$
$$(x + 2)(x - 2) = 0$$

Thus, the critical values are $x = -2$ and $x = 2$; $x = -2$ and $x = 2$ are also partition numbers for f'.

The sign chart for f' is:

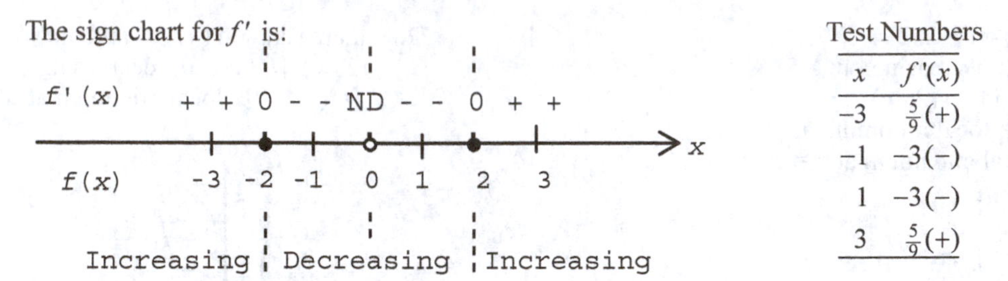

Test Numbers

x	$f'(x)$
-3	$\frac{5}{9}(+)$
-1	$-3(-)$
1	$-3(-)$
3	$\frac{5}{9}(+)$

Therefore, f is increasing on $(-\infty, -2)$ and on $(2, \infty)$, f is decreasing on $(-2, 0)$ and on $(0, 2)$; f has a local maximum at $x = -2$ and a local minimum at $x = 2$.

87. $f(x) = 1 + \frac{1}{x} + \frac{1}{x^2}$ [Note: f is not defined at $x = 0$.]

$f'(x) = -\frac{1}{x^2} - \frac{2}{x^3}$

Critical values: $x = 0$ is not a critical value of f since 0 is not in the domain of f; $x = 0$ is a partition number for f'.

$$f'(x) = -\frac{1}{x^2} - \frac{2}{x^3} = 0$$

$$-x - 2 = 0$$

$$x = -2$$

Thus, the critical value is $x = -2$; -2 is also a partition number for f'.

The sign chart for f' is:

f '(x) - - - - - 0 + + + + + ND - - - - - -

x: -3 -2 -1 0 1

f(x) Decreasing | Increasing | Decreasing

Test Numbers

x	$f'(x)$
-3	$-\frac{1}{27}(-)$
-1	$1(+)$
1	$-3(-)$

Therefore, f is increasing on $(-2, 0)$ and f is decreasing on $(-\infty, -2)$ and on $(0, \infty)$; f has a local minimum at $x = -2$.

89. $f(x) = \frac{x^2}{x-2}$ [Note: f is not defined at $x = 2$.]

$$f'(x) = \frac{(x-2)(2x) - x^2(1)}{(x-2)^2} = \frac{x^2 - 4x}{(x-2)^2}$$

Critical values: $x = 2$ is *not* a critical value of f since 2 is not in the domain of f; $x = 2$ is a partition number for f'.

$$f'(x) = \frac{x^2 - 4x}{(x-2)^2} = 0$$

$$x^2 - 4x = 0$$

$$x(x-4) = 0$$

Thus, the critical values are $x = 0$ and $x = 4$; 0 and 4 are also partition numbers for f'.

The sign chart for f' is:

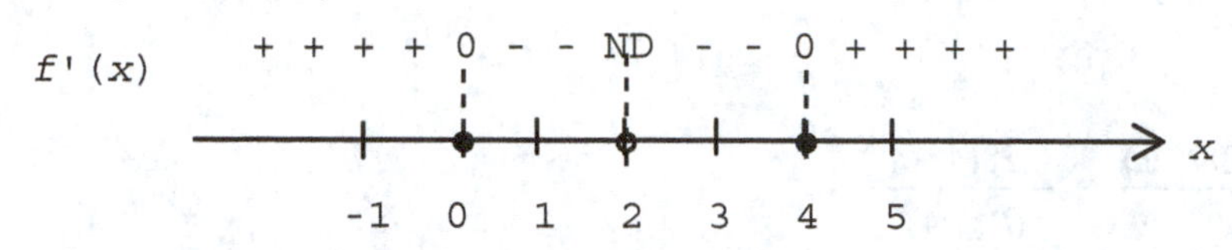

Test Numbers

x	$f'(x)$
-1	$\frac{5}{9}(+)$
1	$-3(-)$
3	$-3(-)$
5	$\frac{5}{9}(+)$

Therefore, f is increasing on $(-\infty, 0)$ and on $(4, \infty)$, f is decreasing on (0, 2) and on (2, 4); f has a local maximum at $x = 0$ and a local minimum at $x = 4$.

91. (A) The marginal profit function, P', is positive on (0, 600), zero at $x = 600$, and negative on (600, 1,000).

(B)

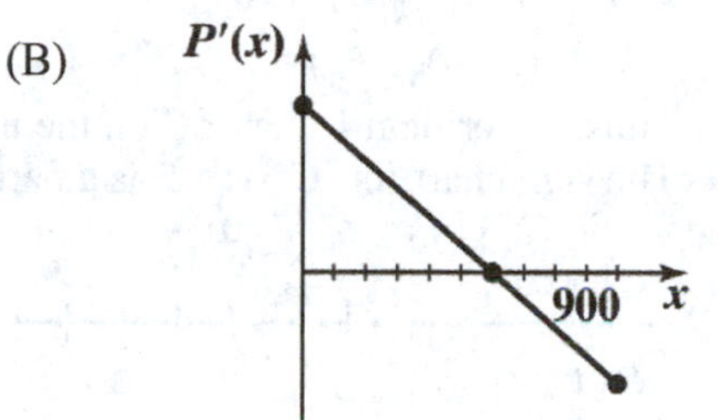

93. (A) The price function, $B(t)$, decreases for the first 15 months to a local minimum, increases for the next 40 months to a local maximum, and then decreases for the remaining 15 months.

(B)

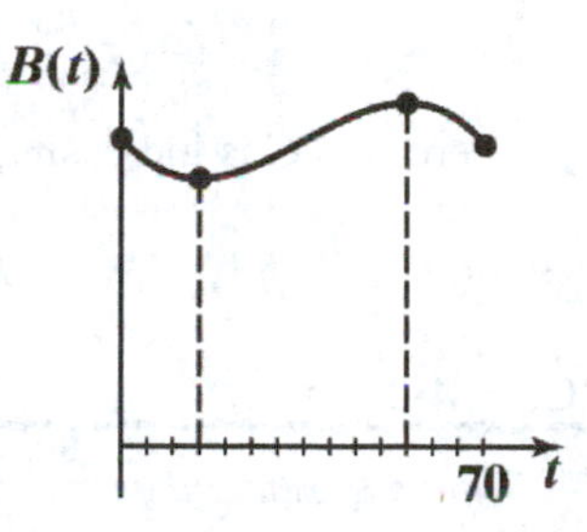

95. $C(x) = \dfrac{x^2}{20} + 20x + 320$

(A) $\overline{C}(x) = \dfrac{C(x)}{x} = \dfrac{x}{20} + 20 + \dfrac{320}{x}$

(B) Critical values:

$$\overline{C}'(x) = \frac{1}{20} - \frac{320}{x^2} = 0$$

$$x^2 - 320(20) = 0$$

$$x^2 - 6400 = 0$$

$$(x - 80)(x + 80) = 0$$

Thus, the critical value of $\overline{C}$ on the interval (0, 150) is $x = 80$.

Next, construct the sign chart for $\overline{C}'$ ($x = 80$ is a partition number for $\overline{C}'$).

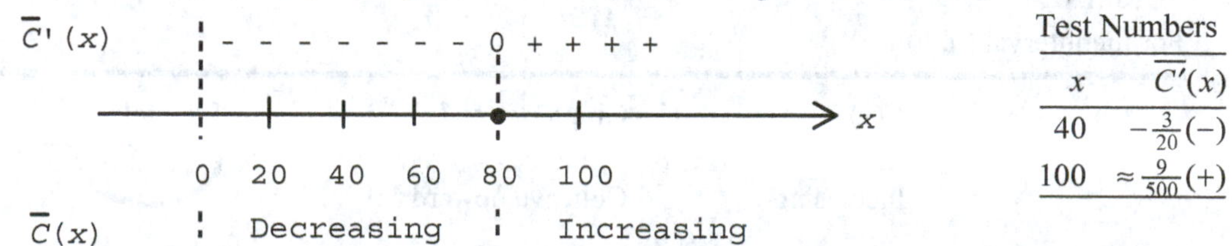

Test Numbers

x	$\overline{C}'(x)$
40	$-\frac{3}{20}(-)$
100	$\approx \frac{9}{500}(+)$

Therefore, $\overline{C}$ is increasing for $80 < x < 150$ and decreasing for $0 < x < 80$; $\overline{C}$ has a local minimum at $x = 80$.

97. $C(t) = \dfrac{0.28t}{t^2+4}$, $0 < t < 24$

$$C'(t) = \frac{(t^2+4)(0.28) - 0.28t(2t)}{(t^2+4)^2} = \frac{0.28(4-t^2)}{(t^2+4)^2}$$

Critical values: C' is continuous for all t on the interval (0, 24).

$$C'(t) = \frac{0.28(4-t^2)}{(t^2+4)^2} = 0$$

$$4 - t^2 = 0$$

$$(2-t)(2+t) = 0$$

Thus, the critical value of C on the interval (0, 24) is $t = 2$.

The sign chart for C' ($t = 2$ is a partition number) is:

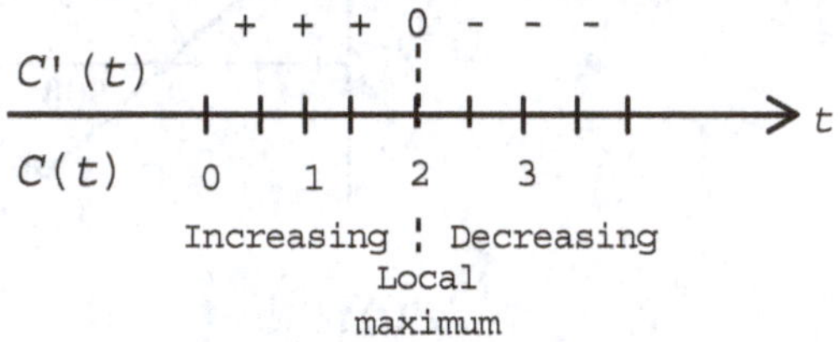

```
           + + + 0 - - -
C'(t)            ¦
------+-+-+-+-+-+-+-+----> t
C(t)  0   1   2   3
   Increasing ¦ Decreasing
            Local
           maximum
```

Test Numbers

t	$C'(t)$
1	(+)
3	(−)

Therefore, C is increasing on (0, 2) and decreasing on (2, 24); $C(2) = 0.07$ is a local maximum.

EXERCISE 4-2

Things to remember:

1. CONCAVITY

   The graph of a function f is CONCAVE UPWARD on the interval (a, b) if $f'(x)$ is *increasing* on (a, b) and is CONCAVE DOWNWARD on the interval (a, b) if $f'(x)$ is *decreasing* on (a, b).

2. SECOND DERIVATIVE

   For $y = f(x)$, the SECOND DERIVATIVE of f, provided it exists, is:

   $$f''(x) = \frac{d}{dx}f'(x)$$

   Other notations for $f''(x)$ are:

   $$\frac{d^2y}{dx^2} \text{ and } y''.$$

3. SUMMARY

   For the interval (a, b):

$f''(x)$	$f'(x)$	Graph of $y = f(x)$	Example
+	Increasing	Concave upward	 ∪
−	Decreasing	Concave downward	 ∩

4. INFLECTION POINT

An INFLECTION POINT is a point on the graph of a function where the concavity changes (from upward to downward, or from downward to upward). If f is continuous on (a, b) and has an inflection point at $x = c$, then either $f''(c) = 0$ or $f''(c)$ does not exist.

5. GRAPHING STRATEGY

Step 1. Analyze $f(x)$.

Find the domain and the intercepts. The x intercepts are the solutions to $f(x) = 0$ and the y intercept is $f(0)$.

Step 2. Analyze $f'(x)$.

Find the partition numbers of $f'(x)$ and the critical values of $f(x)$. Construct a sign chart for $f'(x)$, determine the intervals where f is increasing and decreasing, and find the local maxima and minima.

Step 3. Analyze $f''(x)$.

Find the partition numbers of $f''(x)$. Construct a sign chart for $f''(x)$, determine the intervals where the graph of f is concave upward and concave downward, and find the inflection points.

Step 4. Sketch the graph of f.

Locate intercepts, local maxima and minima, and inflection points. Sketch in what you know from steps 1 – 3. Plot additional points as needed and complete the sketch.

---

1. $f(x) = x^2$ on $(-\infty, \infty)$; concave up

3. $m(x) = x^3$ on $(-\infty, 0)$; concave down

5. $p(x) = \sqrt{x}$ on $(0, \infty)$; concave down

7. $g(x) = |x|$ on $(-\infty, 0)$; neither

9. (A) $(a,c), (c,d), (e,g)$ (B) $(d,e), (g,h)$ (C) $(d,e), (g,h)$

(D) $(a,c), (c,d), (e,g)$ (E) $(a,c), (c,d), (e,g)$ (F) $(d,e), (g,h)$

11. (A) $f(-2) = 3$ is a local maximum; $f(2) = -1$ is a local minimum .

(B) $(0,1)$ is an inflection point.

(C) $f'(x)$ has local extremum at $x = 0$.

13. $f'(x) > 0$, $f''(x) > 0$; (C)

15. $f'(x) < 0, f''(x) > 0$; (D)

17. $f(x) = 2x^3 - 4x^2 + 5x - 6$

$f'(x) = 6x^2 - 8x + 5$

$f''(x) = 12x - 8$

19. $h(x) = 2x^{-1} - 3x^{-2}$

$h'(x) = -2x^{-2} + 6x^{-3}$

$h''(x) = 4x^{-3} - 18x^{-4}$

21. $y = x^2 - 18x^{1/2}$

$\frac{dy}{dx} = 2x - 9x^{-1/2}$

$\frac{d^2y}{dx^2} = 2 + \frac{9}{2}x^{-3/2}$

23. $y = (x^2 + 9)^4$

$y' = 4(x^2 + 9)^3(2x) = 8x(x^2 + 9)^3$

$y'' = 24x(x^2 + 9)^2(2x) + 8(x^2 + 9)^3$

$= 48x^2(x^2+9)^2+8(x^2+9)^3=8(x^2+9)^2(7x^2+9)$

25. $f(x) = x^3 + 30x^2$; $f'(x) = 3x^2 + 60x$; $f''(x) = 6x + 60$

$f''(x) = 0$: $6x + 60 = 0$

$x = -10$

$f(-10) = (-10)^3 + 30(-10)^2 = -1,000 + 3,000 = 2,000$. Inflection point: $(-10,\ 2,000)$

27. $f(x) = x^{5/3} + 2$; $f'(x) = \frac{5}{3}x^{2/3}$; $f''(x) = \frac{10}{9}x^{-1/3} = \frac{10}{9x^{1/3}}$.

$f''(x)$ does not exist at $x = 0$; $f(0) = 2$. Inflection point: $(0, 2)$.

29. $f(x) = 1 + x + x^{2/5}$; $f'(x) = 1 + \frac{2}{5}x^{-3/5}$; $f''(x) = -\frac{6}{25}x^{-8/5} = -\frac{6}{25x^{8/5}} < 0$ for all $x \neq 0$.

The graph is concave downward; no inflection points.

31. $f(x) = x^4 + 6x^2$; $f'(x) = 4x^3 + 12x$; $f''(x) = 12x^2 + 12 \geq 12 > 0$

The graph of f is concave upward for all x; there are no inflection points.

33. $f(x) = x^3 - 4x^2 + 5x - 2$; $f'(x) = 3x^2 - 8x + 5$; $f''(x) = 6x - 8$

$f''(x) = 0$: $6x - 8 = 0$

$x = \frac{4}{3}$

Sign chart for f'' $\left(\text{partition number is } \frac{4}{3}\right)$:

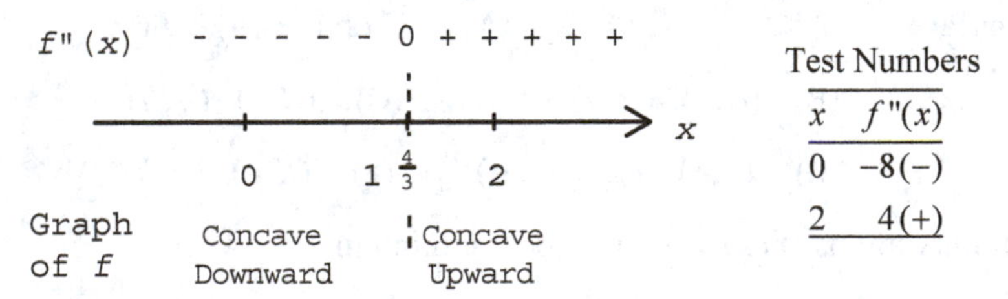

Therefore, the graph of f is concave downward on $\left(-\infty, \frac{4}{3}\right)$ and concave upward on $\left(\frac{4}{3}, \infty\right)$; there is an inflection point at $x = \frac{4}{3}$.

35. $f(x) = -x^4 + 12x^3 - 12x + 24$; $f'(x) = -4x^3 + 36x^2 - 12$; $f''(x) = -12x^2 + 72x$

$f''(x) = 0$: $-12x^2 + 72x = 0$

$-12x(x - 6) = 0$

$x = 0, 6$

Sign chart for f'' (partition numbers 0, 6):

f" (x) - - - 0 + + + + + 0 - - -

-1 0 1 6 7 → x

Graph of f: Concave Downward | Concave Upward | Concave Downward

Test Numbers

x	$f''(x)$
-1	$-84(-)$
1	$60(+)$
7	$-84(-)$

Therefore, the graph of f is concave downward on $(-\infty, 0)$ and $(6, \infty)$; concave upward on $(0, 6)$; there are inflection points at $x = 0$ and $x = 6$.

37. $f(x) = \ln(x^2 - 2x + 10)$ (Note: $x^2 - 2x + 10 = (x-1)^2 + 9 > 0$ for all x)

$$f'(x) = \frac{1}{x^2 - 2x + 10}(2x - 2) = \frac{2x - 2}{x^2 - 2x + 10}$$

$$f''(x) = \frac{(x^2 - 2x + 10)(2) - (2x - 2)(2x - 2)}{(x^2 - 2x + 10)^2} = \frac{2x^2 - 4x + 20 - [4x^2 - 8x + 4]}{(x^2 - 2x + 10)^2} = \frac{-2x^2 + 4x + 16}{(x^2 - 2x + 10)^2}$$

$$= \frac{-2(x^2 - 2x - 8)}{(x^2 - 2x + 10)^2};$$

$$f''(x) = \frac{-2(x - 4)(x + 2)}{(x^2 - 2x + 10)^2}$$

$f''(x) = 0$: $-2(x - 4)(x + 2) = 0$

$x = 4, -2$

Sign chart for f'' (partition numbers 4, –2):

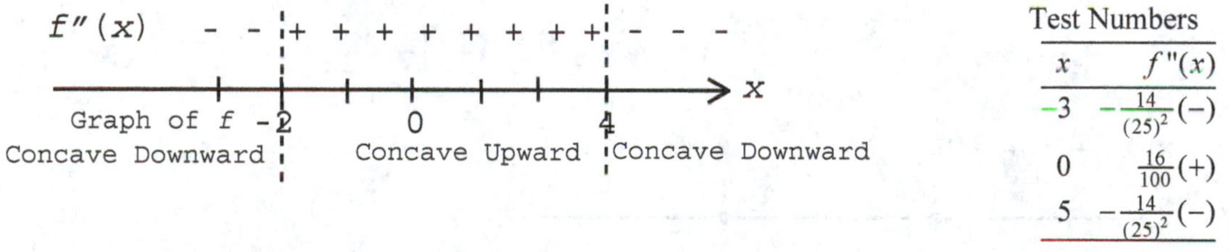

Test Numbers

x	$f''(x)$
-3	$-\frac{14}{(25)^2}(-)$
0	$\frac{16}{100}(+)$
5	$-\frac{14}{(25)^2}(-)$

The graph of f is concave downward on $(-\infty, -2)$ and $(4, \infty)$; the graph of f is concave upward on $(-2, 4)$; there are inflection points at $x = -2$ and $x = 4$.

39. $f(x) = 8e^x - e^{2x}$; $f'(x) = 8e^x - 2e^{2x}$; $f''(x) = 8e^x - 4e^{2x}$

$f''(x) = 0$: $8e^x - 4e^{2x} = 0$

$4e^x(2 - e^x) = 0$

$e^x = 2$

$x = \ln 2$

Sign chart for f'' (partition number $\ln 2 \approx 0.69$):

f″(x) + + + + | - - - -

0 0.69 1 → x

Graph of f: Concave Upward | Concave Downward

Test Numbers

x	$f''(x)$
0	$4(+)$
1	$8e - 4e^2(-)$

The graph of f is concave upward on $(-\infty, \ln 2)$ and concave downward on $(\ln 2, \infty)$; there is an inflection point at $x = \ln 2$.

41.

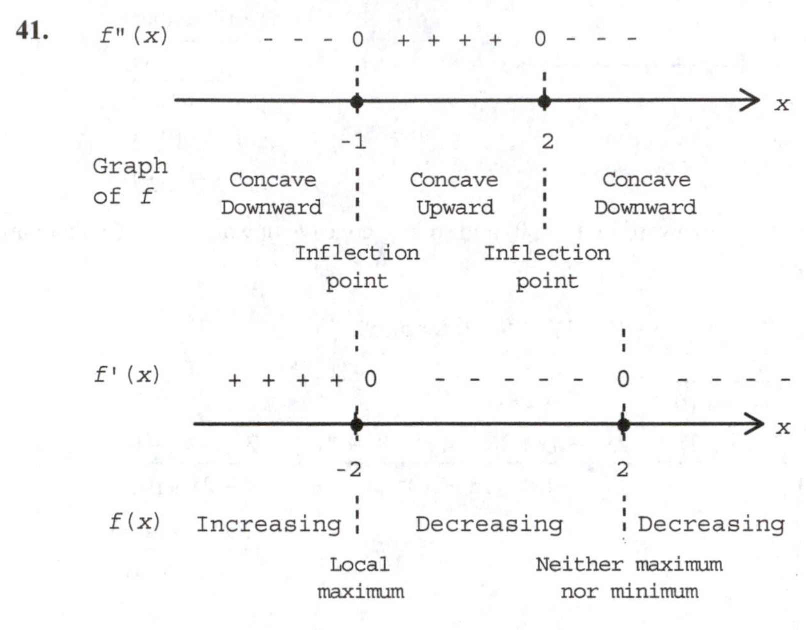

Using this information together with the points (–4, 0), (–2, 3), (–1, 1.5), (0, 0), (2, –1), (4, –3) on the graph,we have

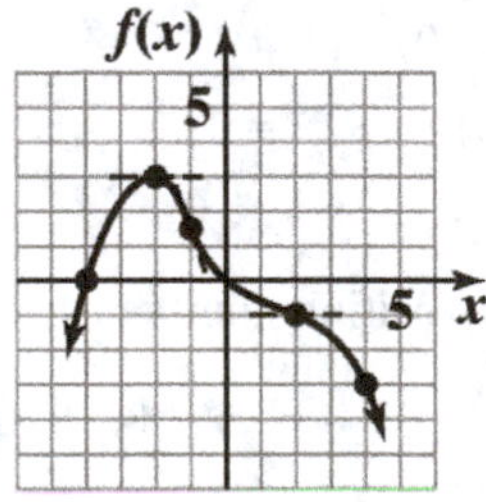

43.

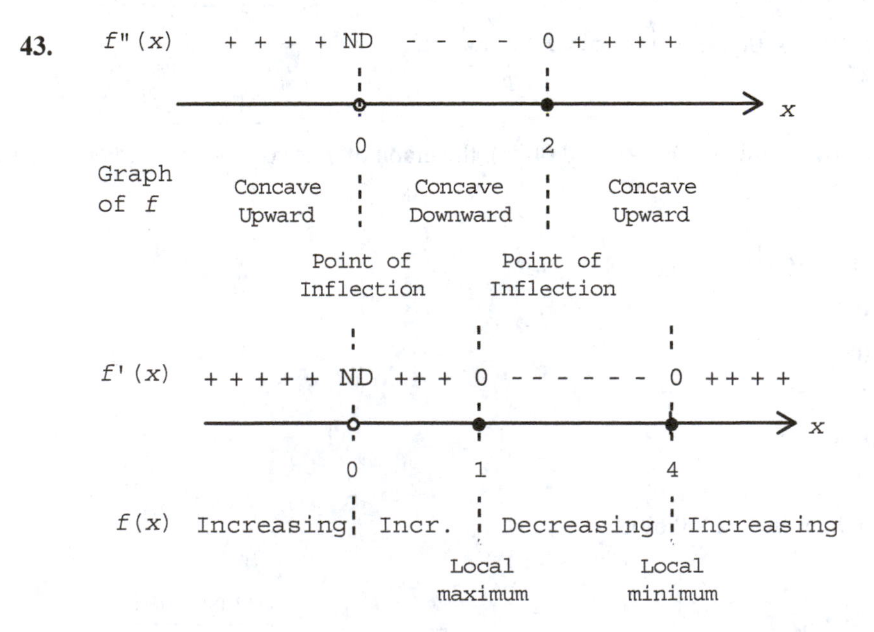

Using this information together with the points (–3, –4), (0, 0), (1, 2), (2, 1), (4, –1), (5, 0) on the graph, we have

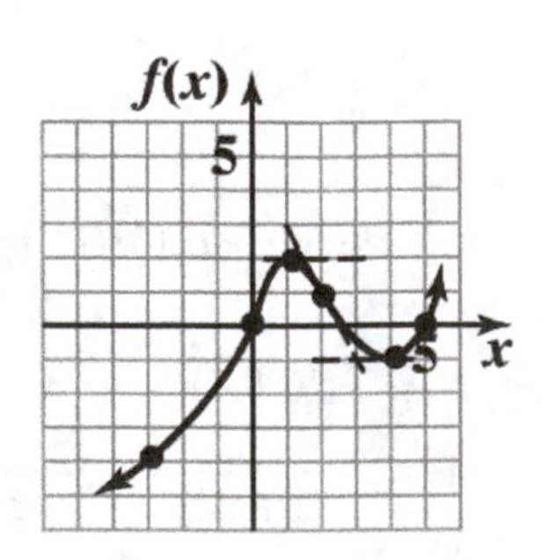

45.

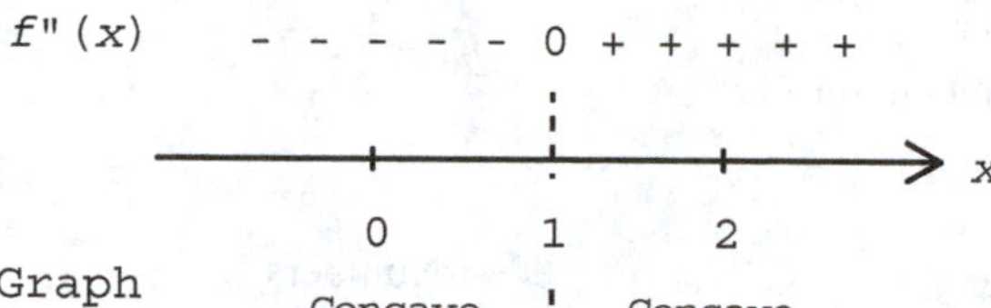

Graph of f

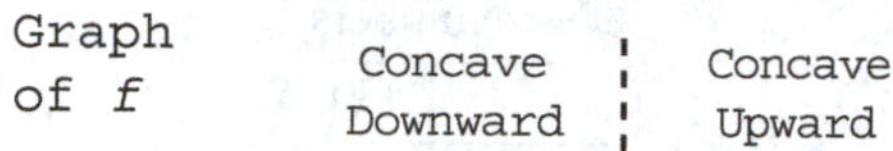

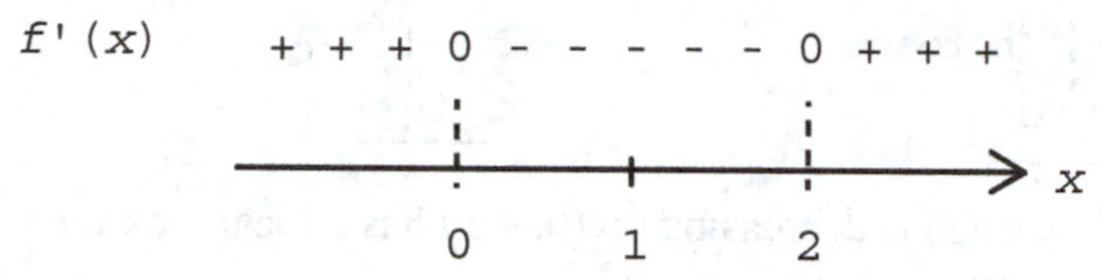

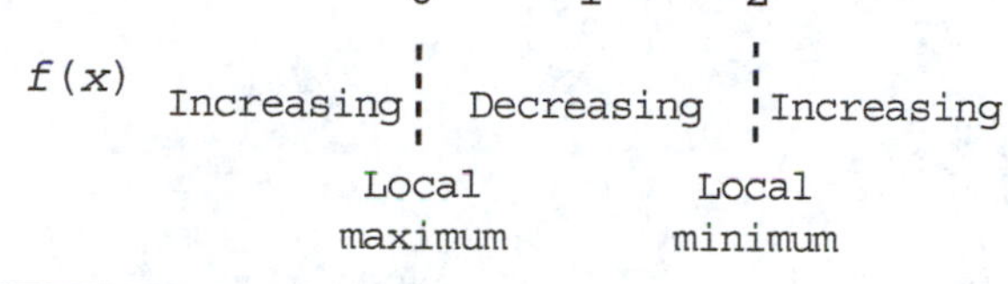

x	0	1	2
$f(x)$	2	0	-2

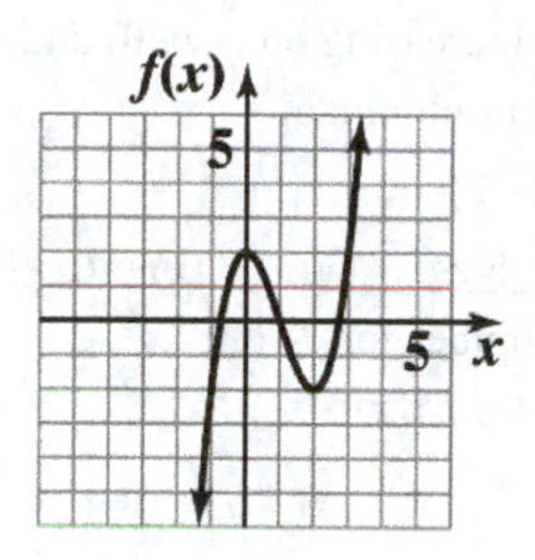

47.

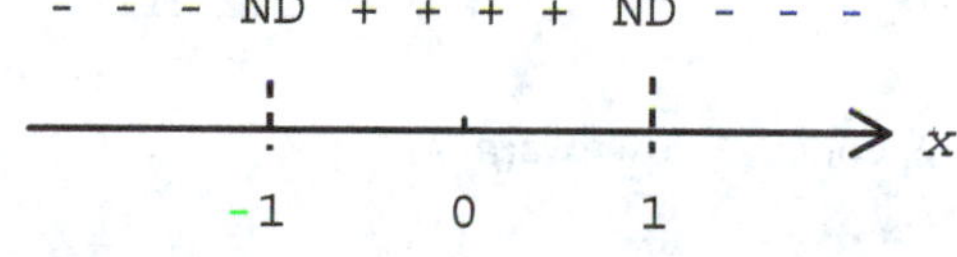

Graph of f: Concave downward | Concave upward | Concave downward

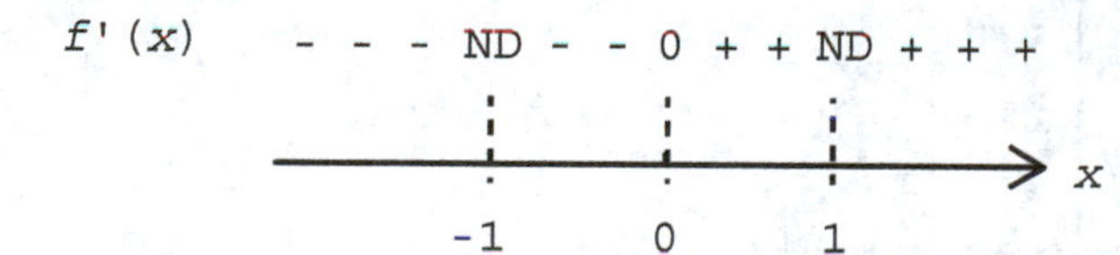

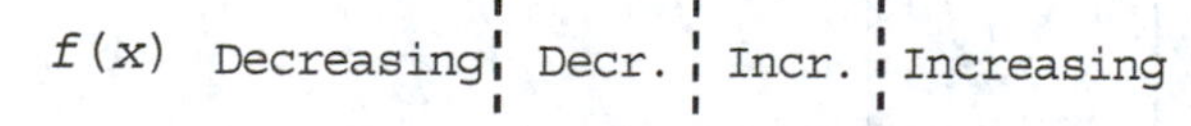

x	-1	0	1
$f(x)$	0	-2	0

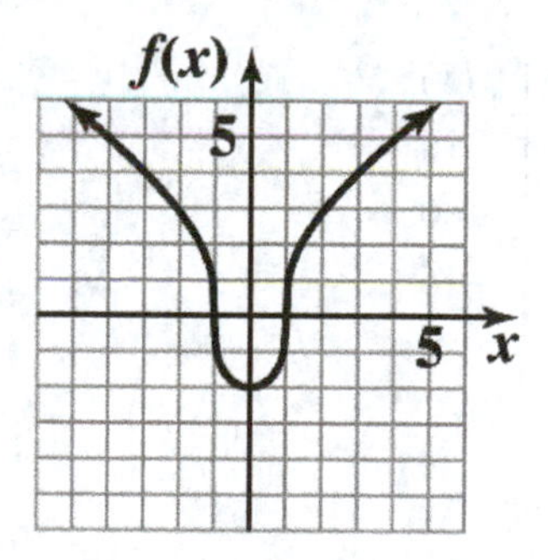

49. $f(x) = (x-2)(x^2 - 4x - 8) = x^3 - 6x^2 + 16$

Step 1. Analyze $f(x)$. Domain of f: $(-\infty, \infty)$

x-intercept(s): $f(x) = 0$:

$$(x-2)(x^2 - 4x - 8) = 0$$

$$x = 2, 2 \pm 2\sqrt{3}$$

y-intercept: $f(0) = 16$

Step 2. Analyze $f'(x)$. $f'(x) = 3x^2 - 12x = 3x(x-4)$
Critical values for f: $x = 0, x = 4$. Partition numbers for f' : 0, 4.

Sign chart for $f'(x)$:

```
f'(x)   + + + 0 - - - - - 0 + + +
        --+---+---+---+---+---+---+--> x
f(x)     -1   0   1   2   3   4   5
  Increasing  |  Decreasing   |  Increasing
            Local           Local
           maximum         minimum
```

Test Numbers

x	$f'(x)$
−1	15(+)
2	−12(−)
5	15(+)

Thus, f is increasing on $(-\infty, 0)$ and on $(4, \infty)$; $f(x)$ is decreasing on $(0, 4)$; f has a local maximum at $x = 0$ and a local minimum at
$x = 4$.

Step 3. Analyze $f''(x)$. $f''(x) = 6x - 12 = 6(x-2)$
Partition number for $f''(x)$: $x = 2$
Sign chart for $f''(x)$:

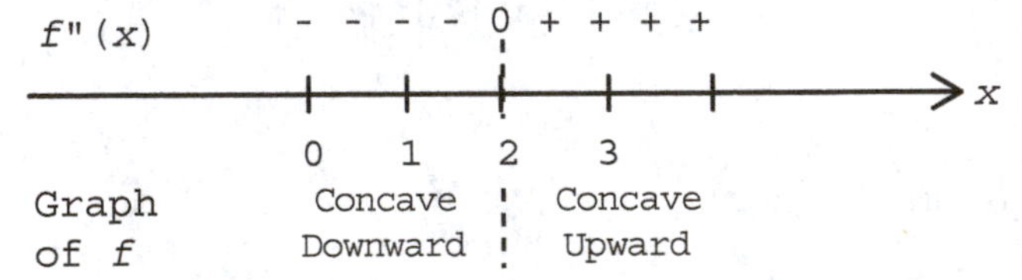

Test Numbers

x	$f''(x)$
0	−12(−)
3	6(+)

The graph of f is concave upward on $(2, \infty)$, concave downward on $(-\infty, 2)$; and has an inflection point at $x = 2$.
Step 4. Sketch the graph of f.

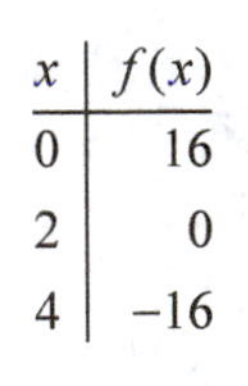

x	$f(x)$
0	16
2	0
4	−16

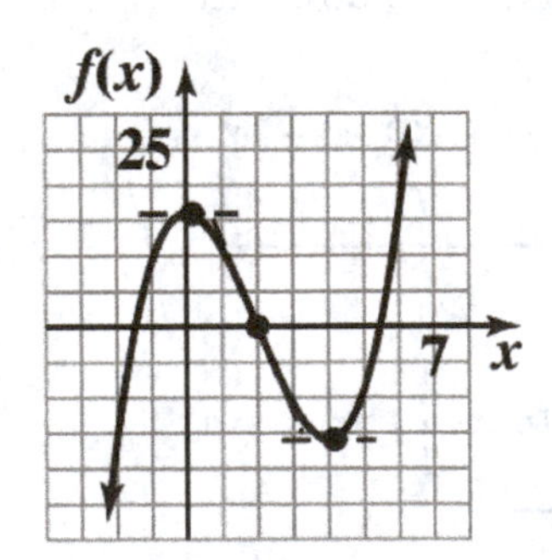

51. $f(x) = (x+1)(x^2 - x + 2) = x^3 + x + 2$
Step 1. Analyze $f(x)$. Domain of f: $(-\infty, \infty)$
x-intercept(s): $f(x) = 0$
$(x+1)(x^2 - x + 2) = 0$
$x = -1$ (the quadratic factor does not have real roots)

y-intercept: $f(0) = 2$

Step 2. Analyze $f'(x)$. $f'(x) = 3x^2 + 1 > 0$ for all x.
Zeros of $f'(x)$: $f'(x)$ does not have any zeros.

Sign chart for $f'(x)$:

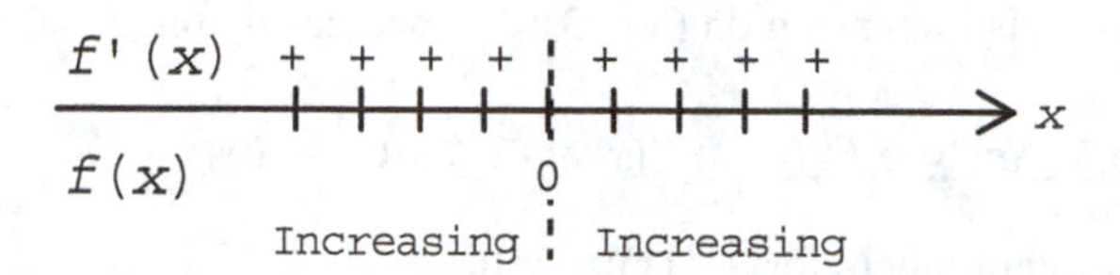

Thus, $f(x)$ is increasing on $(-\infty, \infty)$.

Step 3. Analyze $f''(x)$: $f''(x) = 6x$
Partition numbers for $f''(x)$: $x = 0$
Sign chart for $f''(x)$:

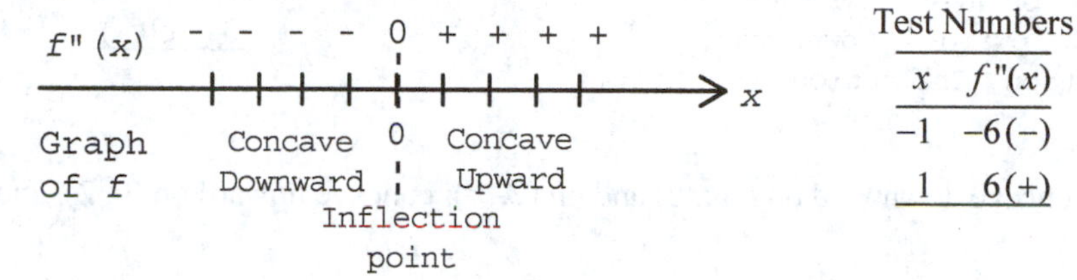

Test Numbers

x	$f''(x)$
-1	$-6(-)$
1	$6(+)$

Thus, the graph of f is concave upward on $(0, \infty)$ and concave downward on $(-\infty, 0)$; the graph has an inflection point at $x = 0$.

Step 4. Sketch the graph of f.

x	$f(x)$
-1	0
0	2
1	4

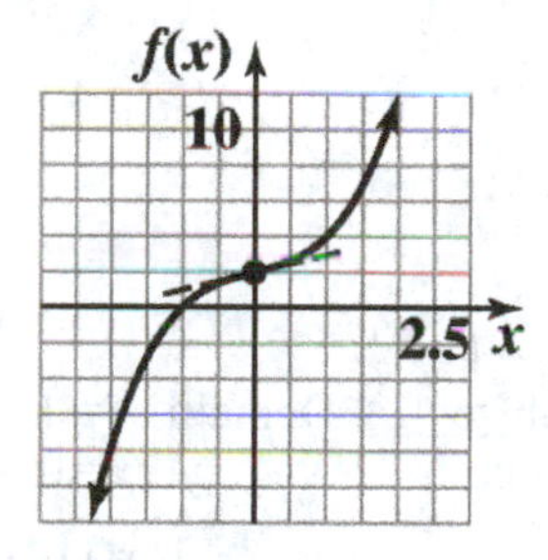

53. $f(x) = -0.25x^4 + x^3 = -\frac{1}{4}x^4 + x^3$

Step 1. Analyze $f(x)$. Domain of f: $(-\infty, \infty)$.
x-intercept(s): $f(x) = 0$:

$$-\frac{1}{4}x^4 + x^3 = 0$$

$$x^3\left(-\frac{1}{4}x + 1\right) = 0$$

$$x = 0, 4$$

y-intercept: $f(0) = 0$

Step 2. Analyze $f'(x)$. $f'(x) = -x^3 + 3x^2 = -x^2(x - 3)$
Critical values for f: $x = 0, 3$. Partition numbers for f': 0, 3.
Sign chart for $f'(x)$:

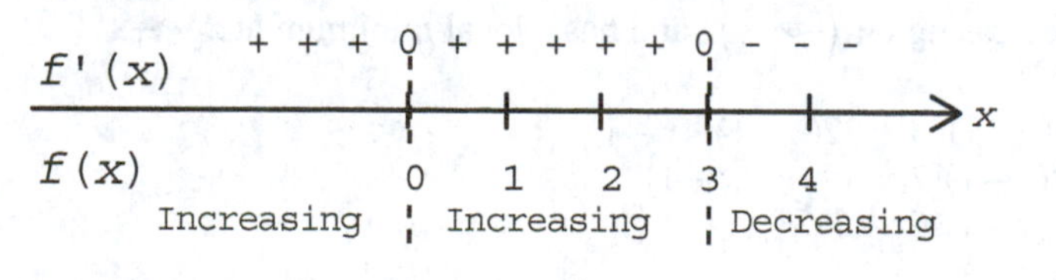

Test Numbers

x	$f'(x)$
-1	$4(+)$
2	$4(+)$
4	$-16(-)$

Thus, f is increasing on $(-\infty, 3)$; f is decreasing on $(3, \infty)$; f has a local maximum at $x = 3$.

Step 3. Analyze $f''(x)$. $f''(x) = -3x^2 + 6x = -3x(x-2)$

Partition numbers for $f''(x)$: $x = 0, 2$
Sign chart for $f''(x)$:

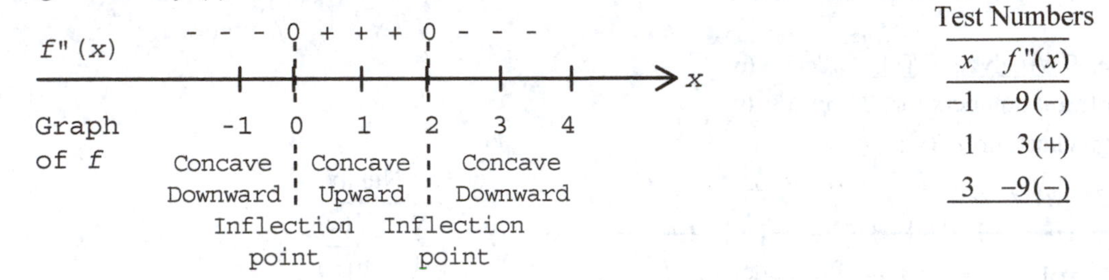

Test Numbers

x	$f''(x)$
-1	$-9(-)$
1	$3(+)$
3	$-9(-)$

Thus, the graph of f is concave downward on $(-\infty, 0)$ and on $(2, \infty)$; concave upward on $(0, 2)$, and has inflection points at $x = 0, 2$.

Step 4. Sketch the graph of f.

x	$f(x)$
0	0
2	4
3	$\frac{27}{4}$
4	0

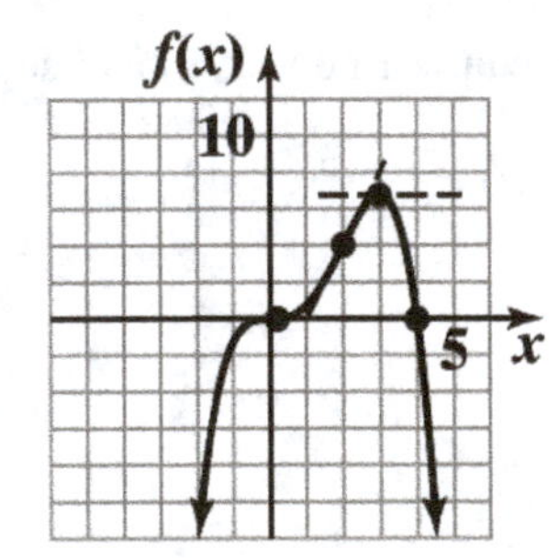

55. $f(x) = 16x(x-1)^3$

Step 1. Analyze $f(x)$. Domain of f: $(-\infty, \infty)$.

x-intercept(s):
$$f(x) = 0$$
$$16x(x-1)^3 = 0$$
$$x = 0, 1$$

y-intercept: $f(0) = 0$

Step 2. Analyze $f'(x)$.
$$f'(x) = 16x(3)(x-1)^2 + 16(x-1)^3$$
$$= 16(x-1)^2(3x + x - 1)$$
$$= 16(x-1)^2(4x-1)$$

Critical values for f: $x = 1$, $x = \frac{1}{4}$. Partition numbers for f': $1, \frac{1}{4}$.

Sign chart for $f'(x)$:

```
f'(x)    - - - 0 + + 0 + + +
        ---+-+-+-+-+-+-+-+-+-+-----> x
f(x)        0 1/4    1
      Decreasing |  Increasing
               Local
              minimum
```

Test Numbers

x	$f'(x)$
0	$-16(-)$
$\frac{1}{2}$	$4(+)$
$\frac{3}{2}$	$20(+)$

Thus, f is increasing on $\left(\frac{1}{4}, \infty\right)$; decreasing on $\left(-\infty, \frac{1}{4}\right)$; and has a local minimum at $x = \frac{1}{4}$.

Step 3. Analyze $f''(x)$.
$$f''(x) = 16(x-1)^2 4 + 32(x-1)(4x-1)$$
$$= 32(x-1)[2(x-1) + 4x - 1]$$
$$= 32(x-1)(6x-3)$$

Partition numbers for f'': $x = 1, \frac{1}{2}$

Sign chart for $f''(x)$:

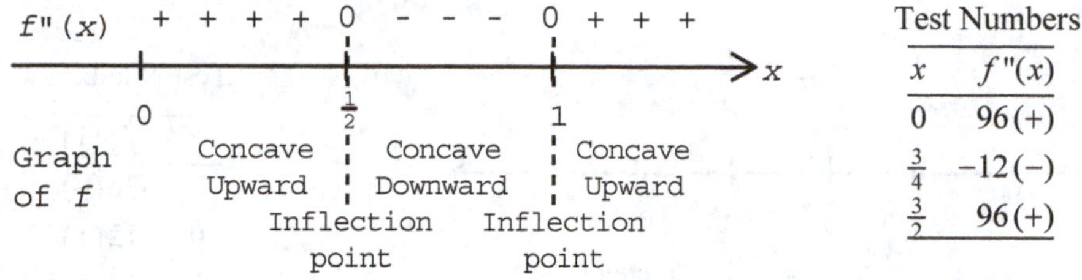

Test Numbers

x	$f''(x)$
0	96(+)
$\frac{3}{4}$	−12(−)
$\frac{3}{2}$	96(+)

Thus, the graph of f is concave upward on $\left(-\infty, \frac{1}{2}\right)$ and on $(1, \infty)$, concave downward on $\left(\frac{1}{2}, 1\right)$, and has inflection points at $x = \frac{1}{2}, 1$.

Step 4. Sketch the graph of f.

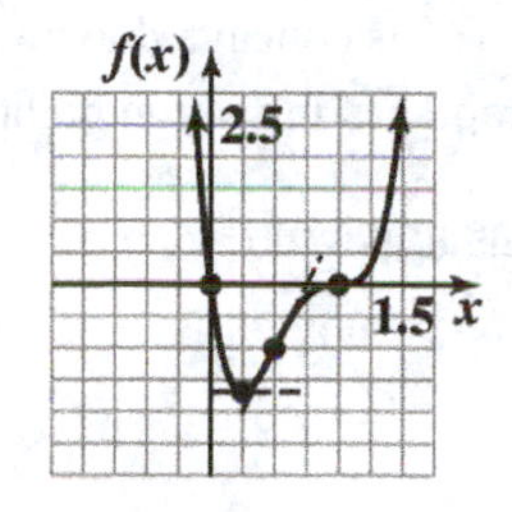

57. $f(x) = (x^2 + 3)(9 - x^2)$

Step 1. Analyze $f(x)$. Domain of f: $(-\infty, \infty)$.

Intercepts: y-intercept: $f(0) = 3(9) = 27$

x-intercepts: $(x^2 + 3)(9 - x^2) = 0$

$(3 - x)(3 + x) = 0$

$x = 3, -3$

Step 2. Analyze $f'(x)$. $f'(x) = (x^2 + 3)(-2x) + (9 - x^2)(2x)$

$= 2x[9 - x^2 - (x^2 + 3)]$

$= 2x(6 - 2x^2)$

$= 4x(\sqrt{3} + x)(\sqrt{3} - x)$

Critical values for f: $x = 0, x = -\sqrt{3}, x = \sqrt{3}$

Partition numbers for f': $0, -\sqrt{3}, \sqrt{3}$.

Sign chart for f':

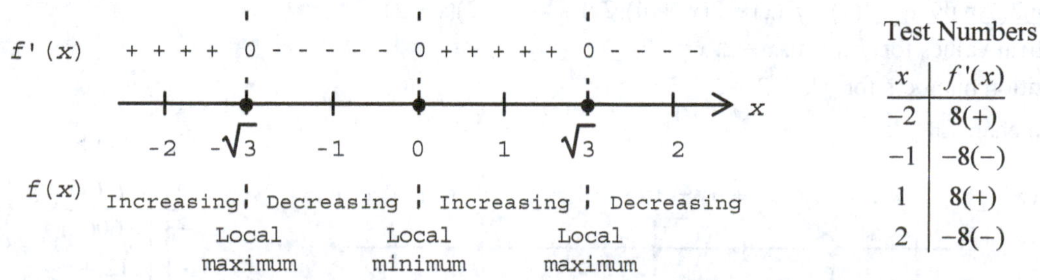

Test Numbers

x	$f'(x)$
−2	8(+)
−1	−8(−)
1	8(+)
2	−8(−)

Thus, f is increasing on $(-\infty, -\sqrt{3})$ and on $(0, \sqrt{3})$; f is decreasing on $(-\sqrt{3}, 0)$ and on $(\sqrt{3}, \infty)$; f has local maxima at $x = -\sqrt{3}$ and $x = \sqrt{3}$ and a local minimum at $x = 0$.

Step 3. Analyze $f''(x)$. $f''(x) = 2x(-4x) + (6 - 2x^2)(2) = 12 - 12x^2 = -12(x - 1)(x + 1)$
Partition numbers for f'': $x = 1, x = -1$
Sign chart for f'':

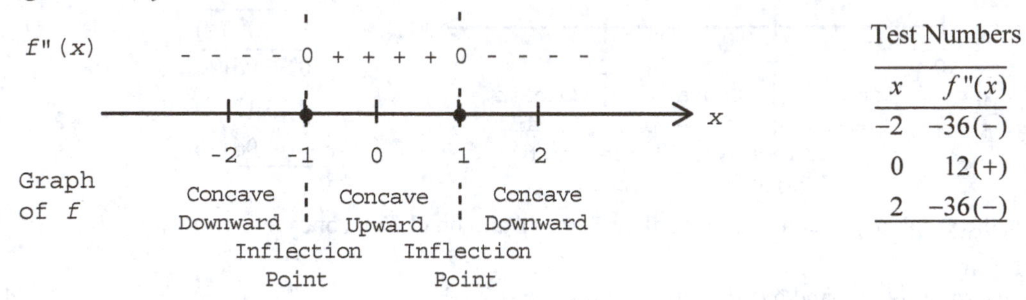

Test Numbers

x	$f''(x)$
−2	−36(−)
0	12(+)
2	−36(−)

Thus, the graph of f is concave downward on $(-\infty, -1)$ and on $(1, \infty)$; the graph of f is concave upward on $(-1, 1)$; the graph has inflection points at $x = -1$ and $x = 1$.

Step 4. Sketch the graph of f:

x	$f(x)$
$-\sqrt{3}$	36
−1	32
0	27
1	32
$\sqrt{3}$	36

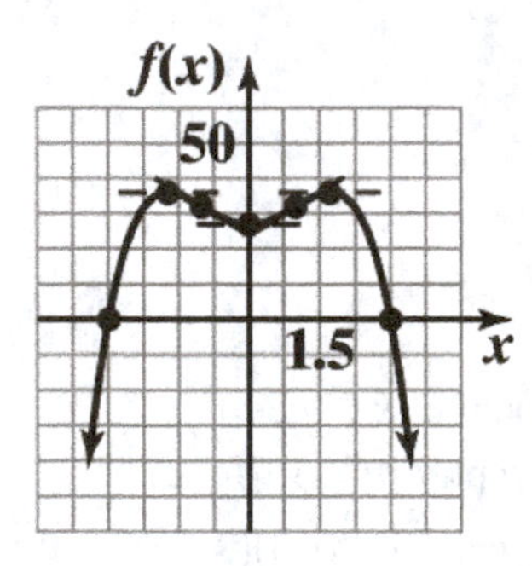

59. $f(x) = (x^2 - 4)^2$

Step 1. Analyze $f(x)$. Domain of f: $(-\infty, \infty)$.

Intercepts: y-intercept: $f(0) = (-4)^2 = 16$

x-intercepts: $(x^2 - 4)^2 = 0$

$$[(x - 2)(x + 2)]^2 = 0$$
$$(x - 2)^2(x + 2)^2 = 0$$
$$x = 2, -2$$

Step 2. Analyze $f'(x)$. $f'(x) = 2(x^2 - 4)(2x) = 4x(x - 2)(x + 2)$
Critical values for f: $x = 0, x = 2, x = -2$
Partition numbers for f': 0, 2, −2
Sign chart for f':

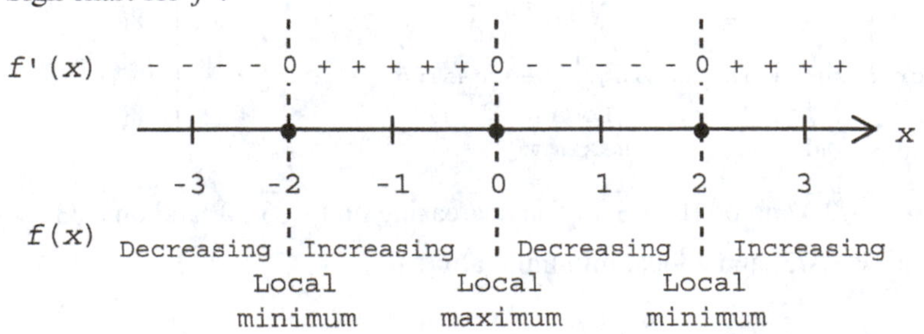

Test Numbers

x	$f'(x)$
−3	−60(−)
−1	12(+)
1	−12(−)
3	60(+)

Thus, f is decreasing on $(-\infty, -2)$ and on $(0, 2)$; f is increasing on $(-2, 0)$ and on $(2, \infty)$; f has local minima at $x = -2$ and $x = 2$ and a local maximum at $x = 0$.

Step 3. Analyze $f''(x)$. $f''(x) = 4x(2x) + (x^2 - 4)(4) = 12x^2 - 16 = 12\left(x^2 - \frac{4}{3}\right)$

$$= 12\left(x - \frac{2\sqrt{3}}{3}\right)\left(x + \frac{2\sqrt{3}}{3}\right)$$

Partition numbers for f'': $x = \frac{2\sqrt{3}}{3}, x = \frac{-2\sqrt{3}}{3}$

Sign chart for f'':

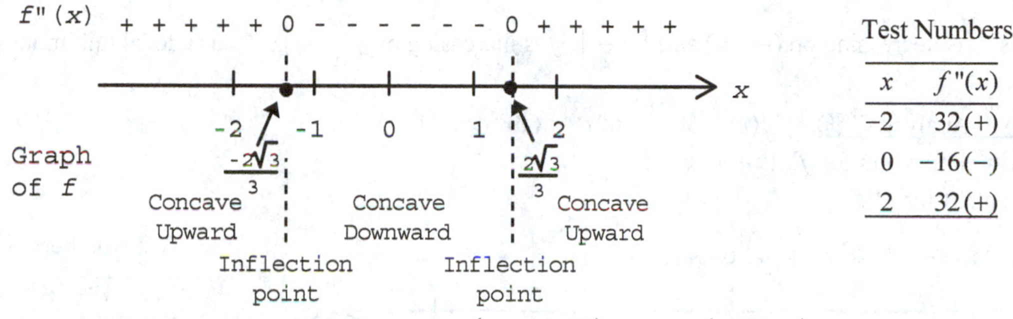

Test Numbers

x	$f''(x)$
-2	$32(+)$
0	$-16(-)$
2	$32(+)$

Thus, the graph of f is concave upward on $\left(-\infty, \frac{-2\sqrt{3}}{3}\right)$ and on $\left(\frac{2\sqrt{3}}{3}, \infty\right)$; the graph of f is concave downward on $\left(\frac{-2\sqrt{3}}{3}, \frac{2\sqrt{3}}{3}\right)$; the graph has inflection points at $x = \frac{-2\sqrt{3}}{3}$ and $x = \frac{2\sqrt{3}}{3}$.

Step 4. Sketch the graph of f.

x	$f(x)$
-2	0
$-\frac{2\sqrt{3}}{3}$	$\frac{64}{9}$
0	16
$\frac{2\sqrt{3}}{3}$	$\frac{64}{9}$
2	0

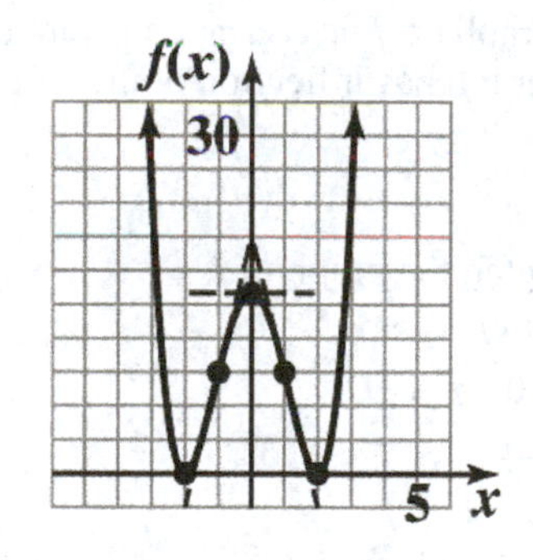

61. $f(x) = 2x^6 - 3x^5$

Step 1. Analyze $f(x)$. Domain of f: $(-\infty, \infty)$.

Intercepts: y-intercept: $f(0) = 2\cdot 0^6 - 3\cdot 0^5 = 0$

x-intercepts: $2x^6 - 3x^5 = 0$

$$x^5(2x - 3) = 0$$

$$x = 0, \frac{3}{2}$$

Step 2. Analyze $f'(x)$. $f'(x) = 12x^5 - 15x^4 = 12x^4\left(x - \frac{5}{4}\right)$

Critical values for f: $x = 0, x = \frac{5}{4}$

Partition numbers for f' : $0, \frac{5}{4}$

Sign chart for f':

f'(x) - - - - 0 - - - - - - - 0 + + + +

-1 0 1 $\frac{5}{4}$ 2 x

f(x) Decreasing ¦ Decreasing ¦ Increasing

Local minimum

Test Numbers

x	$f'(x)$
-1	$-27(-)$
1	$-3(-)$
2	$144(+)$

Thus, f is decreasing on $(-\infty, 0)$ and $\left(0, \frac{5}{4}\right)$; f is increasing on $\left(\frac{5}{4}, \infty\right)$; f has a local minimum at $x = \frac{5}{4}$.

Step 3. Analyze $f''(x)$. $f''(x) = 60x^4 - 60x^3 = 60x^3(x - 1)$

Partition numbers for f'': $x = 0, x = 1$

Sign chart for f'':

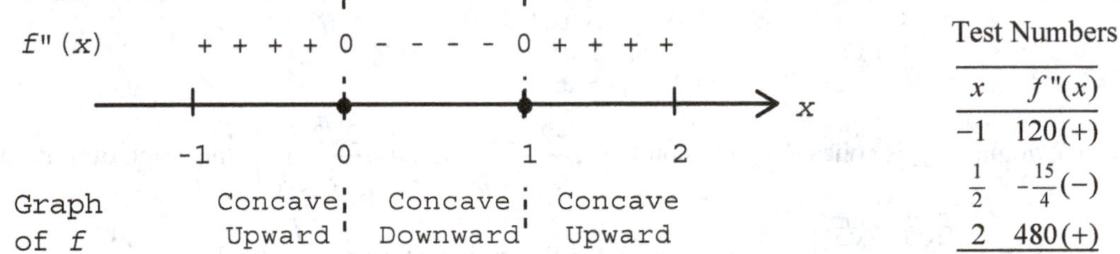

Test Numbers

x	$f''(x)$
-1	$120(+)$
$\frac{1}{2}$	$-\frac{15}{4}(-)$
2	$480(+)$

Thus, the graph of f is concave upward on $(-\infty, 0)$ and on $(1, \infty)$; the graph of f is concave downward on $(0, 1)$; the graph has inflection points at $x = 0$ and $x = 1$.

Step 4. Sketch the graph of f.

x	$f(x)$
0	0
1	-1
$\frac{5}{4}$	≈ -1.5

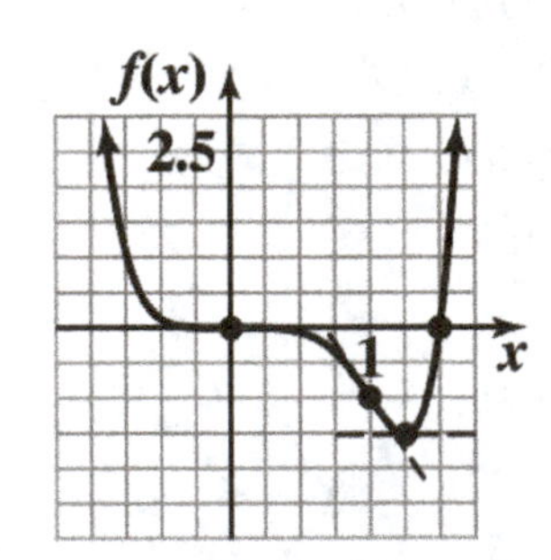

63. $f(x) = 1 - e^{-x}$

Step 1. Analyze $f(x)$. Domain of f: $(-\infty, \infty)$

Intercepts: x-intercepts: $f(x) = 0$

$$1 - e^{-x} = 0$$
$$e^{-x} = 1$$
$$-x = \ln 1 = 0$$
$$x = 0$$

y-intercept: $f(0) = 0$

Step 2. Analyze $f'(x)$. $f'(x) = e^{-x} > 0$ for all x.

Sign chart for $f(x)$:

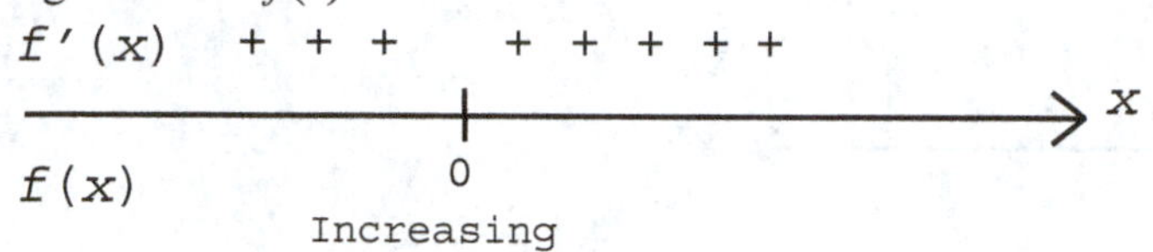

Thus, f is increasing on $(-\infty, \infty)$.

Step 3. Analyze $f''(x)$. $f''(x) = -e^{-x} < 0$ for all x.
Sign chart for $f''(x)$:

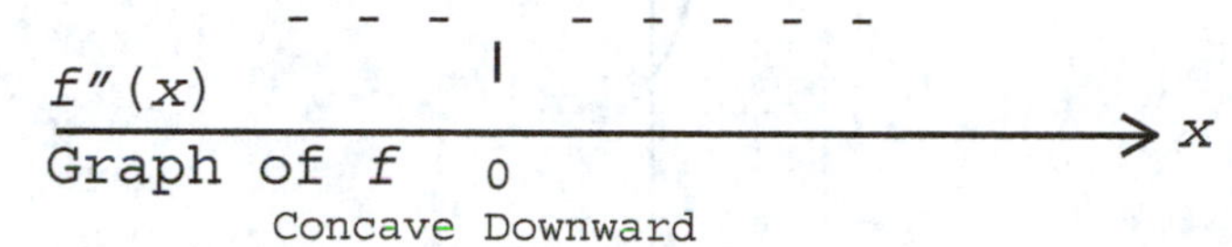

Thus, the graph of f is concave downward on $(-\infty, \infty)$.

Step 4. Sketch the graph of f:

x	$f(x)$
0	0

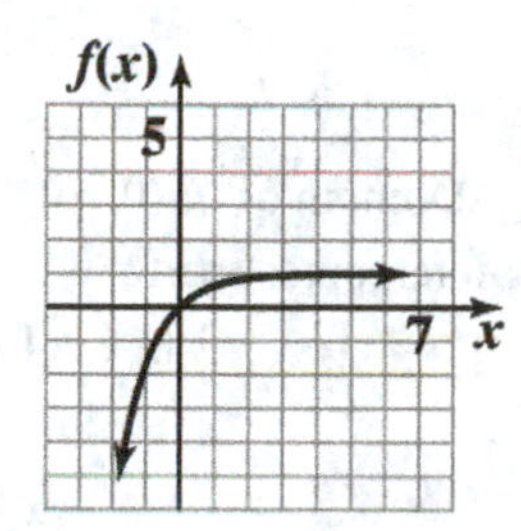

65. $f(x) = e^{0.5x} + 4e^{-0.5x}$

Step 1. Analyze $f(x)$. Domain of f: $(-\infty, \infty)$

Intercepts: x-intercepts:

$$f(x) = 0$$
$$e^{0.5x} + 4e^{-0.5x} = 0$$
$$e^{-0.5x}[e^{x} + 4] = 0$$

This equation has no solutions; $e^{-0.5x} > 0$ and $e^{x} > 0$ for all x; there are no x-intercepts.

y-intercept: $f(0) = 5$

Step 2. Analyze $f'(x)$. $f'(x) = 0.5e^{0.5x} - 2e^{-0.5x}$; $f'(x) = 0$:

$$0.5e^{0.5x} - 2e^{-0.5x} = 0$$
$$e^{-0.5x}[0.5e^{x} - 2] = 0$$
$$e^{x} = 4$$
$$x = \ln 4$$

Critical values for f: $x = \ln 4$
Partition numbers for f': $\ln 4$
Sign chart for $f(x)$:

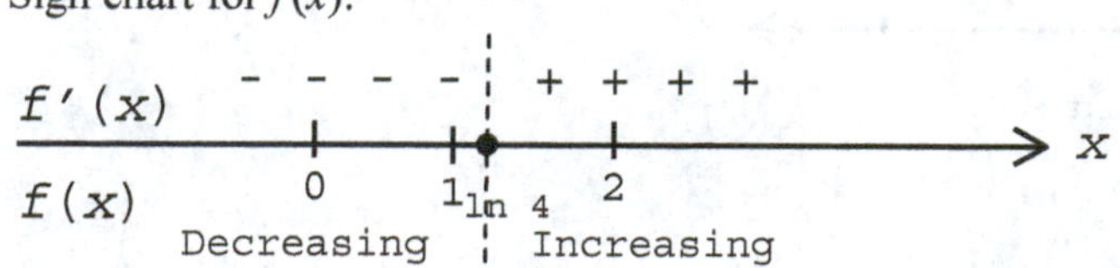

Test Numbers

x	$f'(x)$
0	$-\frac{3}{2}(-)$
2	$\frac{e^2-4}{2e}(+)$

Thus, f is decreasing on $(-\infty, \ln 4)$; f is increasing on $(\ln 4, \infty)$; f has a local minimum at $x = \ln 4$.

Step 3. Analyze $f''(x)$. $f''(x) = 0.25e^{0.5x} + e^{-0.5x} > 0$ for all x.

Sign chart for $f''(x)$:

```
f″(x)    + + + +   + + + +
─────────────────┼──────────────────→ x
Graph of f       0
            Concave Upward
```

Thus, the graph of f is concave upward on $(-\infty, \infty)$.

Step 4. Sketch the graph of f:

x	$f(x)$
0	5
ln 4	4

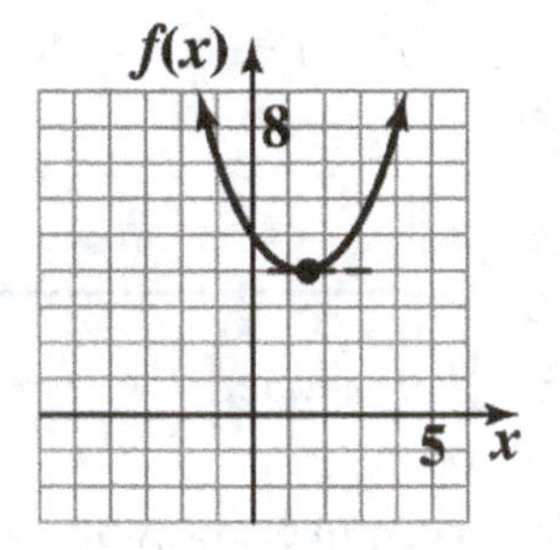

67. $f(x) = 2 \ln x - 4$

Step 1. Analyze $f(x)$. Domain of f: $(0, \infty)$

Intercepts: x-intercepts: $f(x) = 0$

$$2 \ln x - 4 = 0$$
$$\ln x = 2$$
$$x = e^2$$

y-intercept: no y-intercept; $f(0)$ is not defined.

Step 2. Analyze $f'(x)$. $f'(x) = \dfrac{2}{x} > 0$ on $(0, \infty)$

Sign chart for $f(x)$:

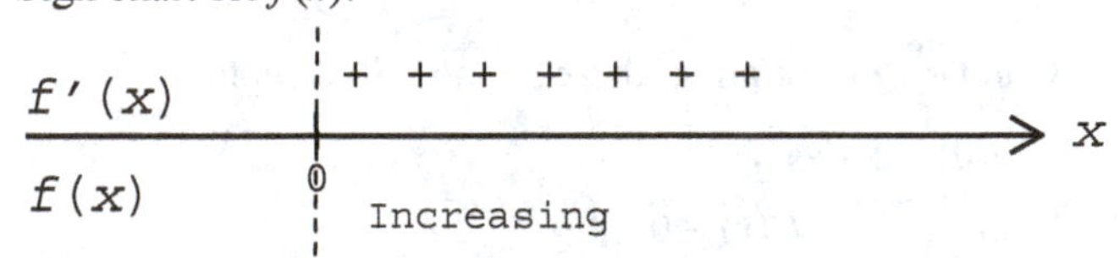

Thus, f is increasing on $(0, \infty)$.

Step 3. Analyze $f''(x)$. $f''(x) = -\dfrac{2}{x^2} < 0$ on $(0, \infty)$.

Sign chart for $f''(x)$:

```
f″(x)          - - - - - - -
───────────┼──────────────────────→ x
Graph of 0 f
             Concave Downward
```

Step 4. Sketch the graph of f:

x	$f(x)$
e^2	0
e^3	2

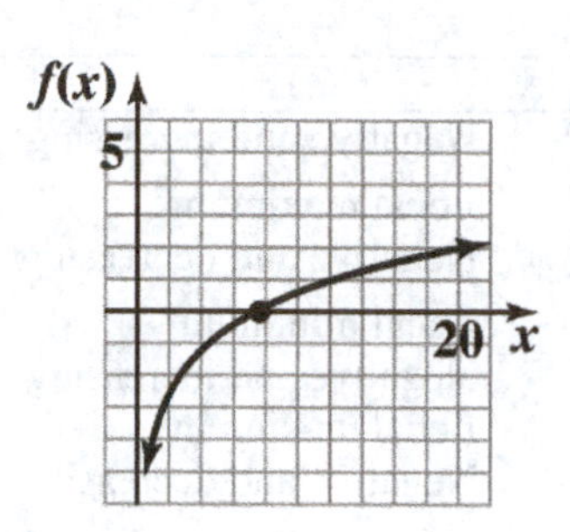

69. $f(x) = \ln(x+4) - 2$

Step 1. Analyze $f(x)$. Domain of f: $(-4, \infty)$

Intercepts: x-intercepts: $f(x) = 0$

$$\ln(x+4) - 2 = 0$$
$$\ln(x+4) = 2$$
$$x + 4 = e^2$$
$$x = e^2 - 4 \approx 3.4$$

y-intercept: $f(0) = \ln 4 - 2 \approx -0.61$

Step 2. Analyze $f'(x)$. $f'(x) = \dfrac{1}{x+4} > 0$ on $(-4, \infty)$

Sign chart for $f'(x)$:

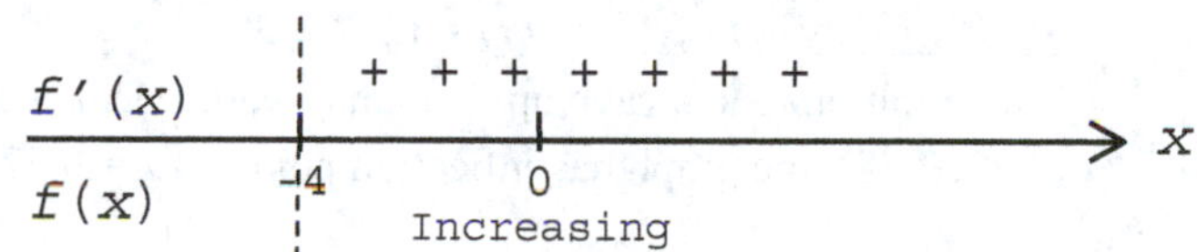

Thus, f is increasing on $(-4, \infty)$.

Step 3. Analyze $f''(x)$. $f''(x) = \dfrac{-1}{(x+4)^2} < 0$ on $(-4, \infty)$.

Sign chart for $f''(x)$:

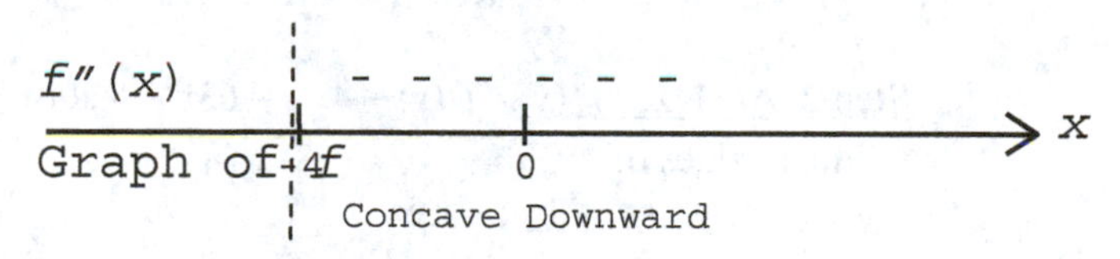

Step 4. Sketch the graph of f:

x	$f(x)$
0	−0.61
$e^2 - 4$	0

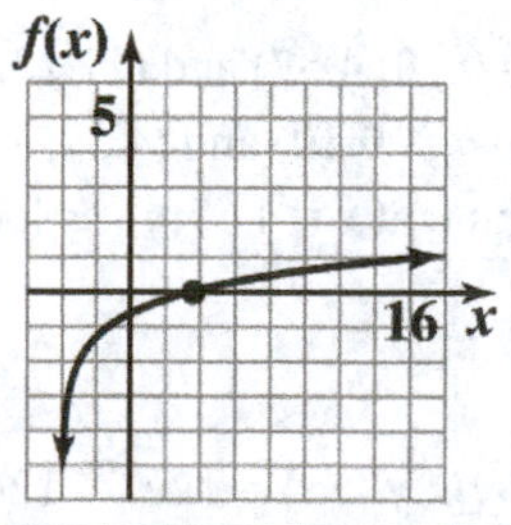

71.

x	$f'(x)$	$f(x)$
$-\infty < x < -1$	Positive and decreasing	Increasing and concave downward
$x = -1$	x-intercept	Local maximum
$-1 < x < 0$	Negative and decreasing	Decreasing and concave downward
$x = 0$	Local minimum	Inflection point
$0 < x < 2$	Negative and increasing	Decreasing and concave upward
$x = 2$	Local maximum	Inflection point
$2 < x < \infty$	Negative and decreasing	Decreasing and concave downward

73.

x	$f'(x)$	$f(x)$
$-\infty < x < -2$	Negative and increasing	Decreasing and concave upward
$x = -2$	Local maximum	Inflection point
$-2 < x < 0$	Negative and decreasing	Decreasing and concave downward
$x = 0$	Local minimum	Inflection point
$0 < x < 2$	Negative and increasing	Decreasing and concave upward
$x = 2$	Local maximum	Inflection point
$2 < x < \infty$	Negative and decreasing	Decreasing and concave downward

75. $f(x) = x^4 - 5x^3 + 3x^2 + 8x - 5$

Step 1. Analyze $f(x)$. Domain of f: $(-\infty, \infty)$.

Intercepts: y-intercept: $f(0) = -5$

x-intercepts: $x \approx -1.18, 0.61, 1.87, 3.71$

Step 2. Analyze $f'(x)$. $f'(x) = 4x^3 - 15x^2 + 6x + 8$

Critical values for f: $x \approx -0.53, 1.24, 3.04$

f is decreasing on $(-\infty, -0.53)$ and $(1.24, 3.04)$; f is increasing on $(-0.53, 1.24)$ and $(3.04, \infty)$; f has local minima at $x = -0.53$ and 3.04; f has a local maximum at $x = 1.24$

Step 3. Analyze $f''(x)$. $f''(x) = 12x^2 - 30x + 6$

The graph of f is concave upward on $(-\infty, 0.22)$ and $(2.28, \infty)$; the graph of f is concave downward on $(0.22, 2.28)$; the graph has inflection points at $x = 0.22$ and 2.28.

77. $f(x) = x^4 - 21x^3 + 100x^2 + 20x + 100$

Step 1. Analyze $f(x)$. Domain of f: $(-\infty, \infty)$.

Intercepts: y-intercept: $f(0) = 100$

x-intercept: $x \approx 8.01, 13.36$

Step 2. Analyze $f'(x)$. $f'(x) = 4x^3 - 63x^2 + 200x + 20$

Critical values of f:

$x \approx -0.10, 4.57, 11.28$

f is increasing on $(-0.10, 4.57)$ and $(11.28, \infty)$;
f is decreasing on $(-\infty, -0.10)$ and $(4.57, 11.28)$;
f has a local maximum at $x = 4.57$; f has local minima at $x = -0.10$ and 11.28.

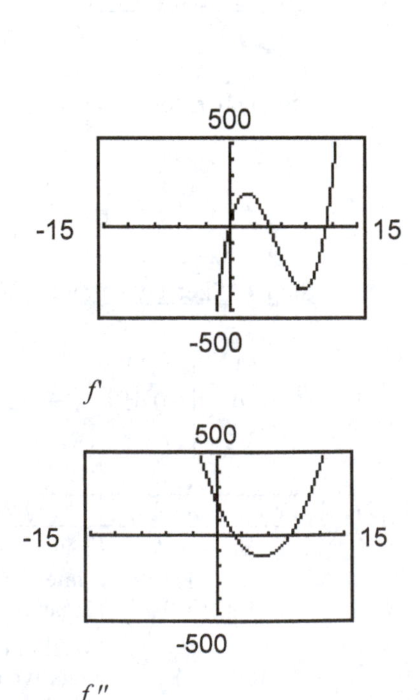

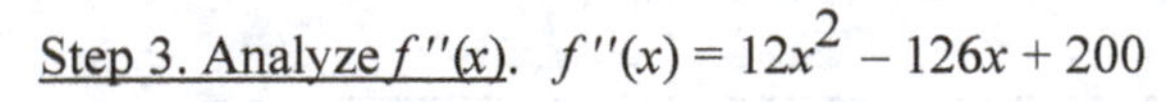

Step 3. Analyze $f''(x)$. $f''(x) = 12x^2 - 126x + 200$

The graph of f is concave upward on $(-\infty, 1.95)$ and $(8.55, \infty)$; the graph of f is concave downward on $(1.95, 8.55)$; the graph has inflection points at $x = 1.95$ and $x = 8.55$.

79. $f(x) = -x^4 - x^3 + 2x^2 - 2x + 3$

Step 1. Analyze $f(x)$. Domain of f: $(-\infty, \infty)$.

Intercepts: y-intercept: $f(0) = 3$
x-intercepts: $x \approx -2.40, 1.16$

Step 2. Analyze $f'(x)$. $f'(x) = -4x^3 - 3x^2 + 4x - 2$
Critical value for f: $x \approx -1.58$
f is increasing on $(-\infty, -1.58)$; f is decreasing on $(-1.58, \infty)$; f has a local maximum at $x = -1.58$

Step 3. Analyze $f''(x)$. $f''(x) = -12x^2 - 6x + 4$
The graph of f is concave downward on $(-\infty, -0.88)$ and $(0.38, \infty)$; the graph of f is concave upward on $(-0.88, 0.38)$; the graph has inflection points at $x = -0.88$ and $x = 0.38$.

81. $f(x) = 0.1x^5 + 0.3x^4 - 4x^3 - 5x^2 + 40x + 30$

Step 1. Analyze $f(x)$. Domain of f: $(-\infty, \infty)$.

Intercepts: y-intercept: $f(0) = 3$
x-intercepts: $x \approx -6.68, -3.64, -0.72$

Step 2. Analyze $f'(x)$. $f'(x) = 0.5x^4 + 1.2x^3 - 12x^2 - 10x + 40$
Critical values for f: $x \approx -5.59, -2.27, 1.65, 3.82$
f is increasing on $(-\infty, -5.59)$, $(-2.27, 1.65)$, and $(3.82, \infty)$; f is decreasing on $(-5.59, -2.27)$ and $(1.65, 3.82)$; f has local minima at $x = -2.27$ and 3.82; f has local maxima at $x = -5.59$ and 1.65

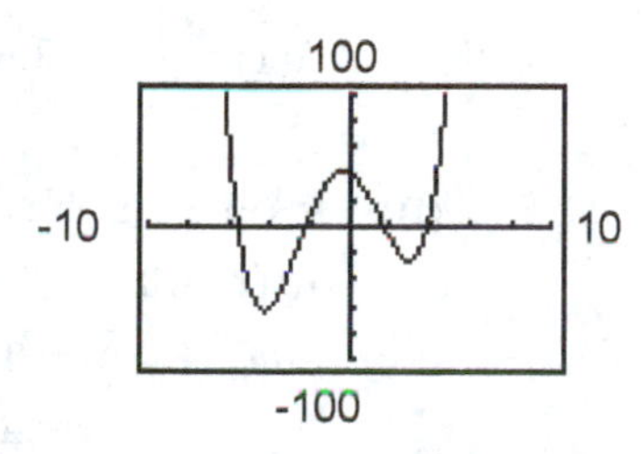

f

Step 3. Analyze $f''(x)$. $f''(x) = 2x^3 + 3.6x^2 - 24x - 10$
The graph of f is concave downward on $(-\infty, -4.31)$ and $(-0.40, 2.91)$; the graph of f is concave upward on $(-4.31, -0.40)$ and $(2.91, \infty)$; the graph has inflection points at $x = -4.31, -0.40$ and 2.91.

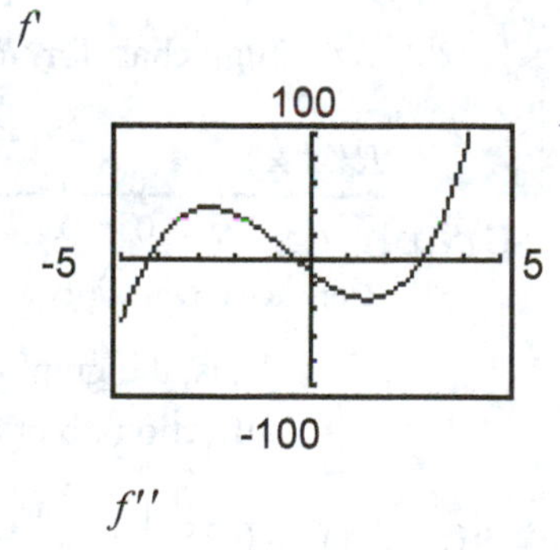

f''

83. The graph of the CPI is concave up.

85. The graph of C is increasing and concave down. Therefore, the graph of C' is positive and decreasing. Since the marginal costs are decreasing, the production process is becoming more efficient.

87. $R(x) = xp = 1296x - 0.12x^3$, $0 < x < 80$
$R'(x) = 1296 - 0.36x^2$
Critical values: $R'(x) = 1296 - 0.36x^2 = 0$

$$x^2 = \frac{1296}{0.36} = 3600$$

$$x = \pm 60$$

Thus, $x = 60$ is the only critical value on the interval $(0, 80)$.

$R''(x) = -0.72x$

$R''(60) = -43.2 < 0$

(A) R has a local maximum at $x = 60$.

(B) Since $R''(x) = -0.72x < 0$ for $0 < x < 80$, R is concave downward on this interval.

89. Demand: $p = 10e^{-x}$, $0 \le x \le 5$

Revenue function: $R(x) = xp(x) = 10xe^{-x}$, $0 \le x \le 5$

(A) $R'(x) = -10xe^{-x} + 10e^{-x} = 10e^{-x}(1 - x)$

$R'(x) = 0$:

$10e^{-x}(1 - x) = 0$

$x = 1$

Sign chart for $R'(x)$:

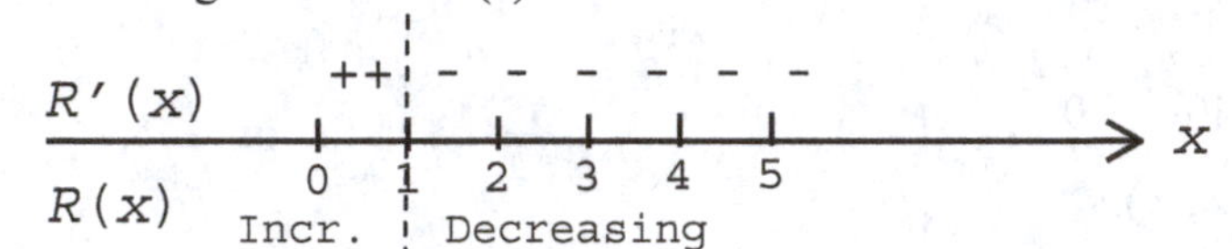

Test Numbers

x	$R'(x)$
0	10 (+)
2	$-\frac{10}{e^2}$ (–)

Thus, R is increasing on [0, 1); R is decreasing on (1, 5]; R has a local maximum at $x = 1$;

$R(1) = \dfrac{10}{e} \approx 3.68$.

(B) $R''(x) = 10xe^{-x} - 10e^{-x} - 10e^{-x} = 10e^{-x}(x - 2)$

$R''(x) = 0$:

$10e^{-x}(x - 2) = 0$

$x = 2$

Sign chart for R'':

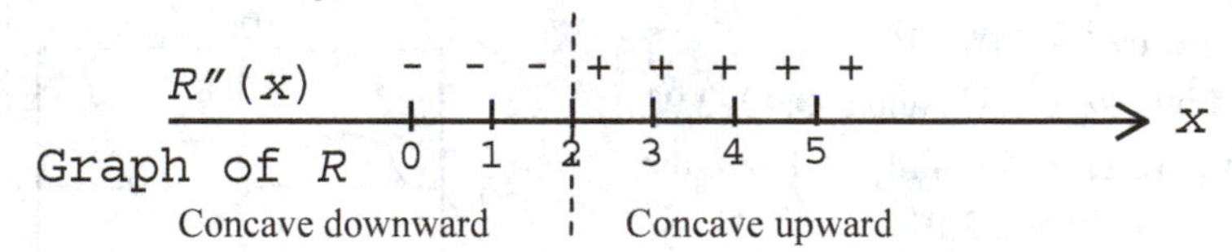

Test Numbers

x	$R''(x)$
0	–20 (–)
3	$\frac{10}{e^3}$ (+)

Thus, the graph of R is concave downward on (0, 2) and concave upward on (2, 5); R has an inflection point at $x = 2$.

91. $T(x) = -0.25x^4 + 5x^3 = -\frac{1}{4}x^4 + 5x^3$, $0 \le x \le 15$

$T'(x) = -x^3 + 15x^2$

$T''(x) = -3x^2 + 30x = -3x(x - 10)$

Partition numbers for $T''(x)$: $x = 10$

Sign chart for $T''(x)$:

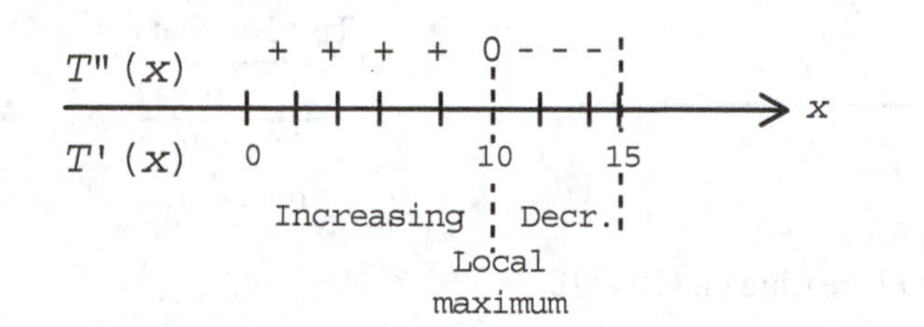

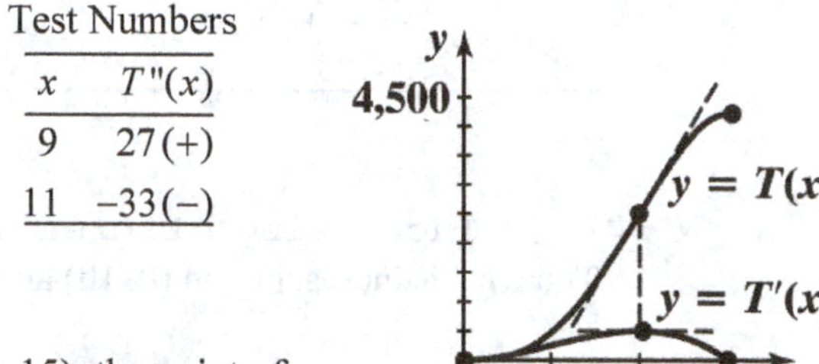

Test Numbers

x	$T''(x)$
9	27(+)
11	−33(−)

Thus, T is increasing on (0, 10) and decreasing on (10, 15); the point of diminishing returns is $x = 10$; the maximum rate of change is $T'(10) = 500$.

93. $N(x) = -0.25x^4 + 23x^3 - 540x^2 + 80{,}000,\ 24 \le x \le 45$

$N'(x) = -x^3 + 69x^2 - 1080x$

$N''(x) = -3x^2 + 138x - 1080 = -3(x^2 - 46x + 360) = -3(x - 10)(x - 36)$

Partition numbers for $N''(x)$: $x = 36$

Sign chart for $N''(x)$:

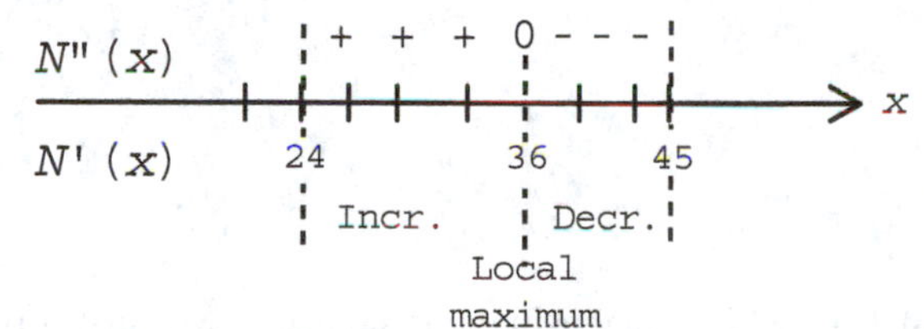

Test Numbers

x	$N''(x)$
35	75(+)
37	−81(−)

Thus, N' is increasing on (24, 36) and decreasing on (36, 45); the point of diminishing returns is $x = 36$; the maximum rate of change is $N'(36) = -(36)^3 + 69(36)^2 - 1080(36) = 3888$.

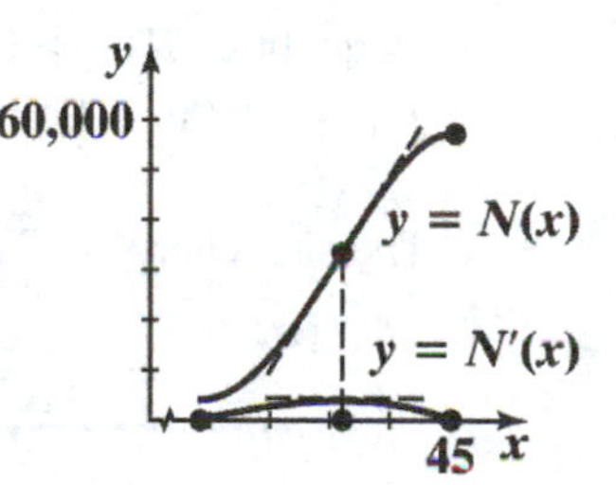

95. (A)

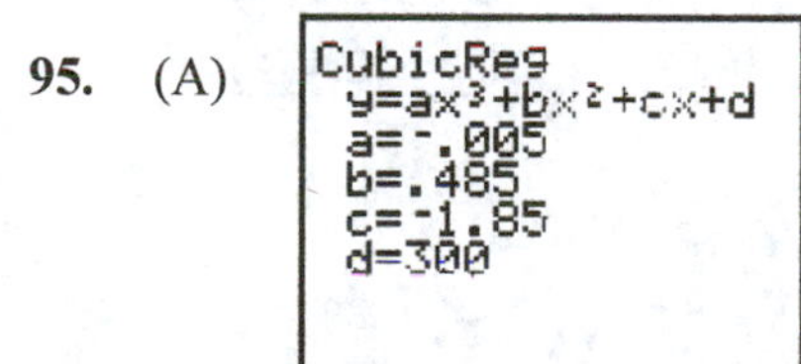

(B) From part (A),

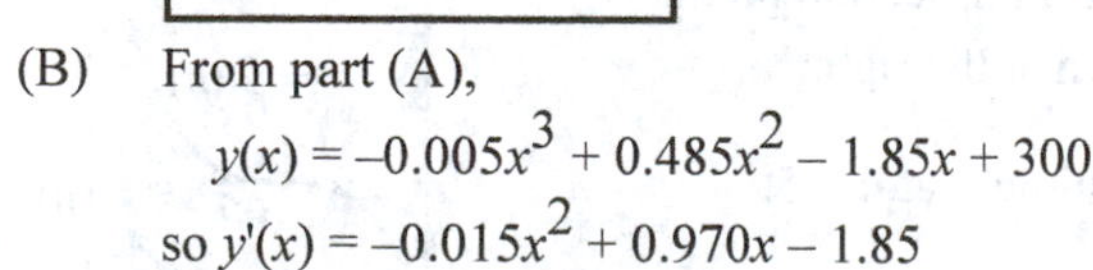

$y(x) = -0.005x^3 + 0.485x^2 - 1.85x + 300$

so $y'(x) = -0.015x^2 + 0.970x - 1.85$

The graph of $y'(x)$ is shown at the right and the maximum value of y' occurs at $x \approx 32$; and $y(32) \approx 574$.

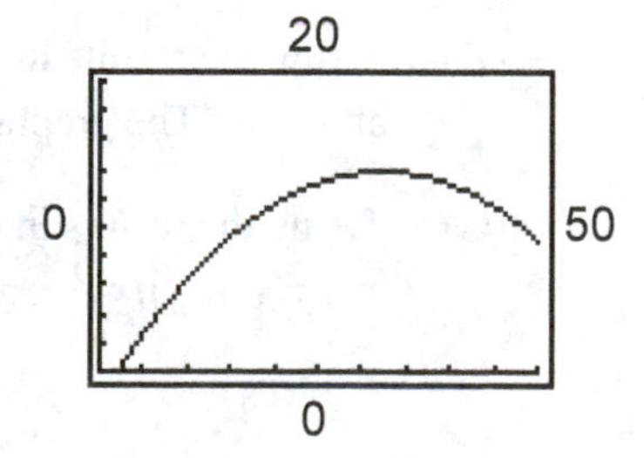

The manager should place 32 ads each month to maximize the rate of change of sales; the manager can expect to sell 574 cars.

97. $N(t) = 1000 + 30t^2 - t^3,\ 0 \le t \le 20$

$N'(t) = 60t - 3t^2$

$N''(t) = 60 - 6t$

(A) To determine when N' is increasing or decreasing, we must solve the inequalities $N''(t) > 0$ and $N''(t) < 0$, respectively. Now

$N''(t) = 60 - 6t = 0$

$t = 10$

The sign chart for N'' (partition number is 10) is:

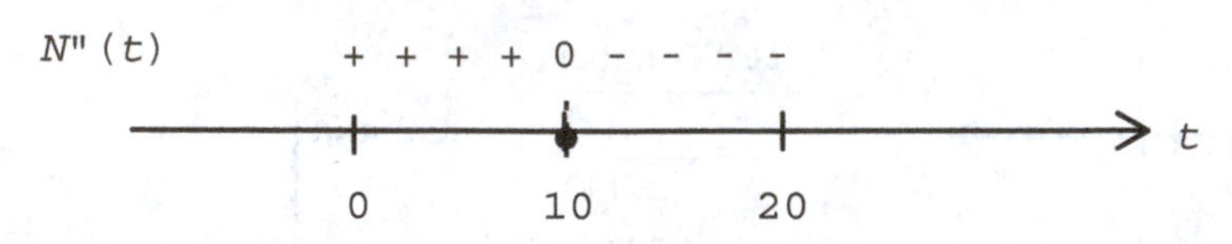

Test Numbers

t	$N''(t)$
0	60(+)
20	−60(−)

Thus, N' is increasing on (0, 10) and decreasing on (10, 20).

(B) From the results in (A), the graph of N has an inflection point at $t = 10$.

(C)

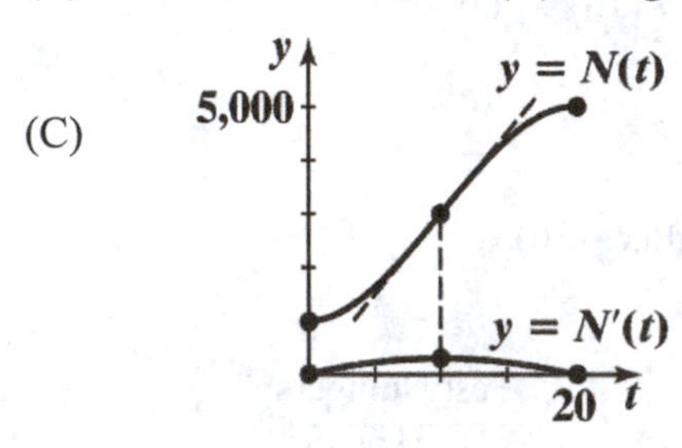

(D) Using the results in (A), N' has a local maximum at $t = 10$: $N'(10) = 300$

99. $T(n) = 0.08n^3 - 1.2n^2 + 6n, n \geq 0$

$T'(n) = 0.24n^2 - 2.4n + 6, \; n \geq 0$

$T''(n) = 0.48n - 2.4$

(A) To determine when the rate of change of T, i.e., T', is increasing or decreasing, we must solve the inequalities $T''(n) > 0$ and $T''(n) < 0$, respectively. Now

$T''(n) = 0.48n - 2.4 = 0$

$n = 5$

The sign chart for T'' (partition number is 5) is:

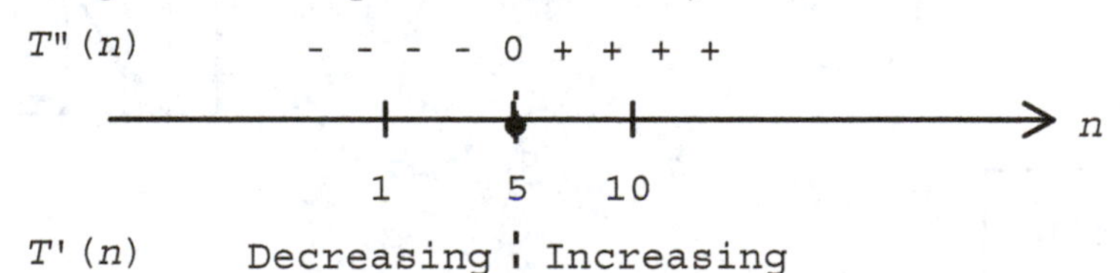

Test Numbers

t	$T''(t)$
1	−1.92(−)
20	2.4(+)

Thus, T is increasing on $(5, \infty)$ and decreasing on (0, 5).

(B) Using the results in (A), the graph of T has an inflection point at $n = 5$. The graphs of T and T' areshown at the right.

(C) Using the results in (A), T' has a local minimum at $n = 5$:
$T'(5) = 0.24(5)^2 - 2.4(5) + 6 = 0$

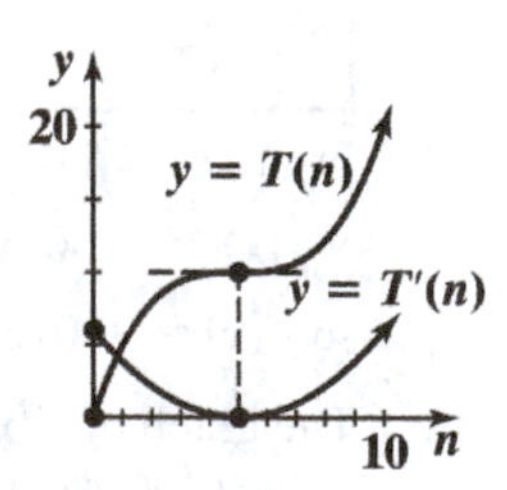

EXERCISE 4-3

Things to remember:

1. L'HÔPITAL'S RULE FOR 0/0 INDETERMINATE FORMS: VERSION 1

For c a real number, if $\lim_{x\to c} f(x) = 0$ and $\lim_{x\to c} g(x) = 0$, then

$$\lim_{x\to c}\frac{f(x)}{g(x)} = \lim_{x\to c}\frac{f'(x)}{g'(x)}$$

provided the second limit exists or is $+\infty$ or $-\infty$.

2. L'HÔPITAL'S RULE FOR 0/0 INDETERMINATE FORMS: VERSION 2
(For One-Sided Limits and Limits at Infinity)
The first version of L'Hôpital's rule remains valid if the symbol $x \to c$ is replaced everywhere it occurs with one of the following symbols:

$$x \to c^{+},\ x \to c^{-},\ x \to \infty,\ \ x \to -\infty.$$

3. L'HÔPITAL'S RULE FOR THE INDETERMINATE FORM ∞/∞: VERSION 3
Versions 1 and 2 of L'Hôpital's rule for the indeterminate form 0/0 are also valid if the limit of f and the limit of g are both infinite; that is both $+\infty$ and $-\infty$ are permissible for either limit.

Note: Throughout this section D_x *denotes* $\frac{d}{dx}$.

1. $\frac{5}{0.01} = 500;\ 500$

3. $\frac{3}{1{,}000} = 0.003;\ 0$

5. $\frac{1}{2(1.01-1)} = \frac{1}{0.02} = 50;\ 50$

7. $\frac{\ln 100}{100} < \frac{5}{100};\ 0$

9. $\lim_{x\to 3}\frac{x^2-9}{x-3} = \lim_{x\to 3}\frac{D_x(x^2-9)}{D_x(x-3)} = \lim_{x\to -3}\frac{2x}{1} = 6$. Therefore $\lim_{x\to 3}\frac{x^2-9}{x-3} = 6$.

11. $\lim_{x\to -5}\frac{x+5}{x^2-25} = \lim_{x\to -5}\frac{D_x(x+5)}{D_x(x^2-25)} = \lim_{x\to -5}\frac{1}{2x} = -\frac{1}{10}$. Therefore, $\lim_{x\to -5}\frac{x+5}{x^2-25} = -\frac{1}{10}$.

13. $\lim_{x\to 1}\frac{x^2+5x-6}{x-1} = \lim_{x\to 1}\frac{D_x(x^2+5x-6)}{D_x(x-1)} = \lim_{x\to 1}\frac{2x+5}{1} = 7$. Therefore $\lim_{x\to 1}\frac{x^2+5x-6}{x-1} = 7$.

15. $\lim_{x\to -9}\frac{x+9}{x^2+13x+36} = \lim_{x\to -9}\frac{D_x(x+9)}{D_x(x^2+13x+36)} = \lim_{x\to -9}\frac{1}{2x+13} = -\frac{1}{5}$. Therefore
$\lim_{x\to -9}\frac{x+9}{x^2+13x+36} = -\frac{1}{5}$.

17. $\lim_{x\to\infty}\frac{2x+3}{5x-1} = \lim_{x\to\infty}\frac{D_x(2x+3)}{D_x(5x-1)} = \lim_{x\to\infty}\frac{2}{5} = \frac{2}{5}$. Therefore $\lim_{x\to\infty}\frac{2x+3}{5x-1} = \frac{2}{5}$.

19. $\lim_{x\to\infty}\frac{3x^2-1}{x^3+4} = \lim_{x\to\infty}\frac{D_x(3x^2-1)}{D_x(x^3+4)} = \lim_{x\to\infty}\frac{6x}{3x^2} = \lim_{x\to\infty}\frac{2}{x} = 0$. Therefore $\lim_{x\to\infty}\frac{3x^2-1}{x^3+4} = 0$.

21. $\lim_{x\to -\infty}\frac{x^2-9}{x-3} = \lim_{x\to -\infty}\frac{D_x(x^2-9)}{D_x(x-3)} = \lim_{x\to -\infty}\frac{2x}{1} = -\infty$. Therefore $\lim_{x\to -\infty}\frac{x^2-9}{x-3} = -\infty$.

23. $\lim_{x\to\infty}\frac{2x^2+3x+1}{3x^2-2x+1}=\lim_{x\to\infty}\frac{D_x(2x^2+3x+1)}{D_x(3x^2-2x+1)}=\lim_{x\to\infty}\frac{4x+3}{6x-2}=\lim_{x\to\infty}\frac{4x}{6x}=\frac{2}{3}$. Therefore

$\lim_{x\to\infty}\frac{2x^2+3x+1}{3x^2-2x+1}=\frac{2}{3}$.

25. $\lim_{x\to0}\frac{e^x-1}{2x}$ (0/0 form)

$\lim_{x\to0}\frac{e^x-1}{2x}=\lim_{x\to0}\frac{D_x(e^x-1)}{D_x(2x)}=\lim_{x\to0}\frac{e^x}{2}=\frac{1}{2}$. Therefore, $\lim_{x\to0}\frac{e^x-1}{2x}=\frac{1}{2}$.

27. $\lim_{x\to1}\frac{x-1}{\ln x}$. (0/0 form)

$\lim_{x\to1}\frac{x-1}{\ln x}=\lim_{x\to1}\frac{D_x(x-1)}{D_x(\ln x)}=\lim_{x\to1}\frac{1}{(1/x)}=\lim_{x\to1}\frac{x}{1}=1$. Therefore $\lim_{x\to1}\frac{x-1}{\ln x}=1$.

29. $\lim_{x\to\infty}\frac{x^2}{e^x}$ (∞/∞ form)

$\lim_{x\to\infty}\frac{x^2}{e^x}=\lim_{x\to\infty}\frac{D_x(x^2)}{D_x(e^x)}=\lim_{x\to\infty}\frac{2x}{e^x}=\lim_{x\to\infty}\frac{2}{e^x}=0$. Therefore $\lim_{x\to\infty}\frac{x^2}{e^x}=0$.

31. $\lim_{x\to0}\frac{e^{4x}-1}{x}$ (0/0 form)

$\lim_{x\to0}\frac{e^{4x}-1}{x}=\lim_{x\to0}\frac{D_x[e^{4x}-1]}{D_x(x)}=\lim_{x\to0}\frac{4e^{4x}}{1}=4$. Therefore $\lim_{x\to0}\frac{e^{4x}-1}{x}=4$.

33. $\lim_{x\to1}\frac{x^2+5x+4}{x^3+1}$; $\lim_{x\to1}x^3+1=2$. Therefore L'Hôpital's Rule does not apply. Use the rule for the limit of a quotient:

$$\lim_{x\to1}\frac{x^2+5x+4}{x^3+1}=\frac{\lim_{x\to1}(x^2+5x+4)}{\lim_{x\to1}(x^3+1)}=\frac{10}{2}=5.$$

35. $\lim_{x\to2}\frac{x+2}{(x-2)^4}$; $\lim_{x\to2}(x-2)^4=0$ and $\lim_{x\to2}(x+2)=4$

Therefore the limit is not an indeterminate form.

$\lim_{x\to2}\frac{x+2}{(x-2)^4}=\infty$

37. $\lim_{x\to0}\frac{e^{4x}-1-4x}{x^2}$

Step 1:

$\lim_{x\to0}(e^{4x}-1-4x)=e^0-1=0$ and $\lim_{x\to0}x^2=0$.

Thus, L'Hôpital's rule 3a applies.

Step 2:

$$\lim_{x\to 0}\frac{D_x(e^{4x}-1-4x)}{D_x x^2}=\lim_{x\to 0}\frac{4e^{4x}-4}{2x}.$$

Since $\lim_{x\to 0}(4e^{4x}-4)=4e^0-4=0$ and $\lim_{x\to 0}2x=0$, $\lim_{x\to 0}\frac{4e^{4x}-4}{2x}$ is a 0/0 indeterminate form and 3a applies.

Step 3: Apply L'Hôpital's rule again.

$$\lim_{x\to 0}\frac{D_x(4e^{4x}-4)}{D_x 2x}=\lim_{x\to 0}\frac{16e^{4x}}{2}=8e^0=8.$$

Thus, $\lim_{x\to 0}\frac{e^{4x}-1-4x}{x^2}=\lim_{x\to 0}\frac{4e^{4x}-4}{2x}=\lim_{x\to 0}\frac{16e^{4x}}{2}=8.$

39. $\lim_{x\to 2}\frac{\ln(x-1)}{x-1}$

Step 1:

$\lim_{x\to 2}\ln(x-1)=\ln(1)=0$ and $\lim_{x\to 2}(x-1)=1$. Thus, L'Hôpital's rule does not apply.

Step 2: Use the quotient property for limits.

$$\lim_{x\to 2}\frac{\ln(x-1)}{x-1}=\frac{\ln 1}{1}=\frac{0}{1}=0.$$

41. $\lim_{x\to 0^+}\frac{\ln(1+x^2)}{x^3}$

Step 1:

$\lim_{x\to 0^+}\ln(1+x^2)=\ln 1=0$ and $\lim_{x\to 0^+}x^3=0$.

Thus, L'Hôpital's rule 3b applies.

Step 2:

$$\lim_{x\to 0^+}\frac{D_x\ln(1+x^2)}{D_x x^3}=\lim_{x\to 0^+}\frac{\frac{2x}{1+x^2}}{3x^2}=\lim_{x\to 0^+}\frac{2}{3x(1+x^2)}=\infty.$$

Thus, $\lim_{x\to 0^+}\frac{\ln(1+x^2)}{x^3}=\infty.$

43. $\lim_{x\to 0^+}\frac{\ln(1+\sqrt{x})}{x}$

Step 1:

$\lim_{x\to 0^+}\ln(1+\sqrt{x})=\ln 1=0$ and $\lim_{x\to 0^+}x=0$.

Thus, L'Hôpital's rule 3b applies.

Step 2:

$$\lim_{x\to 0^+}\frac{D_x \ln(1+\sqrt{x})}{D_x x} = \lim_{x\to 0^+}\frac{\frac{1}{1+\sqrt{x}}\cdot\frac{1}{2}x^{-1/2}}{1} = \lim_{x\to 0^+}\frac{1}{2\sqrt{x}(1+\sqrt{x})} = \infty.$$

Thus, $\lim_{x\to 0^+}\frac{\ln(1+\sqrt{x})}{x} = \infty.$

45. $\lim_{x\to -2}\frac{x^2+2x+1}{x^2+x+1}$

Step 1:

Since $\lim_{x\to -2}(x^2+x+1) = 4 - 2 + 1 = 3$, L'Hôpital's rule does not apply.

Step 2:

Using the limit properties, we have:

$$\lim_{x\to -2}\frac{x^2+2x+1}{x^2+x+1} = \frac{(-2)^2+2(-2)+1}{(-2)^2+(-2)+1} = \frac{4-4+1}{4-2+1} = \frac{1}{3}.$$

47. $\lim_{x\to -1}\frac{x^3+x^2-x-1}{x^3+4x^2+5x+2}$

Step 1:

$\lim_{x\to -1}(x^3+x^2-x-1) = -1 + 1 + 1 - 1 = 0$ and $\lim_{x\to -1}(x^3+4x^2+5x+2) = -1 + 4 - 5 + 2 = 0.$

Thus, L'Hôpital's rule 3a applies.

Step 2:

$$\lim_{x\to -1}\frac{D_x(x^3+x^2-x-1)}{D_x(x^3+4x^2+5x+2)} = \lim_{x\to -1}\frac{3x^2+2x-1}{3x^2+8x+5}.$$

Since, $\lim_{x\to -1}(3x^2+2x-1) = 3 - 2 - 1 = 0$ and $\lim_{x\to -1}(3x^2+8x+5) = 3 - 8 + 5 = 0$, $\lim_{x\to -1}\frac{3x^2+2x-1}{3x^2+8x+5}$ is a 0/0 indeterminate form and 3a applies again.

Step 3:

$$\lim_{x\to -1}\frac{D_x(3x^2+2x-1)}{D_x(3x^2+8x+5)} = \lim_{x\to -1}\frac{6x+2}{6x+8} = \frac{-4}{2} = -2.$$

Thus,

$$\lim_{x\to -1}\frac{x^3+x^2-x-1}{x^3+4x^2+5x+2} = \lim_{x\to -1}\frac{3x^2+2x-1}{3x^2+8x+5} = \lim_{x\to -1}\frac{6x+2}{6x+8} = -2.$$

49. $\lim_{x\to 2^-}\frac{x^3-12x+16}{x^3-6x^2+12x-8}$

Step 1:

$\lim_{x\to 2^-}(x^3-12x+16) = 8 - 24 + 16 = 0$ and $\lim_{x\to 2^-}(x^3-6x^2+12x-8) = 8 - 24 + 24 - 8 = 0.$

Thus L'Hôpital's 3b applies.

Step 2:

$$\lim_{x\to 2^-} \frac{D_x(x^3-12x+16)}{D_x(x^3-6x^2+12x-8)} = \lim_{x\to 2^-} \frac{3x^2-12}{3x^2-12x+12}.$$

Since $\lim_{x\to 2^-}(3x^2-12) = 12-12=0$ and $\lim_{x\to 2^-}(3x^2-12x+12) = 12-24+12=0$, $\lim_{x\to 2^-}\frac{3x^2-12}{3x^2-12x+12}$ is a 0/0 indeterminate form and 3b applies again.

Step 3:

$$\lim_{x\to 2^-} \frac{D_x(3x^2-12)}{D_x(3x^2-12x+12)} = \lim_{x\to 2^-} \frac{6x}{6x-12} = \lim_{x\to 2^-} \frac{x}{x-2} = -\infty.$$

Thus,

$$\lim_{x\to 2^-} \frac{x^3-12x+16}{x^3-6x^2+12x-8} = \lim_{x\to 2^-} \frac{3x^2-12}{3x^2-12x+12} = \lim_{x\to 2^-} \frac{x}{x-2} = -\infty.$$

51. $\lim_{x\to\infty} \frac{3x^2+5x}{4x^3+7}$

Step 1:

$\lim_{x\to\infty}(3x^2+5x) = \infty$ and $\lim_{x\to\infty}(4x^3+7) = \infty$.

Thus, L'Hôpital's rule 4 applies.

Step 2:

$$\lim_{x\to\infty} \frac{D_x(3x^2+5x)}{D_x(4x^3+7)} = \lim_{x\to\infty} \frac{6x+5}{12x^2}.$$

Since $\lim_{x\to\infty}(6x+5) = \infty$ and $\lim_{x\to\infty} 12x^2 = \infty$, $\lim_{x\to\infty}\frac{6x+5}{12x^2}$ is an ∞/∞ indeterminate form and 4 applies again.

Step 3:

$$\lim_{x\to\infty} \frac{D_x(6x+5)}{D_x(12x^2)} = \lim_{x\to\infty} \frac{6}{24x} = \frac{1}{4}\lim_{x\to\infty}\frac{1}{x} = 0.$$

Thus, $\lim_{x\to\infty} \frac{3x^2+5x}{4x^3+7} = \lim_{x\to\infty} \frac{6x+5}{12x^2} = \frac{1}{4}\lim_{x\to\infty}\frac{1}{x} = 0.$

An alternative approach is:

$$\lim_{x\to\infty} \frac{3x^2+5x}{4x^3+7} = \lim_{x\to\infty} \frac{x^2\left(3+\frac{5}{x}\right)}{x^2\left(4x+\frac{7}{x^2}\right)} = \lim_{x\to\infty} \frac{3+\frac{5}{x}}{4x+\frac{7}{x^2}} = 0.$$

53. $\lim_{x\to\infty} \frac{x^2}{e^{2x}}$

Step 1

$\lim_{x\to\infty} x^2 = \infty$ and $\lim_{x\to\infty} e^{2x} = \infty$. Thus, L'Hôpital's rule 4 applies.

Step 2:

$$\lim_{x\to\infty} \frac{D_x x^2}{D_x e^{2x}} = \lim_{x\to\infty} \frac{2x}{2e^{2x}} = \lim_{x\to\infty} \frac{x}{e^{2x}}.$$

Since $\lim_{x\to\infty} x = \infty$ and $\lim_{x\to\infty} e^{2x} = \infty$, $\lim_{x\to\infty} \frac{x}{e^{2x}}$ is an ∞/∞ indeterminate form and 4 applies again.

Step 3:

$$\lim_{x\to\infty} \frac{D_x x}{D_x e^{2x}} = \lim_{x\to\infty} \frac{1}{2e^{2x}} = 0. \text{ Thus, } \lim_{x\to\infty} \frac{x^2}{e^{2x}} = \lim_{x\to\infty} \frac{x}{e^{2x}} = \lim_{x\to\infty} \frac{1}{2e^{2x}} = 0.$$

55. $\lim_{x\to\infty} \frac{1+e^{-x}}{1+x^2}$

Step 1:

$\lim_{x\to\infty} (1+e^{-x}) = 1$ and $\lim_{x\to\infty} (1+x^2) = \infty$. Thus, this limit is *not* an indeterminate form.

Step 2:

Since $\lim_{x\to\infty} (1+e^{-x}) = 1$ and $\lim_{x\to\infty} (1+x^2) = \infty$, $\lim_{x\to\infty} \frac{1+e^{-x}}{1+x^2} = 0$.

57. $\lim_{x\to\infty} \frac{e^{-x}}{\ln(1+4e^{-x})}$

Step 1:

$\lim_{x\to\infty} e^{-x} = 0$ and $\lim_{x\to\infty} \ln(1 + 4e^{-x}) = \ln 1 = 0$.

Thus, L'Hôpital's rule 3b applies.

Step 2:

$$\lim_{x\to\infty} \frac{D_x(e^{-x})}{D_x \ln(1+4e^{-x})} = \lim_{x\to\infty} \frac{-e^{-x}}{\frac{1}{1+4e^{-x}}\cdot(-4e^{-x})} = \lim_{x\to\infty} \frac{1+4e^{-x}}{4} = \frac{1}{4}.$$

Thus, $\lim_{x\to\infty} \frac{e^{-x}}{\ln(1+4e^{-x})} = \frac{1}{4}$.

59. $\lim_{x\to 0} \frac{e^x - e^{-x} - 2x}{x^3}$

Step 1:

$\lim_{x\to 0} (e^x - e^{-x} - 2x) = e^0 - e^0 = 0$ and $\lim_{x\to 0} x^3 = 0$.

Thus, L'Hôpital's rule 3a applies.

Step 2:

$$\lim_{x\to 0}\frac{D_x(e^x - e^{-x} - 2x)}{D_x x^3} = \lim_{x\to 0}\frac{(e^x + e^{-x} - 2)}{3x^2}.$$

Since $\lim_{x\to 0}(e^x + e^{-x} - 2) = e^0 + e^0 - 2 = 0$ and $\lim_{x\to 0} 3x^2 = 0$, L'Hôpital's rule 3a applies again.

Step 3:

$$\lim_{x\to 0}\frac{D_x(e^x + e^{-x} - 2)}{D_x 3x^2} = \lim_{x\to 0}\frac{e^x - e^{-x}}{6x}.$$ Since $\lim_{x\to 0}(e^x - e^{-x}) = e^0 - e^0 = 0$

and $\lim_{x\to 0} 6x = 0$, we apply 3a a third time.

Step 4:

$$\lim_{x\to 0}\frac{D_x(e^x - e^{-x})}{D_x 6x} = \lim_{x\to 0}\frac{e^x + e^{-x}}{6} = \frac{e^0 + e^0}{6} = \frac{2}{6} = \frac{1}{3}.$$

Therefore, $\lim_{x\to 0}\frac{e^x - e^{-x} - 2x}{x^3} = \frac{1}{3}$.

61. $\lim_{x\to 0^+} x\ln x = \lim_{x\to 0^+}\frac{\ln x}{\frac{1}{x}}$

Step 1:

$\lim_{x\to 0^+}\ln x = -\infty$ and $\lim_{x\to 0^+}\frac{1}{x} = \infty$.

Therefore, L'Hôpital's rule 4 applies.

Step 2:

$$\lim_{x\to 0^+}\frac{D_x \ln x}{D_x \frac{1}{x}} = \lim_{x\to 0^+}\frac{\frac{1}{x}}{-\frac{1}{x^2}} = \lim_{x\to 0^+}(-x) = 0.$$ Thus, $\lim_{x\to 0^+} x\ln x = 0$.

63. $\lim_{x\to\infty}\frac{\ln x}{x^n}$, n a positive integer.

Step 1:

$\lim_{x\to\infty}\ln x = \infty$ and $\lim_{x\to\infty} x^n = \infty$. Therefore, L'Hôpital's rule 4 applies.

Step 2:

$$\lim_{x\to\infty}\frac{D_x \ln x}{D_x x^n} = \lim_{x\to\infty}\frac{\frac{1}{x}}{nx^{n-1}} = \lim_{x\to\infty}\frac{1}{nx^n} = 0.$$ Thus, $\lim_{x\to\infty}\frac{\ln x}{x^n} = 0$.

65. $\lim_{x\to\infty}\frac{e^x}{x^n}$, n a positive integer.

Step 1:

$\lim_{x\to\infty} e^x = \infty$ and $\lim_{x\to\infty} x^n = \infty$. Therefore, L'Hôpital's rule 4 applies.

Step 2:

$$\lim_{x\to\infty}\frac{D_x e^x}{D_x x^n}=\lim_{x\to\infty}\frac{e^x}{nx^{n-1}}$$

If $n=1$, this limit is $\lim_{x\to\infty}\frac{e^x}{1}=\infty$. If $n>1$, then L'Hôpital's rule 4 applies again.

Step 3:

$$\lim_{x\to\infty}\frac{D_x e^x}{D_x nx^{n-1}}=\lim_{x\to\infty}\frac{e^x}{n(n-1)x^{n-2}}.$$

This limit is ∞ if $n=2$ and has the indeterminate form ∞/∞ if $n>2$. Applying L'Hôpital's rule 4 n-times, we have $\lim_{x\to\infty}\frac{e^x}{n!}=\infty$. Thus $\lim_{x\to\infty}\frac{e^x}{x^n}=\infty$.

67. $\lim_{x\to\infty}\frac{\sqrt{1+x^2}}{x}$

Step 1:

$\lim_{x\to\infty}\sqrt{1+x^2}=\infty$ and $\lim_{x\to\infty}x=\infty$.

Thus, L'Hôpital's rule applies.

Step 2:

$$\lim_{x\to\infty}\frac{D_x(\sqrt{1+x^2})}{D_x(x)}=\lim_{x\to\infty}\frac{\frac{x}{\sqrt{1+x^2}}}{1}=\lim_{x\to\infty}\frac{x}{\sqrt{1+x^2}}$$

Since $\lim_{x\to\infty}x=\lim_{x\to\infty}\sqrt{1+x^2}=\infty$, L'Hôpital's rule applies again.

Step 3.

$$\lim_{x\to\infty}\frac{D_x(x)}{D_x\sqrt{1+x^2}}=\lim_{x\to\infty}\frac{1}{\frac{x}{\sqrt{1+x^2}}}=\lim_{x\to\infty}\frac{\sqrt{1+x^2}}{x}$$, the original limit.

Algebraic manipulation: for $x>0$

$$\frac{\sqrt{1+x^2}}{x}=\sqrt{\frac{1+x^2}{x^2}}=\sqrt{\frac{1}{x^2}+1}.$$

Therefore,

$$\lim_{x\to\infty}\frac{\sqrt{1+x^2}}{x}=\lim_{x\to\infty}\sqrt{\frac{1}{x^2}+1}=1$$

69. $\lim_{x\to-\infty}\frac{\sqrt[3]{x^3+1}}{x}=\lim_{x\to-\infty}\frac{(x^3+1)^{1/3}}{x}$

Step 1:

$\lim_{x\to-\infty}(x^3+1)^{1/3}=-\infty$ and $\lim_{x\to-\infty}x=-\infty$

Thus, L'Hôpital's rule applies.

Step 2:

$$\lim_{x\to-\infty}\frac{D_x(x^3+1)^{1/3}}{D_x(x)} = \lim_{x\to-\infty}\frac{\frac{1}{3}(x^3+1)^{-2/3}\cdot 3x^2}{1} = \lim_{x\to-\infty}\frac{x^2}{(x^3+1)^{2/3}}$$

Since $\lim_{x\to-\infty} x^2 = \lim_{x\to-\infty} (x^3+1)^{2/3} = \infty$, L'Hôpital's rule applies again.

Step 3:

$$\lim_{x\to-\infty}\frac{D_x(x^2)}{D_x(x^3+1)^{2/3}} = \lim_{x\to-\infty}\frac{2x}{\frac{2}{3}(x^3+1)^{-1/3}(3x^2)} = \lim_{x\to-\infty}\frac{(x^3+1)^{1/3}}{x}\text{, the original limit.}$$

Algebraic manipulation: for $x < 0$,

$$\frac{\sqrt[3]{x^3+1}}{x} = \sqrt[3]{\frac{x^3+1}{x^3}} = \sqrt[3]{1+\frac{1}{x^3}}.$$

Therefore,

$$\lim_{x\to-\infty}\frac{\sqrt[3]{x^3+1}}{x} = \lim_{x\to-\infty}\sqrt{1+\frac{1}{x^3}} = 1.$$

EXERCISE 4-4

Things to remember:

1. GRAPHING STRATEGY

Step 1. Analyze $f(x)$. $f''(x)$

(A) Find the domain of f.

(B) Find the intercepts.

(C) Find asymptotes.

Step 2. Analyze $f'(x)$. Find the partition numbers and critical values of $f'(x)$. Construct a sign chart for $f'(x)$, determine the intervals where f is increasing and decreasing, and find local maxima and minima.

Step 3. Analyze $f''(x)$. Find the partition numbers of $f''(x)$. Construct a sign chart for $f''(x)$, determine the intervals where the graph of f is concave upward and concave downward, and find inflection points.

Step 4. Sketch the graph of f. Draw asymptotes and locate intercepts, local maxima and minima, and inflection points. Sketch in what you know from steps 1—3. Plot additional points as needed and complete the sketch.

1. $f(x)=3x+36$. Domain: All real numbers; x-intercept: $3x+36=0,\ x=-12$;
y-intercept: $f(0)=36$.

3. $f(x)=\sqrt{25-x}$. Domain: $(-\infty,5]$; x-intercepts: $\sqrt{25-x}=0,\ x=25$;
y-intercept: $f(0)=5$.

5. $f(x)=\dfrac{x+1}{x-2}$. Domain: All real numbers except $x=2$; x-intercepts: $f(x)=\dfrac{x+1}{x-2}=0,\ x=-1$; y-intercept: $f(0)=-\dfrac{1}{2}$.

7. $f(x)=\dfrac{3}{x^2-1}$. Domain: All real numbers except $x=-1, 1$; x-intercepts: none, y-intercept: $f(0)=-3$.

9. (A) $f'(x)<0$ on $(-\infty, b)$, $(0, e)$, (e, g)
(B) $f'(x)>0$ on (b, d), $(d, 0)$, (g, ∞)
(C) $f(x)$ is increasing on (b, d), $(d, 0)$, (g, ∞)
(D) $f(x)$ is decreasing on $(-\infty, b)$, $(0, e)$, (e, g)
(E) $f(x)$ has a local maximum at $x=0$
(F) $f(x)$ has local minima at $x=b$ and $x=g$
(G) $f''(x)<0$ on $(-\infty, a)$, (d, e), (h, ∞)
(H) $f''(x)>0$ on (a, d), (e, h)
(I) The graph of f is concave upward on (a, d) and (e, h).
(J) The graph of f is concave downward on $(-\infty, a)$, (d, e), and (h, ∞).
(K) Inflection points at $x=a$, $x=h$
(L) Horizontal asymptote: $y=L$
(M) Vertical asymptotes: $x=d$, $x=e$

11. Step 1. Analyze $f(x)$:
(A) Domain: All real numbers
(B) Intercepts: y-intercept: 0
x-intercepts: –4, 0, 4
(C) Asymptotes: Horizontal asymptote: $y=2$

Step 2. Analyze $f'(x)$:

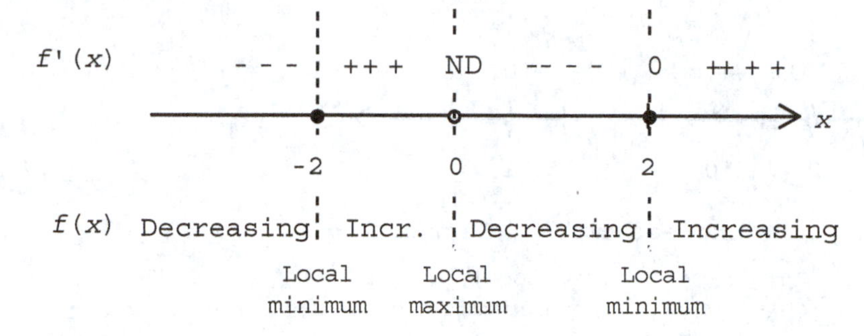

Step 3. Analyze $f''(x)$:

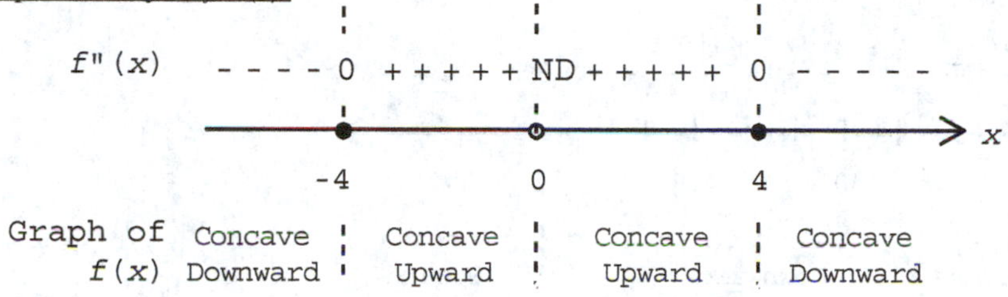

Step 4. Sketch the graph of f:

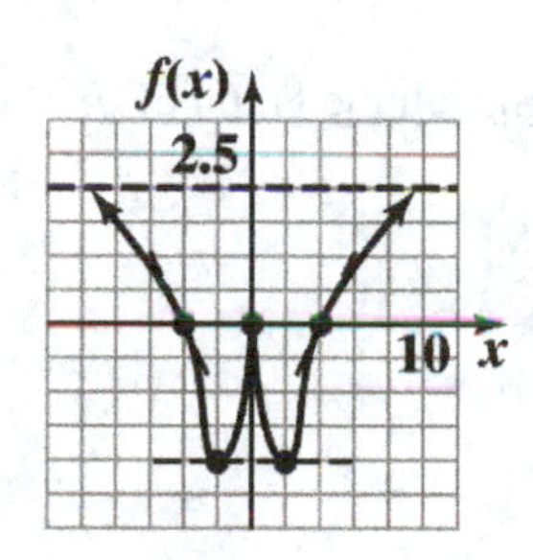

13. Step 1. Analyze $f(x)$:

(A) Domain: All real numbers except $x = -2$

(B) Intercepts: y-intercept: 0

x-intercepts: -4, 0

(C) Asymptotes: Horizontal asymptote: $y = 1$

Vertical asymptote: $x = -2$

Step 2. Analyze $f'(x)$:

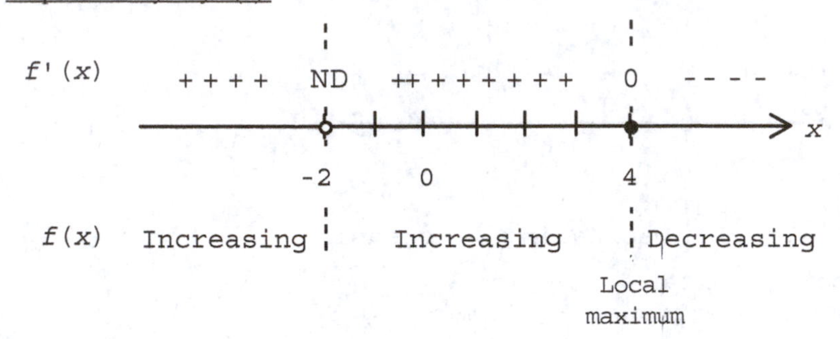

Step 3. Analyze $f''(x)$:

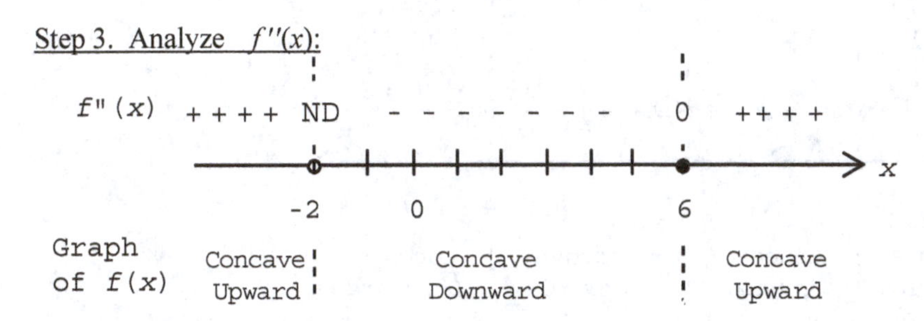

Step 4. Sketch the graph of f:

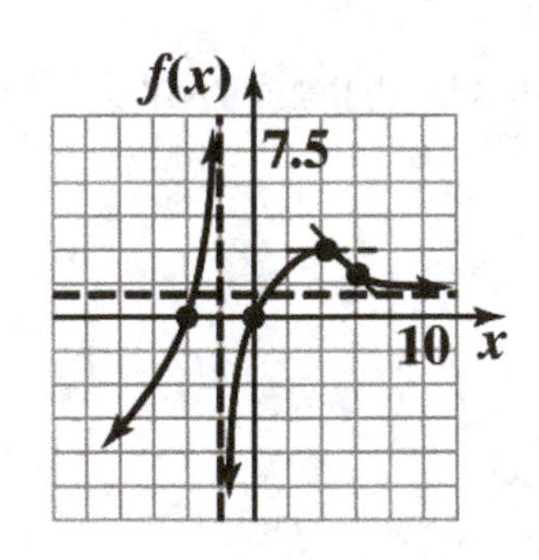

15. Step 1. Analyze $f(x)$:

(A) Domain: All real numbers except $x = -1$

(B) Intercepts: y-intercept: -1

x-intercept: 1

(C) Asymptotes: Horizontal asymptote: $y = 1$

Vertical asymptote: $x = -1$

Step 2. Analyze $f'(x)$:

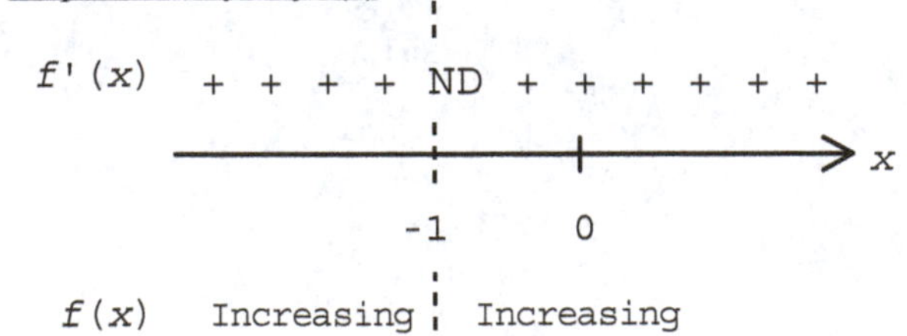

Step 3. Analyze $f''(x)$:

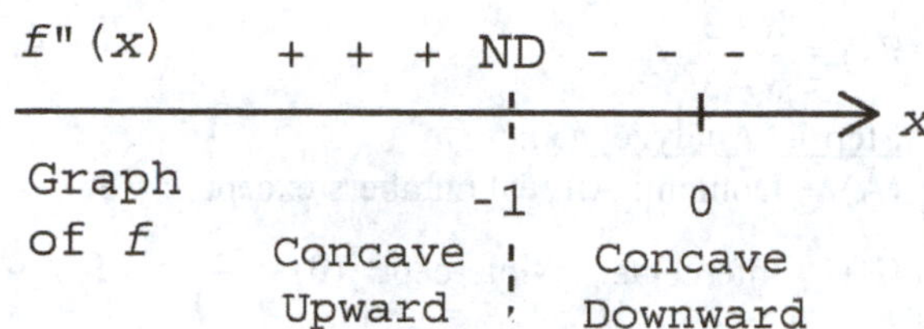

Step 4. Sketch the graph of f:

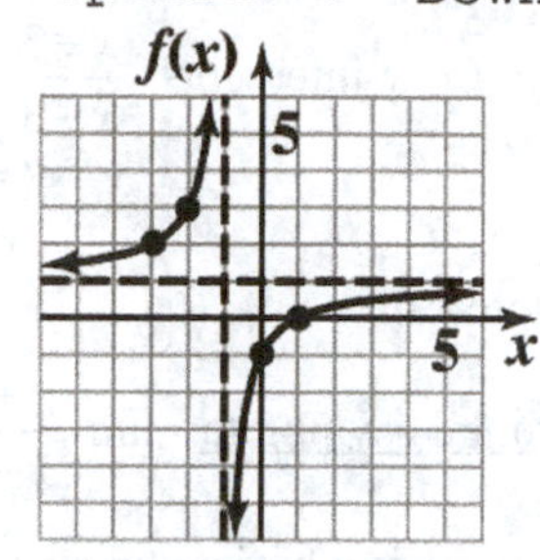

17. Step 1. Analyze $f(x)$:

(A) Domain: All real numbers except $x = -2$, $x = 2$

(B) Intercepts: y-intercept: 0
x-intercept: 0

(C) Asymptotes: Horizontal asymptote: $y = 0$
Vertical asymptotes: $x = -2$, $x = 2$

Step 2. Analyze $f'(x)$:

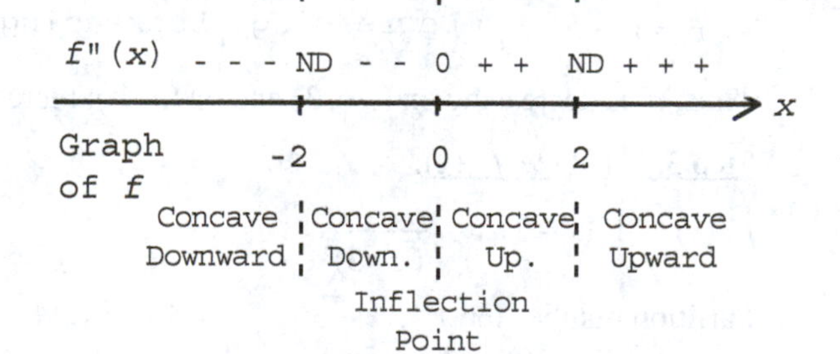

Step 3. Analyze $f''(x)$:

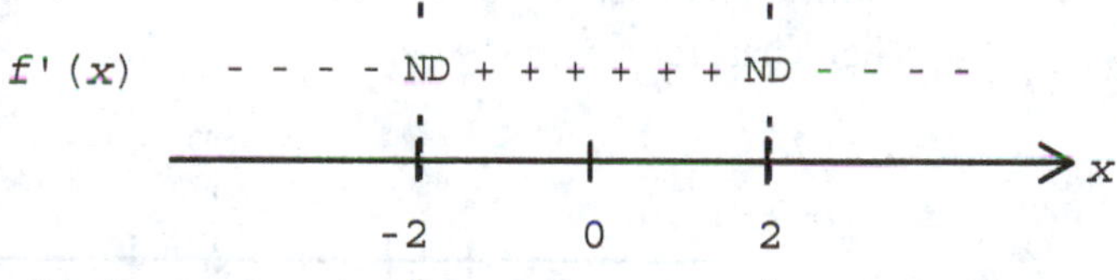

Step 4. Sketch the graph of f:

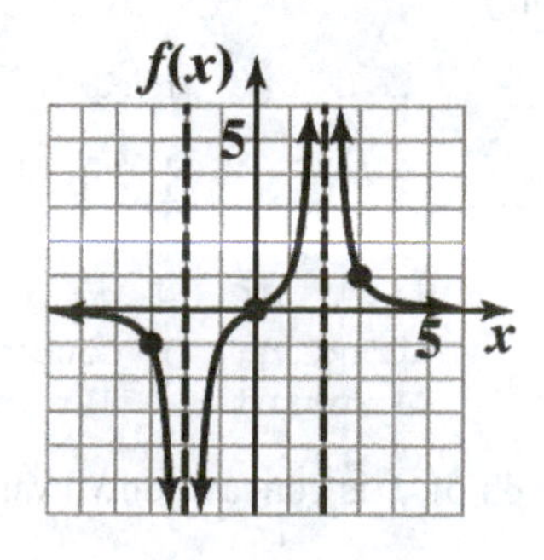

19. $f(x) = \dfrac{x+3}{x-3}$

Step 1. Analyze $f(x)$:

(A) Domain: All real numbers except $x = 3$.

(B) Intercepts: y-intercept: $f(0) = \dfrac{3}{-3} = -1$

x-intercepts: $\dfrac{x+3}{x-3} = 0$

$x + 3 = 0$

$x = -3$

(C) Asymptotes:

Horizontal asymptote: $\lim\limits_{x\to\infty} \dfrac{x+3}{x-3} = \lim\limits_{x\to\infty} \dfrac{x\left(1+\frac{3}{x}\right)}{x\left(1-\frac{3}{x}\right)} = 1.$

Thus, $y = 1$ is a horizontal asymptote.

Vertical asymptote: The denominator is 0 at $x = 3$ and the numerator is not 0 at $x = 3$. Thus, $x = 3$ is a vertical asymptote.

Step 2. Analyze $f'(x)$:

$$f'(x) = \frac{(x-3)(1)-(x+3)(1)}{(x-3)^2} = \frac{-6}{(x-3)^2} = -6(x-3)^{-2}$$

Critical values: None

Partition number: $x = 3$

Sign chart for f':

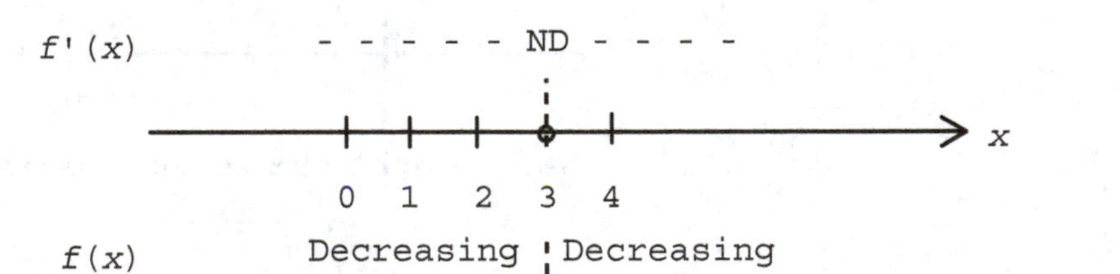

Test Numbers

x	$f'(x)$
2	$-6(-)$
4	$-6(-)$

Thus, f is decreasing on $(-\infty, 3)$ and on $(3, \infty)$; there are no local extrema.

Step 3. Analyze $f''(x)$:

$$f''(x) = 12(x-3)^{-3} = \frac{12}{(x-3)^3}$$

Partition number for f'': $x = 3$

Sign chart for f'':

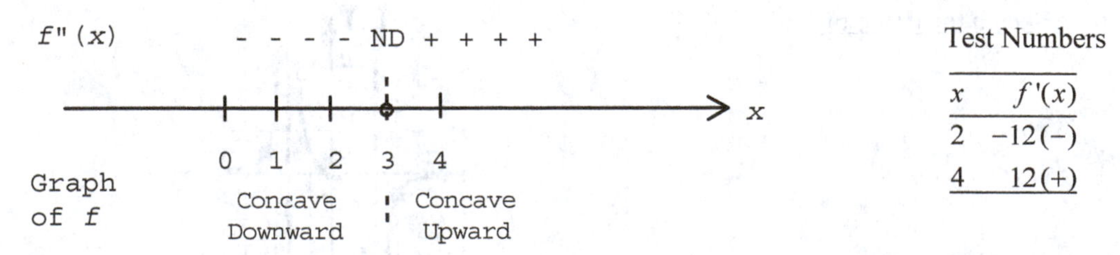

Test Numbers

x	$f'(x)$
2	$-12(-)$
4	$12(+)$

Thus, the graph of f is concave downward on $(-\infty, 3)$ and concave upward on $(3, \infty)$.

Step 4. Sketch the graph of f:

x	$f(x)$
–3	0
0	–1
5	4

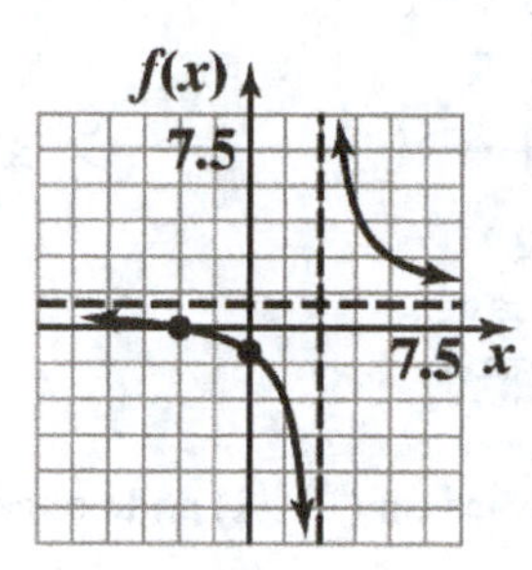

21. $f(x) = \dfrac{x}{x-2}$

Step 1. Analyze $f(x)$:

(A) Domain: All real numbers except $x = 2$.

(B) Intercepts: y-intercept: $f(0) = \dfrac{0}{-2} = 0$

x-intercepts: $\dfrac{x}{x-2} = 0$

$x = 0$

(C) Asymptotes:

Horizontal asymptote: $\lim\limits_{x\to\infty} \dfrac{x}{x-2} = \lim\limits_{x\to\infty} \dfrac{x}{x\left(1-\frac{2}{x}\right)} = 1.$

Thus, $y = 1$ is a horizontal asymptote.

Vertical asymptote: The denominator is 0 at $x = 2$ and the numerator is not 0 at $x = 2$. Thus, $x = 2$ is a vertical asymptote.

Step 2. Analyze $f'(x)$:

$$f'(x) = \frac{(x-2)(1) - x(1)}{(x-2)^2} = \frac{-2}{(x-2)^2} = -2(x-2)^{-2}$$

Critical values: None

Partition number: $x = 2$

Sign chart for f':

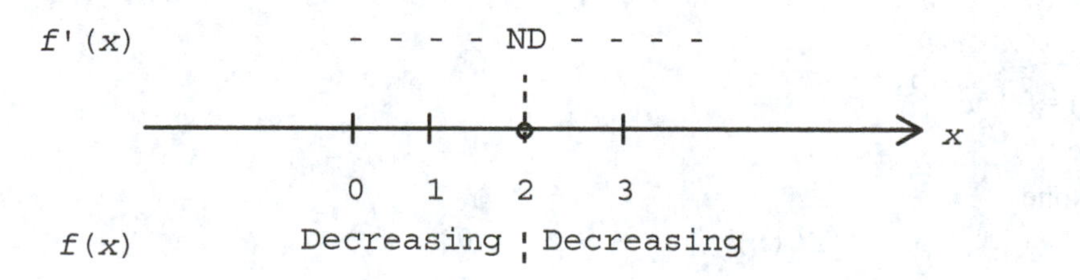

Test Numbers

x	$f'(x)$
0	$-\frac{1}{2}(-)$
3	$-2(-)$

Thus, f is decreasing on $(-\infty, 2)$ and on $(2, \infty)$; there are no local extrema.

Step 3. Analyze $f''(x)$:

$$f''(x) = 4(x-2)^{-3} = \frac{4}{(x-2)^3}$$

Partition number for f'': $x = 2$

Sign chart for f'':

$f''(x)$ - - - - ND + + + +

Graph of f: 0 1 2 3 — Concave Downward | Concave Upward

Test Numbers

x	$f''(x)$
0	$-\frac{1}{2}(-)$
3	$4(+)$

Thus, the graph of f is concave downward on $(-\infty, 2)$ and concave upward on $(2, \infty)$.

Step 4. Sketch the graph of f:

x	$f(x)$
0	0
4	2

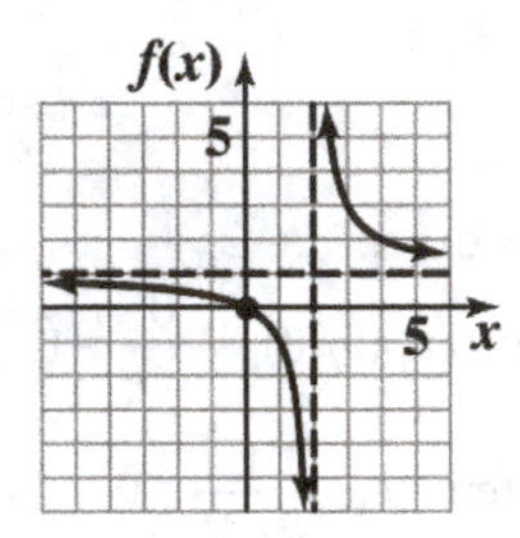

23. $f(x) = 5 + 5e^{-0.1x}$

Step 1. Analyze $f(x)$:

(A) Domain: All real numbers.

(B) Intercepts: y-intercept: $f(0) = 5 + 5e^{0} = 10$

x-intercept: $5 + 5e^{-0.1x} = 0$

$e^{-0.1x} = -1$; no solutions

$e^{-0.1x} > 0$ for all x

(C) Asymptotes:

Vertical asymptotes: None

Horizontal asymptotes: $\lim_{x\to\infty} (5 + 5e^{-0.1x}) = \lim_{x\to\infty} \left(5 + \frac{5}{e^{0.1x}}\right) = 5$

$\lim_{x\to-\infty} (5 + 5e^{-0.1x})$ does not exist

$y = 5$ is a horizontal asymptote.

Step 2. Analyze $f'(x)$:

$f'(x) = 5e^{-0.1x}(-0.1) = -0.5e^{-0.1x}$

Critical values: None

Partition numbers: None

Sign chart for f':

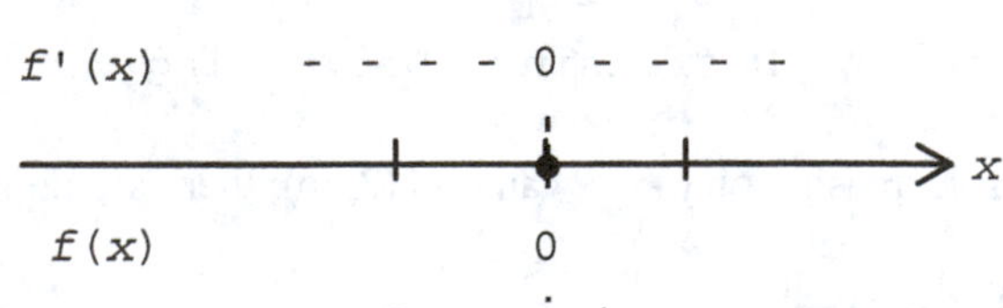

Thus, f decreases on $(-\infty, \infty)$.

Step 3. Analyze $f''(x)$:

$f''(x) = -0.5e^{-0.1x}(-0.1) = 0.05e^{-0.1x}$

Partition numbers for $f''(x)$: None

Sign chart for f'':

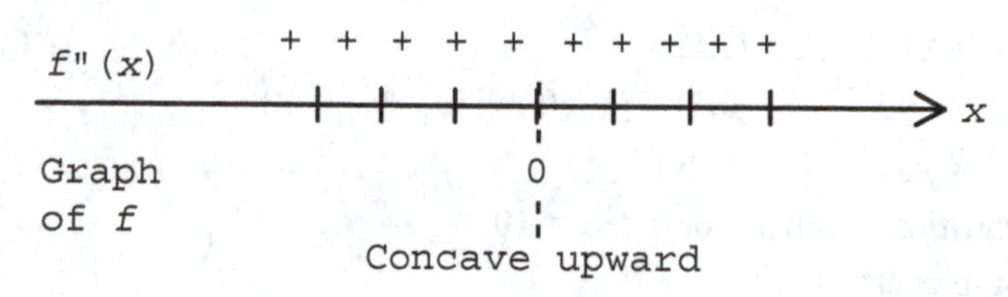

Thus, the graph of f is concave upward on $(-\infty, \infty)$.

Step 4. Sketch the graph of f:

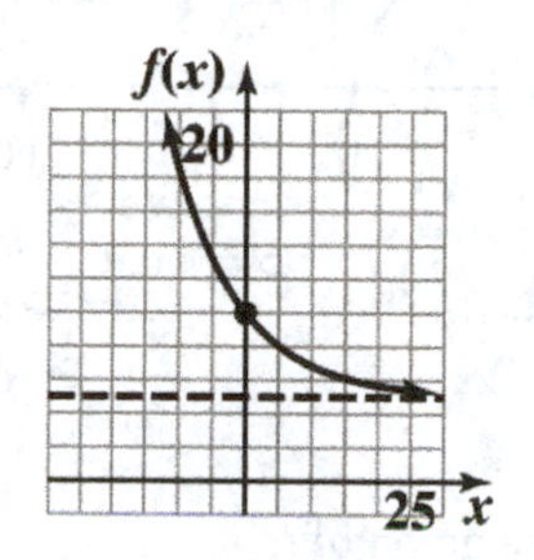

25. $f(x) = 5xe^{-0.2x}$

Step 1. Analyze $f(x)$:

(A) Domain: All real numbers.

(B) Intercepts: y-intercept: $f(0) = 5(0)e^{0} = 0$

x-intercept: $5xe^{-0.2x} = 0$

$x = 0$

(C) Asymptotes:

Vertical asymptotes: None

Horizontal asymptotes:

x	10	20	30	$40 \to \infty$
$f(x)$	6.77	1.83	0.37	$0.067 \to 0$

x	-10	-20	$\to -\infty$
$f(x)$	-369.45	-5458.01	$\to -\infty$

$y = 0$ is a horizontal asymptote

Step 2. Analyze $f'(x)$:

$f'(x) = 5xe^{-0.2x}(-0.2) + e^{-0.2x}\,5 = 5e^{-0.2x}[1 - 0.2x]$

Critical values: $x = 5$

Partition numbers: $x = 5$

Sign chart for f':

```
          + + + + + + + 0 - - - -
f'(x)  ---+-+-+-+-+-+-+-●-+-+----> x
f(x)          0         5
             Increasing ¦ Decreasing
                      Local
                     maximum
```

Test Numbers

x	$f'(x)$
0	$5(+)$
6	$-e^{-1.2}(-)$

Thus, $f(x)$ increases on $(-\infty, 5)$, has a local maximum at $x = 5$, and decreases on $(5, \infty)$.

Step 3. Analyze $f''(x)$:

$f''(x) = 5e^{-0.2x}(-0.2) + [1 - 0.2x]5e^{-0.2x}(-0.2)$

$= -e^{-0.2x}[2 - 0.2x]$

Partition numbers for f'': $x = 10$

Sign chart for f'':

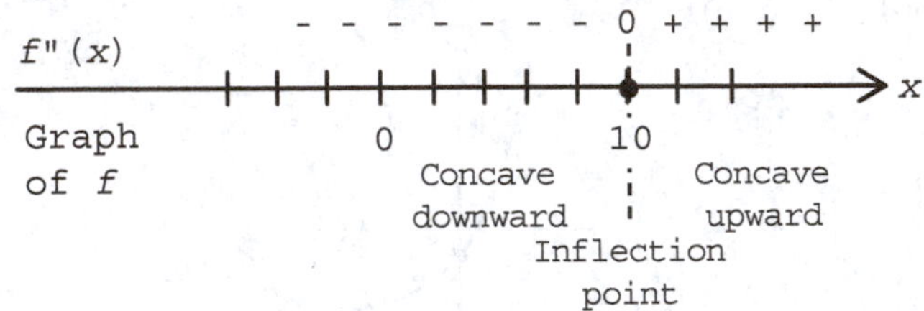

Test Numbers

x	$f''(x)$
0	$-2(-)$
20	$2e^{-4}\ (+)$

Step 4. Sketch the graph of f:

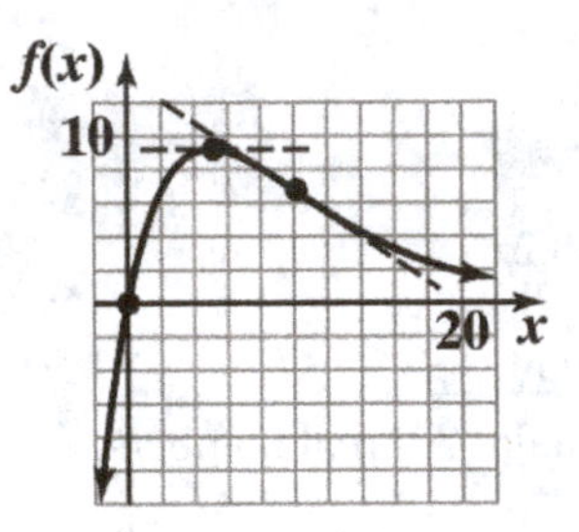

27. $f(x) = \ln(1 - x)$

Step 1. Analyze $f(x)$:

(A) Domain: All real numbers x such that $1 - x > 0$, i.e., $x < 1$ or $(-\infty, 1)$.

(B) Intercepts: y-intercept: $f(0) = \ln(1 - 0) = \ln 1 = 0$

x-intercepts: $\ln(1 - x) = 0$

$1 - x = 1$

$x = 0$

(C) Asymptotes:

Horizontal asymptote: $\lim_{x\to-\infty} f(x) = \lim_{x\to-\infty} \ln(1 - x)$ does not exist. Thus, there are no horizontal asymptotes.

Vertical asymptote: From the table,

x	0.9	0.99	0.99999	0.9999999	$\to 1$
$f(x)$	–2.30	–4.61	–11.51	–16.12	$\to -\infty$

We conclude that $x = 1$ is a vertical asymptote.

Step 2. Analyze $f'(x)$:

$f'(x) = \frac{1}{1-x}(-1), \quad x < 1$

$= \frac{1}{x-1}$

Now, $f'(x) = \frac{1}{x-1} < 0$ on $(-\infty, 1)$.

Thus, f is decreasing on $(-\infty, 1)$; there are no critical values and no local extrema.

Step 3. Analyze $f''(x)$:

$f'(x) = (x-1)^{-1}$

$f''(x) = -1(x-1)^{-2} = \dfrac{-1}{(x-1)^2}$

Since $f''(x) = \dfrac{-1}{(1-x)^2} < 0$ on $(-\infty, 1)$, the graph of f is concave downward on $(-\infty, 1)$; there are no inflection points.

Step 4. Sketch the graph of f:

x	$f(x)$
0	0
−2	≈ 1.10
.9	≈ −2.30

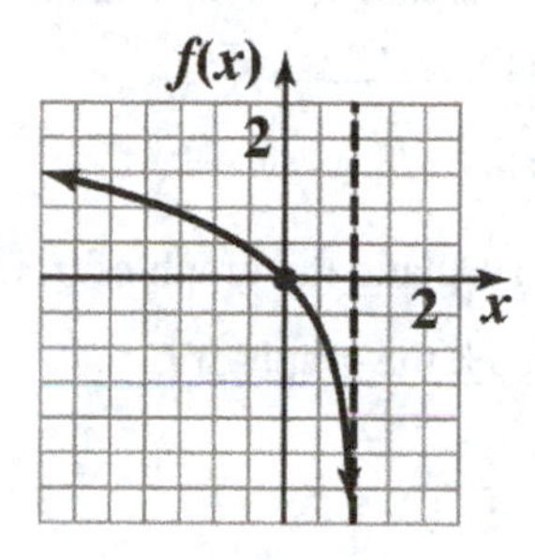

29. $f(x) = x - \ln x$

Step 1. Analyze $f(x)$:

(A) Domain: All positive real numbers, $(0, \infty)$.
[Note: $\ln x$ is defined only for positive numbers.]

(B) Intercepts: y-intercept: There is no y intercept; $f(0) = 0 - \ln(0)$ is not defined.
x-intercept: $x - \ln x = 0$
$\ln x = x$
Since the graph of $y = \ln x$ is below the graph of $y = x$, there are no solutions to this equation; there are no x-intercepts.

(C) Asymptotes:

Horizontal asymptote: None

Vertical asymptotes: Since $\lim_{x\to 0^+} \ln x = -\infty$, $\lim_{x\to 0^+} (x - \ln x) = \infty$. $x = 0$ is a vertical asymptote.

Step 2. Analyze $f'(x)$:

$f'(x) = 1 - \dfrac{1}{x} = \dfrac{x-1}{x}, x > 0$

Critical values: $\dfrac{x-1}{x} = 0, \quad x = 1$

Partition numbers: $x = 1$

Sign chart for $f'(x) = \frac{x-1}{x}$:

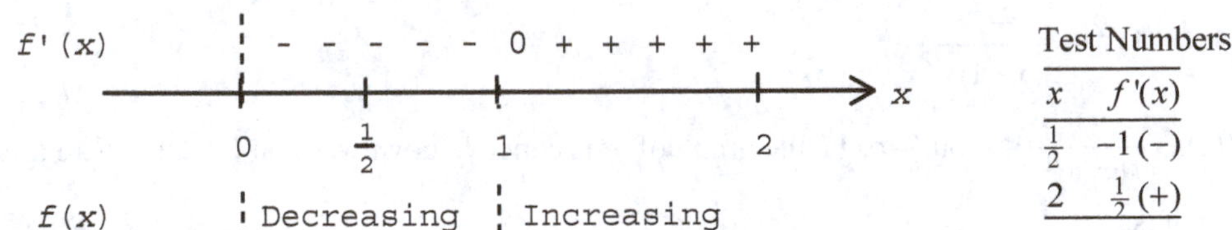

Test Numbers

x	$f'(x)$
$\frac{1}{2}$	$-1(-)$
2	$\frac{1}{2}(+)$

Thus, f is decreasing on (0, 1) and increasing on $(1, \infty)$; f has a local minimum at $x = 1$.

Step 3. Analyze $f''(x)$:

$f''(x) = \frac{1}{x^2}, x > 0$

Thus, $f''(x) > 0$ and the graph of f is concave upward on $(0, \infty)$.

Step 4. Sketch the graph of f:

x	$f(x)$
0.1	≈ 2.4
1	1
10	≈ 7.7

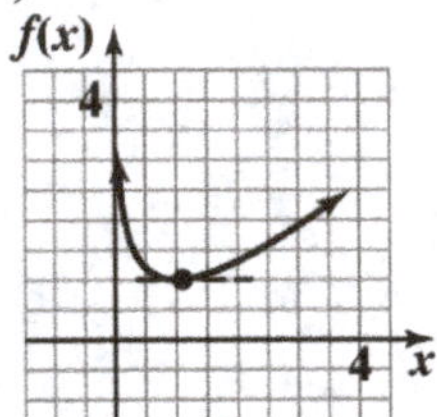

31. $f(x) = \frac{x}{x^2 - 4} = \frac{x}{(x-2)(x+2)}$

Step 1. Analyze $f(x)$:

(A) Domain: All real numbers except $x = 2$, $x = -2$.

(B) Intercepts: y-intercept: $f(0) = \frac{0}{-4} = 0$

x-intercept: $\frac{x}{x^2 - 4} = 0$

$x = 0$

(C) Asymptotes:

Horizontal asymptote:

$$\lim_{x\to\infty} \frac{x}{x^2 - 4} = \lim_{x\to\infty} \frac{x}{x^2\left(1 - \frac{4}{x^2}\right)} = \lim_{x\to\infty} \frac{1}{x}\left(\frac{1}{1 - \frac{4}{x^2}}\right) = 0$$

Thus, $y = 0$ (the x axis) is a horizontal asymptote.

Vertical asymptotes: The denominator is 0 at $x = 2$ and $x = -2$. The numerator is nonzero at each of these points. Thus, $x = 2$ and $x = -2$ are vertical asymptotes.

Step 2. Analyze $f'(x)$:

$$f'(x) = \frac{(x^2 - 4)(1) - x(2x)}{(x^2 - 4)^2} = \frac{-(x^2 + 4)}{(x^2 - 4)^2}$$

Critical values: None ($x^2 + 4 \neq 0$ for all x)

Partition numbers: $x = 2$, $x = -2$

Sign chart for f':

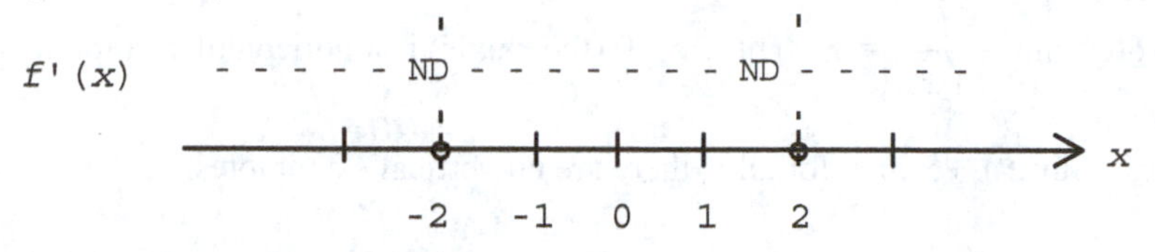

Thus, f is decreasing on $(-\infty, -2)$, on $(-2, 2)$, and on $(2, \infty)$; f has no local extrema.

Step 3. Analyze $f''(x)$:

$$f''(x) = \frac{(x^2-4)^2(-2x) - [-(x^2+4)](2)(x^2-4)(2x)}{(x^2-4)^4}$$

$$= \frac{(x^2-4)(-2x) + 4x(x^2+4)}{(x^2-4)^3} = \frac{2x^3+24x}{(x^2-4)^3} = \frac{2x(x^2+12)}{(x^2-4)^3}$$

Partition numbers for f'': $x = 0, x = 2, x = -2$

Sign chart for f'':

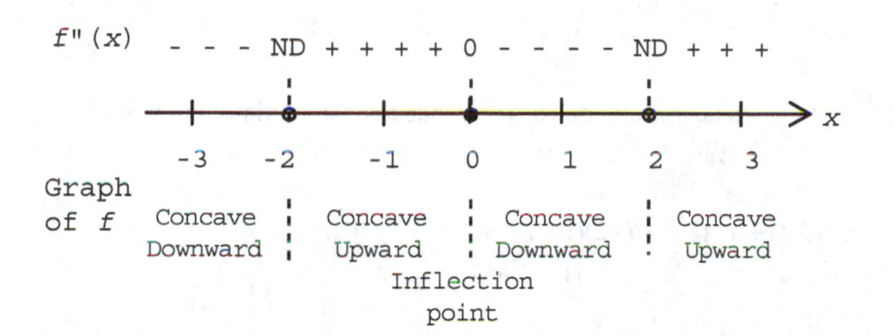

Test Numbers

x	$f''(x)$
-3	$-\frac{126}{125}(-)$
-1	$\frac{26}{27}(+)$
1	$-\frac{26}{27}(-)$
3	$\frac{126}{125}(+)$

Thus, the graph of f is concave downward on $(-\infty, -2)$ and on $(0, 2)$; the graph of f is concave upward on $(-2, 0)$ and on $(2, \infty)$; the graph has an inflection point at $x = 0$.

Step 4. Sketch the graph of f:

x	$f(x)$
0	0
1	$-\frac{1}{3}$
-1	$\frac{1}{3}$
3	$\frac{3}{5}$
-3	$-\frac{3}{5}$

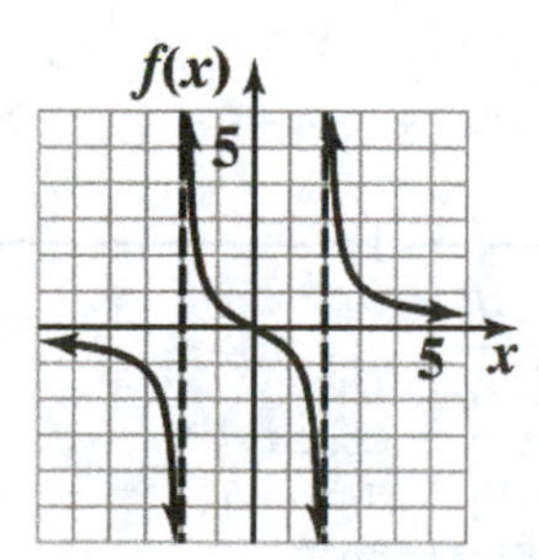

33. $f(x) = \dfrac{1}{1+x^2}$

Step 1. Analyze $f(x)$:

(A) Domain: All real numbers ($1 + x^2 \neq 0$ for all x).

(B) Intercepts: y-intercept: $f(0) = 1$

x-intercept: $\dfrac{1}{1+x^2} \neq 0$ for all x; no x intercepts

(C) Asymptotes:

Horizontal asymptote: $\lim\limits_{x\to\infty}\dfrac{1}{1+x^2}=0$. Thus, $y=0$ (the x-axis) is a horizontal asymptote.

Vertical asymptotes: Since $1+x^2\neq 0$ for all x, there are no vertical asymptotes.

Step 2. Analyze $f'(x)$:

$$f'(x)=\frac{(1+x^2)(0)-1(2x)}{(1+x^2)^2}=\frac{-2x}{(1+x^2)^2}$$

Critical values: $x=0$
Partition numbers: $x=0$

Sign chart for f':

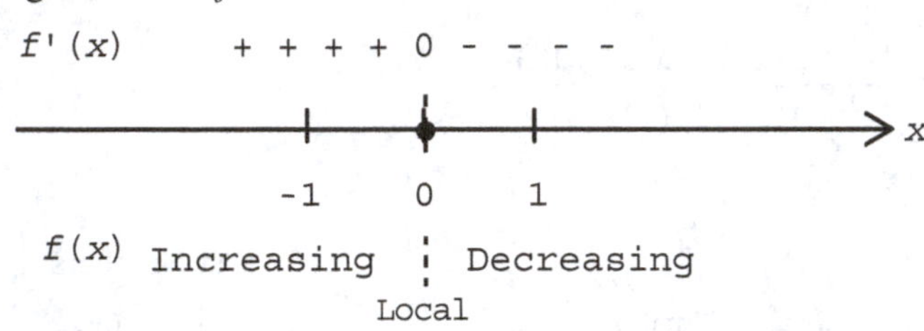

Test Numbers

x	$f'(x)$
-1	$\frac{1}{2}(+)$
1	$-\frac{1}{2}(-)$

Thus, f is increasing on $(-\infty, 0)$; f is decreasing on $(0, \infty)$; f has a local maximum at $x=0$.

Step 3. Analyze $f''(x)$:

$$f''(x)=\frac{(1+x^2)^2(-2)-(-2x)(2)(1+x^2)2x}{(1+x^2)^4}=\frac{(-2)(1+x^2)+8x^2}{(1+x^2)^3}$$

$$=\frac{6x^2-2}{(1+x^2)^3}=\frac{6\left(x+\frac{\sqrt{3}}{3}\right)\left(x-\frac{\sqrt{3}}{3}\right)}{(1+x^2)^3}$$

Partition numbers for f'': $x=-\dfrac{\sqrt{3}}{3}$, $x=\dfrac{\sqrt{3}}{3}$

Sign chart for f'':

f"(x) + + + + 0 - - - 0 + + + +

x

-2 -1 $-\frac{\sqrt{3}}{3}$ 0 $\frac{\sqrt{3}}{3}$ 1 2

Graph of f: Concave Upward | Concave Downward | Concave Upward

Inflection point Inflection point

Test Numbers

x	$f''(x)$
-1	$\frac{1}{2}(+)$
0	$-2(-)$
1	$\frac{1}{2}(+)$

Thus, the graph of f is concave upward on $\left(-\infty, \dfrac{-\sqrt{3}}{3}\right)$ and on $\left(\dfrac{\sqrt{3}}{3}, \infty\right)$; the graph of f is concave downward on $\left(\dfrac{-\sqrt{3}}{3}, \dfrac{\sqrt{3}}{3}\right)$; the graph has inflection points at $x=\dfrac{-\sqrt{3}}{3}$ and $x=\dfrac{\sqrt{3}}{3}$.

Step 4. Sketch the graph of f:

x	$f(x)$
$-\frac{\sqrt{3}}{3}$	$\frac{3}{4}$
0	1
$\frac{\sqrt{3}}{3}$	$\frac{3}{4}$

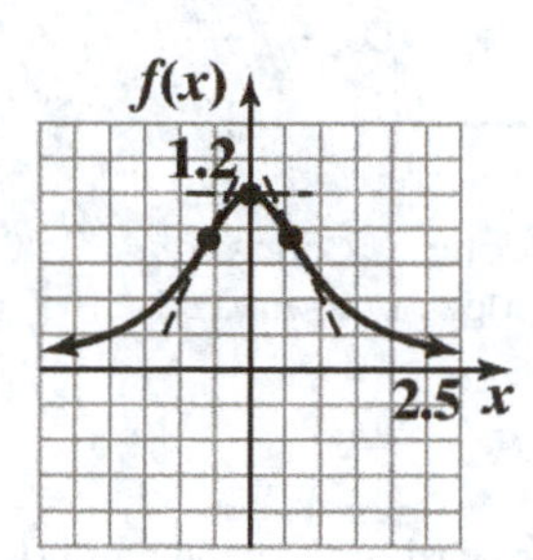

35. $f(x) = \dfrac{2x}{1-x^2}$

Step 1. Analyze $f(x)$:

(A) Domain: All real numbers except $x = -1$ and $x = 1$.

(B) Intercepts: y-intercept: $f(0) = \dfrac{0}{1} = 0$

x-intercepts: $\dfrac{2x}{1-x^2} = 0$

$x = 0$

(C) Asymptotes:

Horizontal asymptote: $\lim\limits_{x\to\infty} \dfrac{2x}{1-x^2} = \lim\limits_{x\to\infty} \dfrac{\frac{2}{x}}{\frac{1}{x^2}-1} = 0$. Thus, $y = 0$ (the x-axis) is a horizontal asymptote.

Vertical asymptotes: The denominator is 0 at $x = \pm 1$ and the numerator is not 0 at $x = \pm 1$. Thus, $x = -1$, $x = 1$ are vertical asymptotes.

Step 2. Analyze $f'(x)$:

$$f'(x) = \frac{(1-x^2)2 - 2x(-2x)}{(1-x^2)^2} = \frac{2x^2+2}{(1-x^2)^2}$$

Critical values: none

Partition numbers: $x = -1, x = 1$

Sign chart for f':

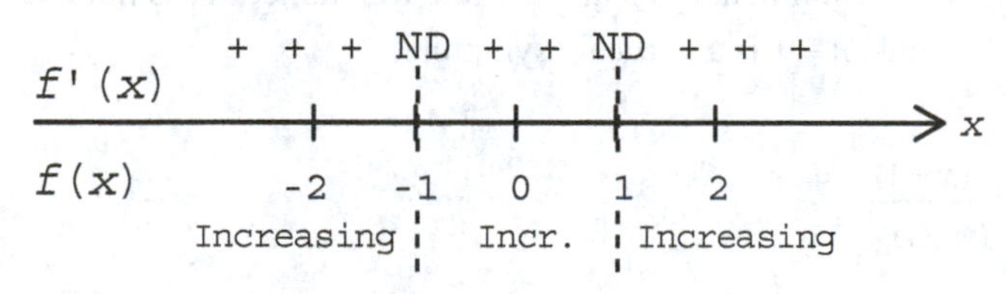

Test Numbers

x	$f'(x)$
-2	$\frac{10}{9}(+)$
0	$2(+)$
2	$\frac{10}{9}(+)$

Thus, f is increasing on $(-\infty, -1)$, $(-1, 1)$, and $(1, \infty)$.

Step 3. Analyze $f''(x)$:

$$f''(x) = \frac{(1-x^2)^2 4x - (2x^2+2)(2)(1-x^2)(-2x)}{(1-x^2)^4} = \frac{4x(x^2+3)}{(1-x^2)^3}$$

Partition numbers for $f''(x)$: $x = -1, x = 0, x = 1$

Sign chart for f'':

```
                + + ND - 0 + ND - -
f" (x)
      ---------+-----+-----+-----+-----+------> x
              -2    -1     0     1     2
Graph     Concave  ¦ Conc.¦ Conc.¦ Concave
of f      Upward   ¦downw.¦ Upw. ¦ downward
                   Inflection
                      point
```

Test Numbers

x	$f''(x)$
-2	$\frac{56}{27}(+)$
$-\frac{1}{2}$	$-\frac{416}{27}(-)$
$\frac{1}{2}$	$\frac{416}{27}(+)$
2	$-\frac{56}{27}(-)$

Thus, the graph of f is concave upward on $(-\infty, -1)$ and $(0, 1)$, concave downward on $(-1, 0)$ and $(1, \infty)$, and has an inflection point at $x = 0$.

Step 4. Sketch the graph of f:

x	$f(x)$
0	0

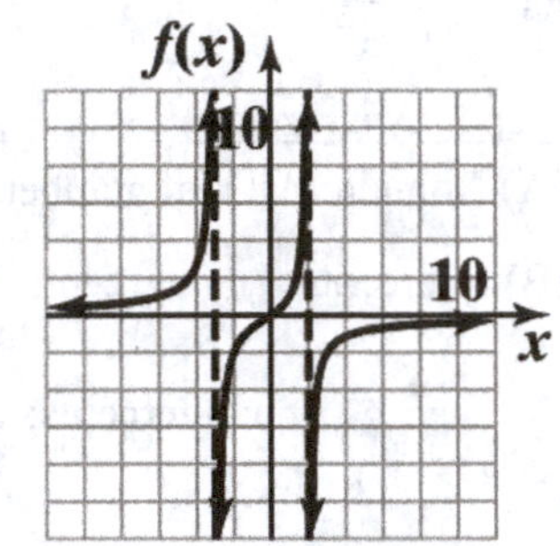

37. $f(x) = \dfrac{-5x}{(x-1)^2} = \dfrac{-5x}{x^2 - 2x + 1}$

Step 1. Analyze $f(x)$:

(A) Domain: All real numbers except $x = 1$.

(B) Intercepts: y-intercept: $f(0) = 0$

x-intercepts: $\dfrac{-5x}{(x-1)^2} = 0$

$x = 0$

(C) Asymptotes:

Horizontal asymptote: $\lim_{x\to\infty} \dfrac{-5x}{x^2 - 2x + 1} = \lim_{x\to\infty} \dfrac{-\frac{5}{x}}{1 - \frac{2}{x} + \frac{1}{x^2}} = 0$. Thus, $y = 0$ (the x-axis) is a horizontal asymptote.

Vertical asymptotes: The denominator is 0 at $x = 1$ and the numerator is not 0 at $x = 1$. Thus, $x = 1$ is a vertical asymptote.

Step 2. Analyze $f'(x)$:

$$f'(x) = \frac{(x-1)^2(-5) + 5x(2)(x-1)}{(x-1)^4} = \frac{5(x+1)}{(x-1)^3}$$

Critical values: $x = -1$

Partition numbers: $x = -1, x = 1$

Sign chart for f':

```
         + + + 0 - - - ND + + +
f'(x)  ----+----+----+----+----+----> x
f(x)      -2   -1    0    1    2
      Increasing | Decr. | Increasing
               Local
              maximum
```

Test Numbers

x	$f'(x)$
-2	$\frac{5}{27}(+)$
0	$-5(-)$
2	$15(+)$

Thus, f is increasing on $(-\infty, -1)$, $(1, \infty)$, decreasing on $(-1, 1)$, and has a local maximum at $x = -1$.

Step 3. Analyze $f''(x)$:

$$f''(x) = \frac{(x-1)^3 5 - 5(x+1)(3)(x-1)^2}{(x-1)^6} = \frac{-10(x+2)}{(x-1)^4}$$

Partition numbers for f'': $x = -2$, $x = 1$

Sign chart for f'':

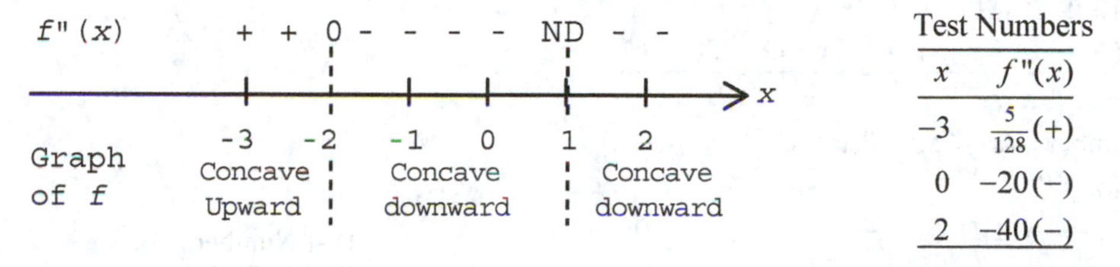

Test Numbers

x	$f''(x)$
-3	$\frac{5}{128}(+)$
0	$-20(-)$
2	$-40(-)$

Thus, the graph of f is concave upward on $(-\infty, -2)$, concave downward on $(-2, 1)$ and $(1, \infty)$, and has an inflection point at $x = -2$.

Step 4. Sketch the graph of f:

x	$f(x)$
-2	$\frac{10}{9}$
-1	$\frac{5}{4}$
0	0

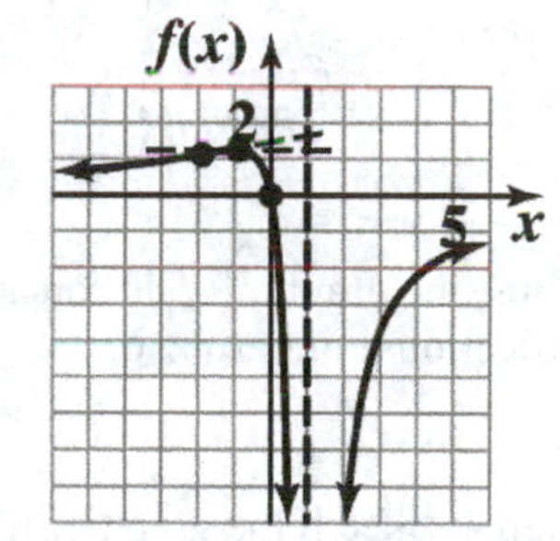

39. $f(x) = \dfrac{x^2 + x - 2}{x^2} = \dfrac{(x+2)(x-1)}{x^2}$

Step 1. Analyze $f(x)$:

(A) Domain: All real numbers except $x = 0$.

(B) Intercepts: y-intercept: $f(0)$ not defined; no y-intercept

x-intercepts: $\dfrac{(x+2)(x-1)}{x^2} = 0$

$x = -2, 1$

(C) Asymptotes:

Horizontal asymptote: $\lim_{x\to\infty} \dfrac{x^2 + x - 2}{x^2} = \lim_{x\to\infty} \dfrac{1 + \frac{1}{x} - \frac{2}{x^2}}{1} = 1$; $y = 1$ is a horizontal asymptote.

Vertical asymptotes: $x = 0$ (the x-axis)

Step 2. Analyze $f'(x)$:

$$f'(x) = \frac{x^2(2x+1)-(x^2+x-2)(2x)}{x^4} = \frac{4-x}{x^3}$$

Critical values: $x = 4$

Partition numbers: $x = 0, x = 4$

Sign chart for f':

$f'(x)$: - - - ND + + + + 0 - - -

x: -1 0 1 2 3 4 5

$f(x)$: Decreasing | Increasing | Decreasing

Local maximum (at 4)

Test Numbers

x	$f'(x)$
−1	−5(−)
1	3(+)
5	$\frac{-1}{125}$(−)

Thus, f is increasing on (0, 4), decreasing on $(-\infty, 0)$ and $(4, \infty)$, and has a local maximum at $x = 4$.

Step 3. Analyze $f''(x)$:

$$f''(x) = \frac{x^3(-1)-(4-x)(3x^2)}{x^6} = \frac{2x-12}{x^4} = \frac{2(x-6)}{x^4}$$

Partition numbers for f'': $x = 0, x = 6$

Sign chart for f'':

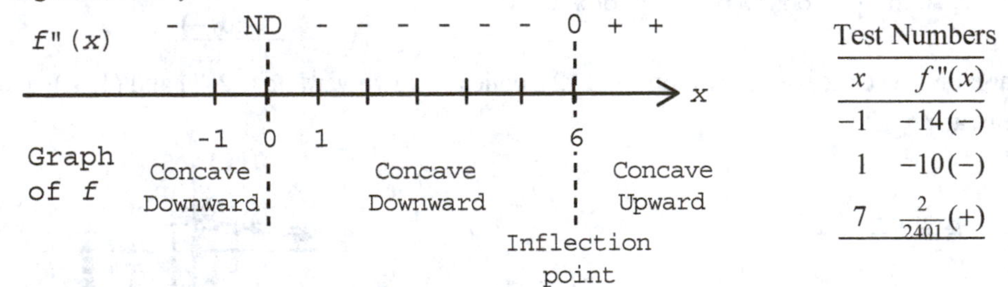

Test Numbers

x	$f''(x)$
−1	−14(−)
1	−10(−)
7	$\frac{2}{2401}$(+)

Thus, the graph of f is concave upward on $(6, \infty)$, concave downward on $(-\infty, 0)$ and (0, 6), and has an inflection point at $x = 6$.

Step 4. Sketch the graph of f:

x	$f(x)$
4	$\frac{9}{8}$
6	$\frac{10}{9}$

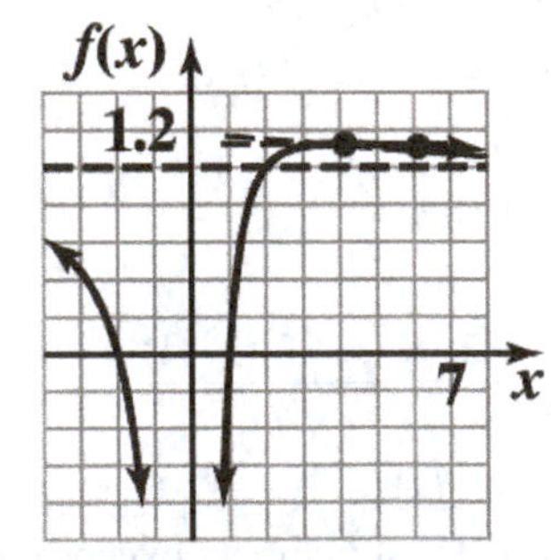

41. $f(x) = \dfrac{x^2}{x-1}$

Step 1. Analyze $f(x)$:

(A) Domain: All real numbers except $x = 1$.

(B) Intercepts: y-intercept: $f(0) = 0$

x-intercepts: $\dfrac{x^2}{x-1} = 0$

$x = 0$

(C) Asymptotes:

Horizontal asymptote: $\frac{x^2}{x} = x$; no horizontal asymptote

Vertical asymptote: $x = 1$

Oblique asymptote: It follows from above that $y = x$ is an oblique asymptote.

Step 2. Analyze $f'(x)$:

$$f'(x) = \frac{(x-1)(2x) - x^2}{(x-1)^2} = \frac{x^2 - 2x}{(x-1)^2} = \frac{x(x-2)}{(x-1)^2}$$

Critical values: $x = 0, x = 2$

Partition numbers: $x = 0, x = 1, x = 2$

Sign chart for f':

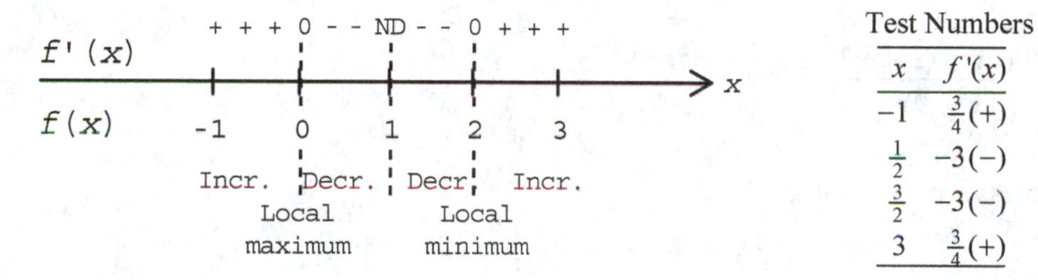

Test Numbers

x	$f'(x)$
-1	$\frac{3}{4}(+)$
$\frac{1}{2}$	$-3(-)$
$\frac{3}{2}$	$-3(-)$
3	$\frac{3}{4}(+)$

Thus, f is increasing on $(-\infty, 0)$ and $(2, \infty)$, decreasing on $(0, 1)$ and $(1, 2)$, and has a local maximum at $x = 0$ and a local minimum at $x = 2$.

Step 3. Analyze $f''(x)$:

$$f''(x) = \frac{(x-1)^2(2x-2) - (x^2 - 2x)(2)(x-1)}{(x-1)^4} = \frac{2}{(x-1)^3}$$

Partition numbers for f'': $x = 1$

Sign chart for f'':

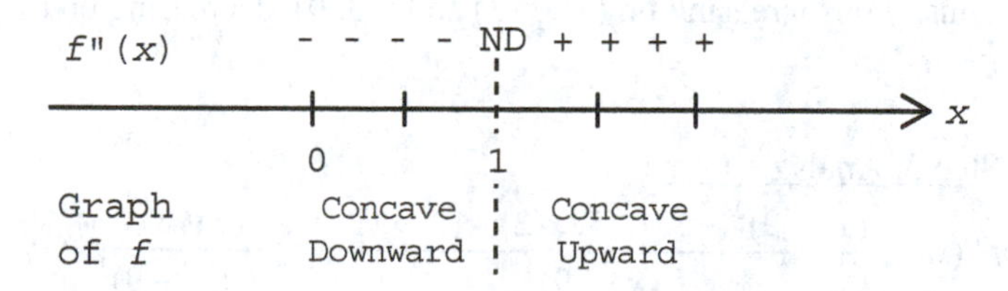

Thus, the graph of f is concave upward on $(1, \infty)$ and concave downward on $(-\infty, 1)$.

Step 4. Sketch the graph of f:

x	$f(x)$
0	0
2	4

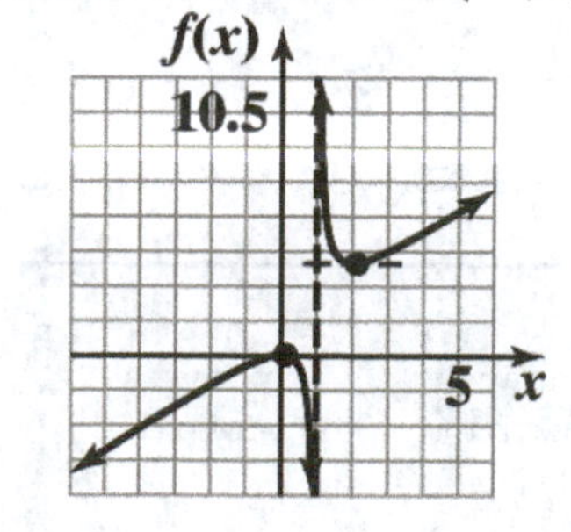

43. $f(x) = \dfrac{3x^2 + 2}{x^2 - 9}$

Step 1. Analyze $f(x)$:

(A) Domain: All real numbers except $x = -3$, $x = 3$.

(B) Intercepts: y-intercept: $f(0) = -\dfrac{2}{9}$

x-intercepts: $3x^2 + 2 \neq 0$ for all x; no x-intercepts

(C) Asymptotes:

Horizontal asymptote: $\dfrac{3x^2}{x^2} = 3$; $y = 3$ is a horizontal asymptote

Vertical asymptotes: $x = -3$, $x = 3$

Step 2. Analyze $f'(x)$:

$$f'(x) = \frac{(x^2-9)(6x) - (3x^2+2)(2x)}{(x^2-9)^2} = \frac{-58x}{(x^2-9)^2}$$

Critical values: $x = 0$

Partition numbers: $x = -3$, $x = 0$, $x = 3$

Sign chart for f':

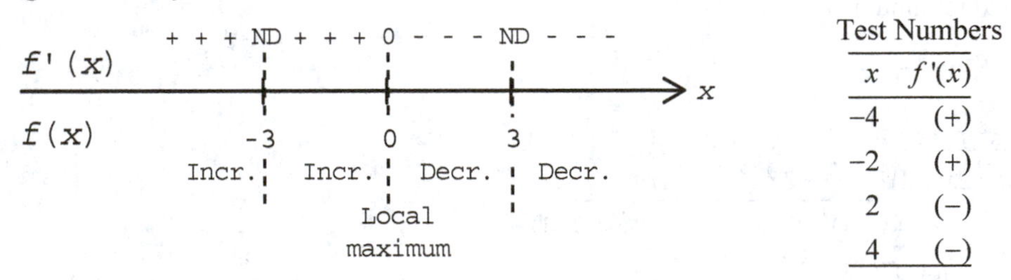

Test Numbers

x	$f'(x)$
–4	(+)
–2	(+)
2	(–)
4	(–)

Thus, f is increasing on $(-\infty, -3)$ and $(-3, 0)$, decreasing on $(0, 3)$ and $(3, \infty)$, and has a local maximum at $x = 0$.

Step 3. Analyze $f''(x)$:

$$f''(x) = \frac{(x^2-9)^2(-58) + 58x(2)(x^2-9)(2x)}{(x^2-9)^4} = \frac{174(x^2+3)}{(x^2-9)^3}$$

Partition numbers for f'': $x = -3$, $x = 3$

Sign chart for f'':

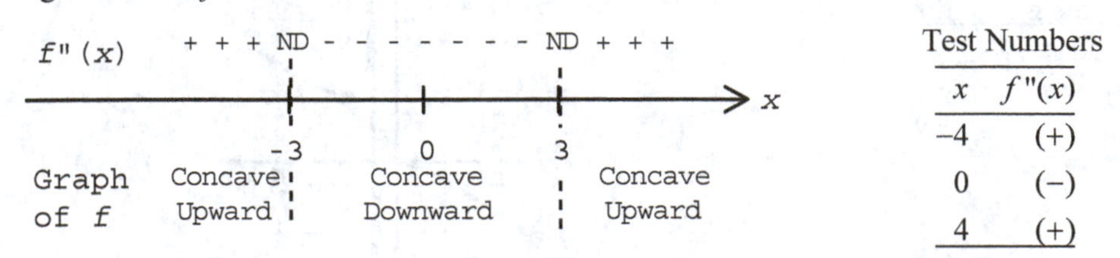

Test Numbers

x	$f''(x)$
–4	(+)
0	(–)
4	(+)

Thus, the graph of f is concave upward on $(-\infty, -3)$ and $(3, \infty)$, and concave downward on $(-3, 3)$.

Step 4. Sketch the graph of f:

x	$f(x)$
0	$-\frac{2}{9}$

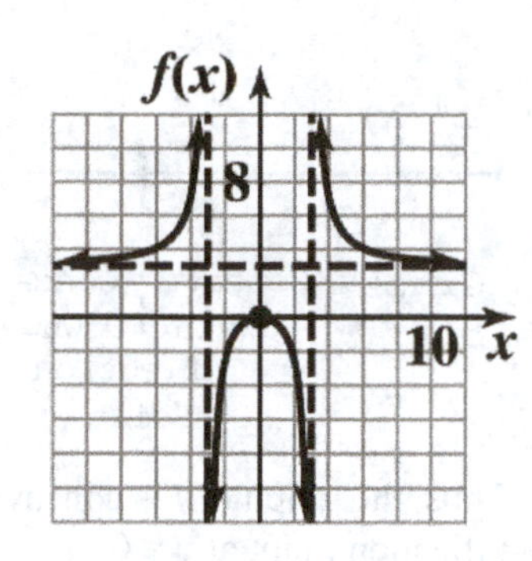

45. $f(x) = \dfrac{x^3}{x-2}$

Step 1. Analyze $f(x)$:

(A) Domain: All real numbers except $x = 2$.

(B) Intercepts: y-intercept: $f(0) = 0$

x-intercepts: $\dfrac{x^3}{x-2} = 0$

$x = 0$

(C) Asymptotes:

Horizontal asymptote: $\dfrac{x^3}{x} = x^2$; no horizontal asymptote

Vertical asymptote: $x = 2$

Step 2. Analyze $f'(x)$:

$$f'(x) = \frac{(x-2)(3x^2) - x^3}{(x-2)^2} = \frac{2x^2(x-3)}{(x-2)^2}$$

Critical values: $x = 0, x = 3$

Partition numbers: $x = 0, x = 2, x = 3$

Sign chart for f':

```
                - - 0 - - - ND - 0 + +
f'(x)
     ----+-----+-----+-----+-----+-----+----> x
f(x)    -1     0     1     2     3     4
       Decreasing ¦ Decr.  ¦ Dec.¦ Increasing
                                 Local
                                 minimum
```

Test Numbers

x	$f'(x)$
-1	$-\frac{8}{9}(-)$
1	$-4(-)$
$\frac{5}{2}$	$-25(-)$
4	$8(+)$

Thus, f is increasing on $(3, \infty)$, decreasing on $(-\infty, 2)$ and $(2, 3)$, and has a local minimum at $x = 3$.

Step 3. Analyze $f''(x)$:

$$f''(x) = \frac{(x-2)^2[2x^2 + 4x(x-3)] - 2x^2(x-3)(2)(x-2)}{(x-2)^4} = \frac{2x(x^2 - 6x + 12)}{(x-2)^3}$$

Partition numbers for f'': $x = 0$, $x = 2$ ($x^2 - 6x + 12$ has no real roots)

Sign chart for f'':

```
f"(x)      + + + 0 - - - - ND + + +
-----------------|--------|------|-------> x
                 0        2      3
Graph    Concave | Concave  | Concave
of f     Upward  | Downward | Upward
        Inflection
          Point
```

Test Numbers

x	$f''(x)$
−1	$\frac{38}{27}(+)$
1	−14(−)
3	18(+)

Thus, the graph of f is concave upward on $(-\infty, 0)$ and $(2, \infty)$, concave downward on $(0, 2)$, and has an inflection point at $x = 0$.

Step 4. Sketch the graph of f:

x	$f(x)$
0	0
3	27

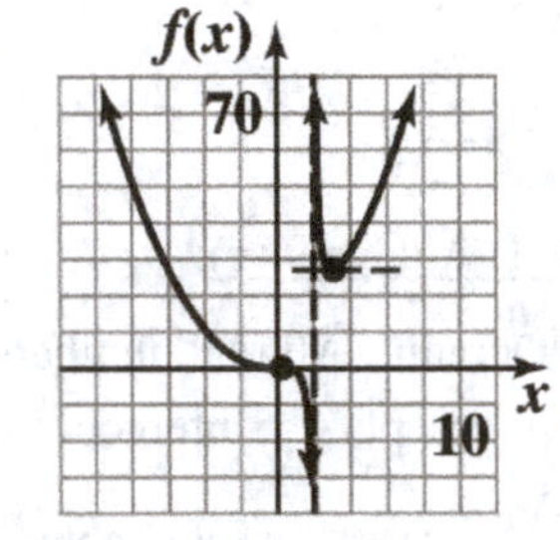

47. $f(x) = (3 - x)e^x$

Step 1. Analyze $f(x)$:

(A) Domain: All real numbers, $(-\infty, \infty)$.

(B) Intercepts: y-intercept: $f(0) = (3 - 0)e^0 = 3$

x-intercept: $(3 - x)e^x = 0$

$3 - x = 0$

$x = 3$

(C) Asymptotes:

Horizontal asymptote: Consider the behavior of f as $x \to \infty$ and as $x \to -\infty$.
Using the following tables,

x	−1	−10	−20
$f(x)$	1.47	0.00059	0.000000047

x	5	10
$f(x)$	−296.83	−154,185.26

we conclude that $\lim_{x \to -\infty} f(x) = 0$ and $\lim_{x \to \infty} f(x)$ does not exist. Because of the first limit, $y = 0$ is a horizontal asymptote.

Vertical asymptotes: There are no vertical asymptotes.

Step 2. Analyze $f'(x)$:

$f'(x) = (3 - x)e^x + e^x(-1) = (2 - x)e^x$

Critical values: $(2 - x)e^x = 0$

$x = 2$ [Note: $e^x > 0$]

Partition numbers: $x = 2$

Sign chart for f':

f'(x)   + + + + + 0 - - - - -

0   1   2   3   x

f(x)   Increasing ¦ Decreasing

Test Numbers

x	$f'(x)$
0	$2(+)$
3	$-e^3(-)$

Thus, f is increasing on $(-\infty, 2)$ and decreasing on $(2, \infty)$; f has a local maximum at $x = 2$.

Step 3. Analyze $f''(x)$:

$f''(x) = (2 - x)e^x + e^x(-1) = (1 - x)e^x$

Partition number for f'': $x = 1$

Sign chart for f'':

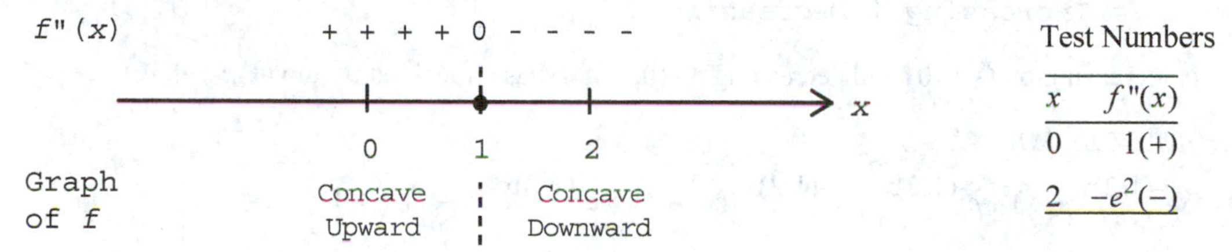

Test Numbers

x	$f''(x)$
0	$1(+)$
2	$-e^2(-)$

Thus, the graph of f is concave upward on $(-\infty, 1)$ and concave downward on $(1, \infty)$; the graph has an inflection point at $x = 1$.

Step 4. Sketch the graph of f:

x	$f(x)$
0	3
2	$e^2 \approx 7.4$
3	0

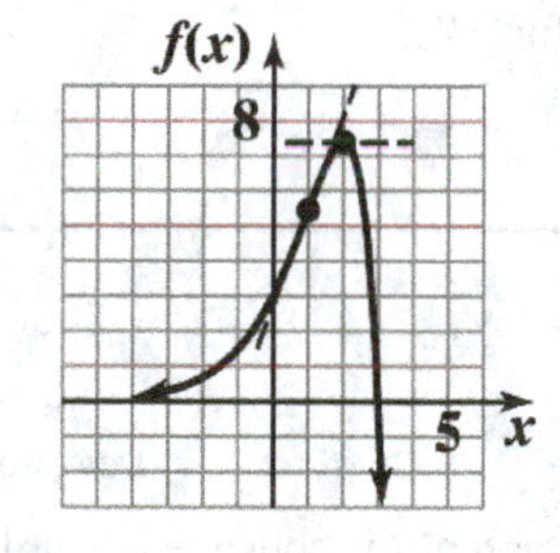

49. $f(x) = e^{-(1/2)x^2}$

Step 1. Analyze $f(x)$

(A) Domain: All real numbers, $(-\infty, \infty)$.

(B) Intercepts: y-intercept: $f(0) = e^{-(1/2)0} = e^0 = 1$

x-intercepts: Since $e^{-(1/2)x^2} \neq 0$ for all x, there are no x-intercepts.

(C) Asymptotes: $\lim_{x\to\infty} f(x) = \lim_{x\to\infty} e^{-(1/2)x^2} = \lim_{x\to\infty} \frac{1}{e^{(1/2)x^2}} = 0$

$\lim_{x\to-\infty} f(x) = \lim_{x\to-\infty} e^{-(1/2)x^2} = \lim_{x\to-\infty} \frac{1}{e^{(1/2)x^2}} = 0$

Thus, $y = 0$ is a horizontal asymptote.

Since $f(x) = e^{-(1/2)x^2} = \frac{1}{e^{(1/2)x^2}}$ and $e^{(1/2)x^2} \neq 0$ for all x, there are no vertical asymptotes.

Step 2. Analyze $f'(x)$:

$f'(x) = e^{-(1/2)x^2}(-x) = -xe^{-(1/2)x^2}$

Critical values: $-xe^{-(1/2)x^2} = 0$
$x = 0$

Partition numbers: $x = 0$
Sign chart for f':

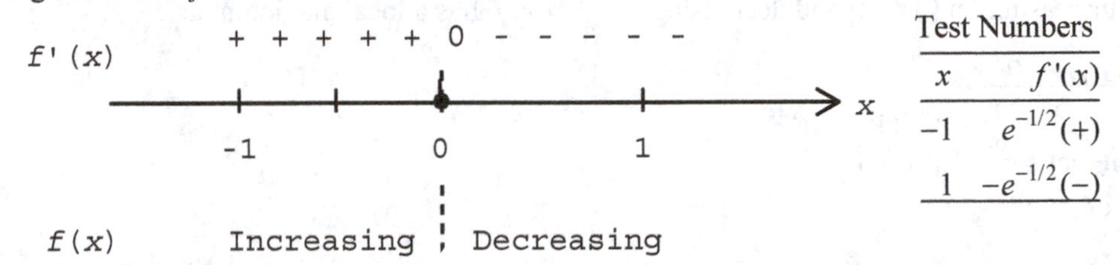

Test Numbers

x	$f'(x)$
-1	$e^{-1/2}(+)$
1	$-e^{-1/2}(-)$

Thus, f is increasing on $(-\infty, 0)$ and decreasing on $(0, \infty)$; f has a local maximum at $x = 0$.

Step 3. Analyze $f''(x)$:

$f''(x) = -xe^{-(1/2)x^2}(-x) - e^{-(1/2)x^2} = e^{-(1/2)x^2}(x^2 - 1) = e^{-(1/2)x^2}(x-1)(x+1)$

Partition numbers for f'': $e^{-(1/2)x^2}(x-1)(x+1) = 0$
$(x-1)(x+1) = 0$
$x = -1, 1$

Sign chart for f'':

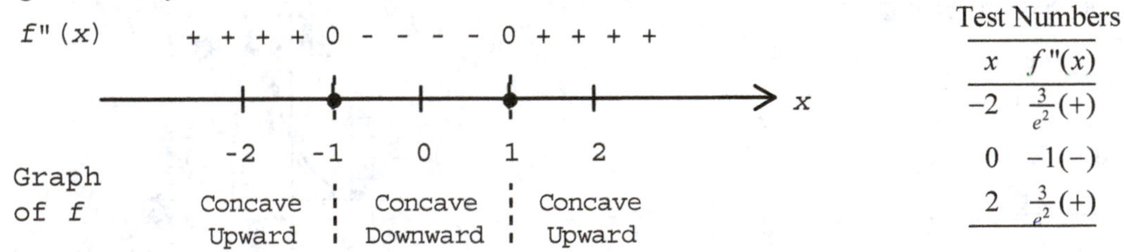

Test Numbers

x	$f''(x)$
-2	$\frac{3}{e^2}(+)$
0	$-1(-)$
2	$\frac{3}{e^2}(+)$

Thus, the graph of f is concave upward on $(-\infty, -1)$ and on $(1, \infty)$; the graph of f is concave downward on $(-1, 1)$; the graph has inflection points at $x = -1$ and at $x = 1$.

Step 4. Sketch the graph of f:

x	$f'(x)$
0	1
-1	≈ 0.61
1	≈ 0.61

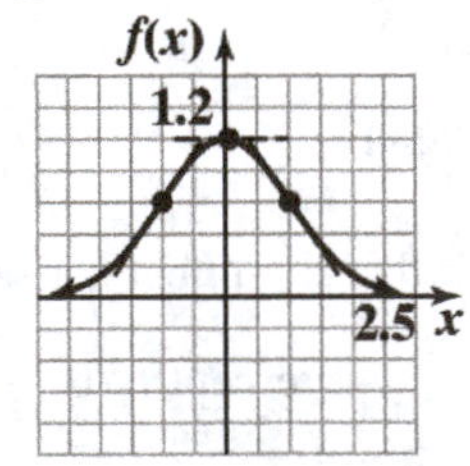

51. $f(x) = x^2 \ln x$.

Step 1. Analyze $f(x)$:

(A) Domain: All positive numbers, $(0, \infty)$.

(B) Intercepts: y-intercept: There is no y intercept.

x-intercept: $x^2 \ln x = 0$
$\ln x = 0$
$x = 1$

(C) Asymptotes: Consider the behavior of f as $x \to \infty$ and as $x \to 0$. It is clear that $\lim_{x\to\infty} f(x)$ does not exist; f is unbounded as x approaches ∞.

The following table indicates that f approaches 0 as x approaches 0.

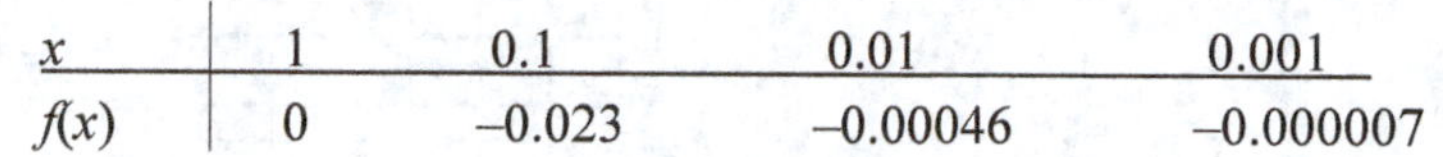

x	1	0.1	0.01	0.001
$f(x)$	0	–0.023	–0.00046	–0.000007

Thus, there are no vertical or horizontal asymptotes.

Step 2. Analyze $f'(x)$:

$$f'(x) = x^2\left(\frac{1}{x}\right) + (\ln x)(2x) = x(1 + 2\ln x)$$

Critical values: $x(1 + 2\ln x) = 0$

$$1 + 2\ln x = 0 \quad [\text{Note: } x > 0]$$

$$\ln x = -\frac{1}{2}$$

$$x = e^{-1/2} = \frac{1}{\sqrt{e}} \approx 0.6065$$

Partition number: $x = \frac{1}{\sqrt{e}} \approx 0.6065$

Sign chart for f':

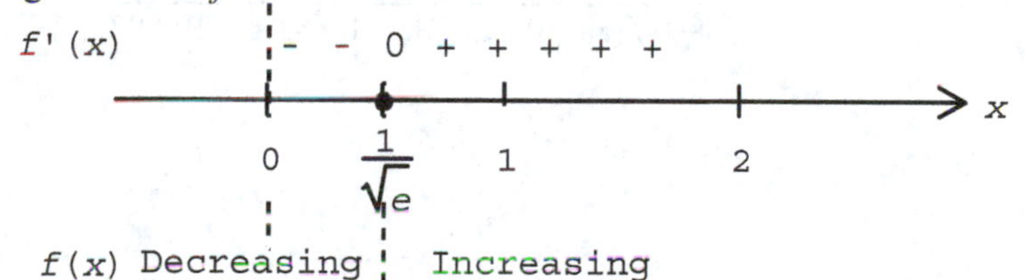

Test Numbers

x	$f'(x)$
$\frac{1}{2}$	$\approx -.19(-)$
1	1(+)

Thus, f is decreasing on $(0, e^{-1/2})$ and increasing on $(e^{-1/2}, \infty)$; f has a local minimum at $x = e^{-1/2}$.

Step 3. Analyze $f''(x)$:

$$f''(x) = x\left(\frac{2}{x}\right) + (1 + 2\ln x) = 3 + 2\ln x$$

Partition number for f'': $3 + 2\ln x = 0$

$$\ln x = -\frac{3}{2}$$

$$x = e^{-3/2} \approx 0.2231$$

Sign chart for f'':

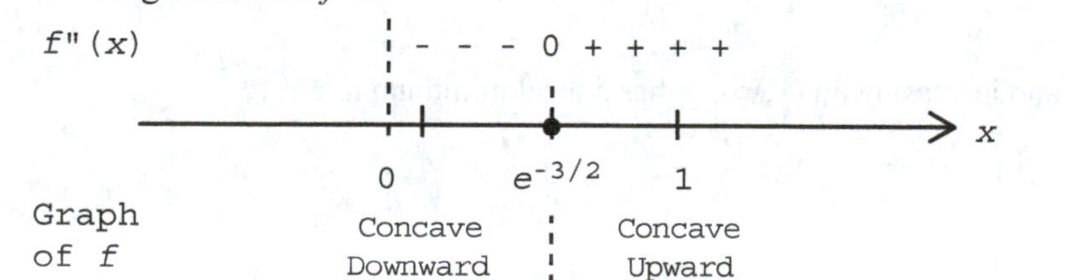

Test Numbers

x	$f''(x)$
$\frac{1}{10}$	$\approx -1.61(-)$
1	3(+)

Thus, the graph of f is concave downward on $(0, e^{-3/2})$ and concave upward on $(e^{-3/2}, \infty)$; the graph has an inflection point at $x = e^{-3/2}$.

Step 4. Sketch the graph of f:

x	$f(x)$
$e^{-3/2}$	≈ -0.075
$e^{-1/2}$	≈ -0.18
1	0

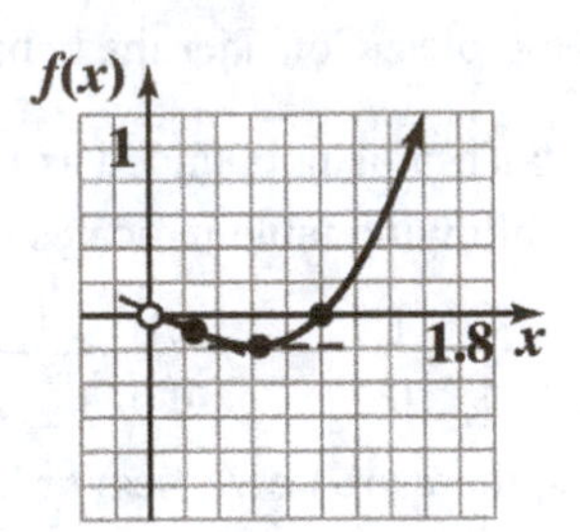

53. $f(x) = (\ln x)^2$

Step 1. Analyze $f(x)$:

(A) Domain: All positive numbers, $(0, \infty)$.

(B) Intercepts: y-intercept: There is no y-intercept.

x-intercept: $(\ln x)^2 = 0$

$\ln x = 0$

$x = 1$

(C) Asymptotes:
Consider the behavior of f as $x \to \infty$ and as $x \to 0$. It is clear that $\lim_{x\to\infty} f(x)$ does not exist; $f(x) \to \infty$ as $x \to \infty$. Thus, there is no horizontal asymptote.
The following table indicates that $f(x) \to \infty$ as $x \to 0$;
$x = 0$ (the y-axis) is a vertical asymptote.

x	1	0.01	0.0001	0.000001
$f(x)$	0	21.21	84.83	190.87

Step 2. Analyze $f'(x)$:

$$f'(x) = 2(\ln x)\frac{d}{dx}\ln x = \frac{2\ln x}{x}$$

Critical values: $\frac{2\ln x}{x} = 0$

$\ln x = 0$

$x = 1$

Partition numbers: $x = 1$

Sign chart for f':

```
f'(x)          - - - - 0 + + + +
                       |
---------------+-------*-------+----------> x
                       |
f(x)           0       1
          Decreasing   |  Increasing
                       |
                     Local
                    minimum
```

Test Numbers

x	$f'(x)$
0.5	$-2.77(-)$
2	$0.69(+)$

Thus, f is decreasing on $(0, 1)$ and increasing on $(1, \infty)$; f has a local minimum at $x = 1$.

Step 3. Analyze $f''(x)$:

$$f''(x) = \frac{x\left(\frac{2}{x}\right) - 2\ln x}{x^2} = \frac{2(1-\ln x)}{x^2}$$

Partition numbers for f'': $\frac{2(1-\ln x)}{x^2} = 0$

$\ln x = 1$

$x = e$

Sign chart for f'':

```
f"(x)      + + + + + + + 0 - - - -
          ──┼──┼──┼──┼──┼──┼──┼──┼──> x
Graph      0     1     2  e 3
of f         Concave         Concave
             upward          downward
                 Inflection
                   point
```

Test Numbers

x	$f''(x)$
1	2(+)
4	−0.048(−)

Thus, the graph of f is concave upward on $(0, e)$ and concave downward on (e, ∞); the graph has an inflection point at $x = e$.

Step 4. Sketch the graph of f:

x	$f(x)$
1	0
e	1

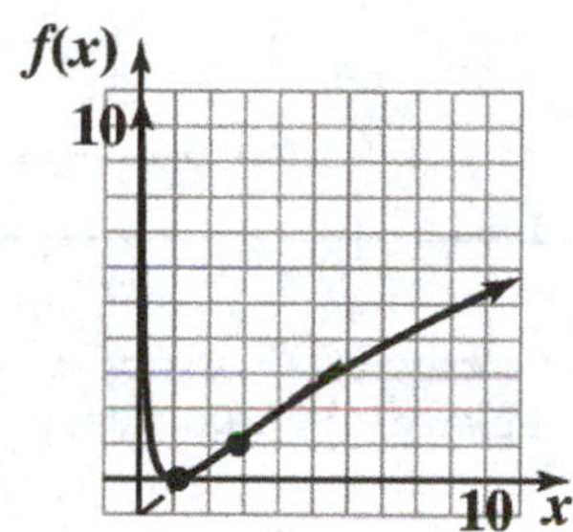

55. $f(x) = \dfrac{1}{x^2 + 2x - 8} = \dfrac{1}{(x+4)(x-2)}$

Step 1. Analyze $f(x)$:

(A) Domain: All real numbers except $x = -4$, $x = 2$.

(B) Intercepts: y-intercept: $f(0) = -\dfrac{1}{8}$

x-intercepts: no x-intercept

(C) Asymptotes:

Horizontal asymptote: $\dfrac{1}{x^2} \to 0$ as $x \to \infty$; $y = 0$ (the x-axis) is a horizontal asymptote.

Vertical asymptote: $x = -4$, $x = 2$

Step 2. Analyze $f'(x)$:

$$f'(x) = \frac{-(2x+2)}{(x^2+2x-8)^2} = \frac{-2(x+1)}{(x^2+2x-8)^2} = \frac{-2(x+1)}{(x-2)^2(x+4)^2}$$

Critical values: $x = -1$

Partition numbers: $x = -4$, $x = -1$, $x = 2$

Sign chart for f':

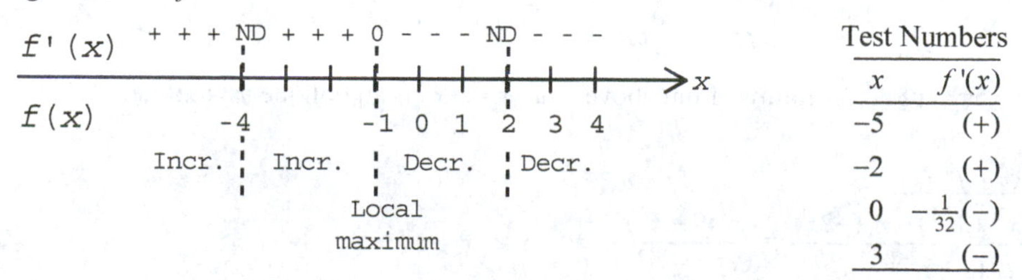

Test Numbers

x	$f'(x)$
−5	(+)
−2	(+)
0	$-\frac{1}{32}$(−)
3	(−)

Thus, f is increasing on $(-\infty, -4)$ and $(-4, -1)$, decreasing on $(-1, 2)$ and $(2, \infty)$, and has a local maximum at $x = -1$.

Step 3. Analyze $f''(x)$:

$$f''(x) = \frac{(x^2+2x-8)^2(-2)+(2x+2)(2)(x^2+2x-8)(2x+2)}{(x^2+2x-8)^4} = \frac{6(x^2+2x+4)}{(x^2+2x-8)^3}$$

Partition numbers for f'': $x = -4$, $x = 2$ ($x^2 + 2x + 4$ has no real roots)

Sign chart for f'':

$f''(x)$	+ + ND	- - - - - - - - -	ND + +
x	-4	0	2
Graph of f	Concave Upward	Concave Downward	Concave Upward

Test Numbers

x	$f''(x)$
-5	$(+)$
0	$(-)$
3	$(+)$

Thus, the graph of f is concave upward on $(-\infty, -4)$ and $(2, \infty)$, and concave downward on $(-4, 2)$.

Step 4. Sketch the graph of f:

x	$f(x)$
-1	$-\frac{1}{9}$
0	$-\frac{1}{8}$

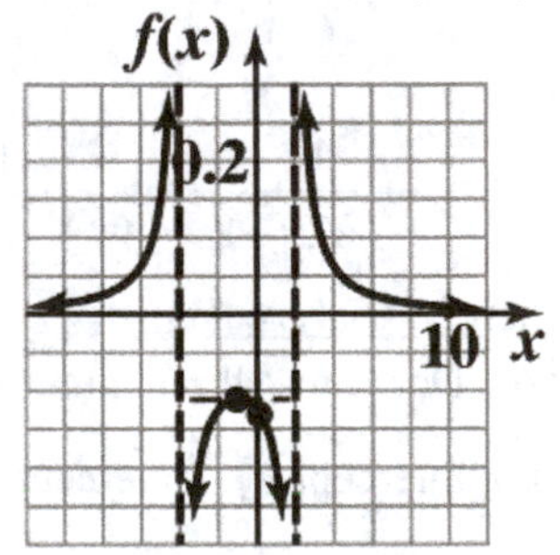

57. $f(x) = \dfrac{x^3}{3-x^2}$

Step 1. Analyze $f(x)$:

(A) Domain: All real numbers except $x = -\sqrt{3}$, $x = \sqrt{3}$.

(B) Intercepts: y-intercept: $f(0) = 0$

x-intercepts: $\dfrac{x^3}{3-x^2} = 0$, $x = 0$

(C) Asymptotes:

Horizontal asymptote: $\dfrac{x^3}{-x^2} = -x$; no horizontal asymptote

Vertical asymptote: $x = -\sqrt{3}$, $x = \sqrt{3}$

Oblique asymptote: It follows from above that $y = -x$ is an oblique asymptote.

Step 2. Analyze $f''(x)$:

$$f'(x) = \frac{(3-x^2)(3x^2)-x^3(-2x)}{(3-x^2)^2} = \frac{x^2(9-x^2)}{(3-x^2)^2}$$

Critical values: $x = -3, x = 0, x = 3$
Partition numbers: $x = -3, x = -\sqrt{3}, x = 0, x = \sqrt{3}, x = 3$
Sign chart for f':

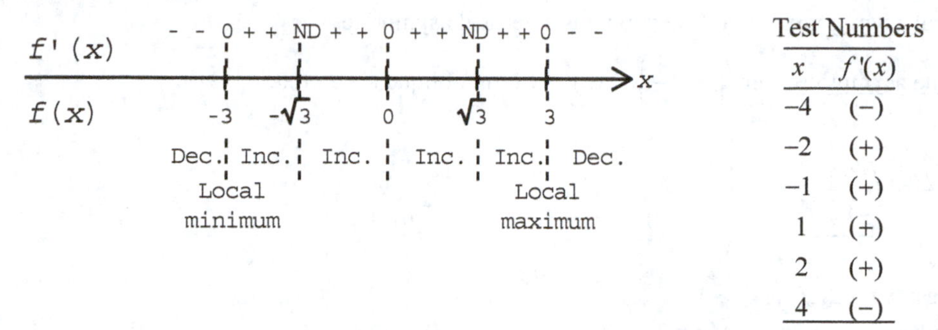

Test Numbers

x	$f'(x)$
–4	(–)
–2	(+)
–1	(+)
1	(+)
2	(+)
4	(–)

Thus, f is increasing on $(-3, -\sqrt{3})$, $(-\sqrt{3}, \sqrt{3})$, $(\sqrt{3}, 3)$, decreasing on $(-\infty, -3)$ and $(3, \infty)$, and has a local minimum at $x = -3$ and a local maximum at $x = 3$.

Step 3. Analyze $f''(x)$:

$$f''(x) = \frac{(3-x^2)^2(18x-4x^3)-x^2(9-x^2)2(3-x^2)(-2x)}{(3-x^2)^4} = \frac{6x(9+x^2)}{(3-x^2)^3}$$

Partition numbers for f'': $x = -\sqrt{3},\ x = 0,\ x = \sqrt{3}$
Sign chart for f'':

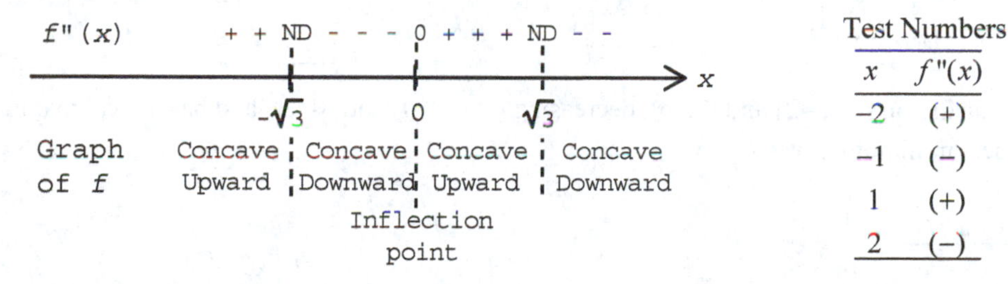

Test Numbers

x	$f''(x)$
–2	(+)
–1	(–)
1	(+)
2	(–)

Thus, the graph of f is concave upward on $(-\infty, -\sqrt{3})$ and $(0, \sqrt{3})$, concave downward on $(-\sqrt{3}, 0)$ and $(\sqrt{3}, \infty)$, and has an inflection point at $x = 0$.

Step 4. Sketch the graph of f:

x	$f(x)$
–3	$\frac{9}{2}$
0	0
3	$-\frac{9}{2}$

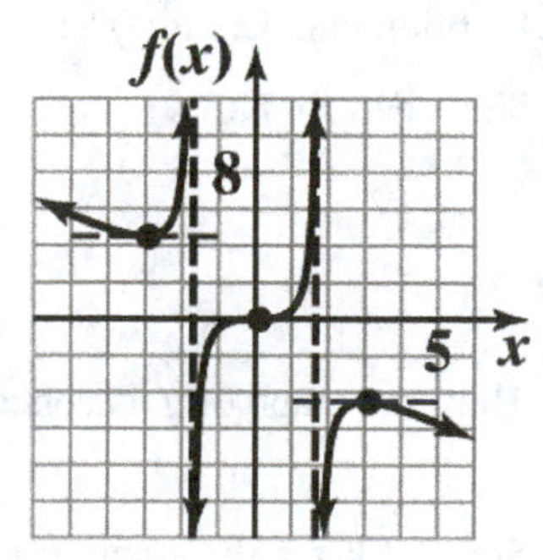

59. $f(x) = x + \dfrac{4}{x} = \dfrac{x^2+4}{x}$

Step 1. Analyze $f(x)$:
(A) Domain: All real numbers except $x = 0$.
(B) Intercepts: y-intercept: no y-intercept
x-intercepts: no x-intercepts

(C) Asymptotes:

Horizontal asymptote: $\frac{x^2}{x} = x$; no horizontal asymptote

Vertical asymptote: $x = 0$ (the y-axis) is a vertical asymptote

Oblique asymptote: $\lim_{x\to\infty}\left(x+\frac{4}{x}\right) = x$; $y = x$ is an oblique asymptote.

Step 2. Analyze $f'(x)$:

$$f'(x) = 1 - \frac{4}{x^2} = \frac{x^2-4}{x^2}$$

Critical values: $x = -2, x = 2$

Partition numbers: $x = -2, x = 0, x = 2$

Sign chart for f':

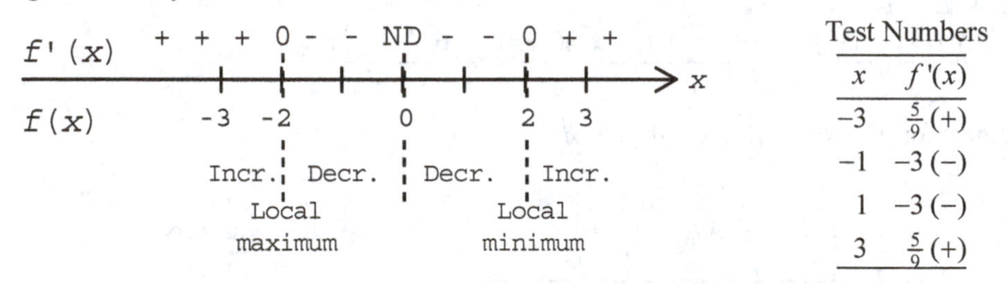

Test Numbers

x	$f'(x)$
-3	$\frac{5}{9}$ (+)
-1	-3 (−)
1	-3 (−)
3	$\frac{5}{9}$ (+)

Thus, f is increasing on $(-\infty, -2)$ and $(2, \infty)$, decreasing on $(-2, 0)$ and $(0, 2)$, and has a local maximum at $x = -2$ and a local minimum at $x = 2$.

Step 3. Analyze $f''(x)$:

$$f''(x) = \frac{8}{x^3}$$

Partition numbers for f'': $x = 0$

Sign chart for f'':

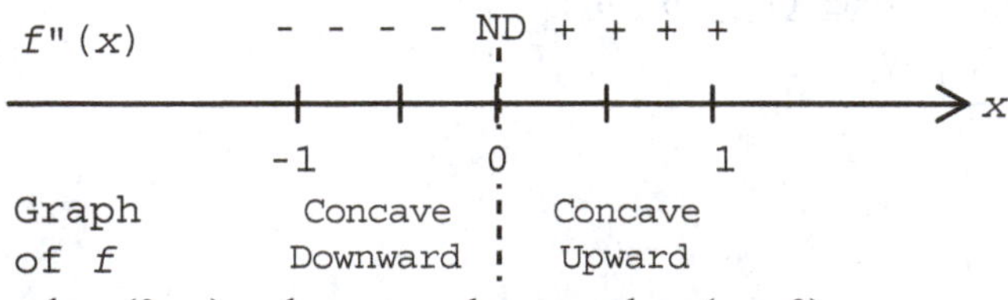

Thus, the graph of f is concave upward on $(0, \infty)$ and concave downward on $(-\infty, 0)$.

Step 4. Sketch the graph of f:

x	$f(x)$
-2	-4
2	4

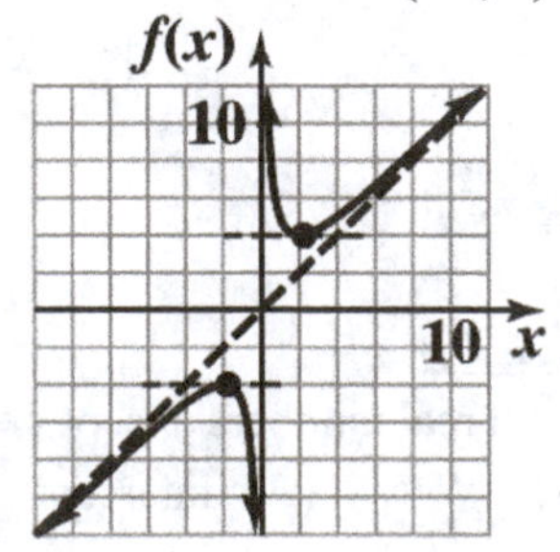

61. $f(x) = x - \dfrac{4}{x^2} = \dfrac{x^3 - 4}{x^2}$

Step 1. Analyze $f(x)$:

(A) Domain: All real numbers except $x = 0$.

(B) Intercepts: y-intercept: no y-intercept

x-intercepts: $\dfrac{x^3 - 4}{x} = 0,\ x = \sqrt[3]{4}$

(C) Asymptotes:

Horizontal asymptote: $\dfrac{x^3}{x^2} = x$; no horizontal asymptote

Vertical asymptote: $x = 0$ (the y-axis) is a vertical asymptote

Oblique asymptote: $\lim\limits_{x\to\infty}\left(x - \dfrac{4}{x^2}\right) = x$; $y = x$ is an oblique asymptote

Step 2. Analyze $f'(x)$:

$f'(x) = 1 + \dfrac{8}{x^3} = \dfrac{x^3 + 8}{x^3}$

Critical values: $x = -2$

Partition numbers: $x = -2, x = 0$

Sign chart for f':

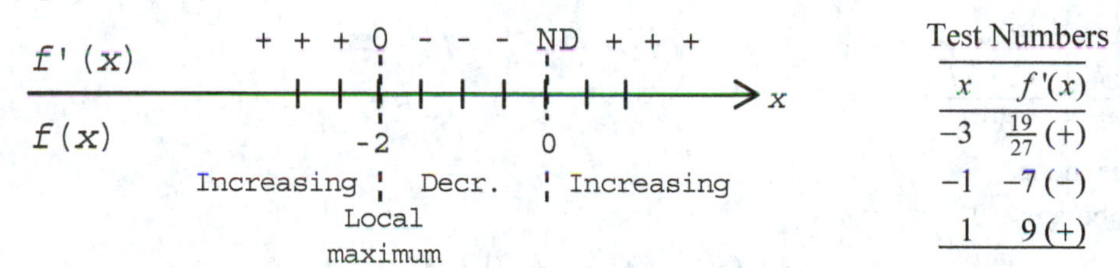

Test Numbers

x	$f'(x)$
-3	$\frac{19}{27}$ (+)
-1	-7 (–)
1	9 (+)

Thus, f is increasing on $(-\infty, -2)$ and $(0, \infty)$, decreasing on $(-2, 0)$; f has a local maximum at $x = -2$.

Step 3. Analyze $f''(x)$:

$f''(x) = -\dfrac{24}{x^4}$

Partition numbers for $f''(x)$: $x = 0$

Sign chart for f'':

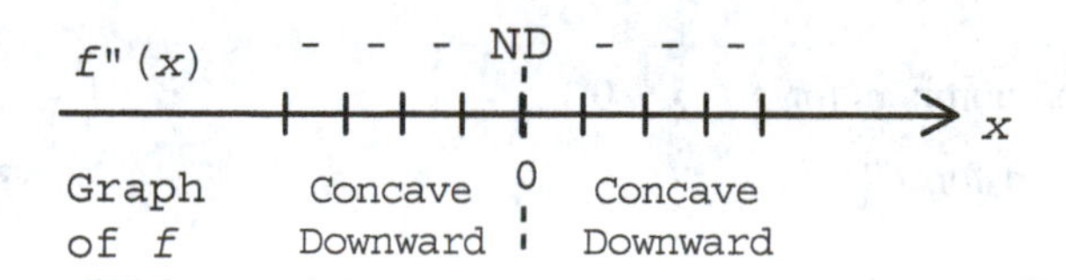

Thus, the graph of f is concave downward on $(-\infty, 0)$ and $(0, \infty)$.

Step 4. Sketch the graph of f:

x	$f(x)$
-2	-3

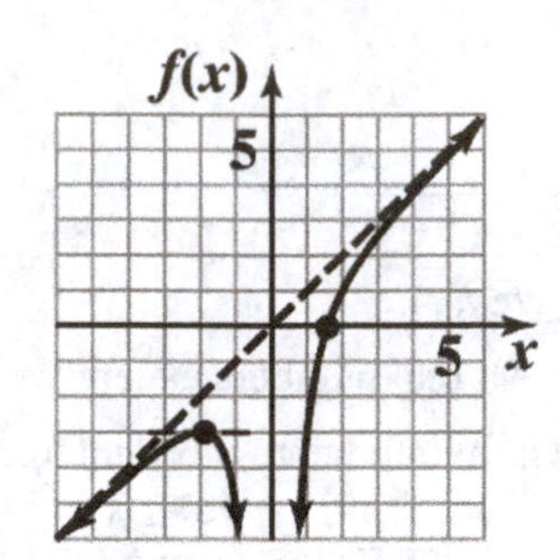

63. $f(x) = x - \frac{9}{x^3} = \frac{x^4 - 9}{x^3} = \frac{(x^2-3)(x^2+3)}{x^3}$

Step 1. Analyze $f(x)$:

(A) Domain: All real numbers except $x = 0$.

(B) Intercepts: y-intercept: no y-intercept

x-intercepts: $x = -\sqrt{3}$, $x = \sqrt{3}$

(C) Asymptotes:

Horizontal asymptote: $\frac{x^4}{x^3} = x$; no horizontal asymptote

Vertical asymptote: $x = 0$ (the y-axis) is a vertical asymptote

Oblique asymptote: $\lim_{x\to\infty}\left(x + \frac{9}{x^3}\right) = x$; $y = x$ is an oblique asymptote

Step 2. Analyze $f'(x)$:

$f'(x) = 1 + \frac{27}{x^4} = \frac{x^4+27}{x^4}$

Critical values: none

Partition numbers: $x = 0$

Sign chart for f':

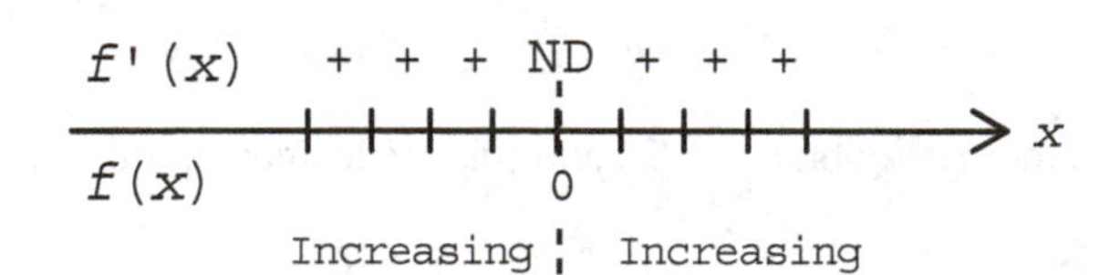

Thus, f is increasing on $(-\infty, 0)$ and $(0, \infty)$.

Step 3. Analyze $f''(x)$:

$f''(x) = -\frac{108}{x^5}$

Partition numbers for f'': $x = 0$

Sign chart for f'':

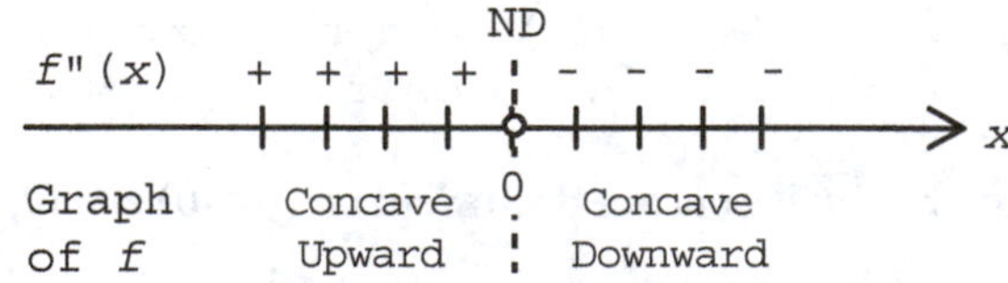

Thus, the graph of f is concave upward on $(-\infty, 0)$ and concave downward on $(0, \infty)$.

Step 4. Sketch the graph of f:

x	$f(x)$
$-\sqrt{3}$	0
$\sqrt{3}$	0

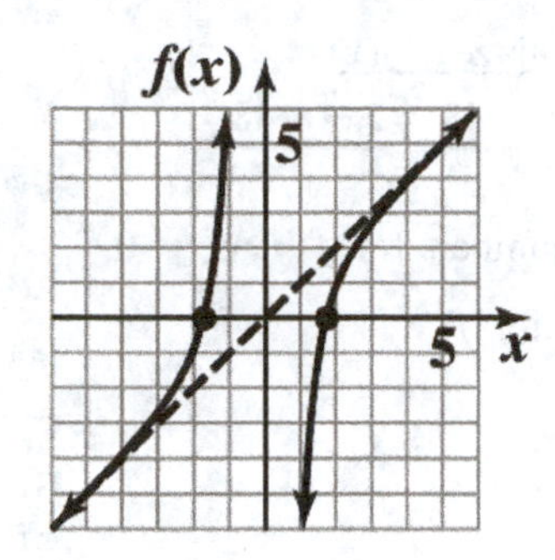

65. $f(x) = x + \frac{1}{x} + \frac{4}{x^3} = \frac{x^4 + x^2 + 4}{x^3}$

Step 1. Analyze $f(x)$:

(A) Domain: All real numbers except $x = 0$.

(B) Intercepts: y-intercept: no y-intercept
x-intercepts: no x-intercepts

(C) Asymptotes:

Horizontal asymptote: $\frac{x^4}{x^3} = x$; no horizontal asymptote

Vertical asymptote: $x = 0$ (the y-axis) is a vertical asymptote

Oblique asymptote: $\lim_{x\to\infty}\left(x + \frac{1}{x} + \frac{4}{x^3}\right) = x$; $y = x$ is an oblique asymptote

Step 2. Analyze $f'(x)$:

$$f'(x) = 1 - \frac{1}{x^2} - \frac{12}{x^4} = \frac{x^4 - x^2 - 12}{x^4} = \frac{(x^2-4)(x^2+3)}{x^4}$$

Critical values: $x = -2, x = 2$

Partition numbers: $x = -2,\ x = 0,\ x = 2$

Sign chart for f':

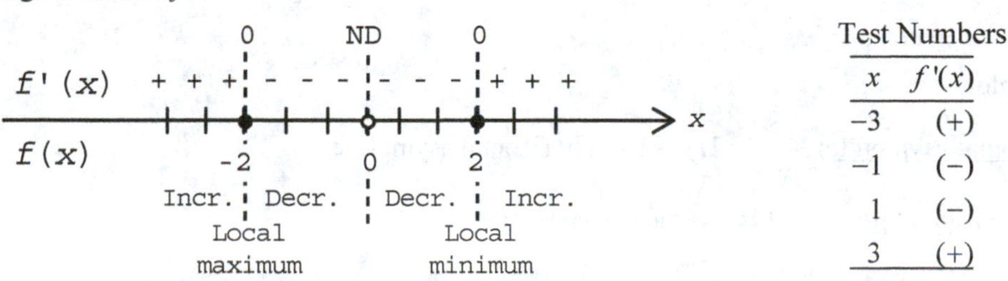

Test Numbers

x	$f'(x)$
−3	(+)
−1	(−)
1	(−)
3	(+)

Thus, f is increasing on $(-\infty, -2)$ and $(2, \infty)$, decreasing on $(-2, 0)$ and $(0, 2)$; f has a local maximum at $x = -2$ and a local minimum at $x = 2$.

Step 3. Analyze $f''(x)$:

$f''(x) = \frac{2}{x^3} + \frac{48}{x^5} = \frac{2x^2 + 48}{x^5}$

Partition numbers for $f''(x)$: $x = 0$

Sign chart for f'':

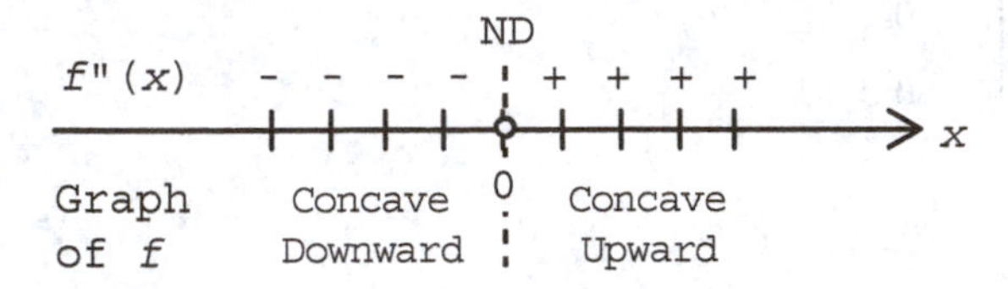

Thus, the graph of f is concave upward on $(0, \infty)$ and concave downward on $(-\infty, 0)$.

Step 4. Sketch the graph of f:

x	$f(x)$
–2	–3
2	3

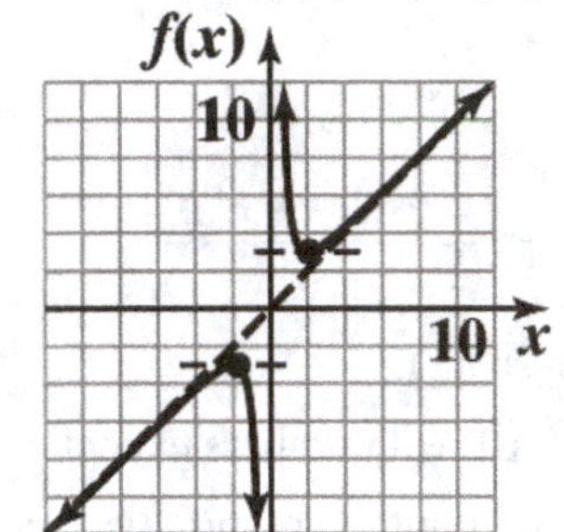

67. $C(x) = 10,000 + 90x + 0.02x^2$.

Average cost function: $\overline{C}(x) = \frac{C(x)}{x} = \frac{10,000}{x} + 90 + 0.02x \approx 90 + 0.02x$ for very large x.

69. $C(x) = 95,000 + 210x + 0.1x^2$.

Average cost function: $\overline{C}(x) = \frac{C(x)}{x} = \frac{95,000}{x} + 210 + 0.1x \approx 210 + 0.1x$ for very large x.

71. $f(x) = \frac{x^2 + x - 6}{x^2 - 6x + 8} = \frac{(x+3)(x-2)}{(x-4)(x-2)} = \frac{x+3}{x-4}, x \neq 2$

Step 1. Analyze $f(x)$:

(A) Domain: All real numbers except $x = 2$, $x = 4$.

(B) Intercepts: y-intercept: $f(0) = -\frac{3}{4}$

x-intercepts: $x = -3$

(C) Asymptotes:

Horizontal asymptote: $\frac{x^2}{x^2} = 1$, $y = 1$ is a horizontal asymptote

Vertical asymptote: $x = 4$ is a vertical asymptote

Step 2. Analyze $f'(x)$:

$f'(x) = \frac{x - 4 - (x+3)}{(x-4)^2} = -\frac{7}{(x-4)^2}$

Critical values: None

Partition numbers: $x = 4$

Sign chart for f':

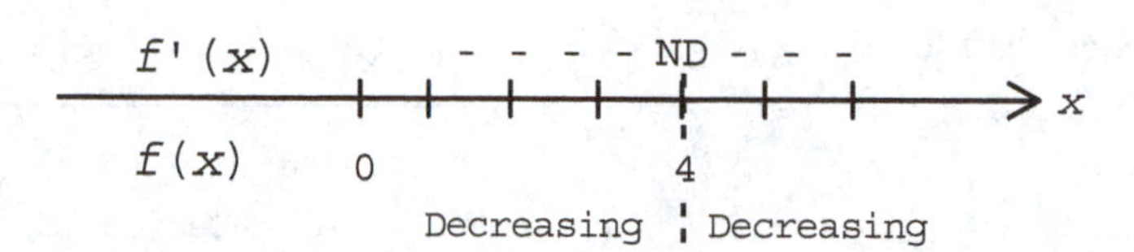

Thus, f is decreasing on $(-\infty, 4)$ and $(4, \infty)$.

Step 3. Analyze $f''(x)$:

$$f''(x) = \frac{14}{(x-4)^3}$$

Partition numbers for f'': $x = 4$

Sign chart for f'':

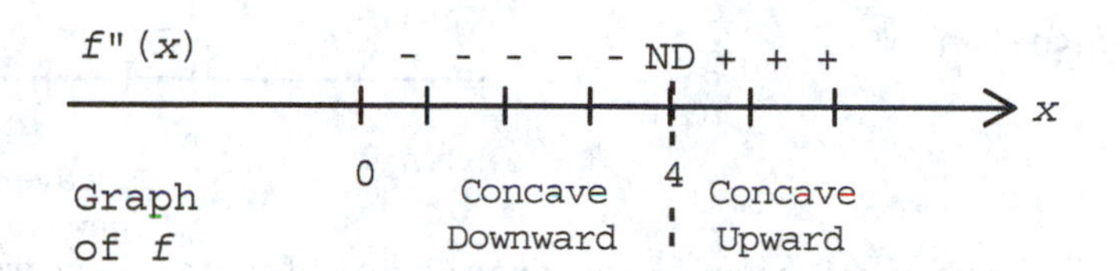

The graph of f is concave upward on $(4, \infty)$ and concave downward on $(-\infty, 4)$.

Step 4. Sketch the graph of f:

x	$f(x)$
-3	0
0	$-\frac{3}{4}$

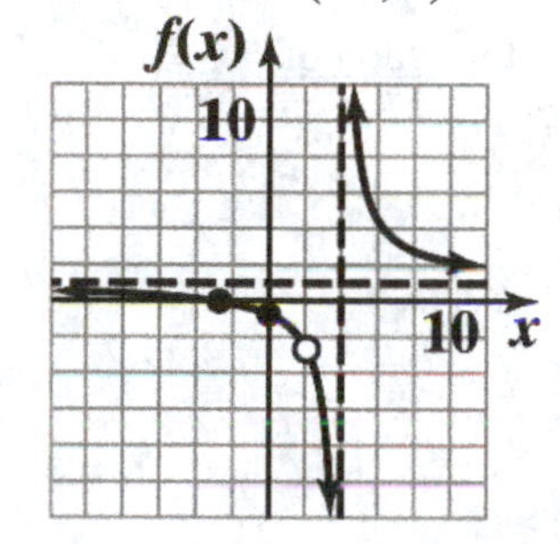

73. $f(x) = \dfrac{2x^2 + x - 15}{x^2 - 9} = \dfrac{(2x-5)(x+3)}{(x-3)(x+3)} = \dfrac{2x-5}{x-3}, x \neq -3$

Step 1. Analyze $f(x)$:

(A) Domain: All real numbers except $x = -3$, $x = 3$.

(B) Intercepts: y-intercept: $f(0) = \frac{5}{3}$

x-intercepts: $x = \frac{5}{2}$

(C) Asymptotes:

Horizontal asymptote: $\frac{2x^2}{x^2} = 2$, $y = 2$ is a horizontal asymptote

Vertical asymptote: $x = 3$ is a vertical asymptote

Step 2. Analyze $f'(x)$:

$$f'(x) = \frac{(x-3)2 - (2x-5)}{(x-3)^2} = \frac{-1}{(x-3)^2}$$

Critical values: None

Partition numbers: $x = 3$

Sign chart for f':

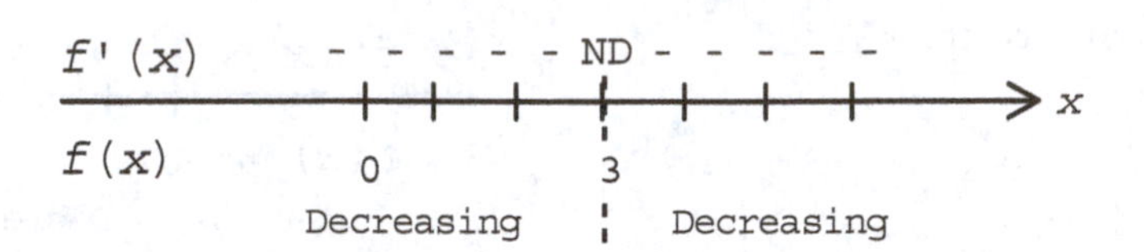

Thus, f is decreasing on $(-\infty, 3)$ and $(3, \infty)$.

Step 3. Analyze $f''(x)$:

$$f''(x) = \frac{2}{(x-3)^3}$$

Partition numbers for f'': $x = 3$

Sign chart for f'':

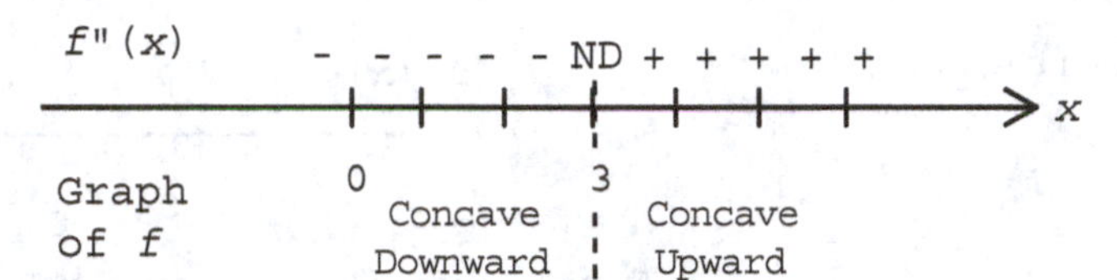

The graph of f is concave upward on $(3, \infty)$ and concave downward on $(-\infty, 3)$.

Step 4. Sketch the graph of f:

x	$f(x)$
$\frac{5}{2}$	0
0	$\frac{5}{3}$

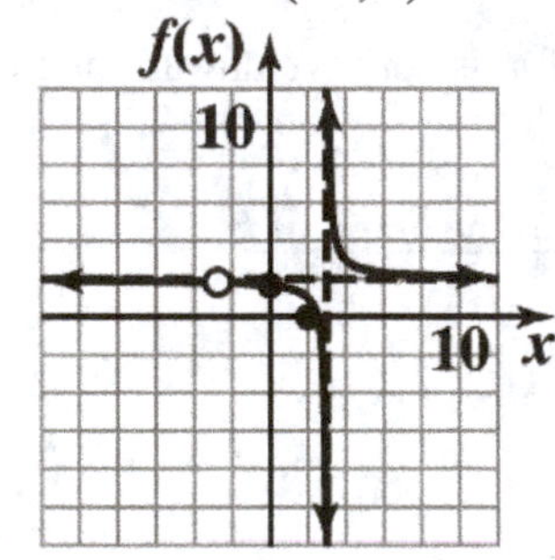

75. $f(x) = \dfrac{x^3 - 5x^2 + 6x}{x^2 - x - 2} = \dfrac{x(x-3)(x-2)}{(x-2)(x+1)} = \dfrac{x(x-3)}{x+1}$, $x \neq 2$

Step 1. Analyze $f(x)$:

(A) Domain: All real numbers except $x = -1$, $x = 2$.

(B) Intercepts: y-intercept: $f(0) = 0$
x-intercepts: $x = 0$, $x = 3$

(C) Asymptotes:

Horizontal asymptote: $\dfrac{x^3}{x^2} = x$; no horizontal asymptote

Vertical asymptote: $x = -1$ is a vertical asymptote

Oblique asymptote: $\dfrac{x^2 - 3x}{x+1} = x - 4 + \dfrac{4}{x+1}$; $y = x - 4$ is an oblique asymptote

Step 2. Analyze $f'(x)$:

$$f'(x) = \frac{(x+1)(2x-3) - (x^2 - 3x)}{(x+1)^2} = \frac{x^2 + 2x - 3}{(x+1)^2} = \frac{(x+3)(x-1)}{(x+1)^2}$$

Critical values: $x = -3$, $x = 1$

Partition numbers: $x = -3$, $x = -1$, $x = 1$

Sign chart for f':

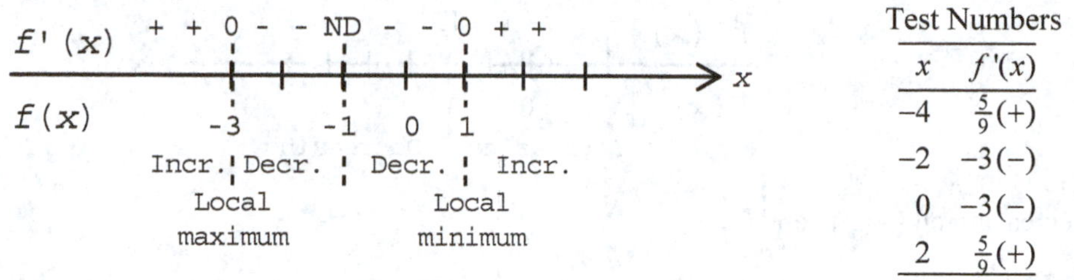

Test Numbers

x	$f'(x)$
−4	$\frac{5}{9}(+)$
−2	$-3(-)$
0	$-3(-)$
2	$\frac{5}{9}(+)$

Thus, f is increasing on $(-\infty, -3)$ and $(1, \infty)$, f is decreasing on $(-3, -1)$ and $(-1, 1)$; f has a local maximum at $x = -3$ and a local minimum at $x = 1$.

Step 3. Analyze $f''(x)$:

$$f''(x) = \frac{(x+1)^2(2x+2)-(x^2+2x-3)(2)(x+1)}{(x+1)^4} = \frac{8}{(x+1)^3}$$

Partition numbers for f'': $x = -1$

Sign chart for f'':

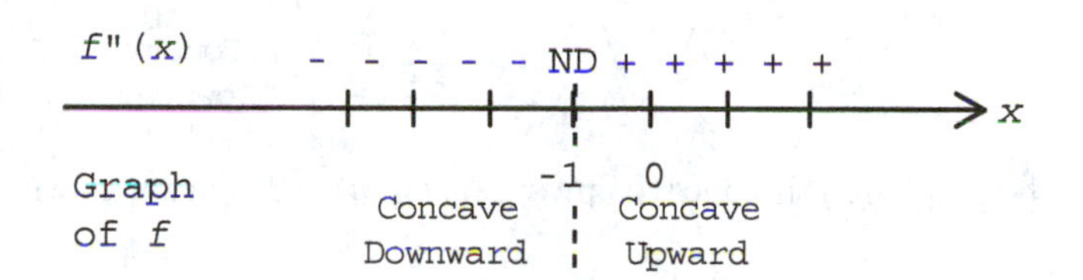

The graph of f is concave upward on $(-1, \infty)$ and concave downward on $(-\infty, -1)$.

Step 4. Sketch the graph of f:

x	$f(x)$
−3	−9
0	0
1	−1
3	0

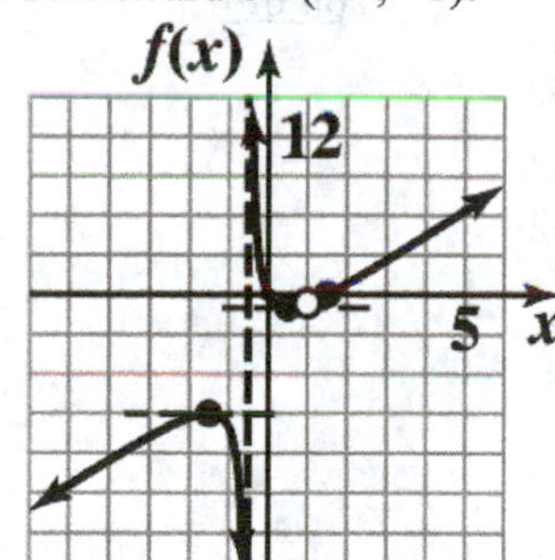

77. $f(x) = \frac{x^2+x-2}{x^2-2x+1} = \frac{(x+2)(x-1)}{(x-1)^2} = \frac{x+2}{x-1}, x \neq 1$

Step 1. Analyze $f(x)$:

(A) Domain: All real numbers except $x = 1$.

(B) Intercepts: y-intercept: $f(0) = -2$

x-intercepts: $x = -2$

(C) Asymptotes:

Horizontal asymptote: $\frac{x^2}{x^2} = 1$; $y = 1$ is a horizontal asymptote

Vertical asymptote: $x = 1$ is a vertical asymptote

Step 2. Analyze $f'(x)$:

$$f'(x) = \frac{(x-1)-(x+2)}{(x-1)^2} = \frac{-3}{(x-1)^2}$$

Critical values: None

Partition numbers: $x = 1$

Sign chart for f':

$f'(x)$ - - - - ND - - - -

$f(x)$ 0 1 x

Decreasing ¦ Decreasing

Thus, f is decreasing on $(-\infty, 1)$ and $(1, \infty)$.

Step 3. Analyze $f''(x)$:

$$f''(x) = \frac{6}{(x-1)^3}$$

Partition numbers for f'': $x = 1$

Sign chart for f'':

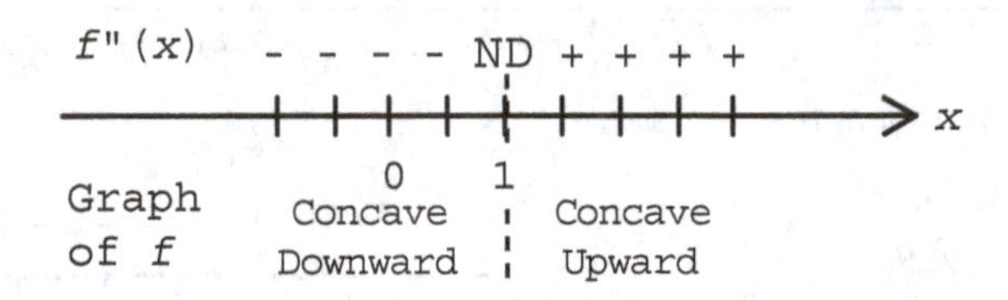

The graph of f is concave upward on $(1, \infty)$ and concave downward on $(-\infty, 1)$.

Step 4. Sketch the graph of f:

x	$f(x)$
–2	0
0	–2

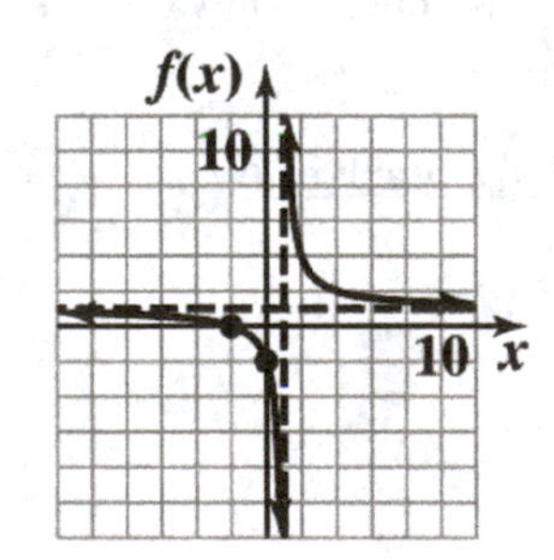

79. $R(x) = 1{,}296x - 0.12x^3, \ 0 \le x \le 80$

$R'(x) = 1{,}296 - 0.36x^2$

$R'(x) = 0$: $\quad 0.36x^2 = 1{,}296$

$x^2 = 3{,}600$

$x = 60$

The critical value or R is $x = 60$.

Sign chart for $R'(x)$:

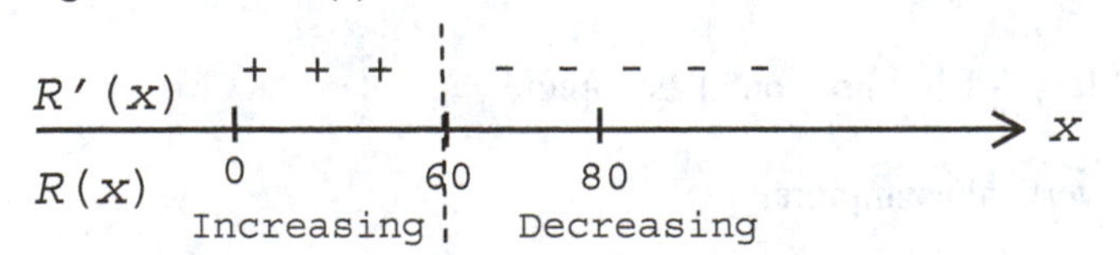

Test Numbers

x	$R'(x)$
30	972 (+)
80	–1008 (–)

R is increasing on (0, 60) and decreasing on (60, 80).
$R''(x) = -0.72x < 0$ on (0, 80).
The graph of R is concave downward on (0, 80).

x	$f(x)$
0	0
60	51,840
80	42,240

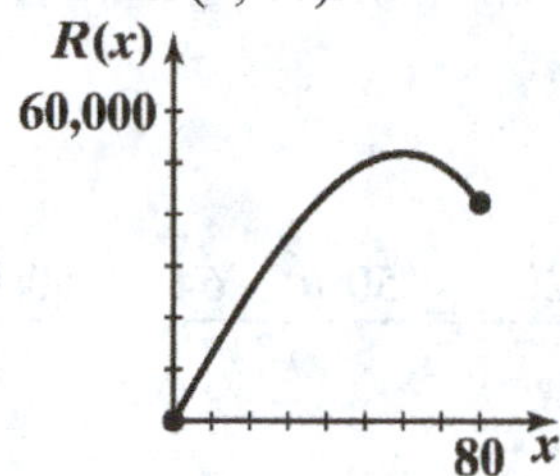

81. $P(x) = \dfrac{2x}{1-x}, 0 \le x < 1$

(A) $P'(x) = \dfrac{(1-x)(2) - 2x(-1)}{(1-x)^2} = \dfrac{2}{(1-x)^2}$

$P'(x) > 0$ for $0 \le x < 1$. Thus, P is increasing on (0, 1).

(B) From (A), $P'(x) = 2(1-x)^{-2}$. Thus,

$P''(x) = -4(1-x)^{-3}(-1) = \dfrac{4}{(1-x)^3}$.

$P''(x) > 0$ for $0 \le x < 1$, and the graph of P is concave upward on (0, 1).

(C) Since the domain of P is [0, 1), there are no horizontal asymptotes. The denominator is 0 at $x = 1$ and the numerator is nonzero there. Thus, $x = 1$ is a vertical asymptote.

(D) $P(0) = \dfrac{2(0)}{1-0} = 0$.

Thus, the origin is both an x and a y intercept of the graph.

(E) The graph of P is:

x	$P(x)$
0	0
$\frac{1}{2}$	2
$\frac{3}{4}$	6

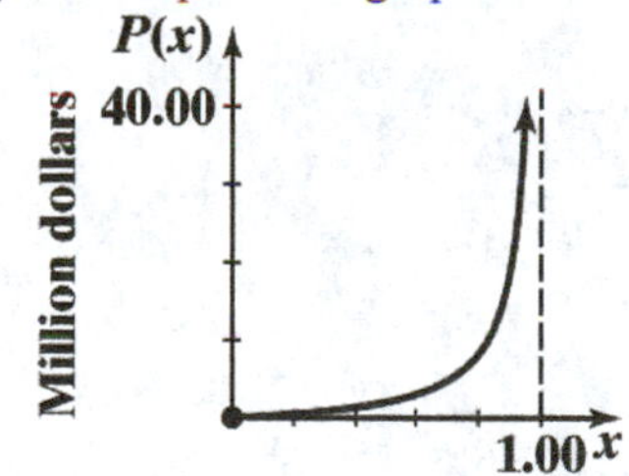

83. $C(n) = 3200 + 250n + 50n^2, 0 < n < \infty$

(A) Average cost per year:

$\overline{C}(n) = \dfrac{C(n)}{n} = \dfrac{3200}{n} + 250 + 50n, \ 0 < n < \infty$

(B) Graph $\overline{C}(n)$:

Step 1. Analyze $\overline{C}(n)$:
Domain: $0 < n < \infty$
Intercepts: C intercept: None ($n > 0$)
n intercepts: $\dfrac{3200}{n} + 250 + 50n > 0$ on $(0, \infty)$; there are no n intercepts.

Asymptotes: For large n, $C(n) = \frac{3200}{n} + 250 + 50n \approx 250 + 50n$. Thus, $y = 250 + 50n$ is an oblique asymptote. As $n \to 0$, $\overline{C} \to \infty$. Thus, $n = 0$ is a vertical asymptote.

Step 2. Analyze $\overline{C}'(n)$:

$$\overline{C}'(n) = -\frac{3200}{n^2} + 50 = \frac{50n^2 - 3200}{n^2} = \frac{50(n^2 - 64)}{n^2} = \frac{50(n-8)(n+8)}{n^2}, \ 0 < n < \infty$$

Critical value: $n = 8$

Sign chart for $\overline{C}'$:

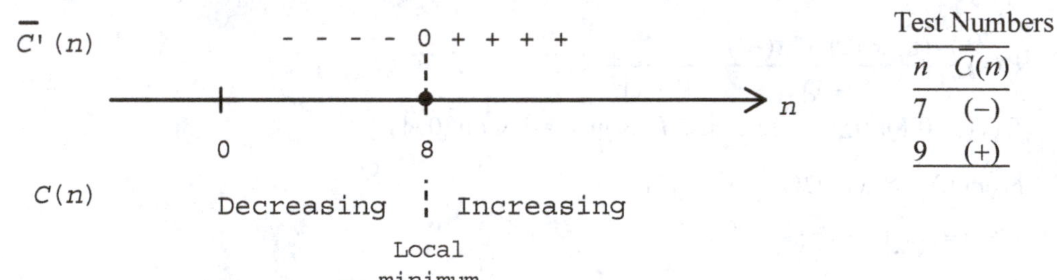

Test Numbers	
n	$\overline{C}(n)$
7	(−)
9	(+)

Thus, $\overline{C}$ is decreasing on (0, 8) and increasing on (8, ∞); $n = 8$ is a local minimum.

Step 3: Analyze $C''(n)$:

$$\overline{C}''(n) = \frac{6400}{n^3}, \ 0 < n < \infty$$

$\overline{C}''(n) > 0$ on (0, ∞). Thus, the graph of $\overline{C}$ is concave upward on (0, ∞).

Step 4. Sketch the graph of $\overline{C}$:

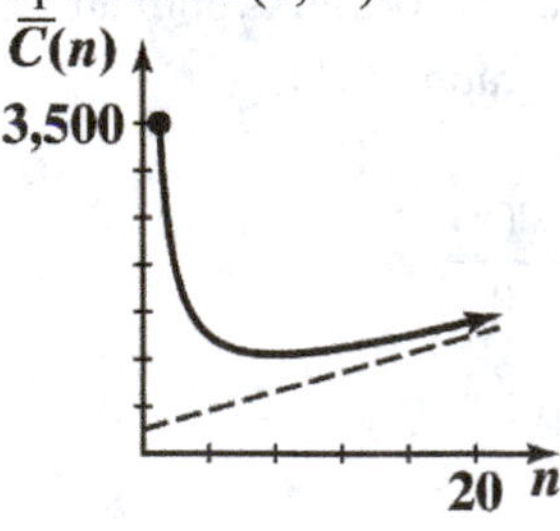

(C) The average cost per year is a minimum when $n = 8$ years.

85. $C(x) = 1000 + 5x + 0.1x^2, 0 < x < \infty$.

(A) The average cost function is: $\overline{C}(x) = \frac{1000}{x} + 5 + 0.1x$.

Now, $\overline{C}'(x) = -\frac{1000}{x^2} + \frac{1}{10} = \frac{x^2 - 10{,}000}{10x^2} = \frac{(x+100)(x-100)}{10x^2}$

Sign chart for $\overline{C}'$:

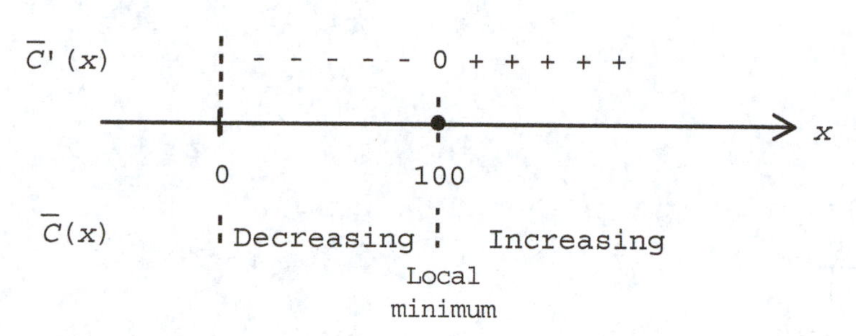

Test Numbers

x		$\overline{C}'(x)$
1	≈	$-1000(-)$
101	≈	$\frac{1}{500}(+)$

Thus, $\overline{C}$ is decreasing on (0, 100) and increasing on (100, ∞); $\overline{C}$ has a minimum at $x = 100$.

Since $\overline{C}''(x) = \dfrac{2000}{x^3} > 0$ for $0 < x < \infty$, the graph of $\overline{C}$ is concave upward on $(0, \infty)$. The line $x = 0$ is a vertical asymptote and the line $y = 5 + 0.1x$ is an oblique asymptote for the graph of $\overline{C}$. The marginal cost function is $C'x) = 5 + 0.2x$.

The graphs of $\overline{C}$ and C' are:

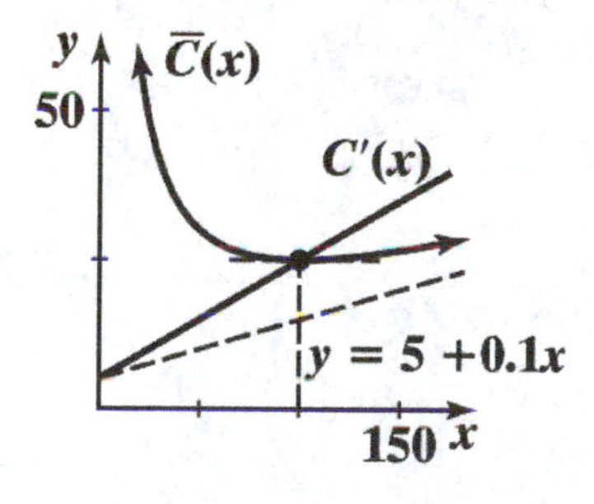

(B) The minimum average cost is:

$$\overline{C}(100) = \frac{1000}{100} + 5 + \frac{1}{10}(100) = 25$$

87. (A)

```
QuadReg
 y=ax²+bx+c
 a=.0100714286
 b=.7835714286
 c=316
```

(B) The average cost function $\overline{y} = \dfrac{y(x)}{x}$ where $y(x)$ is the regression equation found in part (A).

The minimum average cost is \$4.35 when 177 pizzas are produced.

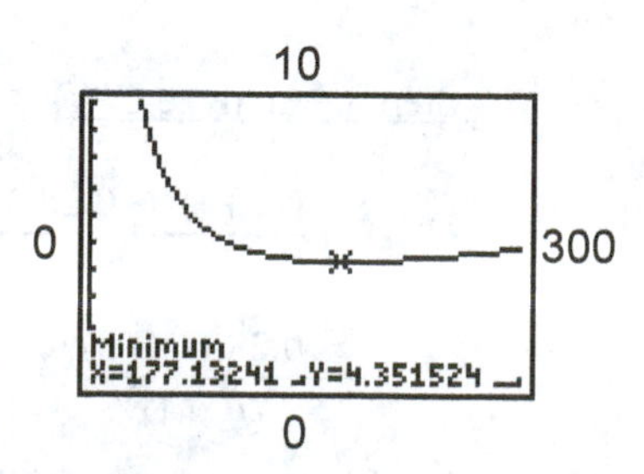

89. $C(t) = \dfrac{0.14t}{t^2+1}$

<u>Step 1. Analyze $C(t)$:</u>

Domain: $t \geq 0$, i.e., $[0, \infty)$

Intercepts: y intercept: $C(0) = 0$

t intercepts: $\dfrac{0.14t}{t^2+1} = 0$

$t = 0$

Asymptotes:

<u>Horizontal asymptote</u>: $\lim_{t\to\infty} \dfrac{0.14t}{t^2+1} = \lim_{t\to\infty} \dfrac{0.14t}{t^2\left(1+\frac{1}{t^2}\right)} = \lim_{t\to\infty} \dfrac{0.14}{t\left(1+\frac{1}{t^2}\right)} = 0$

Thus, $y = 0$ (the t axis) is a horizontal asymptote.

<u>Vertical asymptotes</u>: Since $t^2 + 1 > 0$ for all t, there are no vertical asymptotes.

<u>Step 2. Analyze $C'(t)$:</u>

$$C'(t) = \frac{(t^2+1)(0.14) - 0.14t(2t)}{(t^2+1)^2} = \frac{0.14(1-t^2)}{(t^2+1)^2} = \frac{0.14(1-t)(1+t)}{(t^2+1)^2}$$

Critical values on $[0, \infty)$: $t = 1$

Sign chart for C':

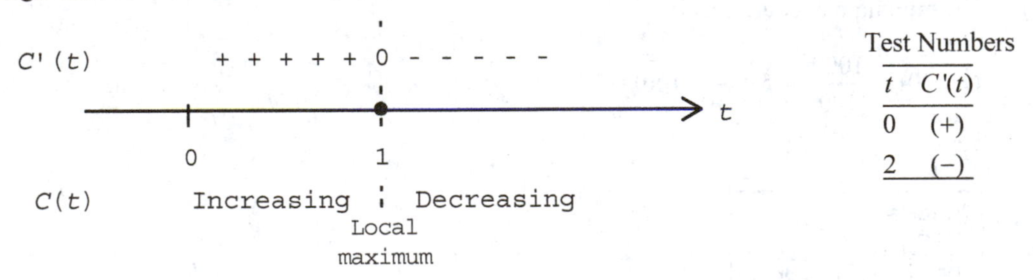

Thus, C is increasing on $(0, 1)$ and decreasing on $(1, \infty)$; C has a maximum value at $t = 1$.

<u>Step 3. Analyze $C''(t)$:</u>

$$C''(t) = \frac{(t^2+1)^2(-0.28t) - 0.14(1-t^2)(2)(t^2+1)(2t)}{(t^2+1)^4} = \frac{(t^2+1)(-0.28t) - 0.56t(1-t^2)}{(t^2+1)^3} = \frac{0.28t^3 - 0.84t}{(t^2+1)^3}$$

$$= \frac{0.28t(t^2-3)}{(t^2+1)^3} = \frac{0.28t(t-\sqrt{3})(t+\sqrt{3})}{(t^2+1)^3}, \quad 0 \leq t < \infty$$

Partition numbers for C'' on $[0, \infty)$: $t = \sqrt{3}$

Sign chart for C'':

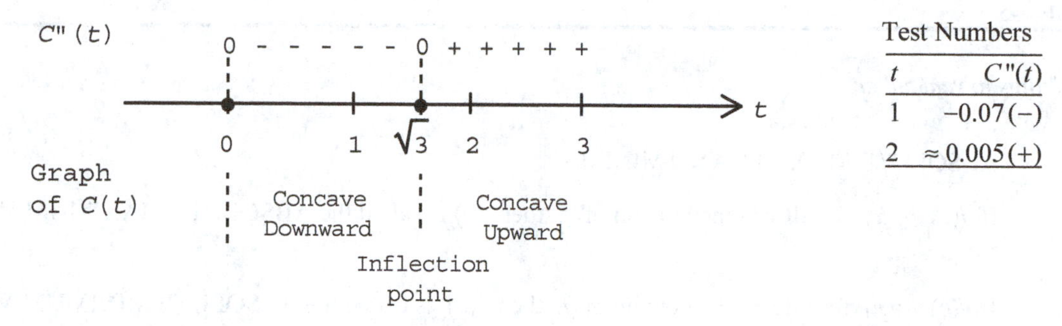

Test Numbers

t	$C''(t)$
1	$-0.07(-)$
2	$\approx 0.005(+)$

Thus, the graph of C is concave downward on $(0, \sqrt{3})$ and concave upward on $(\sqrt{3}, \infty)$; the graph has an inflection point at $t = \sqrt{3}$.

Step 4. Sketch the graph of $C(t)$:

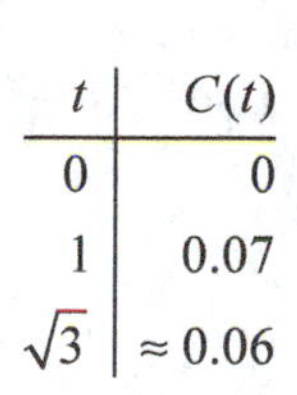

t	$C(t)$
0	0
1	0.07
$\sqrt{3}$	≈ 0.06

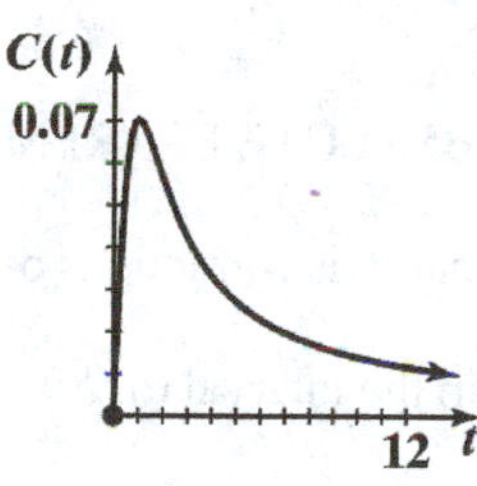

91. $N(t) = \dfrac{5t+20}{t} = 5 + 20t^{-1}, \ 1 \le t \le 30$

Step 1. Analyze $N(t)$:
Domain: $1 \le t \le 30$, or $[1, 30]$.
Intercepts: There are no t or N intercepts.
Asymptotes: Since N is defined only for $1 \le t \le 30$, there are no horizontal asymptotes. Also, since $t \ne 0$ on $[1, 30]$, there are no vertical asymptotes.

Step 2. Analyze $N'(t)$:
$N'(t) = -20t^{-2} = \dfrac{-20}{t^2}, \ 1 \le t \le 30$
Since $N'(t) < 0$ for $1 \le t \le 30$, N is decreasing on $(1, 30)$; N has no local extrema.

Step 3. Analyze $N''(t)$:
$N''(t) = \dfrac{40}{t^3}, \ 1 \le t \le 30$
Since $N''(t) > 0$ for $1 \le t \le 30$, the graph of N is concave upward on $(1, 30)$.

Step 4. Sketch the graph of N:

t	$N(t)$
1	25
5	9
10	7
30	5.67

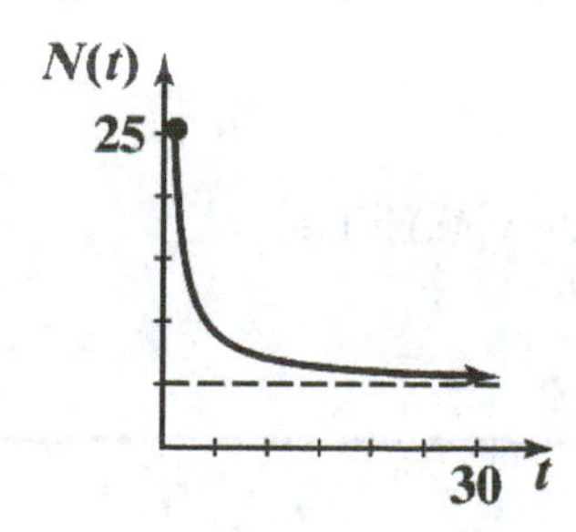

EXERCISE 4-5

Things to remember:

1. ABSOLUTE MAXIMA AND MINIMA

   If $f(c) \geq f(x)$ for all x in the domain of f, then $f(c)$ is called the ABSOLUTE MAXIMUM VALUE of f.

   If $f(c) \leq f(x)$ for all x in the domain of f, then $f(c)$ is called the ABSOLUTE MINIMUM VALUE of f.

2. A function f continuous on a closed interval $[a, b]$ has both an absolute maximum and an absolute minimum on that interval. Absolute extrema (if they exist) must always occur at critical values or at endpoints.

3. PROCEDURE FOR FINDING ABSOLUTE EXTREMA ON A CLOSED INTERVAL

   Step 1. Check to make certain that f is continuous over $[a, b]$.

   Step 2. Find the critical values in the interval (a, b).

   Step 3. Evaluate f at the endpoints a and b and at the critical values found in Step 2.

   Step 4. The absolute maximum $f(x)$ on $[a, b]$ is the largest of the values found in Step 3.

   Step 5. The absolute minimum $f(x)$ on $[a, b]$ is the smallest of the values found in Step 3.

4. SECOND DERIVATIVE TEST

   Let c be a critical value for $f(x)$.

$f'(c)$	$f''(c)$	GRAPH OF f IS:	$f(c)$	EXAMPLE
0	+	Concave upward	Local minimum	∪
0	–	Concave downward	Local maximum	∩
0	0	?	Test does not apply	

5. SECOND DERIVATIVE TEST FOR ABSOLUTE EXTREMUM

   Let f be continuous on an interval I with only one critical value c on I:

   If $f'(c) = 0$ and $f''(c) > 0$, then $f(c)$ is the absolute minimum of f on I.

   If $f'(c) = 0$ and $f''(c) < 0$, then $f(c)$ is the absolute maximum of f on I.

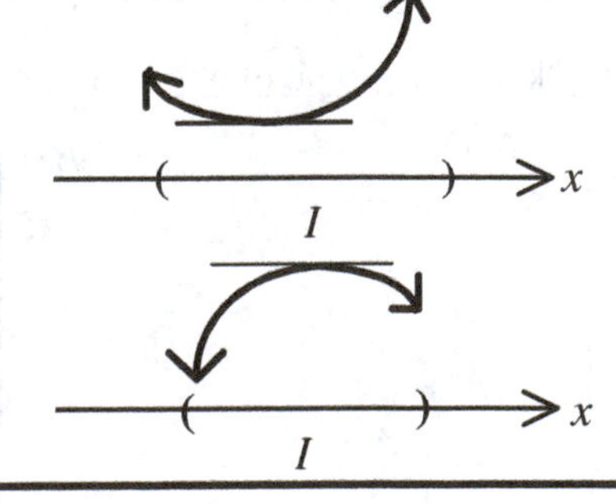

1. Max $f(x) = f(3) = 3$; min $f(x) = f(-2) = -2$.

3. Max $h(x) = h(-5) = 25$; min $h(x) = h(0) = 0$.

5. Max $n(x) = n(4) = 2$; min $n(x) = n(3) = \sqrt{3}$.

7. Max $q(x) = q(27) = -3$; min $q(x) = q(64) = -4$.

9. Interval [0, 10]; absolute minimum: $f(0) = 0$;
absolute maximum: $f(10) = 14$

11. Interval [0, 8]; absolute minimum: $f(0) = 0$;
absolute maximum: $f(3) = 9$

13. Interval [1, 10]; absolute minimum: $f(1) = f(7) = 5$;
absolute maximum: $f(10) = 14$

15. Interval [1, 9]; absolute minimum: $f(1) = f(7) = 5$;
absolute maximum: $f(3) = f(9) = 9$

17. Interval [2, 5]; absolute minimum: $f(5) = 7$;
absolute maximum: $f(3) = 9$

19. $f(x) = 2x - 5$.

(A) On [0,4]: Max $f(x) = f(4) = 3$; min $f(x) = f(0) = -5$.

(B) On [0,10], Max $f(x) = f(10) = 15$; min $f(x) = f(0) = -5$.

(C) On [−5,10], Max $f(x) = f(10) = 15$; min $f(x) = f(-5) = -15$.

21. $f(x) = x^2$.

(A) On [−1,1], Max $f(x) = f(-1) = f(1) = 1$; min $f(x) = f(0) = 0$.

(B) On [1,5], Max $f(x) = f(5) = 25$; min $f(x) = f(1) = 1$.

(C) On [−5,5], Max $f(x) = f(-5) = f(5) = 25$; min $f(x) = f(0) = 0$.

23. $f(x) = e^{-x}$; $f'(x) = -e^{-x} < 0$; f is decreasing on [−1,1].
Absolute maximum: $f(-1) = e \approx 2.718$; absolute minimum: $f(1) = e^{-1} \approx 0.368$.

25. $f(x) = 9 - x^2$ on $[-4,4]$; $f'(x) = -2x$; critical value: $-2x = 0$, $x = 0$.
$f(-4) = -7$, $f(0) = 9$, $f(4) = -7$;
absolute maximum: $f(0) = 9$; absolute minimum: $f(-4) = f(4) = -7$.

27. $f(x) = x^2 - 2x + 3$, $I = (-\infty, \infty)$
$f'(x) = 2x - 2 = 2(x - 1)$
$f'(x) = 0$: $2(x - 1) = 0$
$x = 1$

$x = 1$ is the ONLY critical value on I, and $f(1) = 1^2 - 2(1) + 3 = 2$.
$f''(x) = 2$ and $f''(1) = 2 > 0$. Therefore, $f(1) = 2$ is the absolute minimum. The function does not have an absolute maximum since $\lim_{x \to \pm\infty} f(x) = \infty$.

29. $f(x) = -x^2 - 6x + 9, I = (-\infty, \infty)$
$f'(x) = -2x - 6 = -2(x + 3)$
$f'(x) = 0: \quad -2(x + 3) = 0$
$x = -3$

$x = -3$ is the ONLY critical value on I, and $f(-3) = -(-3)^2 - 6(-3) + 9 = 18$.
$f''(x) = -2$ and $f''(-3) = -2 < 0$. Therefore, $f(-3) = 18$ is the absolute maximum. The function does not have an absolute minimum since $\lim_{x\to\pm\infty} f(x) = -\infty$.

31. $f(x) = x^3 + x, I = (-\infty, \infty)$
$f'(x) = 3x^2 + 1 \geq 1$ on I; f is increasing on I and $\lim_{x\to-\infty} f(x) = -\infty$, $\lim_{x\to\infty} f(x) = \infty$. Therefore, f does not have any absolute extrema.

33. $f(x) = 8x^3 - 2x^4$; domain: all real numbers
$f'(x) = 24x^2 - 8x^3 = 8x^2(3 - x)$
$f''(x) = 48x - 24x^2 = 24x(2 - x)$
Critical values: $x = 0$, $x = 3$
$f''(0) = 0$ (second derivative test fails)
$f''(3) = -72$ f has a local maximum at $x = 3$.

Sign chart for $f'(x) = 8x^2(3 - x)$
(0 and 3 are partition numbers)

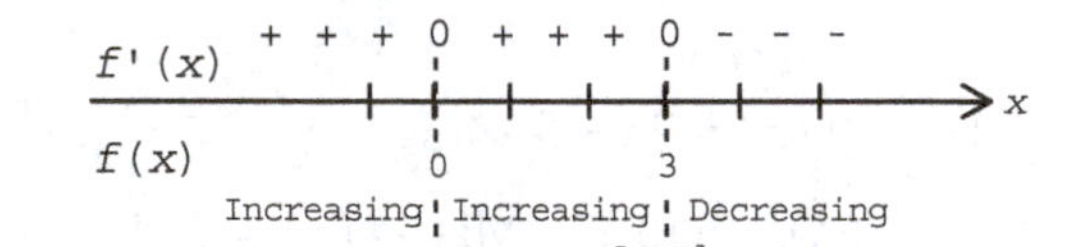

From the sign chart, f does not have a local extremum at $x = 0$;
f has a local maximum at $x = 3$ which must be an absolute maximum since f is increasing on $(-\infty, 3)$ and decreasing on $(3, \infty)$; $f(3) = 54$ is the absolute maximum of f. f does not have an absolute minimum since $\lim_{x\to\infty} f(x) = \lim_{x\to-\infty} f(x) = -\infty$.

35. $f(x) = x + \frac{16}{x}$; domain: all real numbers except $x = 0$.

$f'(x) = 1 - \frac{16}{x^2} = \frac{x^2 - 16}{x^2} = \frac{(x-4)(x+4)}{x^2}$

$f''(x) = \frac{32}{x^3}$

Critical values: $x = -4$, $x = 4$

$f''(-4) = -\frac{1}{2} < 0$; f has a local maximum at $x = -4$

$f''(4) = \frac{1}{2} > 0$; f has a local minimum at $x = 4$

$\lim_{x\to\infty} f(x) = \lim_{x\to\infty}\left(x + \frac{16}{x}\right) = \infty$; $\lim_{x\to-\infty} f(x) = \lim_{x\to-\infty}\left(x + \frac{16}{x}\right) = -\infty$; f has no absolute extrema.

37. $f(x) = \frac{x^2}{x^2 + 1}$; domain: all real numbers

$$f'(x) = \frac{(x^2+1)2x - x^2(2x)}{(x^2+1)^2} = \frac{2x}{(x^2+1)^2}$$

$$f''(x) = \frac{(x^2+1)^2(2) - 2x(2)(x^2+1)(2x)}{(x^2+1)^4} = \frac{2 - 6x^2}{(x^2+1)^3}$$

Critical value: $x = 0$
Since f has only one critical value and $f''(0) = 2 > 0$, $f(0) = 0$ is the absolute minimum of f. Since $\lim_{x\to\infty} f(x) = \lim_{x\to\infty} \frac{x^2}{x^2+1} = 1$, f has no absolute maximum; $y = 1$ is a horizontal asymptote for the graph of f.

39. $f(x) = \frac{2x}{x^2+1}$; domain: all real numbers

$$f'(x) = \frac{(x^2+1)2 - 2x(2x)}{(x^2+1)^2} = \frac{2-2x^2}{(x^2+1)^2} = \frac{2(1-x^2)}{(x^2+1)^2}$$

$$f''(x) = \frac{(x^2+1)^2(-4x) - 2(1-x^2)(2)(x^2+1)(2x)}{(x^2+1)^4} = \frac{4x(x^2-3)}{(x^2+1)^3}$$

Critical values: $x = -1$, $x = 1$
$f''(-1) = 1 > 0$; f has a local minimum at $x = -1$
$f''(1) = -1 < 0$; f has a local maximum at $x = 1$

Sign chart for $f'(x)$
(partition numbers are -1 and 1)

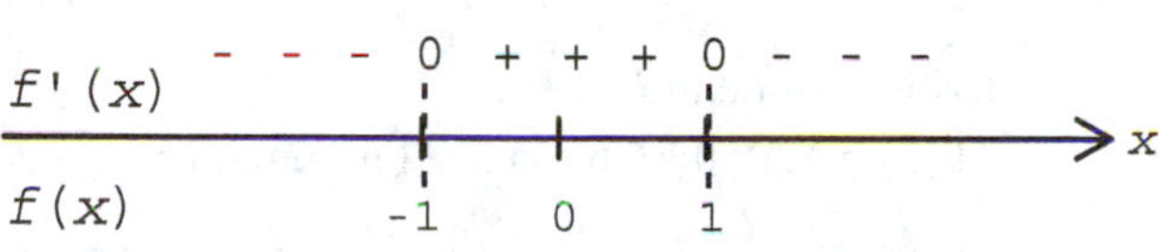

$\lim_{x\to\pm\infty} f(x) = \lim_{x\to\pm\infty} \frac{2x}{x^2+1} = 0$
(the x-axis is a horizontal asymptote)

We can now conclude that $f(1) = 1$ is the absolute maximum of f and $f(-1) = -1$ is the absolute minimum of f.

41. $f(x) = \frac{x^2-1}{x^2+1}$; domain: all real numbers

$$f'(x) = \frac{(x^2+1)(2x) - (x^2-1)(2x)}{(x^2+1)^2} = \frac{4x}{(x^2+1)^2}$$

$$f''(x) = \frac{(x^2+1)^2(4) - 4x(2)(x^2+1)2x}{(x^2+1)^4} = \frac{4(1-3x^2)}{(x^2+1)^3}$$

Critical value: $x = 0$
$f''(0) = 4 > 0$; f has a local minimum at $x = 0$

Sign chart for $f'(x) = \frac{4x}{(x^2+1)^2}$ (0 is the partition number)

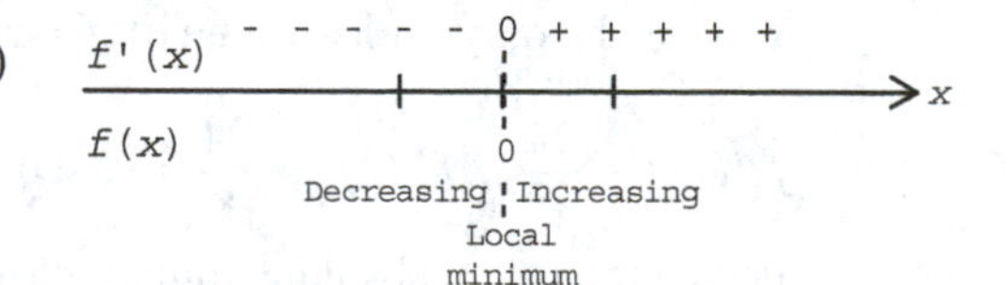

$\lim_{x\to\pm\infty} f(x) = \lim_{x\to\pm\infty} \frac{x^2-1}{x^2+1} = 1$; ($y = 1$ is a horizontal asymptote)

We can now conclude that $f(0) = -1$ is the absolute minimum and f does not have an absolute maximum.

43. $f(x) = 2x^2 - 8x + 6$ on $I = [0, \infty)$
$f'(x) = 4x - 8 = 4(x-2)$
$f''(x) = 4$
Critical value: $x = 2$

$f''(2) = 4 > 0$; f has a local minimum at $x = 2$
Since $x = 2$ is the only critical value of f on I, $f(2) = -2$ is the absolute minimum of f on I.

45. $f(x) = 3x^2 - x^3$ on $I = [0, \infty)$
$f'(x) = 6x - 3x^2 = 3x(2 - x)$
$f''(x) = 6 - 6x$

Critical value (in $(0, \infty)$): $x = 2$
$f''(2) = -6 < 0$; f has a local maximum at $x = 2$

Since $f(0) = 0$ and $x = 2$ is the only critical value of f in $(0, \infty)$, $f(2) = 4$ is the absolute maximum value of f on I.

47. $f(x) = (x + 4)(x - 2)^2$ on $I = [0, \infty)$
$f'(x) = (x + 4)(2)(x - 2) + (x - 2)^2 = (x - 2)[2x + 8 + x - 2] = (x - 2)(3x + 6) = 3x^2 - 12$
$f''(x) = 6x$

Critical value in I: $x = 2$
$f''(2) = 12 > 0$; f has a local minimum at $x = 2$

Since $f(0) = 16$ and $x = 2$ is the only critical value of f in $(0, \infty)$, $f(2) = 0$ is the absolute minimum of f on I.

49. $f(x) = 2x^4 - 8x^3$ on $I = (0, \infty)$
Since $\lim_{x\to\infty} f(x) = \lim_{x\to\infty} (2x^4 - 8x^3) = \infty$, f does not have an absolute maximum on I.

51. $f(x) = 20 - 3x - \dfrac{12}{x}$, $x > 0$; $I = (0, \infty)$

$f'(x) = -3 + \dfrac{12}{x^2}$

$f'(x) = 0$: $-3 + \dfrac{12}{x^2} = 0$

$3x^2 = 12$
$x^2 = 4$
$x = 2$ (-2 is not in I)

$x = 2$ is the only critical value of f on I, and $f(2) = 20 - 3(2) - \dfrac{12}{2} = 8$.

$f''(x) = -\dfrac{24}{x^3}$; $f''(2) = -\dfrac{24}{8} = -3 < 0$. Therefore, $f(2) = 8$ is the absolute maximum of f. The function does not have an absolute minimum since $\lim_{x\to\infty} f(x) = -\infty$. (Also, $\lim_{x\to 0^+} f(x) = -\infty$.)

53. $f(x) = 10 + 2x + \dfrac{64}{x^2}$, $x > 0$; $I = (0, \infty)$

$f'(x) = 2 - \dfrac{128}{x^3}$

$f'(x) = 0$: $2 - \dfrac{128}{x^3} = 0$

$2x^3 = 128$

$$x^3 = 64$$
$$x = 4$$

$x = 4$ is the only critical value of f on I and $f(4) = 10 + 2(4) + \frac{64}{4^2} = 22$.

$f''(x) = \frac{384}{x^4}$; $f''(4) = \frac{384}{4^4} = \frac{3}{2} > 0$. Therefore, $f(4) = 22$ is the absolute minimum of f. The function does not have an absolute maximum since $\lim_{x\to\infty} f(x) = \infty$. (Also, $\lim_{x\to 0^+} f(x) = \infty$.)

55. $f(x) = x + \frac{1}{x} + \frac{30}{x^3}$ on $I = (0, \infty)$

$$f'(x) = 1 - \frac{1}{x^2} - \frac{90}{x^4} = \frac{x^4 - x^2 - 90}{x^4} = \frac{(x^2 - 10)(x^2 + 9)}{x^4}$$

$$f''(x) = \frac{2}{x^3} + \frac{360}{x^5}$$

Critical value (in $(0, \infty)$): $x = \sqrt{10}$

$f''(\sqrt{10}) = \frac{2}{(10)^{3/2}} + \frac{360}{(10)^{5/2}} > 0$; f has a local minimum at $x = \sqrt{10}$.

Since $\sqrt{10}$ is the only critical value of f on I, $f(\sqrt{10}) = \frac{14}{\sqrt{10}}$ is the absolute minimum of f on I.

57. $f(x) = \frac{e^x}{x^2}$, $x > 0$

$$f'(x) = \frac{x^2 \frac{d}{dx} e^x - e^x \frac{d}{dx} x^2}{x^4} = \frac{x^2 e^x - 2xe^x}{x^4} = \frac{xe^x(x-2)}{x^4} = \frac{e^x(x-2)}{x^3}$$

Critical values: $f'(x) = \frac{e^x(x-2)}{x^3} = 0$

$$e^x(x-2) = 0$$

$x = 2$ [Note: $e^x \neq 0$ for all x.]

Thus, $x = 2$ is the only critical value of f on $(0, \infty)$.

Sign chart for f': [Note: This approach is a little easier than calculating $f''(x)$]

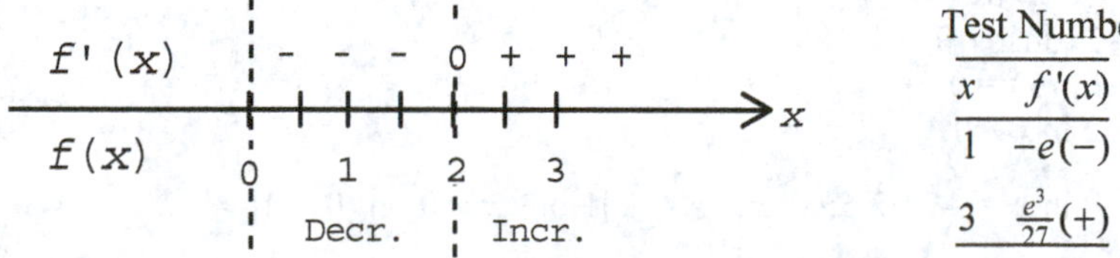

Test Numbers

x	$f'(x)$
1	$-e(-)$
3	$\frac{e^3}{27}(+)$

By the first derivative test, f has a minimum value at $x = 2$;

$f(2) = \frac{e^2}{2^2} = \frac{e^2}{4} \approx 1.847$ is the absolute minimum value of f.

59. $f(x) = \frac{x^3}{e^x}$

$$f'(x) = \frac{\left(\frac{d}{dx}x^3\right)e^x - \left(\frac{d}{dx}e^x\right)x^3}{(e^x)^2} = \frac{3x^2e^x - x^3e^x}{e^{2x}} = \frac{x^2(3-x)e^x}{e^{2x}} = \frac{x^2(3-x)}{e^x}$$

Critical values: $f'(x) = \dfrac{x^2(3-x)}{e^x} = 0$

$$x^2(3-x) = 0$$
$$x = 0 \text{ and } x = 3$$

Sign chart for f': [Note: This approach is a little easier than calculating $f''(x)$]:

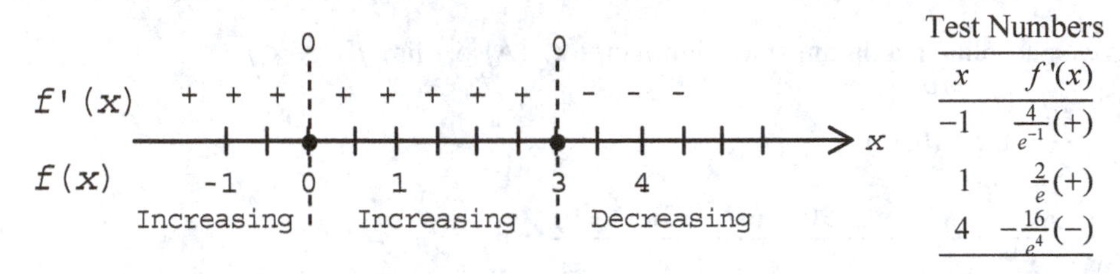

By the first derivative test, f has a maximum value at $x = 3$; $f(3) = \dfrac{27}{e^3} \approx 1.344$ is the absolute maximum value of f.

61. $f(x) = 5x - 2x \ln x, x > 0$

$f'(x) = 5 - 2x\dfrac{d}{dx}(\ln x) - \ln x\dfrac{d}{dx}(2x) = 5 - 2x\left(\dfrac{1}{x}\right) - 2\ln x = 3 - 2\ln x, x > 0$

Critical values: $f'(x) = 3 - 2\ln x = 0$

$$\ln x = \frac{3}{2} = 1.5; \quad x = e^{1.5}$$

Thus, $x = e^{1.5}$ is the only critical value of f on $(0, \infty)$.

Now, $f''(x) = \dfrac{d}{dx}(3 - 2\ln x) = -\dfrac{2}{x}$, and $f''(e^{1.5}) = -\dfrac{2}{e^{1.5}} < 0$.

Therefore, f has a maximum value at $x = e^{1.5}$, and

$f(e^{1.5}) = 5e^{1.5} - 2e^{1.5}\ln(e^{1.5}) = 5e^{1.5} - 2(1.5)e^{1.5} = 2e^{1.5} \approx 8.963$

is the absolute maximum of f.

63. $f(x) = x^2(3 - \ln x), x > 0$

$f'(x) = x^2\dfrac{d}{dx}(3 - \ln x) + (3 - \ln x)\dfrac{d}{dx}x^2 = x^2\left(-\frac{1}{x}\right) + (3 - \ln x)2x = -x + 6x - 2x\ln x = 5x - 2x\ln x$

Critical values: $f'(x) = 5x - 2x\ln x = 0$

$$x(5 - 2\ln x) = 0$$
$$5 - 2\ln x = 0$$
$$\ln x = \frac{5}{2} = 2.5; \quad x = e^{2.5} \quad [\text{Note: } x \neq 0 \text{ on } (0, \infty)]$$

Now $f''(x) = 5 - 2x\left(\dfrac{1}{x}\right) - 2\ln x = 3 - 2\ln x$

and $f''(e^{2.5}) = 3 - 2\cdot\ln(e^{2.5}) = 3 - 2(2.5) = 3 - 5 = -2 < 0$

Therefore, f has a maximum value at $x = e^{2.5}$ and

$f(e^{2.5}) = (e^{2.5})^2(3 - \ln e^{2.5}) = e^5(3 - 2.5) = \dfrac{e^5}{2} \approx 74.207$

is the absolute maximum value of f.

65. $f(x) = \ln(xe^{-x}), x > 0$

$$f'(x) = \frac{1}{xe^{-x}} \frac{d}{dx}(xe^{-x}) = \frac{1}{xe^{-x}}[e^{-x} - xe^{-x}] = \frac{1-x}{x}$$

Critical values: $f'(x) = \dfrac{1-x}{x} = 0$; $x = 1$

Sign chart for $f'(x)$:

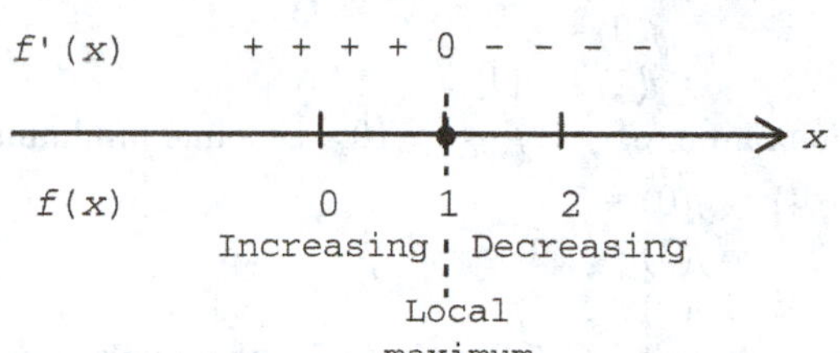

Test Numbers

x	$f(x)$
$\frac{1}{2}$	$1(+)$
2	$-\frac{1}{2}(-)$

By the first derivative test, f has a maximum value at $x = 1$; $f(1) = \ln(e^{-1}) = -1$ is the absolute maximum value of f.

67. $f(x) = x^3 - 6x^2 + 9x - 6$

$f'(x) = 3x^2 - 12x + 9 = 3(x^2 - 4x + 3) = 3(x - 3)(x - 1)$

Critical values: $x = 1, 3$

(A) On the interval $[-1, 5]$:
$f(-1) = -1 - 6 - 9 - 6 = -22$
$f(1) = 1 - 6 + 9 - 6 = -2$
$f(3) = 27 - 54 + 27 - 6 = -6$
$f(5) = 125 - 150 + 45 - 6 = 14$

Thus, the absolute maximum of f is $f(5) = 14$, and the absolute minimum of f is $f(-1) = -22$.

(B) On the interval $[-1, 3]$:
$f(-1) = -22$
$f(1) = -2$
$f(3) = -6$

Absolute maximum of f: $f(1) = -2$; absolute minimum of f: $f(-1) = -22$

(C) On the interval $[2, 5]$:
$f(2) = 8 - 24 + 18 - 6 = -4$
$f(3) = -6$
$f(5) = 14$

Absolute maximum of f: $f(5) = 14$; absolute minimum of f: $f(3) = -6$

69. $f(x) = (x - 1)(x - 5)^3 + 1$

$f'(x) = (x - 1)3(x - 5)^2 + (x - 5)^3 = (x - 5)^2(3x - 3 + x - 5) = (x - 5)^2(4x - 8)$

Critical values: $x = 2, 5$

(A) Interval $[0, 3]$:
$f(0) = (-1)(-5)^3 + 1 = 126$
$f(2) = (2 - 1)(2 - 5)^3 + 1 = -26$
$f(3) = (3 - 1)(3 - 5)^3 + 1 = -15$

Absolute maximum of f: $f(0) = 126$; absolute minimum of f: $f(2) = -26$

(B) Interval $[1, 7]$:
$f(1) = 1$
$f(2) = -26$
$f(5) = 1$
$f(7) = (7 - 1)(7 - 5)^3 + 1 = 6 \cdot 8 + 1 = 49$

Absolute maximum of f: $f(7) = 49$; absolute minimum of f: $f(2) = -26$

(C) Interval $[3, 6]$:
$f(3) = (3 - 1)(3 - 5)^3 + 1 = -15$
$f(5) = 1$
$f(6) = (6 - 1)(6 - 5)^3 + 1 = 6$

Absolute maximum of f: $f(6) = 6$; absolute minimum of f: $f(3) = -15$

71. $f(x) = x^4 - 4x^3 + 5$
$f'(x) = 4x^3 - 12x^2 = 4x^2(x-3)$
Critical values: $x = 0,\ x = 3$

(A) Interval $[-1, 2]$: $f(-1) = 10$
$f(0) = 5$
$f(2) = -11$
Absolute maximum of f: $f(-1) = 10$; absolute minimum of f: $f(2) = -11$.

(B) Interval $[0, 4]$: $f(0) = 5$
$f(3) = -22$
$f(4) = 5$
Absolute maximum of f: $f(0) = f(4) = 5$; absolute minimum of f: $f(3) = -22$.

(C) Interval $[-1, 1]$: $f(-1) = 10$
$f(0) = 5$
$f(1) = 2$
Absolute maximum of f: $f(-1) = 10$; absolute minimum of f: $f(1) = 2$

73. f has a local minimum at $x = 2$.

75. Unable to determine from the given information $(f'(-3) = f''(-3) = 0)$.

77. Neither a local maximum nor a local minimum at $x = 6$; $x = 6$ is not a critical value of f.

79. f has a local maximum at $x = 2$.

EXERCISE 4-6

Things to remember:

STRATEGY FOR SOLVING OPTIMIZATION PROBLEMS

Step 1. Introduce variables, look for relationships among these variables, and construct a mathematical model of the form: Maximize (or minimize) $f(x)$ on the interval I

Step 2. Find the critical values of $f(x)$.

Step 3. Use the procedures developed in Section 5-5 to find the absolute maximum (or minimum) value of $f(x)$ on the interval I and the value(s) of x where this occurs.

Step 4. Use the solution to the mathematical model to answer all the questions asked in the problem.

1. $f = xy$ where $x + y = 28$. Since $y = 28 - x$, $f(x) = x(28 - x)$.

3. Diameter x implies radius $r = \frac{x}{2}$. Therefore, $f(x) = \pi\left(\frac{x}{2}\right)^2 = \frac{\pi x^2}{4}$.

5. Volume of a right circular cylinder of radius r and height h: $V = \pi r^2 h$. We have $h = x$. Therefore,
$f(x) = \pi(x)^2(x) = \pi x^3$.

7. $f = xy$ where $2x+2y=120$, or $y=60-x$. Therefore, $f(x)=x(60-x)$.

9. Let x be one of the numbers and let y be the other.
Maximize $P = xy$ subject to $x + y = 15$.
$x + y = 15$ implies $y = 15 - x$ and $P(x) = x(15-x) = 15x - x^2$.

Critical values:

$P'(x) = 15 - 2x$

$15 - 2x = 0;\ x = \frac{15}{2} = 7.5$

$P''(x) = -2 < 0$

P has a local maximum at $x = 7.5$. Since 7.5 is the only critical value, P has an absolute maximum at $x = 7.5$. The numbers are $x = 7.5,\ y = -7.5$.

11. Let x be one of the numbers and let y be the other.
Minimize $P = xy$ subject to $x - y = 15$.
$x - y = 15$ implies $y = x - 15$ and $P(x) = x(x-15) = x^2 - 15x$.

Critical values:

$P'(x) = 2x - 15$

$2x - 15 = 0;\ x = \frac{15}{2} = 7.5$

$P''(x) = 2 > 0$

P has a local minimum at $x = 7.5$. Since 7.5 is the only critical value, P has an absolute minimum at $x = 7.5$. The numbers are $x = y = 7.5$

13. Let x be one of the numbers and let y be the other.
Minimize $P = x + y$ subject to $xy = 15,\ x \geq 0,\ y \geq 0$.

$xy = 15$ implies $y = \frac{15}{x}$ and $P(x) = x + \frac{15}{x}$

Critical values:

$P'(x) = 1 - \frac{15}{x^2}$

$1 - \frac{15}{x^2} = 0$

$x^2 = 15$

$x = \sqrt{15}$

$P''(x) = \frac{30}{x^3};\ P''(\sqrt{15}) = \frac{30}{15\sqrt{15}} = \frac{2}{\sqrt{15}} > 0$

P has a local minimum at $x = \sqrt{15}$. Since $\sqrt{15}$ is the only critical value, P has an absolute minimum at $x = \sqrt{15}$. The numbers are $x = y = \sqrt{15}$.

15. Let x be the length and y the width of the rectangle.
Minimize $P = 2x + 2y$ subject to $xy = 200,\ x \geq 0,\ y \geq 0$.

$xy = 200$ implies $y = \frac{200}{x}$ and $P(x) = 2x + 2\left(\frac{200}{x}\right) = 2x + \frac{400}{x}$

Critical values:

$P'(x) = 2 - \frac{400}{x^2}$

$2 - \frac{400}{x^2} = 0$

$2x^2 = 400$

$x^2 = 200$

$x = \sqrt{200} = 10\sqrt{2}$

$P''(x) = \frac{800}{x^3};\ P''(10\sqrt{2}) = \frac{800}{(10\sqrt{2})^3} = \frac{2}{5\sqrt{2}} > 0$

P has a local minimum at $x = 10\sqrt{2}$ and since this is the only critical value, P has an absolute minimum at $x = 10\sqrt{2}$.

At $x = 10\sqrt{2}$, $y = \frac{200}{10\sqrt{2}} = \frac{20}{\sqrt{2}} = 10\sqrt{2}$. The dimensions of the rectangle are: length $= 10\sqrt{2}$, width = $10\sqrt{2}$.

17. Let x be the length and y the width of the rectangle.
Maximize $A = xy$ subject to $2x + 2y = 148$, $x \geq 0$, $y \geq 0$.
$2x + 2y = 148$ implies $y = 74 - x$ and $A(x) = x(74 - x) = 74x - x^2$.

Critical values:

$A'(x) = 74 - 2x$

$74 - 2x = 0;\ x = \frac{74}{2} = 37$

$A''(x) = -2 < 0$

A has a local maximum at $x = 37$. Since 37 is the only critical value, A has an absolute maximum at $x = 37$.

At $x = 37$, $y = 74 - 37 = 37$. The dimensions of the rectangle are: length = 37, width = 37.

19. Price-demand: $p(x) = 500 - 0.5x$; cost: $C(x) = 20{,}000 + 135x$

(A) Revenue: $R(x) = x \cdot p(x) = 500x - 0.5x^2,\ 0 \leq x < \infty$

$R'(x) = 500 - x$

$R'(x) = 500 - x = 0$ implies $x = 500$

$R''(x) = -1;\ R''(500) = -1 < 0$

R has an absolute maximum at $x = 500$.

$p(500) = 500 - 0.5(500) = 250;\ R(500) = (500)^2 - 0.5(500)^2 = 125{,}000$

The company should produce 500 phones each week at a price of $250 per phone to maximize their revenue. The maximum revenue is $125,000

(B) Profit: $P(x) = R(x) - C(x) = 500x - 0.5x^2 - (20{,}000 + 135x) = 365x - 0.5x^2 - 20{,}000$

$P'(x) = 365 - x$

$P'(x) = 365 - x = 0$ implies $x = 365$

$P''(x) = -1;\ P''(365) = -1 < 0$

P has an absolute maximum at $x = 365$

$p(365) = 500 - 0.5(365) = 317.50$;

$P(365) = (365)^2 - 0.5(365)^2 - 20{,}000 = 46{,}612.50$

To maximize profit, the company should produce 365 phones each week at a price of \$317.50 per phone. The maximum profit is \$46,612.50.

21. (A) Revenue $R(x) = x \cdot p(x) = x\left(200 - \frac{x}{30}\right) = 200x - \frac{x^2}{30}$, $0 \le x \le 6{,}000$

$R'(x) = 200 - \frac{2x}{30} = 200 - \frac{x}{15}$

Now $R'(x) = 200 - \frac{x}{15} = 0$ implies $x = 3000$.

$R''(x) = -\frac{1}{15} < 0$.

Thus, $R''(3000) = -\frac{1}{15} < 0$ and we conclude that R has an absolute maximum at $x = 3000$. The maximum revenue is $R(3000) = 200(3000) - \frac{(3000)^2}{30} = \$300{,}000$

(B) Profit $P(x) = R(x) - C(x) = 200x - \frac{x^2}{30} - (72{,}000 + 60x)$

$$= 140x - \frac{x^2}{30} - 72{,}000$$

$P'(x) = 140 - \frac{x}{15}$

Now $140 - \frac{x}{15} = 0$ implies $x = 2{,}100$. $P''(x) = -\frac{1}{15}$ and $P''(2{,}100) = -\frac{1}{15} < 0$. Thus, the maximum profit occurs when 2,100 television sets are produced. The maximum profit is

$P(2{,}100) = 140(2{,}100) - \frac{(2{,}100)^2}{30} - 72{,}000 = \$75{,}000$

the price that the company should charge is $p(2{,}100) = 200 - \frac{2{,}100}{30} = \$130$ for each set.

(C) If the government taxes the company \$5 for each set, then the profit $P(x)$ is given by

$P(x) = 200x - \frac{x^2}{30} - (72{,}000 + 60x) - 5x = 135x - \frac{x^2}{30} - 72{,}000$.

$P'(x) = 135 - \frac{x}{15}$.

Now $135 - \frac{x}{15} = 0$ implies $x = 2{,}025$.

$P''(x) = -\frac{1}{15}$ and $P''(2{,}025) = -\frac{1}{15} < 0$. Thus, the maximum profit in this case occurs when 2,025 television sets are produced. The maximum profit is

$$P(2{,}025) = 135(2{,}025) - \frac{(2{,}025)^2}{30} - 72{,}000 = \$64{,}687.50$$

and the company should charge $p(2{,}025) = 200 - \frac{2{,}025}{30} = \$132.50$/set.

23. (A)

(B)

LinReg
y=ax+b
a=53.50318471
b=82245.22293

(C) The revenue at the demand level x is:
$R(x) = xp(x)$
where $p(x)$ is the quadratic regression equation in (A).

The cost at the demand level x is $C(x)$ given by the linear regression equation in (B). The profit $P(x) = R(x) - C(x)$.

The maximum profit is \$118,996 at the demand level $x = 1422$.

The price per sleeping bag at the demand level $x = 1422$ is \$195.

25. (A) Let x = number of 10¢ reductions in price. Then
$640 + 40x$ = number of sandwiches sold at x 10¢ reductions
$8 - 0.1x$ = price per sandwich, $0 \le x \le 80$
Revenue: $R(x) = (640 + 40x)(8 - 0.1x) = 5120 + 256x - 4x^2$, $0 \le x \le 80$
$R'(x) = 256 - 8x$
$R'(x) = 256 - 8x = 0$ implies $x = 32$
$R(0) = 5120$, $R(32) = 9216$, $R(80) = 0$

Thus, the deli should charge $8 - 3.20 = \$4.80$ per sandwich to realize a maximum revenue of \$9216.

(B) Let x = number of 20¢ reductions in price. Then
$640 + 15x$ = number of sandwiches sold
$8 - 0.2x$ = price per sandwich $0 \le x \le 40$
Revenue: $R(x) = (640 + 15x)(8 - 0.2x) = 5120 - 8x - 3x^2$, $0 \le x \le 40$
$R'(x) = -8 - 6x$
$R'(x) = -8 - 6x = 0$ has no solutions in $(0, 40)$
Now, $R(0) = 640 \cdot 8 = \$5120$, $R(40) = 0$

Thus, the deli should charge \$8 per sandwich to maximize their revenue under these conditions.

27. Let x = number of dollar increases in the rate per day. Then
$200 - 5x$ = total number of cars rented and $30 + x$ = rate per day.
Total income = (total number of cars rented)(rate)
$y(x) = (200 - 5x)(30 + x)$, $0 \le x \le 40$
$y'(x) = (200 - 5x)(1) + (30 + x)(-5) = 200 - 5x - 150 - 5x = 50 - 10x = 10(5 - x)$
Thus, $x = 5$ is the only critical value and
$y(5) = (200 - 25)(30 + 5) = 6125$.
$y''(x) = -10$
$y''(5) = -10 < 0$
Therefore, the absolute maximum income is $y(5) = \$6125$ when the rate is \$35 per day.

29. Let x = number of additional trees planted per acre. Then
$30 + x$ = total number of trees per acre and $50 - x$ = yield per tree.
Yield per acre = (total number of trees per acre)(yield per tree)

$$y(x) = (30 + x)(50 - x), \ 0 \le x \le 20$$
$$y'(x) = (30 + x)(-1) + (50 - x) = 20 - 2x = 2(10 - x)$$

The only critical value is $x = 10$.

$y(10) = 40(40) = 1600$ pounds per acre.
$y''(x) = -2$
$y''(10) = -2 < 0$
Therefore, the absolute maximum yield is $y(10) = 1600$ pounds per acre when the number of trees per acre is 40.

31. Volume = $V(x) = (12 - 2x)(8 - 2x)x, \ 0 \le x \le 4$
$= 96x - 40x^2 + 4x^3$

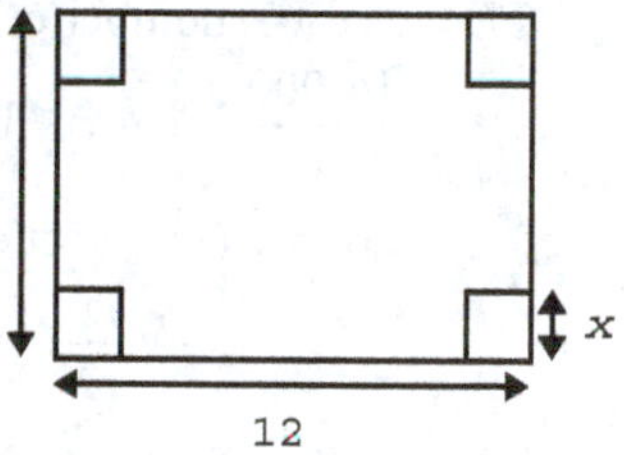

$V'(x) = 96 - 80x + 12x^2 = 4(24 - 20x + 3x^2)$
We solve $24 - 20x + 3x^2 = 0$ by using the quadratic formula:

$$x = \frac{20 \pm \sqrt{400 - 4 \cdot 24 \cdot 3}}{6} = \frac{10 \pm 2\sqrt{7}}{3}$$

Thus, $x = \frac{10 - 2\sqrt{7}}{3} \approx 1.57$ is the only critical value on the interval [0, 4].
$V''(x) = -80 + 24x$
$V''(1.57) = -80 + 24(1.57) < 0$
Therefore, a square with a side of length $x = 1.57$ inches should be cut from each corner to obtain the maximum volume.

33. Area = 800 square feet = xy (1)
Cost = $18x + 6(2y + x)$

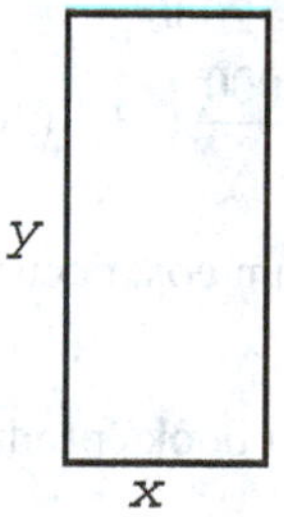

From (1), we have $y = \frac{800}{x}$.

Hence, cost $C(x) = 18x + 6\left(\frac{1600}{x} + x\right)$, or

$C(x) = 24x + \frac{9600}{x}, \ x > 0,$

$C'(x) = 24 - \frac{9600}{x^2} = \frac{24(x^2 - 400)}{x^2} = \frac{24(x - 20)(x + 20)}{x^2}.$

Therefore, $x = 20$ is the only critical value.

$C''(x) = \frac{19,200}{x^3}$

$C''(20) = \frac{19,200}{8000} > 0.$ Therefore, $x = 20$ for the minimum cost.

The dimensions of the fence are shown in the diagram at the right.

35. (A) Let x and y be the width and the length of the rectangle respectively. Then we have
$2x + y + (y - 100) = 240$ or $2x + 2y = 340$ and $x = 170 - y$ where $100 \le y \le 170$.

The Area $= xy = (170 - y)y$.

Let $f(y) = y(170 - y),\ 100 \le y \le 170$.

$f'(y) = 170 - 2y$ and $f''(y) = -2 < 0$. .

$f'(y) = 0$ implies $y = 85$ which is not in the domain of f. We note that $f(100) = 7{,}000,\ f(170) = 0$.

Thus, the maximum of f occurs when $y = 100$ and $x = 170 - y = 70$.

(B) In this case, $2x + 2y - 100 = 400$ or $x + y = 250$ and $x = 250 - y$.

$f(y) = y(250 - y) = 250y - y^2,\ 100 \le y \le 250;\ f'(y) = 250 - 2y;\ f''(y) = -2 < 0.$

$f'(y) = 0$ implies $y = 125$

Thus, f has an absolute maximum at $y = 125,\ x = 250 - y = 125$.

37. Let x = number of cans of paint produced in each production run. Then, number of production runs:
$\frac{16{,}000}{x},\ 1 \le x \le 16{,}000$

Cost: $C(x)$ = cost of storage + cost of set up

$$= \frac{x}{2}(4) + \frac{16{,}000}{x}(500)$$

[Note: $\frac{x}{2}$ is the average number of cans of paint in storage per day.]

Thus,

$$C(x) = 2x + \frac{8{,}000{,}000}{x},\ 1 \le x \le 16{,}000$$

$$C'(x) = 2 - \frac{8{,}000{,}000}{x^2} = \frac{2x^2 - 8{,}000{,}000}{x^2} = \frac{2(x^2 - 4{,}000{,}000)}{x^2}$$

Critical value: $x = 2000$

$$C''(x) = \frac{16{,}000{,}000}{x^3};\quad C''(2000) > 0.$$

Thus, the minimum cost occurs when $x = 2000$ and the number of production runs is $\frac{16{,}000}{2{,}000} = 8$.

39. Let x = number of books produced each printing. Then, the number of printings $= \frac{50{,}000}{x}$.

Cost $= C(x)$ = cost of storage + cost of printing

$$= \frac{x}{2} + \frac{50{,}000}{x}(1000),\quad x > 0$$

[Note: $\frac{x}{2}$ is the average number in storage each day.]

$$C'(x) = \frac{1}{2} - \frac{50{,}000{,}000}{x^2} = \frac{x^2 - 100{,}000{,}000}{2x^2} = \frac{(x + 10{,}000)(x - 10{,}000)}{2x^2}$$

Critical value: $x = 10{,}000$

$$C''(x) = \frac{100{,}000{,}000}{x^3}$$

$$C''(10{,}000) = \frac{100{,}000{,}000}{(10{,}000)^3} > 0$$

Thus, the minimum cost occurs when $x = 10{,}000$ and the number of printings is $\frac{50{,}000}{10{,}000} = 5$.

41. Let x = number of hours it takes the train to travel 360 miles.

Then $360 = xv$ or $x = \frac{360}{v}$.

$$\text{Cost} = \left(300 + \frac{v^2}{4}\right)x = \left(300 + \frac{v^2}{4}\right)\left(\frac{360}{v}\right) = \frac{108,000}{v} + 90v$$

Let $C(v) = \frac{108,000}{v} + 90v,\ v > 0$. We want to minimize $C(v)$.

$$C'(v) = \frac{-108,000}{v^2} + 90 = \frac{-108,000 + 90v^2}{v^2}.$$

$C'(v) = 0$ implies $90v^2 = 108,000$ or $v = 34.64$

$$C''(v) = \frac{216,000}{v^3} > 0 \text{ for } v > 0$$

So, $C(v)$ has an absolute minimum at $v = 34.64$ miles per hour.

43. (A) Let the cost to lay the pipe on the land be 1 unit; then the cost to lay the pipe in the lake is 1.4 units.

$$C(x) = \text{total cost} = (1.4)\sqrt{x^2+25} + (1)(10-x),\ 0 \le x \le 10 = (1.4)(x^2+25)^{1/2} + 10 - x$$

$$C'(x) = (1.4)\frac{1}{2}(x^2+25)^{-1/2}(2x) - 1 = (1.4)x(x^2+25)^{-1/2} - 1 = \frac{1.4x - \sqrt{x^2+25}}{\sqrt{x^2+25}}$$

$C'(x) = 0$ when $1.4x - \sqrt{x^2+25} = 0$ or $1.96x^2 = x^2 + 25$

$$.96x^2 = 25$$

$$x^2 = \frac{25}{.96} = 26.04$$

$$x = \pm 5.1$$

Thus, the critical value is $x = 5.1$.

$$C''(x) = (1.4)(x^2+25)^{-1/2} + (1.4)x\left(-\frac{1}{2}\right)(x^2+25)^{-3/2}2x = \frac{1.4}{(x^2+25)^{1/2}} - \frac{(1.4)x^2}{(x^2+25)^{3/2}}$$

$$= \frac{35}{(x^2+25)^{3/2}}$$

$$C''(5.1) = \frac{35}{[(5.1)^2+25]^{3/2}} > 0$$

Thus, the cost will be a minimum when $x = 5.1$.

Note that: $C(0) = (1.4)\sqrt{25} + 10 = 17$

$C(5.1) = (1.4)\sqrt{51.01} + (10 - 5.1) = 14.9$

$C(10) = (1.4)\sqrt{125} = 15.65$

Thus, the absolute minimum occurs when $x = 5.1$ miles.

(B) $C(x) = (1.1)\sqrt{x^2+25} + (1)(10-x),\ 0 \le x \le 10$

$$C'(x) = \frac{(1.1)x - \sqrt{x^2 + 25}}{\sqrt{x^2 + 25}}$$

$C'(x) = 0$ when $1.1x - \sqrt{x^2 + 25} = 0$ or $(1.21)x^2 = x^2 + 25$

$$.21x^2 = 25$$

$$x^2 = \frac{25}{.21} = 119.05$$

$$x = \pm 10.91$$

Critical value: $x = 10.91 > 10$, i.e., there are no critical values on the interval [0, 10]. Now,

$C(0) = (1.1)\sqrt{25} + 10 = 15.5$,

$C(10) = (1.1)\sqrt{125} \approx 12.30$.

Therefore, the absolute minimum occurs when $x = 10$ miles.

45. $C(t) = 30t^2 - 240t + 500, 0 \le t \le 8$

$C'(t) = 60t - 240$; $t = 4$ is the only critical value.

$C''(t) = 60$

$C''(4) = 60 > 0$

Now, $C(0) = 500$

$C(4) = 30(4)^2 - 240(4) + 500 = 20$,

$C(8) = 30(8)^2 - 240(8) + 500 = 500$.

Thus, 4 days after a treatment, the concentration will be minimum; the minimum concentration is 20 bacteria per cm^3.

47. $H(t) = 4t^{1/2} - 2t, \ 0 \le t \le 2$

$H'(t) = 2t^{-1/2} - 2$

Thus, $t = 1$ is the only critical value.

Now, $H(0) = 4 \cdot 0^{1/2} - 2(0) = 0$,

$H(1) = 4 \cdot 1^{1/2} - 2(1) = 2$,

$H(2) = 4 \cdot 2^{1/2} - 4 \approx 1.66$.

Therefore, $H(1)$ is the absolute maximum, and after one month the maximum height will be 2 feet.

49. $N(t) = 30 + 12t^2 - t^3, 0 \le t \le 8$

The rate of increase $= R(t) = N'(t) = 24t - 3t^2$, and

$R'(t) = N''(t) = 24 - 6t$.

Thus, $t = 4$ is the only critical value of $R(t)$.

Now, $R(0) = 0$,

$R(4) = 24 \cdot 4 - 3 \cdot 4^2 = 48$,

$R(8) = 24 \cdot 8 - 3 \cdot 8^2 = 0$.

Therefore, the absolute maximum value of R occurs when $t = 4$; the maximum rate of increase will occur four years from now.

CHAPTER 4 REVIEW

1. The function f is increasing on (a, c_1), (c_3, c_6). (4-1, 4-2)

2. $f'(x) < 0$ on (c_1, c_3), (c_6, b). (4-1, 4-2)

3. The graph of f is concave downward on (a, c_2), (c_4, c_5), (c_7, b). (4-1, 4-2)

4. A local minimum occurs at $x = c_3$. (4-1)

5. The absolute maximum occurs at $x = c_1, c_6$. (4-1, 4-5)

6. $f'(x)$ appears to be zero at $x = c_1, c_3, c_5$. (4-1)

7. $f'(x)$ does not exist at $x = c_4, c_6$. (4-1)

8. $x = c_2, c_4, c_5, c_7$ are inflection points. (4-2)

9. 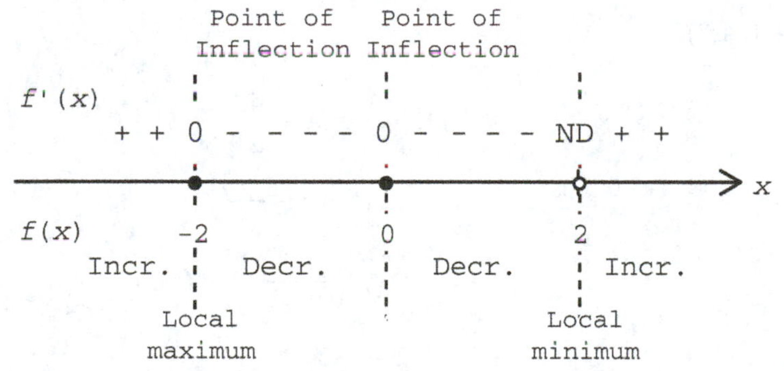

Using this information together with the points (–3, 0), (–2, 3), (–1, 2), (0, 0), (2, –3), (3, 0) on the graph, we have

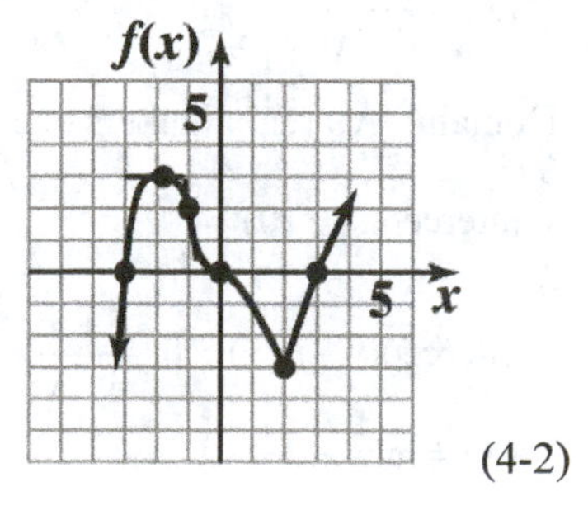

(4-2)

10. Domain: all real numbers
Intercepts: y-intercept: $f(0) = 0$
x-intercepts: $x = 0$
Asymptotes: Horizontal asymptote: $y = 2$
no vertical asymptotes
Critical values: $x = 0$

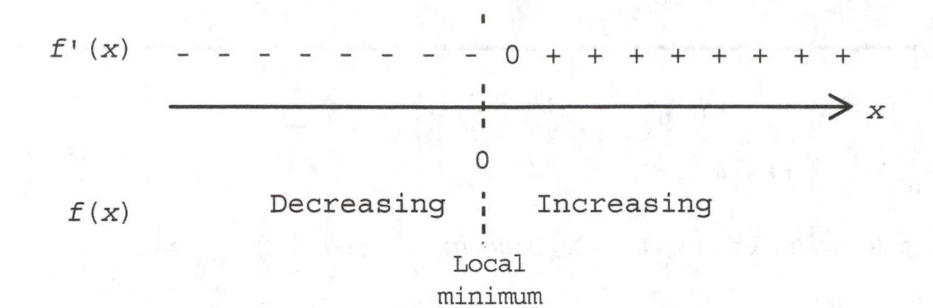

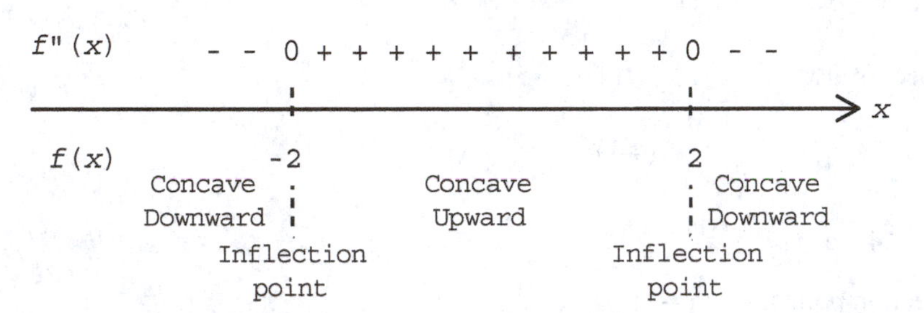

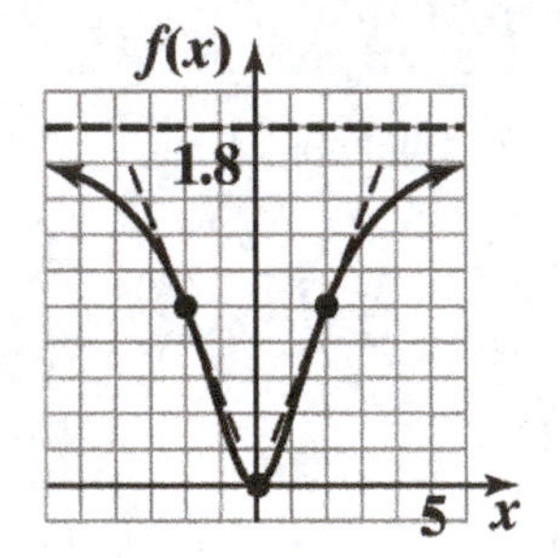

(4-1, 4-2)

11. $f(x) = x^4 + 5x^3$

$f'(x) = 4x^3 + 15x^2$

$f''(x) = 12x^2 + 30x$ (4-2)

12. $y = 3x + \dfrac{4}{x}$

$y' = 3 - \dfrac{4}{x^2}$

$y'' = \dfrac{8}{x^3}$ (4-2)

13. $f(x) = \dfrac{5+x}{4-x}$

Domain: All real numbers except $x = 4$.

y-intercept: $f(0) = \dfrac{5}{4}$

x-intercept: $f(x) = 0;\ \dfrac{5+x}{4-x} = 0,\ \ x = -5$ (4-1)

14. $f(x) = \ln(x+2)$

Domain: $x > -2,\ \ (-2, \infty)$

y-intercept: $f(0) = \ln 2$

x-intercept: $f(x) = 0;\ \ln(x+2) = 0,\ \ x+2 = 1,\ \ x = -1.$ (4-1)

15. $f(x)=\frac{x+3}{x^2-4}$

Horizontal asymptote: $\lim_{x\to\infty}\frac{x+3}{x^2-4}=0$, $y=0$ is a horizontal asymptote.

Vertical asymptotes: $x^2-4=0$; $x=2$, $x=-2$ are vertical asymptotes. (4-4)

16. $f(x)=\frac{2x-7}{3x+10}$

Horizontal asymptote: $\lim_{x\to\infty}\frac{2x-7}{3x+10}=\frac{2}{3}$, $y=\frac{2}{3}$ is a horizontal asymptote.

Vertical asymptotes: $3x+10=0$; $x=-\frac{10}{3}$ is a vertical asymptote. (4-4)

17. $f(x)=x^4-12x^2$, $f'(x)=4x^3-24x$, $f''(x)=12x^2-24$

$$f''(x)=0$$
$$12x^2-24=0$$
$$x^2=2$$
$$x=\pm\sqrt{2}$$

Inflection points: $(-\sqrt{2},-20)$, $(\sqrt{2},-20)$ (4-2)

18. $f(x)=(2x+1)^{1/3}-6$, $f'(x)=\frac{1}{3}(2x+1)^{-2/3}(2)=\frac{2}{3}(2x+1)^{-2/3}$

$$f''(x)=-\frac{4}{9}(2x+1)^{-5/3}(2)=\frac{-8}{9(2x+1)^{5/3}}$$

$f''(x)$ does not exist at $x=-\frac{1}{2}$; inflection point $\left(-\frac{1}{2},-6\right)$ (4-2)

19. $f(x)=x^{1/5}$

(A) $f'(x)=\frac{1}{5}x^{-4/5}=\frac{1}{5x^{4/5}}$

(B) Critical value: $x=0$ ($f'(0)$ does not exist)

(C) Partition number: $x=0$ (4-1)

20. $f(x)=x^{-1/5}$

(A) $f'(x)=-\frac{1}{5}x^{-6/5}=\frac{-1}{5x^{6/5}}$

(B) Critical value: none (0 is not in the domain of f).

(C) Partition number: $x=0$ (4-1)

21. $f(x)=x^3-18x^2+81x$

Step 1. Analyze $f(x)$:

(A) Domain: All real numbers, $(-\infty,\infty)$

(B) Intercepts: y-intercept: $f(0)=0^3-18(0)^2+81(0)=0$

x-intercepts: $x^3-18x^2+81x=0$

$$x(x^2-18x+81)=0$$

$$x(x-9)^2 = 0$$
$$x = 0, 9$$

(C) Asymptotes: No horizontal or vertical asymptotes.

Step 2. Analyze $f'(x)$:

$f'(x) = 3x^2 - 36x + 81 = 3(x^2 - 12x + 27) = 3(x-3)(x-9)$

Critical values: $x = 3, x = 9$

Partition numbers: $x = 3, x = 9$

Sign chart for f':

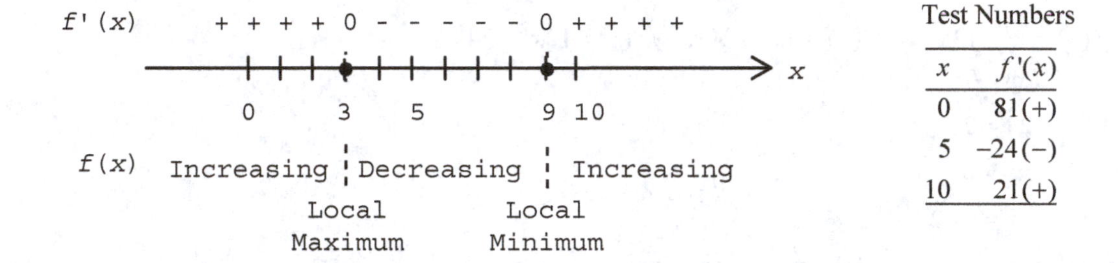

Test Numbers

x	$f'(x)$
0	81(+)
5	−24(−)
10	21(+)

Thus, f is increasing on $(-\infty, 3)$ and on $(9, \infty)$; f is decreasing on $(3, 9)$. There is a local maximum at $x = 3$ and a local minimum at $x = 9$.

Step 3. Analyze $f''(x)$:

$f''(x) = 6x - 36 = 6(x-6)$

Thus, $x = 6$ is a partition number for $f''(x)$.

Sign chart for f'':

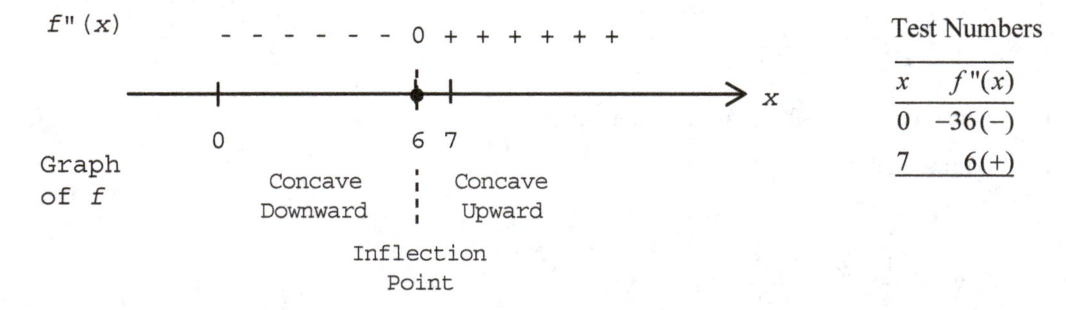

Test Numbers

x	$f''(x)$
0	−36(−)
7	6(+)

Thus, the graph of f is concave downward on $(-\infty, 6)$ and concave upward on $(6, \infty)$. The point $x = 6$ is an inflection point.

Step 4. Sketch the graph of f:

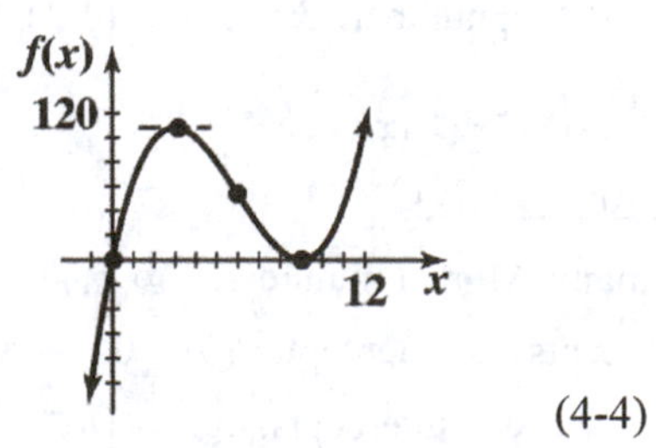

(4-4)

22. $f(x) = (x + 4)(x - 2)^2$

Step 1. Analyze $f(x)$:

(A) Domain: All real numbers, $(-\infty, \infty)$.

(B) Intercepts: y-intercept: $f(0) = 4(-2)^2 = 16$

x-intercepts: $(x + 4)(x - 2)^2 = 0$

$x = -4, 2$

(C) Asymptotes: Since f is a polynomial, there are no horizontal or vertical asymptotes.

Step 2. Analyze $f'(x)$:

$$\begin{aligned} f'(x) &= (x + 4)2(x - 2)(1) + (x - 2)^2(1) \\ &= (x - 2)[2(x + 4) + (x - 2)] \\ &= (x - 2)(3x + 6) \\ &= 3(x - 2)(x + 2) \end{aligned}$$

Critical values: $x = -2, x = 2$

Partition numbers: $x = -2, x = 2$

Sign chart for f':

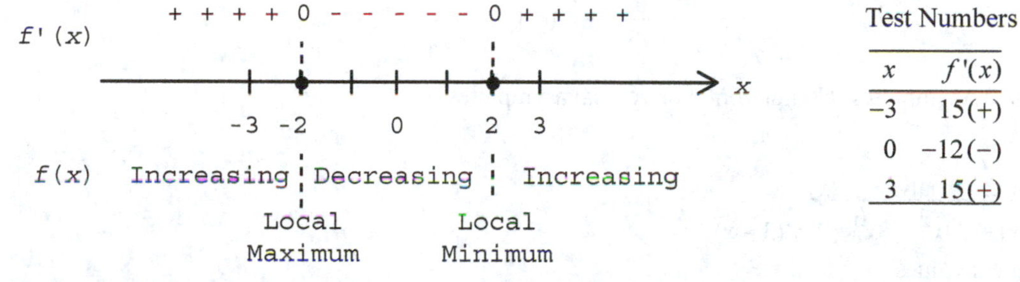

Test Numbers

x	$f'(x)$
-3	$15(+)$
0	$-12(-)$
3	$15(+)$

Thus, f is increasing on $(-\infty, -2)$ and on $(2, \infty)$; f is decreasing on $(-2, 2)$; f has a local maximum at $x = -2$ and a local minimum at $x = 2$.

Step 3. Analyze $f''(x)$:

$f''(x) = 3(x + 2)(1) + 3(x - 2)(1) = 6x$

Partition number for f'': $x = 0$

Sign chart for f'':

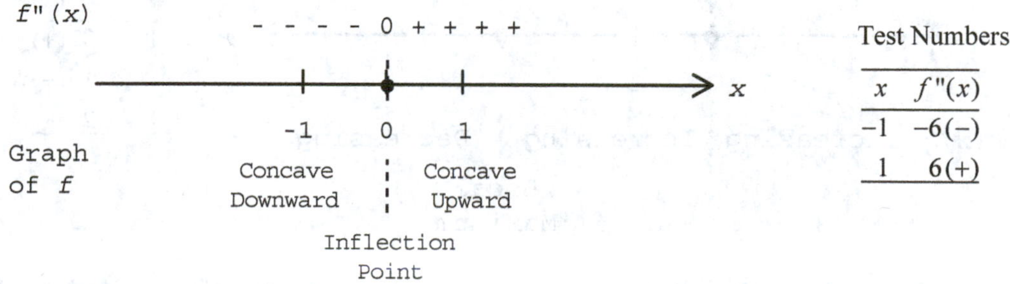

Test Numbers

x	$f''(x)$
-1	$-6(-)$
1	$6(+)$

Thus, the graph of f is concave downward on $(-\infty, 0)$ and concave upward on $(0, \infty)$; there is an inflection point at $x = 0$.

Step 4. Sketch the graph of f:

x	$f(x)$
–2	32
0	16
2	0

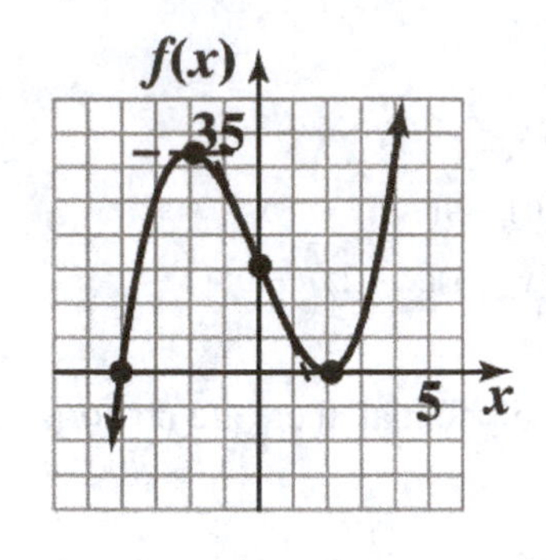

(4-4)

23. $f(x) = 8x^3 - 2x^4$

Step 1. Analyze $f(x)$:

(A) Domain: All real numbers, $(-\infty, \infty)$.

(B) Intercepts: y-intercept: $f(0) = 0$

x-intercepts:
$$8x^3 - 2x^4 = 0$$
$$2x^3(4 - x) = 0$$
$$x = 0, 4$$

(C) Asymptotes: No horizontal or vertical asymptotes.

Step 2. Analyze $f'(x)$:

$f'(x) = 24x^2 - 8x^3 = 8x^2(3 - x)$

Critical values: $x = 0,\ x = 3$

Partition numbers: $x = 0,\ x = 3$

Sign chart for f':

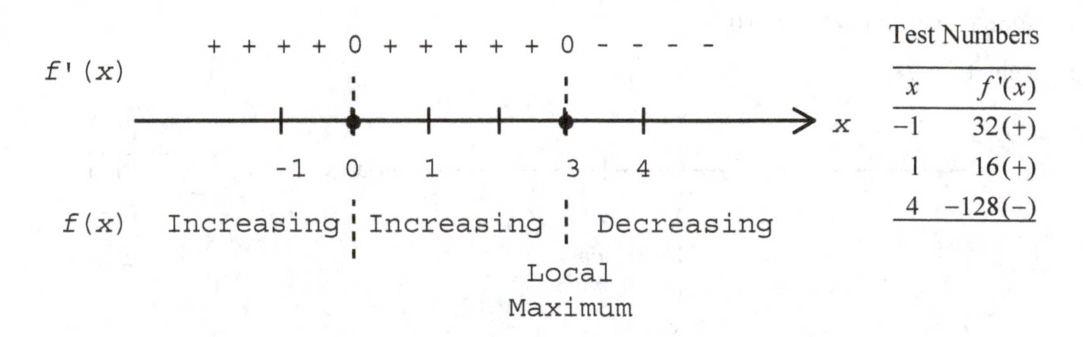

Test Numbers

x	$f'(x)$
–1	32(+)
1	16(+)
4	–128(–)

Thus, f is increasing on $(-\infty, 3)$ and decreasing on $(3, \infty)$; f has a local maximum at $x = 3$.

Step 3. Analyze $f''(x)$:

$f''(x) = 48x - 24x^2 = 24x(2 - x)$

Partition numbers for $f'' = 0$, $x = 2$

Sign chart for f'':

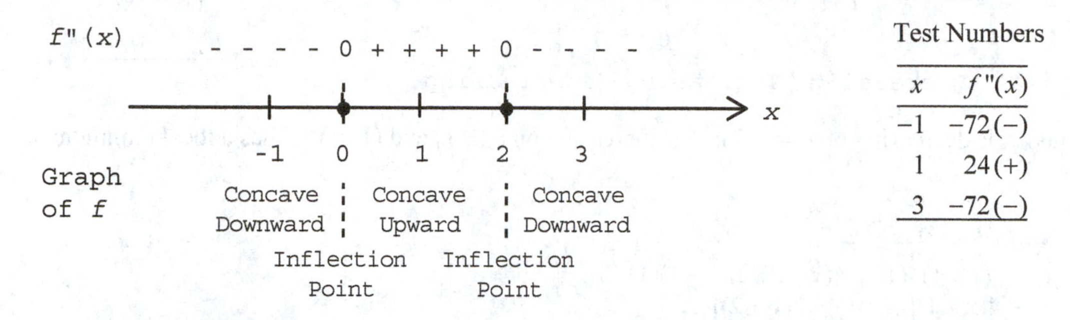

Test Numbers

x	$f''(x)$
-1	$-72(-)$
1	$24(+)$
3	$-72(-)$

Thus, the graph of f is concave downward on $(-\infty, 0)$ and on $(2, \infty)$; the graph is concave upward on $(0, 2)$; there are inflection points at $x = 0$ and $x = 2$.

Step 4. Sketch the graph of f:

x	$f(x)$
0	0
2	32
3	54

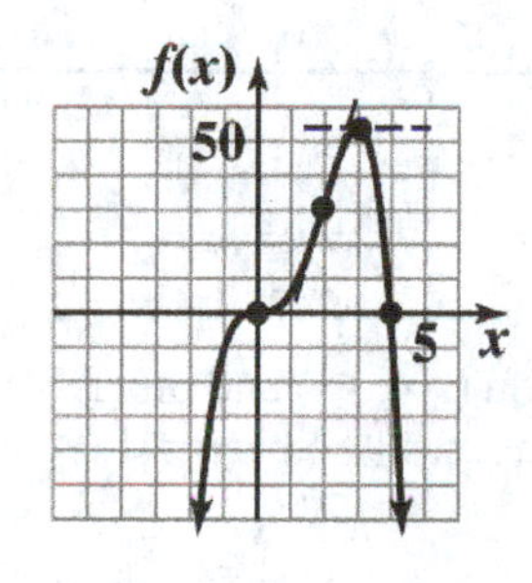

(4-4)

24. $f(x) = (x - 1)^3(x + 3)$

Step 1. Analyze $f(x)$:

(A) Domain: All real numbers.

(B) Intercepts: y-intercept: $f(0) = (-1)^3(3) = -3$

x-intercepts: $(x - 1)^3(x + 3) = 0$

$x = 1, -3$

(C) Asymptotes: Since f is a polynomial (of degree 4), the graph of f has no asymptotes.

Step 2. Analyze $f'(x)$:

$$f'(x) = (x - 1)^3(1) + (x + 3)(3)(x - 1)^2(1)$$
$$= (x - 1)^2[(x - 1) + 3(x + 3)]$$
$$= 4(x - 1)^2(x + 2)$$

Critical values: $x = -2$, $x = 1$

Partition numbers: $x = -2$, $x = 1$

Sign chart for f':

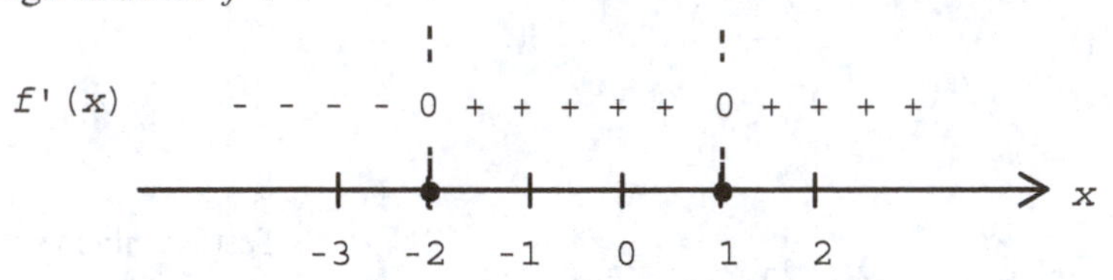

Test Numbers

x	$f'(x)$
−3	−64(−)
0	8(+)
2	16(+)

Thus, f is decreasing on $(-\infty, -2)$; f is increasing on $(-2, 1)$ and $(1, \infty)$; f has a local minimum at $x = -2$.

Step 3. Analyze $f''(x)$:

$$\begin{aligned} f''(x) &= 4(x-1)^2(1) + 4(x+2)(2)(x-1)(1) \\ &= 4(x-1)[(x-1) + 2(x+2)] \\ &= 12(x-1)(x+1) \end{aligned}$$

Partition numbers for f'': $x = -1, x = 1$.

Sign chart for f'':

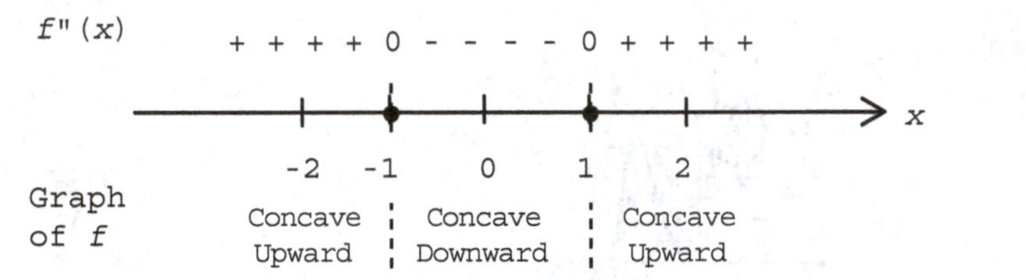

Test Numbers

x	$f''(x)$
−2	36(+)
0	−12(−)
2	36(+)

Thus, the graph of f is concave upward on $(-\infty, -1)$ and on $(1, \infty)$; the graph of f is concave downward on $(-1, 1)$; the graph has inflection points at $x = -1$ and at $x = 1$.

Step 4. Sketch the graph of f:

x	$f(x)$
−2	−27
0	−3
1	0

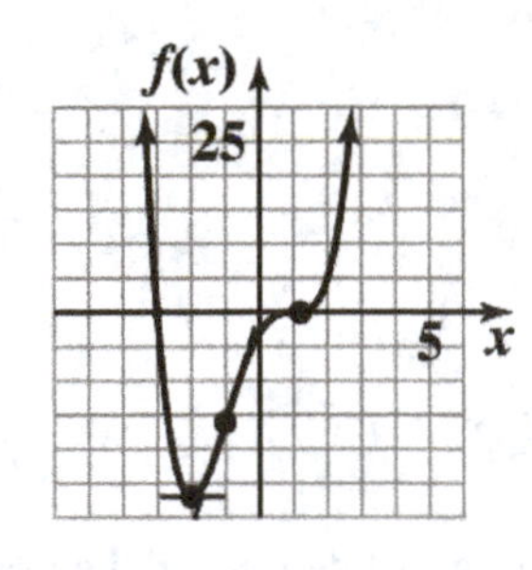

(4-4)

25. $f(x) = \dfrac{3x}{x+2}$

Step 1. Analyze $f(x)$:

The domain of f is all real numbers except $x = -2$.

Intercepts: y-intercept: $f(0) = \dfrac{3(0)}{0+2} = 0$

x-intercepts: $\dfrac{3x}{x+2} = 0$

$$3x = 0$$
$$x = 0$$

Asymptotes:

Horizontal asymptotes: $\dfrac{3x}{x} = 3$. Thus, the line $y = 3$ is a horizontal asymptote.

Vertical asymptote(s): The denominator is 0 at $x = -2$ and the numerator is nonzero at $x = -2$. Thus, the line $x = -2$ is a vertical asymptote.

Step 2. Analyze $f'(x)$:

$$f'(x) = \frac{(x+2)(3) - 3x(1)}{(x+2)^2} = \frac{6}{(x+2)^2}$$

Critical values: $f'(x) = \dfrac{6}{(x+2)^2} \neq 0$ for all x ($x \neq -2$).

Thus, f does not have any critical values.

Partition numbers: $x = -2$ is a partition number for f'.

Sign chart for f':

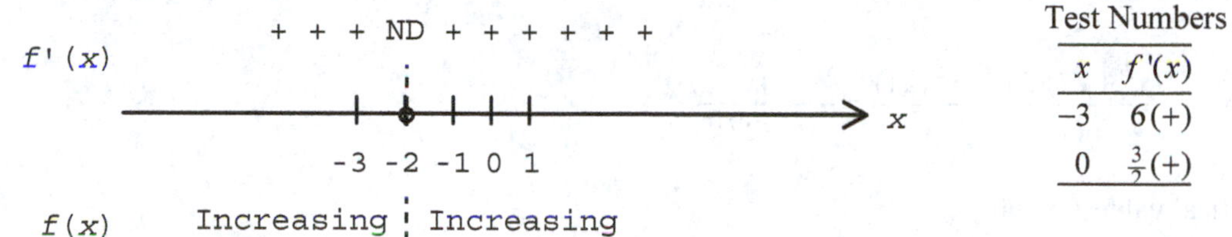

Test Numbers

x	$f'(x)$
-3	$6(+)$
0	$\frac{3}{2}(+)$

Thus, f is increasing on $(-\infty, -2)$ and on $(-2, \infty)$; f does not have any local extrema.

Step 3. Analyze $f''(x)$:

$$f''(x) = -12(x+2)^{-3} = \frac{-12}{(x+2)^3}$$

Partition numbers for f'': $x = -2$

Sign chart for f'':

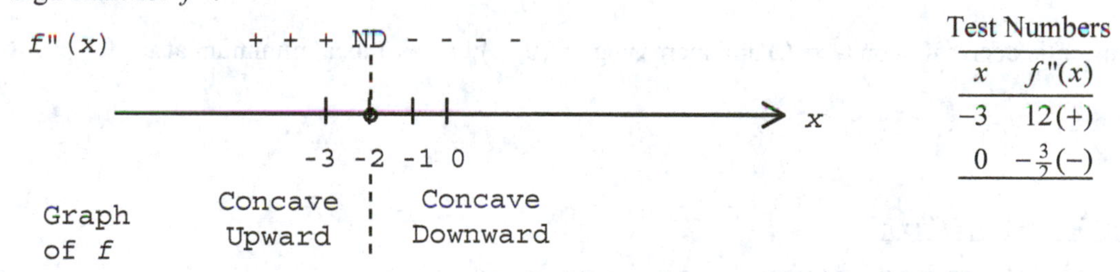

Test Numbers

x	$f''(x)$
-3	$12(+)$
0	$-\frac{3}{2}(-)$

The graph of f is concave upward on $(-\infty, -2)$ and concave downward on $(-2, \infty)$. The graph of f does not have any inflection points.

Step 4. Sketch the graph of f:

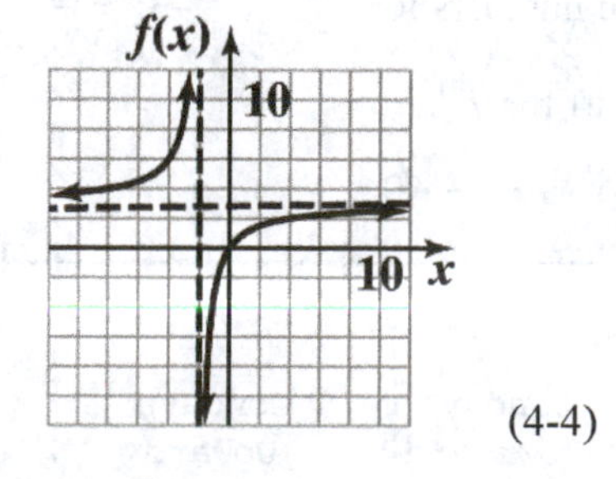

(4-4)

26. $f(x) = \dfrac{x^2}{x^2 + 27}$

Step 1. Analyze $f(x)$:

(A) Domain: All real numbers.

(B) Intercepts: y-intercepts: $f(0) = 0$

x-intercepts: $\frac{x^2}{x^2+27} = 0, \quad x = 0$

(C) Asymptotes:

Horizontal asymptote: $\frac{x^2}{x^2} = 1$; $y = 1$ is a horizontal asymptote

Vertical asymptote: no vertical asymptotes

Step 2. Analyze $f'(x)$:

$$f'(x) = \frac{(x^2+27)(2x) - x^2(2x)}{(x^2+27)^2} = \frac{54x}{(x^2+27)^2}$$

Critical values: $x = 0$

Partition numbers: $x = 0$

Sign chart for f':

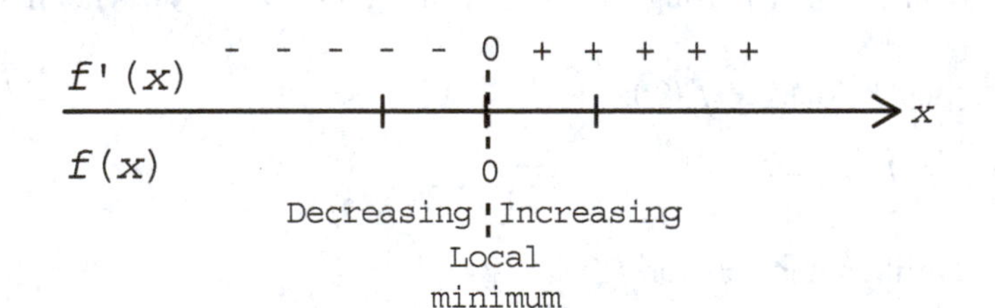

Thus, f is decreasing on $(-\infty, 0)$ and increasing on $(0, \infty)$; f has a local minimum at $x = 0$.

Step 3. Analyze $f''(x)$:

$$f''(x) = \frac{(x^2+27)^2(54) - 54x(2)(x^2+27)2x}{(x^2+27)^4} = \frac{162(9-x^2)}{(x^2+27)^3}$$

Partition numbers for f'': $x = -3, \ x = 3$

Sign chart for f'':

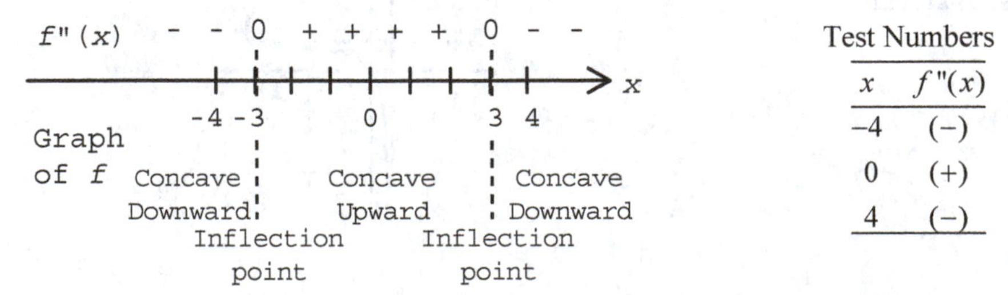

Test Numbers

x	$f''(x)$
-4	$(-)$
0	$(+)$
4	$(-)$

The graph of f is concave upward on $(-3, 3)$ and concave downward on $(-\infty, -3)$ and $(3, \infty)$; the graph has inflection points at $x = -3$ and $x = 3$.

Step 4. Sketch the graph of f:

x	$f(x)$
–3	$\frac{1}{4}$
0	0
3	$\frac{1}{4}$

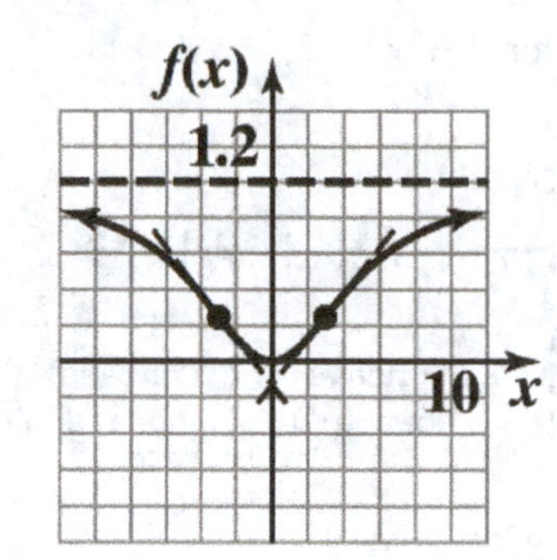

(4-4)

27. $f(x) = \dfrac{x}{(x+2)^2}$

Step 1. Analyze $f(x)$:

(A) Domain: All real numbers except $x = -2$.

(B) Intercepts: y-intercepts: $f(0) = 0$

x-intercepts: $\dfrac{x}{(x+2)^2} = 0, \quad x = 0$

(C) Asymptotes:

Horizontal asymptote: $\dfrac{x}{x^2} = \dfrac{1}{x}$; $y = 0$ (the x-axis) is a horizontal asymptote.

Vertical asymptote: $x = -2$ is a vertical asymptote

Step 2. Analyze $f'(x)$:

$$f'(x) = \frac{(x+2)^2 - x(2)(x+2)}{(x+2)^4} = \frac{2-x}{(x+2)^3}$$

Critical values: $x = 2$

Partition numbers: $x = -2, x = 2$

Sign chart for f':

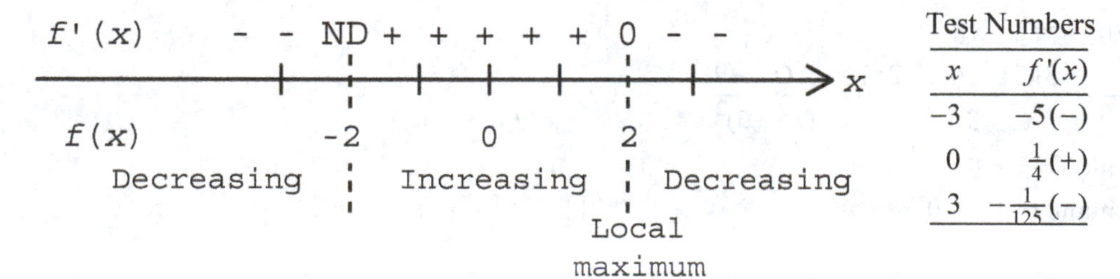

Test Numbers

x	$f'(x)$
–3	–5(–)
0	$\frac{1}{4}$(+)
3	$-\frac{1}{125}$(–)

Thus, f is increasing on $(-2, 2)$ and decreasing on $(-\infty, -2)$ and $(2, \infty)$; f has a local maximum at $x = 2$.

Step 3. Analyze $f''(x)$:

$$f''(x) = \frac{(x+2)^3(-1) - (2-x)(3)(x+2)^2}{(x+2)^6} = \frac{2(x-4)}{(x+2)^4}$$

Partition numbers for f'': $x = -2$, $x = 4$

Sign chart for f'':

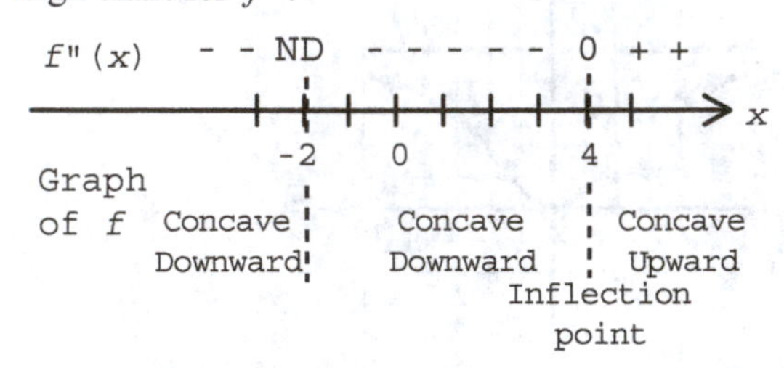

Test Numbers

x	$f''(x)$
-3	$-14(-)$
0	$-\frac{1}{2}(-)$
5	$(+)$

The graph of f is concave upward on $(4, \infty)$ and concave downward on $(-\infty, -2)$ and $(-2, 4)$; the graph has an inflection point at $x = 4$.

Step 4. Sketch the graph of f:

x	$f(x)$
0	0
2	$\frac{1}{8}$
4	$\frac{1}{9}$

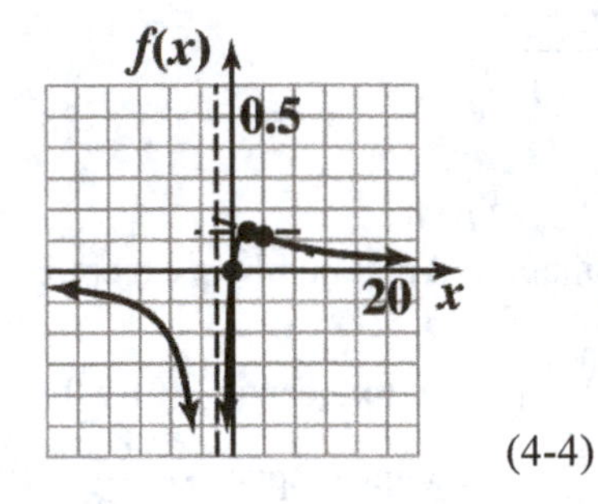

(4-4)

28. $f(x) = \dfrac{x^3}{x^2 + 3}$

Step 1. Analyze $f(x)$:

(A) Domain: All real numbers.

(B) Intercepts: y-intercepts: $f(0) = 0$

x-intercepts: $\dfrac{x^3}{x^2 + 3} = 0, \quad x = 0$

(C) Asymptotes:

Horizontal asymptote: $\dfrac{x^3}{x^2} = x$; no horizontal asymptote.

Vertical asymptote: no vertical asymptotes.

Oblique asymptote: It follows from above that $y = x$ is an oblique asymptote.

Step 2. Analyze $f'(x)$:

$$f'(x) = \frac{(x^2 + 3)(3x^2) - x^3(2x)}{(x^2 + 3)^2} = \frac{x^2(x^2 + 9)}{(x^2 + 3)^2}$$

Critical values: $x = 0$

Partition numbers: $x = 0$

Sign chart for f':

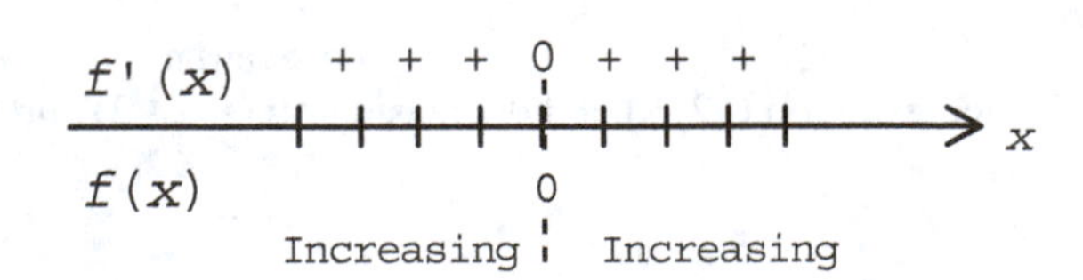

f is increasing on $(-\infty, \infty)$.

Step 3. Analyze $f''(x)$:

$$f''(x) = \frac{(x^2+3)^2(4x^3+18x) - x^2(x^2+9)(2)(x^2+3)2x}{(x^2+3)^4} = \frac{6x(9-x^2)}{(x^2+3)^3}$$

Partition numbers for f'': $x = -3,\ x = 0,\ x = 3$

Sign chart for f'':

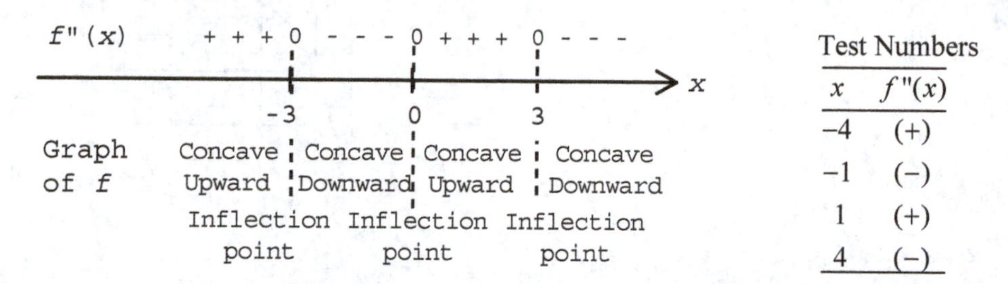

Test Numbers

x	$f''(x)$
-4	(+)
-1	(–)
1	(+)
4	(–)

The graph of f is concave upward on $(-\infty, -3)$ and $(0, 3)$, and concave downward on $(-3, 0)$ and $(3, \infty)$; the graph has inflection points at $x = -3,\ x = 0,\ x = 3$.

Step 4. Sketch the graph of f:

x	$f(x)$
-3	$-\frac{9}{4}$
0	0
3	$\frac{9}{4}$

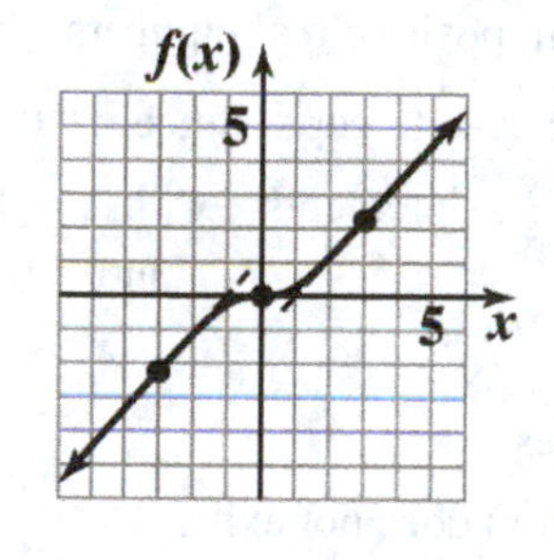

(4-4)

29. $f(x) = 5 - 5e^{-x}$

Step 1. Analyze $f(x)$:

(A) Domain: All real numbers, $(-\infty, \infty)$.

(B) Intercepts: y-intercept: $f(0) = 5 - 5e^{-0} = 0$

x-intercepts: $5 - 5e^{-x} = 0$

$$e^{-x} = 1$$
$$x = 0$$

(C) Asymptotes:

$$\lim_{x\to\infty}(5 - 5e^{-x}) = \lim_{x\to\infty}\left(5 - \frac{5}{e^x}\right) = 5$$

$\lim_{x\to-\infty}(5 - 5e^{-x})$ does not exist.

Thus, $y = 5$ is a horizontal asymptote.

Since $f(x) = 5 - \frac{5}{e^x} = \frac{5e^x - 5}{e^x}$ and $e^x \neq 0$ for all x, there are no vertical asymptotes.

Step 2. Analyze $f'(x)$:

$f'(x) = -5e^{-x}(-1) = 5e^{-x} > 0$ on $(-\infty, \infty)$

Thus, f is increasing on $(-\infty, \infty)$; there are no local extrema.

Step 3. Analyze $f''(x)$:

$f''(x) = -5e^{-x} < 0$ on $(-\infty, \infty)$.

Thus, the graph of f is concave downward on $(-\infty, \infty)$; there are no inflection points.

Step 4. Sketch the graph of f:

x	$f(x)$
0	0
−1	−8.59
2	4.32

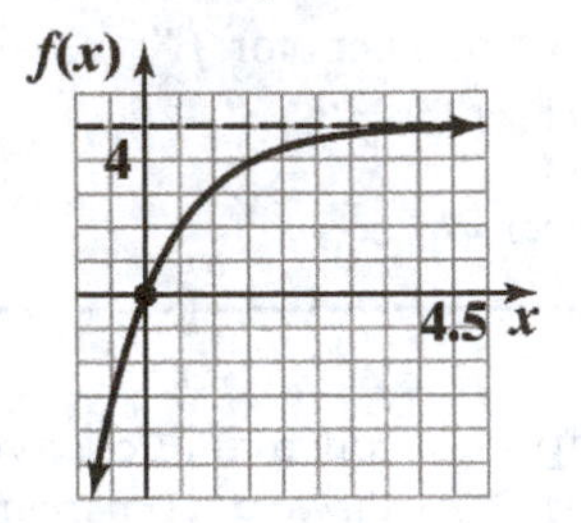

(4-4)

30. $f(x) = x^3 \ln x$

Step 1. Analyze $f(x)$:

(A) Domain: all positive real numbers, $(0, \infty)$.

(B) Intercepts: y-intercept: Since $x = 0$ is not in the domain, there is no y-intercept.

x-intercepts: $x^3 \ln x = 0$

$\ln x = 0$

$x = 1$

(C) Asymptotes:

$\lim_{x\to\infty} (x^3 \ln x)$ does not exist.

It can be shown that $\lim_{x\to 0^+} (x^3 \ln x) = 0$. Thus, there are no horizontal or vertical asymptotes.

Step 2. Analyze $f'(x)$:

$$f'(x) = x^3\left(\frac{1}{x}\right) + (\ln x)3x^2 = x^2[1 + 3\ln x], \quad x > 0$$

Critical values: $x^2[1 + 3\ln x] = 0$

$1 + 3\ln x = 0$ (since $x > 0$)

$\ln x = -\frac{1}{3}$

$x = e^{-1/3} \approx 0.72$

Partition numbers: $x = e^{-1/3}$

Sign chart for f':

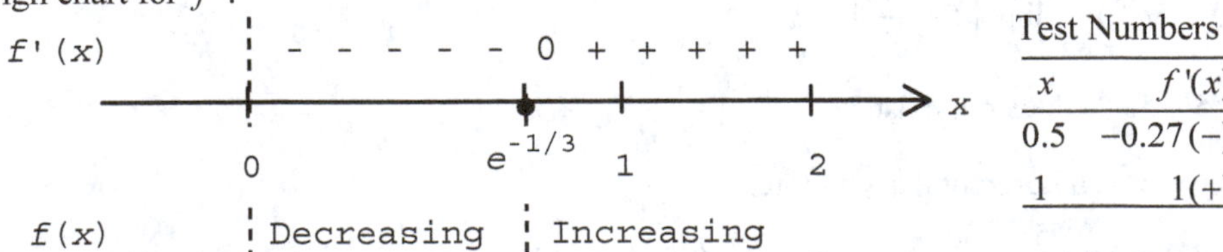

Test Numbers

x	$f'(x)$
0.5	−0.27(−)
1	1(+)

Thus, f is decreasing on $(0, e^{-1/3})$ and increasing on $(e^{-1/3}, \infty)$; f has a local minimum at $x = e^{-1/3}$.

Step 3. Analyze $f''(x)$:

$$f''(x) = x^2\left(\frac{3}{x}\right) + (1 + 3\ln x)2x = x(5 + 6\ln x), \quad x > 0$$

Partition numbers: $x(5 + 6 \ln x) = 0$

$$5 + 6 \ln x = 0$$

$$\ln x = -\frac{5}{6}$$

$$x = e^{-5/6} \approx 0.43$$

Sign chart for f'':

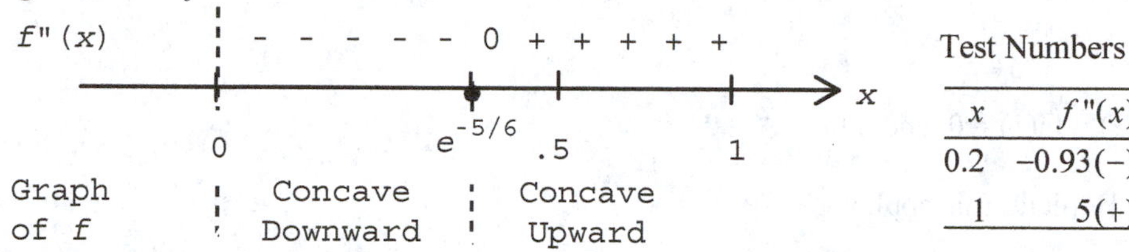

Test Numbers

x	$f''(x)$
0.2	−0.93(−)
1	5(+)

Thus, the graph of f is concave downward on $(0, e^{-5/6})$ and concave upward on $(e^{-5/6}, \infty)$; the graph has an inflection point at $x = e^{-5/6}$.

Step 4. Sketch the graph of f:

x	$f(x)$
$e^{-5/6}$	−0.07
$e^{-1/3}$	−0.12
1	0

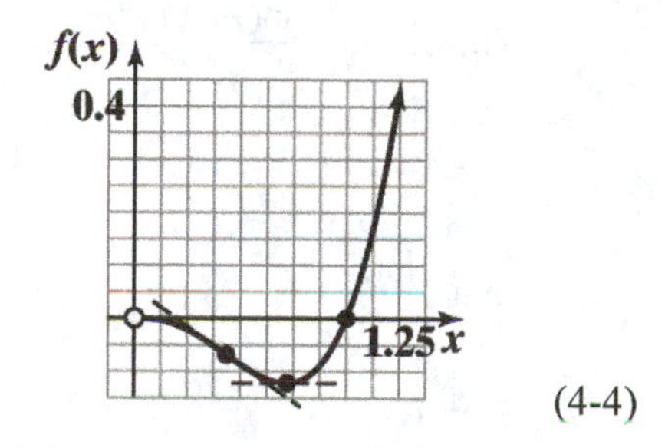

(4-4)

In Problems 31 – 40, D_x denotes $\frac{d}{dx}$.

31. $\lim_{x\to 0} \frac{e^{3x} - 1}{x}$

Step 1:

$\lim_{x\to 0} (e^{3x} - 1) = e^0 - 1 = 0$ and $\lim_{x\to 0} x = 0$.

Therefore, L'Hôpital's rule applies.

Step 2:

$\lim_{x\to 0} \frac{D_x(e^{3x} - 1)}{D_x x} = \lim_{x\to 0} \frac{3e^{3x}}{1} = 3$. Thus, $\lim_{x\to 0} \frac{e^{3x} - 1}{x} = 3$. (4-3)

32. $\lim_{x\to 2} \frac{x^2 - 5x + 6}{x^2 + x - 6}$

Step 1:

$\lim_{x\to 2} (x^2 - 5x + 6) = 2^2 - 10 + 6 = 0$ and $\lim_{x\to 2} (x^2 + x - 6) = 4 + 2 - 6 = 0$.

Therefore, L'Hôpital's rule applies.

Step 2:

$$\lim_{x\to 2}\frac{D_x(x^2-5x+6)}{D_x(x^2+x-6)}=\lim_{x\to 2}\frac{(2x-5)}{(2x+1)}=\frac{-1}{5}.\text{ Thus, }\lim_{x\to 2}\frac{x^2-5x+6}{x^2+x-6}=\frac{-1}{5}.\qquad (4\text{-}3)$$

33. $\lim_{x\to 0^-}\frac{\ln(1+x)}{x^2}$

Step 1:

$\lim_{x\to 0^-}\ln(1+x)=\ln(1)=0$ and $\lim_{x\to 0^-}x^2=0.$

Therefore, L'Hôpital's rule applies.

Step 2:

$$\lim_{x\to 0^-}\frac{D_x\ln(1+x)}{D_x x^2}=\lim_{x\to 0^-}\frac{\frac{1}{1+x}}{2x}=\lim_{x\to 0^-}\frac{1}{2x(1+x)}=-\infty.$$

Thus, $\lim_{x\to 0^-}\frac{\ln(1+x)}{x^2}=-\infty.$ (4-3)

34. $\lim_{x\to 0}\frac{\ln(1+x)}{1+x}$

Step 1:

$\lim_{x\to 0}\ln(1+x)=\ln(1)=0$ and $\lim_{x\to 0}(1+x)=1.$

Therefore, L'Hôpital's rule does not apply.

Step 2:

Using the quotient property for limits

$$\lim_{x\to 0}\frac{\ln(1+x)}{1+x}=\frac{0}{1}=0\qquad (4\text{-}3)$$

35. $\lim_{x\to\infty}\frac{e^{4x}}{x^2}$

Step 1:

$\lim_{x\to\infty}e^{4x}=\infty$ and $\lim_{x\to\infty}x^2=\infty.$ Therefore, L'Hôpital's rule applies.

Step 2:

$$\lim_{x\to\infty}\frac{D_x e^{4x}}{D_x x^2}=\lim_{x\to\infty}\frac{4e^{4x}}{2x}=\lim_{x\to\infty}\frac{2e^{4x}}{x}.$$

Since $\lim_{x\to\infty}2e^{x}=\infty$ and $\lim_{x\to\infty}x=\infty$, we apply L'Hôpital's rule again.

Step 3:

$$\lim_{x\to\infty} \frac{D_x 2e^{4x}}{D_x x} = \lim_{x\to\infty} \frac{8e^{4x}}{1} = \infty. \text{ Thus, } \lim_{x\to\infty} \frac{e^{4x}}{x^2} = \infty \qquad (4\text{-}3)$$

36. $\lim_{x\to 0} \frac{e^x + e^{-x} - 2}{x^2}$

Step 1:

$\lim_{x\to 0} (e^x + e^{-x} - 2) = e^0 + e^0 - 2 = 0$ and $\lim_{x\to 0} x^2 = 0$.

Therefore, L'Hôpital's rule applies.

Step 2:

$\lim_{x\to 0} \frac{D_x(e^x + e^{-x} - 2)}{D_x x^2} = \lim_{x\to 0} \frac{e^x - e^{-x}}{2x}$. Since $\lim_{x\to 0} (e^x - e^{-x}) = e^0 - e^0 = 0$ and $\lim_{x\to 0} 2x = 0$, we apply L'Hôpital's again.

Step 3:

$$\lim_{x\to 0} \frac{D_x(e^x + e^{-x})}{D_x 2x} = \lim_{x\to 0} \frac{e^x + e^{-x}}{2} = \frac{e^0 + e^0}{2} = 1.$$

Thus, $\lim_{x\to 0} \frac{e^x + e^{-x} - 2}{x^2} = 1.$ (4-3)

37. $\lim_{x\to 0^+} \frac{\sqrt{1+x} - 1}{\sqrt{x}}$

Step 1:

$\lim_{x\to 0^+} \left(\sqrt{1+x} - 1\right) = 0$ and $\lim_{x\to 0^+} \sqrt{x} = 0$.

Therefore, L'Hôpital's rule applies.

Step 2:

$$\lim_{x\to 0^+} \frac{D_x\left[\sqrt{1+x} - 1\right]}{D_x\sqrt{x}} = \lim_{x\to 0^+} \frac{\frac{1}{2}(1+x)^{-1/2}}{\frac{1}{2}x^{-1/2}} = \lim_{x\to 0^+} \frac{\sqrt{x}}{\sqrt{1+x}} = 0.$$

Thus, $\lim_{x\to 0^+} \frac{\sqrt{1+x} - 1}{\sqrt{x}} = 0.$ (4-3)

38. $\lim_{x\to\infty} \frac{\ln x}{x^5}$

Step 1:

$\lim_{x\to\infty} \ln x = \infty$ and $\lim_{x\to\infty} x^5 = \infty$. Therefore, L'Hôpital's rule applies.

Step 2:

$$\lim_{x\to\infty} \frac{D_x \ln x}{D_x x^5} = \lim_{x\to\infty} \frac{\frac{1}{x}}{5x^4} = \lim_{x\to\infty} \frac{1}{5x^5} = 0. \text{ Thus, } \lim_{x\to\infty} \frac{\ln x}{x^5} = 0. \qquad (4\text{-}3)$$

39. $\lim\limits_{x\to\infty} \dfrac{\ln(1+6x)}{\ln(1+3x)}$

Step 1:

$\lim\limits_{x\to\infty} \ln(1+6x) = \infty$ and $\lim\limits_{x\to\infty} \ln(1+3x) = \infty$.

Therefore, L'Hôpital's rule applies.

Step 2:

$$\lim_{x\to\infty} \frac{D_x \ln(1+6x)}{D_x \ln(1+3x)} = \lim_{x\to\infty} \frac{\dfrac{6}{1+6x}}{\dfrac{3}{1+3x}} = \lim_{x\to\infty} \frac{2(1+3x)}{1+6x}.$$

Since $\lim\limits_{x\to\infty} 2(1+3x) = \infty$ and $\lim\limits_{x\to\infty} (1+6x) = \infty$, we apply L'Hôpital's rule again.

Step 3:

$$\lim_{x\to\infty} \frac{D_x(2[1+3x])}{D_x(1+6x)} = \lim_{x\to\infty} \frac{D_x(2+6x)}{D_x(1+6x)} = \lim_{x\to\infty} \frac{6}{6} = 1.$$

Thus, $\lim\limits_{x\to\infty} \dfrac{\ln(1+6x)}{\ln(1+3x)} = 1.$ (4-3)

40. $\lim\limits_{x\to 0} \dfrac{\ln(1+6x)}{\ln(1+3x)}$

Step 1:

$\lim\limits_{x\to 0} \ln(1+6x) = \ln 1 = 0$ and $\lim\limits_{x\to 0} \ln(1+3x) = \ln 1 = 0$.

Therefore, L'Hôpital's rule applies.

Step 2:

$$\lim_{x\to 0} \frac{D_x \ln(1+6x)}{D_x \ln(1+3x)} = \lim_{x\to 0} \frac{\dfrac{6}{1+6x}}{\dfrac{3}{1+3x}} = \lim_{x\to 0} \frac{2(1+3x)}{1+6x} = 2.$$

Thus, $\lim\limits_{x\to 0} \dfrac{\ln(1+6x)}{\ln(1+3x)} = 2.$ (4-3)

41.

x	$f'(x)$	$f(x)$
$-\infty < x < -2$	Negative and increasing	Decreasing and concave upward
$x = -2$	x-intercept	Local minimum
$-2 < x < -1$	Positive and increasing	Increasing and concave upward
$x = -1$	Local maximum	Inflection point
$-1 < x < 1$	Positive and decreasing	Increasing and concave downward
$x = 1$	Local minimum	Inflection point
$1 < x < \infty$	Positive and increasing	Increasing and concave upward

(4-2)

42. The graph in (C) could be the graph of $y = f'(x)$. (4-2)

43. $f(x) = x^3 - 6x^2 - 15x + 12$

$f'(x) = 3x^2 - 12x - 15$

Critical values: f' is defined for all x:

$3x^2 - 12x - 15 = 0$

$3(x^2 - 4x - 5) = 0$

$3(x-5)(x+1)=0$
Thus, $x=-1$ and $x=5$ are critical values of f.
$f''(x)=6x-12$
Now, $f''(-1)=6(-1)-12=-18<0$.
Thus, f has a local maximum at $x=-1$.
Also, $f''(5)=6(5)-12=18>0$ and f has a local minimum at $x=5$. (4-1, 4-2)

44. $y=f(x)=x^3-12x+12,\ -3\le x\le 5$
$f'(x)=3x^2-12$
Critical values: f' is defined for all x:
$3x^2-12=0$
$3(x^2-4)=0$
$3(x-2)(x+2)=0$
Thus, the critical values of f are: $x=-2,\ x=2$.
$f(-3)=(-3)^3-12(-3)+12=21$
$f(-2)=(-2)^3-12(-2)+12=28$
$f(2)=2^3-12(2)+12=-4$ Absolute minimum
$f(5)=5^3-12(5)+12=77$ Absolute maximum (4-5)

45. $y=f(x)=x^2+\dfrac{16}{x^2},\ x>0$

$$f'(x)=2x-\frac{32}{x^3}=\frac{2x^4-32}{x^3}=\frac{2(x^4-16)}{x^3}=\frac{2(x-2)(x+2)(x^2+4)}{x^3}$$

$$f''(x)=2+\frac{96}{x^4}$$

The only critical value of f in the interval $(0,\infty)$ is $x=2$.
Since

$$f''(2)=2+\frac{96}{2^4}=8>0,$$

$f(2)=8$ is the absolute minimum of f on $(0,\infty)$. (4-5)

46. $f(x)=11x-2x\ln x,\ x>0$

$$f'(x)=11-2x\left(\frac{1}{x}\right)-(\ln x)(2)=11-2-2\ln x=9-2\ln x,\quad x>0$$

Critical value(s):
$9-2\ln x=0$
$2\ln x=9$
$\ln x=\dfrac{9}{2}$
$x=e^{9/2}$

$$f''(x)=-\frac{2}{x}\ \text{ and }\ f''(e^{9/2})=-\frac{2}{e^{9/2}}<0$$

Since $x=e^{9/2}$ is the only critical value, and $f''(e^{9/2})<0$, f has an absolute maximum at $x=e^{9/2}$. The absolute maximum is:

$$f(e^{9/2})=11e^{9/2}-2e^{9/2}\ln(e^{9/2})=11e^{9/2}-9e^{9/2}=2e^{9/2}\approx 180.03$$ (4-5)

47. $f(x) = 10xe^{-2x}, \quad x > 0$

$f'(x) = 10xe^{-2x}(-2) + 10e^{-2x}(1) = 10e^{-2x}(1 - 2x), \quad x > 0$

Critical value(s):

$10e^{-2x}(1 - 2x) = 0$

$1 - 2x = 0$

$x = \frac{1}{2}$

$f''(x) = 10e^{-2x}(-2) + 10(1 - 2x)e^{-2x}(-2) = -20e^{-2x}(1 + 1 - 2x) = -40e^{-2x}(1 - x)$

$f''\left(\frac{1}{2}\right) = -20e^{-1} < 0$

Since $x = \frac{1}{2}$ is the only critical value, and $f''\left(\frac{1}{2}\right) = -20e^{-1} < 0$, f has an absolute maximum at $x = \frac{1}{2}$.

The absolute maximum of f is: $f\left(\frac{1}{2}\right) = 10\left(\frac{1}{2}\right)e^{-2(1/2)} = 5e^{-1} \approx 1.84$ (4-5)

48. Yes. Consider f on the interval $[a, b]$. Since f is a polynomial, f is continuous on $[a, b]$. Therefore, f has an absolute maximum on $[a, b]$. Since f has a local minimum at $x = a$ and $x = b$, the absolute maximum of f on $[a, b]$ must occur at some point c in (a, b); f has a local maximum at $x = c$. (4-5)

49. No, increasing/decreasing properties are stated in terms of intervals in the domain of f. A correct statement is: $f(x)$ is decreasing on $(-\infty, 0)$ and $(0, \infty)$. (4-1)

50. A critical value for f is a partition number for f' that is also in the domain of f. However, f' may have partition numbers that are not in the domain of f and hence are not critical values for f. For example, let $f(x) = \frac{1}{x}$. Then $f'(x) = -\frac{1}{x^2}$ and 0 is a partition number for f', but 0 is NOT a critical value for f since it is not in the domain of f. (4-1)

51. $f(x) = 6x^2 - x^3 + 8, \ 0 \le x \le 4$

$f'(x) = 12x - 3x^2$

$f''(x) = 12 - 6x$

Now, $f''(x)$ is defined for all x and $f''(x) = 12 - 6x = 0$ implies $x = 2$. Thus, f' has a critical value at $x = 2$. Since this is the only critical value of f' and $[f'(x)]'' = f'''(x) = -6$ so that $f'''(2) = -6 < 0$, it follows that $f'(2) = 12$ is the absolute maximum of f'. The graph is shown at the right.

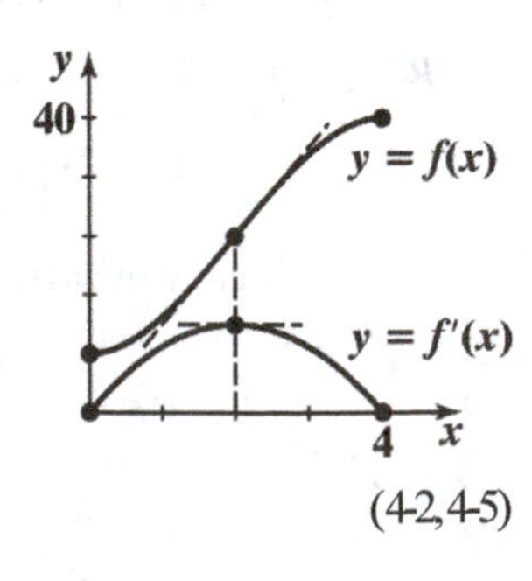

(4-2,4-5)

52. Let $x > 0$ be one of the numbers. Then $y = \frac{400}{x}$ is the other number. Now, we have:

$S(x) = x + \frac{400}{x}, \ x > 0,$

$S'(x) = 1 - \frac{400}{x^2} = \frac{x^2 - 400}{x^2} = \frac{(x-20)(x+20)}{x^2}$

Thus, $x = 20$ is the only critical value of S on $(0, \infty)$.

$S''(x) = \dfrac{800}{x^3}$ and $S''(20) = \dfrac{800}{8000} = \dfrac{1}{10} > 0$

Therefore, $S(20) = 20 + \dfrac{400}{20} = 40$ is the absolute minimum sum, and this occurs when each number is 20. (4-6)

53. $f(x) = x^4 + x^3 - 4x^2 - 3x + 4.$

Step 1. Analyze $f(x)$:

(A) Domain: All real numbers (f is a polynomial function)

(B) Intercepts: y-intercept: $f(0) = 4$
x-intercepts: $x \approx 0.79,\ 1.64$

(C) Asymptotes: Since f is a polynomial function (of degree 4), the graph of f has no asymptotes.

Step 2. Analyze $f'(x)$:

$= 4x^3 + 3x^2 - 8x - 3$

Critical values: $x \approx -1.68,\ -0.35,\ 1.28$;

f is increasing on $(-1.68, -0.35)$ and $(1.28, \infty)$; f is decreasing on $(-\infty, -1.68)$ and $(-0.35, 1.28)$. f has local minima at $x = -1.68$ and $x = 1.28$. f has a local maximum at $x = -0.35$.

Step 3. Analyze $f''(x)$:

$f''(x) = 12x^2 + 6x - 8$

The graph of f is concave downward on $(-1.10, 0.60)$; the graph of f is concave upward on $(-\infty, -1.10)$ and $(0.60, \infty)$; the graph has inflection points at $x \approx -1.10$ and 0.60. (4-4)

54. $f(x) = 0.25x^4 - 5x^3 + 31x^2 - 70x$

Step 1. Analyze $f(x)$:

(A) Domain: all real numbers

(B) Intercepts: y-intercept: $f(0) = 0$
x-intercepts: $x = 0,\ 11.10$

(C) Asymptotes: since f is a polynomial function, the graph of f has no asymptotes; $\lim\limits_{x \to \pm\infty} f(x) = \infty$

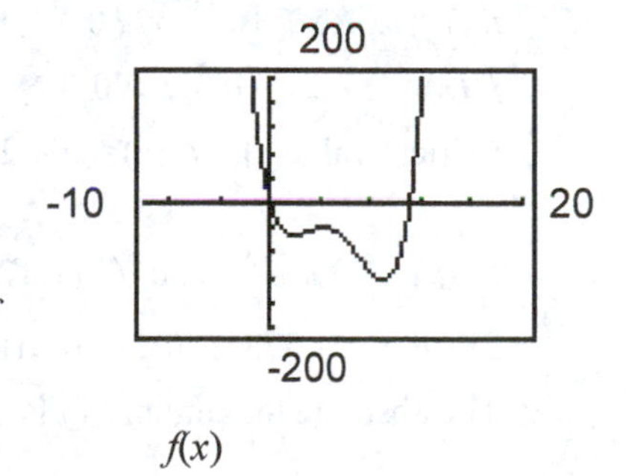

$f(x)$

Step 2. Analyze $f'(x)$:

$f'(x) = x^3 - 15x^2 + 62x - 70$

Critical values: $x \approx 1.87,\ 4.19,\ 8.94$

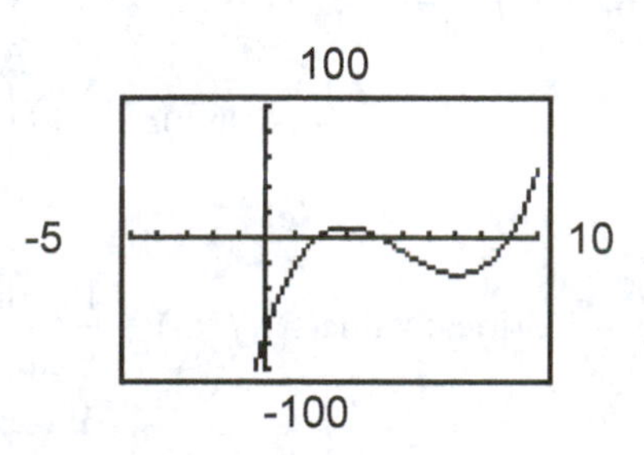

Sign chart for f' :

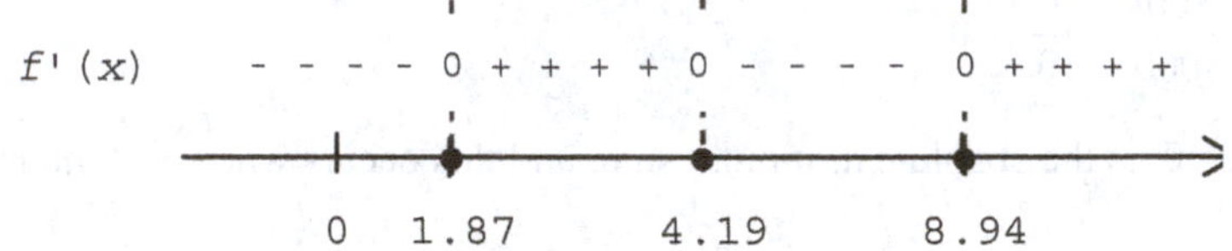

f (x) Decreasing | Increasing | Decreasing | Increasing

Local minimum | Local maximum | Local minimum

f is increasing on (1.87, 4.19) and (8.94, ∞); f is decreasing on ($-\infty$, 1.87) and (4.19, 8.94); f has local minima at $x = 1.87$ and $x = 8.94$; f has a local maximum at $x = 4.19$

Step 3. Analyze $f''(x)$:

$f''(x) = 3x^2 - 30x + 62$

Partition numbers for f'': $x \approx 2.92, 7.08$

Sign chart for f'':

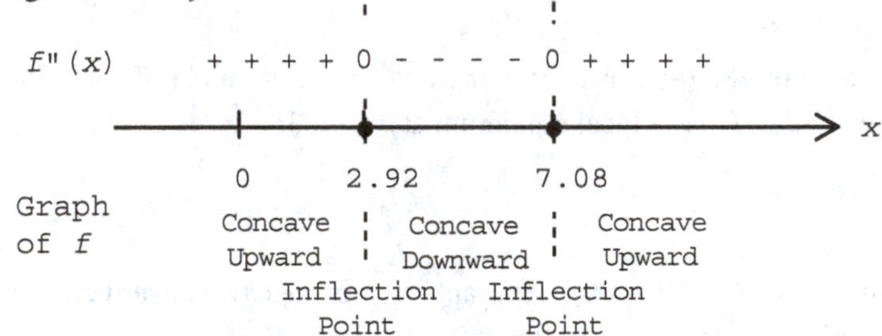

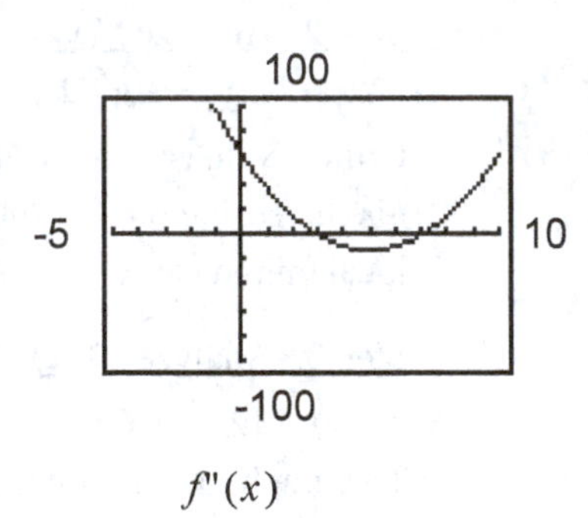

$f''(x)$

The graph of f is concave downward on (2.92, 7.08) and concave upward on ($-\infty$, 2.92) and (7.08, ∞); the graph has inflection points at $x = 2.92$ and $x = 7.08$. (4-4)

55. $f(x) = 3x - x^2 + e^{-x}, x > 0$

$f'(x) = 3 - 2x - e^{-x}, x > 0$

Critical value(s): $f'(x) = 3 - 2x - e^{-x} = 0$

$x \approx 1.373$

$f''(x) = -2 + e^{-x}$ and $f''(1.373) = -2 + e^{-1.373} < 0$

Since $x \approx 1.373$ is the only critical value, and $f''(1.373) < 0$, f has an absolute maximum at $x = 1.373$. The absolute maximum of f is: $f(1.373) = 3(1.373) - (1.373)^2 + e^{-1.373} \approx 2.487$. (4-5)

56. $f(x) = \dfrac{\ln x}{e^x}, x > 0$

$$f'(x) = \frac{e^x\left(\frac{1}{x}\right) - (\ln x)e^x}{(e^x)^2} = \frac{e^x\left(\frac{1}{x} - \ln x\right)}{e^{2x}} = \frac{1 - x\ln x}{xe^x}, \quad x > 0$$

Critical value(s): $f'(x) = \dfrac{1 - x\ln x}{xe^x} = 0$

$1 - x\ln x = 0$

$x\ln x = 1$

$x \approx 1.763$

$$f''(x) = \frac{xe^x[-1-\ln x]-(1-x\ln x)(xe^x+e^x)}{x^2e^{2x}} = \frac{-x(1+\ln x)-(x+1)(1-x\ln x)}{x^2e^x};$$

$$f''(1.763) \approx \frac{-1.763(1.567)-(2.763)(0.000349)}{(1.763)^2e^{1.763}} < 0$$

Since $x = 1.763$ is the only critical value, and $f''(1.763) < 0$, f has an absolute maximum at $x = 1.763$.
The absolute maximum of f is: $f(1.763) = \frac{\ln(1.763)}{e^{1.763}} \approx 0.097$. (4-5)

57. (A) For the first 15 months, the price is increasing and concave down, with a local maximum at $t = 15$. For the next 15 months, the price is decreasing and concave down, with an inflection point at $t = 30$. For the next 15 months, the price is decreasing and concave up, with a local minimum at $t = 45$. For the remaining 15 months, the price is increasing and concave up.

(B)

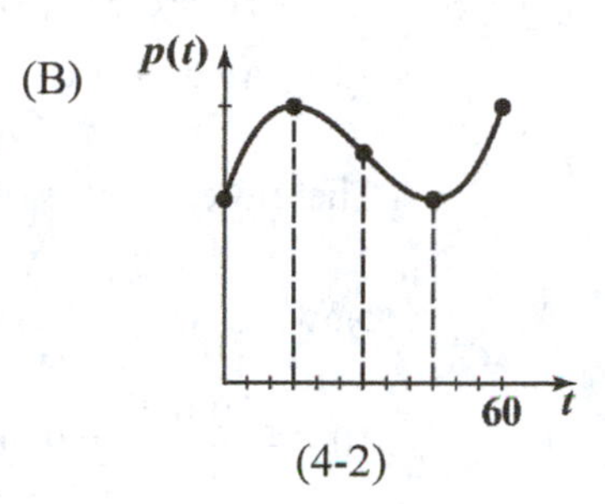

(4-2)

58. (A) $R(x) = xp(x) = 500x - 0.025x^2,\ 0 \le x \le 20{,}000$
$R'(x) = 500 - 0.05x;$
$500 - 0.05x = 0, \quad x = 10{,}000$
Thus, $x = 10{,}000$ is a critical value.

Now, $R(0) = 0$
$R(10{,}000) = 2{,}500{,}000$
$R(20{,}000) = 0$

Thus, $R(10{,}000) = \$2{,}500{,}000$ is the absolute maximum of R.

(B) $P(x) = R(x) - C(x) = 500x - 0.025x^2 - (350x + 50{,}000)$
$= 150x - 0.025x^2 - 50{,}000, \quad 0 \le x \le 20{,}000$
$P'(x) = 150 - 0.05x;$
$150 - 0.05x = 0, \quad x = 3{,}000$

Now, $P(0) = -50{,}000$
$P(3{,}000) = 175{,}000$
$P(20{,}000) = -7{,}050{,}000$

Thus, the maximum profit is \$175,000 when 3000 readers are manufactured and sold at $p(3{,}000) = \$425$ each.

(C) If the government taxes the company \$20 per reader, then the cost equation is:
$C(x) = 370x + 50{,}000$ and
$P(x) = 500x - 0.025x^2 - (370x + 50{,}000) = 130x - 0.025x^2 - 50{,}000, \quad 0 \le x \le 20{,}000$
$P'(x) = 130 - 0.05x;$
$130 - 0.05x = 0, \quad x = 2{,}600$
The maximum profit is $P(2{,}600) = \$119{,}000$ when 2,600 readers are produced and sold for $p(2{,}600) = \$435$ each. (4-6)

59.

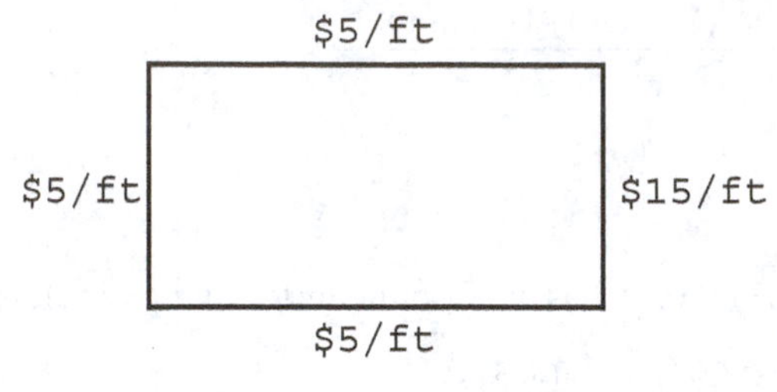

Let x be the length and y the width of the rectangle.

(A) $C(x, y) = 5x + 5x + 5y + 15y = 10x + 20y$

Also, Area $A = xy = 5000$, so $y = \dfrac{5000}{x}$.

Therefore, $C(x) = 10x + \dfrac{100,000}{x}, \quad x > 0.$

Now, $C'(x) = 10 - \dfrac{100,000}{x^2}$ and

$10 - \dfrac{100,000}{x^2} = 0$ implies $10x^2 = 100,000$

$$x^2 = 10,000$$
$$x = \pm 100$$

Thus, $x = 100$ is the critical value. [Note: $x > 0$, so $x = -100$ is not a critical value.]

Now, $C''(x) = \dfrac{200,000}{x^3}$ and $C''(100) = \dfrac{200,000}{1,000,000} = 0.2 > 0$. Thus, C has an absolute minimum when $x = 100$.

The most economical (i.e. least cost) fence will have dimensions: length $x = 100$ feet and width $y = \dfrac{5000}{100} = 50$ feet.

(B) We want to maximize $A = xy$ subject to

$C(x, y) = 10x + 20y = 3000$ or $x = 300 - 2y$

Thus, $A = y(300 - 2y) = 300y - 2y^2, \quad 0 \le y \le 150.$

Now, $A'(y) = 300 - 4y$ and

$300 - 4y = 0$ implies $y = 75$.

Therefore, $y = 75$ is the critical value.

Now, $A''(y) = -4$ and $A''(75) = -4 < 0$. Thus, A has an absolute maximum when $y = 75$.

The dimensions of the rectangle that will enclose maximum area are: length $x = 300 - 2(75) = 150$ feet and width $y = 75$ feet. (4-6)

60. Let x = the number of dollars increase in the nightly rate, $x \ge 0$. Then $200 - 4x$ rooms will be rented at $(40 + x)$ dollars per room. [Note: Since $200 - 4x \ge 0$, $x \le 50$.] The cost of service for $200 - 4x$ rooms at \$8 per room is $8(200 - 4x)$. Thus:

Gross profit: $P(x) = (200 - 4x)(40 + x) - 8(200 - 4x) = (200 - 4x)(32 + x) = 6400 + 72x - 4x^2, \quad 0 \le x \le 50$

$P'(x) = 72 - 8x$

Critical value:

$72 - 8x = 0, \quad x = 9$

Now, $P(0) = 6400$

$P(9) = 6724$ (absolute maximum)

$P(50) = 0$

Thus, the maximum gross profit is \$6724 and this occurs at $x = 9$, i.e., the rooms should be rented at \$49 per night. (4-6)

61. Let x = number of times the company should order. Then, the number of disks per order = $\frac{7200}{x}$. The average number of unsold disks is given by:

$$\frac{7200}{2x} = \frac{3600}{x}$$

Total cost: $C(x) = 5x + 0.2\left(\frac{3600}{x}\right) = 5x + \frac{720}{x}, \quad x > 0.$

$$C'(x) = 5 - \frac{720}{x^2} = \frac{5x^2 - 720}{x^2} = \frac{5(x^2 - 144)}{x^2} = \frac{5(x+12)(x-12)}{x^2}$$

Critical value: $x = 12$ [Note: $x > 0$, so $x = -12$ is not a critical value.]

$C''(x) = \frac{1440}{x^3}$ and $C''(12) = \frac{1440}{12^3} > 0$

Therefore, $C(x)$ is a minimum when $x = 12$. (4-6)

62. $C(x) = 4000 + 10x + 0.1x^2, x > 0$

Average cost = $\bar{C}(x) = \frac{4000}{x} + 10 + 0.1x$

Marginal cost = $C'(x) = 10 + \frac{2}{10}x = 10 + 0.2x$

The graph of C' is a straight line with slope $\frac{1}{5}$ and y intercept 10.

$$\bar{C}'(x) = \frac{-4000}{x^2} + \frac{1}{10} = \frac{-40,000 + x^2}{10x^2} = \frac{(x+200)(x-200)}{10x^2}$$

Thus, $\bar{C}'(x) < 0$ on (0, 200) and $\bar{C}'(x) > 0$ on $(200, \infty)$. Therefore, $\bar{C}(x)$ is decreasing on (0, 200), increasing on $(200, \infty)$, and a minimum occurs at $x = 200$.

Min = $\bar{C}(200) = \frac{4000}{200} + 10 + \frac{1}{10}(200) = 50$

$\bar{C}''(x) = \frac{8000}{x^3} > 0$ on $(0, \infty)$.

Therefore, the graph of $\bar{C}(x)$ is concave upward on $(0, \infty)$.

Using this information and point-by-point plotting (use a calculator), the graphs of $C(x)$ and $\bar{C}(x)$ are as shown in the diagram at the right. The line $y = 0.1x + 10$ is an oblique asymptote for $y = \bar{C}(x)$.

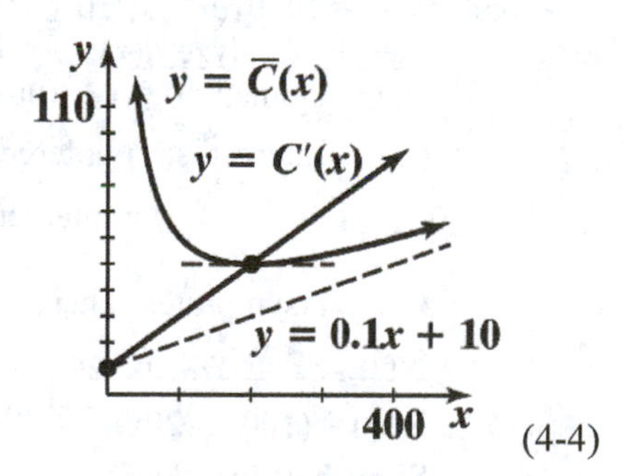

(4-4)

63. Cost: $C(x) = 200 + 50x - 50\ln x, \quad x \ge 1$

Average cost: $\bar{C} = \frac{C(x)}{x} = \frac{200}{x} + 50 - \frac{50}{x}\ln x, \quad x \ge 1$

$$\bar{C}'(x) = \frac{-200}{x^2} - \frac{50}{x}\left(\frac{1}{x}\right) + (\ln x)\frac{50}{x^2} = \frac{50(\ln x - 5)}{x^2}, x \ge 1$$

Critical value(s): $\bar{C}'(x) = \frac{50(\ln x - 5)}{x^2} = 0$

$$\ln x = 5$$
$$x = e^5$$

Sign chart for $\overline{C}\,'$:

$\overline{C}'(x)$: - - - - - - 0 + + + + +

(x-axis, mark at 1)

$\overline{C}(x)$: Decreasing | Increasing

Test Numbers

x	$\overline{C}'(x)$
1	$-250(-)$
e^6	$\frac{50}{e^{12}}(+)$

By the first derivative test, $\overline{C}$ has a local minimum at $x = e^5$. Since this is the only critical value of $\overline{C}$, $\overline{C}$ has as absolute minimum at $x = e^5$. Thus, the minimal average cost is:

$\overline{C}(e^5) = \frac{200}{e^5} + 50 - \frac{50}{e^5}\ln(e^5) = 50 - \frac{50}{e^5} \approx 49.66$ or \$49.66 (4-4)

64. $R(x) = xp(x) = 1000xe^{-0.02x}$

$R'(x) = 1000x \cdot e^{-0.02x}(-0.02) + 1000e^{-0.02x} = (1000 - 20x)e^{-0.02x}$

$R'(x) = 0$: $(1000 - 20x)e^{-0.02x} = 0$

$1000 - 20x = 0$

$x = 50$

Critical value of R: $x = 50$

$R''(x) = (1000 - 20x)e^{-0.02x}(-0.02) + e^{-0.02x}(-20) = e^{-0.02x}[0.4x - 20 - 20] = e^{-0.02x}[0.4x - 40]$;

$R''(50) = -20e^{-1} < 0$.

Since $x = 50$ is the only critical value and $R''(50) < 0$, R has an absolute maximum at a production level of 50 units. The maximum revenue is $R(50) = 1000(50)e^{-0.02(50)} = 50{,}000e^{-1} \approx \$18{,}394$.

The price per unit at the production level of 50 units is:

$p(50) = 1000e^{-0.02(50)} = 1000e^{-1} \approx \$367.88$. (4-6)

65. $R(x) = 1000xe^{-0.02x}$, $0 \le x \le 100$

Step 1. Analyze $R(x)$:

(A) Domain: $0 \le x \le 100$ or $[0, 100]$

(B) Intercepts: y-intercept: $R(0) = 0$

x-intercepts: $100xe^{-0.02x} = 0$

$x = 0$

(C) Asymptotes: There are no horizontal or vertical asymptotes.

Step 2. Analyze $R'(x)$:

$R'(x) = (1000 - 20x)e^{-0.02x}$ and $x = 50$ is a critical value.

Sign chart for R':

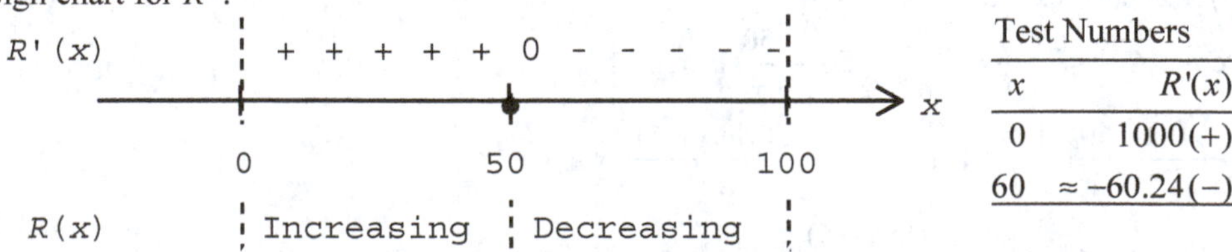

Thus, R is increasing on (0, 50) and decreasing on (50, 100); R has a maximum at $x = 50$.

Step 3. Analyze $R''(x)$:

$R''(x) = (0.4x - 40)e^{-0.02x} < 0$ on (0, 100)

Thus, the graph of R is concave downward on (0, 100).

Step 4. Sketch the graph of R:

x	$R(x)$
0	0
50	18,394
100	13,534

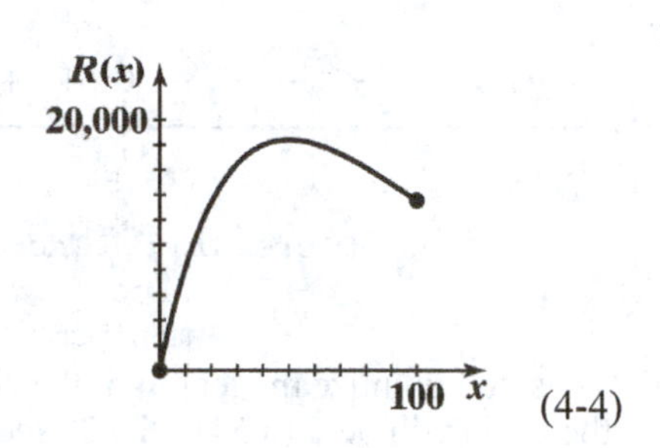

(4-4)

66. Cost: $C(x) = 220x$

Price-demand equation: $p(x) = 1{,}000e^{-0.02x}$

Revenue: $R(x) = xp(x) = 1{,}000xe^{-0.02x}$

Profit: $P(x) = R(x) - C(x) = 1{,}000xe^{-0.02x} - 220x$

On a graphing utility, graph $P(x)$ and calculate its maximum value. The maximum value is \$9,864 at a demand level of 29.969082 ($\approx$ 30). The price at this demand level is: p = \$549.15. (4-6)

67. Let x = the number of cream puffs.

Daily cost: $C(x) = x$ (dollars)

Daily revenue: $R(x) = xp(x)$, where

$p(x) = a + b \ln x$ is the logarithmic regression model for the given data.

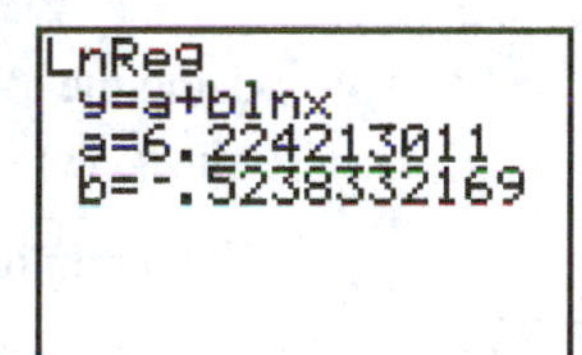

Profit: $P(x) = R(x) - C(x)$

Using a graphing utility, we find that the maximum profit is achieved at the demand level $x \approx 7888$. The price at this demand level is: $p(7888)$ = \$1.52 (to the nearest cent). (4-6)

68. Let x be the length of the vertical portion of the chain. Then the length of each of the "arms" of the "Y" is $\sqrt{(10-x)^2+36} = \sqrt{x^2-20x+136}$. Thus, the total length is given by:

$$L(x) = x + 2\sqrt{x^2-20x+136}, \quad 0 \le x \le 10$$

Now, $L'(x) = 1 + 2\left(\frac{1}{2}\right)(x^2 - 20x + 136)^{-1/2}(2x - 20) = 1 + \dfrac{2x-20}{(x^2-20x+136)^{1/2}}$

$L'(x) = 0$: $\quad 1 + \dfrac{2x-20}{(x^2-20x+136)^{1/2}} = 0$

$$(x^2 - 20x + 136)^{1/2} + 2x - 20 = 0$$

$$(x^2 - 20x + 136)^{1/2} = 2(10 - x)$$

$$x^2 - 20x + 136 = 4(100 - 20x + x^2)$$

$$-3x^2 + 60x - 264 = 0$$

$$x = \frac{20 \pm \sqrt{48}}{2} = 10 \pm 2\sqrt{3}$$

Critical value (in (0, 10)): $x = 10 - 2\sqrt{3} \approx 6.54$

Sign chart for $L'(x)$:

```
L'(x)      - - - - 0 + + + +
      ----+--------+--------+----> x
L(x)      0       6.54      10
          Decreasing : Increasing
                   Local
                  minimum
```

Test Numbers

x	$L'(x)$
0	(–)
10	1(+)

Thus, to minimize the length of the chain, the vertical portion should be 6.54 feet long. The total length of the chain will be $L(6.54) = 20.39$ feet. (4-6)

69. (A)

```
QuadReg
 y=ax²+bx+c
 a=.0061285714
 b=.1224285714
 c=102.2
```

(B) Let $C(x)$ be the regression equation from part (A). The average cost function $\overline{C}(x) = \frac{C(x)}{x}$.

Using the "find the minimum" routine on the graphing utility, we find that $\min \overline{C}(x) = \overline{C}(129) = 1.71$

The minimum average cost is \$1.71 at a production level of 129 dozen cookies. (4-4)

70. $N(x) = -0.25x^4 + 11x^3 - 108x^2 + 3{,}000,\ 9 \le x \le 24$

$N'(x) = -x^3 + 33x^2 - 216x$

$N''(x) = -3x^2 + 66x - 216 = -3(x^2 - 22x + 72) = -3(x-4)(x-18)$

Partition numbers for N'': $x = 18$

Sign chart for $N''(x)$:

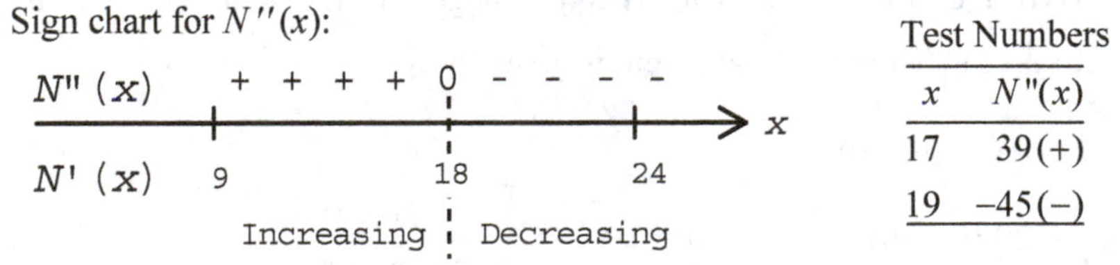

Test Numbers

x	$N''(x)$
17	39(+)
19	–45(–)

Thus, N' is increasing on (9, 18) and decreasing on (18, 24); the point of diminishing returns is $x = 18$; the maximum rate of change is $N'(18) = 972$.

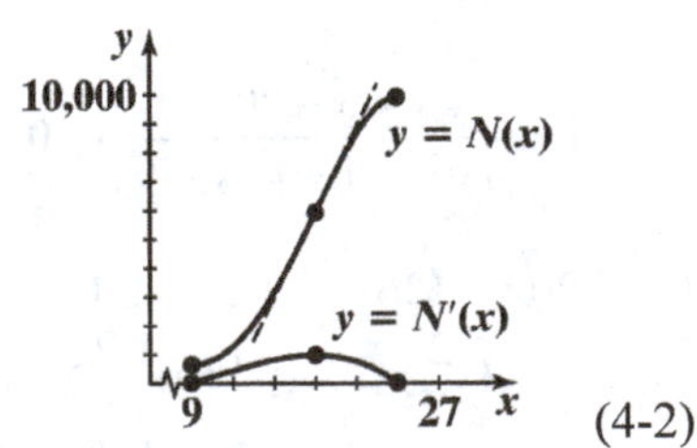

(4-2)

71. (A)

```
CubicReg
 y=ax³+bx²+cx+d
 a=-.01
 b=.83
 c=-2.3
 d=221
```

(B) The regression equation found in (A) is:
$y(x) = -0.01x^3 + 0.83x^2 - 2.3x + 221$

The rate of change of sales with respect to the number of ads is:

$y'(x) = -0.03x^2 + 1.66x - 2.3$
$y''(x) = -0.06x + 1.66$

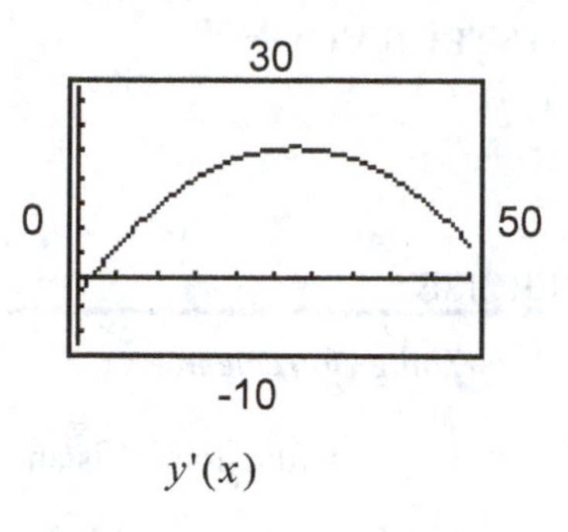

$y'(x)$

Critical value: $-0.06x + 1.66 = 0$
$x \approx 27.667$

From the graph, the absolute maximum of y' occurs at $x \approx 27.667$. Thus, 28 ads should be placed
each month. The expected number of sales is: $y(28) \approx 588$. (4-6)

72. $C(t) = 20t^2 - 120t + 800, 0 \le t \le 9$
$C'(t) = 40t - 120 = 40(t - 3)$
Critical value: $t = 3$
$C''(t) = 40$ and $C''(3) = 40 > 0$
Therefore, a local minimum occurs at $t = 3$.
$C(0) = 800$
$C(3) = 20(3^2) - 120(3) + 800 = 620$ Absolute minimum
$C(9) = 20(81) - 120(9) + 800 = 1340$
Therefore, the bacteria count will be at a minimum three days after a treatment. (4-6)

73. $N = 10 + 6t^2 - t^3, 0 \le t \le 5$
$\frac{dN}{dt} = 12t - 3t^2$

Now, find the critical values of the rate function $R(t)$:
$R(t) = \frac{dN}{dt} = 12t - 3t^2$

$R'(t) = \frac{dR}{dt} = \frac{d^2N}{dt^2} = 12 - 6t$

Critical value: $t = 2$
$R''(t) = -6$ and $R''(2) = -6 < 0$
$R(0) = 0$
$R(2) = 12$ Absolute maximum
$R(5) = -15$

Therefore, $R(t)$ has an absolute maximum at $t = 2$. The rate of increase will be a maximum after two years. (4-6)

5 INTEGRATION

EXERCISE 5-1

Things to remember:

1. A function F is an ANTIDERIVATIVE of f if $F'(x) = f(x)$.

2. THEOREM ON ANTIDERIVATIVES

   If the derivatives of two functions are equal on an open interval (a, b), then the functions can differ by at most a constant. Symbolically: If F and G are differentiable functions on the interval (a, b) and $F'(x) = G'(x)$ for all x in (a, b), then $F(x) = G(x) + k$ for some constant k.

3. The INDEFINITE INTEGRAL of $f(x)$, denoted

   $$\int f(x)dx,$$

   represents all antiderivatives of $f(x)$ and is given by

   $$\int f(x)dx = F(x) + C$$

   where $F(x)$ is any antiderivative of $f(x)$ and C is an arbitrary constant. The symbol $\int$ is called an INTEGRAL SIGN, the function $f(x)$ is called the INTEGRAND, and C is called the CONSTANT OF INTEGRATION.

4. Indefinite integration and differentiation are reverse operations (except for the addition of the constant of integration). This is expressed symbolically by:

   (a) $\frac{d}{dx}\left(\int f(x)dx\right) = f(x)$

   (b) $\int F'(x)dx = F(x) + C$

5. INDEFINITE INTEGRAL FORMULAS:

   (a) $\int x^n\,dx = \frac{x^{n+1}}{n+1} + C, n \neq -1$

   (b) $\int e^x\,dx = e^x + C$

   (c) $\int \frac{dx}{x} = \ln|x| + C, x \neq 0$

6. PROPERTIES OF INDEFINITE INTEGRALS:

   (a) $\int k\,f(x)dx = k\int f(x)dx$, k constant

   (b) $\int [f(x) \pm g(x)]\,dx = \int f(x)dx \pm \int g(x)dx$

1. $\frac{5}{x^4} = 5x^{-4}$.

3. $\frac{3x-2}{x^5} = 3x^{-4} - 2x^{-5}$.

5. $\sqrt{x} + \frac{5}{\sqrt{x}} = x^{1/2} + \frac{5}{x^{1/2}} = x^{1/2} + 5x^{-1/2}$.

7. $\sqrt[3]{x}(4 + x - 3x^2) = 4x^{1/3} + x^{4/3} - 3x^{7/3}$.

9. $\int 7\, dx = 7x + C$

Check: $\frac{d}{dx}(7x + C) = 7$

11. $\int 8x\, dx = 8\int x\, dx = 8\frac{x^2}{2} + C = 4x^2 + C$

Check: $\frac{d}{dx}\left(4x^2 + C\right) = 8x$

13. $\int 9x^2\, dx = 9\int x^2\, dx = 9\frac{x^3}{3} + C = 3x^3 + C$

Check: $\frac{d}{dx}(3x^3 + C) = 9x^2$

15. $\int x^5\, dx = \frac{1}{6}x^6 + C$

Check: $\frac{d}{dx}\left(\frac{1}{6}x^6 + C\right) = x^5$

17. $\int x^{-3}\, dx = \frac{x^{-2}}{-2} + C$

Check: $\frac{d}{dx}\left(\frac{x^{-2}}{-2} + C\right) = x^{-3}$

19. $\int 10x^{3/2}\, dx = 10\int x^{3/2}\, dx = 10\frac{x^{5/2}}{5/2} + C = 4x^{5/2} + C$

Check: $\frac{d}{dx}\left(4x^{5/2} + C\right) = 10x^{3/2}$

21. $\int \frac{3}{z}\, dz = 3\int \frac{1}{z}\, dz = 3\ln|z| + C$

Check: $\frac{d}{dz}(3\ln|z| + C) = \frac{3}{z}$

23. $\int 16e^u\, du = 16\int e^u\, du = 16e^u + C$

Check: $\frac{d}{du}\left(16e^u + C\right) = 16e^u$

25. $F(x) = (x+1)(x+2) = x^2 + 3x + 2,\ F'(x) = 2x + 3 = f(x);$ yes.

27. $F(x) = 1 + x\ln x,\ F'(x) = 1 + \ln x = f(x);$ yes.

29. $F(x) = \frac{(2x+1)^3}{3},\ F'(x) = (2x+1)^2(2) \neq (2x+1)^2 = f(x);$ no.

31. $F(x) = e^{x^3/3},\ F'(x) = e^{x^3/3}(x^2) = x^2e^{x^3/3} \neq e^{x^2} = f(x);$ no.

33. True: $f(x) = \pi,\ f'(x) = 0 = k(x)$

35. False: $f(x) = x^{-1}$ is a counter-example

37. True: $h(x) = 5e^x,\ h'(x) = 5e^x = h(x)$

39. The graphs in this set ARE NOT graphs from a family of antiderivative functions since the graphs are not vertical translations of each other.

41. The graphs in this set could be graphs from a family of antiderivative functions since they appear to be vertical translations of each other.

43. $\int 5x(1-x)\,dx = 5\int x(1-x)\,dx = 5\int (x-x^2)\,dx = 5\int x\,dx - 5\int x^2\,dx = \frac{5x^2}{2} - \frac{5x^3}{3} + C$

<u>Check:</u> $\frac{d}{dx}\left(\frac{5x^2}{2} - \frac{5x^3}{3} + C\right) = 5x - 5x^2 = 5x(1-x)$

45. $\int \frac{du}{\sqrt{u}} = \int \frac{du}{u^{1/2}} = \int u^{-1/2}\,du = \frac{u^{(-1/2)+1}}{-1/2+1} + C = \frac{u^{1/2}}{1/2} + C = 2u^{1/2} + C$ or $2\sqrt{u} + C$

<u>Check:</u> $\frac{d}{du}\left(2u^{1/2} + C\right) = 2\left(\frac{1}{2}\right)u^{-1/2} = \frac{1}{u^{1/2}} = \frac{1}{\sqrt{u}}$

47. $\int \frac{dx}{4x^3} = \frac{1}{4}\int x^{-3}\,dx = \frac{1}{4}\cdot\frac{x^{-2}}{-2} + C = \frac{-x^{-2}}{8} + C$

<u>Check:</u> $\frac{d}{dx}\left(\frac{-x^{-2}}{8} + C\right) = \frac{1}{8}(-2)(-x^{-3}) = \frac{1}{4}x^{-3} = \frac{1}{4x^3}$

49. $\int \frac{4+u}{u}\,du = \int\left(\frac{4}{u} + 1\right)du = 4\int \frac{1}{u}\,du + \int 1\,du = 4\ln|u| + u + C$

<u>Check:</u> $\frac{d}{du}(4\ln|u| + u + C) = \frac{4}{u} + 1 = \frac{4+u}{u}$

51. $\int (5e^z + 4)\,dz = 5\int e^z\,dz + 4\int dz = 5e^z + 4z + C$

<u>Check:</u> $\frac{d}{dz}(5e^z + 4z + C) = 5e^z + 4$

53. $\int\left(3x^2 - \frac{2}{x^2}\right)dx = \int 3x^2\,dx - \int \frac{2}{x^2}\,dx = 3\int x^2\,dx - 2\int x^{-2}\,dx = 3\cdot\frac{x^3}{3} - \frac{2x^{-1}}{-1} + C = x^3 + 2x^{-1} + C$

<u>Check:</u> $\frac{d}{dx}(x^3 + 2x^{-1} + C) = 3x^2 - 2x^{-2} = 3x^2 - \frac{2}{x^2}$

55. $C'(x) = 6x^2 - 4x$

$C(x) = \int (6x^2 - 4x)\,dx = 6\int x^2\,dx - 4\int x\,dx = \frac{6x^3}{3} - \frac{4x^2}{2} + C = 2x^3 - 2x^2 + C$

Given $C(0) = 3000$: $3000 = 2(0^3) - 2(0^2) + C$. Hence, $C = 3000$ and $C(x) = 2x^3 - 2x^2 + 3000$.

57. $\frac{dx}{dt} = \frac{20}{\sqrt{t}}$

$x = \int \frac{20}{\sqrt{t}}\,dt = 20\int t^{-1/2}\,dt = 20\frac{t^{1/2}}{1/2} + C = 40\sqrt{t} + C$

Given $x(1) = 40$: $40 = 40\sqrt{1} + C$ or $40 = 40 + C$. Hence, $C = 0$ and $x = 40\sqrt{t}$.

59. $\frac{dy}{dx} = 2x^{-2} + 3x^{-1} - 1$

$$y = \int (2x^{-2} + 3x^{-1} - 1)\,dx = 2\int x^{-2}\,dx + 3\int x^{-1}\,dx - \int dx = \frac{2x^{-1}}{-1} + 3\ln|x| - x + C$$
$$= \frac{-2}{x} + 3\ln|x| - x + C$$

Given $y(1) = 0$: $0 = -\frac{2}{1} + 3\ln|1| - 1 + C$. Hence, $C = 3$ and $y = -\frac{2}{x} + 3\ln|x| - x + 3$.

61. $\frac{dx}{dt} = 4e^t - 2$

$$x = \int (4e^t - 2)\,dt = 4\int e^t\,dt - 2\int dt = 4e^t - 2t + C$$

Given $x(0) = 1$: $1 = 4e^0 - 2(0) + C = 4 + C$. Hence, $C = -3$ and $x = 4e^t - 2t - 3$.

63. $\frac{dy}{dx} = 4x - 3$

$$y = \int (4x - 3)\,dx = 4\int x\,dx - 3\int dx = \frac{4x^2}{2} - 3x + C = 2x^2 - 3x + C$$

Given $y(2) = 3$: $3 = 2\cdot 2^2 - 3\cdot 2 + C$. Hence, $C = 1$ and $y = 2x^2 - 3x + 1$.

65. $\int \frac{2x^4 - x}{x^3}\,dx = \int \left(\frac{2x^4}{x^3} - \frac{x}{x^3}\right)dx = 2\int x\,dx - \int x^{-2}\,dx = \frac{2x^2}{2} - \frac{x^{-1}}{-1} + C = x^2 + x^{-1} + C$

67. $\int \frac{x^5 - 2x}{x^4}\,dx = \int \left(\frac{x^5}{x^4} - \frac{2x}{x^4}\right)dx = \int x\,dx - 2\int x^{-3}\,dx = \frac{x^2}{2} - \frac{2x^{-2}}{-2} + C = \frac{x^2}{2} + x^{-2} + C$

69. $\int \frac{x^2e^x - 2x}{x^2}\,dx = \int \left(\frac{x^2e^x}{x^2} - \frac{2x}{x^2}\right)dx = \int e^x\,dx - 2\int x^{-1}\,dx = e^x - 2\ln|x| + C$

71. $\frac{d}{dx}\left[\int x^3\,dx\right] = x^3$ [by 4(a)]

73. $\int \frac{d}{dx}(x^4 + 3x^2 + 1)\,dx = x^4 + 3x^2 + 1 + C = x^4 + 3x^2 + C_1$ [by 4(b)]
($C_1 = 1 + C$ is an arbitrary constant since C is arbitrary)

75. $\frac{d}{dx}\left(\frac{x^{n+1}}{n+1} + C\right) = x^n$

77. Assume $x > 0$. Then $|x| = x$ and $\ln|x| = \ln x$.
Therefore, $\frac{d}{dx}(\ln|x| + C) = \frac{d}{dx}(\ln x + C) = \frac{1}{x}$.

79. Assume $\int f(x)\,dx = F(x) + C_1$ and $\int g(x)\,dx = G(x) + C_2$.
Then, $\frac{d}{dx}(F(x) + C_1) = f(x)$, $\frac{d}{dx}(G(x) + C_2) = g(x)$, and

$$\frac{d}{dx}(F(x)+C_1+G(x)+C_2)=\frac{d}{dx}(F(x)+C_1)+\frac{d}{dx}(G(x)+C_2)=f(x)+g(x).$$

81. $\overline{C}'(x)=-\frac{1,000}{x^2}$

$$\overline{C}(x)=\int \overline{C}'(x)\,dx=\int -\frac{1,000}{x^2}\,dx=-1,000\int x^{-2}\,dx=-1,000\frac{x^{-1}}{-1}+C=\frac{1,000}{x}+C$$

Given $\overline{C}(100)=25$: $\frac{1,000}{100}+C=25$

$C=15$

Thus, $\overline{C}(x)=\frac{1,000}{x}+15.$

Cost function: $C(x)=x\overline{C}(x)=15x+1,000$
Fixed costs: $C(0)=\$1,000$

83. (A) The cost function increases from 0 to 8. The graph is concave downward from 0 to 4 and concave upward from 4 to 8. There is an inflection point at $x = 4$.

(B) $C(x)=\int C'(x)\,dx=\int(3x^2-24x+53)\,dx=3\int x^2\,dx-24\int x\,dx+53\int dx=x^3-12x^2+53x+K$

Since $C(0)=30$, we have $K=30$ and $C(x)=x^3-12x^2+53x+30$.

$C(4)=4^3-12(4)^2+53(4)+30=\$114$ thousand

$C(8)=8^3-12(8)^2+53(8)+30=\$198$ thousand

(C)

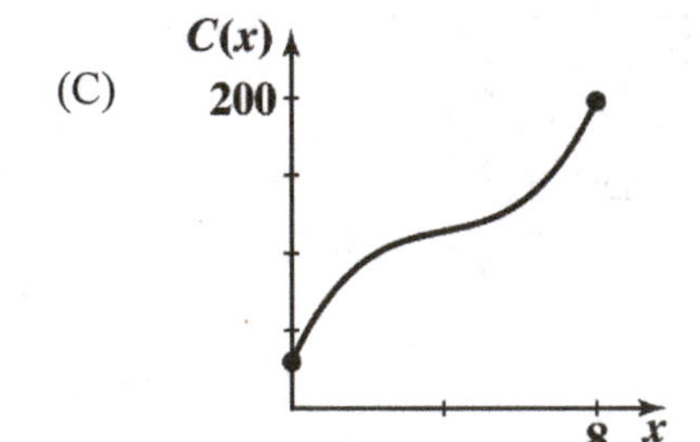

(D) Manufacturing plants are often inefficient at low and high levels of production.

85. $S'(t)=-24t^{1/3}$

$$S(t)=\int -24t^{1/3}\,dt=-24\int t^{1/3}\,dt=-24\frac{t^{4/3}}{4/3}+C=-18t^{4/3}+C$$

Given $S(0)=1200=-18(0)+C$. Hence, $C=1200$ and $S(t)=1,200-18t^{4/3}$.
Now, we want to find t such that $S(t) = 300$, that is:

$$1,200-18t^{4/3}=300$$
$$-18t^{4/3}=-900$$
$$t^{4/3}=50$$
$$t=50^{3/4}\approx 18.803,$$

Thus, the company should manufacture SUV's for 19 months.

87. $S'(t) = -24t^{1/3} - 70$

$$S(t) = \int S'(t)\,dt = \int (-24t^{1/3} - 70)\,dt = -24\frac{t^{4/3}}{4/3} - 70t + C = -18t^{4/3} - 70t + C$$

Given $S(0) = 1{,}200$ implies $C = 1{,}200$ and $S(t) = 1{,}200 - 18t^{4/3} - 70t$

Graphing $y_1 = 1{,}200 - 18t^{4/3} - 70t$, $y_2 = 800$ on $0 \le x \le 10$, $0 \le y \le 1000$, we see that the point of intersection is $t \approx 4.0527553$, $y = 800$. So we get $t \approx 4.05$ months.

89. $L'(x) = g(x) = 2400x^{-1/2}$

$$L(x) = \int g(x)\,dx = \int 2{,}400x^{-1/2}\,dx = 2{,}400\int x^{-1/2}\,dx = 2{,}400\,\frac{x^{1/2}}{1/2} + C = 48{,}000\,x^{1/2} + C$$

Given $L(16) = 19{,}200$: $19{,}200 = 4800(16)^{1/2} + C = 19{,}200 + C$. Hence, $C = 0$ and $L(x) = 4800x^{1/2}$.
$L(25) = 4800(25)^{1/2} = 4800(5) = 24{,}000$ labor hours.

91. $\frac{dW}{dh} = 0.0015h^2$

$$W = \int 0.0015h^2\,dh = 0.0015\int h^2\,dh = 0.0015\frac{h^3}{3} + C = 0.0005h^3 + C$$

Given $W(60) = 108$: $108 = 0.0005(60)^3 + C$ or $108 = 108 + C$.
Hence, $C = 0$ and $W(h) = 0.0005h^3$. Now 5'10" = 70" and $W(70) = 0.0005(70)^3 = 171.5$ lb.

93. $\frac{dN}{dt} = 400 + 600\sqrt{t}$, $0 \le t \le 9$

$$N = \int\left(400 + 600\sqrt{t}\right)dt = 400\int dt + 600\int t^{1/2}\,dt = 400t + 600\frac{t^{3/2}}{3/2} + C = 400t + 400t^{3/2} + C$$

Given $N(0) = 5000$: $5000 = 400(0) + 400(0)^{3/2} + C$. Hence, $C = 5000$ and
$N(t) = 400t + 400t^{3/2} + 5000$.
$N(9) = 400(9) + 400(9)^{3/2} + 5000 = 3600 + 10{,}800 + 5000 = 19{,}400$

EXERCISE 5-2

Things to remember:

1. REVERSING THE CHAIN RULE

   The chain rule formula for differentiating a composite function:

   $$\frac{d}{dx}f[g(x)] = f'[g(x)]g'(x),$$

   yields the integral formula

   $$\int f'[g(x)]\,g'(x)\,dx = f[g(x)] + C$$

2. GENERAL INDEFINITE INTEGRAL FORMULAS (Version 1)

(a) $\int [f(x)]^n f'(x)\,dx = \frac{[f(x)]^{n+1}}{n+1} + C, n \neq -1$

(b) $\int e^{f(x)} f'(x)\,dx = e^{f(x)} + C$

(c) $\int \frac{1}{f(x)} f'(x)\,dx = \ln|f(x)| + C$

3. DIFFERENTIALS

If $y = f(x)$ defines a differentiable function, then:

(a) The DIFFERENTIAL dx of the independent variable x is an arbitrary real number.

(b) The DIFFERENTIAL dy of the dependent variable y is defined as the product of $f'(x)$ and dx; that is: $dy = f'(x)dx$.

4. GENERAL INDEFINITE INTEGRAL FORMULAS (Version 2)

(a) $\int u^n du = \frac{u^{n+1}}{n+1} + C, n \neq -1$

(b) $\int e^u du = e^u + C$

(c) $\int \frac{1}{u} du = \ln|u| + C$

These formulas are valid if u is an independent variable or if u is a function of another variable and du is its differential with respect to that variable.

5. INTEGRATION BY SUBSTITUTION

Step 1. Select a substitution that appears to simplify the integrand. In particular, try to select u so that du is a factor in the integrand.

Step 2. Express the integrand entirely in terms of u and du, completely eliminating the original variable and its differential.

Step 3. Evaluate the new integral, if possible.

Step 4. Express the antiderivative found in Step 3 in terms of the original variable.

---

1. $f(x) = (5x+1)^{10}$, $f'(x) = 10(5x+1)^9(5) = 50(5x+1)^9$.

3. $f(x) = (x^2+1)^7$, $f'(x) = 7(x^2+1)^6(2x) = 14x(x^2+1)^6$.

5. $f(x) = e^{x^2}$, $f'(x) = e^{x^2}(2x) = 2xe^{x^2}$.

7. $f(x) = \ln(x^4 - 10)$, $f'(x) = \frac{1}{x^4 - 10}(4x^3) = \frac{4x^3}{x^4 - 10}$.

9. $\int (3x+5)^2(3)\,dx = \int u^2\,du = \frac{1}{3}u^3 + C = \frac{1}{3}(3x+5)^3 + C$ [Formula 4a]

Let $u = 3x + 5$
Then $du = 3\,dx$

Check: $\frac{d}{dx}\left[\frac{1}{3}(3x+5)^3 + C\right] = \frac{1}{3}\cdot 3(3x+5)^2\frac{d}{dx}(3x+5) = (3x+5)^2(3)$

11. $\int (x^2-1)^5(2x)\,dx = \int u^5\,du = \frac{1}{6}u^6 + C = \frac{1}{6}(x^2-1)^6 + C$ [Formula 4a]

Let $u = x^2 - 1$
Then $du = 2x\,dx$

Check: $\frac{d}{dx}\left[\frac{1}{6}(x^2-1)^6 + C\right] = \frac{1}{6}\cdot 6(x^2-1)^5\frac{d}{dx}(x^2-1) = (x^2-1)^5(2x)$

13. $\int (5x^3+1)^{-3}(15x^2)\,dx = \int u^{-3}\,du = \frac{u^{-2}}{-2} + C = -\frac{1}{2}(5x^3+1)^{-2} + C$ [Formula 4a]

Let $u = 5x^3 + 1$
Then $du = 15x^2\,dx$

Check: $\frac{d}{dx}\left[-\frac{1}{2}(5x^3+1)^{-2} + C\right] = -\frac{1}{2}(-2)(5x^3+1)^{-3}\frac{d}{dx}(5x^3+1) = (5x^3+1)^{-3}(15x^2)$

15. $\int e^{5x}(5)\,dx = \int e^u\,du = e^u + C = e^{5x} + C$ [Formula 4b]

Let $u = 5x$
Then $du = 5\,dx$

Check: $\frac{d}{dx}(e^{5x} + C) = e^{5x}\frac{d}{dx}(5x) = e^{5x}(5)$

17. $\int \frac{1}{1+x^2}(2x)\,dx = \int \frac{1}{u}\,du = \ln|u| + C = \ln|1+x^2| + C = \ln(1+x^2) + C \quad (1+x^2 > 0)$ [Formula 4c]

Let $u = 1 + x^2$.
Then $du = 2x\,dx$

Check: $\frac{d}{dx}(\ln(1+x^2) + C) = \frac{1}{1+x^2}\frac{d}{dx}(1+x^2) = \frac{1}{1+x^2}(2x)$

19. $\int \sqrt{1+x^4}\,(4x^3)\,dx = \int \sqrt{u}\,du = \int u^{1/2}\,du = \frac{u^{3/2}}{3/2} + C = \frac{2}{3}u^{3/2} + C = \frac{2}{3}(1+x^4)^{3/2} + C$

Let $u = 1 + x^4$.
Then $du = 4x^3\,dx$

Check: $\frac{d}{dx}\left[\frac{2}{3}(1+x^4)^{3/2} + C\right] = \frac{3}{2}\cdot\frac{2}{3}(1+x^4)^{1/2}\frac{d}{dx}(1+x^4) = (1+x^4)^{1/2}(4x^3) = \sqrt{1+x^4}\,(4x^3)$

21. $\int (x+3)^{10}\,dx = \int u^{10}\,du = \frac{1}{11}u^{11} + C = \frac{1}{11}(x+3)^{11} + C$

Let $u = x + 3$
Then $du = dx$

Check: $\frac{d}{dx}\left[\frac{1}{11}(x+3)^{11}+C\right] = \frac{1}{11}\cdot 11(x+3)^{10}\frac{d}{dx}(x+3) = (x+3)^{10}$

23. $\int(6t-7)^{-2}\,dt = \int(6t-7)^{-2}\frac{6}{6}dt = \frac{1}{6}\int(6t-7)^{-2}\,6\,dt = \frac{1}{6}\int u^{-2}\,du = \frac{1}{6}\cdot\frac{u^{-1}}{-1}+C = -\frac{1}{6}(6t-7)^{-1}+C$

Let $u = 6t-7$
Then $du = 6\,dt$

Check: $\frac{d}{dt}\left[-\frac{1}{6}(6t-7)^{-1}+C\right] = -\frac{1}{6}(-1)(6t-7)^{-2}\frac{d}{dt}(6t-7) = \frac{1}{6}(6t-7)^{-2}(6) = (6t-7)^{-2}$

25. $\int(t^2+1)^5\,t\,dt = \int(t^2+1)^5\frac{2}{2}t\,dt = \frac{1}{2}\int(t^2+1)^5\,2t\,dt = \frac{1}{2}\int u^5\,du = \frac{1}{2}\cdot\frac{1}{6}u^6+C = \frac{1}{12}(t^2+1)^6+C$

Let $u = t^2+1$
Then $du = 2t\,dt$

Check: $\frac{d}{dt}\left[\frac{1}{12}(t^2+1)^6+C\right] = \frac{1}{12}\cdot 6(t^2+1)^5\frac{d}{dt}(t^2+1) = \frac{1}{2}(t^2+1)^5(2t) = (t^2+1)^5 t$

27. $\int x e^{x^2}\,dx = \int e^{x^2}\frac{2}{2}x\,dx = \frac{1}{2}\int e^{x^2}(2x)\,dx = \frac{1}{2}\int e^u\,du = \frac{1}{2}e^u+C = \frac{1}{2}e^{x^2}+C$

Let $u = x^2$
Then $du = 2x\,dx$

Check: $\frac{d}{dx}\left(\frac{1}{2}e^{x^2}+C\right) = \frac{1}{2}e^{x^2}\frac{d}{dx}(x^2) = \frac{1}{2}e^{x^2}(2x) = xe^{x^2}$

29. $\int\frac{1}{5x+4}dx = \int\frac{1}{5x+4}\cdot\frac{5}{5}dx = \frac{1}{5}\int\frac{1}{5x+4}5\,dx = \frac{1}{5}\int\frac{1}{u}du = \frac{1}{5}\ln|u|+C = \frac{1}{5}\ln|5x+4|+C$

Let $u = 5x+4$
Then $du = 5\,dx$

Check: $\frac{d}{dx}\left[\frac{1}{5}\ln|5x+4|+C\right] = \frac{1}{5}\cdot\frac{1}{5x+4}\frac{d}{dx}(5x+4) = \frac{1}{5}\cdot\frac{1}{5x+4}\cdot 5 = \frac{1}{5x+4}$

31. $\int e^{1-t}\,dt = \int e^{1-t}\left(\frac{-1}{-1}\right)dt = \frac{1}{-1}\int e^{1-t}(-1)\,dt = -\int e^u\,du = -e^u+C = -e^{1-t}+C$

Let $u = 1-t$
Then $du = -dt$

Check: $\frac{d}{dt}[-e^{1-t}+C] = -e^{1-t}\frac{d}{dt}(1-t) = -e^{1-t}(-1) = e^{1-t}$

33. $\int\frac{t}{(3t^2+1)^4}dt = \int(3t^2+1)^{-4}\,t\,dt = \int(3t^2+1)^{-4}\frac{6}{6}t\,dt = \frac{1}{6}\int(3t^2+1)^{-4}\,6t\,dt = \frac{1}{6}\int u^{-4}\,du$

Let $u = 3t^2+1$. Then $du = 6t\,dt$. $\quad = \frac{1}{6}\cdot\frac{u^{-3}}{-3}+C = \frac{-1}{18}(3t^2+1)^{-3}+C$

Check: $\frac{d}{dt}\left[\frac{-1}{18}(3t^2+1)^{-3}+C\right] = \left(\frac{-1}{18}\right)(-3)(3t^2+1)^{-4}(6t) = \frac{t}{(3t^2+1)^4}$

35. $\int x\sqrt{x+4}\,dx$

Let $u = x + 4$. Then $du = dx$ and $x = u - 4$.

$$\int x\sqrt{x+4}\,dx = \int (u-4)u^{1/2}\,du = \int \left(u^{3/2} - 4u^{1/2}\right)du = \frac{u^{5/2}}{5/2} - \frac{4u^{3/2}}{3/2} + C = \frac{2}{5}u^{5/2} - \frac{8}{3}u^{3/2} + C$$

$$= \frac{2}{5}(x+4)^{5/2} - \frac{8}{3}(x+4)^{3/2} + C \text{ (since } u = x+4)$$

Check: $\frac{d}{dx}\left[\frac{2}{5}(x+4)^{5/2} - \frac{8}{3}(x+4)^{3/2} + C\right] = \frac{2}{5}\left(\frac{5}{2}\right)(x+4)^{3/2}(1) - \frac{8}{3}\left(\frac{3}{2}\right)(x+4)^{1/2}(1)$

$$= (x+4)^{3/2} - 4(x+4)^{1/2} = (x+4)^{1/2}[(x+4) - 4]$$
$$= x\sqrt{x+4}$$

37. $\int \frac{x}{\sqrt{x-3}}\,dx$

Let $u = x - 3$. Then $du = dx$ and $x = u + 3$.

$$\int \frac{x}{\sqrt{x-3}}\,dx = \int \frac{u+3}{u^{1/2}}\,du = \int \left(u^{1/2} + 3u^{-1/2}\right)du = \frac{u^{3/2}}{3/2} + \frac{3u^{1/2}}{1/2} + C = \frac{2}{3}u^{3/2} + 6u^{1/2} + C$$

$$= \frac{2}{3}(x-3)^{3/2} + 6(x-3)^{1/2} + C \text{ (since } u = x-3)$$

Check: $\frac{d}{dx}\left[\frac{2}{3}(x-3)^{3/2} + 6(x-3)^{1/2} + C\right] = \frac{2}{3}\left(\frac{3}{2}\right)(x-3)^{1/2}(1) + 6\left(\frac{1}{2}\right)(x-3)^{-1/2}(1)$

$$= (x-3)^{1/2} + \frac{3}{(x-3)^{1/2}} = \frac{x-3+3}{(x-3)^{1/2}} = \frac{x}{\sqrt{x-3}}$$

39. $\int x(x-4)^9\,dx$

Let $u = x - 4$. Then $du = dx$ and $x = u + 4$.

$$\int x(x-4)^9\,dx = \int (u+4)u^9\,du = \int (u^{10} + 4u^9)\,du = \frac{u^{11}}{11} + \frac{4u^{10}}{10} + C = \frac{(x-4)^{11}}{11} + \frac{2}{5}(x-4)^{10} + C$$

Check: $\frac{d}{dx}\left[\frac{(x-4)^{11}}{11} + \frac{2}{5}(x-4)^{10} + C\right] = \frac{1}{11}(11)(x-4)^{10}(1) + \frac{2}{5}(10)(x-4)^9(1)$

$$= (x-4)^9[(x-4) + 4] = x(x-4)^9$$

41. $\int e^{2x}(1+e^{2x})^3\,dx$

Let $u = 1 + e^{2x}$. Then $du = 2e^{2x}dx$.

$$\int e^{2x}(1+e^{2x})^3\,dx = \int \left(1+e^{2x}\right)^3 \frac{2}{2}e^{2x}\,dx = \frac{1}{2}\int \left(1+e^{2x}\right)^3 2e^{2x}\,dx = \frac{1}{2}\int u^3\,du = \frac{1}{2}\cdot\frac{u^4}{4} + C$$

$$= \frac{1}{8}(1+e^{2x})^4 + C$$

Check: $\frac{d}{dx}\left[\frac{1}{8}(1+e^{2x})^4 + C\right] = \left(\frac{1}{8}\right)(4)(1+e^{2x})^3 e^{2x}(2) = e^{2x}(1+e^{2x})^3$

43. $\int \frac{1+x}{4+2x+x^2}dx$ Let $u = 4 + 2x + x^2$. Then $du = (2 + 2x)\,dx = 2(1 + x)\,dx$.

$$\int \frac{1+x}{4+2x+x^2}dx = \int \frac{1}{4+2x+x^2}\cdot\frac{2(1+x)}{2}dx = \frac{1}{2}\int \frac{1}{4+2x+x^2}2(1+x)\,dx = \frac{1}{2}\int \frac{1}{u}du = \frac{1}{2}\ln|u| + C$$

$$= \frac{1}{2}\ln|4 + 2x + x^2| + C$$

Check: $\frac{d}{dx}\left[\frac{1}{2}\ln\left|4+2x+x^2\right|+C\right] = \left(\frac{1}{2}\right)\frac{1}{4+2x+x^2}(2+2x) = \frac{1+x}{4+2x+x^2}$

45. $\int 5(5x+3)\,dx$

(1) By substitution: Let $u = 5x+3$. Then $du = 5\,dx$.

$$\int 5(5x+3)\,dx = \int u\,du = \frac{u^2}{2}+C = \frac{1}{2}(5x+3)^2+C = \frac{1}{2}(25x^2+30x+9)+C$$

$$= \frac{25}{2}x^2+15x+K \quad (K = C+9/2)$$

(2) Expanding the integrand: $\int 5(5x+3)\,dx = \int (25x+15)\,dx = \frac{25}{2}x^2+15x+C$.

47. $\int 2x(x^2-1)\,dx$

(1) By substitution: Let $u = x^2-1$. Then $du = 2x\,dx$.

$$\int 2x(x^2-1)\,dx = \int u\,du = \frac{u^2}{2}+C = \frac{1}{2}(x^2-1)^2+C = \frac{1}{2}(x^4-2x^2+1)+C$$

$$= \frac{1}{2}x^4-x^2+K \quad (K = C+1/2).$$

(2) Expanding the integrand: $\int 2x(x^2-1)\,dx = \int (2x^3-2x)\,dx = 2\frac{x^4}{4}-x^2+C = \frac{1}{2}x^4-x^2+C$.

49. $\int 5x^4(x^5)^4\,dx$

(1) By substitution: Let $u = x^5$. Then $du = 5x^4\,dx$.

$$\int 5x^4(x^5)^4\,dx = \int u^4\,du = \frac{u^5}{5}+C = \frac{1}{5}(x^5)^5+C = \frac{1}{5}x^{25}+C.$$

(2) Expanding the integrand: $\int 5x^4(x^5)^4\,dx = \int 5x^{24}\,dx = 5\frac{x^{25}}{25}+C = \frac{1}{5}x^{25}+C.$

51. $F(x) = x^2e^x,\ F'(x) = x^2e^x + e^x(2x) = x^2e^x + 2xe^x \neq 2xe^x = f(x);$ no.

53. $F(x) = (x^2+4)^6,\ F'(x) = 6(x^2+4)^5(2x) = 12x(x^2+4) = f(x);$ yes.

55. $F(x) = e^{2x}+4,\ F'(x) = e^{2x}(2) = 2e^{2x} \neq e^{2x} = f(x);$ no.

57. $F(x) = 0.5(\ln x)^2+10,\ F'(x) = 2(0.5)(\ln x)^1\left(\frac{1}{x}\right) = \frac{\ln x}{x} = f(x);$ yes.

59. $\int x\sqrt{3x^2+7}\,dx$

Let $u = 3x^2 + 7$. Then $du = 6x\,dx$.

$$\int x\sqrt{3x^2+7}\,dx = \int (3x^2+7)^{1/2}x\,dx = \int (3x^2+7)^{1/2}\frac{6}{6}x\,dx = \frac{1}{6}\int u^{1/2}du = \frac{1}{6}\cdot\frac{u^{3/2}}{3/2} + C$$
$$= \frac{1}{9}(3x^2+7)^{3/2} + C$$

Check: $\frac{d}{dx}\left[\frac{1}{9}(3x^2+7)^{3/2} + C\right] = \frac{1}{9}\left(\frac{3}{2}\right)(3x^2+7)^{1/2}(6x) = x(3x^2+7)^{1/2}$

61. $\int x(x^3+2)^2\,dx = \int x(x^6+4x^3+4)\,dx = \int (x^7+4x^4+4x)\,dx = \frac{x^8}{8} + \frac{4}{5}x^5 + 2x^2 + C$

Check: $\frac{d}{dx}\left[\frac{x^8}{8} + \frac{4}{5}x^5 + 2x^2 + C\right] = x^7 + 4x^4 + 4x = x(x^6+4x^3+4) = x(x^3+2)^2$

63. $\int x^2(x^3+2)^2\,dx$

Let $u = x^3 + 2$. Then $du = 3x^2\,dx$.

$$\int x^2(x^3+2)^2\,dx = \int (x^3+2)^2\frac{3x^2}{3}dx = \frac{1}{3}\int (x^3+2)^2 3x^2\,dx = \frac{1}{3}\int u^2\,du = \frac{1}{3}\cdot\frac{u^3}{3} + C$$
$$= \frac{1}{9}u^3 + C = \frac{1}{9}(x^3+2)^3 + C$$

Check: $\frac{d}{dx}\left[\frac{1}{9}(x^3+2)^3 + C\right] = \frac{1}{9}(3)(x^3+2)^2(3x^2) = x^2(x^3+2)^2$

65. $\int \frac{x^3}{\sqrt{2x^4+3}}dx$

Let $u = 2x^4 + 3$. Then $du = 8x^3dx$.

$$\int \frac{x^3}{\sqrt{2x^4+3}}dx = \int (2x^4+3)^{-1/2}x^3dx = \int (2x^4+3)^{-1/2}\frac{8}{8}x^3dx = \frac{1}{8}\int u^{-1/2}du = \frac{1}{8}\cdot\frac{u^{1/2}}{1/2} + C$$
$$= \frac{1}{4}(2x^4+3)^{1/2} + C$$

Check: $\frac{d}{dx}\left[\frac{1}{4}(2x^4+3)^{1/2} + C\right] = \frac{1}{4}\left(\frac{1}{2}\right)(2x^4+3)^{-1/2}(8x^3) = \frac{x^3}{(2x^4+3)^{1/2}}$

67. $\int \frac{(\ln x)^3}{x}dx$

Let $u = \ln x$. Then $du = \frac{1}{x}\,dx$.

$$\int \frac{(\ln x)^3}{x}dx = \int u^3du = \frac{u^4}{4} + C = \frac{(\ln x)^4}{4} + C$$

Check: $\frac{d}{dx}\left[\frac{(\ln x)^4}{4} + C\right] = \frac{1}{4}(4)(\ln x)^3\cdot\frac{1}{x} = \frac{(\ln x)^3}{x}$

69. $\int \frac{1}{x^2} e^{-1/x}\, dx$

Let $u = \frac{-1}{x} = -x^{-1}$. Then $du = \frac{1}{x^2}\, dx$.

$\int \frac{1}{x^2} e^{-1/x}\, dx = \int e^u du = e^u + C = e^{-1/x} + C$

Check: $\frac{d}{dx}[e^{-1/x} + C] = e^{-1/x}\left(\frac{1}{x^2}\right) = \frac{1}{x^2} e^{-1/x}$

71. $\frac{dx}{dt} = 7t^2(t^3 + 5)^6$

Let $u = t^3 + 5$. Then $du = 3t^2 dt$.

$x = \int 7t^2(t^3+5)^6\, dt = \frac{7}{3}\int (t^3+5)^6\, 3t^2\, dt = \frac{7}{3}\int u^6\, du = \frac{1}{3}u^7 + C = \frac{1}{3}(t^3+5)^7 + C$

73. $\frac{dy}{dt} = \frac{3t}{\sqrt{t^2 - 4}}$

Let $u = t^2 - 4$. Then $du = 2t\, dt$.

$y = \int \frac{3t}{(t^2-4)^{1/2}}\, dt = 3\int (t^2-4)^{-1/2}\, t\, dt = 3\int (t^2-4)^{-1/2}\frac{2}{2}t\, dt = \frac{3}{2}\int u^{-1/2}\, du = \frac{3}{2} \cdot \frac{u^{1/2}}{1/2} + C$

$= 3(t^2 - 4)^{1/2} + C$

75. $\frac{dp}{dx} = \frac{e^x + e^{-x}}{(e^x - e^{-x})^2}$

Let $u = e^x - e^{-x}$. Then $du = (e^x + e^{-x})\, dx$.

$p = \int \frac{e^x + e^{-x}}{(e^x - e^{-x})^2}\, dx = \int (e^x - e^{-x})^{-2}(e^x + e^{-x})\, dx = \int u^{-2}\, du = \frac{u^{-1}}{-1} + C = -(e^x - e^{-x})^{-1} + C$

77. $p'(x) = \frac{-6000}{(3x+50)^2}$

Let $u = 3x + 50$. Then $du = 3\, dx$.

$p(x) = \int \frac{-6000}{(3x+50)^2}\, dx = -6000\int (3x+50)^{-2}\, dx = -6000\int (3x+50)^{-2}\frac{3}{3}\, dx = -2000\int u^{-2}\, du$

$= -2000 \cdot \frac{u^{-1}}{-1} + C = \frac{2000}{3x+50} + C$

Given $p(150) = 8$:

$8 = \frac{2000}{(3 \cdot 150 + 50)} + C = \frac{2000}{500} + C = 4 + C$. Therefore, $C = 4$.

Thus, $p(x) = \frac{2000}{3x+50} + 4$.

Now

$$6.50 = \frac{2000}{3x+50} + 4$$
$$2.50 = \frac{2000}{3x+50}$$
$$2.50(3x+50) = 2000$$
$$7.50x + 125 = 2000$$
$$7.50x = 1875$$
$$x = 250$$

Thus, the demand is 250 bottles when the price is \$6.50.

79. $C'(x) = 12 + \frac{500}{x+1}, x > 0$

$$C(x) = \int\left(12 + \frac{500}{x+1}\right)dx = \int 12\,dx + 500\int\frac{1}{x+1}dx \quad (u = x+1,\ du = dx)$$
$$= 12x + 500\ln(x+1) + C$$

Now, $C(0) = 2000$. Thus, $C(x) = 12x + 500\ln(x+1) + 2000$. The average cost is:

$$\overline{C}(x) = 12 + \frac{500}{x}\ln(x+1) + \frac{2000}{x}$$

and

$$\overline{C}(1000) = 12 + \frac{500}{1000}\ln(1001) + \frac{2000}{1000} = 12 + \frac{1}{2}\ln(1001) + 2 \approx 17.45$$ or \$17.45 per pair of shoes

81. $S'(t) = 10 - 10e^{-0.1t}, 0 \le t \le 24$

(A) $S(t) = \int\left(10 - 10e^{-0.1t}\right)dt = \int 10\,dt - 10\int e^{-0.1t}\,dt = 10t - \frac{10}{-0.1}e^{-0.1t} + C = 10t + 100e^{-0.1t} + C$

Given $S(0) = 0$: $0 + 100e^{0} + C = 0$

$$100 + C = 0$$
$$C = -100$$

Total sales at time t:

$$S(t) = 10t + 100e^{-0.1t} - 100, \quad 0 \le t \le 24.$$

(B) $S(12) = 10(12) + 100e^{-0.1(12)} - 100$

$$= 20 + 100e^{-1.2} \approx 50$$

Total estimated sales for the first twelve months: \$50 million.

(C) On a graphing utility, solve

$$10t + 100e^{-0.1t} - 100 = 100$$

or

$$10t + 100e^{-0.1t} = 200$$

The result is: $t \approx 18.41$ months.

83. $Q(t) = \int R(t)\,dt = \int\left(\frac{100}{t+1} + 5\right)dt = 100\int\frac{1}{t+1}dt + \int 5\,dt = 100\ln(t+1) + 5t + C$

Given $Q(0) = 0$:

$0 = 100\ln(1) + 0 + C$

Thus, $C = 0$ and $Q(t) = 100 \ln(t + 1) + 5t,\ 0 \le t \le 20$.
$Q(9) = 100 \ln(9 + 1) + 5(9) = 100 \ln 10 + 45 \approx 275$ thousand barrels.

85. $W(t) = \int w(t)\,dt = \int 0.2\,e^{0.1t}\,dt = \frac{0.2}{0.1}\int e^{0.1t}(0.1)\,dt = 2e^{0.1t} + C$

Given $W(0) = 2$:

$2 = 2e^0 + C$.

Thus, $C = 0$ and $W(t) = 2e^{0.1t}$.

The weight of the culture after 8 hours is given by:

$W(8) = 2e^{0.1(8)} = 2e^{0.8} \approx 4.45$ grams.

87. $\frac{dN}{dt} = -\frac{2000t}{1+t^2},\ 0 \le t \le 10$

(A) To find the minimum value of $\frac{dN}{dt}$, calculate

$$\frac{d}{dt}\left(\frac{dN}{dt}\right) = \frac{d^2N}{dt^2} = -\frac{(1+t^2)(2000) - 2000t\ (2t)}{(1+t^2)^2} = -\frac{2000[1-t^2]}{(1+t^2)^2} = \frac{-2000(1-t)(1+t)}{(1+t^2)^2}$$

critical value: $t = 1$

Now $\left.\frac{dN}{dt}\right|_{t=0} = 0$

$\left.\frac{dN}{dt}\right|_{t=1} = -1{,}000$

$\left.\frac{dN}{dt}\right|_{t=10} = \frac{-20{,}000}{101} \approx -198.02$

Thus, the minimum value of $\frac{dN}{dt}$ is $-1{,}000$ bacteria/ml per day.

(B) $N(t) = \int \frac{-2{,}000t}{1+t^2}\,dt$

Let $u = 1 + t^2$. Then $du = 2t\,dt$

$$N(t) = \int \frac{-2{,}000t}{1+t^2}\,dt = -1{,}000\int \frac{2t}{1+t^2}\,dt = -1{,}000\int \frac{1}{u}\,du = -1{,}000 \ln|u| + C$$

$$= -1{,}000 \ln(1 + t^2) + C$$

Given $N(0) = 5{,}000$:
$5{,}000 = -1{,}000 \ln(1) + C = C$ $(\ln 1 = 0)$

Thus, $C = 5{,}000$ and $N(t) = 5{,}000 - 1{,}000 \ln(1 + t^2)$

Now, $N(10) = 5{,}000 - 1{,}000 \ln(1 + 10^2) = 5{,}000 - 1{,}000 \ln(101) \approx 385$ bacteria/ml

(C) Set $N(t) = 1{,}000$ and solve for t:

$1{,}000 = 5{,}000 - 1{,}000 \ln(1 + t^2)$

$\ln(1 + t^2) = 4$

$1 + t^2 = e^4$

$t^2 = e^4 - 1$

$t = \sqrt{e^4 - 1} \approx 7.32$ days

89. $N'(t) = 6e^{-0.1t}, 0 \le t \le 15$

$N(t) = \int N'(t)\,dt = \int 6e^{-0.1t}\,dt = 6\int e^{-0.1t}\,dt = \frac{6}{-0.1}\int e^{-0.1t}(-0.1)\,dt = -60e^{-0.1t} + C$

Given $N(0) = 40$:

$40 = -60e^{0} + C$

Hence, $C = 100$ and $N(t) = 100 - 60e^{-0.1t}, 0 \le t \le 15$.
The number of words per minute after completing the course is:

$N(15) = 100 - 60e^{-0.1(15)} = 100 - 60e^{-1.5} \approx 87$ words per minute.

91. $\frac{dE}{dt} = 5000(t+1)^{-3/2}, t \ge 0$

Let $u = t + 1$, then $du = dt$

$E = \int 5000(t+1)^{-3/2}\,dt = 5000\int (t+1)^{-3/2}\,dt = 5000\int u^{-3/2}\,du = 5000\frac{u^{-1/2}}{-1/2} + C$

$= -10{,}000(t+1)^{-1/2} + C = \frac{-10{,}000}{\sqrt{t+1}} + C$

Given $E(0) = 2000$:

$2000 = \frac{-10{,}000}{\sqrt{1}} + C$

Hence, $C = 12{,}000$ and $E(t) = 12{,}000 - \frac{10{,}000}{\sqrt{t+1}}$.

The projected enrollment 15 years from now is:

$E(15) = 12{,}000 - \frac{10{,}000}{\sqrt{15+1}} = 12{,}000 - \frac{10{,}000}{\sqrt{16}} = 12{,}000 - \frac{10{,}000}{4} = 9500$ students.

EXERCISE 5-3

Things to remember:

1. A DIFFERENTIAL EQUATION is an equation that involves an unknown function and one or more of its derivatives. The ORDER of a differential equation is the order of the highest derivative of the unknown function.

2. A SLOPE FIELD for a first-order differential equation is obtained by drawing tangent line segments determined by the equation at each point in a grid.

3. EXPONENTIAL GROWTH LAW

   If $\frac{dQ}{dt} = rQ$ and $Q(0) = Q_0$, then $Q(t) = Q_0e^{rt}$, where

   Q_0 = Amount at $t = 0$
   r = Relative growth rate (expressed as a decimal)
   t = Time
   Q = Quantity at time t

4. COMPARISON OF EXPONENTIAL GROWTH PHENOMENA

DESCRIPTION	MODEL	SOLUTION	GRAPH	USES
Unlimited growth: Rate of growth is proportional to the amount present	$\frac{dy}{dt} = ky$ $k, t > 0$ $y(0) = c$	$y = ce^{kt}$		•Short-term population growth (people, bacteria, etc.) •Growth of money at continuous compound interest •Price-supply curves •Depletion of natural resources
Exponential decay: Rate of growth is proportional to the amount present	$\frac{dy}{dt} = -ky$ $k, t > 0$ $y(0) = c$	$y = ce^{-kt}$		•Radioactive decay •Light absorption in water •Price-demand curves •Atmospheric pressure (t is altitude)
Limited growth: Rate of growth is proportional to the difference between the amount present and a fixed limit	$\frac{dy}{dt} = k(M - y)$ $k, t > 0$ $y(0) = 0$	$y = M(1 - e^{-kt})$		•Sales fads (e.g., skateboards) •Depreciation of equipment •Company growth •Learning
Logistic growth: Rate of growth is proportional to the amount present and to the difference between the amount present and a fixed limit	$\frac{dy}{dt} = ky(M - y)$ $k, t > 0$ $y(0) = \frac{M}{1 + c}$	$y = \frac{M}{1 - ce^{-kMt}}$		•Long-term population growth •Epidemics •Sales of new products •Rumor spread •Company growth

---

1. The derivative of $f(x) = e^{5x}$ is 5 times f; $y' = 5y$.

3. The derivative of $f(x) = 10e^{-x}$ is $-f$; $y' = -y$.

5. The derivative of $f(x) = 3.2e^{x^2}$ is $2x$ times f; $y' = 2xy$.

7. The derivative of $f(x) = 1 - e^{-x}$ is 1 minus f; $y' = 1 - y$.

9. $\frac{dy}{dx} = 6x$

$$\int \frac{dy}{dx}\,dx = \int 6x\,dx = 6\int x\,dx$$

$$\int dy = 6\int x\,dx$$

$$y = 6 \cdot \frac{x^2}{2} + C = 3x^2 + C$$

General solution: $y = 3x^2 + C$

11. $\frac{dy}{dx} = \frac{7}{x}$

$$\int \frac{dy}{dx}\,dx = 7\int \frac{1}{x}\,dx$$

$$\int dy = 7\int \frac{1}{x}\,dx$$

General solution: $y = 7\ln|x| + C$

13. $\frac{dy}{dx} = e^{0.02x}$

$$\int \frac{dy}{dx}\,dx = \int e^{0.02x}\,dx$$

$$\int dy = \int e^{0.02x}\,dx \quad (u = 0.02x,\ du = 0.02dx)$$

$$y = \int e^{u}\frac{1}{0.02}\,du = \frac{1}{0.02}\int e^{u}\,du = \frac{1}{0.02}e^{u} + C = 50e^{0.02x} + C$$

General solution: $y = 50e^{0.02x} + C$

15. $\frac{dy}{dx} = x^2 - x;\ y(0) = 0$

$$\int \frac{dy}{dx}\,dx = \int (x^2 - x)\,dx$$

$$y = \frac{1}{3}x^3 - \frac{1}{2}x^2 + C$$

Given $y(0) = 0$: $\frac{1}{3}(0)^3 - \frac{1}{2}(0)^2 + C = 0,\quad C = 0$

Particular solution: $y = \frac{1}{3}x^3 - \frac{1}{2}x^2$

17. $\frac{dy}{dx} = -2xe^{-x^2};\ y(0) = 3$

$$\int \frac{dy}{dx}\,dx = \int -2x\,e^{-x^2}\,dx$$

$$y = \int -2xe^{-x^2}\,dx$$

Let $u = -x^2$. Then $du = -2x\ dx$ and

$$\int -2x\,e^{-x^2}\,dx = \int e^{u}\,du = e^{u} + C = e^{-x^2} + C$$

Thus, $y = e^{-x^2} + C.$

Given $y(0) = 3$: $3 = e^{0} + C$

$3 = 1 + C,\quad C = 2$

Particular solution: $y = e^{-x^2} + 2$

19. $\frac{dy}{dx} = \frac{2}{1+x};\ y(0) = 5$

$$\int \frac{dy}{dx}\,dx = \int \frac{2}{1+x}\,dx = 2\int \frac{1}{1+x}\,dx$$

$$\int dy = 2\int \frac{1}{1+x}\,dx \quad (u = 1 + x,\ du = dx)$$

$$y = 2\int \frac{1}{u}\,du = 2\ln|u| + C = 2\ln|1 + x| + C$$

Given $y(0) = 5$: $5 = 2\ln 1 + C$

$5 = C$

Particular solution: $y = 2\ln|1 + x| + 5.$

21. Second order

23. Third order

25. $y = 5x,\ \dfrac{dy}{dx} = 5 = \dfrac{5x}{x} = \dfrac{y}{x}$; yes.

27. $y = \sqrt{9+x^2},\ y' = \dfrac{1}{2\sqrt{9+x^2}}(2x) = \dfrac{x}{\sqrt{9+x^2}} = \dfrac{x}{y}$; yes.

29. $y = e^{3x},\ y' = 3e^{3x},\ y'' = 9e^{3x};\ 9e^{3x} - 4(3e^{3x}) + 3(e^{3x}) = 12e^{3x} - 12e^{3x} = 0$; yes.

31. $y = 100e^{3x},\ y' = 300e^{3x},\ y'' = 900e^{3x};\ 900e^{3x} - 4(300e^{3x}) + 3(100e^{3x}) = 1200e^{3x} - 1200e^{3x} = 0$; yes.

33. Figure (B). When $x = 1$, $\dfrac{dy}{dx} = 1 - 1 = 0$ for any y. When $x = 0$, $\dfrac{dy}{dx} = 0 - 1 = -1$ for any y. When $x = 2$, $\dfrac{dy}{dx} = 2 - 1 = 1$ for any y; and so on. These facts are consistent with the slope-field in Figure (B); they are not consistent with the slope-field in Figure (A).

35. $\dfrac{dy}{dx} = x - 1$

$$\int \frac{dy}{dx}\,dx = \int (x-1)\,dx$$

General solution: $y = \dfrac{1}{2}x^2 - x + C$

Given $y(0) = -2$: $\dfrac{1}{2}(0)^2 - 0 + C = -2,\quad C = -2$

Particular solution: $y = \dfrac{1}{2}x^2 - x - 2$

37.

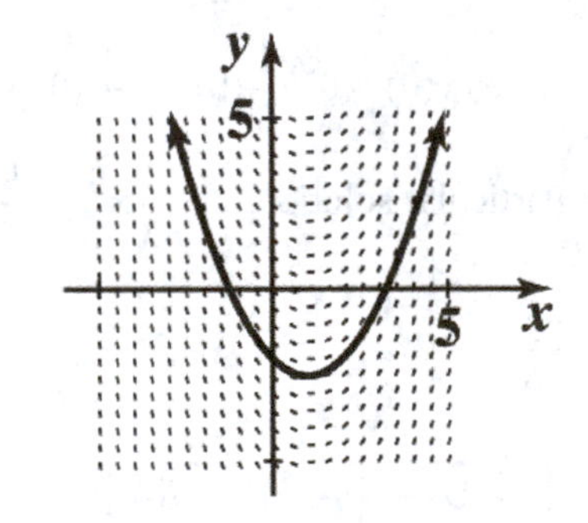

39. $\dfrac{dy}{dt} = 2y$

$$\frac{1}{y}\frac{dy}{dt} = 2$$

$$\int \frac{1}{y}\frac{dy}{dt}\,dt = \int 2\,dt$$

$$\int \frac{1}{u}\,du = \int 2\,dt \qquad [u = y,\ du = dy = \frac{dy}{dt}\cdot dt]$$

$\ln|u| = 2t + K$ [K an arbitrary constant]

$|u| = e^{2t+K} = e^K e^{2t}$

$|u| = Ce^{2t}$ $[C = e^K, C > 0]$

so $|y| = Ce^{2t}$

Now, if we set $y(t) = Ce^{2t}$, C ANY constant, then

$$y'(t) = 2Ce^{2t} = 2y(t),$$

So $y = Ce^{2t}$ satisfies the differential equation where C is any constant. This is the general solution. Note, the differential equation is the model for exponential growth with growth rate 2.

41. $\frac{dy}{dx} = -0.5y, \quad y(0) = 100$

$$\frac{1}{y}\frac{dy}{dx} = -0.5$$

$$\int \frac{1}{y}\frac{dy}{dx}\,dx = \int -0.5\,dx$$

$$\int \frac{1}{u}\,du = \int -0.5\,dx \quad [u = y,\ du = dy = \frac{dy}{dx}\,dx]$$

$$\ln|u| = -0.5x + K$$
$$|u| = e^{-0.5x+K} = e^{K}e^{-0.5x}$$
$$|y| = Ce^{-0.5x},\ C = e^{K} > 0.$$

So, general solution: $y = Ce^{-0.5x}$, C any constant.

Given $y(0) = 100$: $100 = Ce^{0} = C$; particular solution: $y = 100e^{-0.5x}$

43. $\frac{dx}{dt} = -5x$

$$\frac{1}{x}\frac{dx}{dt} = -5$$

$$\int \frac{1}{x}\frac{dx}{dt}\,dt = \int -5\,dt$$

$$\int \frac{1}{x}\,dx = -5\int dt$$

$$\ln|x| = -5t + K$$
$$|x| = e^{-5t+K} = e^{K}e^{-5t} = Ce^{-5t}, \quad C = e^{K} > 0.$$

General solution: $x = Ce^{-5t}$, C any constant.

45. $\frac{dx}{dt} = -5t$

$$\int \frac{dx}{dt}\,dx = \int -5t\,dt = -5\int t\,dt$$

General solution: $x = -\frac{5t^2}{2} + C$

47. $y' = 2.5y(300 - y)$; logistic growth

49. $y' = 0.43y$; exponential growth

51. Figure (A). When $y = 1$, $\frac{dy}{dx} = 1 - 1 = 0$ for any x.

When $y = 2$, $\frac{dy}{dx} = 1 - 2 = -1$ for any x; and so on. This is consistent with the slope-field in Figure (A); it is not consistent with the slope-field in Figure (B).

53. $y = 1 - Ce^{-x}$

$\frac{dy}{dx} = \frac{d}{dx}[1 - Ce^{-x}] = Ce^{-x}$

From the original equation, $Ce^{-x} = 1 - y$
Thus, we have

$\frac{dy}{dx} = 1 - y$

and $y = 1 - Ce^{-x}$ is a solution of the differential equation for any number C.

Given $y(0) = 0$: $0 = 1 - Ce^{0} = 1 - C$
$C = 1$

Particular solution: $y = 1 - e^{-x}$

55.

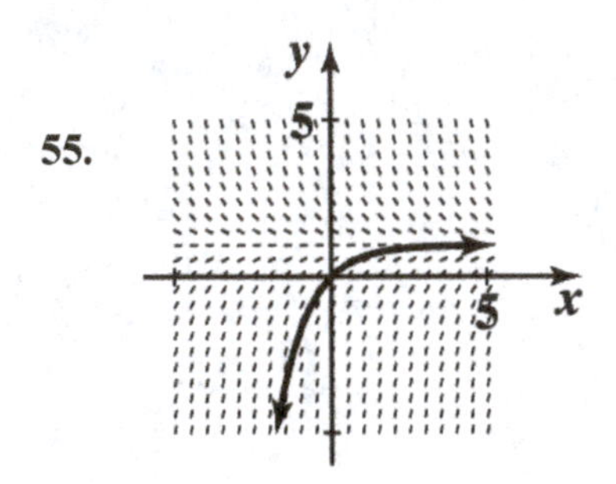

57.

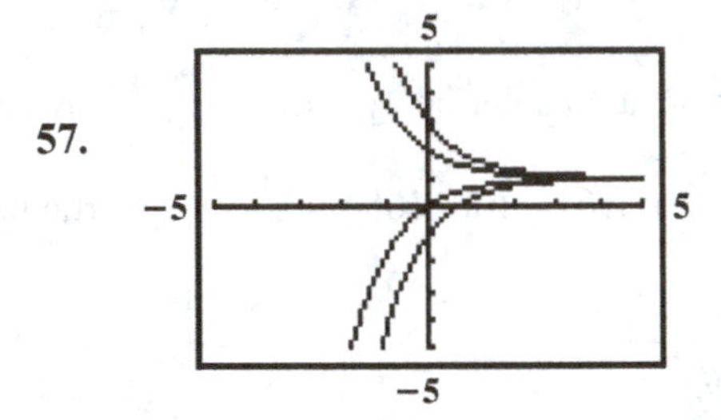

59. $y = \sqrt{C - x^2} = (C - x^2)^{1/2}$

$\frac{dy}{dx} = \frac{1}{2}(C - x^2)^{-1/2}(-2x) = \frac{-x}{\sqrt{C - x^2}} = -\frac{x}{y}$

Thus, $y = \sqrt{C - x^2}$ satisfies the differential equation $\frac{dy}{dx} = -\frac{x}{y}$.

Setting $x = 3$, $y = 4$, gives $4 = \sqrt{C - 9}$, $16 = C - 9$, $C = 25$.

The solution that passes through (3, 4) is: $y = \sqrt{25 - x^2}$.

61. $y = Cx$; $C = \frac{y}{x}$

$\frac{dy}{dx} = C = \frac{y}{x}$

Therefore $y = Cx$ satisfies the differential equation $\frac{dy}{dx} = \frac{y}{x}$.

Setting $x = -8, y = 24$ gives $24 = -8C$, $C = -3$
The solution that passes through (–8, 24) is $y = -3x$

63. $y = \frac{1}{1 + ce^{-t}} = \frac{e^t}{e^t + c}$ (multiply numerator and denominator by e^t)

$\frac{dy}{dt} = \frac{(e^t + c)e^t - e^t(e^t)}{(e^t + c)^2} = \frac{ce^t}{(e^t + c)^2}$

$$y(1-y) = \left(\frac{e^t}{e^t+c}\right)\left(1-\frac{e^t}{e^t+c}\right) = \frac{e^t}{e^t+c}\cdot\frac{c}{e^t+c} = \frac{ce^t}{(e^t+c)^2}$$

Thus $\frac{dy}{dt} = y(1-y)$ and $y = \frac{1}{1+ce^{-t}}$ satisfies the differential equation.

Setting $t=0,\ y=-1$ gives

$$-1 = \frac{1}{1+c},\ -1-c=1,\quad c=-2$$

The solution that passes through (0, –1) is

$$y = \frac{1}{1-2e^{-t}}$$

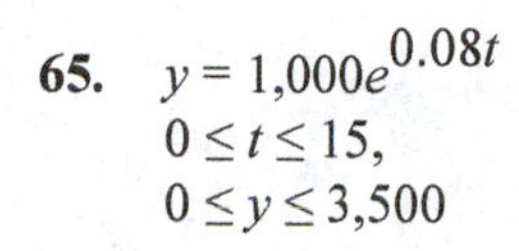

65. $y = 1{,}000e^{0.08t}$
$0 \le t \le 15$,
$0 \le y \le 3{,}500$

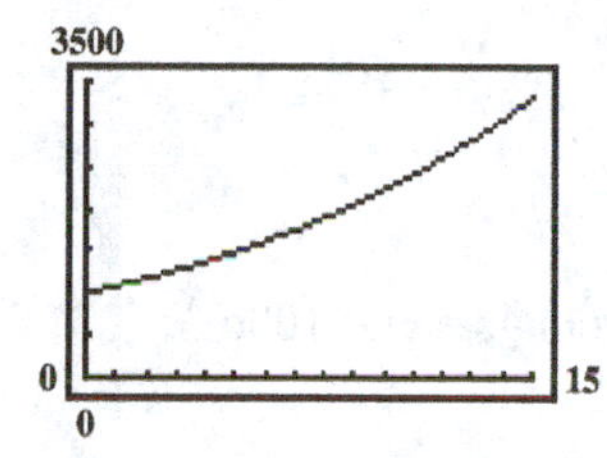

67. $p = 100e^{-0.05x}$
$0 \le x \le 30$,
$0 \le p \le 100$

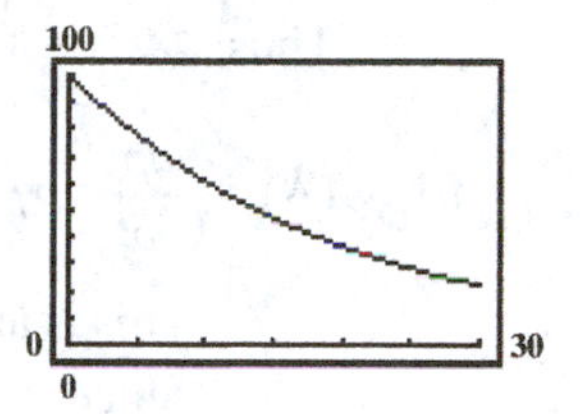

69. $N = 100(1-e^{-0.05t})$
$0 \le t \le 100,\ 0 \le N \le 100$

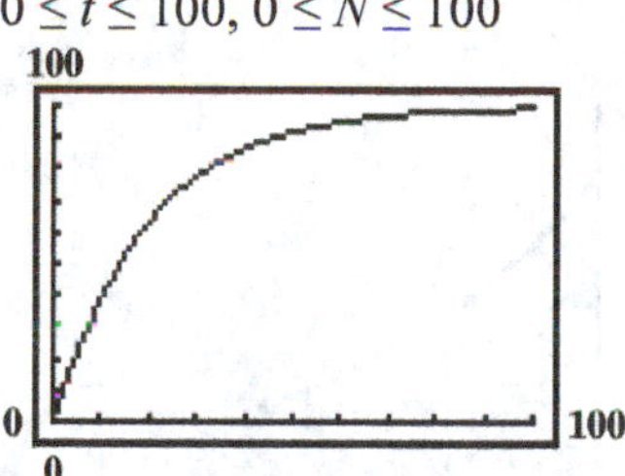

71. $N = \frac{1{,}000}{1+999e^{-0.4t}}$
$0 \le t \le 40,\ 0 \le N \le 1{,}000$

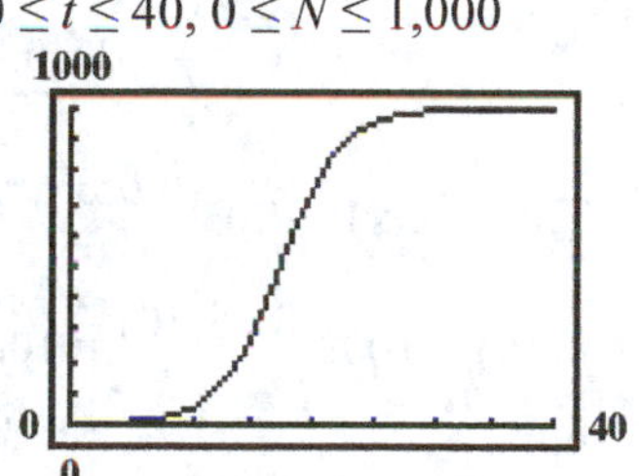

73. $\frac{dy}{dt} = ky(M-y)$, k, M positive constants. Set $f(y) = ky(M-y) = kMy - ky^2$.
This is a quadratic function which opens downward; it has a maximum value. Now
$f'(y) = kM - 2ky$

Critical value: $kM - 2ky = 0$ and $y = \frac{M}{2}$

$f''(y) = -2k < 0$. Thus, f has a maximum value at $y = \frac{M}{2}$.

75. In 1999: $\frac{dQ}{dt} = 6e^{0.013} \approx 6.079$

In 2009: $\frac{dQ}{dt} = 6.8e^{0.012} \approx 6.882$

The rate of growth in 2009 was greater than the rate of growth in 1999.

77. $\frac{dA}{dt} = 0.03A$ and $A(0) = 1{,}000$ is an unlimited growth model. From 4, the amount in the account after t years is: $A(t) = 1000e^{0.03t}$.

79. $\frac{dA}{dt} = rA,\ A(0) = 8{,}000$

is an unlimited growth model. From 4, $A(t) = 8{,}000e^{rt}$.

Since $A(2) = 8{,}260.14$, we solve $8{,}000e^{2r} = 8{,}260.14$ for r:

$$\begin{aligned} 8000e^{2r} &= 8{,}260.14 \\ e^{2r} &= \frac{8{,}260.14}{8{,}000} \\ 2r &= \ln(8{,}260.14/8{,}000) \\ r &= \frac{\ln(8{,}260.14/8{,}000)}{2} \approx 0.016\,. \end{aligned}$$

Thus, $A(t) = 8{,}000e^{0.016t}$.

81. (A) $\frac{dp}{dx} = rp,\ p(0) = 100$

This is an Unlimited Growth Model. From 4, $p(x) = 100e^{rx}$.

Since $p(5) = 77.88$, we have

$$\begin{aligned} 77.88 &= 100e^{5r} \\ e^{5r} &= 0.7788 \\ 5r &= \ln(0.7788) \\ r &= \frac{\ln(0.7788)}{5} \approx -0.05 \end{aligned}$$

Thus, $p(x) = 100e^{-0.05x}$.

(B) $p(10) = 100e^{-0.05(10)} = 100e^{-0.5} \approx \$60.65$ per unit.

(C)

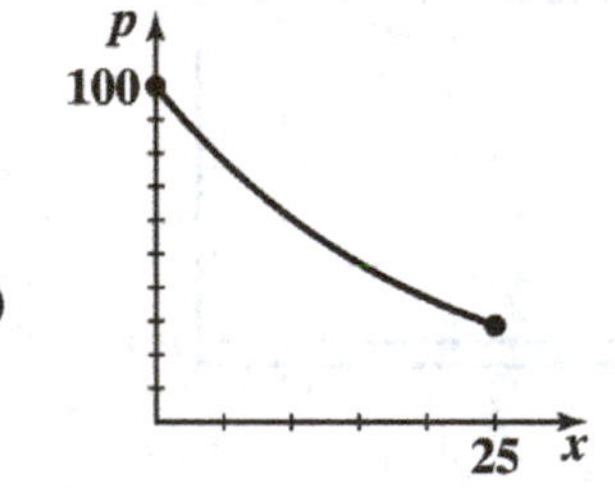

83. (A) $\frac{dN}{dt} = k(L - N);\ N(0) = 0$

This is a Limited Growth Model. From 4, $N(t) = L(1 - e^{-kt})$.

Since $N(10) = 0.4L$, we have

$$\begin{aligned} 0.4L &= L(1 - e^{-10k}) \\ 1 - e^{-10k} &= 0.4 \\ e^{-10k} &= 0.6 \\ -10k &= \ln(0.6) \\ k &= \frac{\ln(0.6)}{-10} \approx 0.051 \end{aligned}$$

Thus, $N(t) = L(1 - e^{-0.051t})$.

(B) $N(5) = L[1 - e^{-0.051(5)}] = L[1 - e^{-0.255}] \approx 0.225L$

Approximately 22.5% of the possible viewers will have been exposed after 5 days.

(C) Solve $L(1 - e^{-0.051t}) = 0.8L$ for t:

$$1 - e^{-0.051t} = 0.8$$
$$e^{-0.051t} = 0.2$$
$$-0.051t = \ln(0.2)$$
$$t = \frac{\ln(0.2)}{-0.051} \approx 31.56$$

It will take 32 days for 80% of the possible viewers to be exposed.

(D)

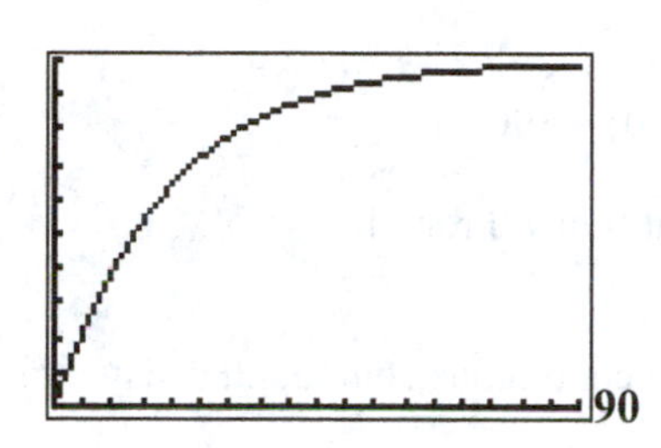

85. $\frac{dI}{dx} = -kI,\ I(0) = I_0$

This is an exponential decay model. From 4, $I(x) = I_0e^{-kx}$ with $k = 0.00942$. We have

$$I(x) = I_0e^{-0.00942x}$$

To find the depth at which the light is reduced to half of that at the surface, solve

$$I_0e^{-0.00942x} = \frac{1}{2}I_0$$

for x:

$$e^{-0.00942x} = 0.5$$
$$-0.00942x = \ln(0.5)$$
$$x = \frac{\ln(0.5)}{-0.00942} \approx 74 \text{ feet}$$

87. $\frac{dQ}{dt} = -0.04Q,\ Q(0) = Q_0.$

(A) This is a model for exponential decay. From 4,

$$Q(t) = Q_0e^{-0.04t}$$

With $Q_0 = 3$, we have

$$Q(t) = 3e^{-0.04t}$$

(B) $Q(10) = 3e^{-0.04(10)} = 3e^{-0.4} \approx 2.01.$

There are approximately 2.01 milliliters in the body after 10 hours.

(C)

$$3e^{-0.04t} = 1$$
$$e^{-0.04t} = \frac{1}{3}$$
$$-0.04t = \ln(1/3)$$
$$t = \frac{\ln(1/3)}{-0.04} \approx 27.47$$

(D)

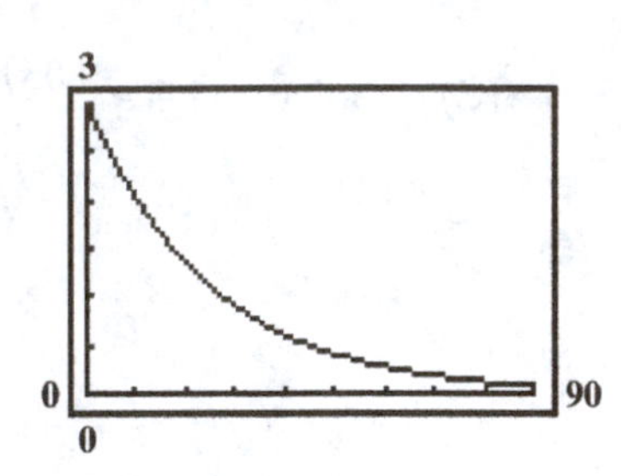

It will take approximately 27.47 hours for Q to decrease to 1 milliliter.

89. Using the exponential decay model, we have $\frac{dy}{dt} = -ky$, $y(0) = 100$, $k > 0$ where $y = y(t)$ is the amount of cesium -137 present at time t. From 4,

$$y(t) = 100e^{-kt}$$

Since $y(3) = 93.3$, we solve $93.3 = 100e^{-3k}$ for k to find the continuous compound decay rate:

$$93.3 = 100e^{-3k}$$
$$e^{-3k} = 0.933$$
$$-3k = \ln(0.933)$$
$$k = \frac{\ln(0.933)}{-3} \approx 0.023117$$

91. From Example 3: $Q = Q_0e^{-0.0001238t}$

Now, the amount of radioactive carbon-14 present is 5% of the original amount. Thus,
$0.05Q_0 = Q_0e^{-0.0001238t}$ or $e^{-0.0001238t} = 0.05$.

Therefore, $-0.0001238t = \ln(0.05) \approx -2.9957$ and $t \approx 24{,}200$ years.

93. $N(k) = 180e^{-0.11(k-1)}$, $1 \le k \le 10$

Thus, $N(6) = 180e^{-0.11(6-1)} = 180e^{-0.55} \approx 104$ times
and $N(10) = 180e^{-0.11(10-1)} = 180e^{-0.99} \approx 67$ times.

95. (A) $x(t) = \dfrac{400}{1+399e^{-0.4t}}$

$$x(5) = \frac{400}{1+399e^{(-0.4)5}} = \frac{400}{1+399e^{-2}} \approx \frac{400}{55} \approx 7 \text{ people}$$

$$x(20) = \frac{400}{1+399e^{(-0.4)20}} = \frac{400}{1+399e^{-8}} \approx 353 \text{ people}$$

(B) $\lim_{t\to\infty} x(t) = 400$.

(C)

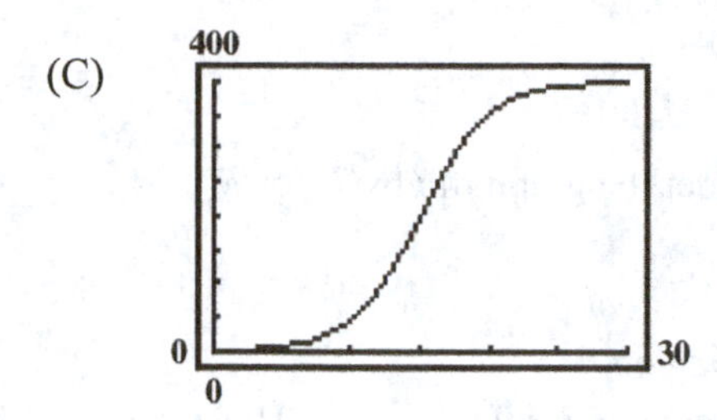

EXERCISE 5-4

Things to remember:

1. APPROXIMATING AREAS BY LEFT AND RIGHT SUMS

   Let $f(x)$ be defined and positive on the interval $[a, b]$. Divide the interval into n subintervals of equal length

   $\Delta x = \dfrac{b-a}{n}$,

   with endpoints $a = x_0 < x_1 < x_2 < \ldots < x_{n-1} < x_n = b$.

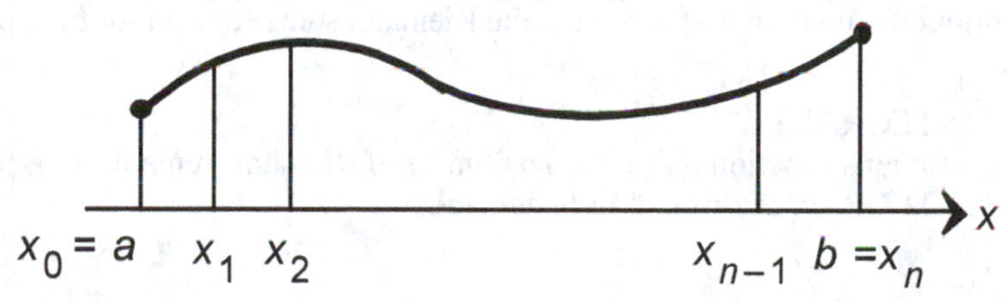

   Then

   $$L_n = f(x_0)\Delta x + f(x_1)\Delta x + f(x_2)\Delta x + \ldots + f(x_{n-1})\Delta x$$

   is called a LEFT SUM;

   $$R_n = f(x_1)\Delta x + f(x_2)\Delta x + \ldots + f(x_{n-1})\Delta x + f(x_n)\Delta x$$

   is called a RIGHT SUM.

   Left and right sums are approximations of the area between the graph of f and the x-axis from $x = a$ to $x = b$.

2. ERROR IN AN APPROXIMATION

   The ERROR IN AN APPROXIMATION is the absolute value of the difference between the approximation and the actual value.

3. ERROR BOUNDS FOR APPROXIMATIONS OF AREA BY LEFT AND RIGHT SUMS

   If $f(x) > 0$ and is either increasing on $[a, b]$ or decreasing on $[a, b]$, then

$|f(b) - f(a)| \cdot \frac{b-a}{n}$

is an error bound for the approximation of the area under the graph of f by L_n or R_n.

4. LIMITS OF LEFT AND RIGHT SUMS

If $f(x) > 0$ and is either increasing on $[a, b]$ or decreasing on $[a, b]$, then its left and right sums approach the same real number I as $n \to \infty$. This number is the area between the graph of f and the x-axis from $x = a$ to $x = b$.

5. RIEMANN SUMS

Let f be defined on the interval $[a, b]$. Divide the interval into n subintervals of equal length $\Delta x = \frac{b-a}{n}$ with endpoints

$a = x_0 < x_1 < x_2 < \ldots < x_{n-1} < x_n = b.$

Choose a point $c_1 \in [x_0, x_1]$, a point $c_2 \in [x_1, x_2]$, ..., and a point $c_n \in [x_{n-1}, x_n]$. Then

$$S_n = f(c_1)\Delta x + f(c_2)\Delta x + \ldots + f(c_n)\Delta x$$

is called a RIEMANN SUM. Note that left sums and right sums are special cases of Riemann Sums.

6. LIMIT OF RIEMANN SUMS

If f is a continuous function on $[a, b]$ then the Riemann sums for f on $[a, b]$ approach a real number I as $n \to \infty$.

7. DEFINITE INTEGRAL

Let f be a continuous function on $[a, b]$. The limit I of Riemann sums for f on $[a, b]$ is called the DEFINITE INTEGRAL of f from a to b, denoted

$$\int_a^b f(x)dx$$

The INTEGRAND is $f(x)$, the LOWER LIMIT OF INTEGRATION is a, and the UPPER LIMIT OF INTEGRATION is b.

8. GEOMETRIC INTERPRETATION OF THE DEFINITE INTEGRAL

If $f(x)$ is positive for some value of x on $[a, b]$ and negative for others, then the DEFINITE INTEGRAL SYMBOL

$$\int_a^b f(x)dx$$

represents the cumulative sum of the signed areas between the curve $y = f(x)$ and the x-axis where the areas above the x-axis are counted positively and the areas below the x-axis are counted negatively (see the figure where A and B are actual areas of the indicated regions).

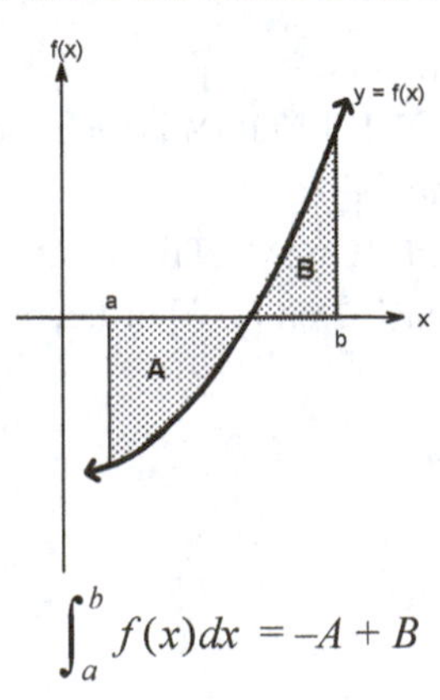

$$\int_a^b f(x)dx = -A + B$$

9. PROPERTIES OF DEFINITE INTEGRALS

(a) $\int_a^a f(x)dx = 0$

(b) $\int_a^b f(x)dx = -\int_b^a f(x)dx$

(c) $\int_a^b Kf(x)dx = K\int_a^b f(x)dx$ K is a constant

(d) $\int_a^b [f(x) \pm g(x)]dx = \int_a^b f(x)dx \pm \int_a^b g(x)dx$

(e) $\int_a^b f(x)dx = \int_a^c f(x)dx + \int_c^b f(x)dx$

---

1. The area of each rectangle is $8\times 2 = 16$ sq. in.; area of 5 such rectangles $5\times 16 = 80$ sq. in.

3. $4[2(3+4+5+6)] = 4[2(18)] = 4(36) = 144$ sq. meters.

5. The square has side length $\sqrt{2}$ and area 2. The circle has area $\pi 1^2 = \pi$. No, the area inside the circle and outside the square is $\pi - 2 \approx 1.14 > 1$.

7. C, E **9.** B **11.** H, I **13.** H

15.

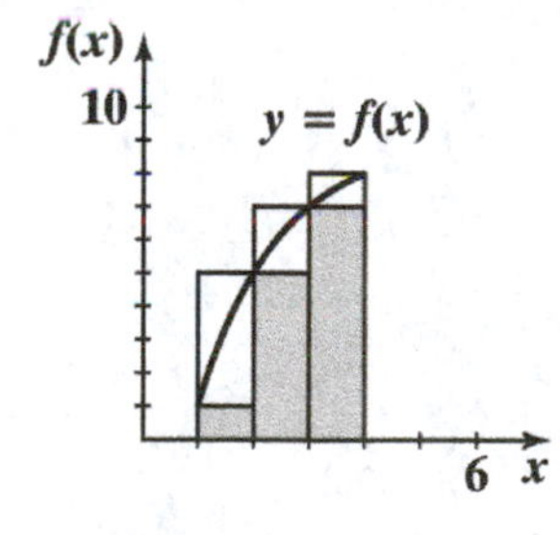

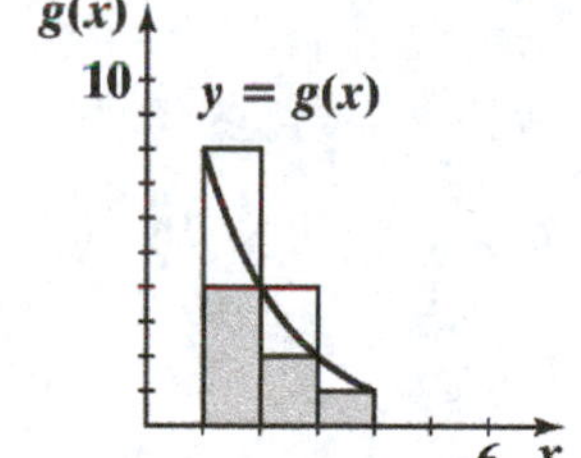

17. For Figure (A):

$$L_3 = f(1)\cdot 1 + f(2)\cdot 1 + f(3)\cdot 1$$
$$= 1 + 5 + 7 = 13$$

$$R_3 = f(2)\cdot 1 + f(3)\cdot 1 + f(4)\cdot 1$$
$$= 5 + 7 + 8 = 20$$

For Figure (B):

$$L_3 = g(1)\cdot 1 + g(2)\cdot 1 + g(3)\cdot 1$$
$$= 8 + 4 + 2 = 14$$

$$R_3 = g(2)\cdot 1 + g(3)\cdot 1 + g(4)\cdot 1$$
$$= 4 + 2 + 1 = 7$$

19. $L_3 \le \int_1^4 f(x)dx \le R_3$, $R_3 \le \int_1^4 g(x)dx \le L_3$; since f is increasing on [1, 4], L_3 underestimates the area and R_3 overestimates the area; since g is decreasing on [1, 4], L_3 overestimates the area and R_3 underestimates the area.

21. For Figure (A).

Error bound for L_3 and R_3:

$$\text{Error} \le |f(4) - f(1)|\left(\frac{4-1}{3}\right) = |8 - 1| = 7$$

For Figure (B).

Error bound for L_3 and R_3:

$$\text{Error} \le |f(4) - f(1)|\left(\frac{4-1}{3}\right) = |1 - 8| = |-7| = 7$$

23. $f(x) = 25 - 3x^2$ on [–2, 8]

$\Delta x = \frac{8-(-2)}{5} = \frac{10}{5} = 2; x_0 = -2, x_1 = 0, x_2 = 2, \ldots, x_5 = 8$

$c_i = \frac{x_{i-1} + x_i}{2}; c_1 = -1, c_2 = 1, c_3 = 3, c_4 = 5, c_5 = 7$

$S_5 = f(-1)2 + f(1)2 + f(3)2 + f(5)2 + f(7)2$

$= [22 + 22 - 2 - 50 - 122]2 = (-130)2 = -260$

25. $f(x) = 25 - 3x^2$ on [0, 12]

$\Delta x = \frac{12-0}{4} = 3; x_0 = 0, x_1 = 3, x_2 = 6, x_3 = 9, x_4 = 12$

$c_i = \frac{2x_{i-1} + x_i}{3}; c_1 = 1, c_2 = 4, c_3 = 7, c_4 = 10$

$S_4 = f(1)3 + f(4)3 + f(7)3 + f(10)3$

$= [22 - 23 - 122 - 275]3 = (-398)3 = -1194$

27. $f(x) = x^2 - 5x - 6$

$\Delta x = \frac{3-0}{3} = 1; x_0 = 0, x_1 = 1, x_2 = 2, x_3 = 3$

$S_3 = f(0.7)1 + f(1.8)1 + f(2.4)1$

$= -9.01 - 11.76 - 12.24 = -33.01$

29. $f(x) = x^2 - 5x - 6$

$\Delta x = \frac{7-1}{6} = 1; x_0 = 1, x_1 = 2, x_3 = 3, \ldots, x_6 = 7$

$S_6 = f(1)1 + f(3)1 + f(3)1 + f(5)1 + f(5)1 + f(7)1$

$= -10 - 12 - 12 - 6 - 6 + 8 = -38$

31. $\int_b^0 f(x)dx = -\text{area } B = -2.475$

33. $\int_a^c f(x)dx = \text{area } A - \text{area } B + \text{area } C = 1.408 - 2.475 + 5.333 = 4.266$

35. $\int_a^d f(x)dx = \text{area } A - \text{area } B + \text{area } C - \text{area } D$

$= 1.408 - 2.475 + 5.333 - 1.792 = 2.474$

37. $\int_c^0 f(x)dx = -\int_0^c f(x)dx = -\text{area } C = -5.333$

39. $\int_0^a f(x)dx = -\int_a^0 f(x)dx = -[\text{area } A - \text{area } B] = -[1.408 - 2.475] = 1.067$

41. $\int_d^b f(x)dx = -\int_b^d f(x)dx = -[\text{area } B + \text{area } C - \text{area } D]$

$= -[-2.475 + 5.333 - 1.792] = -1.066$

43. $\int_1^4 2x\,dx = 2\int_1^4 x\,dx = 2(7.5) = 15$

45. $\int_1^4 (5x + x^2)dx = 5\int_1^4 x\,dx + \int_1^4 x^2dx = 5(7.5) + 21 = 58.5$

47. $\int_1^4 (x^2 - 10x)dx = \int_1^4 x^2dx - 10\int_1^4 x\,dx = 21 - 10(7.5) = -54$

49. $\int_1^5 6x^2dx = 6\int_1^5 x^2dx = 6\left[\int_1^4 x^2dx + \int_4^5 x^2dx\right] = 6\left[21 + \frac{61}{3}\right] = 126 + 122 = 248$

51. $\int_4^4 (7x - 2)^2dx = 0$

53. $\int_5^4 9x^2dx = -\int_4^5 9x^2dx = -9\int_4^5 x^2dx = -9\left(\frac{61}{3}\right) = -183$

55. 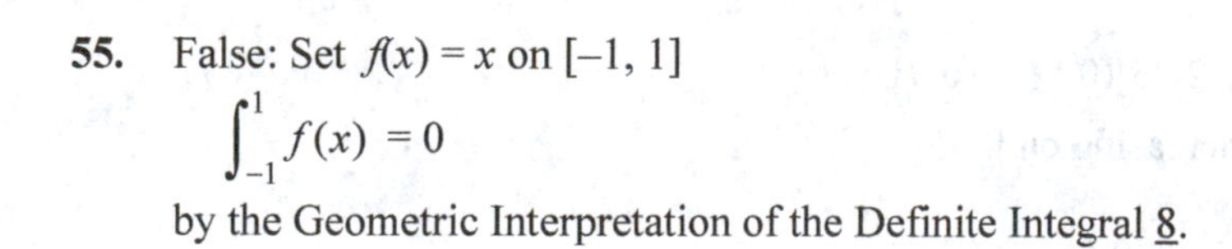
False: Set $f(x) = x$ on $[-1, 1]$

$\int_{-1}^1 f(x) = 0$

by the Geometric Interpretation of the Definite Integral 8.

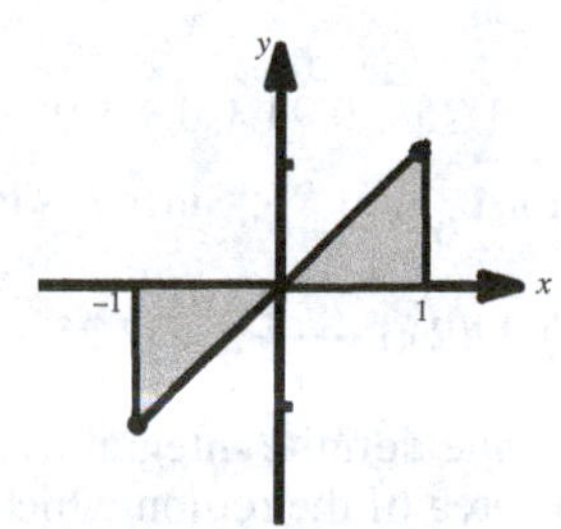

57. False: $f(x) = 2x$ is increasing on $[0, 10]$ and so L_n is less than $\int_0^{10} f(x)\,dx$ for every n.

59. False: Consider $f(x) = 1 - x^2$ on $[-1, 1]$ and let $n = 2$

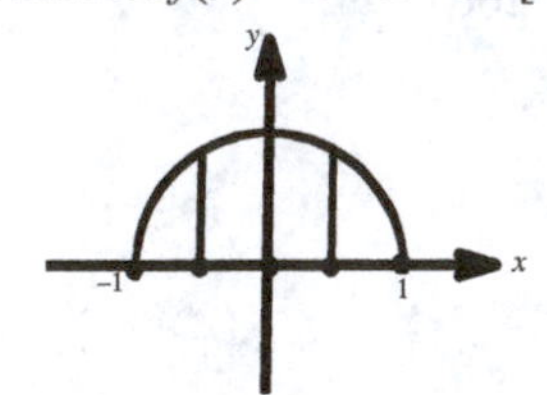

$L_2 = f(-1)1 + f(0)1 = 0 + 1 = 1$
$R_2 = f(0)1 + f(1)1 = 1 + 0 = 1$
(Note: In this case $L_n = R_n$ for every n.)

61. $h(x)$ is an increasing function; $\Delta x = 100$
$L_{10} = h(0)100 + h(100)100 + h(200)100 + \ldots + h(900)(100)$

$= [0 + 183 + 235 + 245 + 260 + 286 + 322 + 388 + 453 + 489]100$

$= (2{,}861)100 = 286{,}100$ sq ft
Error bound for L_{10}:

$$\text{Error} \le |h(1{,}000) - h(0)|\left(\frac{1000-0}{10}\right) = 500(100) = 50{,}000 \text{ sq ft}$$

We want to find n such that $|I - L_n| \le 2{,}500$:

$$|h(1000) - h(0)|\left(\frac{1000-0}{n}\right) \le 2{,}500$$

$$500\left(\frac{1000}{n}\right) \le 2{,}500$$

$$500{,}000 \le 2{,}500n$$

$$n \ge 200$$

63. $f(x) = 0.25x^2 - 4$ on $[2, 5]$
$L_6 = f(2)\Delta x + f(2.5)\Delta x + f(3)\Delta x + f(3.5)\Delta x + f(4)\Delta x + f(4.5)\Delta x,$ where $\Delta x = 0.5$
Thus,

$L_6 = [-3 - 2.44 - 1.75 - 0.94 + 0 + 1.06](0.5) = -3.53$

$R_6 = f(2.5)\Delta x + f(3)\Delta x + f(3.5)\Delta x + f(4)\Delta x + f(4.5)\Delta x + f(5)\Delta x,$ where $\Delta x = 0.5$
Thus,
$R_6 = [-2.44 - 1.75 - 0.94 + 0 + 1.06 + 2.25](0.5) = -0.91$
Error bound for L_6 and R_6: Since f is increasing on $[2, 5]$,

$$\text{Error} \le |f(5) - f(2)|\left(\frac{5-2}{6}\right) = |2.25 - (-3)|(0.5) = 2.63$$

Geometrically, the definite integral over the interval $[2, 5]$ is the area of the region which lies above the x-axis minus the area of the region which lies below the x-axis. From the figure, if R_1 represents the region bounded by the graph of f and the x-axis for $2 \le x \le 4$ and
R_2 represents the region bounded by the graph of f and the x-axis for
$4 \le x \le 5$, then

$$\int_2^5 f(x)dx = \text{area}(R_2) - \text{area}(R_1)$$

65. $f(x) = e^{-x^2}$

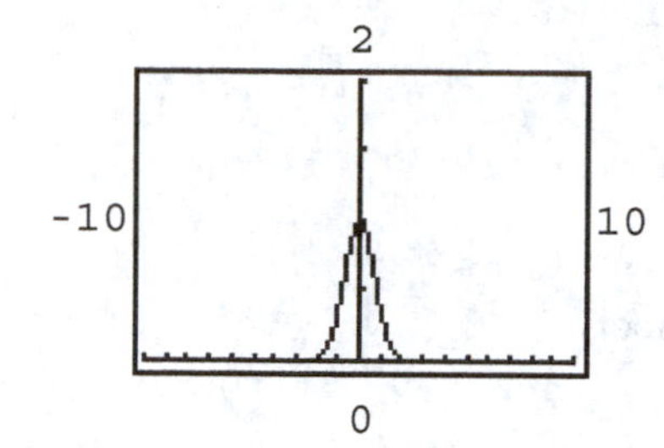

Thus, f is increasing on $(-\infty, 0]$ and decreasing on $[0, \infty)$.

67. $f(x) = x^4 - 2x^2 + 3$

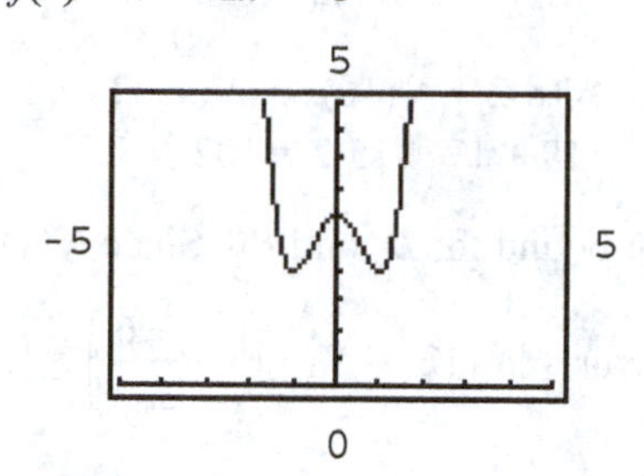

Thus, f is decreasing on $(-\infty, -1]$ and on $[0, 1]$, and increasing on $[-1, 0]$ and on $[1, \infty)$.

69. $\int_1^3 \ln x\, dx$

$|I - R_n| \le |\ln 3 - \ln 1| \dfrac{3-1}{n} \approx \dfrac{(1.0986)2}{n} = \dfrac{2.1972}{n}$

Now $\dfrac{2.1972}{n} \le 0.1$ implies $n \ge \dfrac{2.1972}{0.1} = 21.972$

so $n \ge 22$.

71. $\int_1^3 x^x\, dx$

$|I - L_n| \le |3^3 - 1^1| \dfrac{3-1}{n} = \dfrac{26 \cdot 2}{n} = \dfrac{52}{n}$

Now $\dfrac{52}{n} \le 0.5$ implies $n \ge \dfrac{52}{0.5} = 104$

73. From $t = 0$ to $t = 60$

$L_3 = N(0)20 + N(20)20 + N(40)20$

$= (10 + 51 + 68)20 = 2580$

$R_3 = N(20)20 + N(40)20 + N(60)20$

$= (51 + 68 + 76)20 = 3900$

Error bound for L_3 and R_3: Since $N(t)$ is increasing,

$$\text{Error} \le |N(60) - N(0)| \left(\frac{60-0}{3}\right) = (76 - 10)20 = 1{,}320 \text{ units}$$

75. (A) $L_5 = A'(0)1 + A'(1)1 + A'(2)1 + A'(3)1 + A'(4)1$

$= 0.90 + 0.81 + 0.74 + 0.67 + 0.60$

$= 3.72$ sq cm

$R_5 = A'(1)1 + A'(2)1 + A'(3)1 + A'(4)1 + A'(5)1$

$= (0.81 + 0.74 + 0.67 + 0.60 + 0.55)$

$= 3.37$ sq cm

(B) Since $A'(t)$ is a decreasing function

$R_5 = 3.37 \le \int_0^5 A'(t)dt \le 3.72 = L_5$

77. $L_3 = N'(6)2 + N'(8)2 + N'(10)2$
$= (21 + 19 + 17)2 = 114$

$R_3 = N'(8)2 + N'(10)2 + N'(12)2$
$= (19 + 17 + 15)2 = 102$

Error bound for L_3 and R_3: Since $N'(x)$ is decreasing

$$\text{Error} \le |N'(12) - N'(6)|\left(\frac{12-6}{3}\right) = |15 - 21|(2) = 12 \text{ code symbols}$$

EXERCISE 5-5

1. FUNDAMENTAL THEOREM OF CALCULUS

If f is a continuous function on the closed interval $[a, b]$ and F is any antiderivative of f, then

$$\int_a^b f(x)dx = F(x)\Big|_a^b = F(b) - F(a);$$
$$F'(x) = f(x)$$

2. AVERAGE VALUE OF A CONTINUOUS FUNCTION OVER $[a, b]$

Let f be continuous on $[a, b]$. Then the AVERAGE VALUE of f over $[a, b]$ is:

$$\frac{1}{b-a}\int_a^b f(x)dx$$

1. $f(x) = 100$ on $[1,6]$. The region bounded by the graph of f and the x-axis is a rectangle of length 100 and width 5; area: $A = 500$.

3. $f(x) = x + 5$ on $[0,4]$. The region bounded by the graph of f and the x-axis is a trapezoid with bases of length 5 and 9, and height 4; area: $A = \frac{1}{2}(5+9)(4) = 28$.

5. $f(x) = 3x$ on $[-4,4]$. The region bounded by the graph of f and the x-axis consists of two congruent right triangles with base 4 and height 12. Area: $2\left(\frac{1}{2}\right)4\cdot 12 = 48$.

7. $f(x) = \sqrt{9-x^2}$ on $[-3,3]$. The region bounded by the graph of f and the x-axis is a semi-circle of radius 3; area $A = \frac{1}{2}\pi(3^2) = 4.5\pi \approx 14.14$.

9. $F(x) = 3x^2 + 160$
(A) $F(15) - F(10) = 3(15)^2 + 160 - [3(10)^2 + 160] = 675 - 300 = 375$

(B)

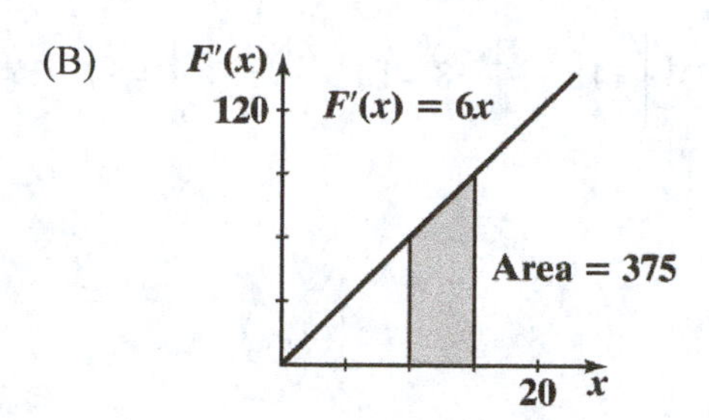

Area of trapezoid:

$$\frac{F'(15)+F'(10)}{2}\cdot 5 = \frac{90+60}{2}\cdot 5$$
$$= 75(5) = 375$$

(C) By the Fundamental Theorem of Calculus:

$$\int_{10}^{15} 6x\,dx = 3x^2\Big|_{10}^{15} = 3(15)^2 - 3(10)^2 = 375$$

11. $F(x) = -x^2 + 42x + 240$

(A) $F(15) - F(10) = -(15)^2 + 42(15) + 240 - [-(10)^2 + 42(10) + 240]$
$= -225 + 630 + 240 - (-100 + 420 + 240) = -225 + 630 + 100 - 420 = 85$

(B)

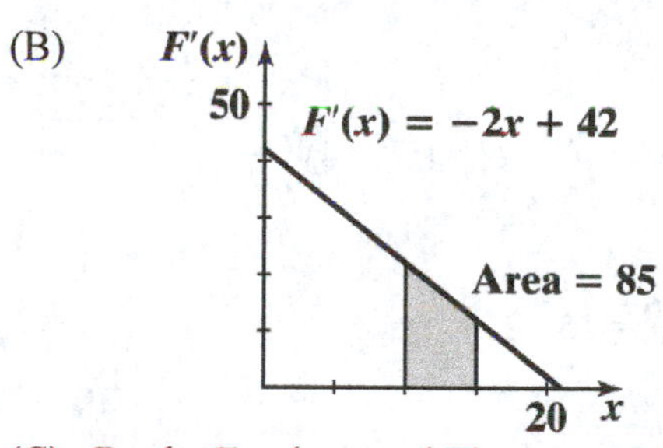

Area of trapezoid:

$$\frac{F'(15)+F'(10)}{2}\cdot 5 = \frac{(-30+42)+(-20+42)}{2}\cdot 5$$
$$= 17(5) = 85$$

(C) By the Fundamental Theorem of Calculus:

$$\int_{10}^{15} (-2x+42)dx = \left[-x^2+42x\right]_{10}^{15} = -(15)^2 + 42(15) - [-(10)^2 + 42(10)]$$
$$= -225 + 630 + 100 - 420 = 85$$

13. $\int_0^{10} 4\,dx = 4x\Big|_0^{10} = 4(10) - 4(0) = 40$

15. $\int_0^6 x^2\,dx = \frac{1}{3}x^3\Big|_0^6 = \frac{1}{3}(6)^3 - \frac{1}{3}(0)^3 = 72$

17. $\int_1^4 (5x+3)\,dx = \left[\frac{5}{2}x^2 + 3x\right]_1^4 = \left(\frac{5}{2}\cdot 4^2 + 3\cdot 4\right) - \left(\frac{5}{2}\cdot 1^2 + 3\cdot 1\right) = 52 - \frac{11}{2} = \frac{93}{2} = 46.5$

19. $\int_0^1 e^x\,dx = e^x\Big|_0^1 = e - 1 \approx 1.718$

21. $\int_1^2 \frac{1}{x}dx = \ln|x|\,\Big|_1^2 = \ln 2 - \ln 1 = \ln 2 \approx 0.693$

23. $\int_{-2}^2 (x^3+7x)\,dx = \left[\frac{1}{4}x^4 + \frac{7}{2}x^2\right]_{-2}^2 = \left[\frac{1}{4}\cdot 2^4 + \frac{7}{2}\cdot 2^2\right] - \left[\frac{1}{2}(-2)^4 + \frac{7}{2}(-2)^2\right] = 18 - 18 = 0$

25. $\int_2^5 (2x+9)\,dx = \left[x^2 + 9x\right]_2^5 = (5^2 + 9\cdot 5) - (2^2 + 9\cdot 2) = 70 - 22 = 48$

27. $\int_5^2 (2x+9)\,dx = -\int_2^5 (2x+9)\,dx = -48$ (Property 2 and Problem 25)

29. $\int_2^3 (6-x^3)\,dx = \left[6x - \frac{1}{4}x^4\right]_2^3 = \left[6\cdot 3 - \frac{1}{4}(3)^4\right] - \left[6\cdot 2 - \frac{1}{4}(2)^4\right] = 18 - \frac{81}{4} - 8 = 10 - \frac{81}{4} = -\frac{41}{4} = -10.25$

31. $\int_6^6 (x^2 - 5x + 1)^{10}\,dx = 0$

33. $\int_1^2 (2x^{-2} - 3)\,dx = \left[-\frac{2}{x} - 3x\right]_1^2 = \left[-\frac{2}{2} - 3\cdot 2\right] - \left[-\frac{2}{1} - 3\cdot 1\right] = -7 + 5 = -2$

35. $\int_1^4 3\sqrt{x}\,dx = 3\int_1^4 x^{1/2}\,dx = 3\cdot\frac{2}{3}x^{3/2}\Big]_1^4 = 2x^{3/2}\Big]_1^4 = 2\cdot 8 - 2\cdot 1 = 14$

37. $\int_2^3 12(x^2-4)^5 x\,dx$. Consider the indefinite integral $\int 12(x^2-4)^5 x\,dx$.

Let $u = x^2 - 4$. Then $du = 2x\,dx$.

$\int 12(x^2-4)^5 x\,dx = 6\int (x^2-4)^5 2x\,dx = 6\int u^5\,du = 6\frac{u^6}{6} + C = u^6 + C = (x^2-4)^6 + C$

Thus,

$\int_2^3 12(x^2-4)^5 x\,dx = (x^2-4)^6\Big|_2^3 = (3^2-4)^6 - (2^2-4)^6 = 5^6 = 15{,}625.$

39. $\int_3^9 \frac{1}{x-1}\,dx$

Let $u = x - 1$. Then $du = dx$ and $u = 8$ when $x = 9$, $u = 2$ when $x = 3$.

Thus,

$\int_3^9 \frac{1}{x-1}\,dx = \int_2^8 \frac{1}{u}\,du = \ln u\Big|_2^8 = \ln 8 - \ln 2 = \ln 4 \approx 1.386.$

41. $\int_{-5}^{10} e^{-0.05x}\,dx$

Let $u = -0.05x$. Then $du = -0.05\,dx$ and $u = -0.5$ when $x = 10$, $u = 0.25$ when $x = -5$. Thus,

$\int_{-5}^{10} e^{-0.05x}\,dx = -\frac{1}{0.05}\int_{-5}^{10} e^{-0.05x}(-0.05)dx = -\frac{1}{0.05}\int_{0.25}^{-0.5} e^u\,du$

$= -\frac{1}{0.05}e^u\Big|_{0.25}^{-0.5} = -\frac{1}{0.05}\left[e^{-0.5} - e^{0.25}\right]$

$= 20(e^{0.25} - e^{-0.5}) \approx 13.550$

43. $\int_1^e \frac{\ln t}{t}\,dt = \int_0^1 u\,du = \frac{1}{2}u^2\Big|_0^1 = \frac{1}{2}$

Let $u = \ln t$. Then $du = \frac{1}{t}\,dt$

$t = 1$ implies $u = \ln 1 = 0$, $t = e$ implies $u = \ln e = 1$

45. $\int_0^1 xe^{-x^2}\,dx = \int_0^1 e^{-x^2}\left(\frac{-2}{-2}\right)x\,dx$

Let $u = -x^2$
Then $du = -2x\,dx$
$x = 0$ implies $u = 0$
$x = 1$ implies $u = -1$

$$= -\frac{1}{2}\int_0^{-1} e^u\,du$$

$$= -\frac{1}{2}e^u\Big|_0^{-1} = -\frac{1}{2}e^{-1} + \frac{1}{2}e^0 = \frac{1}{2}(1 - e^{-1}) \approx 0.316$$

47. $\int_1^1 e^{x^2}\,dx = 0$

49. $f(x) = 500 - 50x$ on $[0, 10]$

(A) $$\text{Avg. } f(x) = \frac{1}{10-0}\int_0^{10}(500 - 50x)dx$$

$$= \frac{1}{10}\left[500x - 25x^2\right]_0^{10}$$

$$= \frac{1}{10}[5{,}000 - 2{,}500] = 250$$

(B)

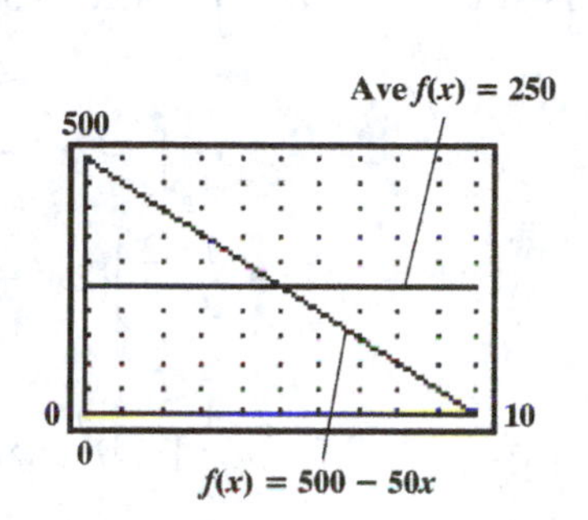

51. $f(t) = 3t^2 - 2t$ on $[-1, 2]$

(A) $$\text{Avg. } f(t) = \frac{1}{2-(-1)}\int_{-1}^{2}(3t^2 - 2t)dt$$

$$= \frac{1}{3}(t^3 - t^2)\Big|_{-1}^{2}$$

$$= \frac{1}{3}[4 - (-2)] = 2$$

(B)

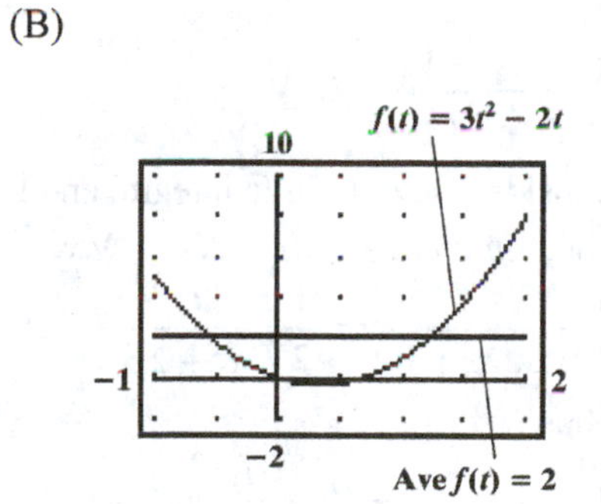

53. $f(x) = \sqrt[3]{x} = x^{1/3}$ on $[1, 8]$

(A) $$\text{Avg. } f(x) = \frac{1}{8-1}\int_1^8 x^{1/3}\,dx$$

$$= \frac{1}{7}\left(\frac{3}{4}x^{4/3}\right)\Bigg|_1^8$$

$$= \frac{3}{28}(16 - 1) = \frac{45}{28} \approx 1.61$$

(B)

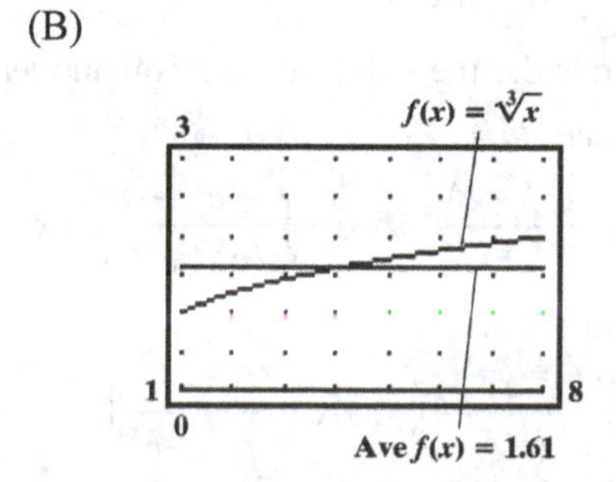

55. $f(x) = 4e^{-0.2x}$ on [0, 10]

(A) $\text{Avg. } f(x) = \dfrac{1}{10-0}\displaystyle\int_0^{10} 4e^{-0.2x}dx$

$= \dfrac{1}{10}\left[-20e^{-0.2x}\right]_0^{10}$

$= \dfrac{1}{10}(20 - 20e^{-2}) \approx 1.73$

(B)

Ave $f(x) = 1.73$

$f(x) = 4e^{-0.2x}$

57. $\displaystyle\int_2^3 x\sqrt{2x^2-3}\,dx = \int_2^3 x(2x^2-3)^{1/2}dx$

$= \dfrac{1}{4}\displaystyle\int_2^3 (2x^2-3)^{1/2}4x\,dx$

$= \dfrac{1}{4}\left(\dfrac{2}{3}\right)(2x^2-3)^{3/2}\Big|_2^3$

[Note: The integrand has the form $u^{1/2}du$; the antiderivative is $\frac{2}{3}u^{3/2} = \frac{2}{3}(2x^2-3)^{3/2}$.]

$= \dfrac{1}{6}[2(3)^2 - 3]^{3/2} - \dfrac{1}{6}[2(2)^2 - 3]^{3/2}$

$= \dfrac{1}{6}(15)^{3/2} - \dfrac{1}{6}(5)^{3/2} = \dfrac{1}{6}[15^{3/2} - 5^{3/2}] \approx 7.819$

59. $\displaystyle\int_0^1 \frac{x-1}{x^2-2x+3}dx$

Consider the indefinite integral and let $u = x^2 - 2x + 3$.

Then $du = (2x - 2)dx = 2(x - 1)dx$.

$$\int \frac{x-1}{x^2-2x+3}dx = \frac{1}{2}\int \frac{2(x-1)}{x^2-2x+3}dx = \frac{1}{2}\int \frac{1}{u}du = \frac{1}{2}\ln|u| + C$$

Thus,

$$\int_0^1 \frac{x-1}{x^2-2x+3}dx = \frac{1}{2}\ln|x^2 - 2x + 3|\,\Big|_0^1 = \frac{1}{2}\ln 2 - \frac{1}{2}\ln 3 = \frac{1}{2}(\ln 2 - \ln 3) \approx -0.203$$

61. $\displaystyle\int_{-1}^1 \frac{e^{-x}-e^x}{(e^{-x}+e^x)^2}dx$

Consider the indefinite integral and let $u = e^{-x} + e^x$.

Then $du = (-e^{-x} + e^x)\,dx = -(e^{-x} - e^x)\,dx$.

$$\int \frac{e^{-x}-e^x}{(e^{-x}+e^x)^2}dx = -\int \frac{-(e^{-x}-e^x)}{(e^{-x}+e^x)^2}dx = -\int u^{-2}\,du = \frac{-u^{-1}}{-1} + C = \frac{1}{u} + C$$

Thus,

$$\int_{-1}^1 \frac{e^{-x}-e^x}{(e^{-x}+e^x)^2}dx = \frac{1}{e^{-x}+e^x}\Big|_{-1}^1 = \frac{1}{e^{-1}+e^1} - \frac{1}{e^{-(-1)}+e^{-1}} = \frac{1}{e^{-1}+e} - \frac{1}{e^{-1}+e} = 0$$

63. $\int_{1.7}^{3.5} x \ln x dx \approx 4.566$

```
fnInt(X*ln X,X,1
.7,3.5)
         4.566415359
```

65. $\int_{-2}^{2} \frac{1}{1+x^2} dx \approx 2.214$

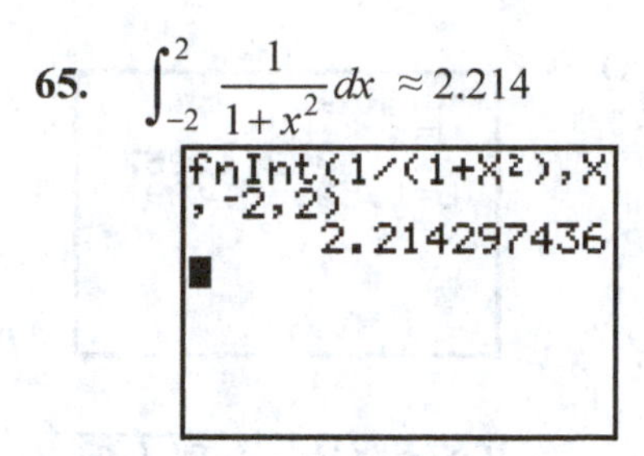

67. If $F(t)$ denotes the position of the car at time t, then the average velocity over the time interval $t = a$ to $t = b$ is given by

$$\frac{F(b) - F(a)}{b - a}.$$

$F'(t)$ gives the instantaneous velocity of the car at time t. By the Mean Value Theorem, there exists at least one time $t = c$ at which

$$\frac{F(b) - F(a)}{b - a} = F'(c).$$

Thus, if $\frac{F(b) - F(a)}{b - a} = 60$, then the instantaneous velocity must equal 60 at least once during the 10 minute time interval.

69. $C'(x) = 500 - \frac{x}{3}$ on [300, 900]

The increase in cost from a production level of 300 bikes per month to a production level of 900 bikes per month is given by:

$$\int_{300}^{900} \left(500 - \frac{x}{3}\right) dx = \left[500x - \frac{1}{6}x^2\right]_{300}^{900} = 315{,}000 - (135{,}000) = \$180{,}000$$

71. Total loss in value in the first 5 years:

$$V(5) - V(0) = \int_0^5 V'(t)\, dt = \int_0^5 500(t - 12) dt = 500\left[\frac{t^2}{2} - 12t\right]_0^5 = 500\left(\frac{25}{2} - 60\right) = -\$23{,}750$$

Total loss in value in the second 5 years:

$$V(10) - V(5) = \int_5^{10} V'(t)\, dt = \int_5^{10} 500(t - 12) dt = 500\left[\frac{t^2}{2} - 12t\right]_5^{10}$$

$$= 500\left[(50 - 120) - \left(\frac{25}{2} - 60\right)\right] = -\$11{,}250$$

73. (A) 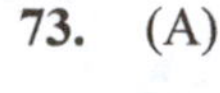

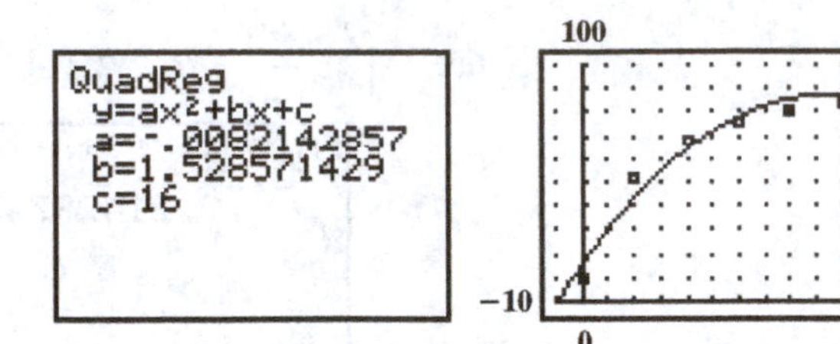

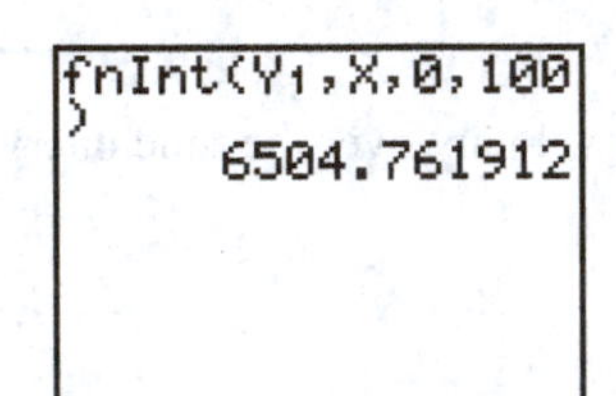

75. To find the useful life, set $C'(t) = R'(t)$ and solve for t.

$$\frac{1}{11}t = 5te^{-t^2}$$

$$e^{t^2} = 55$$

$$t^2 = \ln 55$$

$$t = \sqrt{\ln 55} \approx 2 \text{ years}$$

The total profit accumulated during the useful life is:

$$P(2) - P(0) = \int_0^2 [R'(t) - C'(t)]dt = \int_0^2 \left(5te^{-t^2} - \frac{1}{11}t\right)dt = \int_0^2 5te^{-t^2}dt - \int_0^2 \frac{1}{11}t\,dt$$

$$= -\frac{5}{2}\int_0^2 e^{-t^2}(-2t)dt - \frac{1}{11}\int_0^2 t\,dt$$

$$= -\frac{5}{2}e^{-t^2}\Big|_0^2 - \frac{1}{22}t^2\Big|_0^2$$

$$= -\frac{5}{2}e^{-4} + \frac{5}{2} - \frac{4}{22} = \frac{51}{22} - \frac{5}{2}e^{-4} \approx 2.272$$

[Note: In the first integral, the integrand has the form $e^u du$, where $u = -t^2$; an antiderivative is $e^u = e^{-t^2}$.]

Thus, the total profit is approximately $2,272.

77. $C(x) = 60{,}000 + 300x$

(A) Average cost per unit:

$$\overline{C}(x) = \frac{C(x)}{x} = \frac{60{,}000}{x} + 300$$

$$\overline{C}(500) = \frac{60{,}000}{500} + 300 = \$420$$

(B) $$\text{Avg. } C(x) = \frac{1}{500}\int_0^{500} (60{,}000 + 300x)dx$$

$$= \frac{1}{500}(60{,}000x + 150x^2)\Big|_0^{500}$$

$$= \frac{1}{500}(30{,}000{,}000 + 37{,}500{,}000) = \$135{,}000$$

(C) $\overline{C}(500)$ is the average cost per unit at a production level of 500 units; Avg. $C(x)$ is the average value of the total cost as production increases from 0 units to 500 units.

79. (A)

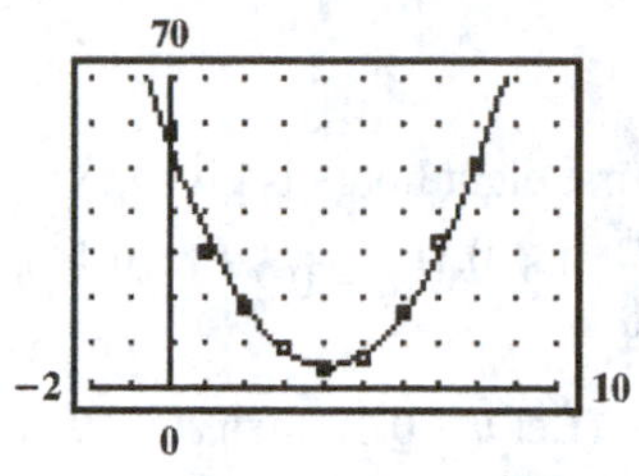

(B) Let $q(x)$ be the quadratic regression model found in part (A). The increase in cost in going from a production level of 2 thousand sunglasses per month to 8 thousand sunglasses per month is given (approximately) by

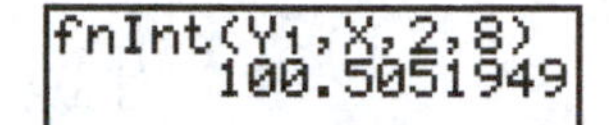

$$\int_2^8 q(x)dx \approx 100.505$$

Therefore, the increase in cost is approximately $100,505.

81. Average price:

$$\text{Avg. } S(x) = \frac{1}{30-20}\int_{20}^{30} 10(e^{0.02x} - 1)dx = \int_{20}^{30}(e^{0.02x} - 1)dx = \int_{20}^{30} e^{0.02x}dx - \int_{20}^{30} dx$$

$$= \frac{1}{0.02}\int_{20}^{30} e^{0.02x}(0.02)dx - x\Big|_{20}^{30} = 50\,e^{0.02x}\Big|_{20}^{30} - (30-20)$$

$$= 50e^{0.6} - 50e^{0.4} - 10 \quad \approx 6.51 \text{ or } \$6.51$$

83. $g(x) = 2400x^{-1/2}$ and $L'(x) = g(x)$.

The number of labor hours to assemble the 17th through the 25th control units is:

$$L(25) - L(16) = \int_{16}^{25} g(x)dx = \int_{16}^{25} 2400x^{-1/2}dx = 2400(2)x^{1/2}\Big|_{16}^{25} = 4800x^{1/2}\Big|_{16}^{25}$$

$$= 4800[25^{1/2} - 16^{1/2}] = 4800 \text{ labor hours.}$$

85. (A) The inventory function is obtained by finding the equation of the line joining (0, 600) and (3, 0).

Slope: $m = \dfrac{0-600}{3-0} = -200$, y intercept: $b = 600$

Thus, the equation of the line is: $I = -200t + 600$

(B) The average of I over [0, 3] is given by:

$$\text{Avg. } I(t) = \frac{1}{3-0}\int_0^3 I(t)dt = \frac{1}{3}\int_0^3 (-200t + 600)dt = \frac{1}{3}(-100t^2 + 600t)\Big|_0^3$$

$$= \frac{1}{3}[-100(3^2) + 600(3) - 0]$$

$$= \frac{900}{3} = 300 \text{ units}$$

87. Rate of production: $R(t) = \dfrac{100}{t+1} + 5,\ 0 \le t \le 20$

Total production from year N to year M is given by:

$$P = \int_N^M R(t)dt = \int_N^M \left(\frac{100}{t+1} + 5\right)dt = 100\int_N^M \frac{1}{t+1}dt + \int_N^M 5\,dt = 100\ln(t+1)\Big]_N^M + 5t\Big]_N^M$$

$$= 100\ln(M+1) - 100\ln(N+1) + 5(M-N)$$

Thus, for total production during the first 10 years, let $M = 10$ and $N = 0$.
$P = 100 \ln 11 - 100 \ln 1 + 5(10 - 0) = 100 \ln 11 + 50 \approx 290$ thousand barrels.
For the total production from the end of the 10th year to the end of the 20th year, let $M = 20$ and $N = 10$.
$P = 100 \ln 21 - 100 \ln 11 + 5(20 - 10) = 100 \ln 21 - 100 \ln 11 + 50 \approx 115$ thousand barrels.

89. $W'(t) = 0.2e^{0.1t}$

The weight increase during the first eight hours is given by:

$$W(8) - W(0) = \int_0^8 W'(t)\,dt = \int_0^8 0.2e^{0.1t}dt = 0.2\int_0^8 e^{0.1t}dt$$

$$= \frac{0.2}{0.1}\int_0^8 e^{0.1t}(0.1)dt \quad \text{(Let } u = 0.1t\text{, then } du = 0.1dt.)$$

$$= 2e^{0.1t}\Big|_0^8 = 2e^{0.8} - 2 \approx 2.45 \text{ grams}$$

The weight increase during the second eight hours, i.e., from the 8th hour through the 16th hour, is given by:

$$W(16) - W(8) = \int_8^{16} W'(t)d = \int_8^{16} 0.2e^{0.1t}dt = 2e^{0.1t}\Big|_8^{16} = 2e^{1.6} - 2e^{0.8} \approx 5.45 \text{ grams.}$$

91. $C(t) = t^3 - 2t + 10,\ \ 0 \le t \le 2.$

Average temperature over time period [0, 2] is given by:

$$\frac{1}{2-0}\int_0^2 C(t)dt = \frac{1}{2}\int_0^2 (t^3 - 2t + 10)dt = \frac{1}{2}\left(\frac{t^4}{4} - \frac{2t^2}{2} + 10t\right)\Bigg|_0^2 = \frac{1}{2}(4 - 4 + 20) = 10° \text{ Celsius}$$

93. $P(t) = \dfrac{8.4t}{t^2 + 49} + 0.1,\ \ 0 \le t \le 24$

(A) Average fraction of people during the first seven months:

$$\frac{1}{7-0}\int_0^7 \left[\frac{8.4t}{t^2+49} + 0.1\right]dt = \frac{4.2}{7}\int_0^7 \frac{2t}{t^2+49}dt + \frac{1}{7}\int_0^7 0.1dt$$

$$= 0.6\ln(t^2 + 49)\Big|_0^7 + \frac{0.1}{7}t\Big|_0^7$$

$$= 0.6[\ln 98 - \ln 49] + 0.1 = 0.6 \ln 2 + 0.1 \approx 0.516$$

(B) Average fraction of people during the first two years:

$$\frac{1}{24-0}\int_0^{24} \left[\frac{8.4t}{t^2+49} + 0.1\right]dt = \frac{4.2}{24}\int_0^{24} \frac{2t}{t^2+49}dt + \frac{1}{24}\int_0^{24} 0.1dt$$

$$= 0.175\ln(t^2 + 49)\Big|_0^{24} + \frac{0.1}{24}t\Big|_0^{24}$$

$$= 0.175[\ln 625 - \ln 49] + 0.1 \approx 0.546$$

CHAPTER 5 REVIEW

1. $\int (6x+3)\,dx = 6\int x\,dx + \int 3\,dx = 6\cdot\frac{x^2}{2} + 3x + C = 3x^2 + 3x + C$ (5-1)

2. $\int_{10}^{20} 5dx = 5x\Big]_{10}^{20} = 5(20) - 5(10) = 50$ (5-5)

3. $\int_0^9 (4-t^2)dt = \int_0^9 4dt - \int_0^9 t^2 dt = 4t\Big|_0^9 - \frac{t^3}{3}\Big|_0^9 = 36 - 243 = -207$ (5-5)

4. $\int (1-t^2)^3 t\,dt = \int (1-t^2)^3 \left(\frac{-2}{-2}\right) t\,dt = -\frac{1}{2}\int (1-t^2)^3(-2t)\,dt = -\frac{1}{2}\int u^3\,du = -\frac{1}{2}\cdot\frac{u^4}{4} + C$

$= -\frac{1}{8}(1-t^2)^4 + C$ (5-2)

5. $\int \frac{1+u^4}{u}du = \int \left(\frac{1}{u} + u^3\right)du = \int \frac{1}{u}du + \int u^3\,du = \ln|u| + \frac{1}{4}u^4 + C$ (5-1)

6. $\int_0^1 xe^{-2x^2}\,dx$

Let $u = -2x^2$. Then $du = -4x\,dx$.

$\int xe^{-2x^2}\,dx = \int e^{-2x^2}\left(\frac{-4}{-4}\right)x\,dx = -\frac{1}{4}\int e^u\,du = -\frac{1}{4}e^u + C = -\frac{1}{4}e^{-2x^2} + C$

$\int_0^1 xe^{-2x^2}\,dx = -\frac{1}{4}e^{-2x^2}\Big|_0^1 = -\frac{1}{4}e^{-2} + \frac{1}{4} \approx 0.216$ (5-5)

7. $F(x) = \ln x^2 = 2\ln x,\quad F'(x) = \frac{2}{x} \neq \ln 2x$; no. (5-1)

8. $F(x) = \ln x^2 = 2\ln x,\quad F'(x) = \frac{2}{x} = f(x)$; yes. (5-1)

9. $F(x) = (\ln x)^2,\quad F'(x) = 2(\ln x)\left(\frac{1}{x}\right) = \frac{2\ln x}{x} \neq 2\ln x = f(x)$; no. (5-1)

10. $F(x) = (\ln x)^2,\quad F'(x) = 2(\ln x)\left(\frac{1}{x}\right) = \frac{2\ln x}{x} = f(x)$; yes. (5-1)

11. $y = 3x + 17,\quad y' = 3;$

$(x+5)y' = (x+5)(3) = 3x + 15 = 3x + 17 - 2 = y - 2$; yes. (5-3)

12. $y = 4x^3 + 7x^2 - 5x + 2,\quad y' = 12x^2 + 14x - 5,\quad y'' = 24x + 14,\quad y''' = 24;$

$(x+2)(24) - 24x = 24x + 48 - 24x = 48$; yes. (5-3)

13. $\frac{d}{dx}\left[\int e^{-x^2}\,dx\right] = e^{-x^2}$ (5-1)

14. $\int \frac{d}{dx}(\sqrt{4+5x})\,dx = \sqrt{4+5x} + C$ (5-1)

15. $\frac{dy}{dx} = 3x^2 - 2$

$y = f(x) = \int 3x^2 - 2)\,dx = f(x) = x^3 - 2x + C$

$f(0) = C = 4;\quad f(x) = x^3 - 2x + 4$ (5-3)

16. (A) $\int (8x^3 - 4x - 1)\,dx = 8\int x^3\,dx - 4\int x\,dx - \int dx = 8 \cdot \frac{1}{4}x^4 - 4\,\frac{1}{2}x^2 - x + C$

$= 2x^4 - 2x^2 - x + C$

(B) $\int \left(e^t - 4t^{-1}\right)dt = \int e^t\,dt - 4\int \frac{1}{t}\,dt = e^t - 4\ln|t| + C$ (5-1)

17. $f(x) = x^2 + 1,\ a = 1,\ b = 5,\ n = 2,\ \Delta x = \frac{5-1}{2} = 2;$

$R_2 = f(3)2 + f(5)2 = 10(2) + 26(2) = 72$

Error bound for R_2: f is increasing on [1, 5], so

$|I - R_2| \le [f(5) - f(1)]\frac{5-1}{2} = (26 - 2)(2) = 48$

Thus, $I = 72 \pm 48$. (5-4)

18. $\int_1^5 (x^2 + 1)\,dx = \left[\frac{1}{3}x^3 + x\right]_1^5 = \frac{125}{3} + 5 - \left(\frac{1}{3} + 1\right) = \frac{136}{3} = 45\tfrac{1}{3}$

$|I - R_2| = \left|45\tfrac{1}{3} - 72\right| = 26\tfrac{2}{3} \approx 26.67$ (5-5)

19. Using the values of f in the table with $a = 1,\ b = 17,\ n = 4$

$\Delta x = \frac{17-1}{4} = 4$, we have

$L_4 = f(1)4 + f(5)4 + f(9)4 + f(13)4 = [1.2 + 3.4 + 2.6 + 0.5]4 = 30.8$ (5-4)

20. $f(x) = 6x^2 + 2x$ on $[-1, 2]$;

$\text{Ave}\,f(x) = \frac{1}{2-(-1)}\int_{-1}^{2} (6x^2 + 2x)\,dx = \frac{1}{3}(2x^3 + x^2)\Big|_{-1}^{2} = \frac{1}{3}[20 - (-1)] = 7$ (5-5)

21. width $= 2 - (-1) = 3$, height = Avg. $f(x) = 7$ (5-5)

22. $f(x) = 100 - x^2$

$\Delta x = \frac{11-3}{4} = \frac{8}{4} = 2;\ c_i = \frac{x_{i-1} + x_i}{2}$ (= midpoint of interval)

$S_4 = f(4)2 + f(6)2 + f(8)\cdot 2 + f(10)\cdot 2 = [84 + 64 + 36 + 0]2 = (184)2 = 368$ (5-4)

23. $f(x) = 100 - x^2$

$\Delta x = \frac{5-(-5)}{5} = \frac{10}{5} = 2$

$S_5 = f(-4)2 + f(-1)2 + f(1)2 + f(2)2 + f(5)2 = [84 + 99 + 99 + 96 + 75]2 = (453)2 = 906$ (5-4)

24. $\int_a^b 5f(x)\,dx = 5\int_a^b f(x)\,dx = 5(-2) = -10$ (5-4, 5-5)

25. $\int_b^c \frac{f(x)}{5}\,dx = \frac{1}{5}\int_b^c f(x)\,dx = \frac{1}{5}(2) = \frac{2}{5} = 0.4$ (5-4, 5-5)

26. $\int_b^d f(x)\,dx = \int_b^c f(x)\,dx + \int_c^d f(x)\,dx = 2 - 0.6 = 1.4$ (5–4, 5-5)

27. $\int_a^c f(x)\,dx = \int_a^b f(x)\,dx + \int_b^c f(x)\,dx = -2+2=0$ (5-4, 5-5)

28. $\int_0^d f(x)\,dx = \int_0^a f(x)\,dx + \int_a^b f(x)\,dx + \int_b^c f(x)\,dx + \int_c^d f(x)\,dx = 1-2+2-0.6=0.4$ (5-4, 5-5)

29. $\int_b^a f(x)\,dx = -\int_a^b f(x)\,dx = -(-2) = 2$ (5-4, 5-5)

30. $\int_c^b f(x)\,dx = -\int_b^c f(x)\,dx = -2$ (5-4, 5-5)

31. $\int_d^0 f(x)\,dx = -\int_0^d f(x)\,dx = -0.4$ (from Problem 28) (5-4, 5-5)

32. (A) $\frac{dy}{dx} = \frac{2y}{x}$; $\left.\frac{dy}{dx}\right|_{(2,1)} = \frac{2(1)}{2} = 1$, $\left.\frac{dy}{dx}\right|_{(-2,-1)} = \frac{2(-1)}{-2} = 1$

(B) $\frac{dy}{dx} = \frac{2x}{y}$; $\left.\frac{dy}{dx}\right|_{(2,1)} = \frac{2(2)}{1} = 4$, $\left.\frac{dy}{dx}\right|_{(-2,-1)} = \frac{2(-2)}{-1} = 4$ (5-3)

33. $\frac{dy}{dx} = \frac{2y}{x}$; from the figure, the slopes at (2, 1) and (–2, –1) are approximately equal to 1 as computed in Problem 32(A), not 4 as computed in Problem 32(B). (5-3)

34. Let $y = Cx^2$. Then $\frac{dy}{dx} = 2Cx$. From the original equation, $C = \frac{y}{x^2}$ so

$$\frac{dy}{dx} = 2x\left(\frac{y}{x^2}\right) = \frac{2y}{x}$$ (5-3)

35. Letting $x=2$ and $y=1$ in $y=Cx^2$, we get

$1 = 4C$ so $C = \frac{1}{4}$ and $y = \frac{1}{4}x^2$

Letting $x=-2$ and $y=-1$ in $y=Cx^2$, we get

$-1 = 4C$ so $C = -\frac{1}{4}$ and $y = -\frac{1}{4}x^2$ (5-3)

36.

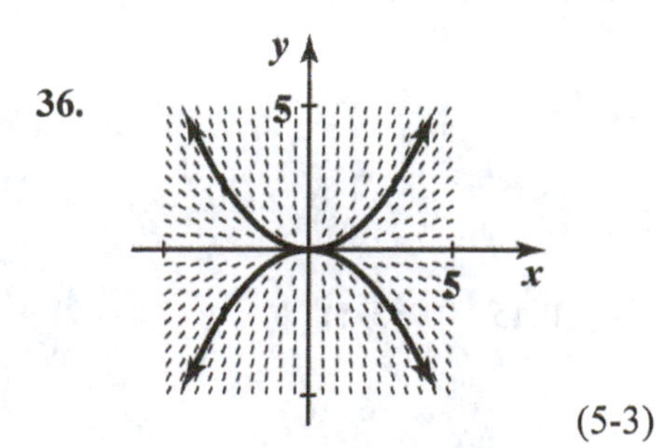

(5-3)

37.

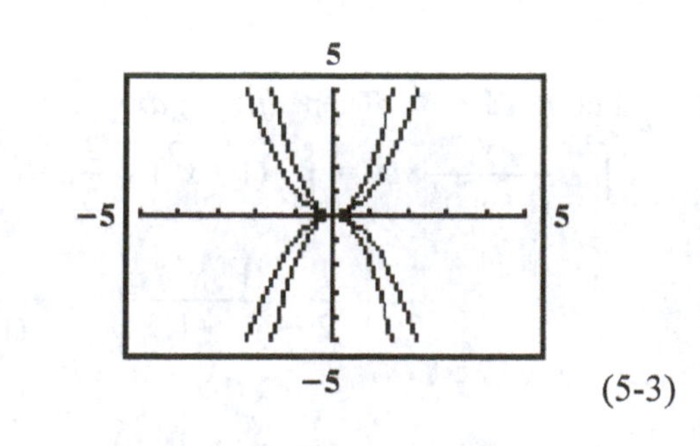

(5-3)

38. $\int_{-1}^{1} \sqrt{1+x}\,dx = \int_{0}^{2} u^{1/2}\,du = \left.\frac{u^{3/2}}{3/2}\right|_{0}^{2} = \frac{2}{3}(2)^{3/2} \approx 1.886$

Let $u = 1 + x$, $du = dx$.
When $x = -1$, $u = 0$, when $x = 1$, $u = 2$. (5-5)

39. $\int_{-1}^{0} x^2(x^3+2)^{-2}\,dx = \int_{-1}^{0} (x^3+2)^{-2}\left(\frac{3}{3}\right)x^2\,dx = \frac{1}{3}\int_{-1}^{0} (x^3+2)^{-2}3x^2\,dx = \frac{1}{3}\int_{1}^{2} u^{-2}\,du$

Let $u = x^3 + 2$. Then $du = 3x^2dx$
When $x = -1$, $u = 1$, when $x = 0$, $u = 2$.

$$= \frac{1}{3}\cdot\frac{u^{-1}}{-1}\Big]_{1}^{2} = -\frac{1}{3u}\Big]_{1}^{2}$$

$$= -\frac{1}{6} + \frac{1}{3} = \frac{1}{6} \quad \text{(5-5)}$$

40. $\int 5e^{-t}\,dt = -5\int e^{-t}(-dt) = -5\int e^{u}\,du = -5e^{u} + C = -5e^{-t} + C$

Let $u = -t$. Then $du = -dt$ (5-2)

41. $\int_{1}^{e} \frac{1+t^2}{t}\,dt = \int_{1}^{e}\left(\frac{1}{t}+t\right)dt = \int_{1}^{e}\frac{1}{t}\,dt + \int_{1}^{e} t\,dt = \ln t\,\Big]_{1}^{e} + \frac{1}{2}t^2\,\Big]_{1}^{e} = \ln e - \ln 1 + \frac{1}{2}e^2 - \frac{1}{2}$

$$= \frac{1}{2} + \frac{1}{2}e^2 \quad \text{(5-5)}$$

42. $\int x e^{3x^2}\,dx = \int e^{3x^2}\left(\frac{6}{6}\right)x\,dx = \frac{1}{6}\int e^{3x^2}\,6x\,dx = \frac{1}{6}\int e^{u}\,du = \frac{1}{6}e^{u} + C = \frac{1}{6}e^{3x^2} + C$

Let $u = 3x^2$ Then $du = 6x\,dx$ (5-2)

43. $\int_{-3}^{1}\frac{1}{\sqrt{2-x}}\,dx = -\int_{-3}^{1}\frac{1}{\sqrt{2-x}}(-dx) = -\int_{5}^{1} u^{-1/2}\,du = \int_{1}^{5} u^{-1/2}\,du = 2u^{1/2}\Big]_{1}^{5} = 2\sqrt{5} - 2 \approx 2.472$

Let $u = 2 - x$. Then $du = -dx$
When $x = -3$, $u = 5$, when $x = 1$, $u = 1$ (5-5)

44. Let $u = 1 + x^2$. Then $du = 2x\,dx$.

$$\int_{0}^{3}\frac{x}{1+x^2}\,dx = \int_{0}^{3}\frac{1}{1+x^2}\frac{2}{2}x\,dx = \frac{1}{2}\int_{0}^{3}\frac{1}{1+x^2}2x\,dx = \frac{1}{2}\ln(1+x^2)\Big|_{0}^{3}$$

$$= \frac{1}{2}\ln 10 - \frac{1}{2}\ln 1 = \frac{1}{2}\ln 10 \approx 1.151 \quad \text{(5-5)}$$

45. Let $u = 1 + x^2$. Then $du = 2x\,dx$.

$$\int_{0}^{3}\frac{x}{(1+x^2)^2}\,dx = \int_{0}^{3}(1+x^2)^{-2}\,\frac{2}{2}x\,dx = \frac{1}{2}\int_{0}^{3}(1+x^2)^{-2}\,2x\,dx$$

$$= \frac{1}{2}\cdot\frac{(1+x^2)^{-1}}{-1}\Big|_{0}^{3} = \frac{-1}{2(1+x^2)}\Big|_{0}^{3} = -\frac{1}{20} + \frac{1}{2} = \frac{9}{20} = 0.45 \quad \text{(5-5)}$$

46. $\int x^3(2x^4+5)^5\,dx$

Let $u = 2x^4 + 5$. Then $du = 8x^3 dx$.

$$\int x^3(2x^4+5)^5\,dx = \int (2x^4+5)^5\left(\frac{8}{8}\right)x^3\,dx = \frac{1}{8}\int (2x^4+5)^5\,8x^3\,dx = \frac{1}{8}\int u^5\,du = \frac{1}{8}\cdot\frac{u^6}{6}+C$$

$$= \frac{(2x^4+5)^6}{48}+C \qquad (5\text{-}2)$$

47. $$\int \frac{e^{-x}}{e^{-x}+3}\,dx = \int \frac{1}{e^{-x}+3}\left(\frac{-1}{-1}\right)e^{-x}\,dx = -\int \frac{1}{u}\,du = -\ln|u| + C = -\ln|e^{-x}+3| + C = -\ln(e^{-x}+3) + C$$

Let $u = e^{-x} + 3$. [Note: Absolute value not needed since $e^{-x} + 3 > 0$.]

Then $du = -e^{-x}dx$. (5-2)

48. $$\int \frac{e^x}{(e^x+2)^2}\,dx = \int (e^x+2)^{-2}\,e^x\,dx = \int u^{-2}\,du\,dx = \frac{u^{-1}}{-1} + C = -(e^x+2)^{-1} + C = \frac{-1}{(e^x+2)} + C \qquad (5\text{-}2)$$

Let $u = e^x + 2$. Then $du = e^x dx$

49. $\frac{dy}{dx} = 3x^{-1} - x^{-2}$

$$y = \int (3x^{-1} - x^{-2})\,dx = 3\int \frac{1}{x}\,dx - \int x^{-2}\,dx = 3\ln|x| - \frac{x^{-1}}{-1} + C = 3\ln|x| + x^{-1} + C$$

Given $y(1) = 5$:

$5 = 3\ln 1 + 1 + C$ and $C = 4$

Thus, $y = 3\ln|x| + x^{-1} + 4$. (5-2, 5-3)

50. $\frac{dy}{dx} = 6x + 1$

$$f(x) = y = \int (6x+1)\,dx = \frac{6x^2}{2} + x + C = 3x^2 + x + C$$

We have $y = 10$ when $x = 2$: $3(2)^2 + 2 + C = 10$, $C = 10 - 12 - 2 = -4$

Thus, the equation of the curve is $y = 3x^2 + x - 4$. (5-3)

51. (A) $f(x) = 3\sqrt{x} = 3x^{1/2}$ on [1, 9]

$$\text{Avg. } f(x) = \frac{1}{9-1}\int_1^9 3x^{1/2}\,dx$$

$$= \frac{3}{8}\cdot\frac{x^{3/2}}{3/2}\Big|_1^9 = \frac{1}{4}x^{3/2}\Big|_1^9$$

$$= \frac{27}{4} - \frac{1}{4} = \frac{26}{4} = 6.5$$

(B)

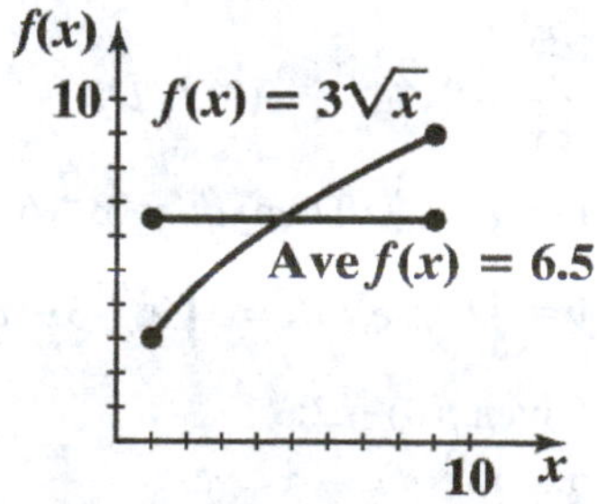

f(x)
10
$f(x) = 3\sqrt{x}$
Ave $f(x)$ = 6.5
10 x

(5-5)

52. Let $u = \ln x$. Then $du = \frac{1}{x}dx$.

$$\int \frac{(\ln x)^2}{x}dx = \int (\ln x)^2 \frac{1}{x}dx = \int u^2\,du = \frac{u^3}{3} + C = \frac{(\ln x)^3}{3} + C \qquad (5\text{-}2)$$

53. $\int x(x^3-1)^2\,dx = \int x(x^6-2x^3+1)\,dx$ (square $x^3 - 1$)

$$= \int (x^7 - 2x^4 + x)\,dx = \frac{x^8}{8} - \frac{2x^5}{5} + \frac{x^2}{2} + C \qquad (5\text{-}2)$$

54. $\int \frac{x}{\sqrt{6-x}}dx$

Let $u = 6 - x$. Then $x = 6 - u$ and $dx = -du$.

$$\int \frac{x}{\sqrt{6-x}}dx = -\int \frac{6-u}{u^{1/2}}du = \int (u^{1/2} - 6u^{-1/2})\,du = \frac{u^{3/2}}{3/2} - \frac{6u^{1/2}}{1/2} + C = \frac{2}{3}u^{3/2} - 12u^{1/2} + C$$

$$= \frac{2}{3}(6-x)^{3/2} - 12(6-x)^{1/2} + C \qquad (5\text{-}2)$$

55. $\int_0^7 x\sqrt{16-x}\,dx$. First consider the indefinite integral:

Let $u = 16 - x$. Then $x = 16 - u$ and $dx = -du$.

$$\int x\sqrt{16-x}\,dx = -\int (16-u)u^{1/2}\,du = \int (u^{3/2} - 16u^{1/2})\,du = \frac{u^{5/2}}{5/2} - \frac{16u^{3/2}}{3/2} + C = \frac{2}{5}u^{5/2} - \frac{32}{3}u^{3/2} + C$$

$$= \frac{2(16-x)^{5/2}}{5} - \frac{32(16-x)^{3/2}}{3} + C$$

$$\int_0^7 x\sqrt{16-x}\,dx = \left[\frac{2(16-x)^{5/2}}{5} - \frac{32(16-x)^{3/2}}{3}\right]_0^7 = \frac{2\cdot 9^{5/2}}{5} - \frac{32\cdot 9^{3/2}}{3} - \left(\frac{2\cdot 16^{5/2}}{5} - \frac{32\cdot 16^{3/2}}{3}\right)$$

$$= \frac{2\cdot 3^5}{5} - \frac{32\cdot 3^3}{3} - \left(\frac{2\cdot 4^5}{5} - \frac{32\cdot 4^3}{3}\right) = \frac{486}{5} - 288 - \left(\frac{2048}{5} - \frac{2048}{3}\right)$$

$$= \frac{1234}{15} \approx 82.267 \qquad (5\text{-}5)$$

56. $\int_1^1 (x+1)^9\,dx = 0$. (Property 9a) (5-4)

57. $\frac{dy}{dx} = 9x^2e^{x^3}$, $f(0) = 2$

Let $u = x^3$. Then $du = 3x^2dx$.

$$y = \int 9x^2e^{x^3}\,dx = 3\int e^{x^3}\cdot 3x^2\,dx = 3\int e^u\,du = 3e^u + C = 3e^{x^3} + C$$

Given $f(0) = 2$:

$2 = 3e^0 + C = 3 + C$

Hence, $C = -1$ and $y = f(x) = 3e^{x^3} - 1$. (5-3)

58. $\frac{dN}{dt} = 0.06N$, $N(0) = 800$, $N > 0$

From the differential equation, $N(t) = Ce^{0.06t}$, where C is an arbitrary constant. Since $N(0) = 800$, we have

$800 = Ce^0 = C$.

Hence, $C = 800$ and $N(t) = 800e^{0.06t}$. (5-3)

59. $N = 50(1 - e^{-0.07t})$,
$0 \le t \le 80$, $0 \le N \le 60$

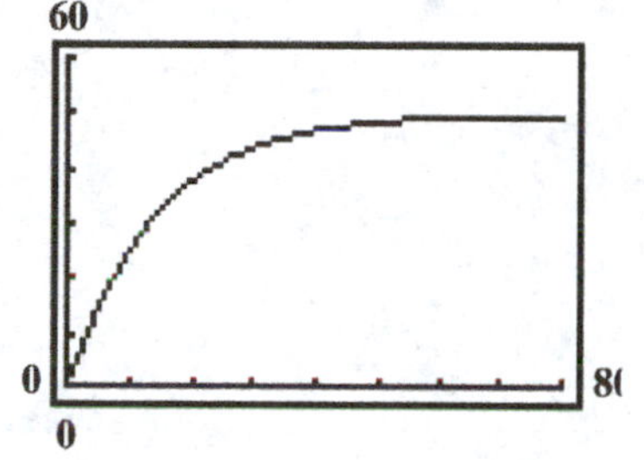

Limited growth

(5-3)

60. $p = 500e^{-0.03x}$,
$0 \le x \le 100$, $0 \le p \le 500$

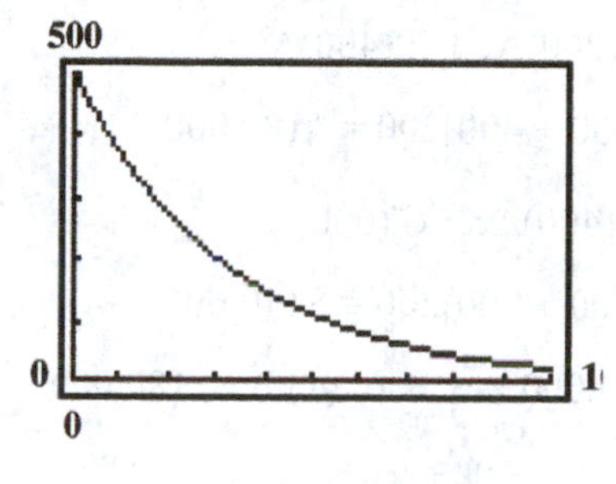

Exponential decay

(5-3)

61. $A = 200e^{0.08t}$,
$0 \le t \le 20$, $0 \le A \le 1{,}000$

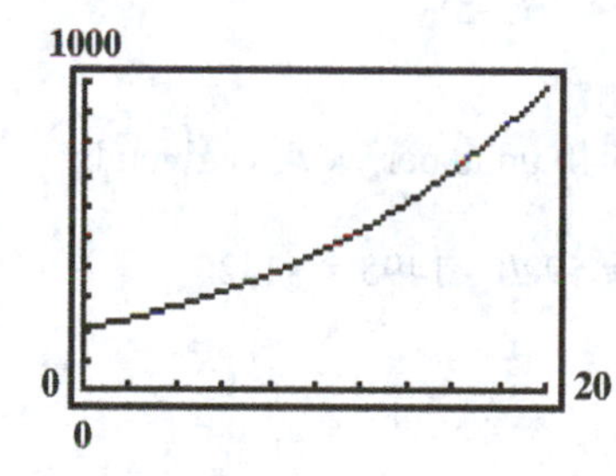

Unlimited growth

(5-3)

62. $N = \frac{100}{1+9e^{-0.3t}}$,
$0 \le t \le 25$, $0 \le N \le 100$

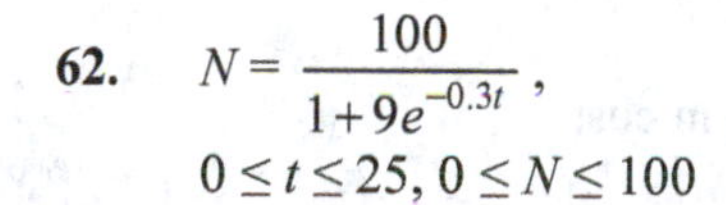

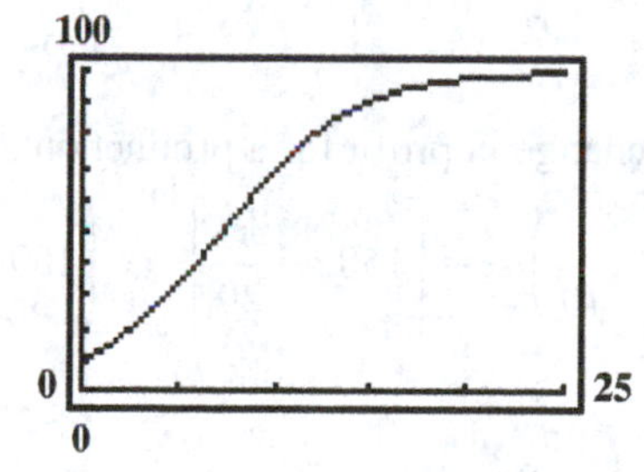

Logistic growth

(5-3)

63. $\int_{-0.5}^{0.6} \frac{1}{\sqrt{1-x^2}}\,dx \approx 1.167$

```
fnInt(Y1,X,-0.5,
0.6)
      1.167099884
```

(5-5)

64. $\int_{-2}^{3} x^2 e^x\,dx \approx 99.074$

```
fnInt(Y1,X,-2,3)
      99.07433178
```

(5-5)

65. $\int_{0.5}^{2.5} \frac{\ln x}{x^2}\,dx \approx -0.153$

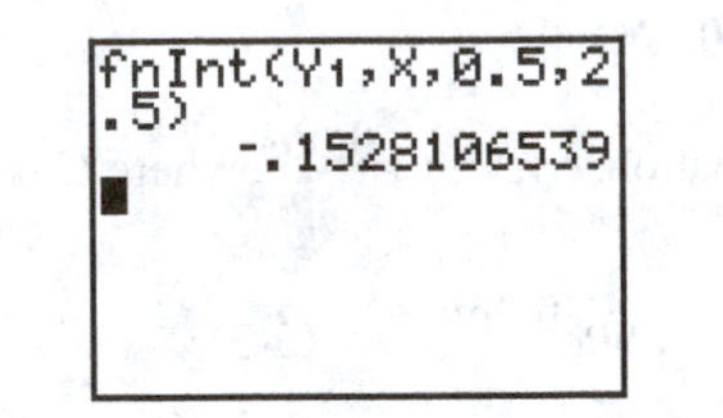

(5-5)

66. $a = 200,\ b = 600,\ n = 2,\ \Delta x = \frac{600-200}{2} = 200$

$L_2 = C'(200)\Delta x + C'(400)\Delta x$

$= [500 + 400]200 = \$180{,}000$

$R_2 = C'(400)\Delta x + C'(600)\Delta x$

$= [400 + 300]200 = \$140{,}000$

$140{,}000 \le \int_{200}^{600} C'(x)dx \le 180{,}000$ (5-4)

67. The graph of $C'(x)$ is a straight line with y-intercept = 600 and slope = $\frac{300-600}{600-0} = -\frac{1}{2}$

Thus, $C'(x) = -\frac{1}{2}x + 600$

Increase in costs:

$\int_{200}^{600}\left(600 - \frac{1}{2}x\right)dx = \left[600x - \frac{1}{4}x^2\right]_{200}^{600} = 270{,}000 - 110{,}000 = \$160{,}000$ (5-5)

68. The total change in profit for a production change from 10 units per week to 40 units per week is given by:

$\int_{10}^{40}\left(150 - \frac{x}{10}\right)dx = \left[150x - \frac{x^2}{20}\right]_{10}^{40} = \left(150(40) - \frac{40^2}{20}\right) - \left(150(10) - \frac{10^2}{20}\right) = 5920 - 1495 = \$4425$

(5-5)

69. $P'(x) = 100 - 0.02x$

$P(x) = \int (100 - 0.02x)\,dx = 100x - 0.02\frac{x^2}{2} + C = 100x - 0.01x^2 + C$

$P(0) = 0 - 0 + C = 0$

$C = 0$

Thus, $P(x) = 100x - 0.01x^2$.

The profit on 10 units of production is given by: $P(10) = 100(10) - 0.01(10)^2 = \$999$ (5-3)

70. The required definite integral is:

$\int_0^{15}(60 - 4t)\,dt = \left[60t - 2t^2\right]_0^{15} = 60(15) - 2(15)^2 = 450$ or 450,000 barrels

The total production in 15 years is 450,000 barrels. (5-5)

71. Average inventory from $t = 3$ to $t = 6$:

Avg. $I(t) = \frac{1}{6-3}\int_3^6 (10 + 36t - 3t^2)\,dt = \frac{1}{3}\left[10t + 18t^2 - t^3\right]_3^6 = \frac{1}{3}[60 + 648 - 216 - (30 + 162 - 27)]$

$= 109$ items (5-5)

72. $S(x) = 8(e^{0.05x} - 1)$
Average price over the interval [40, 50]:

$$\text{Avg. } S(x) = \frac{1}{50-40}\int_{40}^{50} 8(e^{0.05x}-1)\,dx = \frac{8}{10}\int_{40}^{50}(e^{0.05x}-1)\,dx = \frac{4}{5}\left[\frac{e^{0.05x}}{0.05} - x\right]_{40}^{50}$$

$$= \frac{4}{5}[20e^{2.5} - 50 - (20e^{2} - 40)] = 16e^{2.5} - 16e^{2} - 8 \approx \$68.70$$

(5-5)

73. To find the useful life, set $R'(t) = C'(t)$:

$20e^{-0.1t} = 3$

$e^{-0.1t} = \frac{3}{20}$

$-0.1t = \ln\left(\frac{3}{20}\right) \approx -1.897;\quad t = 18.97$ or 19 years

$$\text{Total profit} = \int_0^{19}[R'(t) - C(t)]\,dt = \int_0^{19}(20e^{-0.1t} - 3)\,dt$$

$$= 20\int_0^{19} e^{-0.1t}\,dt - \int_0^{19} 3\,dt = \frac{20}{-0.1}\int_0^{19} e^{-0.1t}(-0.1)\,dt - \int_0^{19} 3dt$$

$$= -200e^{-0.1t}\Big|_0^{19} - 3t\Big|_0^{19} = -200e^{-1.9} + 200 - 57 \approx 113.086 \text{ or } \$113,086$$ (5-5)

74. $S'(t) = 4e^{-0.08t}$, $0 \le t \le 24$. Therefore,

$S(t) = \int 4e^{-0.08t}dt = \frac{4e^{-0.08t}}{-0.08} + C = -50e^{-0.08t} + C.$

Now, $S(0) = 0$, so

$0 = -50e^{-0.08(0)} + C = -50 + C.$

Thus, $C = 50$, and $S(t) = 50(1 - e^{-0.08t})$ gives the total sales after t months.
Estimated sales after 12 months:

$S(12) = 50(1 - e^{-0.08(12)}) = 50(1 - e^{-0.96}) \approx 31$ or \$31 million.

To find the time to reach \$40 million in sales, solve $40 = 50(1 - e^{-0.08t})$ for t.

$0.8 = 1 - e^{-0.08t}$

$e^{-0.08t} = 0.2$

$-0.08t = \ln(0.2)$

$t = \frac{\ln(0.2)}{-0.08} \approx 20$ months (5-3)

75. $\frac{dA}{dt} = -5t^{-2}$, $1 \le t \le 5$

$A = \int -5t^{-2}\,dt = -5\int t^{-2}\,dt = -5 \cdot \frac{t^{-1}}{-1} + C = \frac{5}{t} + C$

Now $A(1) = \frac{5}{1} + C = 5$. Therefore, $C = 0$ and

$A(t) = \frac{5}{t}$

$A(5) = \frac{5}{5} = 1$

The area of the wound after 5 days is 1 cm^2. (5-3)

76. The total amount of seepage during the first four years is given by:

$$T = \int_0^4 R(t)dt = \int_0^4 \frac{1000}{(1+t)^2}dt = 1000\int_0^4 (1+t)^{-2}\,dt = 1000\frac{(1+t)^{-1}}{-1}\Bigg|_0^4$$

[Let $u = 1 + t$. Then $du = dt$.] $= \frac{-1000}{1+t}\Bigg|_0^4 = \frac{-1000}{5} + 1000 = 800$ gallons

(5-5)

77. (A) The exponential growth law applies and we have:

$\frac{dP}{dt} = 0.0107P,\ \ P(0) = 116$ (million)

Thus $P(t) = 116e^{0.0107t}$

The year 2025: $t = 12$, and $P(12) = 116e^{0.0107(12)} = 116e^{0.1284} \approx 132$

Assuming that the population continues to grow at the rate 1.07% per year, the population in 2025 will be approximately 132 million.

(B) Time to double:

$116e^{0.0107t} = 232$

$e^{0.0107t} = 2$

$0.0107t = \ln 2$

$t = \frac{\ln 2}{0.0107} \approx 65$

At the current growth rate it will take approximately 65 years for the population to double.

(5-3)

78. Let $Q = Q(t)$ be the amount of carbon-14 present in the bone at time t. Then,

$\frac{dQ}{dt} = -0.0001238Q$ and $Q(t) = Q_0e^{-0.0001238t}$,

where Q_0 is the amount present originally (i.e., at the time the animal died). We want to find t such that $Q(t) = 0.04Q_0$:

$0.04\,Q_0 = Q_0e^{-0.0001238t}$

$e^{-0.0001238t} = 0.04$

$-0.0001238t = \ln 0.04$

$t = \frac{\ln 0.04}{-0.0001238} \approx 26{,}000$ years (5-3)

79. $N'(t) = 7e^{-0.1t}$ and $N(0) = 25$.

$N(t) = \int 7e^{-0.1t}\,dt = 7\int e^{-0.1t}\,dt = \frac{7}{-0.1}\int e^{-0.1t}(-0.1)\,dt = -70e^{-0.1t} + C, \quad 0 \le t \le 15$

Given $N(0) = 25$: $25 = -70e^{0} + C = -70 + C$

Hence, $C = 95$ and $N(t) = 95 - 70e^{-0.1t}$. The student would be expected to type $N(15) = 95 - 70e^{-0.1(15)}$ $= 95 - 70e^{-1.5} \approx 79$ words per minute after completing the course. (5-3)

6 ADDITIONAL INTEGRATION TOPICS

EXERCISE 6-1

Things to remember:

1. AREA BETWEEN TWO CURVES

   If f and g are continuous and $f(x) \geq g(x)$ over the interval $[a, b]$, then the area bounded by $y = f(x)$ and $y = g(x)$, for $a \leq x \leq b$, is given exactly by:

$$A = \int_a^b [f(x) - g(x)]dx.$$

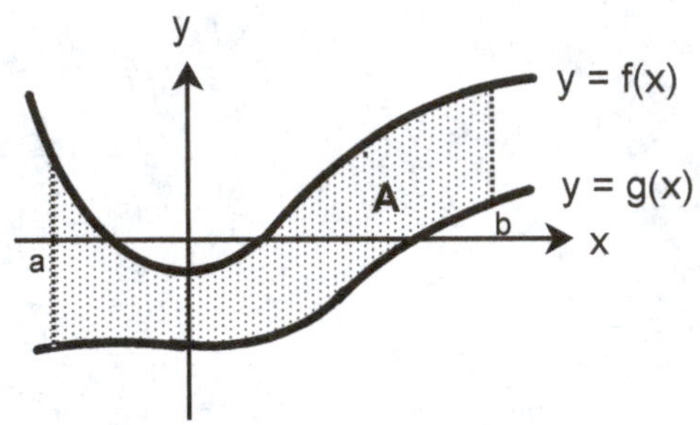

2. GINI INDEX OF INCOME CONCENTRATION

   If $y = f(x)$ is the equation of a Lorenz curve, then the

$$\text{Gini Index} = 2\int_0^1 [x - f(x)]dx.$$

---

1. The region bounded by the graphs of f and g is a rectangle with dimensions 15×10: area = 150.

3. The region bounded by the graphs of f and g is a triangle with vertices (0,6), (5, 16), (5,1). The base has length 15, the height is 5: area $\frac{1}{2}(15\times5) = 37.5$.

5. The region bounded by the graphs of f and g is a trapezoid with vertices $(-1,2)$, $(2,8)$, $(2,-5)$, $(-1,-2)$. The bases have lengths 4 and 13, and height 3. Area: $\frac{1}{2}(4+13)3 = 25.5$.

7. The region bounded by the graphs of f and g is one-eighth of the disk centered at the origin with radius 4: area: $\frac{1}{8}(\pi 4^2) = 2\pi$.

9. $A = \int_a^b g(x)dx$

11. $A = \int_a^b [-h(x)]dx$

13. Since the shaded region in Figure (c) is below the x-axis, $h(x) \leq 0$. Thus, $\int_a^b h(x)dx$ represents the negative of the area of the region.

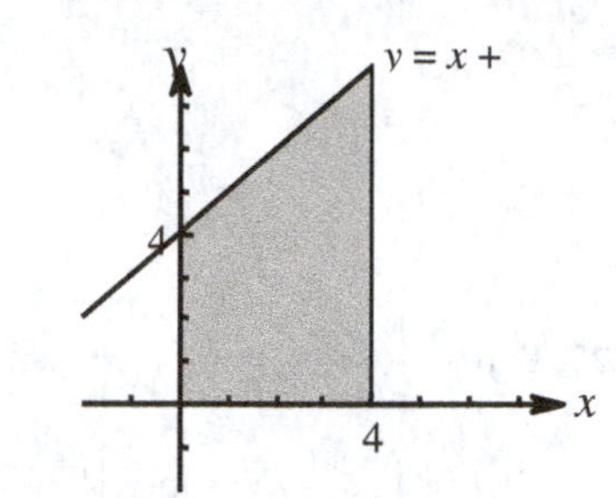

15. $y = x + 4$; $y = 0$ on $[0, 4]$

$$A = \int_0^4 (x+4)dx = \left(\frac{1}{2}x^2 + 4x\right)\Big|_0^4 = (8+16) - 0 = 24$$

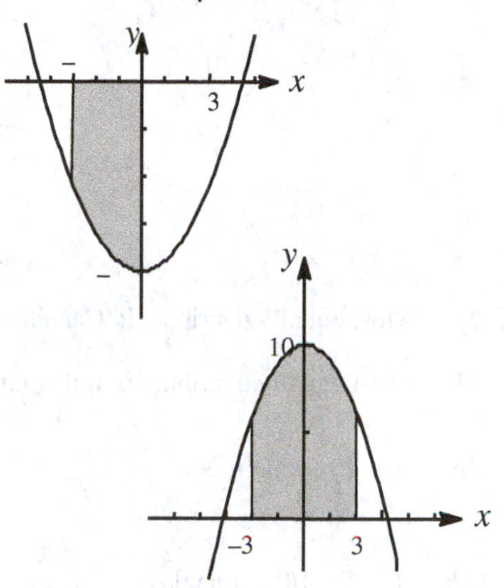

17. $y = x^2 - 20$; $y = 0$ on $[-3, 0]$

$$A = -\int_{-3}^0 (x^2 - 20)dx$$

$$= \int_{-3}^0 (20 - x^2)dx = \left(20x - \frac{1}{3}x^3\right)\Big|_{-3}^0$$

$$= 0 - (-60 + 9) = 51$$

19. $y = -x^2 + 10$; $y = 0$ on $[-3, 3]$

$$A = \int_{-3}^3 (-x^2 + 10)\,dx = \left[-\frac{1}{3}x^3 + 10x\right]_{-3}^3$$

$$= (-9 + 30) - (9 - 30) = 42$$

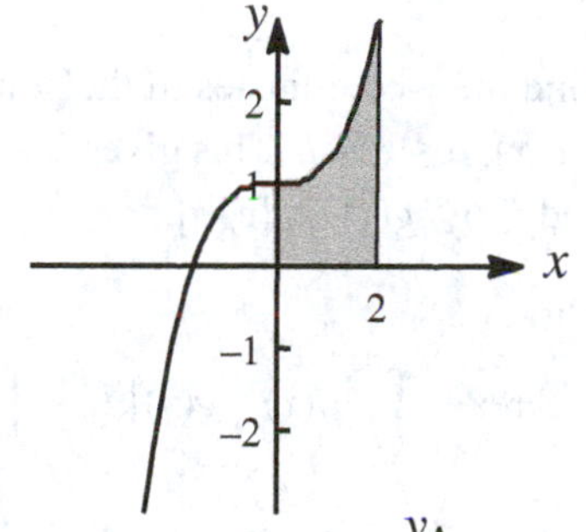

21. $y = x^3 + 1$; $y = 0$ on $[0, 2]$

$$A = \int_0^2 (x^3 + 1)dx = \left(-\frac{1}{4}x^4 + x\right)\Big|_0^2$$

$$= 4 + 2 = 6$$

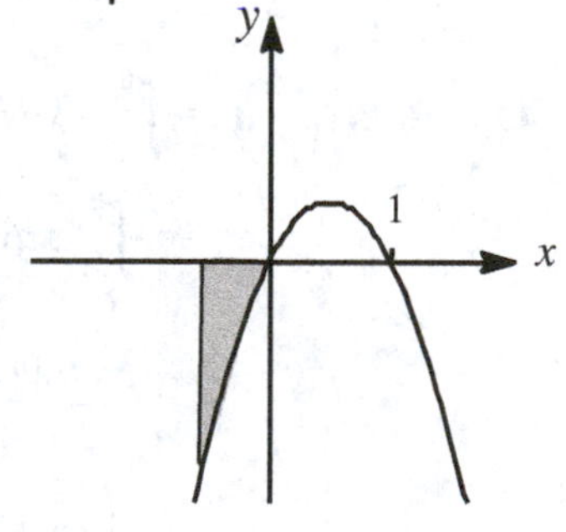

23. $y = x(1 - x)$; $y = 0$ on $[-1, 0]$

$$A = -\int_{-1}^0 x(1-x)\,dx = -\int_{-1}^0 (x - x^2)dx$$

$$= \int_{-1}^0 (x^2 - x)dx$$

$$= \left[\frac{1}{3}x^3 - \frac{1}{2}x^2\right]_{-1}^0$$

$$= 0 - \left(-\frac{1}{3} - \frac{1}{26}\right) = -\frac{5}{6} \approx 0.833$$

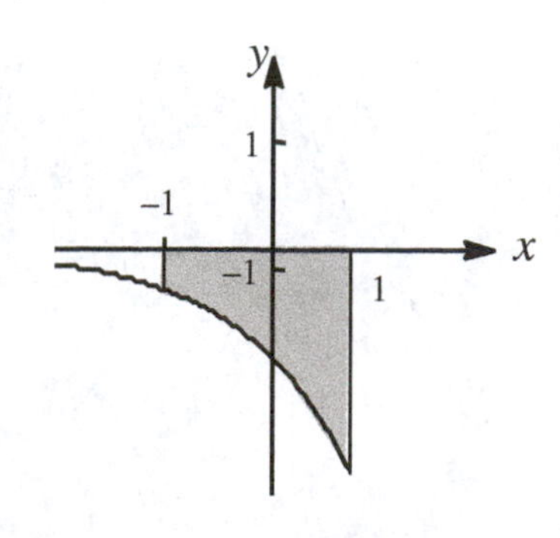

25. $y = -e^x$; $y = 0$ on $[-1, 1]$

$$A = -\int_{-1}^1 -e^x\,dx \qquad = \int_{-1}^1 e^x\,dx$$

$$= e^x\Big|_{-1}^1 = e - e^{-1}$$

$$\approx 2.350$$

27. $y = \frac{1}{x}$; $y = 0$ on $[1, e]$

$$A = \int_1^e \frac{1}{x}\,dx = \ln x \Big|_1^e$$
$$= \ln e - \ln 1 = 1$$

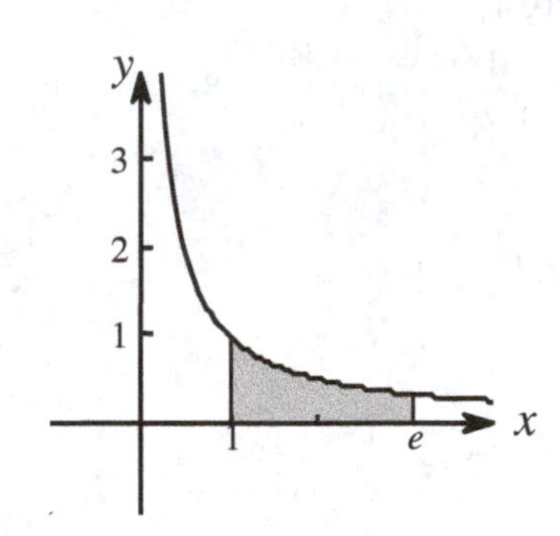

29. Most equally distributed: Canada, Gini index 0.32; least equally distributed: Mexico, Gini index 0.48.

31. Most equally distributed: India, Gini index 0.37; least equally distributed: Brazil, Gini index 0.52.

33. $A = \int_a^b [-f(x)]dx$

35. $A = \int_b^c f(x)dx + \int_c^d [-f(x)]dx$

37. $A = \int_c^d [f(x) - g(x)]dx$

39. $A = \int_a^b [f(x) - g(x)]dx + \int_b^c [g(x) - f(x)]dx$

41. Find the x-coordinates of the points of intersection of the two curves on $[a, d]$ by solving the equation $f(x) = g(x)$, $a \le x \le d$. This gives $x = b$ and $x = c$. Then note that $f(x) \ge g(x)$ on $[a, b]$, $g(x) \ge f(x)$ on $[b, c]$ and $f(x) \ge g(x)$ on $[c, d]$.

Thus,

$$\text{Area} = \int_a^b [f(x) - g(x)]dx + \int_b^c [g(x) - f(x)]dx + \int_c^d [f(x) - g(x)]dx$$

43. $A = A_1 + A_2 = \int_{-2}^0 -x dx + \int_0^1 -(-x)dx$

$$= -\int_{-2}^0 x dx + \int_0^1 x dx$$
$$= -\frac{x^2}{2}\Big|_{-2}^0 + \frac{x^2}{2}\Big|_0^1$$
$$= -\left(0 - \frac{(-2)^2}{2}\right) + \left(\frac{1^2}{2} - 0\right) = 2 + \frac{1}{2} = \frac{5}{2} = 2.5$$

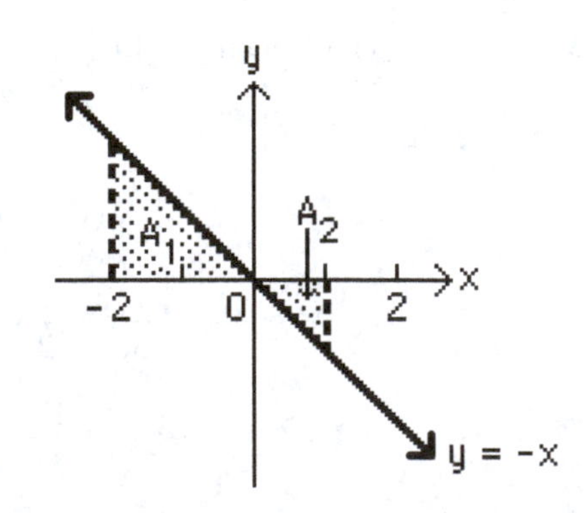

45. $A = A_1 + A_2 = \int_0^2 -(x^2-4)dx + \int_2^3 (x^2-4)dx$

$= \int_0^2 (4-x^2)dx + \int_2^3 (x^2-4)dx$

$= \left(4x - \frac{x^3}{3}\right)\Big|_0^2 + \left(\frac{x^3}{3} - 4x\right)\Big|_2^3$

$= \left(8 - \frac{8}{3}\right) + \left(\frac{27}{3} - 12\right) - \left(\frac{8}{3} - 8\right)$

$= 13 - \frac{16}{3} = \frac{39}{3} - \frac{16}{3} = \frac{23}{3} \approx 7.667$

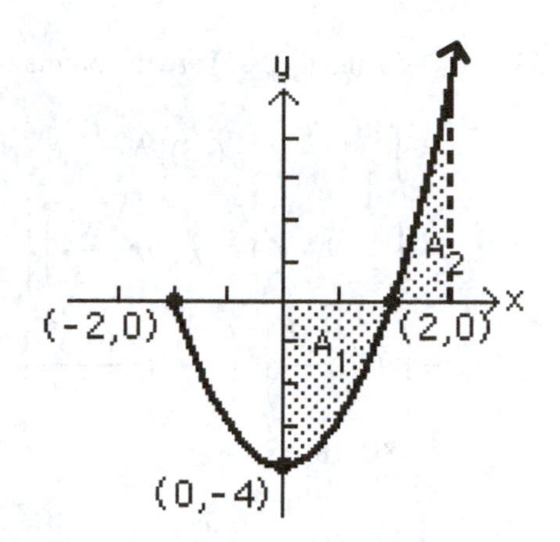

47. $A = A_1 + A_2 = \int_{-2}^0 (x^2-3x)dx + \int_0^2 -(x^2-3x)dx$

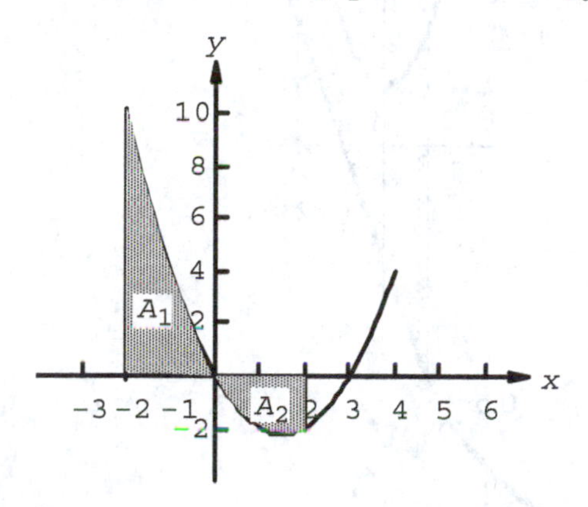

$= \int_{-2}^0 (x^2-3x)dx + \int_0^2 (3x-x^2)dx$

$= \left(\frac{1}{3}x^3 - \frac{3}{2}x^2\right)\Big|_{-2}^0 + \left(\frac{3}{2}x^2 - \frac{1}{3}x^3\right)\Big|_0^2$

$= 0 - \left(-\frac{8}{3} - 6\right) + \left(6 - \frac{8}{3}\right) - 0 = 12$

49. $A = \int_{-1}^2 [12 - (-2x+8)]dx = \int_{-1}^2 (2x+4)dx$

$= \left(\frac{2x^2}{2} + 4x\right)\Big|_{-1}^2 = (x^2+4x)\Big|_{-1}^2$

$= (4+8) - (1-4) = 12 + 3 = 15$

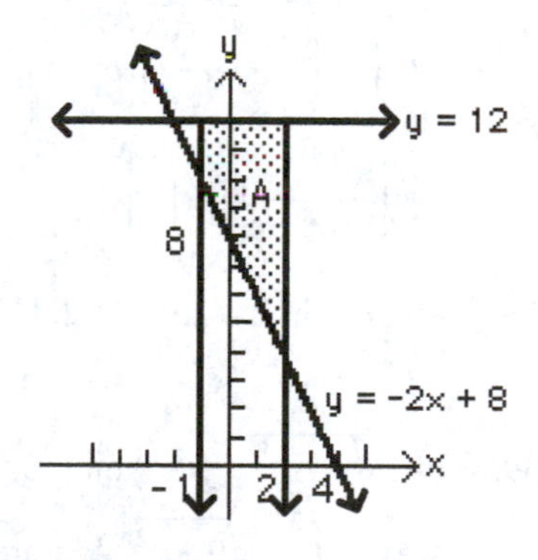

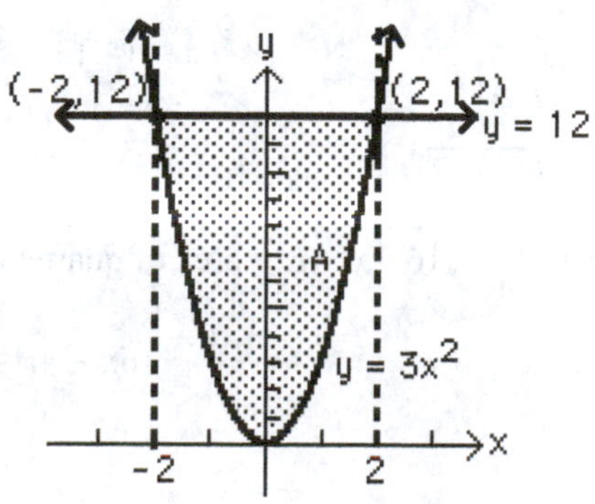

51. $A = \int_{-2}^2 (12-3x^2)dx = \left(12x - \frac{3x^3}{3}\right)\Big|_{-2}^2 = (12x - x^3)\Big|_{-2}^2$

$= (12\cdot 2 - 2^3) - [12\cdot(-2) - (-2)^3]$

$= 16 - (-16) = 32$

53. (3, –5) and (–3, –5) are the points of intersection.

$$A = \int_{-3}^{3} [4 - x^2 - (-5)]dx$$
$$= \int_{-3}^{3} (9 - x^2)dx = \left(9x - \frac{x^3}{3}\right)\Big|_{-3}^{3}$$
$$= \left(9\cdot 3 - \frac{3^3}{3}\right) - \left(9(-3) - \frac{(-3)^3}{3}\right)$$
$$= 18 + 18 = 36$$

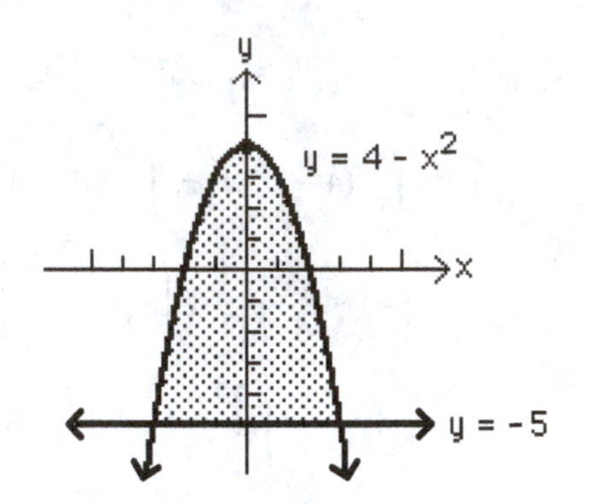

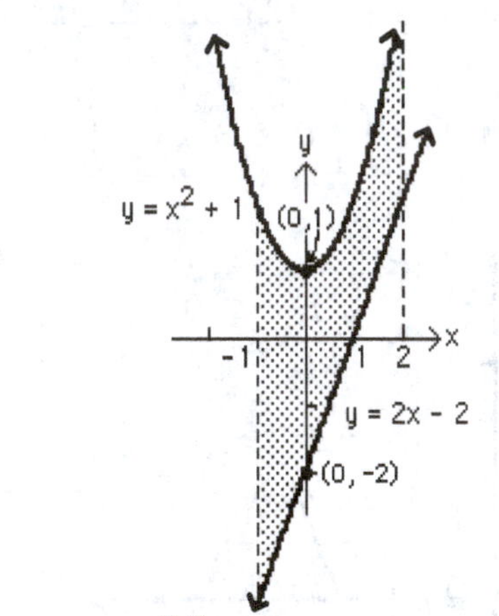

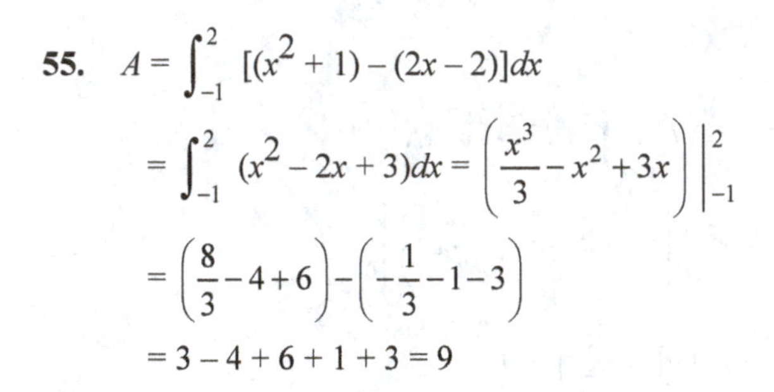

55. $A = \int_{-1}^{2} [(x^2 + 1) - (2x - 2)]dx$
$$= \int_{-1}^{2} (x^2 - 2x + 3)dx = \left(\frac{x^3}{3} - x^2 + 3x\right)\Big|_{-1}^{2}$$
$$= \left(\frac{8}{3} - 4 + 6\right) - \left(-\frac{1}{3} - 1 - 3\right)$$
$$= 3 - 4 + 6 + 1 + 3 = 9$$

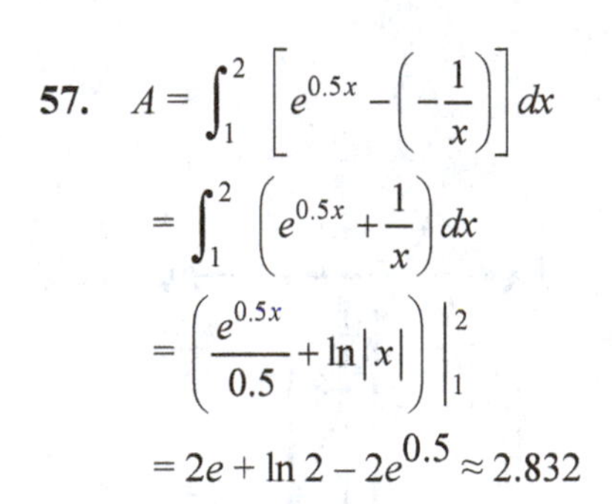

57. $A = \int_{1}^{2} \left[e^{0.5x} - \left(-\frac{1}{x}\right)\right]dx$
$$= \int_{1}^{2} \left(e^{0.5x} + \frac{1}{x}\right)dx$$
$$= \left(\frac{e^{0.5x}}{0.5} + \ln|x|\right)\Big|_{1}^{2}$$
$$= 2e + \ln 2 - 2e^{0.5} \approx 2.832$$

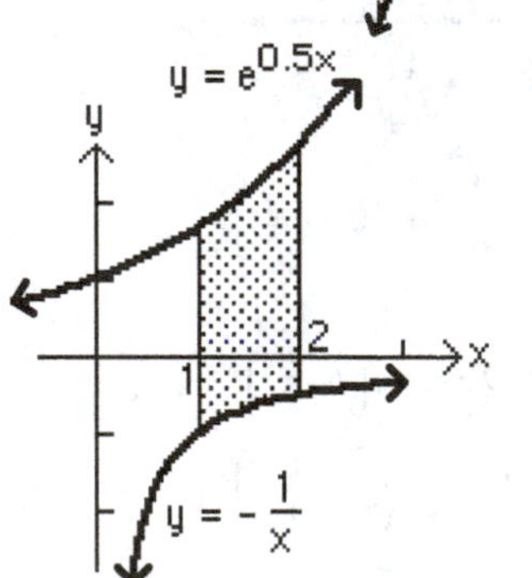

59. $y = \sqrt{9 - x^2}$; $y = 0$ on [–3, 3]

area = $\int_{-3}^{3} \sqrt{9 - x^2}\, dx$ = area of semicircle of radius 3

$$= \frac{9}{2}\pi \approx 14.137$$

61. $y = -\sqrt{16 - x^2}$; $y = 0$ on [0, 4]

area = $\int_{0}^{4} \sqrt{16 - x^2}\, dx$ = area of quarter circle of radius 4

$$= \frac{1}{4} \cdot 16\pi = 4\pi \approx 12.566$$

63. $y = -\sqrt{4-x^2}$; $y = \sqrt{4-x^2}$, on [–2, 2]

$$\text{area} = \int_{-2}^{2} [\sqrt{4-x^2} - (-\sqrt{4-x^2})]dx = \int_{-2}^{2} 2\sqrt{4-x^2}\,dx = 2\int_{-2}^{2} \sqrt{4-x^2}\,dx$$

= 2 area of semi- circle of radius 2 = area of circle of radius 2 = $4\pi \approx 12.566$

65. The graphs of $y = e^x$ and $y = e^{-x}$, $0 \le x \le 4$, are shown at the right.

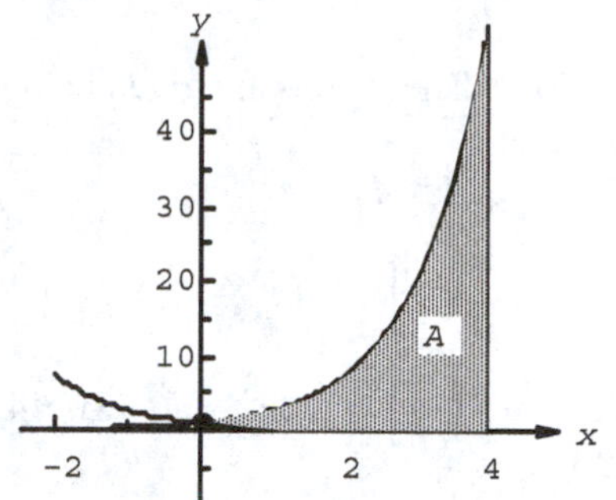

$$A = \int_0^4 (e^x - e^{-x})dx = (e^x + e^{-x})\Big|_0^4$$
$$= e^4 + e^{-4} - (1 + 1)$$
$$\approx 52.616$$

67. The graphs are given at the right. To find the points of intersection, solve:

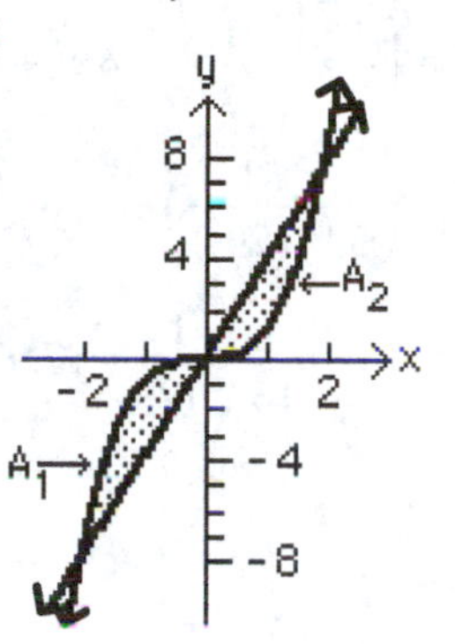

$$x^3 = 4x$$
$$x^3 - 4x = 0$$
$$x(x^2 - 4) = 0$$
$$x(x + 2)(x - 2) = 0$$

Thus, the points of intersection are (–2, –8), (0, 0), and (2, 8).

$$A = A_1 + A_2 = \int_{-2}^{0} (x^3 - 4x)dx + \int_0^2 (4x - x^3)dx$$
$$= \left(\frac{x^4}{4} - 2x^2\right)\Big|_{-2}^{0} + \left(2x^2 - \frac{x^4}{4}\right)\Big|_0^2$$
$$= 0 - \left[\frac{(-2)^4}{4} - 2(-2)^2\right] + \left[2(2^2) - \frac{2^4}{4}\right] - 0$$
$$= -4 + 8 + 8 - 4 = 8$$

69. The graphs are given at the right.
To find the points of intersection, solve:

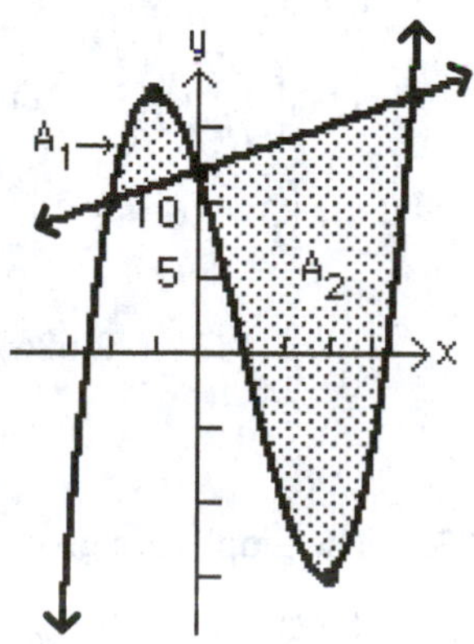

$$x^3 - 3x^2 - 9x + 12 = x + 12$$
$$x^3 - 3x^2 - 10x = 0$$
$$x(x^2 - 3x - 10) = 0$$
$$x(x - 5)(x + 2) = 0$$
$$x = -2, x = 0, x = 5$$

Thus, (–2, 10), (0, 12), and (5, 17) are the points of intersection.

$$A = A_1 + A_2$$
$$= \int_{-2}^{0} [x^3 - 3x^2 - 9x + 12 - (x + 12)]dx + \int_0^5 [x + 12 - (x^3 - 3x^2 - 9x + 12)]dx$$
$$= \int_{-2}^{0} (x^3 - 3x^2 - 10x)dx + \int_0^5 (-x^3 + 3x^2 + 10x)dx$$

$$= \left(\frac{x^4}{4} - x^3 - 5x^2\right)\Big|_{-2}^{0} + \left(-\frac{x^4}{4} + x^3 + 5x^2\right)\Big|_{0}^{5} = -\left[\frac{(-2)^4}{4} - (-2)^3 - 5(-2)^2\right] + \left(\frac{-5^4}{4} + 5^3 + 5\cdot 5^2\right)$$

$$= 8 + \frac{375}{4} = \frac{407}{4} = 101.75$$

71. The graphs are given below. The x-coordinates of the points of intersection are: $x_1 = -2$, $x_2 = 0.5$, $x_3 = 2$

$$A = A_1 + A_2$$

$$= \int_{-2}^{0.5} [(x^3 - x^2 + 2) - (-x^3 + 8x - 2)]\, dx + \int_{0.5}^{2} [(-x^3 + 8x - 2) - (x^3 - x^2 + 2)]dx$$

$$= \int_{-2}^{0.5} (2x^3 - x^2 - 8x + 4)dx + \int_{0.5}^{2} (-2x^3 + x^2 + 8x - 4)dx$$

$$= \left(\frac{1}{2}x^4 - \frac{1}{3}x^3 - 4x^2 + 4x\right)\Big|_{-2}^{0.5} + \left(-\frac{1}{2}x^4 + \frac{1}{3}x^3 + 4x^2 - 4x\right)\Big|_{0.5}^{2}$$

$$= \left(\frac{1}{32} - \frac{1}{24} - 1 + 2\right) - \left(8 + \frac{8}{3} - 16 - 8\right) + \left(-8 + \frac{8}{3} + 16 - 8\right) - \left(-\frac{1}{32} + \frac{1}{24} + 1 - 2\right)$$

$$= 18 + \frac{1}{16} - \frac{1}{12} \approx 17.979$$

73. The graphs are given at the right. The x-coordinates of the points of intersection are: $x_1 \approx -1.924$, $x_2 \approx 1.373$

$$A = \int_{-1.924}^{1.373} [(3 - 2x) - e^{-x}]\, dx$$

$$= \left[3x - x^2 + e^{-x}\right]_{-1.924}^{1.373}$$

$$\approx 2.487 - (-2.626) = 5.113$$

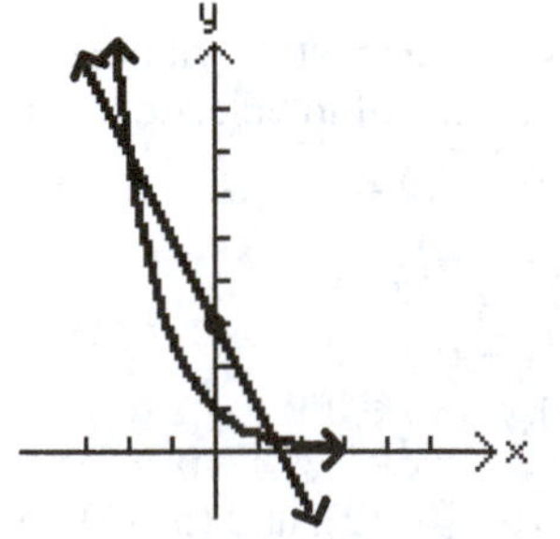

75. The graphs are given at the right. The x-coordinates of the points of intersection are: $x_1 \approx -2.247$, $x_2 \approx 0.264$, $x_3 \approx 1.439$

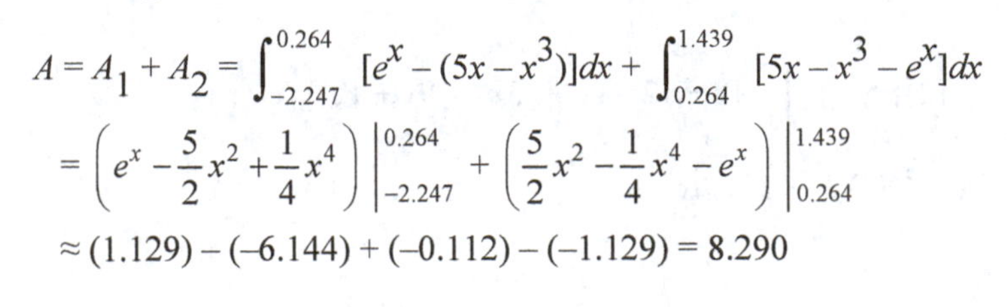

$$A = A_1 + A_2 = \int_{-2.247}^{0.264} [e^x - (5x - x^3)]dx + \int_{0.264}^{1.439} [5x - x^3 - e^x]dx$$

$$= \left(e^x - \frac{5}{2}x^2 + \frac{1}{4}x^4\right)\Big|_{-2.247}^{0.264} + \left(\frac{5}{2}x^2 - \frac{1}{4}x^4 - e^x\right)\Big|_{0.264}^{1.439}$$

$$\approx (1.129) - (-6.144) + (-0.112) - (-1.129) = 8.290$$

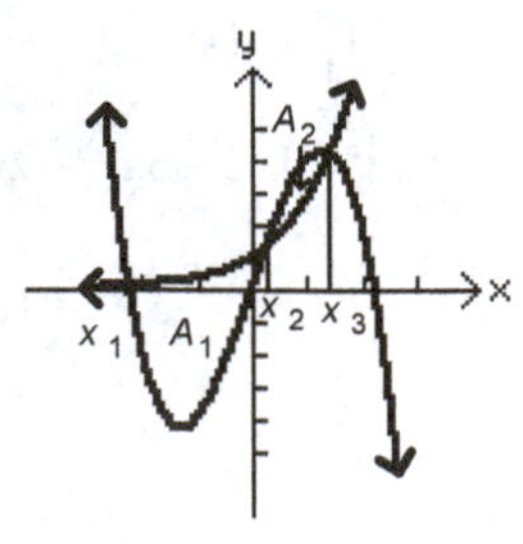

77. $y = e^{-x}; y = \sqrt{\ln x}; 2 \le x \le 5$

The graphs of $y_1 = e^{-x}$ and $y_2 = \sqrt{\ln x}$ are

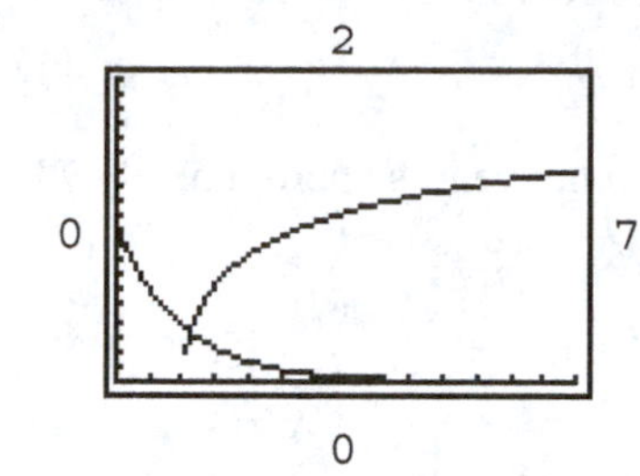

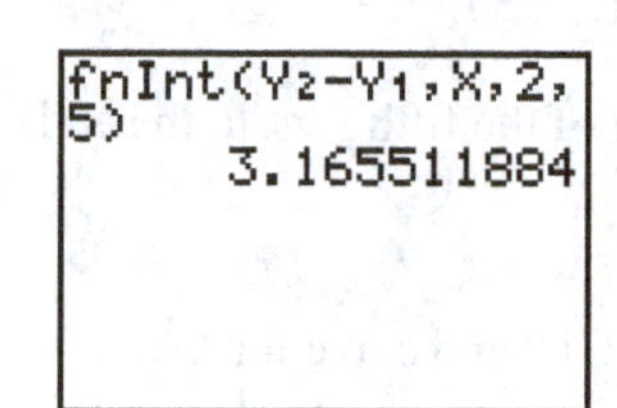

Thus, $A = \int_2^5 (\sqrt{\ln x} - e^{-x})dx \approx 3.166$

79. $y = e^{x^2}; y = x + 2$

The graphs of $y_1 = e^{x^2}$ and $y_2 = x + 2$ are

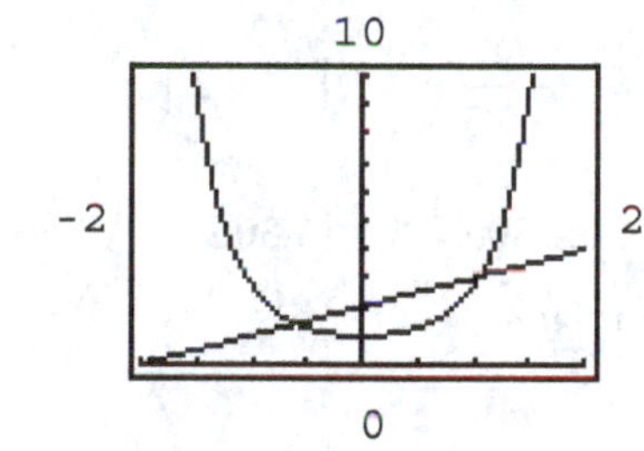

The curves intersect at

$x \approx -0.588$ and $x \approx 1.057$ $\int_{-0.588}^{1.057} (x + 2 - e^{x^2})dx \approx 1.385$

$y \approx 1.412$ $\quad$ $y \approx 3.057$

81. Solve $2\int_0^1 (x - x^c)dx = 0.52$ for c.

$$2\int_0^1 (x - x^c)dx = 2\left[\frac{x^2}{2} - \frac{x^{c+1}}{c+1}\right]_0^1 = 2\left[\frac{1}{2} - \frac{1}{c+1}\right] = 1 - \frac{2}{c+1}$$

$$1 - \frac{2}{c+1} = 0.52, \quad \frac{2}{c+1} = 1 - 0.52 = 0.48, \quad c + 1 = \frac{2}{0.48} \approx 4.17, \quad c = 3.17$$

83. Solve $2\int_0^1 (x - x^c)dx = 0.29$ for c.

$$2\int_0^1 (x - x^c)dx = 2\left[\frac{x^2}{2} - \frac{x^{c+1}}{c+1}\right]_0^1 = 2\left[\frac{1}{2} - \frac{1}{c+1}\right] = 1 - \frac{2}{c+1}$$

$$1 - \frac{2}{c+1} = 0.29, \quad \frac{2}{c+1} = 1 - 0.29 = 0.71, \quad c + 1 = \frac{2}{0.71} \approx 2.82, \quad c = 1.82$$

85. $\int_5^{10} R(t)dt = \int_5^{10} \left(\frac{100}{t+10}+10\right)dt = 100\int_5^{10} \frac{1}{t+10}\,dt + \int_5^{10} 10dt$

$= 100\ln(t+10)\Big|_5^{10} + 10t\Big|_5^{10} = 100\ln 20 - 100\ln 15 + 10(10-5)$

$= 100\ln 20 - 100\ln 15 + 50 \approx 79$

The total production from the end of the fifth year to the end of the tenth year is approximately 79 thousand barrels.

87. To find the useful life, set $R'(t) = C'(t)$ and solve for t:

$9e^{-0.3t} = 2$

$e^{-0.3t} = \frac{2}{9}$

$-0.3t = \ln\frac{2}{9}$

$-0.3t \approx -1.5$

$t \approx 5$ years

$\int_0^5 [R'(t) - C'(t)]\,dt = \int_0^5 [9e^{-0.3t} - 2]\,dt = 9\int_0^5 e^{-0.3t}dt - \int_0^5 2dt = \frac{9}{-0.3}e^{-0.3t}\Big|_0^5 - 2t\Big|_0^5$

$= -30e^{-1.5} + 30 - 10 = 20 - 30e^{-1.5} \approx 13.306$

The total profit over the useful life of the game is approximately $13,306.

89. For 1935: $f(x) = x^{2.4}$

Gini Index $= 2\int_0^1 [x - f(x)]\,dx = 2\int_0^1 (x - x^{2.4})\,dx = 2\left(\frac{x^2}{2} - \frac{x^{3.4}}{3.4}\right)\Big|_0^1 = 2\left(\frac{1}{2} - \frac{1}{3.4}\right) \approx 0.412$

For 1947: $g(x) = x^{1.6}$

Gini Index $= 2\int_0^1 [x - g(x)]\,dx = 2\int_0^1 (x - x^{1.6})dx = 2\left(\frac{x^2}{2} - \frac{x^{2.6}}{2.6}\right)\Big|_0^1 = 2\left(\frac{1}{2} - \frac{1}{2.6}\right) \approx 0.231$

Interpretation: Income was more equally distributed in 1947.

91. For 1963: $f(x) = x^{10}$

Gini Index $= 2\int_0^1 [x - f(x)]\,dx = 2\int_0^1 (x - x^{10})\,dx = 2\left(\frac{x^2}{2} - \frac{x^{11}}{11}\right)\Big|_0^1 = 2\left(\frac{1}{2} - \frac{1}{11}\right) \approx 0.818$

For 1983: $g(x) = x^{12}$

Gini Index $= 2\int_0^1 [x - g(x)]\,dx = 2\int_0^1 (x - x^{12})\,dx = 2\left(\frac{x^2}{2} - \frac{x^{13}}{13}\right)\Big|_0^1 = 2\left(\frac{1}{2} - \frac{1}{13}\right) \approx 0.846$

Interpretation: Total assets were less equally distributed in 1983.

93. (A)

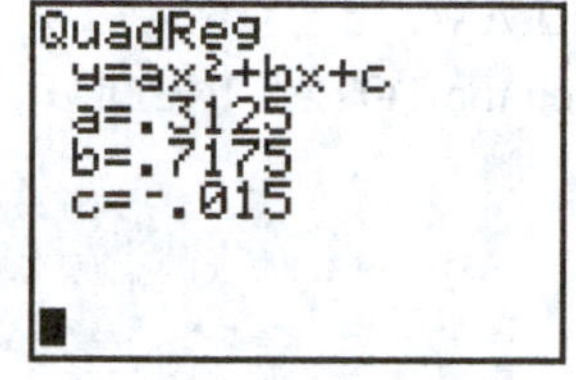

Lorenz curve:

$f(x) = 0.3125x^2 + 0.7175x - 0.015.$

(B) Gini Index:

$$2\int_0^1 [x - f(x)]dx \approx 0.104$$

```
2*fnInt(X-Y1,X,0
,1)
       .1041666667
```

95. $W(t) = \int_0^{10} W'(t)\,dt = \int_0^{10} 0.3e^{0.1t}dt = 0.3\int_0^{10} e^{0.1t}\,dt = \frac{0.3}{0.1}e^{0.1t}\Big|_0^{10} = 3e^{0.1t}\Big|_0^{10} = 3e - 3 \approx 5.15$

Total weight gain during the first 10 hours is approximately 5.15 grams.

97. $V = \int_2^4 \frac{15}{t}\,dt \quad = 15\int_2^4 \frac{1}{t}dt = 15\ln t\Big|_2^4 = 15\ln 4 - 15\ln 2 = 15\ln\left(\frac{4}{2}\right) = 15\ln 2 \approx 10$

Average number of words learned from $t = 2$ to $t = 4$ is 10.

EXERCISE 6-2

Things to remember:

1. PROBABILITY DENSITY FUNCTION

A function f which satisfies the following three conditions:

a. $f(x) \geq 0$ for all real x.

b. The area under the graph of f over the interval $(-\infty, \infty)$ is exactly 1.

c. If $[c, d]$ is a subinterval of $(-\infty, \infty)$, then the probability that the outcome x of an experiment will be in the interval $[c, d]$, denoted Probability $(c \leq x \leq d)$, is given by

$$\text{Probability } (c \leq x \leq d) = \int_c^d f(x)dx$$

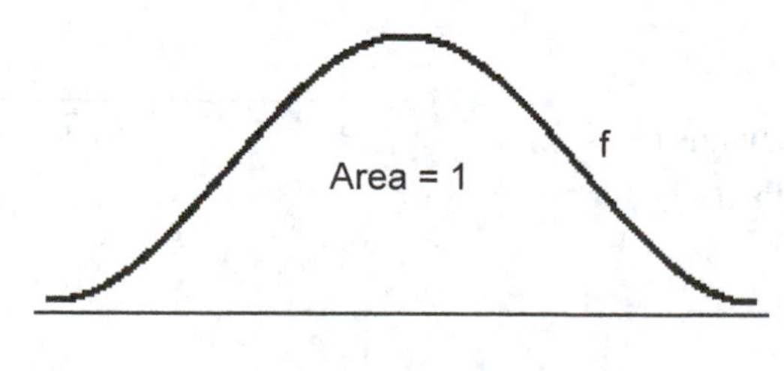

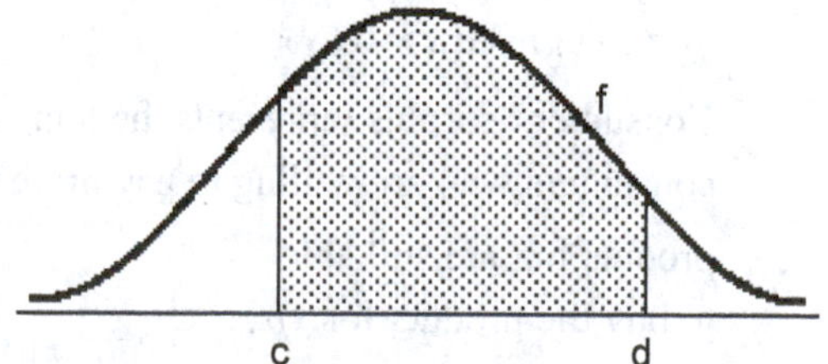

$$\int_c^d f(x)dx = \text{Probability } (c \leq x \leq d)$$

2. TOTAL INCOME FOR A CONTINUOUS INCOME STREAM
If $f(t)$ is the rate of flow of a continuous income stream, then the TOTAL INCOME produced during the time period from
$t = a$ to $t = b$ is:

$$\text{Total income} = \int_a^b f(t)dt$$

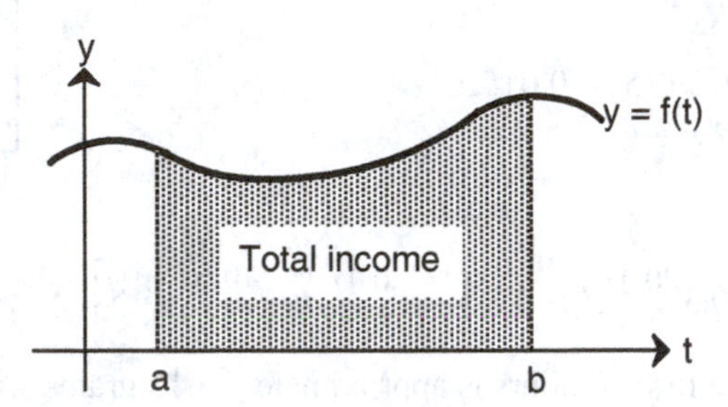

3. FUTURE VALUE OF A CONTINUOUS INCOME STREAM
If $f(t)$ is the rate of flow of a continuous income stream,
$0 \le t \le T$, and if the income is continuously invested at a rate r, compounded continuously, then the FUTURE VALUE, FV, at the end of T years is given by:

$$FV = \int_0^T f(t)e^{r(T-t)}dt = e^{rT}\int_0^T f(t)e^{-rt}dt$$

The future value of a continuous income stream is the total value of all money produced by the continuous income stream (income and interest) at the end of T years.

4. CONSUMERS' SURPLUS
If $(\bar{x}, \bar{p})$ is a point on the graph of the price-demand equation $p = D(x)$ for a particular product, then the CONSUMERS' SURPLUS, CS, at a price level of $\bar{p}$ is

$$CS = \int_0^{\bar{x}} [D(x) - \bar{p}]dx$$

which is the area between $p = \bar{p}$ and $p = D(x)$ from $x = 0$ to $x = \bar{x}$.
Consumers' surplus represents the total savings to consumers who are willing to pay more than $\bar{p}$ for the product but are still able to buy the product for $\bar{p}$.

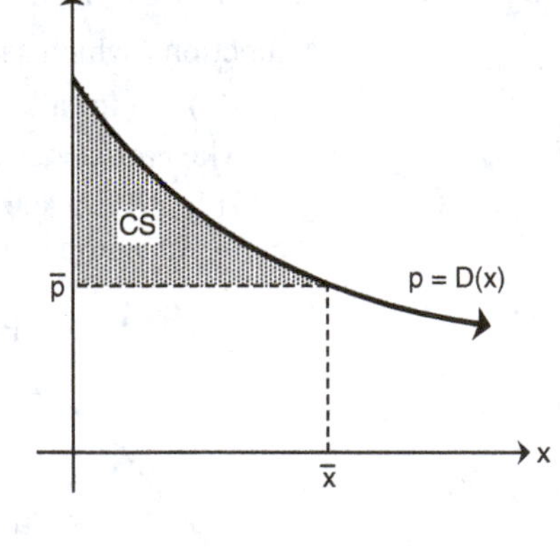

5. PRODUCERS' SURPLUS
If $(\bar{x}, \bar{p})$ is a point on the graph of the price-supply equation $p = S(x)$, then the PRODUCERS' SURPLUS, PS, at a price level of $\bar{p}$ is

$$PS = \int_0^{\bar{x}} [\bar{p} - S(x)]dx$$

which is the area between $p = \bar{p}$ and $p = S(x)$ from $x = 0$ to $x = \bar{x}$

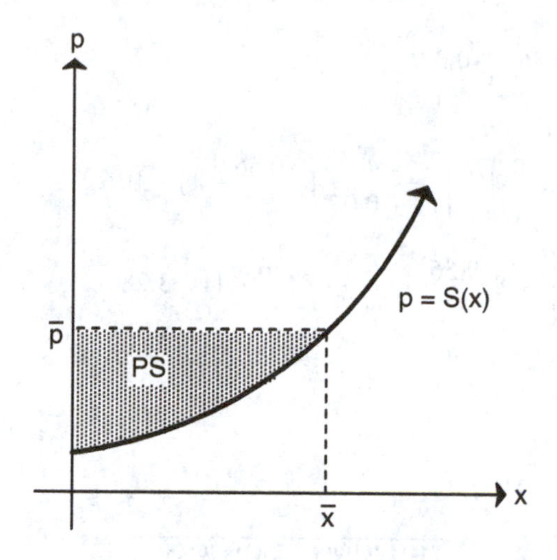

Producers' surplus represents the total gain to producers who are willing to supply units at a lower price than $\overline{p}$ but are still able to supply units at $\overline{p}$.

6. EQUILIBRIUM PRICE AND EQUILIBRIUM QUANTITY

If $p = D(x)$ and $p = S(x)$ are the price-demand and the price-supply equations, respectively, for a product and if $(\overline{x}, \overline{p})$ is the point of intersection of these equations, then $\overline{p}$ is called the EQUILIBRIUM PRICE and $\overline{x}$ is called the EQUILIBRIUM QUANTITY.

---

1. $f(t) = e^{5(4-t)} = e^{20-5t} = e^{20}e^{-5t}, \quad b = 20, \quad c = -5$

3. $f(t) = e^{0.04(8-t)} = e^{0.32-0.04t} = e^{0.32}e^{-0.04t}, \quad b = 0.32, \quad c = -0.04$

5. $f(t) = e^{0.05t}e^{0.08(20-t)} = e^{0.05t}e^{1.6-0.08t} = e^{0.05t}e^{1.6}e^{-0.08t} = e^{1.6}e^{-0.03t}, \quad b = 1.6, \quad c = -0.03$

7. $f(t) = e^{0.09t}e^{0.07(25-t)} = e^{0.09t}e^{1.75-0.07t} = e^{0.09t}e^{1.75}e^{-0.07t} = e^{1.75}e^{0.02t}, \quad b = 1.75, \quad c = 0.02$

9. $\int_0^8 e^{0.06(8-t)}dt = \int_0^8 e^{0.48} \cdot e^{-0.06t}\,dt = e^{0.48}\int_0^8 e^{-0.06t}\,dt = e^{0.48}\left(-\frac{1}{0.06}e^{-0.06t}\right)\Big|_0^8$

$$= -\frac{e^{0.48}}{0.06}[e^{-0.48} - 1] = \frac{e^{0.48} - 1}{0.06} \approx 10.27$$

11. $\int_0^{20} e^{0.08t}e^{0.12(20-t)}dt = \int_0^{20} e^{0.08t} \cdot e^{2.4} \cdot e^{-0.12t}\,dt = e^{2.4}\int_0^{20} e^{-0.04t}\,dt$

$$= e^{2.4}\left(-\frac{1}{0.04}e^{-0.04t}\right)\Big|_0^{20} = -\frac{e^{2.4}}{0.04}(e^{-0.8} - 1) = \frac{-e^{1.6} + e^{2.4}}{0.04} \approx 151.75$$

13. $\int_0^{30} 500e^{0.02t}e^{0.09(30-t)}dt = 500\int_0^{30} e^{0.02t} \cdot e^{2.7} \cdot e^{-0.09t}\,dt = 500e^{2.7}\int_0^{30} e^{-0.07t}\,dt$

$$= 500e^{2.7}\left(-\frac{1}{0.07}e^{-0.07t}\right)\Big|_0^{30} = -\frac{500e^{2.7}}{0.07}(e^{-2.1} - 1)$$

$$= \frac{500(e^{2.7} - e^{0.6})}{0.07} \approx 93{,}268.66$$

15. (A) $\int_0^8 e^{0.07(8-t)}dt = \int_0^8 e^{0.56-0.07t}\,dt = \int_0^8 e^{0.56}\cdot e^{-0.07t}\,dt$

$= e^{0.56}\int_0^8 e^{-0.07t}\,dt = \frac{e^{0.56}}{-0.07}e^{-0.07t}\Big|_0^8 = -\frac{e^{0.56}}{0.07}[e^{-0.56}-1] \approx 10.72$

(B) $\int_0^8 (e^{0.56} - e^{0.07t})\,dt = (e^{0.56})t\Big|_0^8 - \frac{e^{0.07t}}{0.07}\Big|_0^8 = 8e^{0.56} - \frac{1}{0.07}[e^{0.56}-1] \approx 3.28$

(C) $e^{0.56}\int_0^8 e^{-0.07t}\,dt \approx 10.72$ as in (A)

17.

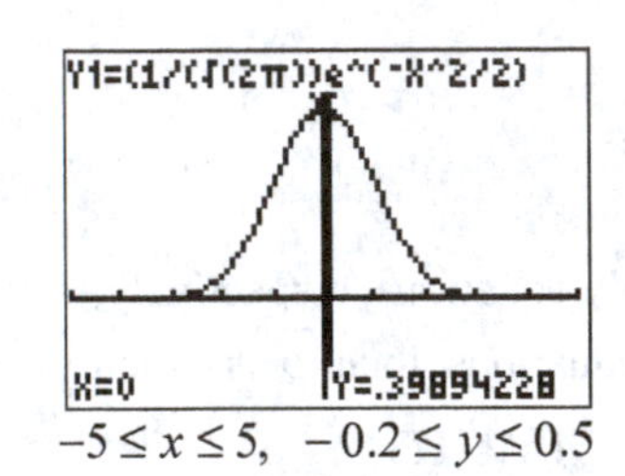

$-5 \le x \le 5,\ -0.2 \le y \le 0.5$

19.

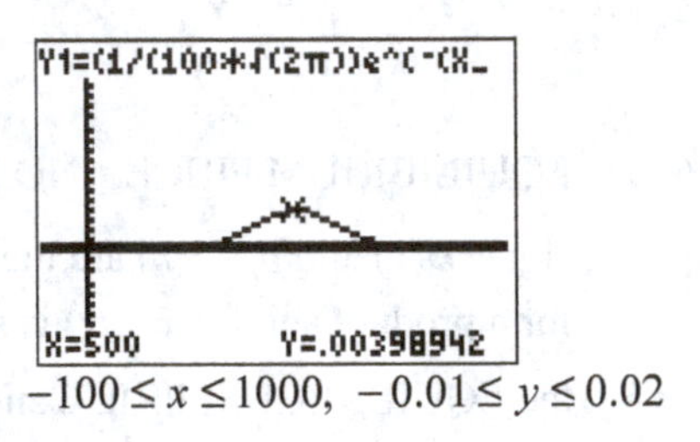

$-100 \le x \le 1000,\ -0.01 \le y \le 0.02$

21. (A) $\int_{-1}^{1} \frac{1}{\sqrt{2\pi}} e^{-x^2/2}\,dx \approx 0.6827$ (B) $\int_{-2}^{2} \frac{1}{\sqrt{2\pi}} e^{-x^2/2}\,dx \approx 0.9545$

(C) $\int_{-3}^{3} \frac{1}{\sqrt{2\pi}} e^{-x^2/2}\,dx \approx 0.9973$

23. (A) $\int_{400}^{600} \frac{1}{100\sqrt{2\pi}} e^{-(x-500)^2/20{,}000}\,dx \approx 0.6827$ (B) $\int_{300}^{700} \frac{1}{100\sqrt{2\pi}} e^{-(x-500)^2/20{,}000}\,dx \approx 0.9545$

(C) $\int_{200}^{800} \frac{1}{100\sqrt{2\pi}} e^{-(x-500)^2/20{,}000}\,dx \approx 0.9973$

25. $f(x) = \begin{cases} \frac{2}{(x+2)^2}, & x \ge 0 \\ 0 & x < 0 \end{cases}$

(A) Probability $(0 \le x \le 6) = \int_0^6 f(x)dx = \int_0^6 \frac{2}{(x+2)^2}dx = 2\frac{(x+2)^{-1}}{-1}\Big|_0^6 = \frac{-2}{(x+2)}\Big|_0^6$

$= -\frac{1}{4} + 1 = \frac{3}{4} = 0.75$

Thus, Probability $(0 \le x \le 6) = 0.75$

(B) Probability $(6 \le x \le 12) = \int_6^{12} f(x)dx = \int_6^{12} \frac{2}{(x+2)^2}dx = \frac{-2}{x+2}\Big|_6^{12} = -\frac{1}{7} + \frac{1}{4} = \frac{3}{28} \approx 0.11$

(C)

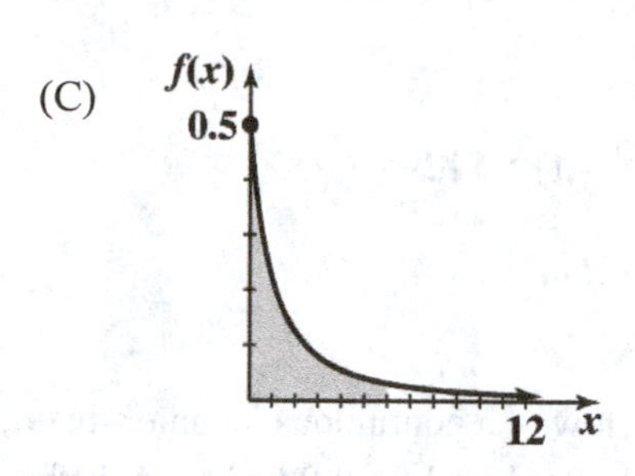

27. We want to find d such that

Probability $(0 \le x \le d) = \int_0^d f(x)\,dx = 0.8$:

$$\int_0^d f(x)\,dx = \int_0^d \frac{2}{(x+2)^2}\,dx = -\frac{2}{x+2}\Big|_0^d = \frac{-2}{d+2} + 1 = \frac{d}{d+2}$$

Now, $\frac{d}{d+2} = 0.8$

$d = 0.8d + 1.6$

$0.2d = 1.6$

$d = 8$ years

29. $f(t) = \begin{cases} 0.01e^{-0.01t} & \text{if } t \ge 0 \\ 0 & \text{otherwise} \end{cases}$

(A) Since t is in months, the probability of failure during the warranty period of the first year is

$$\text{Probability } (0 \le t \le 12) = \int_0^{12} f(t)\,dt = \int_0^{12} 0.01e^{-0.01t}dt = \frac{0.01}{-0.01}e^{-0.01t}\Big|_0^{12}$$

$$= -1(e^{-0.12} - 1) \approx 0.11$$

(B) $\text{Probability } (12 \le t \le 24) = \int_{12}^{24} 0.01e^{-0.01t}dt = -1e^{-0.01t}\Big|_{12}^{24} = -1(e^{-0.24} - e^{-0.12}) \approx 0.10$

31. $\text{Probability } (0 \le t \le \infty) = 1 = \int_0^{\infty} f(t)\,dt.$ But, $\int_0^{\infty} f(t)\,dt = \int_0^{12} f(t)\,dt + \int_{12}^{\infty} f(t)\,dt$

Thus, Probability $(t \ge 12) = 1 - \text{Probability } (0 \le t \le 12) \approx 1 - 0.11 = 0.89$

33. $f(t) = 2500$

Total income $= \int_0^5 2500\,dt = 2500t\Big|_0^5 = \$12{,}500$

35.

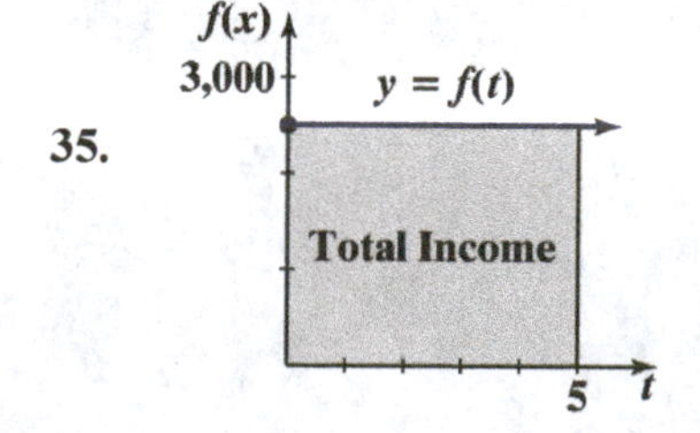

If $f(t)$ is the rate of flow of a continuous income stream, then the total income produced from 0 to 5 years is the area under the curve $y = f(t)$ from $t = 0$ to $t = 5$.

37. $f(t) = 400e^{0.05t}$

Total income $= \int_0^3 400e^{0.05t}\,dt = \frac{400}{0.05}e^{0.05t}\Big|_0^3 = 8000(e^{0.15} - 1) \approx \$1295$

39.

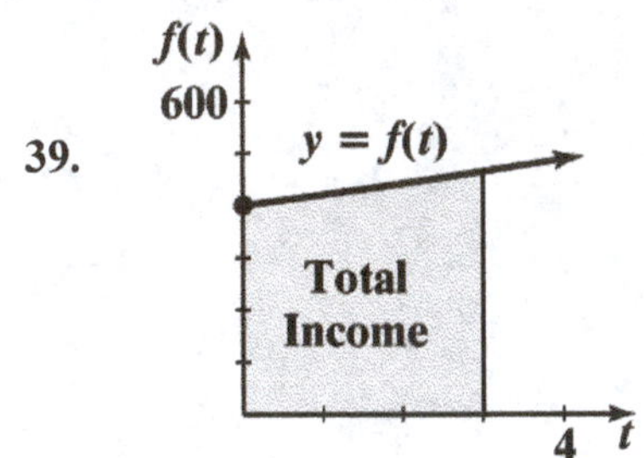

If $f(t)$ is the rate of flow of a continuous income stream, then the total income produced from 0 to 3 years is the area under the curve $y = f(t)$ from $t = 0$ to $t = 3$.

41. $f(t) = 2{,}000e^{0.05t}$

The amount in the account after 40 years is given by:

$$\int_0^{40} 2{,}000e^{0.05t}dt = 40{,}000e^{0.05t}\Big|_0^{40} = 295{,}562.24 - 40{,}000 \approx \$255{,}562$$

Since \$2,000 × 40 = \$80,000 was deposited into the account, the interest earned is:

\$255,562 – \$80,000 = \$175,562

43. $f(t) = 1{,}650e^{-0.02t}$, $r = 0.0325$, $T = 4$.

$$FV = e^{0.0325(4)}\int_0^4 1{,}650e^{-0.02t}e^{-0.0325t}\,dt = 1{,}650e^{0.13}\int_0^4 e^{-0.0525t}dt$$

$$= 1{,}650e^{0.13}\left[\frac{e^{-0.0525t}}{-0.0525}\right]_0^4 = 31{,}428.57(e^{0.13} - e^{-0.08}) \approx \$6{,}779.52\ .$$

45. Total Income $= \int_0^4 1{,}650e^{-0.02t}\,dt = \frac{1650}{-0.02}e^{-0.02t}\Big]_0^4 = -82{,}500(e^{-0.08} - 1) \approx \$6{,}342.90$

From Problem 43,

Interest earned = \$6,779.52 – \$6,342.90 = \$436.62

47. Clothing store: $f(t) = 12{,}000$, $r = 0.04$, $T = 5$.

$$FV = e^{0.04(5)}\int_0^5 12{,}000e^{-0.04t}\,dt = 12{,}000e^{0.2}\int_0^5 e^{-0.04t}\,dt$$

$$= 12{,}000e^{0.2}\frac{e^{-0.04t}}{-0.04}\Bigg]_0^5 = -300{,}000e^{0.2}\left(e^{-0.2} - 1\right) \approx \$66{,}420.83$$

Computer store: $g(t) = 10{,}000e^{0.05t}$, $r = 0.04$, $T = 5$.

$$FV = e^{0.04(5)}\int_0^5 10{,}000e^{0.05t}e^{-0.04t}\,dt = 10{,}000e^{0.2}\int_0^5 e^{0.01t}\,dt$$

$$= 10{,}000\,e^{0.2}\left[\frac{e^{0.01t}}{0.01}\right]_0^5 = 1{,}000{,}000\,e^{0.2}\left(e^{0.05} - 1\right) \approx \$62{,}622.66.$$

The clothing store is the better investment.

49. Bond: $P = \$10{,}000$, $r = 0.0375$, $t = 5$.

$FV = 10{,}000e^{0.0375(5)} = 10{,}000e^{0.1875} \approx \$12{,}062.30$

Business: $f(t) = 2150$, $r = 0.0375$, $T = 5$.

$$FV = e^{0.0375(5)} \int_0^5 2150e^{-0.0375t} dt = 2150e^{0.1875} \int_0^5 e^{-0.0375t}\, dt$$

$$= 2150\,e^{0.1875}\left[\frac{e^{-0.0375t}}{-0.0375}\right]_0^5 = -57,333.33\,e^{0.1875}\left(e^{-0.1875} - 1\right) = 57,333,33\left(e^{0.1875} - 1\right) \approx \$11,823.87.$$

The bond is the better investment.

51. $f(t) = 9,000,\ r = 0.0695,\ T = 8.$

$$FV = e^{0.0695(8)} \int_0^8 9000e^{-0.0695t}\, dx \ = 9000e^{0.556} \int_0^8 e^{-0.0695t} = \frac{9000e^{0.556}}{-0.0695} e^{-0.0695t}\Bigg]_0^8$$

$$\approx -225,800.78(e^{-0.556} - 1) \approx \$96,304.$$

The relationship between present value (PV) and future value (FV) at a continuously compounded interest rate r (expressed as a decimal) for t years is:

$$FV = PVe^{rt} \text{ or } PV = FVe^{-rt}$$

Thus, we have:

$$PV = 96,304e^{-0.0695(8)} = 96,304e^{-0.556} \approx \$55,230$$

53. $f(t) = k$, rate r (expressed as a decimal), years T:

$$FV = e^{rT} \int_0^T ke^{-rt} dt = ke^{rT} \int_0^T e^{-rt} dt = \frac{ke^{rT}}{-r} e^{-rt}\Big|_0^T \ = -\frac{k}{r} e^{rT}(e^{-rT} - 1) \ = \frac{k}{r}(e^{rT} - 1)$$

55. $D(x) = 400 - \frac{1}{20}x,\ \overline{p} = 150$

First, find $\overline{x}$: $150 = 400 - \frac{1}{20}\overline{x}$

$$\overline{x} = 5000$$

$$CS = \int_0^{5000} \left[400 - \frac{1}{20}x - 150\right] dx = \int_0^{5000} \left(250 - \frac{1}{20}x\right) dx = \left(250x - \frac{1}{40}x^2\right)\Bigg|_0^{5000} = \$625,000$$

57.

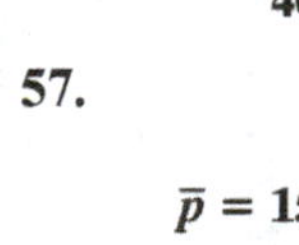

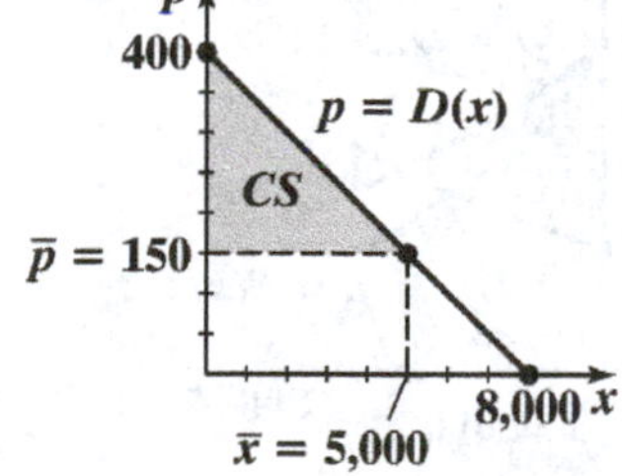

The shaded area is the consumers' surplus and represents the total savings to consumers who are willing to pay more than $150 for a product but are still able to buy the product for $150.

59. $p = S(x) = 10 + 0.1x + 0.0003x^2,\ \overline{p} = 67.$

First find $\overline{x}$: $67 = 10 + 0.1\,\overline{x} + 0.0003\,\overline{x}^2$

$$0.0003\,\overline{x}^2 + 0.1\,\overline{x} - 57 = 0$$

$$\overline{x} = \frac{-0.1 + \sqrt{0.01 + 0.0684}}{0.0006} = \frac{-0.1 + 0.28}{0.0006} = 300$$

$$PS = \int_0^{300} [67 - (10 + 0.1x + 0.0003x^2)]\, dx = \int_0^{300} (57 - 0.1x - 0.0003x^2)\, dx$$

$$= (57x - 0.05x^2 - 0.0001x^3)\Big|_0^{300} = \$9,900$$

61.

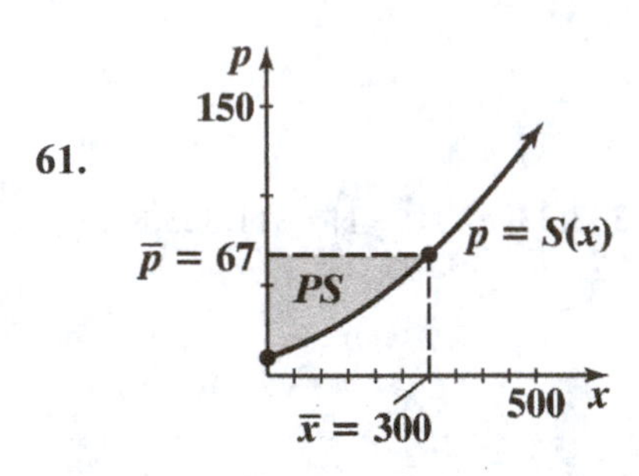

The area of the region *PS* is the producers' surplus and represents the total gain to producers who are willing to supply units at a lower price than \$67 but are still able to supply the product at \$67.

63. $p = D(x) = 50 - 0.1x;\ p = S(x) = 11 + 0.05x$

Equilibrium price: $D(x) = S(x)$

$$\begin{aligned} 50 - 0.1x &= 11 + 0.05x \\ 39 &= 0.15x \\ x &= 260 \end{aligned}$$

Thus, $\bar{x} = 260$ and $\bar{p} = 50 - 0.1(260) = 24$.

$$CS = \int_0^{260} [(50 - 0.1x) - 24]\, dx = \int_0^{260} (26 - 0.1x) dx = (26x - 0.05x^2)\Big|_0^{260} = \$3{,}380$$

$$PS = \int_0^{260} [24 - (11 + 0.05x)]\, dx = \int_0^{260} [13 - 0.05x]\, dx = (13x - 0.025x^2)\Big|_0^{260} = \$1{,}690$$

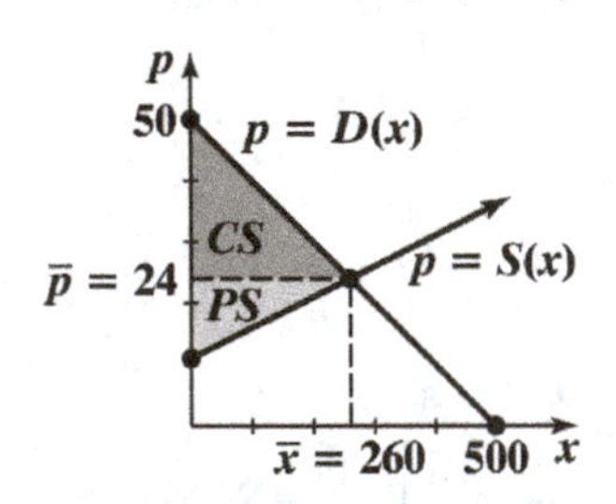

65. $D(x) = 80e^{-0.001x}$ and $S(x) = 30e^{0.001x}$

Equilibrium price: $D(x) = S(x)$

$$\begin{aligned} 80e^{-0.001x} &= 30e^{0.001x} \\ e^{0.002x} &= \frac{8}{3} \\ 0.002x &= \ln\left(\frac{8}{3}\right) \\ \bar{x} &= \frac{\ln\left(\frac{8}{3}\right)}{0.002} \approx 490 \end{aligned}$$

Thus, $\bar{p} = 30e^{0.001(490)} \approx 49$.

$$CS = \int_0^{490} [80e^{-0.001x} - 49]\, dx = \left(\frac{80e^{-0.001x}}{-0.001} - 49x\right)\Bigg|_0^{490} = -80{,}000e^{-0.49} + 80{,}000 - 24{,}010 \approx \$6{,}980$$

$$PS = \int_0^{490} [49 - 30e^{0.001x}] dx = \left(49x - \frac{30e^{0.001x}}{0.001}\right)\Bigg|_0^{490} = 24{,}010 - 30{,}000(e^{0.49} - 1) \approx \$5{,}041$$

67. $D(x) = 80 - 0.04x;\ S(x) = 30e^{0.001x}$

Equilibrium price: $D(x) = S(x)$

$$80 - 0.04x = 30e^{0.001x}$$

Using a graphing utility, we find that

$$\bar{x} \approx 614$$

Thus, $\bar{p} = 80 - (0.04)614 \approx 55$

$$CS = \int_0^{614} [80 - 0.04x - 55]\, dx = \int_0^{614} (25 - 0.04x)\, dx = (25x - 0.02x^2)\Big|_0^{614} \approx \$7{,}810$$

$$PS = \int_0^{614} [55 - (30e^{0.001x})]\, dx = \int_0^{614} (55 - 30e^{0.001x})\, dx = (55x - 30{,}000e^{0.001x})\Big|_0^{614}$$

$$\approx \$8{,}336$$

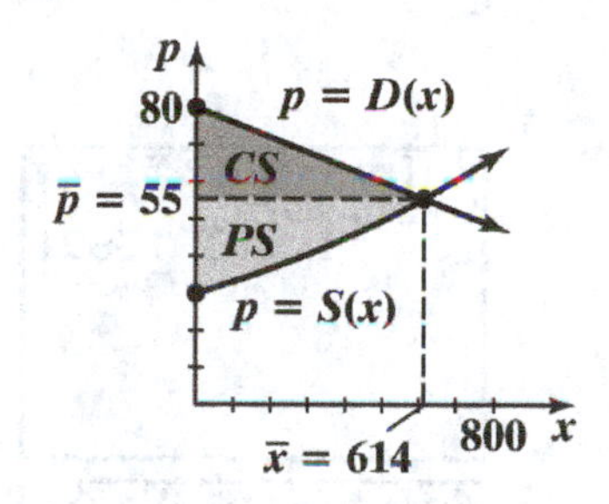

69. $D(x) = 80e^{-0.001x};\ S(x) = 15 + 0.0001x^2$

Equilibrium price: $D(x) = S(x)$

Using a graphing utility, we find that $\bar{x} \approx 556$

Thus, $\bar{p} = 15 + 0.0001(556)^2 \approx 46$

$$CS = \int_0^{556} [80e^{-0.001x} - 46]\, dx = (-80{,}000e^{-0.001x} - 46x)\Big|_0^{556} \approx \$8{,}544$$

$$PS = \int_0^{556} [46 - (15 + 0.0001x^2)]\, dx = \int_0^{556} (31 - 0.0001x^2)\, dx = \left(31x - \frac{0.0001}{3}x^3\right)\Big|_0^{556} \approx \$11{,}507$$

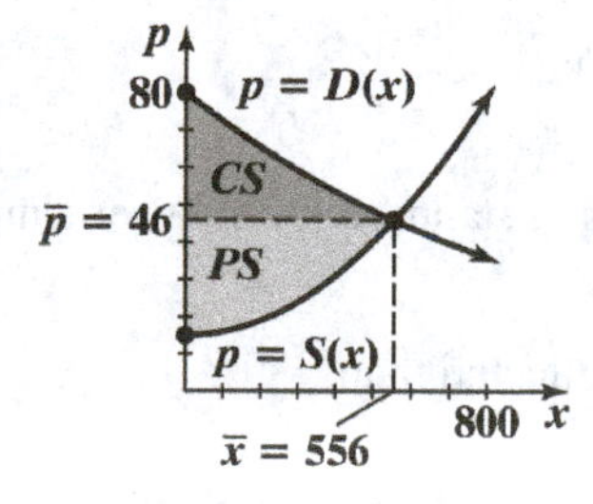

71. (A) Price-Demand

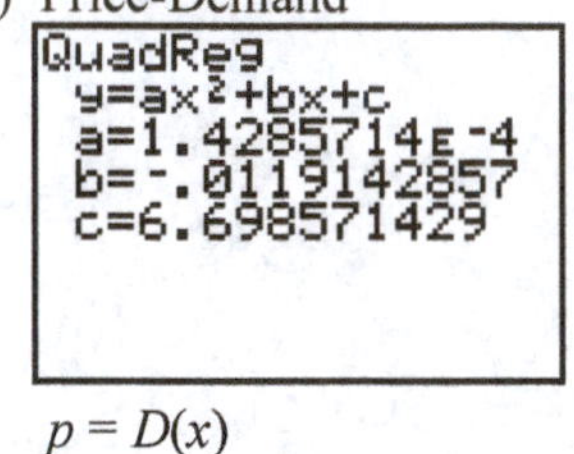

$p = D(x)$

Price-Supply

LinReg
y=ax+b
a=.0046
b=6.41

$p = S(x)$

Graph the price-demand and price-supply models and find their point of intersection.

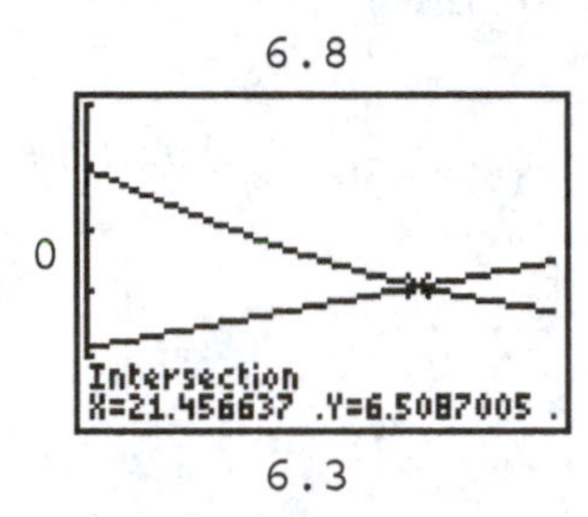

Equilibrium quantity $\bar{x} = 21.457$
Equilibrium price $\bar{p} = 6.51$

(B) Let $D(x)$ be the quadratic regression model in part (A).
Consumers' surplus:

$$CS = \int_0^{21.457} [D(x) - 6.51]dx \approx \$1{,}774.$$

fnInt(Y₁-6.51,X,
0,21.457)
1.773913691

Let $S(x)$ be the linear regression model in part (A).
Producers' surplus

$$PS = \int_0^{21.457} [6.51 - S(x)]dx \approx \$1{,}087$$

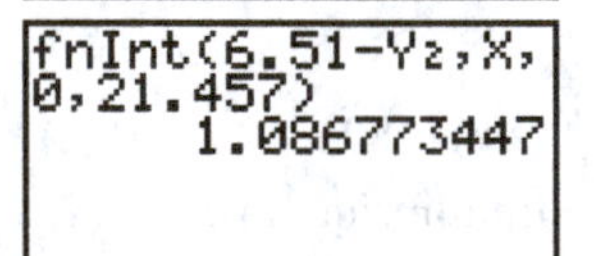

EXERCISE 6-3

Things to remember:

1. INTEGRATION-BY-PARTS FORMULA

$$\int udv = uv - \int vdu$$

2. INTEGRATION-BY-PARTS: SELECTION OF u AND dv

   (a) The product udv must equal the original integrand.

   (b) It must be possible to integrate dv (preferably by using standard formulas or simple substitutions.)

   (c) The new integral, $\int vdu$, should not be more complicated than the original integral $\int udv$.

   (d) For integrals involving $x^p e^{ax}$, try
   $u = x^p$; $dv = e^{ax}dx$.

   (e) For integrals involving $x^p(\ln x)^q$, try
   $u = (\ln x)^q$; $dv = x^p dx$.

1. $f(x)=5x,\ \ f'(x)=5;\ \ g(x)=x^3,\ \ \int x^3\,dx=\frac{x^4}{4}+C$

3. $f(x)=x^3,\ \ f'(x)=3x^2;\ \ g(x)=5x,\ \ \int 5x\,dx=\frac{5}{2}x^2+C$

5 $f(x)=e^{4x},\ \ \ f'(x)=4e^{4x};\ \ g(x)=\frac{1}{x},\ \ \int\frac{1}{x}dx=\ln|x|+C$

7. $f(x)=\frac{1}{x},\ \ f'(x)=-x^{-2}=\frac{-1}{x^2};\ \ g(x)=e^{4x},\ \ \int e^{4x}\,dx=\frac{1}{4}e^{4x}+C$

9. $\int xe^{3x}dx$

Let $u=x$ and $dv=e^{3x}dx$. Then $du=dx$ and $v=\frac{e^{3x}}{3}$.

$$\int xe^{3x}dx=\frac{xe^{3x}}{3}-\int\frac{e^{3x}}{3}dx=\frac{1}{3}xe^{3x}-\frac{1}{3}\int e^{3x}dx=\frac{1}{3}xe^{3x}-\frac{1}{9}e^{3x}+C$$

11. $\int x^2\ln x dx$

Let $u=\ln x$ and $dv=x^2dx$. Then $du=\frac{dx}{x}$ and $v=\frac{x^3}{3}$.

$$\int x^2\ln x\,dx=(\ln x)\left(\frac{x^3}{3}\right)-\int\frac{x^3}{3}\cdot\frac{dx}{x}=\frac{1}{3}x^3\ln x-\frac{1}{3}\int x^2dx$$

$$=\frac{x^3\ln x}{3}-\frac{1}{3}\cdot\frac{x^3}{3}+C=\frac{x^3\ln x}{3}-\frac{x^3}{9}+C$$

13. $\int (x+1)^5(x+2)\,dx$

The better choice is $u=x+2,\ dv=(x+1)^5dx$

The alternative is $u=(x+1)^5,\ dv=(x+2)\,dx$, which will lead to an integral of the form $\int (x+1)^4(x+2)^2dx$.

Let $u=x+2$ and $dv=(x+1)^5dx$. Then $du=dx$ and $v=\frac{1}{6}(x+1)^6$.

Substitute into the integration by parts formula:

$$\int (x+1)^5(x+2)\,dx=\frac{1}{6}(x+1)^6(x+2)-\int\frac{1}{6}(x+1)^6dx=\frac{1}{6}(x+1)^6(x+2)-\frac{1}{42}(x+1)^7+C$$

15. $\int xe^{-x}dx$

Let $u=x$ and $dv=e^{-x}dx$. Then $du=dx$ and $v=-e^{-x}$.

$$\int xe^{-x}dx=x(-e^{-x})-\int(-e^{-x})dx=-xe^{-x}+\int e^{-x}dx=-xe^{-x}-e^{-x}+C$$

17. $\int xe^{x^2}\,dx$

Let $u = x^2$. Then $du = 2x\,dx$.

$\int xe^{x^2}\,dx = \frac{1}{2}\int e^{x^2}2x\,dx = \frac{1}{2}\int e^u\,du = \frac{1}{2}e^u + C = \frac{1}{2}e^{x^2} + C$

19. $\int_0^1 (x-3)e^x dx$

Let $u = (x-3)$ and $dv = e^x dx$. Then $du = dx$ and $v = e^x$.

$\int (x-3)e^x dx = (x-3)e^x - \int e^x dx = (x-3)e^x - e^x + C = xe^x - 4e^x + C.$

Thus, $\int_0^1 (x-3)e^x dx = (xe^x - 4e^x)\Big|_0^1 = (e - 4e) - (-4) = -3e + 4 \approx -4.1548.$

21. $\int_1^3 \ln 2x dx$

Let $u = \ln 2x$ and $dv = dx$. Then $du = \dfrac{dx}{x}$ and $v = x$.

$\int \ln 2x dx = (\ln 2x)(x) - \int x\cdot\dfrac{dx}{x} = x\ln 2x - x + C$

Thus, $\int_1^3 \ln 2x dx = (x\ln 2x - x)\Big|_1^3 = (3\ln 6 - 3) \quad -(\ln 2 - 1) \approx 2.6821.$

23. $\int \dfrac{2x}{x^2+1}dx = \int \dfrac{1}{u}du = \ln|u| + C = \ln(x^2+1) + C$

Substitution: $u = x^2 + 1$, $du = 2xdx$

[Note: Absolute value not needed, since $x^2 + 1 \ge 0$.]

25. $\int \dfrac{\ln x}{x}dx = \int udu = \dfrac{u^2}{2} + C = \dfrac{(\ln x)^2}{2} + C$

Substitution: $u = \ln x$, $du = \dfrac{1}{x}dx$

27. $\int \sqrt{x}\ln x\,dx = \int x^{1/2}\ln x\,dx$

Let $u = \ln x$ and $dv = x^{1/2}dx$. Then $du = \dfrac{dx}{x}$ and $v = \dfrac{2}{3}x^{3/2}$.

$\int x^{1/2}\ln x\,dx = \dfrac{2}{3}x^{3/2}\ln x - \int \dfrac{2}{3}x^{3/2}\dfrac{dx}{x} = \dfrac{2}{3}x^{3/2}\ln x - \dfrac{2}{3}\int x^{1/2}dx = \dfrac{2}{3}x^{3/2}\ln x - \dfrac{4}{9}x^{3/2} + C$

29. $\int (x-3)(x+1)^2\,dx$

Let $u = x - 3$ and $dv = (x+1)^2\,dx$. Then $du = dx$ and $v = \dfrac{1}{3}(x+1)^3$.

$$\int (x-3)(x+1)^2\,dx = \frac{1}{3}(x-3)(x+1)^3 - \int \frac{1}{3}(x+1)^3\,dx = \frac{1}{3}(x-3)(x+1)^3 - \frac{1}{12}(x+1)^4 + C$$

OR

$$\int (x-3)(x+1)^2\,dx = \int (x-3)(x^2+2x+1)\,dx = \int (x^3-x^2-5x-3)\,dx = \frac{1}{4}x^4 - \frac{1}{3}x^3 - \frac{5}{2}x^2 - 3x + C$$

31. $\int (2x+1)(x-2)^2\,dx$

Let $u = 2x+1$ and $dv = (x-2)^2\,dx$. Then $du = 2dx$ and $v = \frac{1}{3}(x-2)^3$.

$$\int (2x+1)(x-2)^2\,dx = \frac{1}{3}(2x+1)(x-2)^3 - \frac{2}{3}\int (x-2)^3\,dx = \frac{1}{3}(2x+1)(x-2)^3 - \frac{1}{6}(x-2)^4 + C$$

OR

$$\int (2x+1)(x-2)^2\,dx = \int (2x+1)(x^2-4x+4)\,dx = \int (2x^3-7x^2+4x+4)dx = \frac{1}{2}x^4 - \frac{7}{3}x^3 + 2x^2 + 4x + C$$

33.

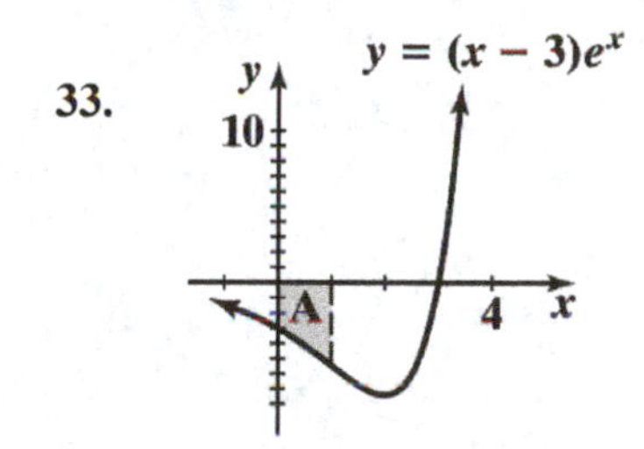

Since $f(x) = (x-3)e^x < 0$ on $[0, 1]$, the integral represents the negative of the area between the graph of f and the x-axis from $x = 0$ to $x = 1$.

35.

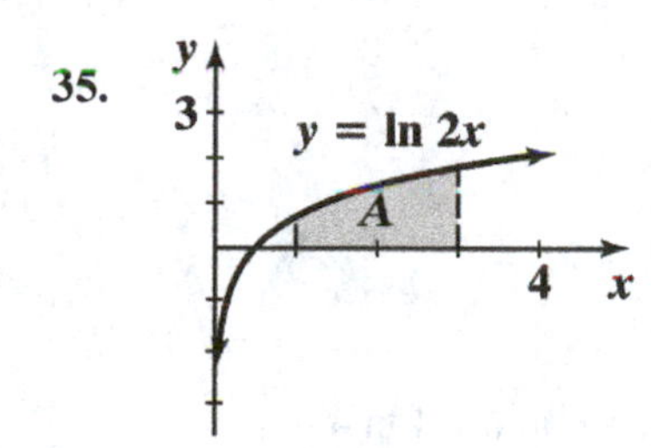

The integral represents the area between the curve $y = \ln 2x$ and the x-axis from $x = 1$ to $x = 3$.

37. $\int x^2e^x dx$

Let $u = x^2$ and $dv = e^x dx$. Then $du = 2x\,dx$ and $v = e^x$.

$$\int x^2e^x dx = x^2e^x - \int e^x(2x)dx = x^2e^x - 2\int xe^x dx$$

$\int xe^x dx$ can be computed by using integration-by-parts again.

Let $u = x$ and $dv = e^x dx$. Then $du = dx$ and $v = e^x$.

$$\int xe^x dx = xe^x - \int e^x dx = xe^x - e^x + C$$

and

$$\int x^2e^x dx = x^2e^x - 2(xe^x - e^x) + C = x^2e^x - 2xe^x + 2e^x + C = (x^2 - 2x + 2)e^x + C$$

39. $\int xe^{ax}dx$

Let $u = x$ and $dv = e^{ax}dx$. Then $du = dx$ and $v = \frac{e^{ax}}{a}$.

$$\int xe^{ax}dx = \frac{xe^{ax}}{a} - \int \frac{e^{ax}}{a}dx = \frac{xe^{ax}}{a} - \frac{e^{ax}}{a^2} + C$$

41. $\int_1^e \frac{\ln x}{x^2}dx$

Let $u = \ln x$ and $dv = \frac{dx}{x^2}$. Then $du = \frac{dx}{x}$ and $v = \frac{-1}{x}$.

$$\int \frac{\ln x}{x^2}dx = (\ln x)\left(-\frac{1}{x}\right) - \int -\frac{1}{x}\cdot\frac{dx}{x} = -\frac{\ln x}{x} + \int \frac{dx}{x^2} = -\frac{\ln x}{x} - \frac{1}{x} + C$$

Thus, $\int_1^e \frac{\ln x}{x^2}dx = \left[-\frac{\ln x}{x} - \frac{1}{x}\right]_1^e = -\frac{\ln e}{e} - \frac{1}{e} - \left(-\frac{\ln 1}{1} - \frac{1}{1}\right) = 1 - \frac{2}{e} \approx 0.2642$.

[Note: $\ln 1 = 0$, $\ln e = 1$.]

43. $\int_0^2 \ln(x+4)\,dx$

Let $t = x + 4$. Then $dt = dx$ and

$\int \ln(x+4)\,dx = \int \ln t\,dt$.

Now, let $u = \ln t$ and $dv = dt$. Then $du = \frac{dt}{t}$ and $v = t$.

$$\int \ln t\,dt = t\ln t - \int t\left(\frac{1}{t}\right)dt = t\ln t - \int dt = t\ln t - t + C$$

Thus, $\int \ln(x+4)\,dx = (x+4)\ln(x+4) - (x+4) + C$

and

$$\int_0^2 \ln(x+4)\,dx = [(x+4)\ln(x+4) - (x+4)]\Big|_0^2 = 6\ln 6 - 6 - (4\ln 4 - 4) = 6\ln 6 - 4\ln 4 - 2$$

$$\approx 3.205.$$

45. $\int xe^{x-2}dx$

Let $u = x$ and $dv = e^{x-2}dx$. Then $du = dx$ and $v = e^{x-2}$.

$$\int xe^{x-2}dx = xe^{x-2} - \int e^{x-2}\,dx = xe^{x-2} - e^{x-2} + C$$

47. $\int x\ln(1+x^2)\,dx$

Let $t = 1 + x^2$. Then $dt = 2x\,dx$ and

$$\int x\ln(1+x^2)\,dx = \int \ln(1+x^2)x\,dx = \int \ln t\frac{dt}{2} = \frac{1}{2}\int \ln t\,dt.$$

Now, for $\int \ln t\,dt$, let $u = \ln t$, $dv = dt$. Then $du = \frac{dt}{t}$ and $v = t$.

$$\int \ln t\,dt = t\ln t - \int t\left(\frac{1}{t}\right)dt = t\ln t - \int dt = t\ln t - t + C$$

Therefore,

$$\int x \ln(1+x^2)\,dx = \frac{1}{2}(1+x^2)\ln(1+x^2) - \frac{1}{2}(1+x^2) + C.$$

49. $\int e^x \ln(1+e^x)\,dx$

Let $t = 1 + e^x$. Then $dt = e^x dx$ and

$$\int e^x \ln(1+e^x)\,dx = \int \ln t\,dt.$$

Now, as shown in Problems 43 and 47,

$$\int \ln t\,dt = t\ln t - t + C.$$

Thus, $\int e^x \ln(1+e^x)\,dx = (1+e^x)\ln(1+e^x) - (1+e^x) + C.$

51. $\int (\ln x)^2 dx$

Let $u = (\ln x)^2$ and $dv = dx$. Then $du = \frac{2\ln x}{x}dx$ and $v = x$.

$$\int (\ln x)^2 dx = x(\ln x)^2 - \int x\cdot\frac{2\ln x}{x}\,dx = x(\ln x)^2 - 2\int \ln x\,dx$$

$\int \ln x\,dx$ can be computed by using integration-by-parts again.

As shown in Problems 43 and 47,

$$\int \ln x\,dx = x\ln x - x + C.$$

Thus, $\int (\ln x)^2 dx = x(\ln x)^2 - 2(x\ln x - x) + C = x(\ln x)^2 - 2x\ln x + 2x + C.$

53. $\int (\ln x)^3\,dx$

Let $u = (\ln x)^3$ and $dv = dx$. Then $du = 3(\ln x)^2\,\frac{1}{x}dx$ and $v = x$.

$$\int (\ln x)^3\,dx = x(\ln x)^3 - \int x\cdot 3(\ln x)^2\cdot\frac{1}{x}dx = x(\ln x)^3 - 3\int (\ln x)^2\,dx$$

Now, using Problem 51,

$$\int (\ln x)^2\,dx = x(\ln x)^2 - 2x\ln x + 2x + C.$$

Therefore, $\int (\ln x^3)dx = x(\ln x)^3 - 3[x(\ln x)^2 - 2x\ln x + 2x] + C$

$$= x(\ln x)^3 - 3x(\ln x)^2 + 6x\ln x - 6x + C.$$

55. $\int_1^e \ln(x^2)\,dx = \int_1^e 2\ln x\,dx = 2\int_1^e \ln x\,dx$

By Example 4, $\int \ln x\,dx = x\ln x - x + C$. Therefore

$$2\int_1^e \ln x\,dx = 2(x\ln x - x)\Big|_1^e = 2[e\ln e - e] - 2[\ln 1 - 1] = 2$$

57. $\int_0^1 \ln(e^{x^2})\,dx = \int_0^1 x^2\,dx = \frac{1}{3}x^3 \Big|_0^1 = \frac{1}{3}$

(Note: $\ln(e^{x^2}) = x^2 \ln e = x^2$.)

59. $y = x - 2 - \ln x,\ 1 \le x \le 4$

$y = 0$ at $x \approx 3.146$

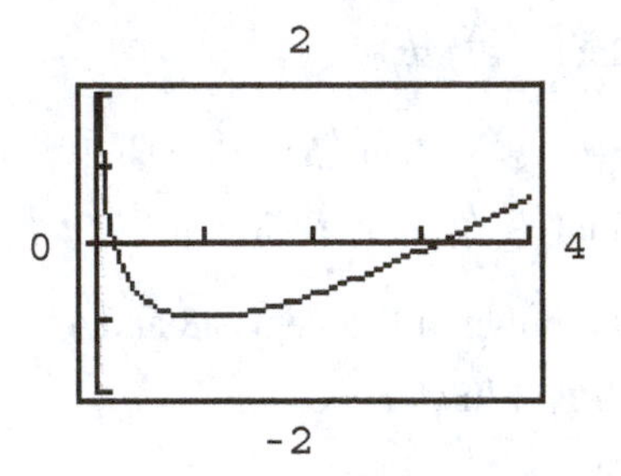

$$A = \int_1^{3.146} [-(x - 2 - \ln x)]\,dx + \int_{3.146}^4 (x - 2 - \ln x)\,dx$$
$$= \int_1^{3.146} (\ln x + 2 - x)\,dx + \int_{3.146}^4 (x - 2 - \ln x)\,dx$$

Now, $\int \ln x\,dx$ is found using integration-by-parts. Let $u = \ln x$ and $dv = dx$. Then $du = \frac{1}{x}dx$ and $v = x$.

$$\int \ln x\,dx = x\ln x - \int x\left(\frac{1}{x}\right)dx = x\ln x - \int dx = x\ln x - x + C$$

Thus,

$$A = \left(x\ln x - x + 2x - \frac{1}{2}x^2\right)\Big|_1^{3.146} + \left(\frac{1}{2}x^2 - 2x - x\ln x + x\right)\Big|_{3.146}^4$$
$$= \left(x\ln x + x - \frac{1}{2}x^2\right)\Big|_1^{3.146} + \left(\frac{1}{2}x^2 - x - x\ln x\right)\Big|_{3.146}^4$$
$$\approx (1.803 - 0.5) + (-1.545 + 1.803) = 1.561$$

61. $y = 5 - xe^x,\ 0 \le x \le 3$

$y = 0$ at $x \approx 1.327$

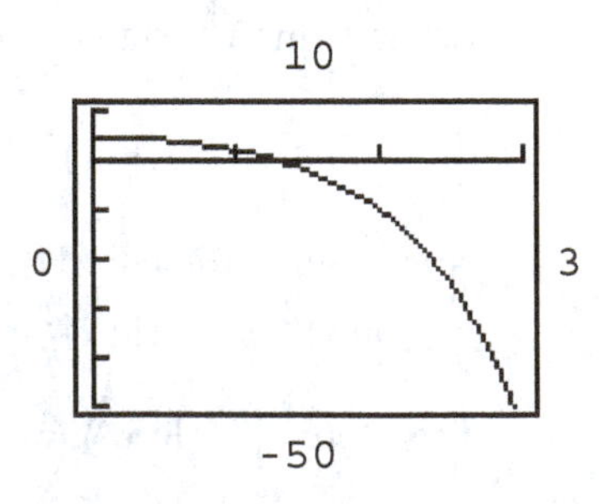

$$A = \int_0^{1.327} (5 - xe^x)\,dx + \int_{1.327}^3 [-(5 - xe^x)]\,dx$$
$$= \int_0^{1.327} (5 - xe^x)\,dx + \int_{1.327}^3 (xe^x - 5)\,dx$$

Now, $\int xe^x\,dx$ is found using integration-by-parts. Let $u = x$ and $dv = e^x\,dx$. Then, $du = dx$ and $v = e^x$.

$$\int xe^x\,dx = xe^x - \int e^x\,dx = xe^x - e^x + C$$

Thus,

$$A = (5x - [xe^x - e^x])\Big|_0^{1.327} + (xe^x - e^x - 5x)\Big|_{1.327}^3 \approx (5.402 - 1) + (25.171 - [-5.402]) \approx 34.98$$

63. Marginal profit: $P'(t) = 2t - te^{-t}$.
The total profit over the first 5 years is given by the definite integral:

$$\int_0^5 (2t - te^{-t})\,dt = \int_0^5 2t\,dt - \int_0^5 te^{-t}dt$$

We calculate the second integral using integration-by-parts.
Let $u = t$ and $dv = e^{-t}\,dt$. Then $du = dt$ and $v = -e^{-t}$

$$\int te^{-t}dt = -te^{-t} - \int -e^{-t}dt = -te^{-t} - e^{-t} + C = -e^{-t}[t+1] + C$$

Thus,

$$\text{Total profit} = t^2\Big|_0^5 + (e^{-t}\ [t+1])\Big|_0^5 \approx 25 + (0.040 - 1) = 24.040$$

To the nearest million, the total profit is \$24 million.

65.

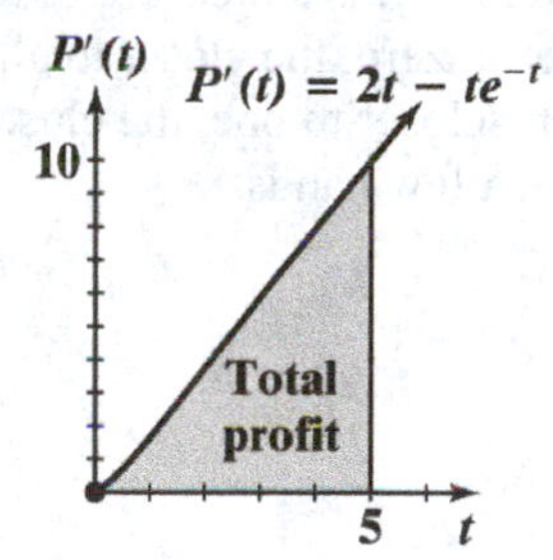

The total profit for the first five years (in millions of dollars) is the same as the area under the marginal profit function, $P'(t) = 2t - te^{-t}$, from $t = 0$ to $t = 5$.

67. From Section 6-2, Future Value $= e^{rT}\int_0^T f(t)e^{-rt}dt$. Now $r = 0.0395$, $T = 5$, $f(t) = 1000 - 200t$. Thus,

$$FV = e^{(0.0395)5}\int_0^5 (1000 - 200t)e^{-0.0395t}dt$$

$$= 1000e^{0.1975}\int_0^5 e^{-0.0395t}dt - 200e^{0.1975}\int_0^5 te^{-0.0395t}dt.$$

We calculate the second integral using integration-by-parts.

Let $u = t$, $dv = e^{-0.0395t}dt$. Then $du = dt$ and $v = \dfrac{e^{-0.0395t}}{-0.0395}$.

$$\int te^{-0.0395t}\,dt = -\frac{te^{-0.0395t}}{0.0395} - \int \frac{e^{-0.0395t}}{-0.0395}dt = -\frac{te^{-0.0395t}}{0.0395} - \frac{e^{-0.0395t}}{(0.0395)^2} + C$$

Thus, we have:

$$FV = 1000\,e^{0.1975}\left[\frac{e^{-0.0395t}}{-0.0395}\right]_0^5 - 200\,e^{0.1975}\left[-\frac{te^{-0.0395t}}{0.0395} - \frac{e^{-0.0395t}}{(0.0395)^2}\right]_0^5$$

$$= 1000\,e^{0.1975}\left[\frac{1}{0.0395} - \frac{e^{-0.1975}}{0.0395}\right] - 200\,e^{0.1975}\left[\frac{1}{(0.0395)^2} - \frac{5e^{-0.1975}}{0.0395} - \frac{e^{-0.1975}}{(0.0395)^2}\right] = \$2854.88$$

69. Gini Index $= 2\int_0^1 (x - xe^{x-1})dx = 2\int_0^1 x dx - 2\int_0^1 xe^{x-1}dx$

We calculate the second integral using integration-by-parts.

Let $u = x,\ dv = e^{x-1}dx$. Then $du = dx,\ v = e^{x-1}$.

$$\int xe^{x-1}dx = xe^{x-1} - \int e^{x-1}dx = xe^{x-1} - e^{x-1} + C$$

Therefore, $2\int_0^1 x dx - 2\int_0^1 xe^{x-1}dx = x^2 \Big|_0^1 - 2[xe^{x-1} - e^{x-1}]\Big|_0^1 = 1 - 2[1 - 1 + (e^{-1})]$

$$= 1 - 2e^{-1} \approx 0.264.$$

71.

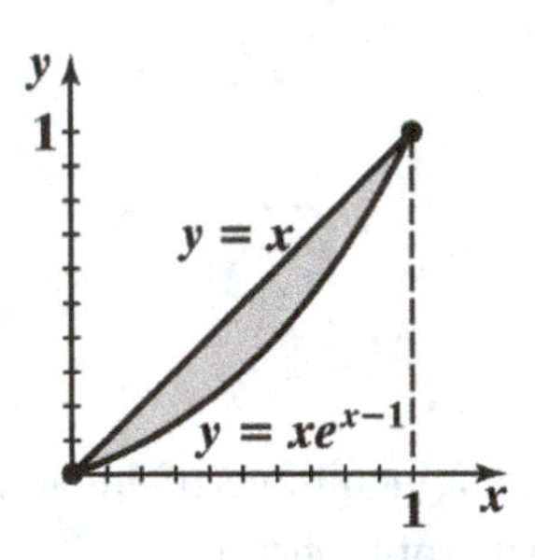

The area bounded by $y = x$ and the Lorenz curve $y = xe^{(x-1)}$ divided by the area under the curve $y = x$ from $x = 0$ to $x = 1$ is the index of income concentration, in this case 0.264. It is a measure of the concentration of income—the closer to zero, the closer to all the income being equally distributed; the closer to one, the closer to all the income being concentrated in a few hands.

73. $S'(t) = -4te^{0.1t}$, $S(0) = 2{,}000$

$$S(t) = \int -4te^{0.1t}\,dt = -4\int te^{0.1t}\,dt$$

Let $u = t$ and $dv = e^{0.1t}\,dt$. Then $du = dt$ and $v = \frac{e^{0.1t}}{0.1} = 10e^{0.1t}$

$$\int te^{0.1t}\,dt = 10te^{0.1t} - \int 10e^{0.1t}\,dt = 10te^{0.1t} - 100e^{0.1t} + C$$

Now, $S(t) = -40te^{0.1t} + 400e^{0.1t} + C$

Since $S(0) = 2{,}000$, we have
$2{,}000 = 400 + C,\ C = 1{,}600$

Thus,
$S(t) = 1{,}600 + 400e^{0.1t} - 40te^{0.1t}$

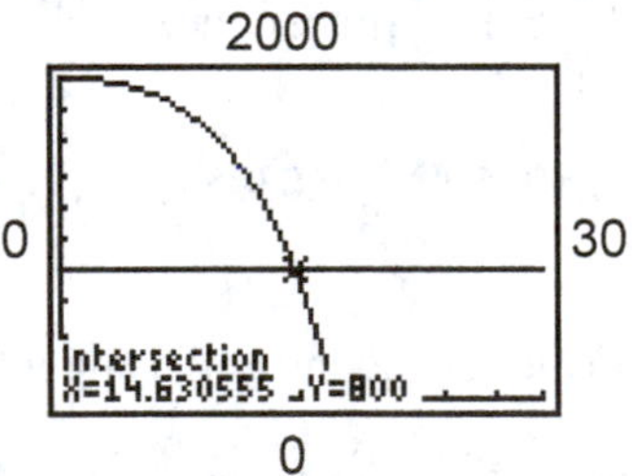

To find how long the company will continue to manufacture this computer, solve $S(t) = 800$ for t.
The company will manufacture the computer for 15 months.

75. $p = D(x) = 9 - \ln(x + 4)$; $\bar{p} = \$2.089$. To find $\bar{x}$, solve

$$9 - \ln(\bar{x} + 4) = 2.089$$
$$\ln(\bar{x} + 4) = 6.911$$
$$\bar{x} + 4 = e^{6.911} \quad \text{(take the exponential of both sides)}$$
$$\bar{x} \approx 1{,}000$$

Now,

$$CS = \int_0^{1{,}000} (D(x) - \bar{p})\,dx = \int_0^{1{,}000} [9 - \ln(x + 4) - 2.089]\,dx = \int_0^{1{,}000} 6.911\,dx - \int_0^{1{,}000} \ln(x + 4)\,dx$$

To calculate the second integral, we first let $z = x + 4$ and $dz = dx$ to get

$$\int \ln(x + 4)\,dx = \int \ln z\,dz$$

Then we use integration-by-parts. Let $u = \ln z$ and $dv = dz$. Then $du = \frac{1}{z}\,dz$ and $v = z$.

$$\int \ln z\,dz = z\ln z - \int z\cdot\frac{1}{z}\,dz = z\ln z - z + C$$

Therefore,

$$\int \ln(x+4)\,dx = (x+4)\ln(x+4) - (x+4) + C$$

and

$$CS = 6.911x\Big|_0^{1{,}000} - [(x+4)\ln(x+4) - (x+4)]\Big|_0^{1{,}000} \approx 6911 - (5935.39 - 1.55) \approx \$977$$

77.

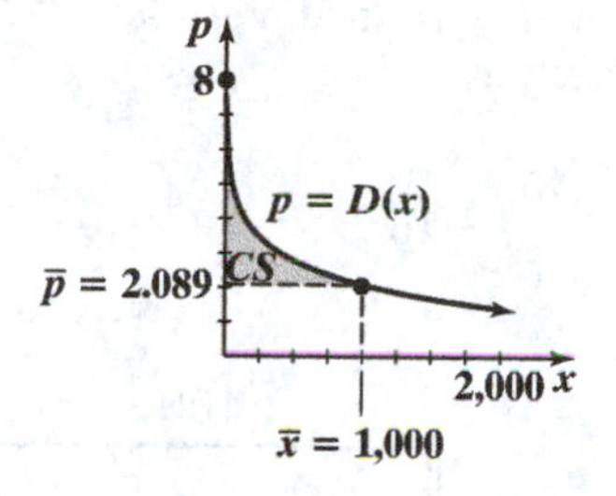

The area bounded by the price-demand equation, $p = 9 - \ln(x+4)$, and the price equation, $y = \bar{p} = 2.089$, from $x = 0$ to $x = \bar{x} = 1{,}000$, represents the consumers' surplus. This is the amount saved by consumers who are willing to pay more than \$2.089.

79. Average concentration: $= \frac{1}{5-0}\int_0^5 \frac{20\ln(t+1)}{(t+1)^2}\,dt = 4\int_0^5 \frac{\ln(t+1)}{(t+1)^2}\,dt$

$\int \frac{\ln(t+1)}{(t+1)^2}\,dt$ is found using integration-by-parts.

Let $u = \ln(t+1)$ and $dv = (t+1)^{-2}dt$.

Then $du = \frac{1}{t+1}\,dt = (t+1)^{-1}dt$ and $v = -(t+1)^{-1}$.

$$\int \frac{\ln(t+1)}{(t+1)^2}\,dt = -\frac{\ln(t+1)}{t+1} - \int -(t+1)^{-1}(t+1)^{-1}dt$$

$$= -\frac{\ln(t+1)}{t+1} + \int (t+1)^{-2}dt = -\frac{\ln(t+1)}{t+1} - \frac{1}{t+1} + C$$

Therefore, the average concentration is:

$$\frac{1}{5}\int_0^5 \frac{20\ln(t+1)}{(t+1)^2}\,dt = 4\left[-\frac{\ln(t+1)}{t+1} - \frac{1}{t+1}\right]_0^5 = 4\left(-\frac{\ln 6}{6} - \frac{1}{6}\right) - 4(-\ln 1 - 1)$$

$$= 4 - \frac{2}{3}\ln 6 - \frac{2}{3} = \frac{1}{3}(10 - 2\ln 6) \approx 2.1388 \text{ ppm}$$

81. $N'(t) = (t + 6)e^{-0.25t}, 0 \le t \le 15; N(0) = 40$

$N(t) - N(0) = \int_0^t N'(x)\,dx;$

$$N(t) = 40 + \int_0^t (x+6)e^{-0.25x}dx = 40 + 6\int_0^t e^{-0.25x}dx + \int_0^t xe^{-0.25x}dx$$

$$= 40 + 6(-4e^{-0.25x})\Big|_0^t + \int_0^t xe^{-0.25x}dx$$

$$= 64 - 24e^{-0.25t} + \int_0^t xe^{-0.25x}dx$$

Let $u = x$ and $dv = e^{-0.25x}dx$. Then $du = dx$ and $v = -4e^{-0.25x}$;

$$\int xe^{-0.25x}dx = -4xe^{-0.25x} - \int -4e^{-0.25x}dx = -4xe^{-0.25x} - 16e^{-0.25x} + C$$

Now, $\int_0^t xe^{-0.25x}dx = (-4xe^{-0.25x} - 16e^{-0.25x})\Big|_0^t = -4te^{-0.25t} - 16e^{-0.25t} + 16$

and

$N(t) = 80 - 40e^{-0.25t} - 4te^{-0.25t}$

To find how long it will take a student to achieve the 70 words per minute level, solve $N(t) = 70$:

100

0 15

Intersection X=7.8665604 Y=70

0

It will take 8 weeks.

By the end of the course, a student should be able to type

$N(15) = 80 - 40e^{-0.25(15)} - 60e^{-0.25(15)} \approx 78$ words per minute.

83. Average number of voters $= \frac{1}{5}\int_0^5 (20 + 4t - 5te^{-0.1t})dt$

$$= \frac{1}{5}\int_0^5 (20 + 4t)dt - \int_0^5 te^{-0.1t}dt$$

$\int te^{-0.1t}dt$ is found using integration-by-parts.

Let $u = t$ and $dv = e^{-0.1t}dt$. Then $du = dt$ and $v = \frac{e^{-0.1t}}{-0.1} = -10e^{-0.1t}$.

$$\int te^{-0.1t}dt = -10te^{-0.1t} - \int -10e^{-0.1t}dt = -10te^{-0.1t} + 10\int e^{-0.1t}dt$$

$$= -10te^{-0.1t} + \frac{10e^{-0.1t}}{-0.1} + C = -10te^{-0.1t} - 100e^{-0.1t} + C$$

Therefore, the average number of voters is:

$$\frac{1}{5}\int_0^5 (20 + 4t)dt - \int_0^5 te^{-0.1t}dt = \frac{1}{5}(20t + 2t^2)\Big|_0^5 - (-10te^{-0.1t} - 100e^{-0.1t})\Big|_0^5$$

$$= \frac{1}{5}(100 + 50) + (10te^{-0.1t} + 100e^{-0.1t})\Big|_0^5$$

$$= 30 + (50e^{-0.5} + 100e^{-0.5}) - 100 = 150e^{-0.5} - 70$$

≈ 20.98 (thousands) or 20,980

EXERCISE 6-4

Things to remember:

1. TRAPEZOIDAL RULE

   Let f be a function defined on an interval $[a,b]$. Partition $[a,b]$ into n subintervals of equal length $\Delta x = (b-a)/n$ with endpoints $a = x_0 < x_1 < x_2 < \cdots < x_n = b$. Then

$$T_n = [f(x_0) + 2f(x_1) + 2f(x_2) + \cdots + 2f(x_{n-1}) + f(x_n)]\frac{\Delta x}{2}$$

   is an approximation of $\int_a^b f(x)dx$.

2. SIMPSON'S RULE

   Let f be a function defined on an interval $[a,b]$. Partition $[a,b]$ into $2n$ subintervals of equal length $\Delta x = (b-a)/n$ with endpoints $a = x_0 < x_1 < x_2 < \cdots < x_n = b$. Then

$$S_n = [f(x_0) + 4f(x_1) + 2f(x_2) + 4f(x_3) + 2f(x_4) + \cdots + 4f(x_{2n-1}) + f(x_{2n})]\frac{\Delta x}{3}$$

   is an approximation of $\int_a^b f(x)dx$.

---

1. $\int_0^6 \sqrt{1+x^4}\,dx$, $f(x) = \sqrt{1+x^4}$. Partition [0,6] into three equal subintervals:

$x_0 = 0,\ x_1 = 2,\ x_2 = 4,\ x_3 = 6,\ \ \Delta x = 6/3 = 2.$

x	$f(x)$
0	1
2	4.1231
4	16.0312
6	36.0139

By the trapezoidal rule:

$T_3 = [f(0) + 2f(2) + 2f(4) + f(6)](2/2) = [1 + 8.2462 + 32.0624 + 36.0139] \approx 77.32$

3. $\int_0^6 \sqrt{1+x^4}\,dx$, $f(x) = \sqrt{1+x^4}$. Partition [0,6] into six equal subintervals:

$x_0 = 0,\ x_1 = 1,\ x_2 = 2,\ x_3 = 3,\ x_4 = 4,\ x_5 = 5,\ x_6 = 6;\ \ \Delta x = 6/6 = 1.$

x	$f(x)$
0	1
1	1.4142
2	4.1231

3	9.0554
4	16.0312
5	25.0200
6	36.0139

By the trapezoidal rule:

$$T_6 = [f(0)+2f(1)+2f(2)+2f(3)+2f(4)+2f(5)+f(6)](1/2)$$
$$= [1+2.8284+8.4262+18.1108+32.0624+50.0400+36.0139](1/2) \approx 74.24$$

5. $\int_1^3 \frac{1}{1+x^2}dx,\ f(x)=\frac{1}{1+x^2}.$ Partition [1,3] into two equal subintervals:

$x_0 = 1,\ x_1 = 2,\ x_2 = 3;\ \Delta x = 1.$

x	$f(x)$
1	0.5000
2	0.2000
3	0.1000

By Simpson's rule:

$$S_2 = [f(1)+4f(2)+f(3)](1/3) = [0.5000+0.8000+0.1000](1/3) \approx 0.47$$

7. $\int_1^3 \frac{1}{1+x^2}dx,\ f(x)=\frac{1}{1+x^2}.$ Partition [1,3] into four equal subintervals:

$x_0 = 1,\ x_1 = 1.5,\ x_2 = 2,\ x_3 = 2.5,\ x_4 = 3;\ \Delta x = 0.5.$

x	$f(x)$
1	0.5000
1.5	0.3077
2	0.2000
2.5	0.1379
3	0.1000

By Simpson's rule:

$$S_4 = [f(1)+4f(1.5)+2f(2)+4f(2.5)+f(3)](1/6)$$
$$= [0.5000+1.2308+0.4000+0.5516+0.1000](1/6)] \approx 0.46.$$

9. Use Formula 9 with $a = b = 1$. $\int \frac{1}{x(1+x)}dx = \frac{1}{1}\ln\left|\frac{x}{1+x}\right| + C = \ln\left|\frac{x}{x+1}\right| + C$

11. Use Formula 18 with $a = 3$, $b = 1$, $c = 5$, $d = 2$:

$$\int \frac{1}{(3+x)^2(5+2x)}\,dx = \frac{1}{3\cdot 2 - 5\cdot 1}\cdot\frac{1}{3+x} + \frac{2}{(3\cdot 2-5\cdot 1)^2}\ln\left|\frac{5+2x}{3+x}\right| + C = \frac{1}{3+x} + 2\ln\left|\frac{5+2x}{3+x}\right| + C$$

13. Use Formula 25 with $a = 16$ and $b = 1$:

$$\int \frac{x}{\sqrt{16+x}}\,dx = \frac{2(x-2\cdot 16)}{3\cdot 1^2}\sqrt{16+x} + C = \frac{2(x-32)}{3}\sqrt{16+x} + C$$

15. Use Formula 29 with $a = 1$: $\displaystyle\int \frac{1}{x\sqrt{1-x^2}}\,dx = -\frac{1}{1}\ln\left|\frac{1+\sqrt{1-x^2}}{x}\right| + C = -\ln\left|\frac{1+\sqrt{1-x^2}}{x}\right| + C$

17. Use Formula 37 with $a = 2$ ($a^2 = 4$): $\displaystyle\int \frac{1}{x\sqrt{x^2+4}}\,dx = \frac{1}{2}\ln\left|\frac{x}{2+\sqrt{x^2+4}}\right| + C$

19. Use Formula 51 with $n = 2$: $\displaystyle\int x^2\ln x\,dx = \frac{x^{2+1}}{2+1}\ln x - \frac{x^{2+1}}{(2+1)^2} + C = \frac{x^3}{3}\ln x - \frac{x^3}{9} + C$

21. Use Formula 48 with $a = c = d = 1$: $\displaystyle\int \frac{1}{1+e^x}\,dx = x - \ln|1+e^x| + C.$

23. First use Formula 5 with $a = 3$ and $b = 1$ to find the indefinite integral.

$$\int \frac{x^2}{3+x}\,dx = \frac{(3+x)^2}{2\cdot 1^3} - \frac{2\cdot 3(3+x)}{1^3} + \frac{3^2}{1^3}\ln|3+x| + C = \frac{(3+x)^2}{2} - 6(3+x) + 9\ln|3+x| + C$$

Thus, $\displaystyle\int_1^3 \frac{x^2}{3+x}\,dx = \left[\frac{(3+x)^2}{2} - 6(3+x) + 9\ln|3+x|\right]_1^3$

$$= \frac{(3+3)^2}{2} - 6(3+3) + 9\ln|3+3| - \left[\frac{(3+1)^2}{2} - 6(3+1) + 9\ln|3+1|\right]$$

$$= 9\ln\frac{3}{2} - 2 \approx 1.6492.$$

25. First use Formula 15 with $a = 3$, $b = c = d = 1$ to find the indefinite integral.

$$\int \frac{1}{(3+x)(1+x)}\,dx = \frac{1}{3\cdot 1 - 1\cdot 1}\ln\left|\frac{1+x}{3+x}\right| + C = \frac{1}{2}\ln\left|\frac{1+x}{3+x}\right| + C$$

Thus, $\displaystyle\int_0^7 \frac{1}{(3+x)(1+x)}\,dx = \frac{1}{2}\ln\left|\frac{1+x}{3+x}\right|\Bigg|_0^7 = \frac{1}{2}\ln\left|\frac{1+7}{3+7}\right| - \frac{1}{2}\ln\left|\frac{1}{3}\right|$

$$= \frac{1}{2}\ln\left|\frac{4}{5}\right| - \frac{1}{2}\ln\left|\frac{1}{3}\right| = \frac{1}{2}\ln\frac{12}{5} \approx 0.4377.$$

27. First use Formula 36 with $a = 3$ ($a^2 = 9$) to find the indefinite integral:

$$\int \frac{1}{\sqrt{x^2+9}}\,dx = \ln\left|x+\sqrt{x^2+9}\right| + C$$

Thus, $\displaystyle\int_0^4 \frac{1}{\sqrt{x^2+9}}\,dx = \ln\left|x+\sqrt{x^2+9}\right|\Bigg|_0^4 = \ln\left|4+\sqrt{16+9}\right| - \ln\left|\sqrt{9}\right| = \ln 9 - \ln 3 = \ln 3 \approx 1.0986.$

29. $\int_3^{13} x^2 dx,\ f(x) = x^2.$ Partition [3,13] into five equal subintervals:

$x_0 = 3,\ x_1 = 5,\ x_2 = 7,\ x_3 = 9,\ x_4 = 11,\ x_5 = 13;\ \Delta x = 2.$

x	$f(x)$
3	9
5	25
7	49
9	81
11	121
13	169

$T_5 = [f(3) + 2f(5) + 2f(7) + 2f(9) + 2f(11) + f(13)](2/2)$

$= [9 + 50 + 98 + 162 + 242 + 169] = 730.$

Exact value: $\int_3^{13} x^2\, dx = \left.\frac{x^3}{3}\right|_3^{13} = \frac{13^3}{3} - 9 = 723.33.$

31. $\int_1^5 \frac{1}{x} dx,\ f(x) = \frac{1}{x}.$ Partition [1,5] into eight equal subintervals:

$x_0 = 1,\ x_1 = 1.5,\ x_2 = 2,\ x_3 = 2.5,\ x_4 = 3,\ x_5 = 3.5,\ x_6 = 4,\ x_7 = 4.5,\ x_8 = 5,\ \Delta x = 0.5.$

x	$f(x)$
1	1
1.5	0.6667
2	0.5000
2.5	0.4000
3	0.3333
3.5	0.2857
4	0.2500
4.5	0.2222
5	0.2000

$$S_8 = [f(1)+4f(1.5)+2f(2)+4f(2.5)+2f(3)+4f(3.5)+2f(4)+4f(4.5)+f(5)](1/6)$$
$$=[1+2.6667+1+1.6000+0.6667+1.1428+0.5000+0.8888+.2000](1/6) \approx 1.61$$

Exact value: $\int_1^5 \frac{1}{x}dx = \ln x\Big|_1^5 = \ln 5 \approx 1.61.$

33. $\int_5^8 (4x-3)dx;\ f(x)=4x-3.$. Partition [5,8] into three equal subintervals:

$x_0 = 5,\ x_1 = 6,\ x_2 = 7,\ x_3 = 8,\ \Delta x = 1.$

x	$f(x)$
5	17
6	21
7	25
8	29

$T_3 = [f(5)+2f(6)+2f(7)+f(8)](1/2) = [17+42+50+29](1/2) = 69.$

Exact value: $\int_5^8 (4x-3)dx = \left[2x^2-3x\right]_5^8 = (128-24)-(50-15) = 69.$

35. $\int_5^9 (3x^2+5x+3)dx;\ f(x)=3x^2+5x+3.$ Partition [5,9] into four equal subintervals:

$x_0 = 5,\ x_1 = 6,\ x_2 = 7,\ x_3 = 8,\ x_4 = 9,\ \Delta x = 1.$

x	$f(x)$
5	103
6	141
7	185
8	235
9	291

$S_4 = [f(5)+4f(6)+2f(7)+4f(8)+f(9)](1/3) = [103+564+370+940+291](1/3) = 756.$

Exact value: $\int_5^9 (3x^2+5x+3)dx = \left[x^3+\frac{5x^2}{2}+3x\right]_5^9 = \left(729+\frac{405}{2}+27\right)-\left(125+\frac{125}{2}+15\right) = 756.$

37. Consider Formula 35. Let $u = 2x$. Then $u^2 = 4x^2$, $x = \frac{u}{2}$, and $dx = \frac{du}{2}$.

$$\int \frac{\sqrt{4x^2+1}}{x^2}dx = \int \frac{\sqrt{u^2+1}}{\frac{u^2}{4}}\frac{du}{2} = 2\int \frac{\sqrt{u^2+1}}{u^2}du = 2\left[-\frac{\sqrt{u^2+1}}{u} + \ln\left|u+\sqrt{u^2+1}\right|\right] + C$$

$$= 2\left[-\frac{\sqrt{4x^2+1}}{2x} + \ln\left|2x+\sqrt{4x^2+1}\right|\right] + C = -\frac{\sqrt{4x^2+1}}{x} + 2\ln\left|2x+\sqrt{4x^2+1}\right| + C$$

39. Let $u = x^2$. Then $du = 2x\,dx$.

$$\int \frac{x}{\sqrt{x^4-16}}dx = \frac{1}{2}\int \frac{1}{\sqrt{u^2-16}}du$$

Now use Formula 43 with $a = 4$ ($a^2 = 16$):

$$\frac{1}{2}\int \frac{1}{\sqrt{u^2-16}}du = \frac{1}{2}\ln\left|u+\sqrt{u^2-16}\right| + C = \frac{1}{2}\ln\left|x^2+\sqrt{x^4-16}\right| + C$$

41. Let $u = x^3$. Then $du = 3x^2dx$.

$$\int x^2\sqrt{x^6+4}\,dx = \frac{1}{3}\int \sqrt{u^2+4}\,du$$

Now use Formula 32 with $a = 2$ ($a^2 = 4$):

$$\frac{1}{3}\int \sqrt{u^2+4}\,du = \frac{1}{3}\cdot\frac{1}{2}\left[u\sqrt{u^2+4} + 4\ln\left|u+\sqrt{u^2+4}\right|\right] + C$$

$$= \frac{1}{6}\left[x^3\sqrt{x^6+4} + 4\ln\left|x^3+\sqrt{x^6+4}\right|\right] + C$$

43. $$\int \frac{1}{x^3\sqrt{4-x^4}}dx = \int \frac{x}{x^4\sqrt{4-x^4}}dx$$

Let $u = x^2$. Then $du = 2x\,dx$.

$$\int \frac{x}{x^4\sqrt{4-x^4}}dx = \frac{1}{2}\int \frac{1}{u^2\sqrt{4-u^2}}du$$

Now use Formula 30 with $a = 2$ ($a^2 = 4$):

$$\frac{1}{2}\int \frac{1}{u^2\sqrt{4-u^2}}du = -\frac{1}{2}\cdot\frac{\sqrt{4-u^2}}{4u} + C = \frac{-\sqrt{4-x^4}}{8x^2} + C$$

45. $$\int \frac{e^x}{(2+e^x)(3+4e^x)}dx = \int \frac{1}{(2+u)(3+4u)}du$$

Substitution: $u = e^x$, $du = e^x\,dx$.
Now use Formula 15 with $a = 2, b = 1, c = 3, d = 4$:

$$\int \frac{1}{(2+u)(3+4u)}du = \frac{1}{2\cdot 4 - 3\cdot 1}\ln\left|\frac{3+4u}{2+u}\right| + C = \frac{1}{5}\ln\left|\frac{3+4e^x}{2+e^x}\right| + C$$

47. $\int \frac{\ln x}{x\sqrt{4+\ln x}}\,dx = \int \frac{u}{\sqrt{4+u}}\,du$

Substitution: $u = \ln x$, $du = \frac{1}{x}\,dx$.

Use Formula 25 with $a = 4$, $b = 1$:

$$\int \frac{u}{\sqrt{4+u}}\,du = \frac{2(u-2\cdot 4)}{3\cdot 1^2}\sqrt{4+u} + C = \frac{2(u-8)}{3}\sqrt{4+u} + C = \frac{2(\ln x - 8)}{3}\sqrt{4+\ln x} + C$$

49. Use Formula 47 with $n = 2$ and $a = 5$:

$$\int x^2e^{5x}dx = \frac{x^2e^{5x}}{5} - \frac{2}{5}\int xe^{5x}dx$$

To find $\int xe^{5x}dx$, use Formula 47 with $n = 1$, $a = 5$:

$$\int xe^{5x}dx = \frac{xe^{5x}}{5} - \frac{1}{5}\int e^{5x}dx = \frac{xe^{5x}}{5} - \frac{1}{5}\cdot\frac{e^{5x}}{5}$$

Thus, $\int x^2e^{5x}dx = \frac{x^2e^{5x}}{5} - \frac{2}{5}\left[\frac{xe^{5x}}{5} - \frac{1}{25}e^{5x}\right] + C = \frac{x^2e^{5x}}{5} - \frac{2xe^{5x}}{25} + \frac{2e^{5x}}{125} + C.$

51. Use Formula 47 with $n = 3$ and $a = -1$.

$$\int x^3e^{-x}dx = \frac{x^3e^{-x}}{-1} - \frac{3}{-1}\int x^2e^{-x}dx = -x^3e^{-x} + 3\int x^2e^{-x}dx$$

Now $\int x^2e^{-x}dx = \frac{x^2e^{-x}}{-1} - \frac{2}{-1}\int xe^{-x}dx = -x^2e^{-x} + 2\int xe^{-x}dx$

and $\int xe^{-x}dx = \frac{xe^{-x}}{-1} - \frac{1}{-1}\int e^{-x}dx = -xe^{-x} - e^{-x}$, using Formula 47.

Thus, $\int x^3e^{-x}dx = -x^3e^{-x} + 3[-x^2e^{-x} + 2(-xe^{-x} - e^{-x})] + C = -x^3e^{-x} - 3x^2e^{-x} - 6xe^{-x} - 6e^{-x} + C.$

53. Use Formula 52 with $n = 3$:

$$\int (\ln x)^3dx = x(\ln x)^3 - 3\int (\ln x)^2dx$$

Now $\int (\ln x)^2dx = x(\ln x)^2 - 2\int \ln x\,dx$ using Formula 52 again, and

$\int \ln x\,dx = x\ln x - x$ by Formula 49.

Thus, $\int (\ln x)^3dx = x(\ln x)^3 - 3[x(\ln x)^2 - 2(x\ln x - x)] + C = x(\ln x)^3 - 3x(\ln x)^2 + 6x\ln x - 6x + C.$

55. $\int_3^5 x\sqrt{x^2-9}\,dx$. First consider the indefinite integral.

Let $u = x^2 - 9$. Then $du = 2x\,dx$ or $x\,dx = \frac{1}{2}du$. Thus,

$$\int x\sqrt{x^2-9}\,dx = \frac{1}{2}\int u^{1/2}du = \frac{1}{2}\cdot\frac{u^{3/2}}{3/2} + C = \frac{1}{3}(x^2-9)^{3/2} + C.$$

Now, $\int_3^5 x\sqrt{x^2-9}\,dx = \frac{1}{3}(x^2-9)^{3/2}\Big|_3^5 = \frac{1}{3}\cdot 16^{3/2} = \frac{64}{3}.$

57. $\int_2^4 \frac{1}{x^2-1}\,dx$. Consider the indefinite integral:

$$\int \frac{1}{x^2-1}\,dx = \frac{1}{2\cdot 1}\ \ln\left|\frac{x-1}{x+1}\right| + C, \text{ using Formula 13 with } a = 1.$$

Thus,

$$\int_2^4 \frac{1}{x^2-1}\,dx = \frac{1}{2}\ \ln\left|\frac{x-1}{x+1}\right|\Big|_2^4 = \frac{1}{2}\ \ln\left|\frac{3}{5}\right| - \frac{1}{2}\ \ln\left|\frac{1}{3}\right| = \frac{1}{2}\ln\frac{9}{5} \approx\ 0.2939.$$

59. $\int \frac{\ln x}{x^2}\,dx = \int x^{-2}\ln x\,dx = \frac{x^{-1}}{-1}\ln x - \frac{x^{-1}}{(-1)^2} + C$ [Formula 51 with $n = -2$]

$$= -\frac{1}{x}\ln x - \frac{1}{x} + C = \frac{-1-\ln x}{x} + C$$

61. $\int \frac{x}{\sqrt{x^2-1}}\,dx = \int \frac{1}{\sqrt{x^2-1}}\left(\frac{2}{2}\right)x dx = \frac{1}{2}\int u^{-1/2}du = u^{1/2} + C = \sqrt{x^2-1} + C$

Let $u = x^2 - 1$. Then $du = 2x\,dx$

63. $\int_{-1}^{1}(ax^2+bx+c)dx$; $f(x) = ax^2+bx+c.$. Partition $[-1,1]$ into two equal subintervals:

$x_0 = -1,\ x_1 = 0,\ x_2 = 1,\ \Delta x = 1.$

x	$f(x)$
-1	$a-b+c$
0	c
1	$a+b+c$

$$S_2 = [f(-1)+4f(0)+f(1)](1/3) = [(a-b+c)+4c+(a+b+c)](1/3) = \frac{2}{3}a+2c.$$

Exact value:

$$\int_{-1}^{1}(ax^2+bx+c)\,dx = \left[\frac{a}{3}x^3+\frac{b}{2}x^2+cx\right]_{-1}^{1} = \left(\frac{a}{3}+\frac{b}{2}+c\right)-\left(-\frac{a}{3}+\frac{b}{2}-c\right) = \frac{2a}{3}+2c.$$

65. $f(x) = \frac{10}{\sqrt{x^2+1}}$, $g(x) = x^2 + 3x$

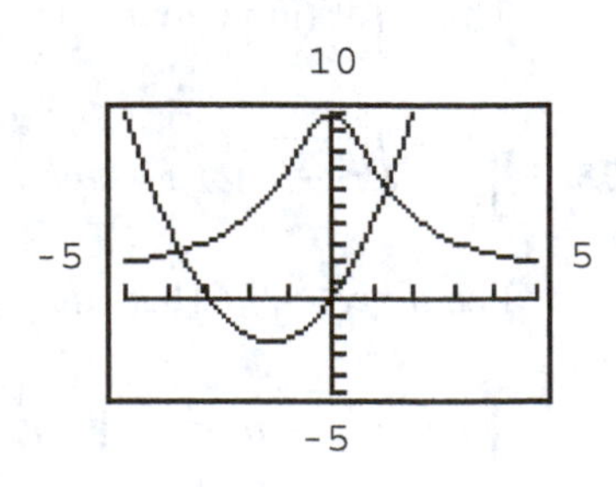

The graphs of f and g are shown at the right. The x-coordinates of the points of intersection are: $x_1 \approx -3.70, x_2 \approx 1.36$

$$A = \int_{-3.70}^{1.36}\left[\frac{10}{\sqrt{x^2+1}} - (x^2+3x)\right]dx$$

$$= 10\int_{-3.70}^{1.36} \frac{1}{\sqrt{x^2+1}}\,dx - \int_{-3.70}^{1.36} (x^2 + 3x)dx$$

For the first integral, use Formula 36 with $a = 1$:

$$A = (10 \ln\ |x + \sqrt{x^2+1}\,|)\Big|_{-3.70}^{1.36} - \left(\frac{1}{3}x^3 + \frac{3}{2}x^2\right)\Big|_{-3.70}^{1.36}$$

$$\approx [11.15 - (-20.19)] - [3.61 - (3.65)] = 31.38$$

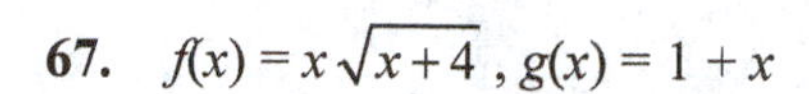

67. $f(x) = x\sqrt{x+4}$, $g(x) = 1 + x$

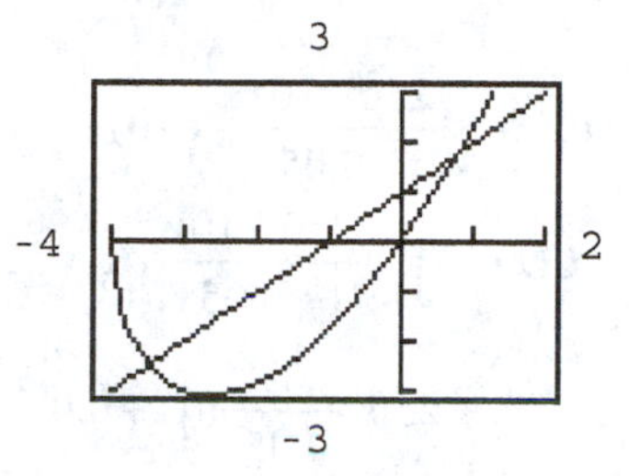

The graphs of f and g are shown at the right. The x-coordinates of the points of intersection are: $x_1 \approx -3.49$, $x_2 \approx 0.83$

$$A = \int_{-3.49}^{0.83} [1 + x - x\sqrt{x+4}\,]dx = \int_{-3.49}^{0.83} (1 + x)dx - \int_{-3.49}^{0.83} x\sqrt{x+4}\,dx$$

For the second integral, use Formula 22 with $a = 4$ and $b = 1$:

$$A = \left(x + \frac{1}{2}x^2\right)\Big|_{-3.49}^{0.83} - \left(\frac{2[3x-8]}{15}\sqrt{(x+4)^3}\right)\Big|_{-3.49}^{0.83}$$

$$\approx (1.17445 - 2.60005) - (-7.79850 + 0.89693) \approx 5.48$$

69. Find $\bar{x}$, the demand when the price $\bar{p} = 15$:

$$15 = \frac{7500 - 30\bar{x}}{300 - \bar{x}}$$

$$4500 - 15\bar{x} = 7500 - 30\bar{x}$$
$$15\bar{x} = 3000$$
$$\bar{x} = 200$$

Consumers' surplus:

$$CS = \int_0^{\bar{x}} [D(x) - \bar{p}\,]dx = \int_0^{200} \left[\frac{7500 - 30x}{300 - x} - 15\right]dx = \int_0^{200} \left[\frac{3000 - 15x}{300 - x}\right]dx$$

Use Formula 20 with $a = 3000$, $b = -15$, $c = 300$, $d = -1$:

$$CS = \left[\frac{-15x}{-1} + \frac{3000(-1) - (-15)(300)}{(-1)^2}\ln|300 - x|\right]_0^{200} = [15x + 1500\ln|300 - x|]\Big|_0^{200}$$

$$= 3000 + 1500\ln(100) - 1500\ln(300) = 3000 + 1500\ln\left(\frac{1}{3}\right) \approx 1352$$

Thus, the consumers' surplus is \$1352.

71. 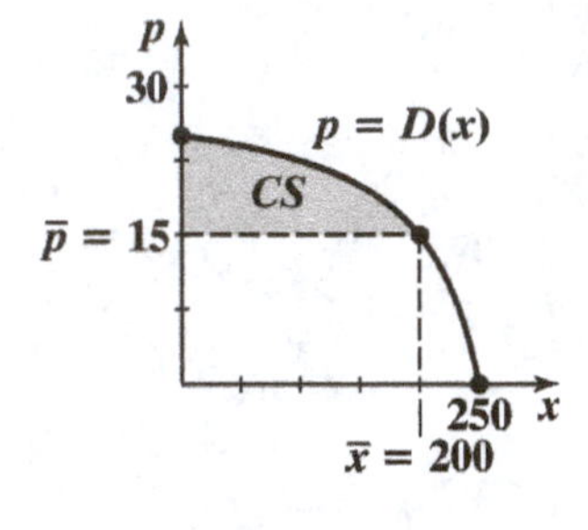

The shaded region represents the consumers' surplus.

73. $C'(x) = \dfrac{250+10x}{1+0.05x}$, $C(0) = 25{,}000$

$$C(x) = \int \frac{250+10x}{1+0.05x}\,dx = 250\int \frac{1}{1+0.05x}\,dx + 10\int \frac{x}{1+0.05x}\,dx$$

$$= 250\left(\frac{1}{0.05}\ln|1+0.05x|\right) + 10\left(\frac{x}{0.05} - \frac{1}{(0.05)^2}\ln|1+0.05x|\right) + K$$

(Formulas 3 and 4)

$$= 5{,}000\ \ln\ |1 + 0.05x| + 200x - 4{,}000\ \ln\ |1 + 0.05x| + K = 1{,}000\ \ln\ |1 + 0.05x| + 200x + K$$

Since $C(0) = 25{,}000$, $K = 25{,}000$ and $C(x) = 1{,}000\ \ln(1 + 0.05x) + 200x + 25{,}000,\ x \geq 0$

To find the production level that produces a cost of \$150,000, solve $C(x) = 150{,}000$ for x:

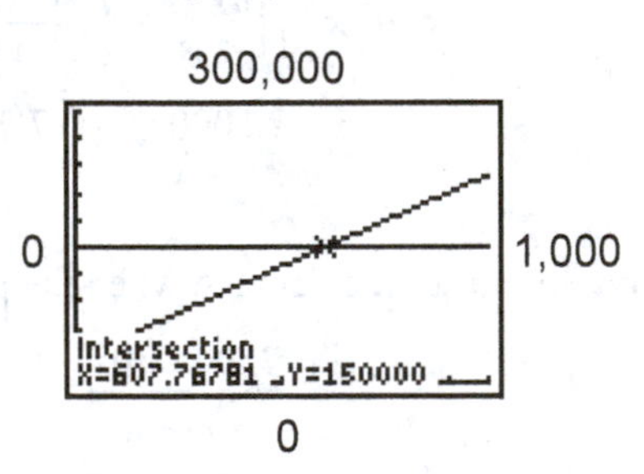

The production level is $x = 608$ pairs of skis.

At a production level of 850 pairs of skis,
$C(850) = 1{,}000\ \ln(1 + 0.05[850]) + 200(850) + 25{,}000 \approx \$198{,}773$.

75. $FV = e^{rT}\int_0^T f(t)e^{-rt}\,dt$

Now, $r = 0.044,\ T = 10,\ f(t) = 50t^2$.

$$FV = e^{(0.044)10}\int_0^{10} 50t^2e^{-0.044t}\,dt = 50\,e^{0.44}\int_0^{10} t^2e^{-0.044t}\,dt$$

To evaluate the integral, use Formula 47 with $n = 2$ and $a = -0.044$:

$$\int t^2e^{-0.044t}\,dt = \frac{t^2e^{-0.044t}}{-0.044} - \frac{2}{-0.044}\int t\,e^{-0.044t}\,dt = \frac{t^2e^{-0.044t}}{-0.044} + \frac{2}{0.044}\int t\,e^{-0.044t}\,dt$$

Now, using Formula 47 again:

$$\int t\,e^{-0.044t}\,dt = \frac{te^{-0.044t}}{-0.044} - \frac{1}{-0.044}\int e^{-0.044t}\,dt = -\frac{te^{-0.044t}}{0.044} - \frac{e^{-0.044t}}{(0.044)^2}$$

Thus

$$\int t^2 e^{-0.044t}\,dt = -\frac{t^2 e^{-0.044t}}{0.044} - \frac{2t\,e^{-0.044t}}{(0.044)^2} - \frac{2\,e^{-0.044t}}{(0.044)^3} + C.$$

Now,

$$FV = 50\,e^{0.44}\left[-\frac{t^2 e^{-0.044t}}{0.044} - \frac{2t\,e^{-0.044t}}{(0.044)^2} - \frac{2\,e^{-0.044t}}{(0.044)^3}\right]_0^{10}$$

$$= 50\,e^{0.44}\left[-\frac{100\,e^{-0.44}}{0.044} - \frac{20\,e^{-0.44}}{(0.044)^2} - \frac{2\,e^{-0.44}}{(0.044)^3} + \frac{2}{(0.044)^3}\right] = \$18,673.95$$

77. Gini Index:

$$2\int_0^1 [x - f(x)]dx = 2\int_0^1 \left[x - \frac{1}{2}x\sqrt{1+3x}\right]dx = \int_0^1 [2x - x\sqrt{1+3x}\,]dx = \int_0^1 2x dx - \int_0^1 x\sqrt{1+3x}\,dx$$

For the second integral, use Formula 22 with $a = 1$ and $b = 3$:

$$= x^2\Big|_0^1 - \frac{2(3\cdot 3x - 2\cdot 1)}{15(3)^2}\sqrt{(1+3x)^3}\,\Big|_0^1 = 1 - \frac{2(9x-2)}{135}\sqrt{(1+3x)^3}\,\Big|_0^1$$

$$= 1 - \frac{14}{135}\sqrt{4^3} - \frac{4}{135}\sqrt{1^3} = 1 - \frac{112}{135} - \frac{4}{135} = \frac{19}{135} \approx 0.1407$$

79.

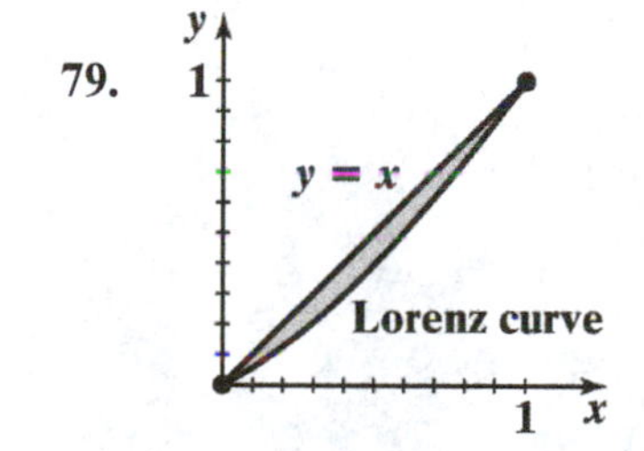

As the area bounded by the two curves gets smaller, the Lorenz curve approaches $y = x$ and the distribution of income approaches perfect equality — all individuals share equally in the income.

81. $S'(t) = \dfrac{t^2}{(1+t)^2}$; $S(t) = \displaystyle\int \frac{t^2}{(1+t)^2}\,dt$

Use Formula 7 with $a = 1$ and $b = 1$:

$$S(t) = \frac{1+t}{1^3} - \frac{1^2}{1^3(1+t)} - \frac{2(1)}{1^3}\ln|1+t| + C = 1 + t - \frac{1}{1+t} - 2\ln|1+t| + C$$

Since $S(0) = 0$, we have $0 = 1 - 1 - 2\ln 1 + C$ and $C = 0$. Thus,

$$S(t) = 1 + t - \frac{1}{1+t} - 2\ln|1+t|.$$

Now, the total sales during the first two years (= 24 months) is given by:

$$S(24) = 1 + 24 - \frac{1}{1+24} - 2\ln|1+24| = 24.96 - 2\ln 25 \approx 18.5$$

Thus, total sales during the first two years is approximately \$18.5 million.

83.

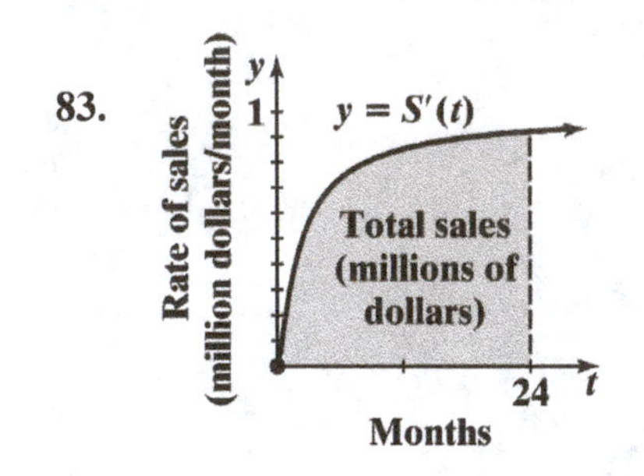

The total sales, in millions of dollars, over the first two years (24 months) is the area under the curve $y = S'(t)$ from $t = 0$ to $t = 24$.

85. $P'(x) = x\sqrt{2+3x}$, $P(1) = -\$2{,}000$

$$P(x) = \int x\sqrt{2+3x}\,dx = \frac{2(9x-4)}{135}(2+3x)^{3/2} + C$$

(Formula 22)

$$P(1) = \frac{2(5)}{135}5^{3/2} + C = -2{,}000$$

$$C = -2{,}000 - \frac{2}{27}5^{3/2} \approx -2{,}000.83$$

Thus, $P(x) = \frac{2(9x-4)}{135}(2+3x)^{3/2} - 2{,}000.83$.

The number of cars that must be sold to have a profit of \$13,000: 54

Profit if 80 cars are sold per week:

$$P(80) = \frac{2(716)}{135}(242)^{3/2} - 2{,}000.83 \approx \$37{,}932.20$$

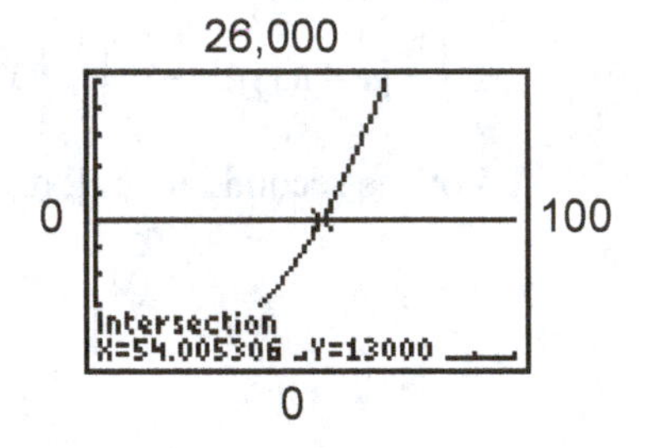

87. $\frac{dR}{dt} = \frac{100}{\sqrt{t^2+9}}$. Therefore,

$$R = \int \frac{100}{\sqrt{t^2+9}}\,dt = 100\int \frac{1}{\sqrt{t^2+9}}\,dt$$

Using Formula 36 with $a = 3$ ($a^2 = 9$), we have:

$$R = 100\ln\left|t + \sqrt{t^2+9}\right| + C$$

Now $R(0) = 0$, so $0 = 100\ln|3| + C$ or $C = -100\ln 3$. Thus,

$$R(t) = 100\ln\left|t + \sqrt{t^2+9}\right| - 100\ln 3$$

and

$$R(4) = 100\ln(4 + \sqrt{4^2+9}) - 100\ln 3 = 100\ln 9 - 100\ln 3 = 100\ln 3 \approx 110 \text{ feet}$$

89. $N'(t) = \frac{60}{\sqrt{t^2+25}}$

The number of items learned in the first twelve hours of study is given by:

$$N = \int_0^{12} \frac{60}{\sqrt{t^2+25}}\,dt = 60\int_0^{12} \frac{1}{\sqrt{t^2+25}}\,dt = 60\left(\ln\left|t+\sqrt{t^2+25}\right|\right)\Big|_0^{12}, \text{ using Formula 36}$$

$$= 60\left[\ln\left|12+\sqrt{12^2+25}\right| - \ln\sqrt{25}\right] = 60(\ln 25 - \ln 5) = 60\ln 5 \approx 96.57 \text{ or } 97 \text{ items}$$

91.

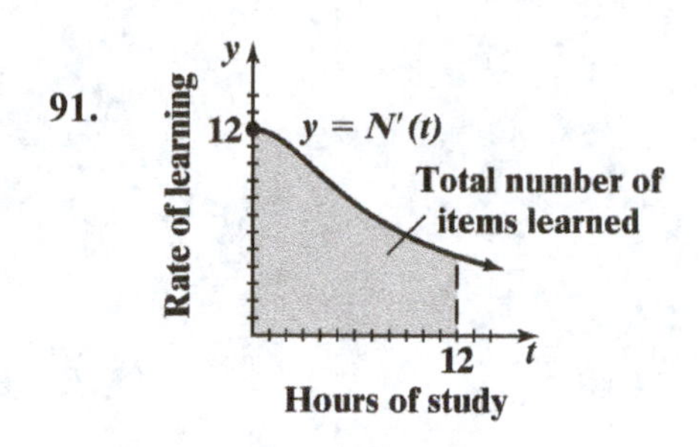

The area under the rate of learning curve, $y = N'(t)$, from $t = 0$ to $t = 12$ represents the total number of items learned in that time interval.

CHAPTER 6 REVIEW

1. $A = \int_a^b f(x)\,dx$ (6-1)

2. $A = \int_b^c [-f(x)]\,dx$ (6-1)

3. $A = \int_a^b f(x)\,dx + \int_b^c [-f(x)]\,dx$ (6-1)

4. $A = \int_{0.5}^{1} [-\ln x]\,dx + \int_1^e \ln x\,dx$

We evaluate the integral using integration-by-parts.
Let $u = \ln x$, $dv = dx$.

Then $du = \frac{1}{x}dx$, $v = x$, and $\int \ln x dx =$

$x \ln x - \int x\left(\frac{1}{x}\right)dx = x \ln x - x + C$

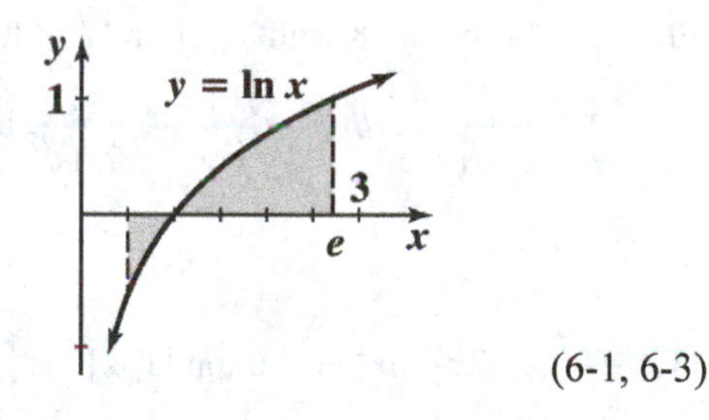

Thus,

$A = -\int_{0.5}^{1} \ln x dx + \int_1^e \ln x dx$

$= (-x \ln x + x)\Big|_{0.5}^{1} + (x \ln x - x)\Big|_1^e$

$\approx (1 - 0.847) + (1) = 1.153$

(6-1, 6-3)

5. $\int xe^{4x}dx$. Use integration-by-parts:

Let $u = x$ and $dv = e^{4x}dx$. Then $du = dx$ and $v = \frac{e^{4x}}{4}$.

$\int xe^{4x}dx = \frac{xe^{4x}}{4} - \int \frac{e^{4x}}{4}dx = \frac{xe^{4x}}{4} - \frac{e^{4x}}{16} + C$ (6-3)

6. $\int x \ln x\,dx$. Use integration-by-parts:

Let $u = \ln x$ and $dv = x\,dx$. Then $du = \frac{1}{x}dx$ and $v = \frac{x^2}{2}$.

$\int x \ln x\,dx = \frac{x^2 \ln x}{2} - \int \frac{1}{x} \cdot \frac{x^2}{2}dx = \frac{x^2 \ln x}{2} - \frac{1}{2}\int x\,dx = \frac{x^2 \ln x}{2} - \frac{x^2}{4} + C$ (6-3)

7. $\int \frac{\ln x}{x}\,dx$

Let $u = \ln x$. Then $du = \frac{1}{x}dx$ and

$$\int \frac{\ln x}{x}\,dx = \int u\,du = \frac{1}{2}u^2 + C = \frac{1}{2}[\ln x]^2 + C \qquad (6\text{-}2)$$

8. $\int \frac{x}{1+x^2}\,dx$

Let $u = 1 + x^2$. Then $du = 2x\,dx$ and

$$\int \frac{x}{1+x^2}\,dx = \int \frac{1/2\,du}{u} = \frac{1}{2}\int \frac{1}{u}\,du = \frac{1}{2}\ln|u| + C = \frac{1}{2}\ln(1+x^2) + C \qquad (6\text{-}2)$$

9. Use Formula 11 with $a = 1$ and $b = 1$.

$$\int \frac{1}{x(1+x)^2}\,dx = \frac{1}{1(1+x)} + \frac{1}{1^2}\ln\left|\frac{x}{1+x}\right| + C = \frac{1}{1+x} + \ln\left|\frac{x}{1+x}\right| + C \qquad (6\text{-}4)$$

10. Use Formula 28 with $a = 1$ and $b = 1$.

$$\int \frac{1}{x^2\sqrt{1+x}}\,dx = -\frac{\sqrt{1+x}}{1\cdot x} - \frac{1}{2\cdot 1\sqrt{1}}\ln\left|\frac{\sqrt{1+x}-\sqrt{1}}{\sqrt{1+x}+\sqrt{1}}\right| + C = -\frac{\sqrt{1+x}}{x} - \frac{1}{2}\ln\left|\frac{\sqrt{1+x}-1}{\sqrt{1+x}+1}\right| + C \qquad (6\text{-}4)$$

11. $y = 5 - 2x - 6x^2;\ y = 0$ on $[1, 2]$

$$A = -\int_1^2 (5 - 2x - 6x^2)\,dx$$

$$= \int_1^2 (6x^2 + 2x - 5)dx = (2x^3 + x^2 - 5x)\Big|_1^2$$

$$= 10 - (-2) = 12 \qquad (6\text{-}1)$$

12. $y = 5x + 7;\ y = 12$ on $[-3, 1]$

$$A = \int_{-3}^{1} [12 - (5x + 7)]\,dx$$

$$= \int_{-3}^{1} (5 - 5x)dx = \left(5x - \frac{5}{2}x^2\right)\Big|_{-3}^{1}$$

$$= \left(5 - \frac{5}{2}\right) - \left(-15 - \frac{45}{2}\right) = 40$$

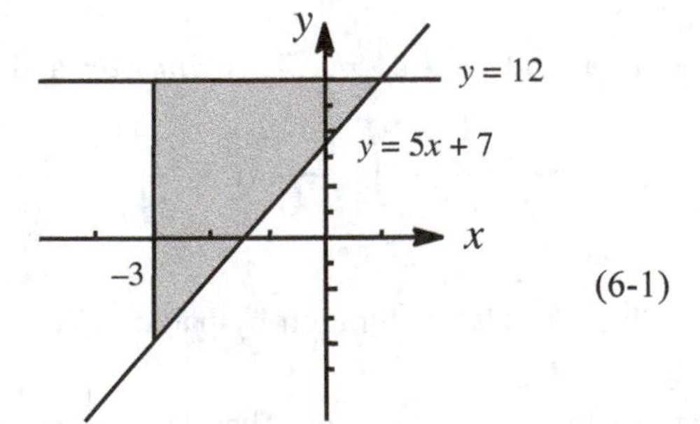

(6-1)

13. $y = -x + 2; y = x^2 + 3$ on $[-1, 4]$

$$A = \int_{-1}^{4} [(x^2 + 3) - (-x + 2)]\, dx$$
$$= \int_{-1}^{4} (x^2 + x + 1)\, dx$$
$$= \left(\frac{1}{3}x^3 + \frac{1}{2}x^2 + x\right)\Big|_{-1}^{4}$$
$$= \frac{64}{3} + 8 + 4 - \left(-\frac{1}{3} + \frac{1}{2} - 1\right)$$
$$= \frac{205}{6} \approx 34.167$$

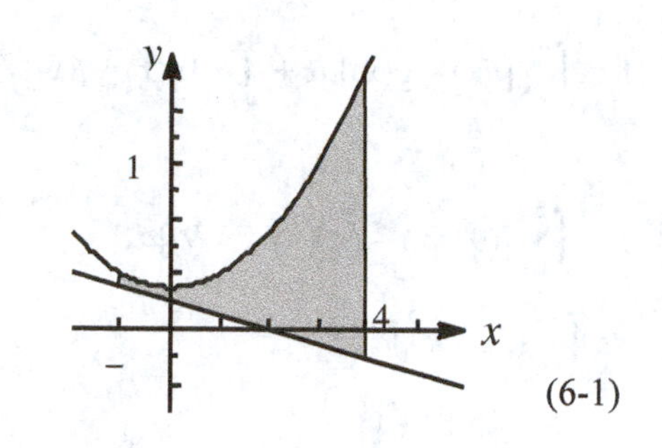

(6-1)

14. $y = \frac{1}{x}; y = -e^{-x}$ on $[1, 2]$

$$A = \int_{1}^{2} \left(\frac{1}{x} - e^{-x}\right) dx = \int_{1}^{2} \left(\frac{1}{x} + e^{-x}\right) dx$$
$$= (\ln x - e^{-x})\Big|_{1}^{2} = (\ln 2 - e^{-2}) + e^{-1} \approx 0.926$$

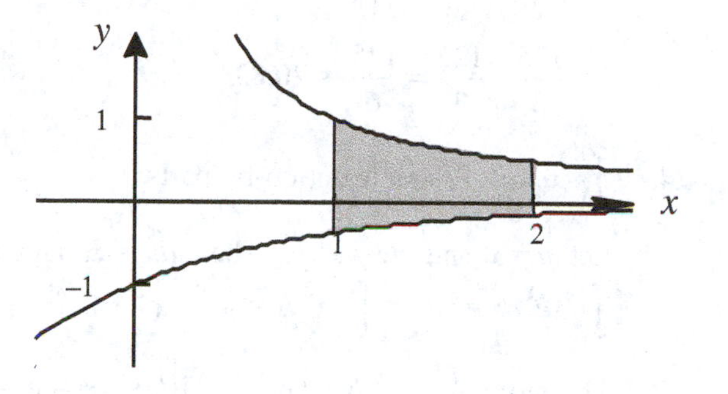

(6-1)

15. $y = x; y = -x^3$ on $[-2, 2]$

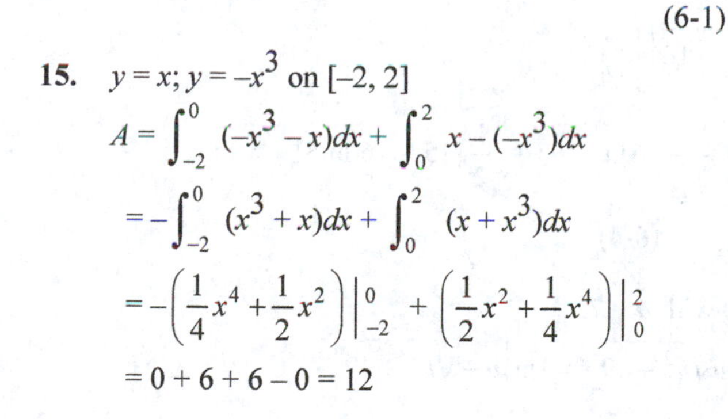

$$A = \int_{-2}^{0} (-x^3 - x)dx + \int_{0}^{2} x - (-x^3)dx$$
$$= -\int_{-2}^{0} (x^3 + x)dx + \int_{0}^{2} (x + x^3)dx$$
$$= -\left(\frac{1}{4}x^4 + \frac{1}{2}x^2\right)\Big|_{-2}^{0} + \left(\frac{1}{2}x^2 + \frac{1}{4}x^4\right)\Big|_{0}^{2}$$
$$= 0 + 6 + 6 - 0 = 12$$

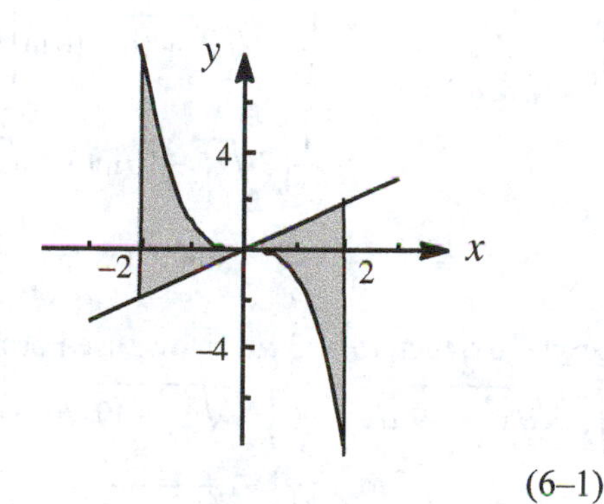

(6–1)

16. $y = x^2; y = -x^4$ on $[-2, 2]$

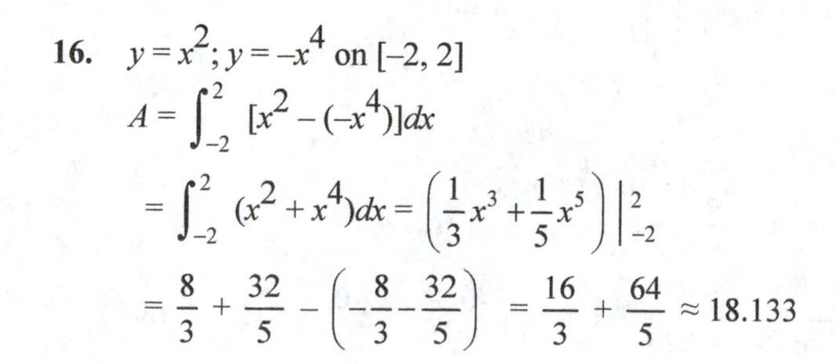

$$A = \int_{-2}^{2} [x^2 - (-x^4)]dx$$
$$= \int_{-2}^{2} (x^2 + x^4)dx = \left(\frac{1}{3}x^3 + \frac{1}{5}x^5\right)\Big|_{-2}^{2}$$
$$= \frac{8}{3} + \frac{32}{5} - \left(-\frac{8}{3} - \frac{32}{5}\right) = \frac{16}{3} + \frac{64}{5} \approx 18.133$$

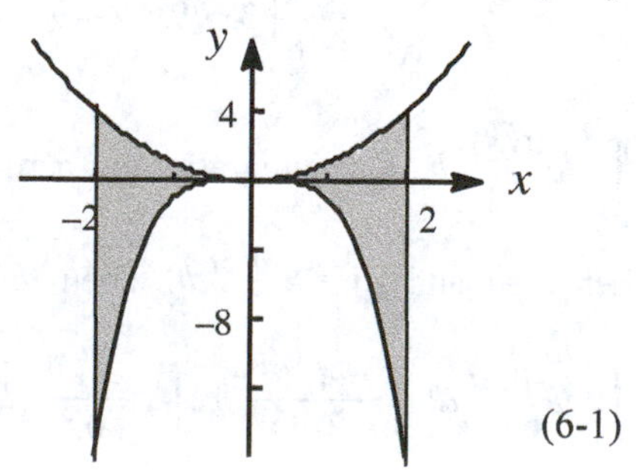

(6-1)

17. Indonesia (6-1)

18. Vietnam (6-1)

19. $A = \int_{a}^{b} [f(x) - g(x)]dx$ (6-1)

20. $A = \int_{b}^{c} [g(x) - f(x)]dx$ (6-1)

21. $A = \int_{b}^{c} [g(x) - f(x)]dx + \int_{c}^{d} [f(x) - g(x)]dx$ (6-1)

22. $A = \int_a^b [f(x) - g(x)]dx + \int_b^c [g(x) - f(x)]dx + \int_c^d [f(x) - g(x)]\,dx$ (6-1)

23. $A = \int_0^5 [(9 - x) - (x^2 - 6x + 9)]dx$

$= \int_0^5 (5x - x^2)dx$

$= \left(\frac{5}{2}x^2 - \frac{1}{3}x^3\right)\Big|_0^5$

$= \frac{125}{2} - \frac{125}{3} = \frac{125}{6} \approx 20.833$

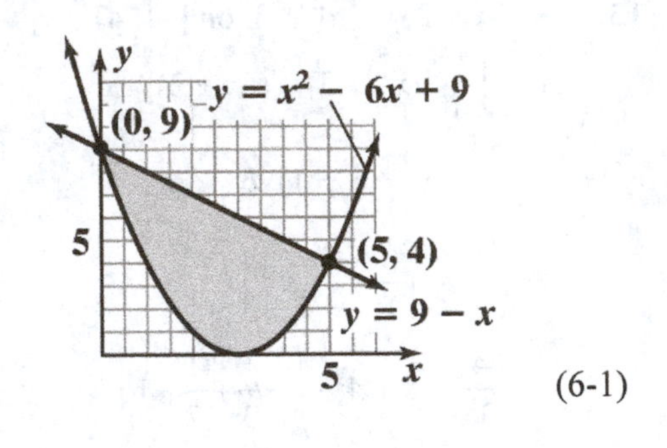

(6-1)

24. $\int_0^1 xe^x dx$. Use integration-by-parts.

Let $u = x$ and $dv = e^x dx$. Then $du = dx$ and $v = e^x$.

$\int xe^x dx = xe^x - \int e^x dx = xe^x - e^x + C$

Therefore, $\int_0^1 xe^x dx = (xe^x - e^x)\Big|_0^1 = 1 \cdot e - e - (0 \cdot 1 - 1) = 1$ (6-3)

25. Use Formula 38 with $a = 4$

$\int_0^3 \frac{x^2}{\sqrt{x^2+16}}dx = \frac{1}{2}\left[x\sqrt{x^2+16} - 16\ln\left|x + \sqrt{x^2+16}\right|\right]\Big|_0^3$

$= \frac{1}{2}\left[3\sqrt{25} - 16\ln(3 + \sqrt{25})\right] - \frac{1}{2}(-16\ln\sqrt{16}) = \frac{1}{2}[15 - 16\ln 8] + 8\ln 4$

$= \frac{15}{2} - 8\ln 8 + 8\ln 4 \approx 1.955$ (6-4)

26. Let $u = 3x$, then $du = 3\,dx$. Now, use Formula 40 with $a = 7$.

$\int \sqrt{9x^2 - 49}\,dx = \frac{1}{3}\int \sqrt{u^2 - 49}\,du = \frac{1}{3} \cdot \frac{1}{2}\left(u\sqrt{u^2 - 49} - 49\ln\left|u + \sqrt{u^2 - 49}\right|\right) + C$

$= \frac{1}{6}\left(3x\sqrt{9x^2 - 49} - 49\ln\left|3x + \sqrt{9x^2 - 49}\right|\right) + C$ (6-4)

27. $\int te^{-0.5t}\,dt$. Use integration-by-parts.

Let $u = t$ and $dv = e^{-0.5t}dt$. Then $du = dt$ and $v = \frac{e^{-0.5t}}{-0.5}$.

$\int te^{-0.5t}\,dt = \frac{-te^{-0.5t}}{0.5} + \int \frac{e^{-0.5t}}{0.5}\,dt = \frac{-te^{-0.5t}}{0.5} + \frac{e^{-0.5t}}{-0.25} + C = -2te^{-0.5t} - 4e^{-0.5t} + C$ (6-3)

28. $\int x^2 \ln x\,dx$. Use integration-by-parts.

Let $u = \ln x$ and $dv = x^2 dx$. Then $du = \frac{1}{x}dx$ and $v = \frac{x^3}{3}$.

$\int x^2 \ln x\,dx = \frac{x^3 \ln x}{3} - \int \frac{1}{x} \cdot \frac{x^3}{3}\,dx = \frac{x^3 \ln x}{3} - \frac{1}{3}\int x^2 dx = \frac{x^3 \ln x}{3} - \frac{x^3}{9} + C$ (6-3)

29. Use Formula 48 with $a = 1$, $c = 1$, and $d = 2$.

$$\int \frac{1}{1+2e^x}\,dx = \frac{x}{1} - \frac{1}{1\cdot 1}\ln|1 + 2e^x| + C = x - \ln|1 + 2e^x| + C \qquad \text{(6-4)}$$

30. (A)

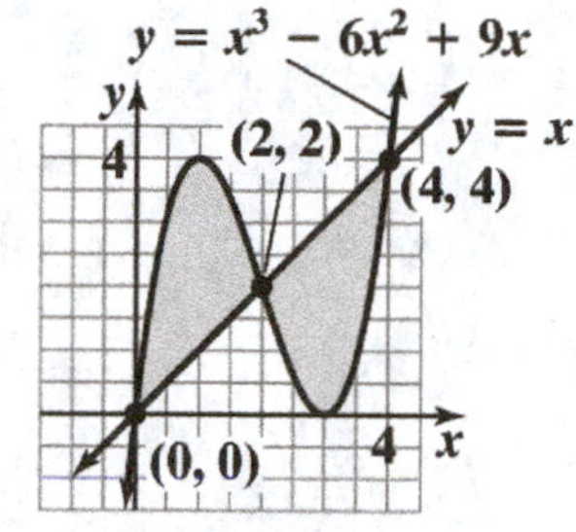

$$A = \int_0^2 [(x^3 - 6x^2 + 9x) - x]\,dx + \int_2^4 [x - (x^3 - 6x^2 + 9x)]\,dx$$

$$= \int_0^2 (x^3 - 6x^2 + 8x)\,dx + \int_2^4 (-x^3 + 6x^2 - 8x)\,dx$$

$$= \left(\frac{1}{4}x^4 - 2x^3 + 4x^2\right)\Big|_0^2 + \left(-\frac{1}{4}x^4 + 2x^3 - 4x^2\right)\Big|_2^4 = 4 + 4 = 8$$

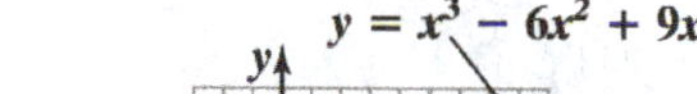

(B)

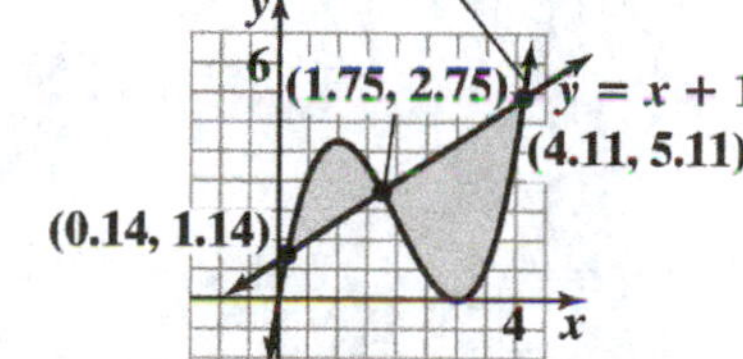

The x-coordinates of the points of intersection are: $x_1 \approx 0.14$, $x_2 \approx 1.75$, $x_3 \approx 4.11$.

$$A = \int_{0.14}^{1.75} [(x^3 - 6x^2 + 9x) - (x + 1)]\,dx + \int_{1.75}^{4.11} [(x + 1) - (x^3 - 6x^2 + 9x)]\,dx$$

$$= \int_{0.14}^{1.75} (x^3 - 6x^2 + 8x - 1)\,dx + \int_{1.75}^{4.11} (1 - x^3 + 6x^2 - 8x)\,dx$$

$$= \left(\frac{1}{4}x^4 - 2x^3 + 4x^2 - x\right)\Big|_{0.14}^{1.75} + \left(x - \frac{1}{4}x^4 + 2x^3 - 4x^2\right)\Big|_{1.75}^{4.11}$$

$$= [2.126 - (-0.066)] + [4.059 - (-2.126)] \approx 8.38 \qquad \text{(6-1)}$$

31. $\int_0^3 e^{x^2}\,dx;\ f(x)=e^{x^2}$. Partition [0,3] into three equal subintervals:

$x_0=0,\ x_1=1,\ x_2=2,\ x_3=3,\ \Delta x=1.$

x	$f(x)$
0	1
1	2.7183
2	54.5982
3	8103.0839

$T_3=[f(0)+2f(1)+2f(2)+f(3)](1/2)=[1+5.4366+109.1964+8103.0839](1/2)\approx 4109.36.$

(6.4)

32. $\int_0^3 e^{x^2}\,dx;\ f(x)=e^{x^2}$. Partition [0,3] into five equal subintervals:

$x_0=0,\ x_1=0.6,\ x_2=1.2,\ x_3=1.8,\ x_4=2.4,\ x_5=3,\ \Delta x=0.6.$

x	$f(x)$
0	1
0.6	1.4333
1.2	4.2207
1.8	25.5337
2.4	317.3483
3	8103.0839

$$T_5=[f(0)+2f(0.6)+2f(1.2)+2f(1.8)+2f(2.4)+f(3)](0.6/2)$$
$$=[1+2.8666+8.4414+51.0674+634.6966+8103.0839](0.3)\approx 2640.35.$$

(6.4)

33. $\int_1^5 (\ln x)^2\, dx,\quad f(x) = (\ln x)^2$. Partition [1,5] into four equal subintervals:

$x_0 = 1,\ x_1 = 2,\ x_2 = 3,\ x_3 = 4,\ x_4 = 5,\ \Delta x = 1.$

x	$f(x)$
1	0
2	0.4805
3	1.2069
4	1.9218
5	2.5903

$S_4 = [f(1) + 4f(2) + 2f(3) + 4f(4) + f(5)](1/3) = [0 + 1.9220 + 2.4138 + 7.6872 + 2.5903](1/3) \approx 4.87$
(6.4)

34. $\int_1^5 (\ln x)^2\, dx,\quad f(x) = (\ln x)^2$. Partition [1,5] into eight equal subintervals:

$x_0 = 1,\ x_1 = 1.5,\ x_2 = 2,\ x_3 = 2.5,\ x_4 = 3,\ x_5 = 3.5,\ x_6 = 4,\ x_7 = 4.5,\ x_8 = 5,\ \Delta x = 0.5.$

x	$f(x)$
1	0
1.5	0.1644
2	0.4805
2.5	0.8396
3	1.2069
3.5	1.5694
4	1.9218
4.5	2.2622
5	2.5903

$S_8 = [f(1) + 4f(1.5) + 2f(2) + 4f(2.5) + 2f(3) + 4f(3.5) + 2f(4) + 4f(4.5) + f(5)](1/6)$

$= [0 + 0.6576 + 0.9610 + 3.3584 + 2.4138 + 6.2776 + 3.8436 + 9.0648 + 2.5903](1/6) \approx 4.86$
(6.4)

35. $\int \frac{(\ln x)^2}{x}dx = \int u^2 du = \frac{u^3}{3} + C = \frac{(\ln x)^3}{3} + C$ (5-2)

Substitution: $u = \ln x$, $du = \frac{1}{x}dx$

36. $\int x(\ln x)^2 dx$. Use integration-by-parts.

Let $u = (\ln x)^2$ and $dv = x\,dx$. Then $du = 2(\ln x)\frac{1}{x}dx$ and $v = \frac{x^2}{2}$.

$$\int x(\ln x)^2 dx = \frac{x^2(\ln x)^2}{2} - \int 2(\ln x)\frac{1}{x}\cdot\frac{x^2}{2}dx = \frac{x^2(\ln x)^2}{2} - \int x \ln x dx$$

Let $u = \ln x$ and $dv = xdx$. Then $du = \frac{1}{x}dx$ and $v = \frac{x^2}{2}$.

$$\int x \ln x\, dx = \frac{x^2 \ln x}{2} - \frac{x^2}{4}$$

Thus, $\int x(\ln x)^2 dx = \frac{x^2(\ln x)^2}{2} - \left[\frac{x^2 \ln x}{2} - \frac{x^2}{4}\right] + C = \frac{x^2(\ln x)^2}{2} - \frac{x^2 \ln x}{2} + \frac{x^2}{4} + C.$ (6-3)

37. Let $u = x^2 - 36$. Then $du = 2xdx$.

$$\int \frac{x}{\sqrt{x^2-36}}dx = \int \frac{x}{(x^2-36)^{1/2}}dx = \frac{1}{2}\int \frac{1}{u^{1/2}}du = \frac{1}{2}\int u^{-1/2}du$$

$$= \frac{1}{2}\cdot\frac{u^{1/2}}{1/2} + C = u^{1/2} + C = \sqrt{x^2-36} + C \quad (6\text{-}2)$$

38. Let $u = x^2$, $du = 2xdx$.

Then use Formula 43 with $a = 6$.

$$\int \frac{x}{\sqrt{x^4-36}}dx = \frac{1}{2}\int \frac{du}{\sqrt{u^2-36}} = \frac{1}{2}\ln\left|u + \sqrt{u^2-36}\right| + C = \frac{1}{2}\ln\left|x^2 + \sqrt{x^4-36}\right| + C \quad (6\text{-}4)$$

39. $\int_0^4 x \ln(10-x)dx$

Consider

$$\int x \ln(10-x)dx = \int (10-t)\ln t\,(-dt) = \int t \ln t\, dt - 10\int \ln t\, dt.$$

Substitution: $t = 10 - x$, $dt = -dx$, $x = 10 - t$

Now use integration-by-parts on the two integrals.

Let $u = \ln t$, $dv = t\,dt$. Then $du = \frac{1}{t}dt$, $v = \frac{t^2}{2}$.

$$\int \ln t\, dt = \frac{t^2}{2}\ln t - \int \frac{t^2}{2}\cdot\frac{1}{t}dt = \frac{t^2 \ln t}{2} - \frac{t^2}{4} + C$$

Let $u = \ln t$, $dv = dt$. Then $du = \frac{1}{t}dt$, $v = t$.

$$\int t \ln t\, dt = t \ln t - \int t\cdot\frac{1}{t}dt = t \ln t - t + C$$

Thus,

$$\int_0^4 x\ln(10-x)dx = \left[\frac{(10-x)^2\ln(10-x)}{2} - \frac{(10-x)^2}{4} - 10(10-x)\ln(10-x) + 10(10-x)\right]_0^4$$

$$= \frac{36\ln 6}{2} - \frac{36}{4} - 10(6)\ln 6 + 10(6) - \left[\frac{100\ln 10}{2} - \frac{100}{4} - 10(10)\ln 10 + 10(10)\right]$$

$$= 18\ln 6 - 9 - 60\ln 6 + 60 - 50\ln 10 + 25 + 100\ln 10 - 100$$

$$= 50\ln 10 - 42\ln 6 - 24 \approx 15.875. \qquad (6\text{-}3)$$

40. Use Formula 52 with $n = 2$.

$$\int (\ln x)^2 dx = x(\ln x)^2 - 2\int \ln x\, dx$$

Now use integration-by-parts to calculate $\int \ln x\, dx$.

Let $u = \ln x$, $dv = dx$. Then $du = \frac{1}{x}dx$, $v = x$.

$$\int \ln x\, dx = x\ln x - \int x \cdot \frac{1}{x}dx = x\ln x - x + C$$

Therefore, $\int (\ln x)^2 dx = x(\ln x)^2 - 2[x\ln x - x] + C = x(\ln x)^2 - 2x\ln x + 2x + C.$ (6-3, 6-4)

41. $\int xe^{-2x^2}dx$

Let $u = -2x^2$. Then $du = -4x\, dx$.

$$\int xe^{-2x^2}dx = -\frac{1}{4}\int e^u\, du = -\frac{1}{4}e^u + C = -\frac{1}{4}e^{-2x^2} + C \qquad (6\text{-}2)$$

42. $\int x^2e^{-2x}dx.$

Use integration-by-parts. Let $u = x^2$ and $dv = e^{-2x}dx$. Then $du = 2x\, dx$ and $v = -\frac{1}{2}e^{-2x}$.

$$\int x^2e^{-2x}dx = -\frac{1}{2}x^2e^{-2x} + \int xe^{-2x}dx$$

Now use integration-by-parts again. Let $u = x$ and $dv = e^{-2x}dx$. Then $du = dx$ and $v = -\frac{1}{2}e^{-2x}$.

$$\int xe^{-2x}dx = -\frac{1}{2}xe^{-2x} + \frac{1}{2}\int e^{-2x}dx = -\frac{1}{2}xe^{-2x} - \frac{1}{4}e^{-2x} + C$$

Thus,

$$\int x^2e^{-2x}dx = -\frac{1}{2}x^2e^{-2x} + \left[-\frac{1}{2}xe^{-2x} - \frac{1}{4}e^{-2x}\right] + C = -\frac{1}{2}x^2e^{-2x} - \frac{1}{2}xe^{-2x} - \frac{1}{4}e^{-2x} + C$$

(6-3)

43. First graph the two functions to find the points of intersection.

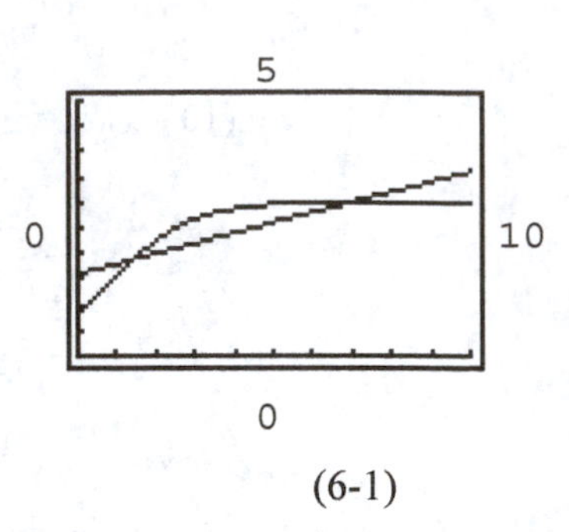

The curves intersect at the points where $x = 1.448$ and $x = 6.965$.

Area $A = \int_{1.448}^{6.965} \left(\frac{6}{2+5e^{-x}} - [0.2x+1.6] \right) dx$
≈ 1.703

(6-1)

44. (A) Probability $(0 \le t \le 1) = \int_0^1 0.21e^{-0.21t}\,dt = -e^{-0.21t}\Big|_0^1 = -e^{-0.21} + 1 \approx 0.189$

(B) Probability $(1 \le t \le 2) = \int_1^2 0.21e^{-0.21t}\,dt = -e^{-0.21t}\Big|_1^2 = e^{-0.21} - e^{-0.42} \approx 0.154$ (6-2)

45.

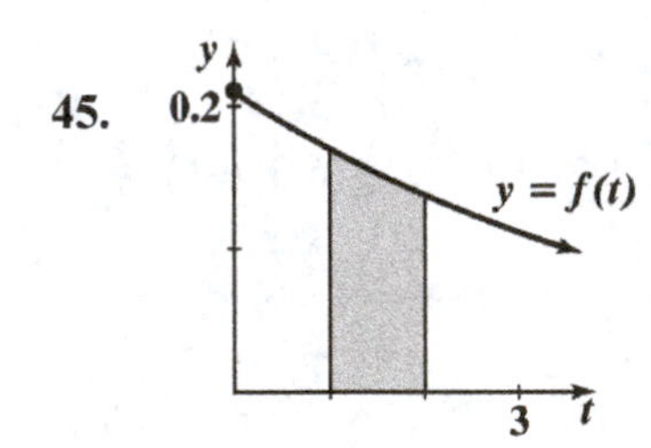

The probability that the product will fail during the second year of warranty is the area under the probability density function $y = f(t)$ from $t = 1$ to $t = 2$. (6-2)

46. $R'(x) = 65 - 6\ln(x + 1)$, $R(0) = 0$

$R(x) = \int [65 - 6\ln(x + 1)]\,dx = 65x - 6\int \ln(x + 1)\,dx$

Let $z = x + 1$. Then $dz = dx$ and $\int \ln(x + 1)\,dx = \int \ln z\,dz$.

Now, let $u = \ln z$ and $dv = dz$. Then $du = \frac{1}{z}dz$ and $v = z$:

$$\int \ln z\,dz = z\ln z - \int z\left(\frac{1}{z}\right)dz = z\ln z - \int dz = z\ln z - z + C$$

Therefore, $\int \ln(x + 1)dx = (x + 1)\ln(x + 1) - (x + 1) + C$ and
$R(x) = 65x - 6[(x + 1)\ln(x + 1) - (x + 1)] + C$.

Since $R(0) = 0$, $C = -6$. Thus, $R(x) = 65x - 6[(x + 1)\ln(x + 1) - x]$

To find the production level for a revenue of \$20,000 per week, solve $R(x) = 20{,}000$ for x.

The production level should be 618 hair dryers per week.

At a production level of 1,000 hair dryers per week, revenue

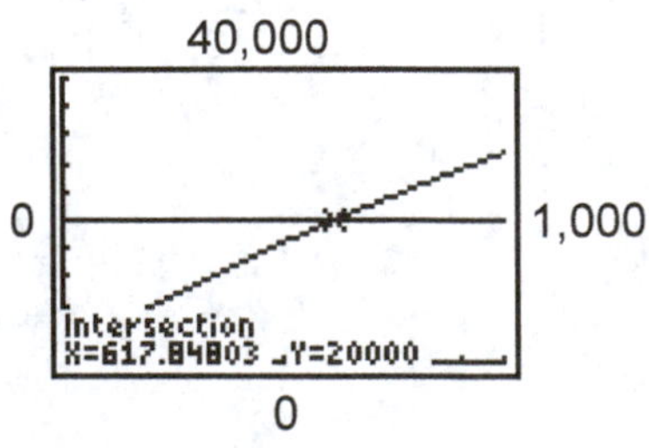

$R(1{,}000) = 65{,}000 - 6[(1{,}001)\ln(1{,}001) - 1{,}000] \approx \$29{,}506$ (6-3)

47. (A)

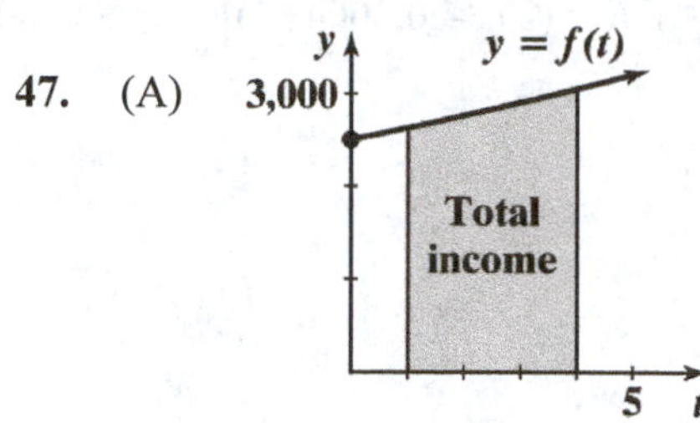

(B) Total income $= \int_1^4 2{,}500e^{0.05t}\,dt = 50{,}000e^{0.05t}\Big|_1^4 = 50{,}000[e^{0.2} - e^{0.05}] \approx \$8{,}507$ (6-2)

48. $f(t) = 2{,}500e^{0.05t}$, $r = 0.04$, $T = 5$

(A) $FV = e^{(0.04)5}\int_0^5 2{,}500e^{0.05t}e^{-0.04t}\,dt = 2{,}500e^{0.2}\int_0^5 e^{0.01t}\,dt = 250{,}000e^{0.2}\;e^{0.01t}\Big|_0^5$

$= 250{,}000[e^{0.25} - e^{0.2}] \approx \$15{,}655.66$

(B) Total income $= \int_0^5 2{,}500e^{0.05t}\,dt = 50{,}000e^{0.05t}\Big|_0^5 = 50{,}000[e^{0.25} - 1] \approx \$14{,}201.27$

Interest $= FV -$ Total income $= \$15{,}655.66 - \$14{,}201.27 = \$1{,}454.39$ (6-2)

49. (A)

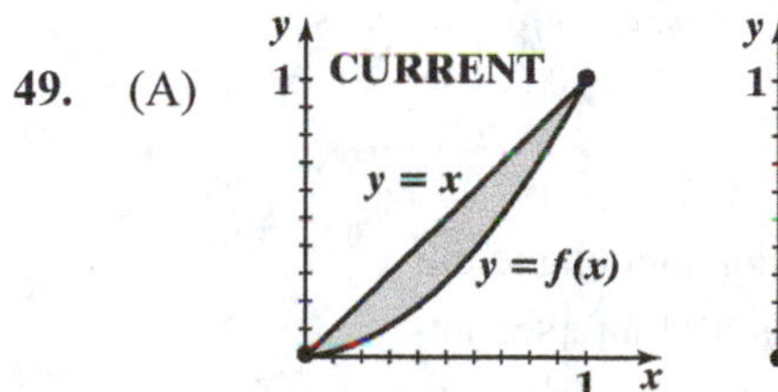

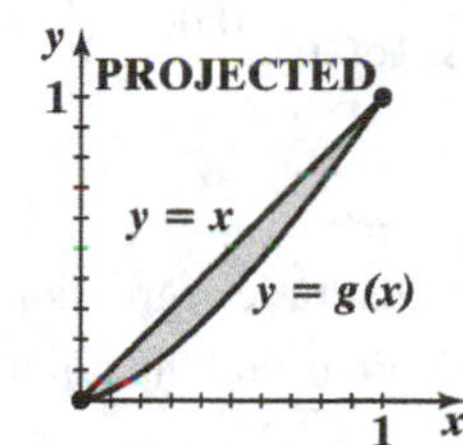

(B) The income will be more equally distributed 10 years from now since the area between $y = x$ and the projected Lorenz curve is less than the area between $y = x$ and the current Lorenz curve.

(C) Current:

Gini Index $= 2\int_0^1 [x - (0.1x + 0.9x^2)]\,dx = 2\int_0^1 (0.9x - 0.9x^2)\,dx = 2(0.45x^2 - 0.3x^3)\Big|_0^1 = 0.30$

Projected:

Gini Index $= 2\int_0^1 (x - x^{1.5})\,dx = 2\int_0^1 (x - x^{3/2})\,dx = 2\left(\frac{1}{2}x^2 - \frac{2}{5}x^{5/2}\right)\Big|_0^1 = 2\left(\frac{1}{10}\right) = 0.2$

Thus, income will be more equally distributed 10 years from now, as indicated in part (B). (6-1)

50. (A) $p = D(x) = 70 - 0.2x$, $p = S(x) = 13 + 0.0012x^2$

Equilibrium price: $D(x) = S(x)$

$$70 - 0.2x = 13 + 0.0012x^2$$

$$0.0012x^2 + 0.2x - 57 = 0$$

$$x = \frac{-0.2 \pm \sqrt{0.04 + 0.2736}}{0.0024} = \frac{-0.2 \pm 0.56}{0.0024}$$

Therefore, $\bar{x} = \frac{-0.2 + 0.56}{0.0024} = 150$, and $\bar{p} = 70 - 0.2(150) = 40$.

$$CS = \int_0^{150} (70 - 0.2x - 40)\,dx = \int_0^{150} (30 - 0.2x)\,dx = (30x - 0.1x^2)\Big|_0^{150} = \$2{,}250$$

$$PS = \int_0^{150} [40 - (13 + 0.0012x^2)]\, dx = \int_0^{150} (27 - 0.0012x^2)\, dx = (27x - 0.0004x^3)\Big|_0^{150} = \$2{,}700$$

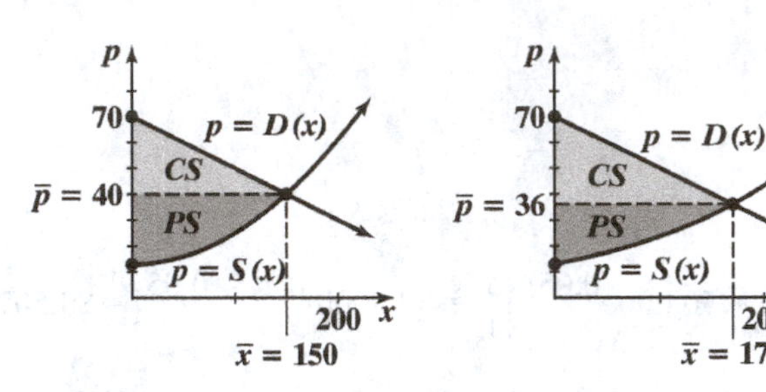

(B) $p = D(x) = 70 - 0.2x$, $p = S(x) = 13e^{0.006x}$

Equilibrium price: $D(x) = S(x)$

$$70 - 0.2x = 13e^{0.006x}$$

Using a graphing utility to solve for x, we get $\bar{x} \approx 170$ and $\bar{p} = 70 - 0.2(170) \approx 36$.

$$CS = \int_0^{170} (70 - 0.2x - 36)\, dx = \int_0^{170} (34 - 0.2x)\, dx = (34x - 0.1x^2)\Big|_0^{170} = \$2{,}890$$

$$PS = \int_0^{170} (36 - 13e^{0.006x})\, dx = (36x - 2{,}166.67e^{0.006x})\Big|_0^{170} \approx \$2{,}278 \qquad (6\text{-}2)$$

51. (A)

Graph the quadratic regression model and the line $p = 52.50$ to find the point of intersection.

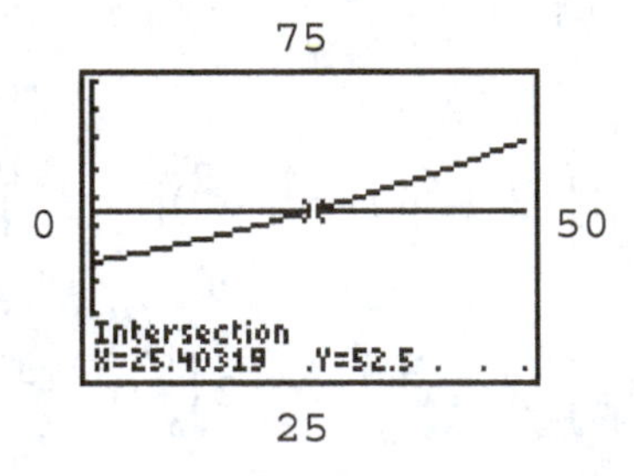

The demand at a price of 52.50 cents per pound is 25,403 lbs.

(B) Let $S(x)$ be the quadratic regression model found in part (A). Then the producers' surplus at the price level of 52.5 cents per pound is given by

$$PS = \int_0^{25.403} [52.5 - S(x)]\, dx \approx \$1{,}216 \qquad (6\text{-}2)$$

52. $R(t) = \dfrac{60t}{(t+1)^2(t+2)}$

The amount of the drug eliminated during the first hour is given by

$$A = \int_0^1 \frac{60t}{(t+1)^2(t+2)}\, dt$$

First use Formula 19 with $a = 1,\ b = 1,\ c = 2,\ d = 1$ to find the indefinite integral:

$$\int \frac{60t}{(t+1)^2(t+2)} dt\ dt = 60\int \int \frac{t}{(t+1)^2 \cdot (t+2)} dt = 60\left[\frac{1}{t+1} - 2\ln\left|\frac{t+2}{t+1}\right|\right] + C = \frac{60}{t+1} - 120\ln\left|\frac{t+2}{t+1}\right| + C$$

Now,

$$A = \int_0^1 \frac{60t}{(t+1)^2(t+2)} dt = \left[\frac{60}{t+1} - 120\ln\left(\frac{t+2}{t+1}\right)\right]_0^1 = 30 - 120\ln\left(\frac{3}{2}\right) - 60 + 120\ln 2$$

$$\approx 4.522 \text{ milliliters}$$

The amount of drug eliminated during the 4th hour is given by:

$$A = \int_3^4 \frac{60t}{(t+1)^2(t+2)} dt = \left[\frac{60}{t+1} - 120\ln\left(\frac{t+2}{t+1}\right)\right]_3^4 = 12 - 120\ln\left(\frac{6}{5}\right) - 15 + 120\ln\left(\frac{5}{4}\right)$$

$$\approx 1.899 \text{ milliliters}$$ (5-5, 6-4)

53.

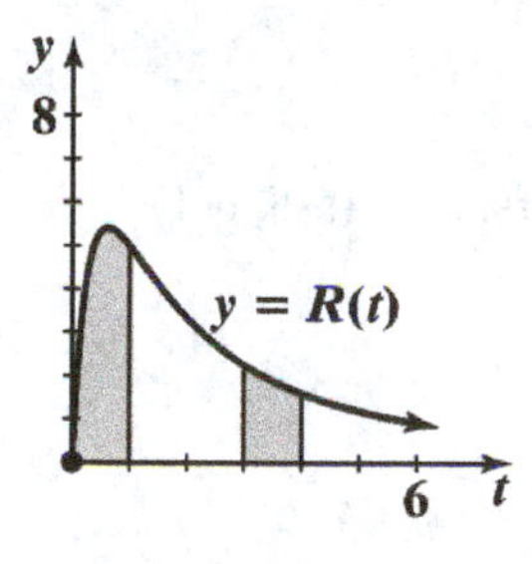

(5-5, 6-1)

54. $f(t) = \begin{cases} \dfrac{4/3}{(t+1)^2} & 0 \le t \le 3 \\ 0 & \text{otherwise} \end{cases}$

(A) Probability $(0 \le t \le 1) = \int_0^1 \frac{4/3}{(t+1)^2} dt$

To calculate the integral, let $u = t + 1$, $du = dt$. Then,

$$\int \frac{4/3}{(t+1)^2} dt = \frac{4}{3}\int u^{-2} du = \frac{4}{3}\frac{u^{-1}}{-1} = -\frac{4}{3u} + C = \frac{-4}{3(t+1)} + C$$

Thus,

$$\int_0^1 \frac{4/3}{(t+1)^2} dt = \left[\frac{-4}{3(t+1)}\right]_0^1 = -\frac{2}{3} + \frac{4}{3} = \frac{2}{3} \approx 0.667$$

(B) Probability $(t \ge 1) = \int_1^3 \frac{4/3}{(t+1)^2} dt = \left[\frac{-4}{3(t+1)}\right]_1^3 = -\frac{1}{3} + \frac{2}{3} = \frac{1}{3} \approx 0.333$ (6-2)

55.

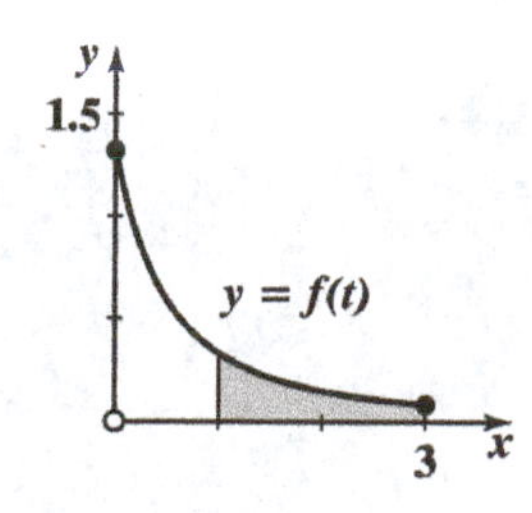

The probability that the doctor will spend more than an hour with a randomly selected patient is the area under the probability density function $y = f(t)$ from $t = 1$ to $t = 3$. (6-2)

56. $N'(t) = \frac{100t}{(1+t^2)^2}$. To find $N(t)$, we calculate

$$\int \frac{100t}{(1+t^2)^2}\,dt$$

Let $u = 1 + t^2$. Then $du = 2t\,dt$, and

$$N(t) = \int \frac{100t}{(1+t^2)^2}\,dt = 50\int \frac{1}{u^2}\,du = 50\int u^{-2}\,du = -50\frac{1}{u} + C = \frac{-50}{1+t^2} + C$$

At $t = 0$, we have

$$N(0) = -50 + C$$

Therefore, $C = N(0) + 50$ and

$$N(t) = \frac{-50}{1+t^2} + 50 + N(0)$$

Now,

$$N(3) = \frac{-50}{1+3^2} + 50 + N(0) = 45 + N(0)$$

Thus, the population will increase by 45 thousand during the next 3 years. (5-5, 6-1)

57. We want to find Probability $(t \geq 2) = \int_2^{\infty} f(t)\,dt$

Since

$$\int_{-\infty}^{\infty} f(t)\,dt = \int_{-\infty}^{2} f(t)\,dt + \int_2^{\infty} f(t)\,dt = 1,$$

$$\int_2^{\infty} f(t)\,dt = 1 - \int_{-\infty}^{2} f(t)\,dt = 1 - \int_0^2 f(t)\,dt \qquad \text{(since } f(t) = 0 \text{ for } t \leq 0\text{)}$$

$$= 1 - \text{Probability } (0 \leq t \leq 2)$$

Now, Probability $(0 \leq t \leq 2) = \int_0^2 0.5e^{-0.5t}\,dt = -e^{-0.5t}\Big]_0^2 = -e^{-1} + 1 \approx 0.632$

Therefore, Probability $(t \geq 2) = 1 - 0.632 = 0.368$. (6-2)

7 MULTIVARIABLE CALCULUS

EXERCISE 7-1

Things to remember:

1. An equation of the form $z = f(x, y)$ describes a FUNCTION OF TWO INDEPENDENT VARIABLES if for each permissible ordered pair (x, y) there is one and only one value of z determined by $f(x, y)$. The variables x and y are INDEPENDENT VARIABLES, and the variable z is a DEPENDENT VARIABLE. The set of all ordered pairs of permissible values of x and y is the DOMAIN of the function, and the set of all corresponding values $f(x, y)$ is the RANGE of the function.

2. CONVENTION ON DOMAINS

   Unless otherwise stated, the domain of a function specified by an equation of the form $z = f(x, y)$ is the set of all ordered pairs of real numbers (x, y) such that $f(x, y)$ is also a real number.

3. Functions of three independent variables $w = f(x, y, z)$, four independent variables $u = f(x, y, z, w)$, and so on, are defined similarly.

---

1. $a = 8,\ b = 5,\ h = 3;\ A = \frac{1}{2}(a+b)h = \frac{1}{2}(8+5)3 = \frac{39}{2} = 19.5\ \text{ft}^2$

3. $V = lwh = 12\times5\times4 = 240\ \text{in}^3$

5. Radius $r = 2,\ h = 8;\ V = \pi r^2 h = \pi 4(8) = 32\pi \approx 100.5\ \text{m}^3$

7. Radius $r = 20,\ h = 48;\ T = \pi r\left(r + \sqrt{r^2 + h^2}\right) = \pi(20)\left(20 + \sqrt{20^2 + 48^2}\right) = \pi(20)(72) = 1{,}440\pi$

$$\approx 4{,}523.9\ \text{cm}^2$$

In Problems 9 -15 , $f(x, y) = 2x + 7y - 5$ and $g(x, y) = \frac{88}{x^2 + 3y}$.

9. $f(4, -1) = 2(4) + 7(-1) - 5 = -4$

11. $f(8, 0) = 2(8) + 7(0) - 5 = 11$

13. $g(1, 7) = \frac{88}{1^2 + 3(7)} = \frac{88}{22} = 4$

15. $g(3, -3)$ not defined; $3^2 + 3(-3) = 0$

In Problems 17 - 20, $f(x, y, z) = 2x - 3y^2 + 5z^3 - 1$.

17. $f(0, 0, 0) = 2(0) - 3(0) + 5(0) - 1 = -1$

19. $f(6, -5, 0) = 2(6) - 3(-5)^2 + 5(0) - 1 = 12 - 75 - 1 = -64$

21. $P(n,r) = \frac{n!}{(n-r)!};\ P(13,5) = \frac{13!}{(13-5)!} = \frac{13!}{8!} = \frac{13\cdot12\cdot11\cdot10\cdot9\cdot8!}{8!} = 13\cdot12\cdot11\cdot10\cdot9 = 154{,}440$

23. $V(R,h)=\pi R^2 h;\ V(4,12)=\pi(4)^2 12=\pi\cdot 16\cdot 12=192\pi\approx 603.2$

25. $S(R,h)=\pi R\sqrt{R^2+h^2};\ S(3,10)=\pi\cdot 3\sqrt{3^2+10^2}=3\pi\sqrt{109}\approx 98.4$

27. $A(P,r,t)=P+Prt$
$A(100, 0.06, 3)=100+100(0.06)3=118$
$(P=100,\ r=0.06,\ \text{and } t=3)$

29. $P(r,T)=\int_0^T 4000e^{-rt}dt,$

$$P(0.05, 12)=\int_0^{12} 4000e^{-0.05t}dt=\frac{4000}{-0.05}e^{-0.05t}\Big|_0^{12}=-80{,}000[e^{-0.6}-1]\approx 36{,}095.07$$

31. $G(x,y)=x^2+3xy+y^2-7;\ f(x)=G(x,0)=x^2+3x(0)+0^2-7=x^2-7$

33. $K(x,y)=10xy+3x-2y+8;\ f(y)=K(4,y)=10(4)y+3(4)-2y+8=38y+20$

35. $M(x,y)=x^2y-3xy^2+5;\ f(y)=M(y,y)=(y)^2y-3(y)y^2+5=-2y^3+5$

37. $F(x,y)=2x+3y-6;\ F(0,y)=2(0)+3y-6=3y-6$
$F(0,y)=0:\ 3y-6=0$
$y=2$

39. $F(x,y)=2xy+3x-4y-1;\ F(x,x)=2x(x)+3x-4x-1=2x^2-x-1$

$F(x,x)=0:\ 2x^2-x-1=0$
$(2x+1)(x-1)=0$
$x=1,-\frac{1}{2}$

41. $F(x,y)=x^2+e^xy-y^2;\ F(x,2)=x^2+2e^x-4.$

We use a graphing utility to solve $F(x,2)=0$. The graph of $u=F(x,2)$ is shown at the right.

The solutions of $F(x,2)=0$ are: $x_1\approx -1.926,\ x_2\approx 0.599$

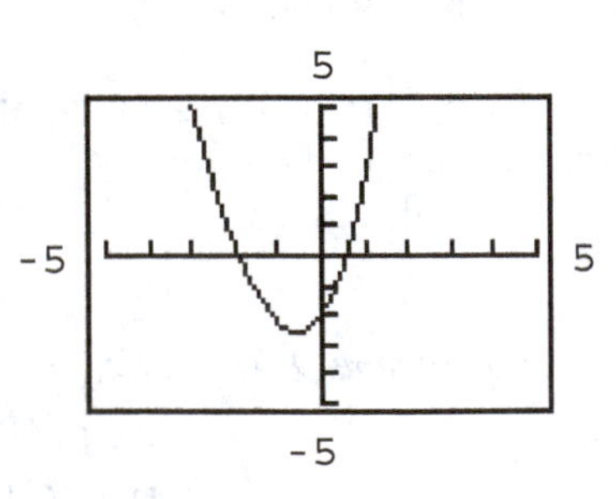

43. $f(x,y)=x^2+2y^2$

$$\frac{f(x+h,y)-f(x,y)}{h}=\frac{(x+h)^2+2y^2-(x^2+2y^2)}{h}=\frac{x^2+2xh+h^2+2y^2-x^2-2y^2}{h}$$

$$=\frac{2xh+h^2}{h}=\frac{h(2x+h)}{h}=2x+h,\ h\neq 0$$

45. $f(x, y) = 2xy^2$

$$\frac{f(x+h,y)-f(x,y)}{h} = \frac{2(x+h)y^2-2xy^2}{h} = \frac{2xy^2+2hy^2-2xy^2}{h} = \frac{2hy^2}{h} = 2y^2, \quad h \neq 0$$

47. Coordinates of point $E = E(0, 0, 3)$.
Coordinates of point $F = F(2, 0, 3)$.

49. $f(x, y) = x^2$

(A) In the plane $y = c$, c any constant, the graph of $z = x^2$ is a parabola.

(B) Cross-section corresponding to $x = 0$: the y-axis

Cross-section corresponding to $x = 1$: the line passing through $(1, 0, 1)$ parallel to the y-axis.

Cross-section corresponding to $x = 2$: the line passing through $(2, 0, 4)$ parallel to the y-axis.

(C) The surface $z = x^2$ is a parabolic trough lying on the y-axis.

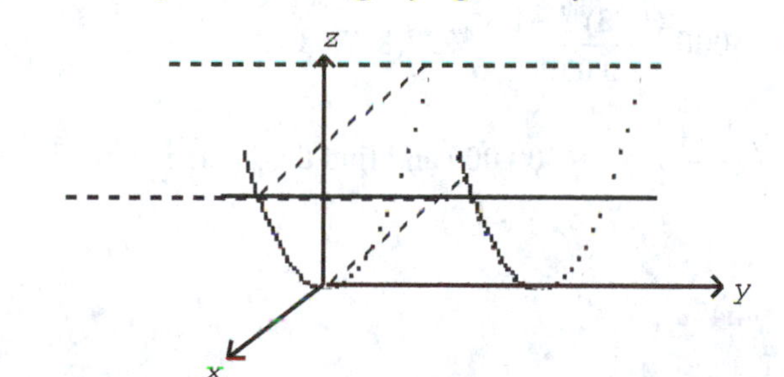

51. $f(x, y) = \sqrt{36 - x^2 - y^2}$

(A) Cross-sections corresponding to $y = 1, y = 2, y = 3, y = 4, y = 5$: Upper semicircles with centers at $(0, 1, 0)$, $(0, 2, 0)$, $(0, 3, 0)$, $(0, 4, 0)$, and $(0, 5, 0)$, respectively.

(B) Cross-sections corresponding to $x = 0, x = 1, x = 2, x = 3, x = 4$, $x = 5$: Upper semicircles with centers at $(0, 0, 0)$, $(1, 0, 0)$, $(2, 0, 0)$, $(3, 0, 0)$, $(4, 0, 0)$ and $(5, 0, 0)$, respectively.

(C) The upper hemisphere of radius 6 with center at the origin.

53. (A) If the points (a, b) and (c, d) both lie on the same circle centered at the origin, then $a^2 + b^2 = r^2 = c^2 + d^2$, where r is the radius of the circle.

(B) The cross-sections are:

(i) $x = 0$, $f(0, y) = e^{-y^2}$

(ii) $y = 0$, $f(x, 0) = e^{-x^2}$

(iii) $x = y$, $f(x, x) = e^{-2x^2}$

These are bell-shaped curves with maximum value 1 at $y = 0$ in (i) and $x = 0$ in (ii) and (iii).

(C) A "bell" with maximum value 1 at the origin, extending infinitely far in all directions, and approaching the x-y plane as $x, y \to \pm\infty$.

55. Monthly cost function = $C(x, y) = 6000 + 210x + 300y$

$$C(20, 10) = 6000 + 210\cdot 20 + 300\cdot 10 = \$13{,}200$$
$$C(50, 5) = 6000 + 210\cdot 50 + 300\cdot 5 = \$18{,}000$$
$$C(30, 30) = 6000 + 210\cdot 30 + 300\cdot 30 = \$21{,}3000$$

57. $R(p, q) = p\cdot x + q\cdot y = 200p - 5p^2 + 4pq + 300q + 2pq - 4q^2$ or

$R(p, q) = -5p^2 + 6pq - 4q^2 + 200p + 300q$

$R(2, 3) = -5\cdot 2^2 + 6\cdot 2\cdot 3 - 4\cdot 3^2 + 200\cdot 2 + 300\cdot 3 = 1280$ or \$1280

$R(3, 2) = -5\cdot 3^2 + 6\cdot 3\cdot 2 - 4\cdot 2^2 + 200\cdot 3 + 300\cdot 2 = 1175$ or \$1175

59. $f(x, y) = 20x^{0.4}y^{0.6}$

$f(1250, 1700) = 20(1250)^{0.4}(1700)^{0.6} \approx 20(17.3286)(86.7500) \approx 30{,}065$ units

61. $F(P, i, n) = P\dfrac{(1+i)^n - 1}{i}$

(A) $P = 5{,}000,\ i = 0.03,\ n = 30$

$$F(5000, 0.03, 30) = 5000\frac{(1.03)^{30} - 1}{0.03} \approx \$237{,}877.08$$

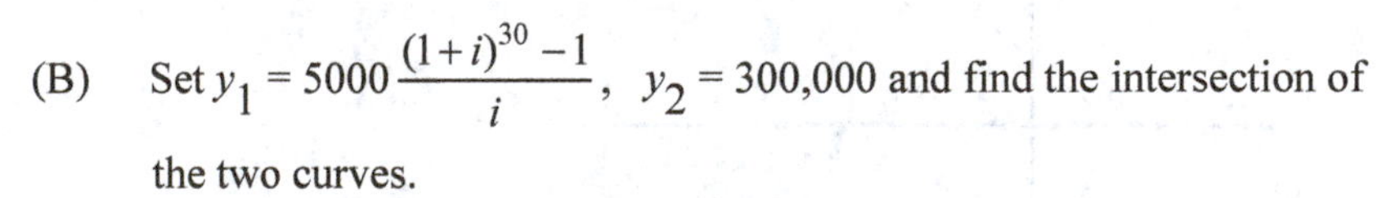

(B) Set $y_1 = 5000\dfrac{(1+i)^{30} - 1}{i}$, $y_2 = 300{,}000$ and find the intersection of the two curves.

Rate of interest: $i = 4.4\%$

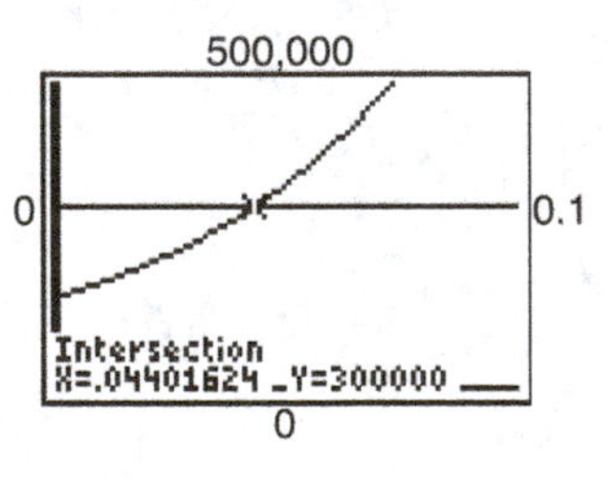

63. $T(V, x) = \dfrac{33V}{x+33}$

$$T(70, 47) = \frac{33\cdot 70}{47+33} = \frac{33\cdot 70}{80} = 28.875 \approx 29 \text{ minutes}$$

$$T(60, 27) = \frac{33\cdot 60}{27+33} = 33 \text{ minutes}$$

65. $C(W, L) = 100\dfrac{W}{L}$

$$C(6, 8) = 100\frac{6}{8} = 75$$

$$C(8.1, 9) = 100\frac{8.1}{9} = 90$$

67. $Q(M, C) = \dfrac{M}{C}100$

$$Q(12, 10) = \frac{12}{10}100 = 120$$

$$Q(10, 12) = \frac{10}{12}100 = 83.33 \approx 83$$

EXERCISE 7-2

Things to remember:

1. Let $z = f(x, y)$ be a function of two independent variables. The PARTIAL DERIVATIVE OF f WITH RESPECT TO x, denoted by

$\frac{\partial z}{\partial x}$, f_x, or $f_x(x, y)$, is given by

$$\frac{\partial z}{\partial x} = \lim_{h\to 0} \frac{f(x+h, y) - f(x, y)}{h}$$

provided this limit exists. Similarly, the PARTIAL DERIVATIVE OF f WITH RESPECT TO y, denoted by $\frac{\partial z}{\partial y}$, f_y, or $f_y(x, y)$, is given by

$$\frac{\partial z}{\partial y} = \lim_{k\to 0} \frac{f(x, y+k) - f(x, y)}{k}$$

provided this limit exists.

2. SECOND-ORDER PARTIAL DERIVATIVES

If $z = f(x, y)$, then:

$$\frac{\partial^2 z}{\partial x^2} = \frac{\partial\left(\frac{\partial z}{\partial x}\right)}{\partial x} = f_{xx}(x, y) = f_{xx}$$

$$\frac{\partial^2 z}{\partial y \partial x} = \frac{\partial\left(\frac{\partial z}{\partial x}\right)}{\partial y} = f_{xy}(x, y) = f_{xy}$$

$$\frac{\partial^2 z}{\partial x \partial y} = \frac{\partial\left(\frac{\partial z}{\partial y}\right)}{\partial x} = f_{yx}(x, y) = f_{yx}$$

$$\frac{\partial^2 z}{\partial y^2} = \frac{\partial\left(\frac{\partial z}{\partial y}\right)}{\partial y} = f_{yy}(x, y) = f_{yy}$$

[Note: For the functions being considered in this text, the mixed partial derivatives f_{xy} and f_{yx} are equal, i.e., $\frac{\partial^2 z}{\partial x \partial y} = \frac{\partial^2 z}{\partial y \partial x}$.]

---

1. $f(x) = \pi x^3 + x\pi^3$, $f'(x) = 3\pi x^2 + \pi^3$

3. $f(x) = x^e + e^x$; $f'(x) = ex^{e-1} + e^x$

5. $z = \frac{x}{e} + \frac{e}{x}$; $\frac{dz}{dx} = \frac{1}{e} - \frac{e}{x^2}$

7. $z = \ln(x^2 + e^2)$; $\frac{dz}{dx} = \frac{1}{x^2 + e^2}(2x) = \frac{2x}{x^2 + e^2}$

9. $f(x, y) = 4x - 3y + 6$

$f_x(x, y) = 4$

11. $f(x, y) = x^2 - 3xy + 2y^2$

$f_y(x, y) = 0 - 3x + 4y = -3x + 4y$

13. $z = x^3 + 4x^2y + 2y^3$

$\frac{\partial z}{\partial x} = 3x^2 + 8xy + 0 = 3x^2 + 8xy$

15. $z = (5x + 2y)^{10}$

$\frac{\partial z}{\partial y} = 10(5x + 2y)^9 \frac{\partial(5x + 2y)}{\partial y}$

$= 10(5x + 2y)^9(2) = 20(5x + 2y)^9$

17. $f(x, y) = 5x^3y - 4xy^2$;
$f_x(x, y) = 15x^2y - 4y^2$
$f_x(1, 3) = 15(1)^2 3 - 4(3)^2 = 45 - 36 = 9$

19. $f(x, y) = 3xe^y$;
$f_y(x, y) = 3xe^y$
$f_y(1, 0) = 3(1)e^0 = 3$

21. $f(x,y) = e^{x^2} - 4y$; $f_y(x,y) = 0 - 4 = -4$; $f_y(2,1) = -4$

23. $f(x, y) = \dfrac{2xy}{1+x^2y^2}$; $f_x(x, y) = \dfrac{(1+x^2y^2)(2y) - 2xy(2xy^2)}{(1+x^2y^2)^2} = \dfrac{2y - 2x^2y^3}{(1+x^2y^2)^2} = \dfrac{2y(1-x^2y^2)}{(1+x^2y^2)^2}$

$$f_x(1, -1) = \frac{2(-1)[1-1^2(-1)^2]}{[1+1^2(-1)^2]^2} = 0$$

25. $M(x,y) = 68 + 0.3x - 0.8y$; $M(32,40) = 68 + 0.3(32) - 0.8(40) = 68 + 9.6 - 32 = 45.6$
Mileage is 45.6 mpg at a tire pressure of 32 psi and a speed of 40 mph.

27. $M(x,y) = 68 + 0.3x - 0.8y$; $M(32,50) = 68 + 0.3(32) - 0.8(50) = 68 + 9.6 - 40 = 37.6$
Mileage is 37.6 mpg at a tire pressure of 32 psi and a speed of 50 mph.

29. $M(x,y) = 68 + 0.3x - 0.8y$; $M_x = 0.3$ mpg per psi; mileage increases at the rate of 0.3 mpg per psi of tire pressure.

31. $f(x,y) = 6x - 5y + 3$; $f_x(x,y) = 6$; $f_{xx}(x,y) = 0$

33. $f(x,y) = 4x^2 + 6y^2 - 10$; $f_x(x,y) = 8x$; $f_{xy}(x,y) = 0$

35. $f(x, y) = e^{xy^2}$
$f_x(x, y) = e^{xy^2}y^2 = y^2e^{xy^2}$
$f_{xy}(x, y) = y^2e^{xy^2}(2xy) + 2ye^{xy^2} = 2xy^3e^{xy^2} + 2ye^{xy^2} = 2y(1 + xy^2)e^{xy^2}$

37. $f(x, y) = \dfrac{\ln x}{y}$; $f_y(x, y) = -\dfrac{\ln x}{y^2}$; $f_{yy}(x, y) = \dfrac{2\ln x}{y^3}$

39. $f(x, y) = (2x + y)^5$; $f_x(x, y) = 5(2x + y)^4 \cdot 2 = 10(2x + y)^4$
$f_{xx}(x, y) = 40(2x + y)^3(2) = 80(2x + y)^3$

41. $f(x, y) = (x^2 + y^4)^{10}$; $f_x(x, y) = 10(x^2 + y^4)^9(2x) = 20x(x^2 + y^4)^9$
$f_{xy}(x, y) = 180x(x^2 + y^4)^8(4y^3) = 720xy^3(x^2 + y^4)^8$

In Problems 43 – 51, $C(x, y) = 3x^2 + 10xy - 8y^2 + 4x - 15y - 120$

43. $C_x(x, y) = 6x + 10y + 4$

45. From Problem 43, $C_x(3, -2) = 6(3) + 10(-2) + 4$
$= 2$

47. From Problem 43, $C_{xx}(x, y) = 6$

49. From Problem 43, $C_{xy}(x, y) = 10$

51. From Problem 47, $C_{xx}(3, -2) = 6$

In Problems 53 –57, $S(T,r) = 50(T-40)(5-r)$.

53. $S(T,r) = 50(T-40)(5-r)$; $S(60,2) = 50(60-40)(5-2) = 50(20)(3) = 3{,}000$
Daily sales are \$3,000 when the temperature is 60° and the rainfall is 2 inches.

55. $S(T,r) = 50(T-40)(5-r)$; $S_r(T,r) = 50(T-40)(-1) = -50(T-40)$
$S_r(90,1) = -50(90-40) = -50(50) = -2{,}500$. Daily sales decrease at a rate of \$2,500 per inch of rain when the temperature is 90° and the rainfall is 1 inch.

57. $S(T,r) = 50(T-40)(5-r)$; $S_T(T,r) = 50(5-r)$; $S_{Tr}(T,r) = -50$
S_r decreases at the rate of \$50 per inch per degree of temperature.

59. (A) $f(x, y) = y^3 + 4y^2 - 5y + 3$
Since f is independent of x, $\frac{\partial f}{\partial x} = 0$

(B) If $g(x, y)$ depends on y only, that is, if $g(x, y) = G(y)$ is independent of x, then
$$\frac{\partial g}{\partial x} = 0$$
Clearly there are an infinite number of such functions.

61. $f(x, y) = x^2y^2 + x^3 + y$

$f_x(x, y) = 2xy^2 + 3x^2$

$f_{xx}(x, y) = 2y^2 + 6x$

$f_{xy}(x, y) = 4xy$

$f_y(x, y) = 2x^2y + 1$

$f_{yx}(x, y) = 4xy$

$f_{yy}(x, y) = 2x^2$

63. $f(x, y) = \frac{x}{y} - \frac{y}{x}$

$f_x(x, y) = \frac{1}{y} + \frac{y}{x^2}$

$f_{xx}(x, y) = -\frac{2y}{x^3}$

$f_{xy}(x, y) = -\frac{1}{y^2} + \frac{1}{x^2}$

$f_y(x, y) = -\frac{x}{y^2} - \frac{1}{x}$

$f_{yx}(x, y) = -\frac{1}{y^2} + \frac{1}{x^2}$

$f_{yy}(x, y) = \frac{2x}{y^3}$

65. $f(x, y) = xe^{xy}$

$f_x(x, y) = xye^{xy} + e^{xy}$

$f_{xx}(x, y) = xy^2e^{xy} + 2ye^{xy}$

$f_{xy}(x, y) = x^2ye^{xy} + 2xe^{xy}$

$f_y(x, y) = x^2e^{xy}$

$f_{yx}(x, y) = x^2ye^{xy} + 2xe^{xy}$

$f_{yy}(x, y) = x^3e^{xy}$

67. $P(x, y) = -x^2 + 2xy - 2y^2 - 4x + 12y - 5$
$P_x(x, y) = -2x + 2y - 0 - 4 + 0 - 0 = -2x + 2y - 4$
$P_y(x, y) = 0 + 2x - 4y - 0 + 12 - 0 = 2x - 4y + 12$

$P_x(x, y) = 0$ and $P_y(x, y) = 0$ when
$-2x + 2y - 4 = 0 \quad (1)$
$2x - 4y + 12 = 0 \quad (2)$

Add equations (1) and (2): $-2y + 8 = 0$
$y = 4$

Substitute $y = 4$ into (1): $-2x + 2 \cdot 4 - 4 = 0$
$-2x + 4 = 0$
$x = 2$

Thus, $P_x(x, y) = 0$ and $P_y(x, y) = 0$ when $x = 2$ and $y = 4$.

69. $F(x, y) = x^3 - 2x^2y^2 - 2x - 4y + 10;$
$F_x(x, y) = 3x^2 - 4xy^2 - 2;\ F_y(x, y) = -4x^2y - 4.$
Set $F_x(x, y) = 0$ and $F_y(x, y) = 0$ and solve simultaneously:

$$3x^2 - 4xy^2 - 2 = 0 \qquad (1)$$
$$-4x^2y - 4 = 0 \qquad (2)$$

From (2), $y = -\frac{1}{x^2}$. Substituting this into (1),

$$3x^2 - 4x\left(-\frac{1}{x^2}\right)^2 - 2 = 0$$
$$3x^2 - 4x\left(\frac{1}{x^4}\right) - 2 = 0$$
$$3x^5 - 2x^3 - 4 = 0$$

Using a graphing utility, we find that $x \approx 1.1996$. Then, $y \approx -0.695$.

71. $f(x, y) = 3x^2 + y^2 - 4x - 6y + 2$

(A) $f(x, 1) = 3x^2 + 1 - 4x - 6 + 2$
$= 3x^2 - 4x - 3$

$\frac{d}{dx}[f(x, 1)] = 6x - 4$; critical values: $6x - 4 = 0,\ x = \frac{2}{3}$

$\frac{d^2}{dx^2}[f(x, 1)] = 6 > 0$

Therefore, $f\left(\frac{2}{3}, 1\right) = 3\left(\frac{2}{3}\right)^2 - 4\left(\frac{2}{3}\right) - 3 = -\frac{13}{3}$ is the minimum value of $f(x, 1)$.

(B) $-\frac{13}{3}$ is the minimum value of $f(x, y)$ on the curve $z = f(x, 1)$; $f(x, y)$ may have smaller values on other curves $z = f(x, k)$, k constant, or $z = f(h, y)$, h constant. For example, the minimum value of

$f(x, 2) = 3x^2 - 4x - 6$ is $f\left(\frac{2}{3}, 2\right) = -\frac{22}{3}$; the minimum value of $f(0, y) = y^2 - 6y + 2$ is $f(0,$

73. $f(x, y) = 4 - x^4y + 3xy^2 + y^5$

(A) Let $y = 2$ and find the maximum value of $f(x, 2) = 4 - 2x^4 + 12x + 32 = -2x^4 + 12x + 36$. Using a graphing utility we find that the maximum value of $f(x, 2)$ is 46.302 at $x = 1.1447152 \approx 1.145$.

(B) $f_x(x, y) = -4x^3y + 3y^2$
$f_x(1.145, 2) = -0.008989 \approx 0$
$f_y(x, y) = -x^4 + 6xy + 5y^4$
$f_y(1.145, 2) = 92.021$

75. $f(x, y) = x^2 + 2y^2$

(A)
$$\lim_{h\to 0} \frac{f(x+h, y) - f(x, y)}{h} = \lim_{h\to 0} \frac{(x+h)^2 + 2y^2 - (x^2 + 2y^2)}{h}$$
$$= \lim_{h\to 0} \frac{x^2 + 2xh + h^2 + 2y^2 - x^2 - 2y^2}{h}$$
$$= \lim_{h\to 0} \frac{h(2x+h)}{h} = \lim_{h\to 0} (2x + h) = 2x$$

(B)
$$\lim_{k\to 0} \frac{f(x, y+k) - f(x, y)}{k} = \lim_{k\to 0} \frac{x^2 + 2(y+k)^2 - (x^2 + 2y^2)}{k}$$
$$= \lim_{k\to 0} \frac{x^2 + 2(y^2 + 2yk + k^2) - x^2 - 2y^2}{k}$$
$$= \lim_{k\to 0} \frac{4yk + 2k^2}{k} = \lim_{k\to 0} (4y + 2k) = 4y$$

77. $R(x, y) = 80x + 90y + 0.04xy - 0.05x^2 - 0.05y^2$
$C(x, y) = 8x + 6y + 20{,}000$
The profit $P(x, y)$ is given by:
$P(x, y) = R(x, y) - C(x, y)$
$= 80x + 90y + 0.04xy - 0.05x^2 - 0.05y^2 - (8x + 6y + 20{,}000)$
$= 72x + 84y + 0.04xy - 0.05x^2 - 0.05y^2 - 20{,}000$
Now
$P_x(x, y) = 72 + 0.04y - 0.1x$
and
$P_x(1200, 1800) = 72 + 0.04(1800) - 0.1(1200) = 72 + 72 - 120 = 24;$
$P_y(x, y) = 84 + 0.04x - 0.1y$
and
$P_y(1200, 1800) = 84 + 0.04(1200) - 0.1(1800) = 84 + 48 - 180 = -48.$
Thus, at the (1200, 1800) output level, profit will increase approximately \$24 per unit increase in production of type A calculators; and profit will decrease \$48 per unit increase in production of type B calculators.

79. $x = 200 - 5p + 4q$
$y = 300 - 4q + 2p$
$\frac{\partial x}{\partial p} = -5, \ \frac{\partial y}{\partial p} = 2$

A $1 increase in the price of brand A will decrease the demand for brand A by 5 pounds at any price level (p, q).

A $1 increase in the price of brand A will increase the demand for brand B by 2 pounds at any price level (p, q).

81. $f(x, y) = 10x^{0.75}y^{0.25}$

(A) $f_x(x, y) = 10(0.75)x^{-0.25}y^{0.25} = 7.5x^{-0.25}y^{0.25}$
$f_y(x, y) = 10(0.25)x^{0.75}y^{-0.75} = 2.5x^{0.75}y^{-0.75}$

(B) Marginal productivity of labor $= f_x(600, 100) = 7.5(600)^{-0.25}(100)^{0.25} \approx 4.79$

Marginal productivity of capital $= f_y(600, 100) = 2.5(600)^{0.75}(100)^{-0.75} \approx 9.58$

(C) The government should encourage the increased use of capital.

83. $x = f(p, q) = 8000 - 0.09p^2 + 0.08q^2$ (Butter)
$y = g(p, q) = 15{,}000 + 0.04p^2 - 0.3q^2$ (Margarine)
$f_q(p, q) = 0.08(2)q = 0.16q > 0$
$g_p(p, q) = 0.04(2)p = 0.08p > 0$
Thus, the products are competitive.

85. $x = f(p, q) = 800 - 0.004p^2 - 0.003q^2$ (Skis)
$y = g(p, q) = 600 - 0.003p^2 - 0.002q^2$ (Ski boots)

$f_q(p, q) = -0.003(2)q = -0.006q < 0$
$g_p(p, q) = -0.003(2)p = -0.006p < 0$
Thus, the products are complementary.

87. $A = f(w, h) = 15.64w^{0.425}h^{0.725}$

(A) $f_w(w, h) = 15.64(0.425)w^{-0.575}h^{0.725} \approx 6.65w^{-0.575}h^{0.725}$
$f_h(w, h) = 15.64(0.725)w^{0.425}h^{-0.275} \approx 11.34w^{0.425}h^{-0.275}$

(B) $f_w(65, 57) = 6.65(65)^{-0.575}(57)^{0.725} \approx 11.31$

For a 65 pound child 57 inches tall, the rate of change of surface area is approximately 11.31 square inches for a one-pound gain in weight, height held fixed.

$f_h(65, 57) = 11.34(65)^{0.425}(57)^{-0.275} \approx 21.99$

For a 65 pound child 57 inches tall, the rate of change of surface area is approximately 21.99 square inches for a one-inch gain in height, weight held fixed.

89. $C(W, L) = 100\frac{W}{L}$

$C_W(W, L) = \frac{100}{L}$

$C_W(6, 8) = \frac{100}{8} = 12.5$

The index increases 12.5 units per 1-inch increase in the width of the head (length held fixed) when $W = 6$ and $L = 8$.

$C_L(W, L) = -\frac{100W}{L^2}$

$C_L(6, 8) = -\frac{100 \times 6}{8^2}$

$= -\frac{600}{64} = -9.38$

The index decreases 9.38 units per 1-inch increase in length (width held fixed) when $W = 6$ and $L = 8$.

EXERCISE 7-3

Things to remember:

1. $f(a, b)$ is a LOCAL MAXIMUM if there exists a circular region in the domain of $f(x, y)$ with (a, b) as the center, such that $f(a, b) \geq f(x, y)$ for all (x, y) in the region. Similarly, $f(a, b)$ is a LOCAL MINIMUM if $f(a, b) \leq f(x, y)$ for all (x, y) in the region.

2. LOCAL EXTREMA AND PARTIAL DERIVATIVES
   Let $f(a, b)$ be a local extremum (a local maximum or a local minimum) for the function f. If both f_x and f_y exist at (a, b) then

   $$f_x(a, b) = 0 \quad \text{and} \quad f_y(a, b) = 0$$

   Points (a, b) such that $f_x(a, b) = f_y(a, b) = 0$ are called CRITICAL POINTS OF f.

3. SECOND-DERIVATIVE TEST FOR LOCAL EXTREMA FOR $z = f(x, y)$
   Given:
   (a) $f_x(a, b) = 0$ and $f_y(a, b) = 0$ [(a, b) is a critical point].
   (b) All second-order partial derivatives of f exist in some circular region containing (a, b) as center.
   (c) $A = f_{xx}(a, b)$, $B = f_{xy}(a, b)$, $C = f_{yy}(a, b)$.
   Then:
   i) If $AC - B^2 > 0$ and $A < 0$, then $f(a, b)$ is a local maximum.
   ii) If $AC - B^2 > 0$ and $A > 0$, then $f(a, b)$ is a local minimum.
   iii) If $AC - B^2 < 0$, then f has a saddle point at (a, b).
   iv) If $AC - B^2 = 0$, then the test fails.

---

1. $f(x) = 2x^3 - 9x^2 + 4, \quad f'(x) = 6x^2 - 18x, \quad f''(x) = 12x - 18$

$f'(0) = 6(0)^2 - 18(0) = 0, \quad f''(0) = 12(0) - 18 = -18 < 0;$ f has a local maximum at $x = 0$.

3. $f(x)=\dfrac{1}{1-x^2},\quad f'(x)=\dfrac{2x}{(1-x^2)^2},$

$$f''(x)=\frac{2(1-x^2)^2-2x[2(1-x^2)(-2x)]}{(1-x^2)^4}=\frac{2(1-x^2)+8x^2}{(1-x^2)^3}=\frac{2+6x^2}{(1-x^2)^3}$$

$f'(0)=\dfrac{2(0)}{(1-0^2)^2}=0,\quad f''(0)=\dfrac{2+6(0)^2}{(1-0^2)^3}=2>0,$ f has a local minimum at $x=0$.

5. $f(x)=e^{-x^2},\quad f'(x)=-2xe^{-x^2},\quad f''(x)=-2e^{-x^2}+4x^2e^{-x^2}.$

$f'(0)=-2(0)e^{-0^2}=0,\quad f''(0)=-2e^{-0^2}+4(0)^2e^{-0^2}=-2<0,$ f has a local maximum at $x=0$.

7. $f(x)=x^3-x^2+x+1;\quad f'(x)=3x^2-2x+1,\quad f''(x)=6x-2$

$f'(0)=3(0)^2-2(0)+1=1,\quad f''(0)=6(0)-2=-2$

$f'(0)=1\neq 0.$ Therefore f has neither a maximum nor a minimum at $x=0$.

9. $f(x, y) = 4x + 5y - 6$

$f_x(x, y) = 4 \neq 0; f_y(x, y) = 5 \neq 0$; the functions $f_x(x, y)$ and $f_y(x, y)$ are nonzero for all (x, y).

11. $f(x, y) = 3.7 - 1.2x + 6.8y + 0.2y^3 + x^4$

$f_x(x, y) = -1.2 + 4x^3; f_y = 6.8 + 0.6y^2$; the function $f_y(x, y)$ is nonzero for all (x, y).

13. $f(x, y) = 6 - x^2 - 4x - y^2$

$f_x(x, y) = -2x - 4 = 0$ implies $x = -2$.

$f_y(x, y) = -2y = 0$ implies $y = 0$.

Thus, $(-2, 0)$ is a critical point.

$f_{xx} = -2,\ f_{xy} = 0,\ f_{yy} = -2,$

$f_{xx}(-2, 0)\cdot f_{yy}(-2, 0) - [f_{xy}(-2, 0)]^2 = (-2)(-2) - 0^2 = 4 > 0$ and $f_{xx}(-2, 0) = -2 < 0.$

Thus, $f(-2, 0) = 6 - (-2)^2 - 4(-2) - 0^2 = 10$ is a local maximum (using 3).

15. $f(x, y) = x^2 + y^2 + 2x - 6y + 14$

$f_x(x, y) = 2x + 2 = 0$ implies $x = -1$

$f_y(x, y) = 2y - 6 = 0$ implies $y = 3$

Thus, $(-1, 3)$ is a critical point.

$f_{xx} = 2,\ f_{xy} = 0,\ f_{yy} = 2$

$f_{xx}(-1, 3) = 2 > 0,\ f_{xy}(-1, 3) = 0,\ f_{yy}(-1, 3) = 2$

$f_{xx}(-1, 3)\cdot f_{yy}(-1, 3) - [f_{xy}(-1, 3)]^2 = 2\cdot 2 - 0^2 = 4 > 0$

Thus, using 3, $f(-1, 3) = 4$ is a local minimum.

17. $f(x, y) = xy + 2x - 3y - 2$

$f_x = y + 2 = 0$ implies $y = -2$

$f_y = x - 3 = 0$ implies $x = 3$

Thus, $(3, -2)$ is a critical point.

$f_{xx} = 0,\ f_{xy} = 1,\ f_{yy} = 0$

$f_{xx}(3, -2) = 0,\ f_{xy}(3, -2) = 1,\ f_{yy}(3, -2) = 0$

$f_{xx}(3, -2)\cdot f_{yy}(3, -2) - [f_{xy}(3, -2)]^2 = 0\cdot 0 - [1]^2 = -1 < 0$

Thus, using 3, f has a saddle point at $(3, -2)$.

19. $f(x, y) = -3x^2 + 2xy - 2y^2 + 14x + 2y + 10$

$f_x = -6x + 2y + 14 = 0$ (1)

$f_y = 2x - 4y + 2 = 0$ (2)

Solving (1) and (2) for x and y, we obtain $x = 3$ and $y = 2$.
Thus, $(3, 2)$ is a critical point.

$f_{xx} = -6,\ f_{xy} = 2,\ f_{yy} = -4$

$f_{xx}(3, 2) = -6 < 0,\ f_{xy}(3, 2) = 2,\ f_{yy}(3, 2) = -4$

$f_{xx}(3, 2)\cdot f_{yy}(3, 2) - [f_{xy}(3, 2)]^2 = (-6)(-4) - 2^2 = 20 > 0$

Thus, using 3, $f(3, 2)$ is a local maximum and $f(3, 2) = -3\cdot 3^2 + 2\cdot 3\cdot 2 - 2\cdot 2^2 + 14\cdot 3 + 2\cdot 2 + 10 = 33$.

21. $f(x, y) = 2x^2 - 2xy + 3y^2 - 4x - 8y + 20$

$f_x = 4x - 2y - 4 = 0$ (1)

$f_y = -2x + 6y - 8 = 0$ (2)

Solving (1) and (2) for x and y, we obtain $x = 2$ and $y = 2$.
Thus, $(2, 2)$ is a critical point.

$f_{xx} = 4,\ f_{xy} = -2,\ f_{yy} = 6$

$f_{xx}(2, 2) = 4 > 0,\ f_{xy}(2, 2) = -2,\ f_{yy}(2, 2) = 6$

$f_{xx}(2, 2)\cdot f_{yy}(2, 2) - [f_{xy}(2, 2)]^2 = 4\cdot 6 - [-2]^2 = 20 > 0$

Thus, using 3, $f(2, 2)$ is a local minimum and $f(2, 2) = 2\cdot 2^2 - 2\cdot 2\cdot 2 + 3\cdot 2^2 - 4\cdot 2 - 8\cdot 2 + 20 = 8$.

23. $f(x, y) = e^{xy}$

$f_x = e^{xy}\dfrac{\partial(xy)}{\partial x}$

$= e^{xy}y = 0$

$y = 0 \quad (e^{xy} \neq 0)$

$f_y = e^{xy}\dfrac{\partial(xy)}{\partial y}$

$= e^{xy}x = 0$

$x = 0 \quad (e^{xy} \neq 0)$

Thus, $(0, 0)$ is a critical point.

$f_{xx} = ye^{xy}\dfrac{\partial(xy)}{\partial x}$,

$= ye^{xy}y$

$= y^2e^{xy}$

$f_{xy} = e^{xy}\cdot 1 + ye^{xy}x$

$= e^{xy} + xye^{xy}$

$f_{yy} = xe^{xy}\dfrac{\partial(xy)}{\partial y}$

$= x^2e^{xy}$

$f_{xx}(0, 0) = 0,\qquad f_{xy}(0, 0) = 1 + 0 = 1,\qquad f_{yy}(0, 0) = 0$

$f_{xx}(0, 0)\cdot f_{yy}(0, 0) - [f_{xy}(0, 0)]^2 = 0 - [1]^2 = -1 < 0$

Thus, using 3, $f(x, y)$ has a saddle point at $(0, 0)$.

25. $f(x, y) = x^3 + y^3 - 3xy$

$f_x = 3x^2 - 3y = 3(x^2 - y) = 0$

Thus, $y = x^2$. (1)

$f_y = 3y^2 - 3x = 3(y^2 - x) = 0$

Thus, $y^2 = x$. (2)

Combining (1) and (2), we obtain $x = x^4$ or $x(x^3 - 1) = 0$. Therefore, $x = 0$ or $x = 1$, and the critical points are (0, 0) and (1, 1).

$f_{xx} = 6x$ $\quad f_{xy} = -3$ $\quad f_{yy} = 6y$

For the critical point (0, 0):

$f_{xx}(0, 0) = 0$ $\quad f_{xy}(0, 0) = -3$ $\quad f_{yy}(0, 0) = 0$

$f_{xx}(0, 0) \cdot f_{yy}(0, 0) - [f_{xy}(0, 0)]^2 = 0 - (-3)^2 = -9 < 0$

Thus, using 3, $f(x, y)$ has a saddle point at (0, 0).

For the critical point (1, 1):

$f_{xx}(1, 1) = 6$ $\quad f_{xy}(1, 1) = -3$ $\quad f_{yy}(1, 1) = 6$

$f_{xx}(1, 1) \cdot f_{yy}(1, 1) - [f_{xy}(1, 1)]^2 = 6 \cdot 6 - (-3)^2 = 27 > 0$

$f_{xx}(1, 1) > 0$

Thus, using 3, $f(1, 1)$ is a local minimum and $f(1, 1) = 1^3 + 1^3 - 3 \cdot 1 \cdot 1 = 2 - 3 = -1$.

27. $f(x, y) = 2x^4 + y^2 - 12xy$

$f_x = 8x^3 - 12y = 0$

Thus, $y = \frac{2}{3}x^3$.

$f_y = 2y - 12x = 0$

Thus, $y = 6x$

Therefore, $6x = \frac{2}{3}x^3$

$x^3 - 9x = 0$

$x(x^2 - 9) = 0$

$x = 0, \ x = 3, \ x = -3$

Thus, the critical points are (0, 0), (3, 18), (–3, –18). Now,

$f_{xx} = 24x^2$ $\quad f_{xy} = -12$ $\quad f_{yy} = 2$

For the critical point (0, 0):

$f_{xx}(0, 0) = 0$ $\quad f_{xy}(0, 0) = -12$ $\quad f_{yy}(0, 0) = 2$

and

$f_{xx}(0, 0) \cdot f_{yy}(0, 0) - [f_{xy}(0, 0)]^2 = 0 \cdot 2 - (-12)^2 = -144 < 0.$

Thus, $f(x, y)$ has a saddle point at (0, 0).

For the critical point (3, 18):

$f_{xx}(3, 18) = 24 \cdot 3^2 = 216 > 0$ $\quad f_{xy}(3, 18) = -12$ $\quad f_{yy}(3, 18) = 2$

and

$f_{xx}(3, 18) \cdot f_{yy}(3, 18) - [f_{xy}(3, 18)]^2 = 216 \cdot 2 - (-12)^2 = 288 > 0$

Thus, $f(3, 18) = -162$ is a local minimum.

For the critical point $(-3, -18)$:

$f_{xx}(-3, -18) = 216 > 0 \qquad f_{xy}(-3, -18) = -12 \qquad f_{yy}(-3, -18) = 2$

and

$f_{xx}(-3, -18) \cdot f_{yy}(-3, -18) - [f_{xy}(-3, -18)]^2 = 288 > 0$

Thus, $f(-3, -18) = -162$ is a local minimum.

29. $f(x, y) = x^3 - 3xy^2 + 6y^2$

$f_x = 3x^2 - 3y^2 = 0$

Thus, $y^2 = x^2$ or $y = \pm x$.

$f_y = -6xy + 12y = 0$ or $-6y(x - 2) = 0$

Thus, $y = 0$ or $x = 2$.

Therefore, the critical points are $(0, 0)$, $(2, 2)$, and $(2, -2)$.

Now,

$f_{xx} = 6x, \qquad f_{xy} = -6y, \qquad f_{yy} = -6x + 12$

For the critical point $(0, 0)$:

$f_{xx}(0, 0) \cdot f_{yy}(0, 0) - [f_{xy}(0, 0]^2 = 0 \cdot 12 - 0^2 = 0$

Thus, the second-derivative test fails.

For the critical point $(2, 2)$:

$f_{xx}(2, 2) \cdot f_{yy}(2, 2) - [f_{xy}(2, 2)]^2 = 12 \cdot 0 - (-12)^2 = -144 < 0$

Thus, $f(x, y)$ has a saddle point at $(2, 2)$.

For the critical point $(2, -2)$:

$f_{xx}(2, -2) \cdot f_{yy}(2, -2) - [f_{xy}(2, -2)]^2 = 12 \cdot 0 - (12)^2 = -144 < 0$

Thus, $f(x, y)$ has a saddle point at $(2, -2)$.

31. $f(x, y) = y^3 + 2x^2y^2 - 3x - 2y + 8;$

$f_x = 4xy^2 - 3; \; f_y = 3y^2 + 4x^2y - 2$

Set $f_x = 0$ and $f_y = 0$ to find the critical points:

$$4xy^2 - 3 = 0 \qquad (1)$$

$$3y^2 + 4x^2y - 2 = 0 \qquad (2)$$

From (1) $x = \dfrac{3}{4y^2}$. Substituting this into (2), we have

$$3y^2 + 4\left(\frac{3}{4y^2}\right)^2 y - 2 = 0$$

$$3y^2 + 4\left(\frac{9}{16y^4}\right)y - 2 = 0$$
$$12y^5 - 8y^3 + 9 = 0$$

Using a graphing utility, we find that $y \approx -1.105$ and $x \approx 0.614$.

Now, $f_{xx} = 4y^2$ and $f_{xx}(0.614, -1.105) \approx 4.884$

$f_{xy} = 8xy$ and $f_{xy}(0.614, -1.105) \approx -5.428$

$f_{yy} = 6y + 4x^2$ and $f_{yy}(0.614, -1.105) \approx -5.122$

$f_{xx}(0.614, -1.105) \cdot f_{yy}(0.614, -1.105) - [f_{xy}(0.614, -1.105)]^2 \approx -54.479 < 0$

Thus, $f(x, y)$ has a saddle point at $(0.614, -1.105)$.

33. $f(x, y) = x^2 \geq 0$ for all (x, y) and $f(x, y) = 0$ when $x = 0$. Thus, f has a local minimum at each point $(0, y, 0)$ on the y–axis.

35. $f(x, y) = x^4e^y + x^2y^4 + 1$

(A) $f_x = 4x^3e^y + 2xy^4 = 0$ (1)

$f_y = x^4e^y + 4x^2y^3 = 0$ (2)

The values $x = 0$, $y = 0$ satisfy (1) and (2) so $(0, 0)$ is a critical point.

$A = f_{xx} = 12x^2e^y + 2y^4 = 0$ at $(0, 0)$,

$B = f_{xy} = 4x^3e^y + 8xy^3 = 0$ at $(0, 0)$,

$C = f_{yy} = x^4e^y + 12x^2y^2 = 0$ at $(0, 0)$.

$AC - B^2 = 0$; the second derivative test fails.

(B) Cross–sections of f by the planes $y = 0$, $x = 0$, $y = x$ and $y = -x$ are shown at the right. The cross-sections indicate that f has a local minimum at $(0, 0)$.

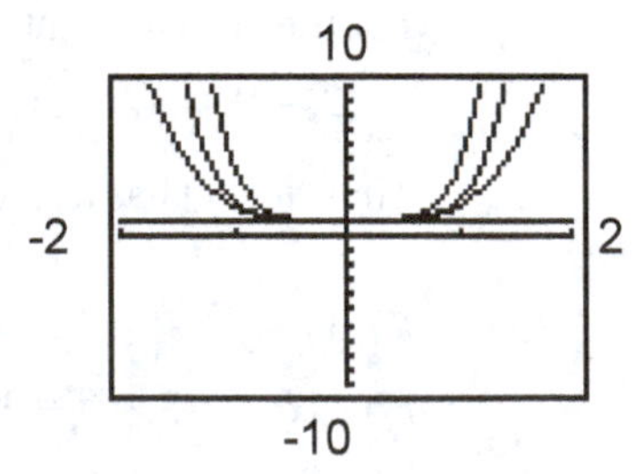

37. $P(x, y) = R(x, y) - C(x, y)$

$= 2x + 3y - (x^2 - 2xy + 2y^2 + 6x - 9y + 5) = -x^2 + 2xy - 2y^2 - 4x + 12y - 5$

$P_x = -2x + 2y - 4 = 0$ (1)

$P_y = 2x - 4y + 12 = 0$ (2)

Solving (1) and (2) for x and y, we obtain $x = 2$ and $y = 4$. Thus, $(2, 4)$ is a critical point.

$P_{xx} = -2$ and $P_{xx}(2, 4) = -2 < 0$

$P_{xy} = 2$ and $P_{xy}(2, 4) = 2$

$P_{yy} = -4$ and $P_{yy}(2, 4) = -4$

$P_{xx}(2, 4) \cdot P_{yy}(2, 4) - [P_{xy}(2, 4)]^2 = (-2)(-4) - [2]^2 = 4 > 0$

The maximum occurs when 2000 type A and 4000 type B earphones are produced. The maximum profit is given by $P(2, 4)$. Hence,

$\max P = P(2, 4) = -(2)^2 + 2\cdot2\cdot4 - 2\cdot4^2 - 4\cdot2 + 12\cdot4 - 5 = -4 + 16 - 32 - 8 + 48 - 5 = \$15$ million.

39. $x = 260 - 3p + q$ (Brand A)

$y = 180 + p - 2q$ (Brand B)

(A)

p	q	x	y
100	120	80	40
110	110	40	70

(B) In terms of p and q, the cost function C is given by:

$C = 60x + 80y = 60(260 - 3p + q) + 80(180 + p - 2q) = 30{,}000 - 100p - 100q$

The revenue function R is given by:

$R = px + qy = p(260 - 3p + q) + q(180 + p - 2q)$

$= -3p^2 + 2pq - 2q^2 + 260p + 180q$

Thus, the profit $P = R - C$ is given by:

$P = -3p^2 + 2pq - 2q^2 + 260p + 180q - (30{,}000 - 100p - 100q)$

$= -3p^2 + 2pq - 2q^2 + 360p + 280q - 30{,}000$

Now, calculating P_p and P_q and setting these equal to 0, we have:

$P_p = -6p + 2q + 360 = 0$ (1)

$P_q = 2p - 4q + 280 = 0$ (2)

Solving (1) and (2) for p and q, we get $p = 100$ and $q = 120$. Thus, (100, 120) is a critical point of the profit function P.

$P_{pp} = -6$ and $P_{pp}(100, 120) = -6$

$P_{pq} = 2$ and $P_{pq}(100, 120) = 2$

$P_{qq} = -4$ and $P_{qq}(100, 120) = -4$

$P_{pp}\cdot P_{qq} - [P_{pq}]^2 = (-6)(-4) - (2)^2 = 20 > 0$

Since $P_{pp}(100, 120) = -6 < 0$, we conclude that the maximum profit occurs when $p = \$100$ and $q = \$120$. The maximum profit is:

$P(100, 120) = -3(100)^2 + 2(100)(120) - 2(120)^2 + 360(100) + 280(120) - 30{,}000 = \$4{,}800$

41. The square of the distance from P to A is: $x^2 + y^2$

The square of the distance from P to B is:

$(x - 2)^2 + (y - 6)^2 = x^2 - 4x + y^2 - 12y + 40$

The square of the distance from P to C is:

$(x - 10)^2 + y^2 = x^2 - 20x + y^2 + 100$

Thus, we have:

$P(x, y) = 3x^2 - 24x + 3y^2 - 12y + 140$

$P_x = 6x - 24 = 0$ $\quad\quad$ $P_y = 6y - 12 = 0$

$x = 4$ $\quad\quad$ $y = 2$

Therefore, (4, 2) is a critical point.

$P_{xx} = 6$ and $P_{xx}(4, 2) = 6 > 0$
$P_{xy} = 0$ and $P_{xy}(4, 2) = 0$
$P_{yy} = 6$ and $P_{yy}(4, 2) = 6$
$P_{xx} \cdot P_{yy} - [P_{xy}]^2 = 6 \cdot 6 - 0 = 36 > 0$
Therefore, P has a minimum at the point (4, 2).

43. Let x = length, y = width, and z = height. Then $V = xyz = 64$ or $z = \dfrac{64}{xy}$. The surface area of the box is:

$S = xy + 2xz + 4yz$ or $S(x, y) = xy + \dfrac{128}{y} + \dfrac{256}{x}, \quad x > 0,\ y > 0$

$S_x = y - \dfrac{256}{x^2} = 0$ or $y = \dfrac{256}{x^2}$ (1)

$S_y = x - \dfrac{128}{y^2} = 0$ or $x = \dfrac{128}{y^2}$

Thus, $y = \dfrac{256}{\dfrac{(128)^2}{y^4}}$ or $y^4 - 64y = 0$

$y(y^3 - 64) = 0$ (Since $y > 0$, $y = 0$ does not yield a critical point.)

$y = 4$

Setting $y = 4$ in (1), we find $x = 8$. Therefore, the critical point is (8, 4).

Now we have:

$S_{xx} = \dfrac{512}{x^3}$ and $S_{xx}(8, 4) = 1 > 0$
$S_{xy} = 1$
$S_{yy} = \dfrac{256}{y^3}$ and $S_{yy}(8, 4) = 4$
$S_{xx}(8, 4) \cdot S_{yy}(8, 4) - [S_{xy}(8, 4)]^2 = 1 \cdot 4 - 1^2 = 3 > 0$

Thus, the dimensions that will require the least amount of material are: length $x = 8$ inches; width $y = 4$ inches; height $z = \dfrac{64}{8(4)} = 2$ inches.

45. Let x = length of the package, y = width, and z = height. Then
$x + 2y + 2z = 120$ (1)
Volume = $V = xyz$.

From (1), $z = \dfrac{120 - x - 2y}{2}$. Thus, we have:

$V(x, y) = xy\left(\dfrac{120 - x - 2y}{2}\right) = 60xy - \dfrac{x^2y}{2} - xy^2, \quad x > 0,\ y > 0$

$V_x = 60y - xy - y^2 = 0$

$y(60 - x - y) = 0$

$60 - x - y = 0$ (2) (Since $y > 0$, $y = 0$ does not yield a critical point.)

$V_y = 60x - \frac{x^2}{2} - 2xy = 0$

$x\left(60 - \frac{x}{2} - 2y\right) = 0$

$120 - x - 4y = 0$ (3) (Since $x > 0$, $x = 0$ does not yield a critical point.)

Solving (2) and (3) for x and y, we obtain $x = 40$ and $y = 20$. Thus, (40, 20) is the critical point.

$V_{xx} = -y$ and $V_{xx}(40, 20) = -20 < 0$

$V_{xy} = 60 - x - 2y$ and $V_{xy}(40, 20) = 60 - 40 - 40 = -20$

$V_{yy} = -2x$ and $V_{yy}(40, 20) = -80$

$V_{xx}(40, 20) \cdot V_{yy}(40, 20) - [V_{xy}(40, 20)]^2 = (-20)(-80) - [-20]^2 = 1600 - 400 = 1200 > 0$

Thus, the maximum volume of the package is obtained when $x = 40$, $y = 20$, and $z = \frac{120 - 40 - 2 \cdot 20}{2} = 20$ inches. The package has dimensions: length $x = 40$ inches; width $y = 20$ inches; height $z = 20$ inches.

EXERCISE 7-4

Things to remember:

1. Any local maxima or minima of the function $z = f(x, y)$ subject to the constraint $g(x, y) = 0$ will be among those points (x_0, y_0) for which (x_0, y_0, λ_0) is a solution to the system:

$$F_x(x, y, \lambda) = 0$$
$$F_y(x, y, \lambda) = 0$$
$$F_\lambda(x, y, \lambda) = 0$$

where $F(x, y, \lambda) = f(x, y) + \lambda g(x, y)$, provided all the partial derivatives exist.

2. METHOD OF LAGRANGE MULTIPLIERS FOR FUNCTIONS OF TWO INDEPENDENT VARIABLES

Step 1. Formulate the problem in the form:
Maximize (or Minimize) $z = f(x, y)$
Subject to: $g(x, y) = 0$

Step 2. Form the function F:
$$F(x, y, \lambda) = f(x, y) + \lambda g(x, y)$$

Step 3. Find the critical points (x_0, y_0, λ_0) for F, that is, solve the system:

$$F_x(x, y, \lambda) = 0$$
$$F_y(x, y, \lambda) = 0$$
$$F_\lambda(x, y, \lambda) = 0$$

Step 4. If (x_0, y_0, λ_0) is the only critical point of F, then assume that (x_0, y_0) is the solution to the problem. If F has more than one critical point, then evaluate $z = f(x, y)$ at (x_0, y_0) for each critical point (x_0, y_0, λ_0) of F. Assume that the largest of these values is the maximum value of $f(x, y)$ subject to the constraint $g(x, y) = 0$, and the smallest is the minimum value of $f(x, y)$ subject to the constraint $g(x, y) = 0$.

3. METHOD OF LAGRANGE MULTIPLIERS FOR FUNCTIONS OF THREE VARIABLES

Any local maxima or minima of the function $w = f(x, y, z)$ subject to the constraint $g(x, y, z) = 0$ will be among the set of points (x_0, y_0, z_0) for which $(x_0, y_0, z_0, \lambda_0)$ is a solution to the system

$$F_x(x,y,z,\lambda)=0$$
$$F_y(x,y,z,\lambda)=0$$
$$F_z(x,y,z,\lambda)=0$$
$$F_\lambda(x,y,z,\lambda)=0$$

where $F(x, y, z, \lambda) = f(x, y, z) + \lambda g(x, y, z)$, provided that all the partial derivatives exist.

---

1. Minimize $f(x,y)=x^2+xy+y^2$, subject to: $y=4$.

$f(x,4)=x^2+4x+16,\quad f'(x,4)=2x+4,\quad f''(x,4)=2$

$f'(x,4)=0:\ 2x+4=0,\quad x=-2;\quad f''(-2,4)=2>0.$

Therefore, f has a minimum at $(-2, 4)$; min $f(x,y)=f(-2,4)=4-8+16=12.$

3. Minimize $f(x,y)=4xy$ subject to: $x-y=2$ or $y=x-2$

$f(x,x-2)=4x(x-2)=4x^2-8x,\quad f'(x,x-2)=8x-8,\quad f''(x,x-2)=8$

$f'(x,x-2)=0:\ 8x-8=0,\quad x=1;\quad f''(1,-1)=8>0.$

Therefore, f has a minimum at $(1, -1)$; min $f(x,y)=f(1,-1)=4(1)(-1)=-4.$

5. Maximize $f(x,y)=2x+y$, subject to: $x^2+y=1$ or $y=1-x^2$.

$f(x,1-x^2)=F(x)=2x+1-x^2=-x^2+2x+1,\quad F'(x)=-2x+2,\quad F''(x)=-2.$

$F'(x)=0:\ -2x+2=0,\quad x=1,\quad y=0;\quad F''(1)=-2<0.$

Therefore, f has a maximum at $(1, 0)$; max $f(x,y)=f(1,0)=2.$

7. Step 1. Maximize $f(x, y) = 2xy$
Subject to: $g(x, y) = x + y - 6 = 0$

Step 2. $F(x, y, \lambda) = f(x, y) + \lambda g(x, y)$
$= 2xy + \lambda (x + y - 6)$

Step 3. $F_x = 2y + \lambda = 0$ (1)
$F_y = 2x + \lambda = 0$ (2)
$F_\lambda = x + y - 6 = 0$ (3)

From (1) and (2), we obtain:

$x=-\frac{\lambda}{2},\ y=-\frac{\lambda}{2}$

Substituting these into (3), we have:

$-\frac{\lambda}{2}-\frac{\lambda}{2}-6=0$

$\lambda=-6.$

Thus, the critical point is $(3, 3, -6)$.

Step 4. Since (3, 3, –6) is the only critical point for F, we conclude that max $f(x, y) = f(3, 3) = 2\cdot3\cdot3 = 18$.

9. Step 1. Minimize $f(x, y) = x^2 + y^2$
Subject to: $g(x, y) = 3x + 4y - 25 = 0$

Step 2. $F(x, y, \lambda) = f(x, y) + \lambda g(x, y) = x^2 + y^2 + \lambda(3x + 4y - 25)$

Step 3.
$F_x = 2x + 3\lambda = 0$ (1)
$F_y = 2y + 4\lambda = 0$ (2)
$F_\lambda = 3x + 4y - 25 = 0$ (3)

From (1) and (2), we obtain:

$$x = -\frac{3\lambda}{2},\ y = -2\lambda$$

Substituting these into (3), we have:

$$3\left(-\frac{3\lambda}{2}\right) + 4(-2\lambda) - 25 = 0$$

$$\frac{25}{2}\lambda = -25$$

$$\lambda = -2$$

The critical point is (3, 4, –2).

Step 4. Since (3, 4, –2) is the only critical point for F, we conclude that min $f(x, y) = f(3, 4) = 3^2 + 4^2 = 25$.

11. Step 1. Maximize $f(x, y) = 4y - 3x$ subject to $2x + 5y - 3 = 0$

Step 2. $F(x, y, \lambda) = f(x, y) + \lambda g(x, y) = 4y - 3x + \lambda(2x + 5y - 3)$

Step 3.
$F_x = -3 + 2\lambda = 0$ (1)
$F_y = 4 + 5\lambda = 0$ (2)
$F_\lambda = 2x + 5y - 3 = 0$ (3)

From (1), $\lambda = \frac{3}{2}$, from (2), $\lambda = -\frac{4}{5}$. Thus, the system (1), (2), (3) does not have a solution; by Theorem 1 there are no maxima or minima.

13. Step 1. Maximize and minimize $f(x, y) = 2xy$
Subject to: $g(x, y) = x^2 + y^2 - 18 = 0$

Step 2. $F(x, y, \lambda) = f(x, y) + \lambda g(x, y) = 2xy + \lambda(x^2 + y^2 - 18)$

Step 3.
$F_x = 2y + 2\lambda x = 0$ (1)
$F_y = 2x + 2\lambda y = 0$ (2)
$F_\lambda = x^2 + y^2 - 18 = 0$ (3)

From (1), (2), and (3), we obtain the critical points
(3, 3, –1), (3, –3, 1), (–3, 3, 1) and (–3, –3, –1).

Step 4. $f(3, 3) = 2 \cdot 3 \cdot 3 = 18$
$f(3, -3) = 2 \cdot 3(-3) = -18$
$f(-3, 3) = 2(-3) \cdot 3 = -18$
$f(-3, -3) = 2(-3)(-3) = 18$
Thus, $\max f(x, y) = f(3, 3) = f(-3, -3) = 18$; $\min f(x, y) = f(3, -3) = f(-3, 3) = -18$.

15. Let x and y be the required numbers.

Step 1. Maximize $f(x, y) = xy$
Subject to: $x + y = 10$ or $g(x, y) = x + y - 10 = 0$

Step 2. $F(x, y, \lambda) = xy + \lambda(x + y - 10)$

Step 3. $F_x = y + \lambda = 0$ (1)
$F_y = x + \lambda = 0$ (2)
$F_\lambda = x + y - 10 = 0$ (3)
From (1) and (2), we obtain:
$x = -\lambda, y = -\lambda$
Substituting these into (3), we have:
$\lambda = -5$
The critical point is $(5, 5, -5)$.

Step 4. Since $(5, 5, -5)$ is the only critical point for F, we conclude that $\max f(x, y) = f(5, 5) = 5 \cdot 5 = 25$. Thus, the maximum product is 25 when $x = 5$ and $y = 5$.

17. Step 1. Minimize $f(x, y, z) = x^2 + y^2 + z^2$
Subject to: $g(x, y) = 2x - y + 3z + 28 = 0$

Step 2. $F(x, y, z, \lambda) = x^2 + y^2 + z^2 + \lambda(2x - y + 3z + 28)$

Step 3. $F_x = 2x + 2\lambda = 0$ (1)
$F_y = 2y - \lambda = 0$ (2)
$F_z = 2z + 3\lambda = 0$ (3)
$F_\lambda = 2x - y + 3z + 28 = 0$ (4)
From (1), (2), and (3), we obtain:
$x = -\lambda, y = \frac{\lambda}{2}, z = -\frac{3}{2}\lambda$
Substituting these into (4), we have:

$$2(-\lambda) - \frac{\lambda}{2} + 3\left(-\frac{3}{2}\lambda\right) + 28 = 0$$

$$-\frac{14}{2}\lambda + 28 = 0$$

$$\lambda = 4$$

The critical point is $(-4, 2, -6, 4)$.

Step 4. Since $(-4, 2, -6, 4)$ is the only critical point for F, we conclude that $\min f(x, y, z) = f(-4, 2, -6) = 56$.

19. Step 1. Maximize and minimize $f(x, y, z) = x + y + z$
Subject to: $g(x, y, z) = x^2 + y^2 + z^2 - 12 = 0$

Step 2. $F(x, y, z, \lambda) = f(x, y, z) + \lambda g(x, y, z) = x + y + z + \lambda(x^2 + y^2 + z^2 - 12)$

Step 3.
$F_x = 1 + 2x\lambda = 0$ (1)
$F_y = 1 + 2y\lambda = 0$ (2)
$F_z = 1 + 2z\lambda = 0$ (3)
$F_\lambda = x^2 + y^2 + z^2 - 12 = 0$ (4)

From (1), (2), and (3), we obtain:

$$x = -\frac{1}{2\lambda},\quad y = -\frac{1}{2\lambda},\quad z = -\frac{1}{2\lambda}$$

Substituting these into (4), we have:

$$\left(-\frac{1}{2\lambda}\right)^2 + \left(-\frac{1}{2\lambda}\right)^2 + \left(-\frac{1}{2\lambda}\right)^2 - 12 = 0$$

$$\frac{3}{4\lambda^2} - 12 = 0$$

$$1 - 16\lambda^2 = 0$$

$$\lambda = \pm\frac{1}{4}$$

Thus, the critical points are $\left(2, 2, 2, -\frac{1}{4}\right)$ and $\left(-2, -2, -2, \frac{1}{4}\right)$.

Step 4. $f(2, 2, 2) = 2 + 2 + 2 = 6$
$f(-2, -2, -2) = -2 - 2 - 2 = -6$
Thus, max $f(x, y, z) = f(2, 2, 2) = 6$; min $f(x, y, z) = f(-2, -2, -2) = -6$.

21. Step 1. Maximize $f(x, y) = y + xy^2$
Subject to: $x + y^2 = 1$ or $g(x, y) = x + y^2 - 1 = 0$

Step 2. $F(x, y, \lambda) = y + xy^2 + \lambda(x + y^2 - 1)$

Step 3.
$F_x = y^2 + \lambda = 0$ (1)
$F_y = 1 + 2xy + 2y\lambda = 0$ (2)
$F_\lambda = x + y^2 - 1 = 0$ (3)

From (1), $\lambda = -y^2$ and from (3), $x = 1 - y^2$. Substituting these values into (2), we have

$$1 + 2(1 - y^2)y - 2y^3 = 0$$

or

$$4y^3 - 2y - 1 = 0$$

Using a graphing utility to solve this equation, we get $y \approx 0.885$. Then $x \approx 0.217$ and max $f(x, y) = f(0.217, 0.885) \approx 1.055$.

23. Step 1. Maximize $f(x, y) = e^x + 3e^y$ subject to $g(x, y) = x - 2y - 6 = 0$

Step 2. $F(x, y, \lambda) = f(x, y) + \lambda g(x, y) = e^x + 3e^y + \lambda(x - 2y - 6)$

Step 3.
$$F_x = e^x + \lambda = 0 \quad (1)$$
$$F_y = 3e^y - 2\lambda = 0 \quad (2)$$
$$F_\lambda = x - 2y - 6 = 0 \quad (3)$$

From (1), $\lambda = -e^x$, which implies λ is negative.

From (2), $\lambda = \frac{3}{2}e^y$ which implies λ is positive.

Thus, (1) and (2) have no simultaneous solution.

25. The constraint $g(x, y) = y - 5 = 0$ implies $y = 5$. Replacing y by 5 in the function f, the problem reduces to maximizing the function $h(x) = f(x, 5)$, a function of one independent variable.

27. Maximize $f(x, y) = e^{-(x^2+y^2)}$

Subject to: $g(x, y) = x^2 + y - 1 = 0$

(A) $x^2 + y - 1 = 0$; $y = 1 - x^2$

Substituting $y = 1 - x^2$ into $f(x, y)$, we get

$$h(x) = f(x, 1 - x^2) = e^{-(x^2 + [1 - x^2]^2)} = e^{-(x^4 - x^2 + 1)}$$

Now, $h'(x) = e^{-(x^4 - x^2 + 1)}(-4x^3 + 2x)$.

Critical numbers: $(2x - 4x^3)e^{-(x^4 - x^2 + 1)} = 0$

$$2x(1 - 2x^2) = 0$$
$$x = 0, \frac{\sqrt{2}}{2}, -\frac{\sqrt{2}}{2}$$

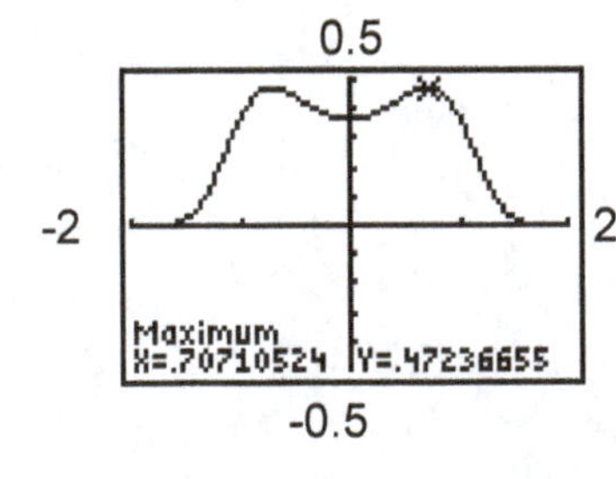

From the constraint equation, $y = \frac{1}{2}$ when $x = \pm\frac{\sqrt{2}}{2}$, and $y = 1$ when $x = 0$.

$f(0,1) = e^{-1} \approx 0.368$ and $f\left(\pm\frac{\sqrt{2}}{2}, \frac{1}{2}\right) \approx 0.472$

Thus, max $f(x, y) = f\left(-\frac{\sqrt{2}}{2}, \frac{1}{2}\right) = f\left(\frac{\sqrt{2}}{2}, \frac{1}{2}\right) \approx 0.472.$

(B) $F(x, y, \lambda) = e^{-(x^2+y^2)} + \lambda(x^2 + y - 1)$

$$F_x = -2xe^{-(x^2+y^2)} + 2x\lambda = 0 \quad (1)$$
$$F_y = -2ye^{-(x^2+y^2)} + \lambda = 0 \quad (2)$$
$$F_\lambda = x^2 + y - 1 = 0 \quad (3)$$

From (3), $y = 1 - x^2$ and from (2)

$\lambda = 2ye^{-(x^2+y^2)} = 2(1 - x^2)e^{-(x^4 - x^2 + 1)}$

Substituting these values into (1), we have

$$-2xe^{-(x^4-x^2+1)} + 2x\,[2(1-x^2)\,e^{-(x^4-x^2+1)}] = 0$$

$$2x\,[2 - 2x^2 - 1] = 0$$

$$2x(1 - 2x^2) = 0$$

$$x = 0, \frac{\sqrt{2}}{2}, -\frac{\sqrt{2}}{2}$$

Now, $y = 1$ when $x = 0$, and $y = \frac{1}{2}$ when $x = \pm\frac{\sqrt{2}}{2}$.

$f(0, 1) = e^{-1} \approx 0.368$

$f\left(\frac{\sqrt{2}}{2}, \frac{1}{2}\right) = f\left(-\frac{\sqrt{2}}{2}, \frac{1}{2}\right) \approx 0.472$

Thus, $\max f(x, y) \approx 0.472$.

29. Step 1. Minimize cost function $C(x, y) = 6x^2 + 12y^2$
Subject to: $x + y = 90$ or $g(x, y) = x + y - 90 = 0$

Step 2. $F(x, y, \lambda) = 6x^2 + 12y^2 + \lambda\,(x + y - 90)$

Step 3. $F_x = 12x + \lambda = 0$ (1)

$F_y = 24y + \lambda = 0$ (2)

$F_\lambda = x + y - 90 = 0$ (3)

From (1) and (2), we obtain

$x = -\frac{\lambda}{12},\ y = -\frac{\lambda}{24}$

Substituting these into (3), we have:

$$-\frac{\lambda}{12} - \frac{\lambda}{24} - 90 = 0$$

$$\frac{3\lambda}{24} = -90$$

$$\lambda = -720$$

The critical point is (60, 30, –720).

Step 4. Since (60, 30, –720) is the only critical point for F, we conclude that:
$\min C(x, y) = C(60, 30) = 6\cdot60^2 + 12\cdot30^2 = 21{,}600 + 10{,}800 = \$32{,}400$
Thus, 60 of model A and 30 of model B will yield a minimum cost of \$32,400 per week.

31. (A) Step 1. Maximize the production function $N(x, y) = 50x^{0.8}y^{0.2}$
Subject to the constraint: $C(x,y) = 40x + 80y = 400{,}000$
i.e., $g(x, y) = 40x + 80y - 400{,}000 = 0$

Step 2. $F(x, y, \lambda) = 50x^{0.8}y^{0.2} + \lambda\,(40x + 80y - 400{,}000)$

Step 3. $F_x = 40x^{-0.2}y^{0.2} + 40\lambda = 0$ (1)

$F_y = 10x^{0.8}y^{-0.8} + 80\lambda = 0$ (2)

$F_\lambda = 40x + 80y - 400{,}000 = 0$ (3)

From (1), $\lambda = -\frac{y^{0.2}}{x^{0.2}}$. From (2), $\lambda = -\frac{x^{0.8}}{8y^{0.8}}$.

Thus, we obtain

$$-\frac{y^{0.2}}{x^{0.2}} = -\frac{x^{0.8}}{8y^{0.8}} \quad \text{or} \quad x = 8y$$

Substituting into (3), we have:

$320y + 80y - 400{,}000 = 0$

$y = 1000$

Therefore, $x = 8000$, $\lambda \approx -0.6598$, and the critical point is (8000, 1000, –0.6598).

Thus, we conclude that:

$\max N(x, y) = N(8000, 1000) = 50(8000)^{0.8}(1000)^{0.2} \approx 263{,}902$ units

and production is maximized when 8000 labor units and 1000 capital units are used.

(B) The marginal productivity of money is $-\lambda \approx 0.6598$. The increase in production if an additional \$50,000 is budgeted for production is: 0.6598(50,000) = 32,990 units

33. Let x = length, y = width, and z = height.

Step 1. Maximize volume $V = xyz$

Subject to: $S(x, y, z) = xy + 3xz + 3yz - 192 = 0$

Step 2. $F(x, y, z, \lambda) = xyz + \lambda(xy + 3xz + 3yz - 192)$

Step 3. $F_x = yz + \lambda(y + 3z) = 0$ (1)

$F_y = xz + \lambda(x + 3z) = 0$ (2)

$F_z = xy + \lambda(3x + 3y) = 0$ (3)

$F_\lambda = xy + 3xz + 3yz - 192 = 0$ (4)

Solving this system of equations, (1)–(4), simultaneously, yields:

$x = 8,\ y = 8,\ z = \frac{8}{3},\ \lambda = -\frac{4}{3}$

Thus, the critical point is $\left(8, 8, \frac{8}{3}, -\frac{4}{3}\right)$.

Step 4. Since $\left(8, 8, \frac{8}{3}, -\frac{4}{3}\right)$ is the only critical point for F:

$\max V(x, y, z) = V\left(8, 8, \frac{8}{3}\right) = \frac{512}{3} \approx 170.67$

Thus, the dimensions that will maximize the volume of the box are: length $x = 8$ inches; width $y = 8$ inches; height $z = \frac{8}{3}$ inches.

35. Step 1. Maximize $A = xy$

Subject to: $P(x, y) = y + 4x - 400 = 0$

Step 2. $F(x, y, \lambda) = xy + \lambda(y + 4x - 400)$

Step 3. $F_x = y + 4\lambda = 0$ (1)

$F_y = x + \lambda = 0$ (2)

$F_\lambda = y + 4x - 400 = 0$ (3)

From (1) and (2), we have:

$y = -4\lambda$ and $x = -\lambda$

Substituting these into (3), we obtain:

$-4\lambda - 4\lambda - 400 = 0$

Thus, $\lambda = -50$ and the critical point is (50, 200, –50).

Step 4. Since (50, 200, -50) is the only critical point for F, max $A(x, y) = A(50, 200) = 10{,}000$.

Therefore, $x = 50$ feet, $y = 200$ feet will produce the maximum area $A(50, 200) = 10{,}000$ square feet.

EXERCISE 7-5

Things to remember:

1. LEAST SQUARES APPROXIMATION FORMULAS

For a set of n points $(x_1, y_1), (x_2, y_2), \ldots, (x_n, y_n)$, the coefficients of the least squares line $y = ax + b$ are the solutions of the system of NORMAL EQUATIONS

$$\left(\sum_{k=1}^{n} x_k^2\right)a + \left(\sum_{k=1}^{n} x_k\right)b = \sum_{k=1}^{n} x_k y_k \quad (1)$$

$$\left(\sum_{k=1}^{n} x_k\right)a + nb = \sum_{k=1}^{n} y_k$$

and are given by the formulas

$$a = \frac{n\left(\sum_{k=1}^{n} x_k y_k\right) - \left(\sum_{k=1}^{n} x_k\right)\left(\sum_{k=1}^{n} y_k\right)}{n\left(\sum_{k=1}^{n} x_k^2\right) - \left(\sum_{k=1}^{n} x_k\right)^2} \quad (2)$$

$$b = \frac{\sum_{k=1}^{n} y_k - a\left(\sum_{k=1}^{n} x_k\right)}{n} \quad (3)$$

[Note: To find a and b, either solve system (1) directly, or use formulas (2) and (3). If the formulas are used, the value of a must be calculated first since it is used in the formula for b.

1. $\sum_{k=1}^{5} x_k = 0+1+2+3+4=10$

3. $\sum_{k=1}^{5} x_k\, y_k = 0(4)+1(5)+2(7)+3(9)+4(13)=98$

5. $\sum_{k=1}^{5} x_k \sum_{k=1}^{5} y_k = 10(4+5+7+9+13)=380$

7.

	x_k	y_k	$x_k y_k$	x_k^2
	1	1	1	1
	2	3	6	4
	3	4	12	9
	4	3	12	16
Totals	10	11	31	30

Thus, $\sum_{k=1}^{4} x_k = 10,\ \sum_{k=1}^{4} y_k = 11,\ \sum_{k=1}^{4} x_k y_k = 31,\ \sum_{k=1}^{4} x_k^2 = 30.$

Substituting these values into formulas (2) and (3) for a and b, respectively, we have:

$$a = \frac{n\left(\sum_{k=1}^{n} x_k y_k\right) - \left(\sum_{k=1}^{n} x_k\right)\left(\sum_{k=1}^{n} y_k\right)}{n\left(\sum_{k=1}^{n} x_k^2\right) - \left(\sum_{k=1}^{n} x_k\right)^2} = \frac{4(31)-(10)(11)}{4(30)-(10)^2} = \frac{14}{20} = 0.7$$

$$b = \frac{\sum_{k=1}^{n} y_k - a\left(\sum_{k=1}^{n} x_k\right)}{n} = \frac{11-0.7(10)}{4} = 1$$

Thus, the least squares line is $y = ax + b = 0.7x + 1$. Refer to the graph at the right.

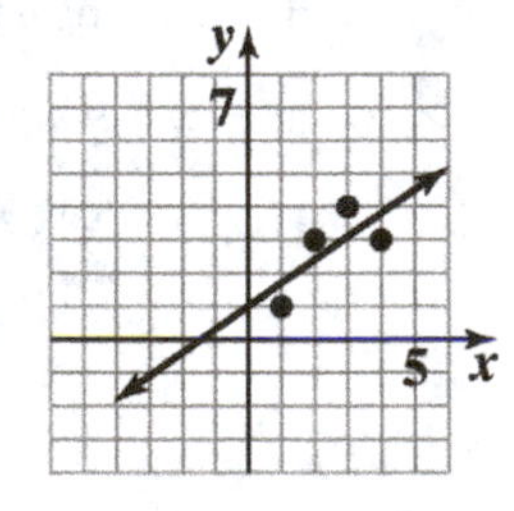

9.

	x_k	y_k	$x_k y_k$	x_k^2
	1	8	8	1
	2	5	10	4
	3	4	12	9
	4	0	0	16
Totals	10	17	30	30

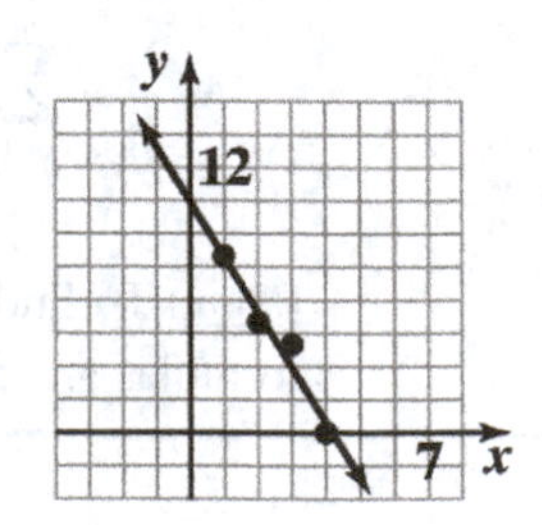

Thus, $\sum_{k=1}^{4} x_k = 10,\ \sum_{k=1}^{4} y_k = 17,\ \sum_{k=1}^{4} x_k y_k = 30,$

$\sum_{k=1}^{4} x_k^2 = 30.$

Substituting these values into system (1), we have:

$30a + 10b = 30$
$10a + \ \ 4b = 17$

The solution of this system is $a = -2.5$, $b = 10.5$. Thus, the least squares line is $y = ax + b = -2.5x + 10.5$. Refer to the graph on the previous page.

11.

	x_k	y_k	$x_k y_k$	x_k^2
	1	3	3	1
	2	4	8	4
	3	5	15	9
	4	6	24	16
Totals	10	18	50	30

Thus, $\sum_{k=1}^{4} x_k = 10$, $\sum_{k=1}^{4} y_k = 18$, $\sum_{k=1}^{4} x_k y_k = 50$, $\sum_{k=1}^{4} x_k^2 = 30$.

Substituting these values into the formulas for a and b [formulas (2) and (3)], we have:

$$a = \frac{4(50) - (10)(18)}{4(30) - (10)^2} = \frac{20}{20} = 1$$

$$b = \frac{18 - 1(10)}{4} = \frac{8}{4} = 2$$

Thus, the least squares line is $y = ax + b = x + 2$. Refer to the graph at the right.

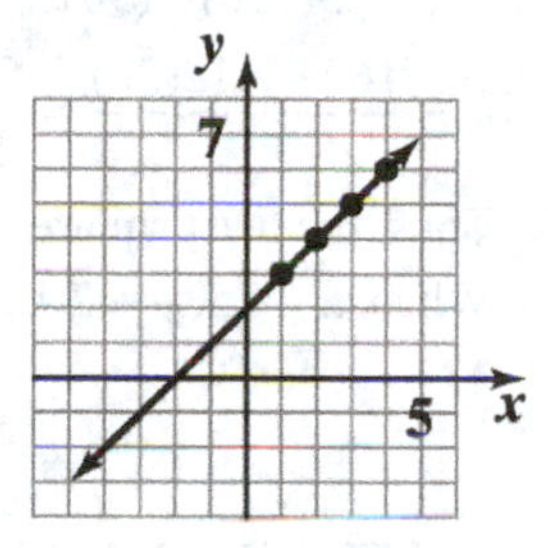

13.

	x_k	y_k	$x_k y_k$	x_k^2
	1	3	3	1
	2	1	2	4
	2	2	4	4
	3	0	0	9
Totals	8	6	9	18

Thus, $\sum_{k=1}^{4} x_k = 8$, $\sum_{k=1}^{4} y_k = 6$, $\sum_{k=1}^{4} x_k y_k = 9$, $\sum_{k=1}^{4} x_k^2 = 18$.

Substituting these values into formulas (2) and (3) for a and b, respectively, we have:

$$a = \frac{4(9) - 8(6)}{4(18) - 8^2} = \frac{36 - 48}{72 - 64} = \frac{-12}{8} = -\frac{3}{2} = -1.5$$

$$b = \frac{6 - (-3/2)(8)}{4} = \frac{6 + 12}{4} = \frac{9}{2} = 4.5$$

Thus, the least squares line is $y = -1.5x + 4.5$.

When $x = 2.5$, $y = -1.5(2.5) + 4.5 = 0.75$.

15.

	x_k	y_k	$x_k y_k$	x_k^2
	0	10	0	0
	5	22	110	25
	10	31	310	100
	15	46	690	225
	20	51	1020	400
Totals	50	160	2130	750

Thus, $\sum_{k=1}^{5} x_k = 50$, $\sum_{k=1}^{5} y_k = 160$, $\sum_{k=1}^{5} x_k y_k = 2130$, $\sum_{k=1}^{5} x_k^2 = 750$.

Substituting these values into formulas (2) and (3) for a and b, respectively, we have:

$$a = \frac{5(2130) - (50)(160)}{5(750) - (50)^2} = \frac{2650}{1250} = 2.12$$

$$b = \frac{160 - 2.12(50)}{5} = \frac{54}{5} = 10.8$$

Thus, the least squares line is $y = 2.12x + 10.8$.
When $x = 25$, $y = 2.12(25) + 10.8 = 63.8$.

17.

	x_k	y_k	$x_k y_k$	x_k^2
	−1	14	−14	1
	1	12	12	1
	3	8	24	9
	5	6	30	25
	7	5	35	49
Totals	15	45	87	85

Thus, $\sum_{k=1}^{5} x_k = 15$, $\sum_{k=1}^{5} y_k = 45$, $\sum_{k=1}^{5} x_k y_k = 87$, $\sum_{k=1}^{5} x_k^2 = 85$.

Substituting these values into formulas (2) and (3) for a and b, respectively, we have:

$$a = \frac{5(87) - (15)(45)}{5(85) - (15)^2} = \frac{-240}{200} = -1.2$$

$$b = \frac{45 - (-1.2)(15)}{5} = 12.6$$

Thus, the least squares line is
$y = -1.2x + 12.6$.
When $x = 2$, $y = -1.2(2) + 12.6 = 10.2$.

19.

	x_k	y_k	$x_k y_k$	x_k^2
	0.5	25	12.5	0.25
	2.0	22	44.0	4.00
	3.5	21	73.5	12.25
	5.0	21	105.0	25.00

	6.5	18	117.0	42.25
	9.5	12	114.0	90.25
	11.0	11	121.0	121.00
	12.5	8	100.0	156.25
	14.0	5	70.0	196.00
	15.5	1	15.5	240.25
Totals	80.0	144	772.5	887.50

Thus, $\sum_{k=1}^{10} x_k = 80$, $\sum_{k=1}^{10} y_k = 144$, $\sum_{k=1}^{10} x_k y_k = 772.5$, $\sum_{k=1}^{10} x_k^2 = 887.5$.

Substituting these values into formulas (2) and (3) for a and b, respectively, we have:

$$a = \frac{10(772.5) - (80)(144)}{10(887.5) - (80)^2} = \frac{-3795}{2475} \approx -1.53$$

$$b = \frac{144 - (-1.53)(80)}{10} = \frac{266.4}{10} = 26.64$$

Thus, the least squares line is

$y = -1.53x + 26.64$.

When $x = 8$, $y = -1.53(8) + 26.64 = 14.4$.

21. Minimize

$$F(a, b, c) = (a + b + c - 2)^2 + (4a + 2b + c - 1)^2 + (9a + 3b + c - 1)^2 + (16a + 4b + c - 3)^2$$

$$F_a(a, b, c) = 2(a + b + c - 2) + 8(4a + 2b + c - 1) + 18(9a+3b+c-1) + 32(16a+4b+c-3)$$
$$= 708a + 200b + 60c - 126$$

$$F_b(a, b, c) = 2(a + b + c - 2) + 4(4a + 2b + c - 1) + 6(9a + 3b + c - 1) + 8(16a + 4b + c - 3)$$
$$= 200a + 60b + 20c - 38$$

$$F_c(a, b, c) = 2(a + b + c - 2) + 2(4a + 2b + c - 1) + 2(9a + 3b + c - 1) + 2(16a + 4b + c - 3)$$
$$= 60a + 20b + 8c - 14$$

The system is:

$$F_a(a, b, c) = 0$$
$$F_b(a, b, c) = 0$$
$$F_c(a, b, c) = 0$$

or:

$$708a + 200b + 60c = 126$$
$$200a + 60b + 20c = 38$$
$$60a + 20b + 8c = 14$$

The solution is $(a, b, c) = (0.75, -3.45, 4.75)$, which gives us the equation for the parabola shown at the right:

$y = ax^2 + bx + c$

or

$y = 0.75x^2 - 3.45x + 4.75$

The given points: (1, 2), (2, 1), (3, 1), (4, 3) also shown.

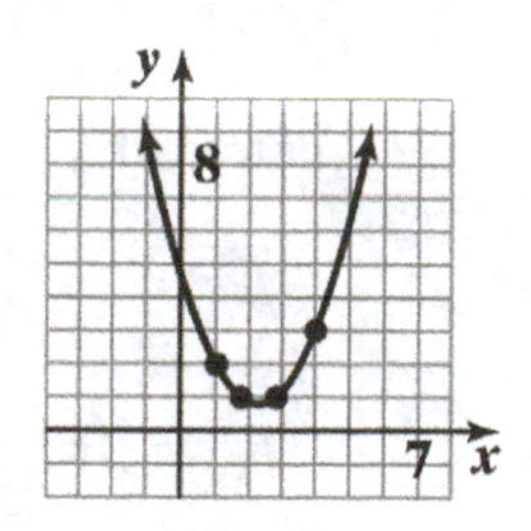

23. System (1) is:

$$\left(\sum_{k=1}^{n} x_k\right)a + nb = \sum_{k=1}^{n} y_k \qquad \text{(a)}$$

$$\left(\sum_{k=1}^{n} x_k^2\right)a + \left(\sum_{k=1}^{n} x_k\right)b = \sum_{k=1}^{n} x_k y_k \qquad \text{(b)}$$

Multiply equation (a) by $-\left(\sum_{k=1}^{n} x_k\right)$, equation (b) by n, and add the resulting equations. This will eliminate b from the system.

$$\left[-\left(\sum_{k=1}^{n} x_k\right)^2 + n\sum_{k=1}^{n} x_k^2\right]a = -\left(\sum_{k=1}^{n} x_k\right)\left(\sum_{k=1}^{n} y_k\right) + n\sum_{k=1}^{n} x_k y_k$$

Thus, $$a = \frac{n\left(\sum_{k=1}^{n} x_k y_k\right) - \left(\sum_{k=1}^{n} x_k\right)\left(\sum_{k=1}^{n} y_k\right)}{n\left(\sum_{k=1}^{n} x_k^2\right) - \left(\sum_{k=1}^{n} x_k\right)^2}$$

which is equation (2). Solving equation (a) for b, we have

$$b = \frac{\sum_{k=1}^{n} y_k - m\left(\sum_{k=1}^{n} x_k\right)}{n},$$ which is equation (3).

25. (A) Suppose that $n = 5$ and $x_1 = -2, x_2 = -1, x_3 = 0, x_4 = 1, x_5 = 2$. Then

$\sum_{k=1}^{5} x_k = -2 - 1 + 0 + 1 + 2 = 0$. Therefore, from formula (2),

$$a = \frac{5\sum_{k=1}^{5} x_k y_k}{5\sum_{k=1}^{5} x_k^2} = \frac{\sum x_k y_k}{\sum x_k^2}$$ From formula (3), $b = \frac{\sum_{k=1}^{5} y_k}{5}$,

which is the average of y_1, y_2, y_3, y_4, and y_5.

(B) If the average of the x-coordinates is 0, then

$$\frac{\sum_{k=1}^{n} x_k}{n} = 0, \quad \text{i.e.,} \quad \sum_{k=1}^{n} x_k = 0.$$

Then all calculations will be the same as in part (A) with "n" instead of 5.

27. (A)

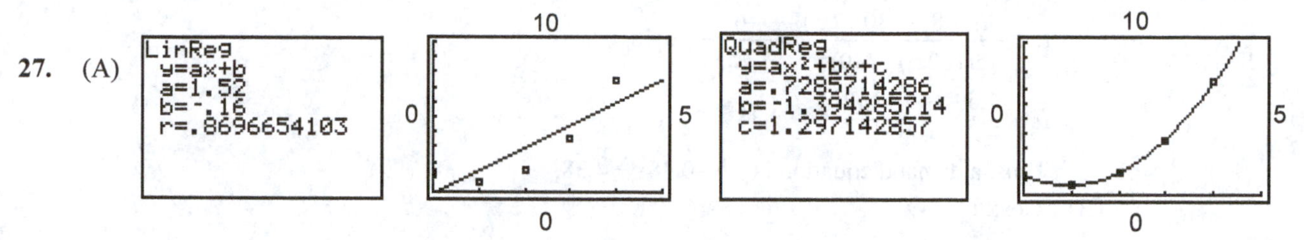

(B) The quadratic regression function best fits the data.

29. The cubic regression function has the form $y = ax^3 + bx^2 + cx + d$. The normal equations form a system of 4 linear equations in the 4 variables *a, b, c* and *d*. The system can be solved using Gauss-Jordan elimination.

31. (A)

	x_k	y_k	$x_k y_k$	x_k^2
	1	3,658	3,658	1
	3	3,591	10,773	9
	5	3,431	17,155	25
	7	3,276	22,932	49
	9	3,041	27,369	81
	11	2,908	31,988	121
Totals	36	19,905	113,875	286

Thus, $\sum_{k=1}^{6} x_k = 36$, $\sum_{k=1}^{6} y_k = 19{,}905$, $\sum_{k=1}^{6} x_k y_k = 113{,}875$, $\sum_{k=1}^{6} x_k^2 = 286$.

Substituting these values in the formulas for a and b, we have:

$$a = \frac{6(113{,}875) - (36)(19{,}905)}{6(286) - (36)^2} \approx -79.36 \text{ and } b = \frac{19{,}905 - (-79.36)(36)}{6} \approx 3{,}793.6$$

Thus, the least squares line is $y = -79.36x + 3{,}793.6$

(B) The property crime rate in 2024 will be (approximately) $-79.36(24) + 3{,}793.6 \approx 1{,}889$ crimes per 100,000 population.

33. (A)

	x_k	y_k	$x_k y_k$	x_k^2
	5.0	2.0	10	25
	5.5	1.8	9.9	30.25
	6.0	1.4	8.4	36
	6.5	1.2	7.8	42.25
	7.0	1.1	7.7	49
Totals	30	7.5	43.8	182.5

Thus, $\sum_{k=1}^{5} x_k = 30$, $\sum_{k=1}^{5} y_k = 7.5$, $\sum_{k=1}^{5} x_k y_k = 43.8$, $\sum_{k=1}^{5} x_k^2 = 182.5$.

Substituting these values into the formulas for *a* and *b*, we have:

$$a = \frac{5(43.8) - (30)(7.5)}{5(182.5) - (30)^2} = \frac{-6}{12.5} = -0.48$$

$$b = \frac{7.5 - (-0.48)(30)}{5} = 4.38$$

Thus, a demand equation is $y = -0.48x + 4.38$.

(B) Cost: $C = 4y$

Revenue: $R = xy = -0.48x^2 + 4.38x$

Profit: $P = R - C = -0.48x^2 + 4.38x - 4(-0.48x + 4.38)$

or $P(x) = -0.48x^2 + 6.3x - 17.52$

Now, $P'(x) = -0.96x + 6.3$.

Critical value: $P'(x) = -0.96x + 6.3 = 0$

$$x = \frac{6.3}{0.96} \approx 6.56$$

$P''(x) = -0.96$ and $P''(6.56) = -0.96 < 0$

Thus, $P(x)$ has a maximum at $x = 6.56$; the price per bottle should be \$6.56 to maximize the monthly profit.

35. (A) We use the linear regression feature on a graphing utility. Thus, the least squares line is: $y = 0.0222x + 18.94$.

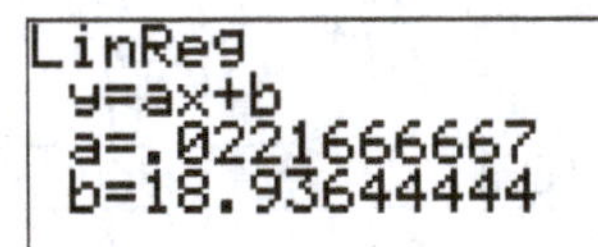

(B) The year 2024 corresponds to $x = 44$; $y(44) = 0.0222(44) + 18.94 \approx 19.92$

An estimate for the winning height in the Olympic Games is 19.92 feet.

37. (A) We use the linear regression feature on a graphing utility with 1885 as $x = 0$, 1895 as $x = 10$, …, etc.

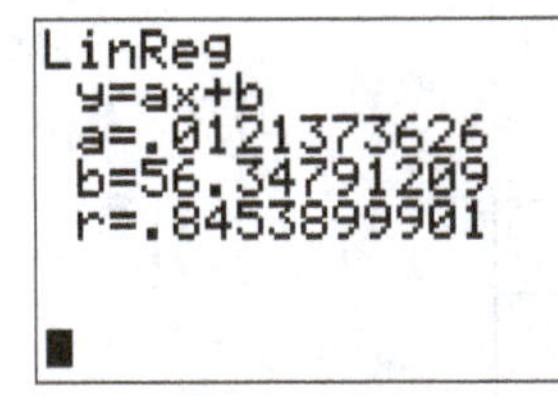

Thus, the least squares line is: $y = 0.00121x + 56.35$

(B) The year 2085 corresponds to $x = 200$;
$y(200) = 0.0121(200) + 56.35 \approx 58.77°$ F.

EXERCISE 7-6

Things to remember:

GIVEN A FUNCTION $z = f(x, y)$:

1. $\int f(x, y)dx$ means antidifferentiate $f(x, y)$ with respect to x, holding y fixed.

   $\int f(x, y)dy$ means antidifferentiate $f(x, y)$ with respect to y, holding x fixed.

2. The DOUBLE INTEGRAL of $f(x, y)$ over the rectangle $R = \{(x, y) \mid a \le x \le b, c \le y \le d\}$ is:

$$\iint_R f(x, y)dA = \int_a^b \left[\int_c^d f(x, y)dy\right]dx$$
$$= \int_c^d \left[\int_a^b f(x, y)dx\right]dy$$

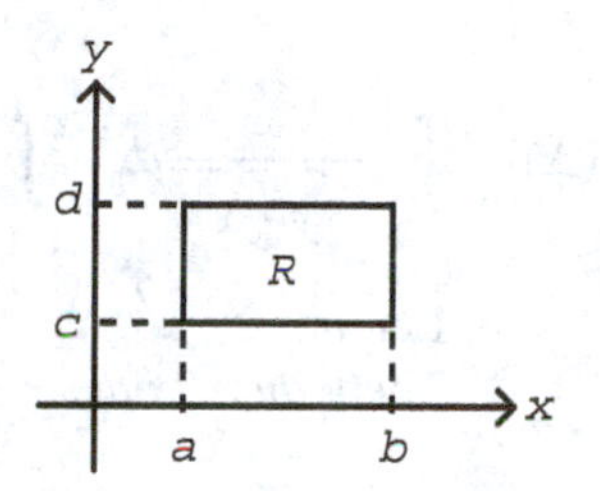

3. The AVERAGE VALUE of $f(x, y)$ over the rectangle $R = \{(x, y) | a \le x \le b, c \le y \le d\}$ is:

$$\frac{1}{(b-a)(d-c)}\iint_R f(x, y)dA$$

4. VOLUME UNDER A SURFACE

   If $f(x, y) \ge 0$ over a rectangle $R = \{(x, y) | a \le x \le b, c \le y \le d\}$, then the volume of the solid formed by graphing f over the rectangle R is given by:

$$V = \iint_R f(x, y)dA$$

---

1. $\int(\pi + x)dx = \pi x + \frac{1}{2}x^2 + C$

3. $\int\left(1 + \frac{\pi}{x}\right)dx = x + \pi \ln x + C$

5. $\int e^{\pi x}dx$ (Let $u = \pi x$. Then $du = \pi dx$.) $\int e^{\pi x}dx = \frac{1}{\pi}\int e^u du = \frac{1}{\pi}e^u + C = \frac{e^{\pi x}}{\pi} + C$

7. (A) $\int 12x^2y^3dy = 12x^2\int y^3dy$ (x is treated as a constant.)

$$= 12x^2\frac{y^4}{4} + C(x)$$ (The "constant" of integration is a function of x.)

$$= 3x^2y^4 + C(x)$$

(B) $\int_0^1 12x^2y^3dy = 3x^2y^4\Big|_0^1 = 3x^2$

9. (A) $\int (4x + 6y + 5)\,dx$

$= \int 4x\,dx + \int (6y + 5)\,dx$ (y is treated as a constant.)

$= 2x^2 + (6y + 5)x + E(y)$ (The "constant" of integration is a function of y.)

$= 2x^2 + 6xy + 5x + E(y)$

(B) $\int_{-2}^{3} (4x + 6y + 5)\,dx = (2x^2 + 6xy + 5x)\Big|_{-2}^{3} = 2\cdot 3^2 + 6\cdot 3y + 5\cdot 3 - [2(-2)^2 + 6(-2)y + 5(-2)]$

$= 30y + 35$

11. (A) $\int \frac{x}{\sqrt{y+x^2}}\,dx = \int (y + x^2)^{-1/2} x\,dx = \frac{1}{2}\int (y + x^2)^{-1/2} 2x\,dx$

Let $u = y + x^2$.
Then $du = 2x\,dx$.

$= \frac{1}{2}\int u^{-1/2}\,du$

$= u^{1/2} + E(y) = \sqrt{y+x^2} + E(y)$

(B) $\int_0^2 \frac{x}{\sqrt{y+x^2}}\,dx = \sqrt{y+x^2}\,\Big|_0^2 = \sqrt{y+4} - \sqrt{y}$

13. (A) $\int \frac{\ln x}{xy}\,dy = \frac{\ln x}{x}\int \frac{1}{y}\,dy = \frac{\ln x}{x}\cdot \ln y + C(x)$

(B) $\int_1^{e^2} \frac{\ln x}{xy}\,dy = \frac{\ln x \ln y}{x}\Big]_1^{e^2} = \frac{\ln x \ln e^2}{x} - \frac{\ln x \ln 1}{x} = \frac{2\ln x}{x}$

15. $\int_{-1}^{2}\int_0^1 12x^2y^3\,dy\,dx = \int_{-1}^{2}\left[\int_0^1 12x^2y^3\,dy\right]dx = \int_{-1}^{2} 3x^2\,dx$ (see Problem 7)

$= x^3\Big|_{-1}^{2} = 8 + 1 = 9$

17. $\int_1^4\int_{-2}^{3} (4x + 6y + 5)\,dx\,dy = \int_1^4\left[\int_{-2}^{3}(4x+6y+5)\,dx\right]dy = \int_1^4 (30y + 35)\,dy$ (see Problem 9)

$= (15y^2 + 35y)\Big|_1^4 = 15\cdot 4^2 + 35\cdot 4 - (15 + 35) = 330$

19. $\int_1^5\int_0^2 \frac{x}{\sqrt{y+x^2}}\,dx\,dy = \int_1^5\left[\int_0^2 \frac{x}{\sqrt{y+x^2}}\,dx\right]dy = \int_1^5 (\sqrt{4+y} - \sqrt{y})\,dy$ (see Problem 11)

$= \left[\frac{2}{3}(4+y)^{3/2} - \frac{2}{3}y^{3/2}\right]_1^5 = \frac{2}{3}(9)^{3/2} - \frac{2}{3}(5)^{3/2} - \left(\frac{2}{3}\cdot 5^{3/2} - \frac{2}{3}\cdot 1^{3/2}\right)$

$= 18 - \frac{4}{3}(5)^{3/2} + \frac{2}{3} = \frac{56 - 20\sqrt{5}}{3}$

21. $\int_1^e \int_1^{e^2} \frac{\ln x}{xy} dydx = \int_1^e \left[\int_1^{e^2} \frac{\ln x}{xy} dy\right] dx = \int_1^e \frac{2\ln x}{x} dx$ (see Problem 13)

$= 2\int_1^e \frac{\ln x}{x} dx = (\ln x)^2 \Big]_1^e = 1$ Substitution: $u = \ln x,\ du = \frac{1}{x} dx$

23. $\iint_R xydA = \int_0^2 \int_0^4 xydydx = \int_0^2 \left[\int_0^4 xy\,dy\right] dx = \int_0^2 \left[\frac{xy^2}{2}\right]_0^4 dx = \int_0^2 8x\,dx = 4x^2 \Big|_0^2 = 16$

$\iint_R xydA = \int_0^4 \int_0^2 xy\,dxdy = \int_0^4 \left[\int_0^2 xy\,dx\right] dy = \int_0^4 \left[\frac{x^2 y}{2}\right]_0^2 dy = \int_0^4 2y\,dy = y^2 \Big|_0^4 = 16$

25. $\iint_R (x+y)^5 dA = \int_{-1}^1 \int_1^2 (x+y)^5 dydx = \int_{-1}^1 \left[\frac{(x+y)^6}{6}\right]_1^2 dx = \int_{-1}^1 \left[\frac{(x+2)^6}{6} - \frac{(x+1)^6}{6}\right] dx$

$= \left[\frac{(x+2)^7}{42} - \frac{(x+1)^7}{42}\right]_{-1}^1 = \frac{3^7}{42} - \frac{2^7}{42} - \frac{1}{42} = 49$

$\iint_R (x+y)^5 dA = \int_1^2 \int_{-1}^1 (x+y)^5 dxdy = \int_1^2 \left[\int_{-1}^1 (x+y)^5 dx\right] dy = \int_1^2 \left[\frac{(x+y)^6}{6}\right]_{-1}^1 dy$

$= \int_1^2 \left[\frac{(y+1)^6}{6} - \frac{(y-1)^6}{6}\right] dy = \left[\frac{(y+1)^7}{42} - \frac{(y-1)^7}{42}\right]_1^2 = \frac{3^7}{42} - \frac{1}{42} - \frac{2^7}{42} = 49$

27. Average value $= \frac{1}{(5-1)[1-(-1)]} \iint_R (x+y)^2 dA = \frac{1}{8} \int_{-1}^1 \int_1^5 (x+y)^2 dxdy = \frac{1}{8} \int_{-1}^1 \left[\frac{(x+y)^3}{3}\right]_1^5 dy$

$= \frac{1}{8} \int_{-1}^1 \left[\frac{(5+y)^3}{3} - \frac{(1+y)^3}{3}\right] dy = \frac{1}{8}\left[\frac{(5+y)^4}{12} - \frac{(1+y)^4}{12}\right]_{-1}^1 = \frac{1}{96}[6^4 - 2^4 - 4^4] = \frac{32}{3}$

29. Average value $= \frac{1}{(4-1)(7-2)} \iint_R \frac{x}{y} dA = \frac{1}{15} \int_1^4 \int_2^7 \frac{x}{y} dy\,dx = \frac{1}{15} \int_1^4 [x \ln y]_2^7 dx$

$= \frac{1}{15} \int_1^4 [x \ln 7 - x \ln 2]\, dx = \frac{\ln 7 - \ln 2}{15} \int_1^4 x\,dx = \left[\frac{\ln 7 - \ln 2}{15} \cdot \frac{x^2}{2}\right]_1^4$

$= \frac{\ln 7 - \ln 2}{15}\left(\frac{4^2}{2} - \frac{1^2}{2}\right) = \frac{1}{2}(\ln 7 - \ln 2) = \frac{1}{2} \ln\left(\frac{7}{2}\right) \approx 0.626$

31. $V = \iint_R (2 - x^2 - y^2) dA = \int_0^1 \int_0^1 \iint_R (2 - x^2 - y^2)\, dy\, dx$

$= \int_0^1 \left[\int_0^1 (2 - x^2 - y^2) dy\right] dx = \int_0^1 \left[2y - x^2 y - \frac{y^3}{3}\right]_0^1 dx$

$= \int_0^1 \left(2 - x^2 - \frac{1}{3}\right) dx = \int_0^1 \left(\frac{5}{3} - x^2\right) dx = \left[\frac{5}{3}x - \frac{x^3}{3}\right]_0^1 = \frac{5}{3} - \frac{1}{3} = \frac{4}{3}$

33. $V = \iint_R (4 - y^2)dA = \int_0^2 \int_0^2 (4 - y^2)dxdy = \int_0^2 \left[\int_0^2 (4 - y^2)dx\right]dy$

$= \int_0^2 \left[(4x - xy^2)\Big|_0^2\right]dy = \int_0^2 (8 - 2y^2)dy = \left[8y - \frac{2}{3}y^3\right]_0^2 = 16 - \frac{16}{3} = \frac{32}{3}$

35. $\iint_R xe^{xy}dA = \int_0^1 \int_1^2 xe^{xy}dydx = \int_0^1 \left[\int_1^2 xe^{xy}dy\right]dx = \int_0^1 \left[x\int_1^2 e^{xy}dy\right]dx = \int_0^1 \left[x \cdot \frac{e^{xy}}{x}\right]_1^2 dx$

$= \int_0^1 (e^{2x} - e^x)\, dx = \left[\frac{e^{2x}}{2} - e^x\right]_0^1 = \frac{e^2}{2} - e - \left(\frac{1}{2} - 1\right) = \frac{e^2}{2} - e + \frac{1}{2}$

37. $\iint_R \frac{2y + 3xy^2}{1 + x^2}dA = \int_0^1 \int_{-1}^1 \frac{2y + 3xy^2}{1 + x^2}dydx = \int_0^1 \left[\int_{-1}^1 \frac{2y + 3xy^2}{1 + x^2}dy\right]dx = \int_0^1 \left[\frac{1}{1 + x^2}(y^2 + xy^3\right]_{-1}^1 dx$

$= \int_0^1 \left[\frac{1}{1 + x^2}(1 + x - [1 - x])\right]dx = \int_0^1 \frac{2x}{1 + x^2}dx = \ln(1 + x^2)\Big|_0^1 = \ln 2$

Substitution: $u = 1 + x^2$, $du = 2x\, dx$

39. $\int_0^2 \int_0^2 (1 - y)dxdy = \int_0^2 \left[\int_0^2 (1 - y)dx\right]dy = \int_0^2 [(x - xy)]_0^2\, dy = \int_0^2 (2 - 2y)\, dy = (2y - y^2)\Big|_0^2 = 0$

Since $f(x, y) = 1 - y$ is NOT positive over the entire rectangle $R = \{(x, y) \mid 0 \le x \le 2, 0 \le y \le 2\}$, the double integral does not represent the volume of a solid.

41. $f(x, y) = x^3 + y^2 - e^{-x} - 1$ on $R = \{(x, y) \mid -2 \le x \le 2, -2 \le y \le 2\}$.

(A) Average value of f:

$\frac{1}{b - a} \cdot \frac{1}{d - c} \iint_R f(x, y)\, dA = \frac{1}{2 - (-2)} \cdot \frac{1}{2 - (-2)} \int_{-2}^2 \int_{-2}^2 (x^3 + y^2 - e^{-x} - 1)\, dx\, dy$

$= \frac{1}{16} \int_{-2}^2 \left[\frac{1}{4}x^4 + xy^2 + e^{-x} - x\right]_{-2}^2 dy = \frac{1}{16} \int_{-2}^2 [4y^2 + e^{-2} - e^2 - 4]\, dy$

$= \frac{1}{16}\left[\frac{4}{3}y^3 + e^{-2}y - e^2y - 4y\right]_{-2}^2 = \frac{1}{16}\left[\frac{64}{3} + 4e^{-2} - 4e^2 - 16\right] = \frac{1}{3} + \frac{1}{4}e^{-2} - \frac{1}{4}e^2$

(B)

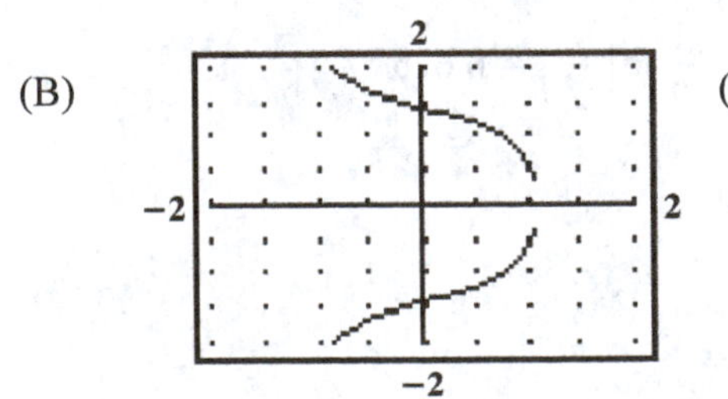

(C) $f(x, y) > 0$ at the points which lie to the right of the curve in part (B);
$f(x, y) < 0$ at the points which lie to the left of the curve in part (B).

43. $S(x, y) = \frac{y}{1-x}$, $0.6 \le x \le 0.8$, $5 \le y \le 7$.

The *average* total amount of spending is given by:

$$T = \frac{1}{(0.8-0.6)(7-5)}\iint_R \frac{y}{1-x}\,dA = \frac{1}{0.4}\int_{0.6}^{0.8}\int_5^7 \frac{y}{1-x}\,dy\,dx = \frac{1}{0.4}\int_{0.6}^{0.8}\left[\frac{1}{1-x}\cdot\frac{y^2}{2}\right]_5^7 dx$$

$$= \frac{1}{0.4}\int_{0.6}^{0.8} \frac{1}{1-x}\left(\frac{49}{2}-\frac{25}{2}\right)dx = \frac{12}{0.4}\int_{0.6}^{0.8} \frac{1}{1-x}\,dx = 30[-\ln(1-x)]\Big|_{0.6}^{0.8}$$

$= 30[-\ln(0.2) + \ln(0.4)] = 30 \ln 2 \approx \$20.8$ billion

45. $N(x, y) = x^{0.75}y^{0.25}$, $10 \le x \le 20$, $1 \le y \le 2$

$$\text{Average value} = \frac{1}{(20-10)(2-1)}\int_{10}^{20}\int_1^2 x^{0.75}y^{0.25}\,dy\,dx = \frac{1}{10}\int_{10}^{20}\left[x^{0.75}\frac{y^{1.25}}{1.25}\right]_1^2 dx$$

$$= \frac{1}{10}\int_{10}^{20}\left[x^{0.75}\frac{2^{1.25}-1}{1.25}\right]dx = \frac{1}{12.5}(2^{1.25}-1)\int_{10}^{20} x^{0.75}dx = \left[\frac{1}{12.5}(2^{1.25}-1)\frac{x^{1.75}}{1.75}\right]_{10}^{20}$$

$$= \frac{1}{21.875}(2^{1.25}-1)(20^{1.75}-10^{1.75}) \approx 8.375 \text{ or } 8{,}375 \text{ items}$$

47. $C = 10 - \frac{1}{10}d^2 = 10 - \frac{1}{10}(x^2+y^2) = C(x, y)$, $-8 \le x \le 8$, $-6 \le y \le 6$

Average concentration

$$= \frac{1}{16(12)}\int_{-8}^{8}\int_{-6}^{6}\left[10-\frac{1}{10}(x^2+y^2)\right]dydx = \frac{1}{192}\int_{-8}^{8}\left[10y-\frac{1}{10}\left(x^2y+\frac{y^3}{3}\right)\right]_{-6}^{6}dx$$

$$= \frac{1}{192}\int_{-8}^{8}\left\{60-\frac{1}{10}\left(6x^2+\frac{216}{3}\right)-\left[-60-\frac{1}{10}\left(-6x^2-\frac{216}{3}\right)\right]\right\}dx$$

$$= \frac{1}{192}\int_{-8}^{8}\left[120-\frac{1}{10}(12x^2+144)\right]dx = \frac{1}{192}\left[120x-\frac{1}{10}(4x^3+144x)\right]_{-8}^{8}$$

$$= \frac{1}{192}(1280) = \frac{20}{3} \approx 6.67 \text{ insects per square foot}$$

49. $C = 100 - 15d^2 = 100 - 15(x^2+y^2) = C(x, y)$, $-2 \le x \le 2$, $-1 \le y \le 1$

$$\text{Average concentration} = \frac{1}{4(2)}\int_{-2}^{2}\int_{-1}^{1}[100-15(x^2+y^2)]\,dy\,dx$$

$$= \frac{1}{8}\int_{-2}^{2}\left[100y-15x^2y-5y^3\right]_{-1}^{1}dx$$

$$= \frac{1}{8}\int_{-2}^{2}(190-30x^2)\,dx = \left[\frac{1}{8}\left(190x-10x^3\right)\right]_{-2}^{2} = \frac{1}{8}(600) = 75 \text{ parts per million}$$

51. $L = 0.0000133xy^2$, $2000 \le x \le 3000$, $50 \le y \le 60$

Average length $= \dfrac{1}{10{,}000}\int_{2000}^{3000}\int_{50}^{60} 0.0000133xy^2\,dy\,dx$

$$= \frac{0.0000133}{10{,}000}\int_{2000}^{3000}\left[\frac{xy^3}{3}\right]_{50}^{60} dx = \frac{0.0000133}{10{,}000}\int_{2000}^{3000}\frac{91{,}000}{3}x\,dx$$

$$= \left[\frac{1.2103}{30{,}000}\cdot\frac{x^2}{2}\right]_{2000}^{3000} = \frac{1.2103}{60{,}000}(5{,}000{,}000) \approx 100.86 \text{ feet}$$

53. $Q(x, y) = 100\left(\dfrac{x}{y}\right)$, $8 \le x \le 16$, $10 \le y \le 12$

Average intelligence $= \dfrac{1}{16}\int_{8}^{16}\int_{10}^{12} 100\left(\dfrac{x}{y}\right) dydx = \dfrac{100}{16}\int_{8}^{16}\left[x\ln y\right]_{10}^{12} dx$

$$= \frac{100}{16}\int_{8}^{16} x\,(\ln 12 - \ln 10)\,dx = \left[\frac{100(\ln 12 - \ln 10)}{16}\cdot\frac{x^2}{2}\right]_{8}^{16}$$

$$= \frac{100(\ln 12 - \ln 10)}{32}(192) = 600\ \ln(1.2) \approx 109.4$$

EXERCISE 7-7

Things to remember:

1. REGULAR REGIONS:

   A region R in the xy plane is a REGULAR x REGION if there exist functions $f(x)$ and $g(x)$ and numbers a and b so that

   $$R = \{(x, y) \mid g(x) \le y \le f(x),\ a \le x \le b\}.$$

   A region R is a REGULAR y REGION if there exist functions $h(y)$ and $k(y)$ and numbers c and d so that

   $$R = \{(x, y) \mid h(y) \le x \le k(y),\ c \le y \le d\}.$$

   See Figure 3 in the text for a geometric interpretation.

2. DOUBLE INTEGRATION OVER REGULAR REGIONS:

   If $R = \{(x, y) \mid g(x) \le y \le f(x),\ a \le x \le b\}$, then

   $$\iint_R F(x, y)\,dA = \int_a^b\left[\int_{g(x)}^{f(x)} F(x, y)dy\right]dx.$$

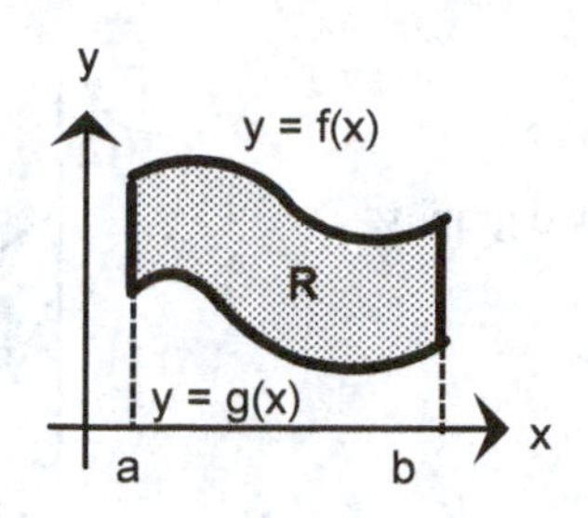

Regular x region

If $R = \{(x, y) \mid h(y) \le x \le k(y), c \le y \le d\}$, then

$$\iint_R F(x, y)\, dA = \int_c^d \left[\int_{h(y)}^{k(y)} F(x, y) dx \right] dy$$

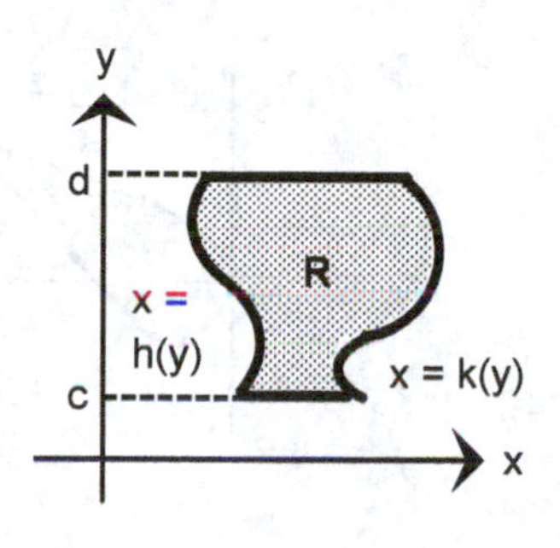

Regular y region

3. If $f(x, y) \ge 0$ over a regular region R, then the double integral of f over R is the VOLUME of the solid formed by the graph f over R.

---

1. $y = 4 - x^2,\ y = 0, 0 \le x \le 2$
The region is
$R = \{(x, y) \mid 0 \le y \le 4 - x^2, 0 \le x \le 2\}$
which is a regular x region.
Also, on the interval $0 \le x \le 2$, we can write the equation $y = 4 - x^2$ as $x = \sqrt{4 - y}$, $0 \le y \le 4$. Thus,
$R = \{(x, y) \mid 0 \le x \le \sqrt{4 - y}, 0 \le y \le 4\}$
which is a regular y region. Hence R is both a regular x region and a regular y region.

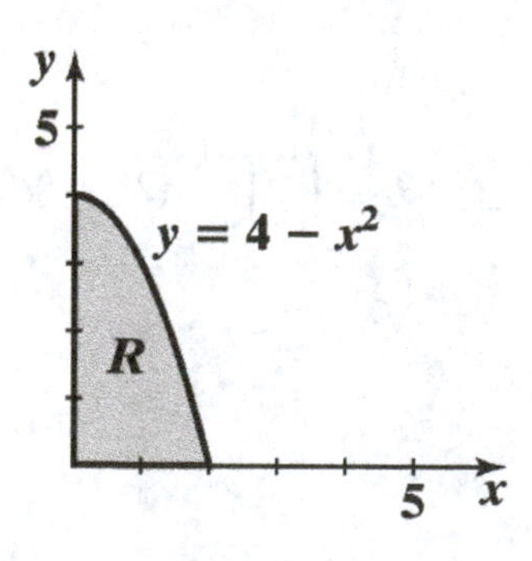

3. $y = x^3$, $y = 12 - 2x$, $x = 0$
or $g(x) = x^3$, $f(x) = 12 - 2x$, $x = 0$
We set $f(x) = g(x)$ to find the points of intersection of the two graphs:
$12 - 2x = x^3$
or $x^3 + 2x - 12 = 0$
$(x - 2)(x^2 + 2x + 6) = 0$
Therefore, $x = 2$ and $(2, 8)$ is the point of intersection. The region is:
$R = \{(x, y) \mid x^3 \le y \le 12 - 2x, 0 \le x \le 2\}$, which is a regular x region.

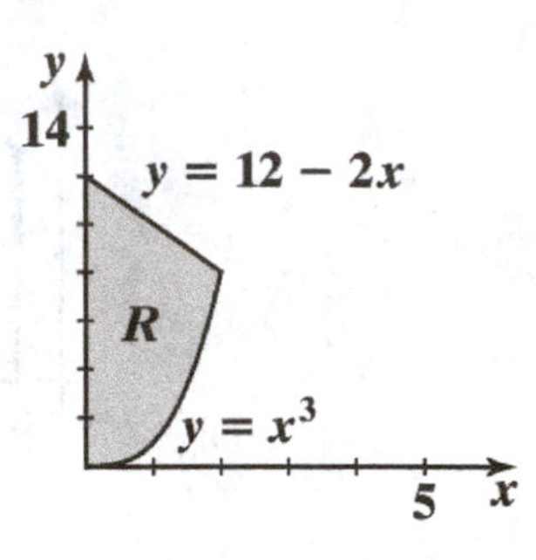

5. $y^2 = 2x$, $y = x - 4$ or $k(y) = \frac{1}{2}y^2$ and $x = h(y) = y + 4$.

To find the point(s) of intersection,
set $h(y) = k(y)$:
$y + 4 = \frac{1}{2}y^2$
or $y^2 - 2y - 8 = 0$
$(y - 4)(y + 2) = 0$
and $y = -2$, $y = 4$. The points of intersection are $(2, -2)$ and $(8, 4)$.
Therefore, the region R is
$R = \left\{(x, y) \middle| \frac{1}{2}y^2 \le x \le y + 4, -2 \le y \le 4\right\}$, which is a regular y region.

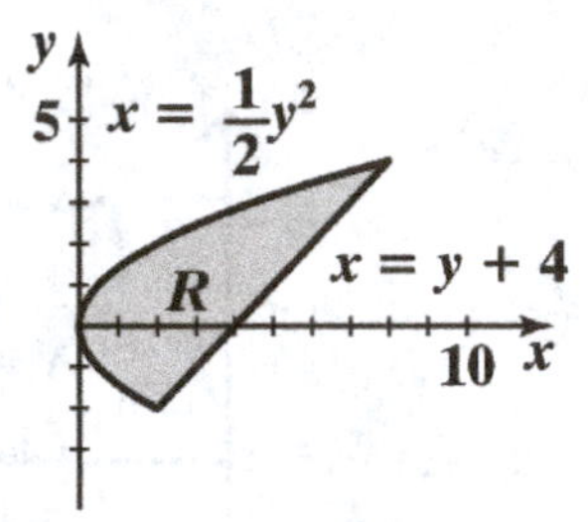

7. $$\int_0^1 \int_0^x (x + y)\,dy\,dx = \int_0^1 \left[\int_0^x (x + y)\,dy\right] dx = \int_0^1 \left[xy + \frac{1}{2}y^2\right]_0^x dx$$
$$= \int_0^1 \left[x^2 + \frac{1}{2}x^2\right] dx = \int_0^1 \frac{3}{2}x^2 dx = \left[\frac{1}{2}x^3\right]_0^1 = \frac{1}{2}(1 - 0) = \frac{1}{2}$$

9. $$\int_0^1 \int_{y^3}^{\sqrt{y}} (2x + y)\,dx\,dy = \int_0^1 \left[\int_{y^3}^{\sqrt{y}} (2x + y)\,dx\right] dy = \int_0^1 \left[(x^2 + xy)\right]_{y^3}^{\sqrt{y}} dy$$
$$= \int_0^1 \{[(\sqrt{y})^2 + y\sqrt{y}] - [(y^3)^2 + y(y^3)]\}\,dy = \int_0^1 (y + y^{3/2} - y^6 - y^4)\,dy$$
$$= \left[\frac{y^2}{2} + \frac{2}{5}y^{5/2} - \frac{1}{7}y^7 - \frac{1}{5}y^5\right]_0^1 = \frac{1}{2} + \frac{2}{5} - \frac{1}{7} - \frac{1}{5} = \frac{39}{70}$$

11. $R = \{(x, y) \mid |x| \le 2, |y| \le 3\}$.
R is the rectangle with vertices $(-2, 3)$, $(2, 3)$, $(2, -3)$, $(-2, -3)$;
R is a regular x region **and** a regular y region.

13. $R = \{(x, y) \mid x^2 + y^2 \geq 1, |x| \leq 2, 0 \leq y \leq 2\}$
R is the region that lies outside the circle of radius 1 centered at the origin and inside the rectangle with vertices $(-2, 2)$, $(2, 2)$, $(2, 0)$, $(-2, 0)$; R is a regular x region but not a regular y region.

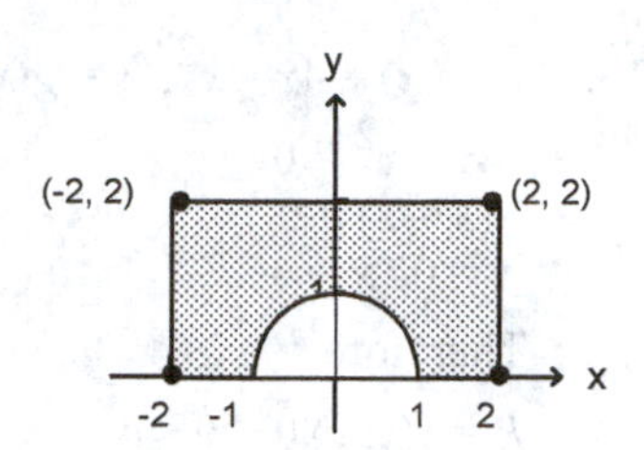

15. $R = \{(x, y) \mid 0 \leq y \leq 2x, 0 \leq x \leq 2\}$

$$\iint_R (x^2 + y^2)dA = \int_0^2 \int_0^{2x} (x^2 + y^2)dy\,dx = \int_0^2 \left[\int_0^{2x} (x^2 + y^2)dy\right]dx = \int_0^2 \left[x^2 y + \frac{1}{3}y^3\right]_0^{2x} dx$$

$$= \int_0^2 \left[x^2(2x) + \frac{1}{3}(2x)^3\right]dx = \int_0^2 \left(2x^3 + \frac{8}{3}x^3\right)dx = \int_0^2 \frac{14}{3}x^3 dx = \frac{7}{6}x^4\Big]_0^2 = \frac{56}{3}$$

17. $R = \{(x, y) \mid 0 \leq x \leq y + 2, 0 \leq y \leq 1\}$

$$\iint_R (x + y - 2)^3 dA = \int_0^1 \int_0^{y+2} (x + y - 2)^3 dx\,dy = \int_0^1 \left[\int_0^{y+2} (x + y - 2)^3 dx\right]dy$$

$$= \int_0^1 \left[\frac{1}{4}(x + y - 2)^4\right]_0^{y+2} dy = \int_0^1 \left[4y^4 - \frac{1}{4}(y - 2)^4\right]dy = \left[\frac{4}{5}y^5 - \frac{1}{20}(y - 2)^5\right]_0^1$$

$$= \frac{4}{5} - \frac{1}{20}(-1)^5 - \left[-\frac{1}{20}(-2)^5\right] = \frac{4}{5} + \frac{1}{20} - \frac{32}{20} = -\frac{3}{4}$$

19. $R = \{(x, y) \mid -x \leq y \leq x, 0 \leq x \leq 2\}$

$$\iint_R e^{x+y} dA = \int_0^2 \int_{-x}^{x} e^{x+y} dy\,dx = \int_0^2 \left[\int_{-x}^{x} e^x e^y dy\right]dx = \int_0^2 e^x \left[e^y\right]_{-x}^{x} dx = \int_0^2 e^x(e^x - e^{-x})\,dx$$

$$= \int_0^2 (e^{2x} - 1)dx = \left[\frac{1}{2}e^{2x} - x\right]_0^2 = \frac{1}{2}e^4 - 2 - \left(\frac{1}{2}e^0\right) = \frac{1}{2}e^4 - \frac{5}{2}$$

21. $R = \{(x, y) \mid 0 \leq y \leq x + 1, 0 \leq x \leq 1\}$

$$\iint_R \sqrt{1 + x + y}\, dA = \int_0^1 \int_0^{x+1} (1 + x + y)^{1/2} dy\,dx$$

$$= \int_0^1 \left[\frac{2}{3}(1 + x + y)^{3/2}\right]_0^{x+1} dx$$

$$= \int_0^1 \left[\frac{2}{3}(2 + 2x)^{3/2} - \frac{2}{3}(1 + x)^{3/2}\right]dx$$

$$= \int_0^1 \left[\frac{4\sqrt{2}}{3}(1 + x)^{3/2} - \frac{2}{3}(1 + x)^{3/2}\right]dx$$

$$= \frac{(4\sqrt{2} - 2)}{3}\int_0^1 (1 + x)^{3/2} dx = \left[\frac{(4\sqrt{2} - 2)}{3} \cdot \frac{2}{5}(1 + x)^{5/2}\right]_0^1$$

$$= \frac{(8\sqrt{2} - 4)}{15} \cdot 2^{5/2} - \frac{(8\sqrt{2} - 4)}{15} = \frac{68 - 24\sqrt{2}}{15}$$

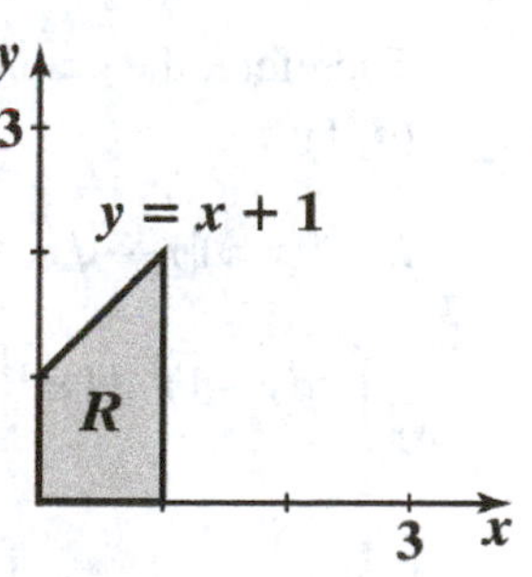

[Note: $(2 + 2x)^{3/2} = 2\sqrt{2}\,(1 + x)^{3/2}$]

23. $y = f(x) = 4x - x^2,\ y = g(x) = 0$

$4x - x^2 = 0$

$x(4 - x) = 0$

$x = 0,\ x = 4$

Therefore,

$R = \{(x, y) | 0 \le y \le 4x - x^2, 0 \le x \le 4.\}$

y

6

y = 4x − x²

R

6 x

$$\iint_R \sqrt{y + x^2}\, dA = \int_0^4 \int_0^{4x-x^2} (y + x^2)^{1/2}\, dy\, dx$$

$$= \int_0^4 \left[\frac{2}{3}(y + x^2)^{3/2}\right]_0^{4x-x^2} dx$$

$$= \int_0^4 \frac{2}{3}[(4x - x^2 + x^2)^{3/2} - (0 + x^2)^{3/2}]\, dx$$

$$= \int_0^4 \left(\frac{16}{3}x^{3/2} - \frac{2}{3}x^3\right) dx \qquad \text{[Note: } (4x)^{3/2} = 4^{3/2}x^{3/2} = 8x^{3/2}.]$$

$$= \left[\frac{16}{3}\cdot\frac{2}{5}x^{5/2} - \frac{2}{3}\cdot\frac{1}{4}x^4\right]_0^4 = \frac{32}{15}(4)^{5/2} - \frac{1}{6}(4)^4 \qquad \text{[Note: } 4^{5/2} = 2^5 = 32.]$$

$$= \frac{32 \cdot 32}{15} - \frac{256}{6} = \frac{128}{5}$$

25. $y = g(x) = 1 - \sqrt{x}\,,\ y = f(x) = 1 + \sqrt{x}\,,\ x = 4$

$f(x) = g(x)$

$1 + \sqrt{x} = 1 - \sqrt{x}$

$2\sqrt{x} = 0$

$x = 0$

Therefore, the graphs intersect at the point (0, 1):

y

4

y = 1 + √x

R

5 x

y = 1 − √x

$R = \{(x, y) | 1 - \sqrt{x} \le y \le 1 + \sqrt{x}\,, 0 \le x \le 4\}$

$$\iint_R x(y - 1)^2 dA = \int_0^4 \int_{1-\sqrt{x}}^{1+\sqrt{x}} x(y - 1)^2 dy\, dx$$

$$= \int_0^4 \left[\frac{1}{3}x(y - 1)^3\right]_{1-\sqrt{x}}^{1+\sqrt{x}} dx$$

$$= \int_0^4 \frac{1}{3}x[(1 + \sqrt{x} - 1)^3 - (1 - \sqrt{x} - 1)^3]\, dx = \int_0^4 \frac{2}{3}x^{5/2}\, dx$$

$$= \left[\frac{2}{3}\cdot\frac{2}{7}x^{7/2}\right]_0^4 = \frac{4}{21}(4)^{7/2} = \frac{4 \cdot 128}{21} = \frac{512}{21}$$

27. $$\int_0^3 \int_0^{3-x} (x + 2y)\, dy\, dx = \int_0^3 \left[\int_0^{3-x} (x + 2y)\, dy\right] dx = \int_0^3 \left[(xy + y^2)\right]_0^{3-x} dx$$

$$= \int_0^3 [x(3 - x) + (3 - x)^2]\, dx = \int_0^3 (9 - 3x)\, dx$$

$$= \left[9x - \frac{3}{2}x^2\right]_0^3 = 27 - \frac{27}{2} = \frac{27}{2}$$

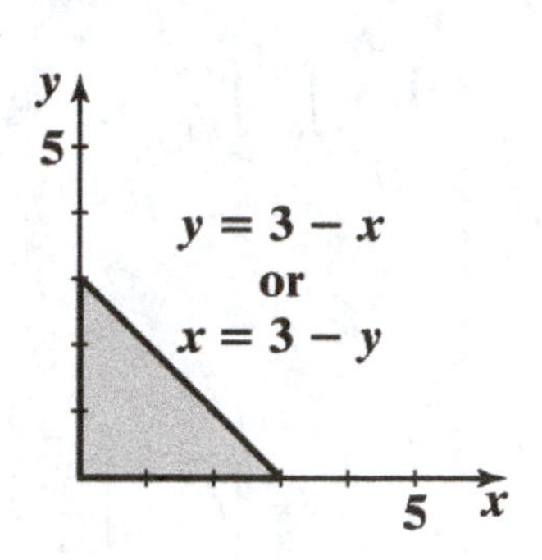

$R = \{(x, y) \mid 0 \le y \le 3 - x, 0 \le x \le 3\}$

Now, $y = 3 - x$ or $x = 3 - y$; and if $x = 0$, then $y = 3$, if $x = 3$, then $y = 0$. Therefore,

$R = \{(x, y) \mid 0 \le x \le 3 - y, 0 \le y \le 3\}$

and integration with the order reversed is:

$$\int_0^3 \int_0^{3-y} (x + 2y)dx\,dy = \int_0^3 \left[\frac{1}{2}x^2 + 2xy\right]_0^{3-y} dy = \int_0^3 \left[\frac{1}{2}(3-y)^2 + 2y(3-y)\right] dy$$

$$= \int_0^3 \left(-\frac{3}{2}y^2 + 3y + \frac{9}{2}\right) dy = \left[-\frac{1}{2}y^3 + \frac{3}{2}y^2 + \frac{9}{2}y\right]_0^3 = -\frac{27}{2} + \frac{27}{2} + \frac{27}{2} = \frac{27}{2}$$

29. $\displaystyle\int_0^1 \int_0^{1-x^2} x\sqrt{y}\,dy\,dx = \int_0^1 \left[\int_0^{1-x^2} xy^{1/2}dy\right]dx = \int_0^1 \left[\frac{2}{3}xy^{3/2}\right]_0^{1-x^2} dx = \int_0^1 \frac{2}{3}x(1-x^2)^{3/2}dx$

$$= \frac{2}{3}\int_0^1 x(1-x^2)^{3/2}dx$$

Let $u = 1 - x^2$. Then $du = -2x\,dx$ or $x\,dx = -\frac{1}{2}du$

$$= -\frac{1}{3}\int_1^0 u^{3/2}du$$

$u = 1$ when $x = 0$, $u = 0$ when $x = 1$.

$$= \frac{1}{3}\int_0^1 u^{3/2}du = \left[\frac{1}{3}\cdot\frac{2}{5}u^{5/2}\right]_0^1 = \frac{2}{15}$$

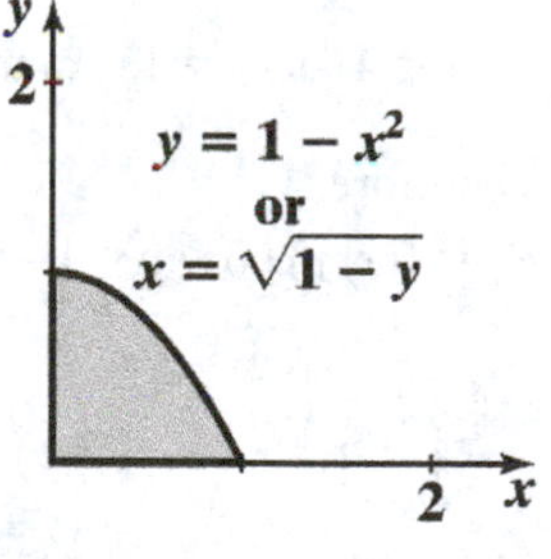

$R = \{(x, y) \mid 0 \le y \le 1 - x^2, 0 \le x \le 1\}$

Now, $y = 1 - x^2$; therefore, $x^2 = 1 - y$ or

$x = \sqrt{1-y}$, and if $x = 0$, then $y = 1$, if $x = 1$, then $y = 0$.

Thus,

$R = \{(x, y) \mid 0 \le x \le \sqrt{1-y}, 0 \le y \le 1\}$, and integration with the order reversed is:

$$\int_0^1 \int_0^{\sqrt{1-y}} x\sqrt{y}\,dx\,dy = \int_0^1 \left[\int_0^{\sqrt{1-y}} x\sqrt{y}dx\right]dy = \int_0^1 \left[\frac{1}{2}x^2\sqrt{y}\right]_0^{\sqrt{1-y}} dy = \int_0^1 \frac{1}{2}(1-y)\sqrt{y}\,dy$$

$$= \frac{1}{2}\int_0^1 (y^{1/2} - y^{3/2})\,dy = \frac{1}{2}\left[\frac{2}{3}y^{3/2} - \frac{2}{5}y^{5/2}\right]_0^1 = \frac{1}{2}\left(\frac{2}{3} - \frac{2}{5}\right) = \frac{2}{15}$$

31. $\int_0^4 \int_{x/4}^{\sqrt{x}/2} x\,dy\,dx = \int_0^4 \left[\int_{x/4}^{\sqrt{x}/2} x\,dy\right] dx = \int_0^4 [xy]_{x/4}^{\sqrt{x}/2}\,dx = \int_0^4 \left(\frac{1}{2}x^{3/2} - \frac{1}{4}x^2\right) dx$

$$= \left[\frac{1}{2}\cdot\frac{2}{5}x^{5/2} - \frac{1}{12}x^3\right]_0^4 = \frac{32}{5} - \frac{16}{3} = \frac{16}{15}$$

$R = \left\{(x, y) \;\middle|\; \frac{x}{4} \le y \le \frac{\sqrt{x}}{2}, 0 \le x \le 4\right\}$

Now, $y = \frac{\sqrt{x}}{2}$ or $x = 4y^2$, and $y = \frac{x}{4}$ or $x = 4y$.

Also, $y = 0$ when $x = 0$ and $y = 1$ when $x = 4$.
Therefore,

$R = \{(x, y) \mid 4y^2 \le x \le 4y, 0 \le y \le 1\}$

$\int_0^4 \int_{x/4}^{\sqrt{x}/2} x\,dy\,dx = \int_0^1 \int_{4y^2}^{4y} x\,dx\,dy = \int_0^1 \left[\frac{1}{2}x^2\right]_{4y^2}^{4y} dy = \int_0^1 \frac{1}{2}[(4y)^2 - (4y^2)^2]\,dy$

$$= \int_0^1 (8y^2 - 8y^4)\,dy = \left[\frac{8}{3}y^3 - \frac{8}{5}y^5\right]_0^1 = \frac{8}{3} - \frac{8}{5} = \frac{16}{15}$$

33. $v = \iint\limits_R (4 - x - y)\,dA$

$x + y = 4$ or $y = 4 - x$, and $x = 0$, $y = 0$.

Therefore,
$R = \{(x, y) \mid 0 \le y \le 4 - x, 0 \le x \le 4\}$

$v = \int_0^4 \int_0^{4-x} (4 - x - y)\,dy\,dx = \int_0^4 \left[-\frac{1}{2}(4 - x - y)^2\right]_0^{4-x} dx$

$= \int_0^4 -\frac{1}{2}[(4 - x - 4 + x)^2 - (4 - x - 0)^2]\,dx = \int_0^4 \frac{1}{2}(4 - x)^2 dx$

$= \left[-\frac{1}{6}(4 - x)^3\right]_0^4 = -\frac{1}{6}[(4-4)^3 - (4-0)^3] = -\frac{1}{6}(-64) = \frac{32}{3}$

35.

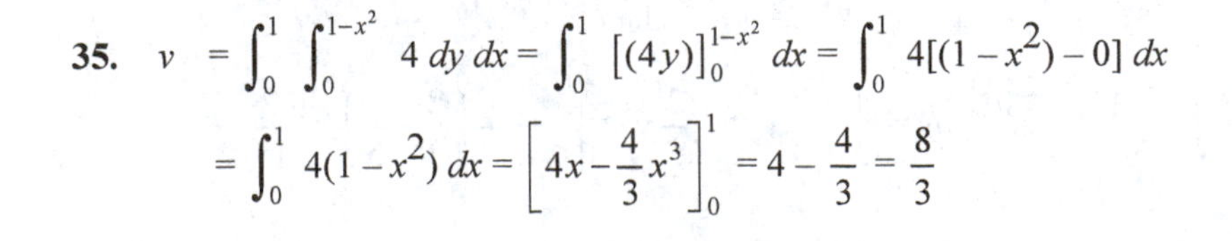

$v = \int_0^1 \int_0^{1-x^2} 4\,dy\,dx = \int_0^1 [(4y)]_0^{1-x^2} dx = \int_0^1 4[(1 - x^2) - 0]\,dx$

$= \int_0^1 4(1 - x^2)\,dx = \left[4x - \frac{4}{3}x^3\right]_0^1 = 4 - \frac{4}{3} = \frac{8}{3}$

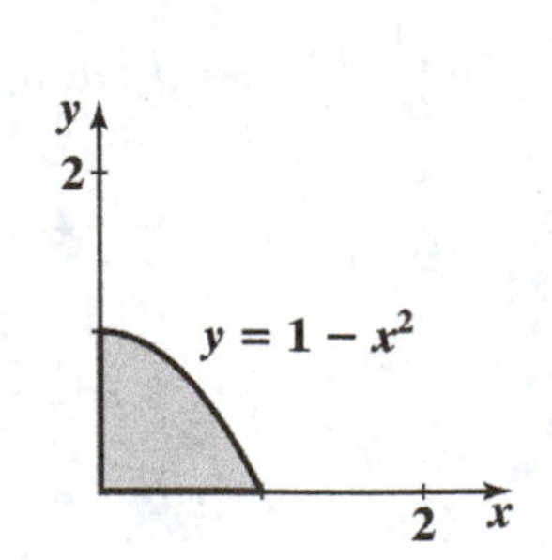

37. $R = \{(x, y) \mid x^2 \le y \le 4, 0 \le x \le 2\}$

Now, $y = x^2$ or $x = \sqrt{y}$; and $y = 0$ when $x = 0$, $y = 4$ when $x = 2$. Therefore,

$R = \{(x, y) \mid 0 \le x \le \sqrt{y}, 0 \le y \le 4\}$

and

$$\int_0^2 \int_{x^2}^4 \frac{4x}{1+y^2}\,dy\,dx = \int_0^4 \int_0^{\sqrt{y}} \frac{4x}{1+y^2}\,dx\,dy = \int_0^4 \left[\frac{2x^2}{1+y^2}\right]_0^{\sqrt{y}} dy = \int_0^4 \frac{2y}{1+y^2}\,dy$$

$$= \ln(1+y^2)\Big]_0^4 = \ln 17 - \ln 1 = \ln 17$$

39. $R = \{(x, y) \mid y^2 \le x \le 1, 0 \le y \le 1\}$

Now, $x = y^2$ or $y = \sqrt{x}$; and $y = 0$ when $x = 0$; $y = 1$ when $x = 1$.

Therefore, $R = \{(x, y) \mid 0 \le y \le \sqrt{x}, 0 \le x \le 1\}$ and

$$\int_0^1 \int_{y^2}^1 4ye^{x^2}\,dx\,dy = \int_0^1 \int_0^{\sqrt{x}} 4ye^{x^2}\,dy\,dx = \int_0^1 \left[2y^2e^{x^2}\right]_0^{\sqrt{x}} dx = \int_0^1 2xe^{x^2}\,dx = e^{x^2}\Big]_0^1 = e - 1.$$

41. $y = 1 + \sqrt{x}$, $y = x^2$, $x = 0$

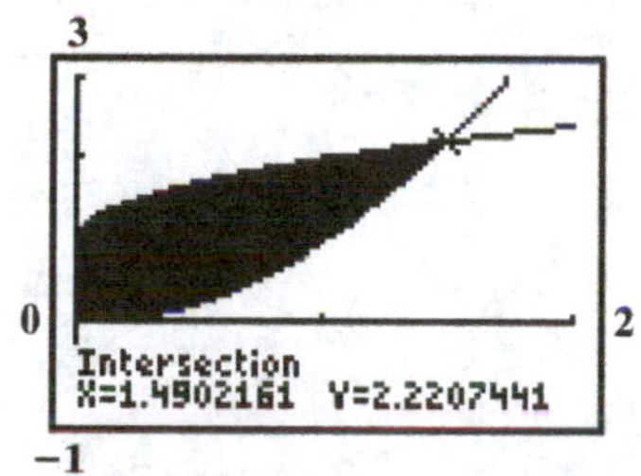

$R = \{(x, y) \mid x^2 \le y \le 1 + \sqrt{x}, 0 \le x \le 1.49\}$

$$\iint x\,dA = \int_0^{1.49} \left[\int_{x^2}^{1+\sqrt{x}} x\,dy\right] dx = \int_0^{1.49} [xy]_{x^2}^{1+\sqrt{x}}\,dx = \int_0^{1.49} (x + x^{3/2} - x^3)dx$$

$$= \left[\frac{x^2}{2} + \frac{2}{5}x^{5/2} - \frac{x^4}{4}\right]_0^{1.49} \approx 0.96$$

43. $y = \sqrt[3]{x}$, $y = 1 - x$, $y = 0$

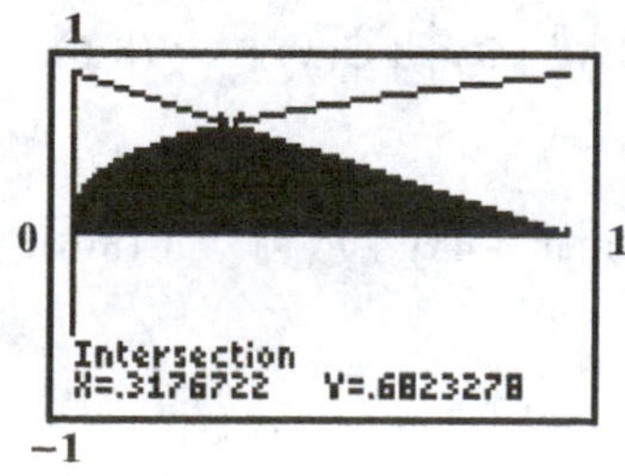

$R = \{(x, y) \mid y^3 \le x \le 1 - y, \ 0 \le y \le 0.68\}$

$$\iint_R 24xy\,dA = \int_0^{0.68} \left[\int_{y^3}^{1-y} 24xy\,dx\right] dy = \int_0^{0.68} \left[12x^2y\right]_{y^3}^{1-y} dy = 12\int_0^{0.68} [y(1-y)^2 - y^7]\,dy$$

$$= 12\int_0^{0.68} (y^3 - 2y^2 + y - y^7)\,dy = 12\left[\frac{1}{4}y^4 - \frac{2}{3}y^3 + \frac{1}{2}y^2 - \frac{1}{8}y^8\right]_0^{0.68} \approx 0.83$$

45. $y = e^{-x}$, $y = 3 - x$

5

−3 3

Intersection
X=-1.505241 Y=4.5052415

−2

$R = \{(x, y) \mid e^{-x} \le y \le 3 - x, -1.51 \le x \le 2.95\}$ Regular x region

$= \{(x, y) \mid -\ln y \le x \le 3 - y, 0.05 \le y \le 4.51\}$ Regular y region

$$\iint_R 4y\,dA = \int_{0.05}^{4.51}\left[\int_{-\ln y}^{3-y} 4y\,dx\right]dy = \int_{0.05}^{4.51}[4xy]_{-\ln y}^{3-y}\,dy = 4\int_{0.05}^{4.51}[y(3-y) + y\ln y]\,dy$$

$$= 4\left[\frac{3}{2}y^2 - \frac{1}{3}y^3 + \frac{1}{2}y^2\ln y - \frac{1}{4}y^2\right]_{0.05}^{4.51} \approx 40.67 \qquad \left(\int y\ln y\,dy = \frac{1}{2}y^2\ln y - \frac{1}{4}y^2\right)$$

CHAPTER 7 REVIEW

1. $f(x, y) = 2000 + 40x + 70y$

$f(5, 10) = 2000 + 40\cdot5 + 70\cdot10 = 2900$

$f_x(x, y) = 40$

$f_y(x, y) = 70$ (8-1, 8-2)

2. $z = x^3y^2$

$$\frac{\partial z}{\partial x} = 3x^2y^2, \qquad \frac{\partial^2 z}{\partial x^2} = \frac{\partial\left(\frac{\partial z}{\partial x}\right)}{\partial x} = \frac{\partial(3x^2y^2)}{\partial x} = 6xy^2$$

$$\frac{\partial z}{\partial y} = 2x^3y, \qquad \frac{\partial^2 z}{\partial x\partial y} = \frac{\partial\left(\frac{\partial z}{\partial y}\right)}{\partial x} = \frac{\partial(2x^3y)}{\partial x} = 6x^2y \qquad \text{(8-2)}$$

3. $\int(6xy^2 + 4y)\,dy = 6x\int y^2\,dy + 4\int y\,dy = 6x\cdot\frac{y^3}{3} + 4\cdot\frac{y^2}{2} + C(x) = 2xy^3 + 2y^2 + C(x)$ (8-6)

4. $\int(6xy^2 + 4y)\,dx = 6y^2\int x\,dx + 4y\int dx = 6y^2\cdot\frac{x^2}{2} + 4yx + E(y) = 3x^2y^2 + 4xy + E(y)$ (8-6)

5. $\int_0^1\int_0^1 4xy\,dy\,dx = \int_0^1\left[\int_0^1 4xy\,dy\right]dx = \int_0^1\left[2xy^2\right]_0^1 dx = \int_0^1 2x\,dx = x^2\Big]_0^1 = 1$ (8-6)

6. $f(x, y) = 6 + 5x - 2y + 3x^2 + x^3$

$f_x(x, y) = 5 + 6x + 3x^2$, $\qquad f_y(x, y) = -2 \ne 0$

The function $f_y(x, y)$ is nonzero for all (x, y). (8-3)

7. $f(x, y) = 3x^2 - 2xy + y^2 - 2x + 3y - 7$
$f(2, 3) = 3\cdot 2^2 - 2\cdot 2\cdot 3 + 3^2 - 2\cdot 2 + 3\cdot 3 - 7 = 7$
$f_y(x, y) = -2x + 2y + 3$
$f_y(2, 3) = -2\cdot 2 + 2\cdot 3 + 3 = 5$ (8-1, 8-2)

8. $f(x, y) = -4x^2 + 4xy - 3y^2 + 4x + 10y + 81$

$f_x(x, y) = -8x + 4y + 4,$ $f_y(x, y) = 4x - 6y + 10$

$f_{xx}(x, y) = -8,$ $f_{yy}(x, y) = -6$

$f_{xy}(x, y) = 4$

Now, $f_{xx}(2, 3)\cdot f_{yy}(2, 3) - [f_{xy}(2, 3)]^2 = (-8)(-6) - 4^2 = 32.$ (8-2)

9. $f(x, y) = x + 3y$ and $g(x, y) = x^2 + y^2 - 10.$
Let $F(x, y, \lambda) = f(x, y) + \lambda g(x, y) = x + 3y + \lambda (x^2 + y^2 - 10)$. Then, we have:
$F_x = 1 + 2x\lambda$
$F_y = 3 + 2y\lambda$
$F_\lambda = x^2 + y^2 - 10$
Setting $F_x = F_y = F_\lambda = 0$, we obtain:
$1 + 2x\lambda = 0$ (1)
$3 + 2y\lambda = 0$ (2)
$x^2 + y^2 - 10 = 0$ (3)
From the first equation, $x = -\frac{1}{2\lambda}$; from the second equation, $y = -\frac{3}{2\lambda}$. Substituting these into the third equation gives:

$$\frac{1}{4\lambda^2} + \frac{9}{4\lambda^2} - 10 = 0$$
$$40\lambda^2 = 10$$
$$\lambda^2 = \frac{1}{4}$$
$$\lambda = \pm\frac{1}{2}$$

Thus, the critical points are $\left(-1, -3, \frac{1}{2}\right)$ and $\left(1, 3, -\frac{1}{2}\right)$. (8-4)

10.

x_k	y_k	$x_k y_k$	x_k^2
2	12	24	4
4	10	40	16
6	7	42	36
8	3	24	64
Totals 20	32	130	120

Thus, $\sum_{k=1}^{4} x_k = 20$, $\sum_{k=1}^{4} y_k = 32$, $\sum_{k=1}^{4} x_k y_k = 130$, $\sum_{k=1}^{4} x_k^2 = 120$.

Substituting these values into the formulas for a and b, we have:

$$a = \frac{4\left(\sum_{k=1}^{4} x_k y_k\right) - \left(\sum_{k=1}^{4} x_k\right)\left(\sum_{k=1}^{4} y_k\right)}{4\left(\sum_{k=1}^{4} x_k^2\right) - \left(\sum_{k=1}^{4} x_k\right)^2} = \frac{4(130)-(20)(32)}{4(120)-(20)^2} = \frac{-120}{80} = -1.5$$

$$b = \frac{\sum_{k=1}^{4} y_k - (-1.5)\sum_{k=1}^{4} x_k}{4} = \frac{32+(1.5)(20)}{4} = \frac{62}{4} = 15.5$$

Thus, the least squares line is: $y = ax + b = -1.5x + 15.5$
When $x = 10$, $y = -1.5(10) + 15.5 = 0.5$. (8-5)

11. $$\iint_R (4x+6y)\,dA = \int_{-1}^{1}\int_{1}^{2} (4x+6y)\,dy\,dx = \int_{-1}^{1}\left[\int_{1}^{2}(4x+6y)\,dy\right]dx = \int_{-1}^{1}\left[(4xy+3y^2)\right]_1^2 dx$$

$$= \int_{-1}^{1} (8x+12-4x-3)\,dx = \int_{-1}^{1} (4x+9)\,dx = (2x^2+9x)\Big|_{-1}^{1} = 2+9-(2-9) = 18$$

$$\iint_R (4x+6y)dA = \int_{1}^{2}\int_{-1}^{1} (4x+6y)\,dx\,dy = \int_{1}^{2}\left[\int_{-1}^{1}(4x+6y)\,dx\right]dy = \int_{1}^{2}\left[(2x^2+6xy)\right]_{-1}^{1} dy$$

$$= \int_{1}^{2} [2+6y-(2-6y)]\,dy = \int_{1}^{2} 12y\,dy = 6y^2\Big]_1^2 = 24-6 = 18 \qquad (8\text{-}6)$$

12. $R = \{(x, y) \mid \sqrt{y} \le x \le 1, 0 \le y \le 1\}$

R is a regular y-region.
(R is also a regular x-region.)

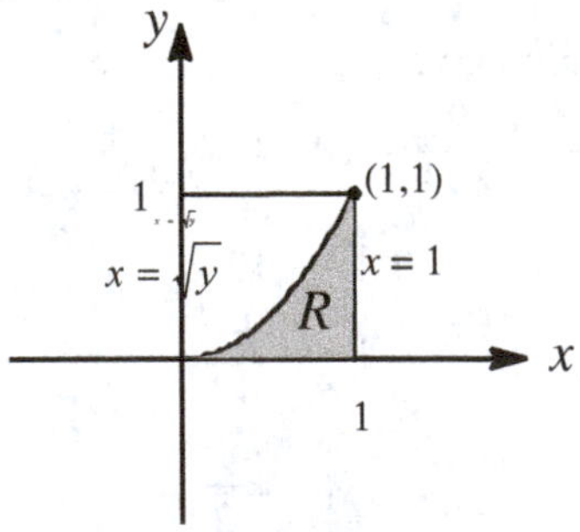

$$\iint_R (6x+y)dA = \int_0^1 \int_{\sqrt{y}}^{1} (6x+y)dx\,dy$$

$$= \int_0^1 \left[3x^2+xy\right]_{\sqrt{y}}^{1} dy$$

$$= \int_0^1 [(3+y)-(3y+y^{3/2})]\,dy$$

$$= \int_0^1 (3-2y-y^{3/2})\,dy = \left[3y-y^2-\frac{2}{5}y^{5/2}\right]_0^1 = 3-1-\frac{2}{5} = \frac{8}{5} \qquad (8\text{-}7)$$

13. $f(x, y) = e^{x^2+2y}$

$f_x(x, y) = e^{x^2+2y}\cdot 2x = 2xe^{x^2+2y}$, $\quad f_y(x, y) = e^{x^2+2y}\cdot 2 = 2e^{x^2+2y}$

$f_{xy}(x, y) = 2xe^{x^2+2y}\cdot 2 = 4xe^{x^2+2y}$ (8-2)

14. $f(x, y) = (x^2 + y^2)^5$

$f_x(x, y) = 5(x^2 + y^2)^4 \cdot 2x = 10x(x^2 + y^2)^4$

$f_{xy}(x, y) = 10x(4)(x^2 + y^2)^3 \cdot 2y = 80xy(x^2 + y^2)^3$ (8-2)

15. $f(x, y) = x^3 - 12x + y^2 - 6y$

$f_x(x, y) = 3x^2 - 12$ $\quad$ $f_y(x, y) = 2y - 6$

$3x^2 - 12 = 0$ $\quad$ $2y - 6 = 0$

$x^2 = 4$ $\quad$ $y = 3$

$x = \pm 2$

Thus, the critical points are $(2, 3)$ and $(-2, 3)$.

$f_{xx}(x, y) = 6x,$ $\quad$ $f_{xy}(x, y) = 0,$ $\quad$ $f_{yy}(x, y) = 2$

For the critical point (2, 3):

$f_{xx}(2, 3) = 12 > 0$

$f_{xy}(2, 3) = 0$

$f_{yy}(2, 3) = 2$

$f_{xx}(2, 3) \cdot f_{yy}(2, 3) - [f_{xy}(2, 3)]^2 = 12 \cdot 2 = 24 > 0$

Therefore, $f(2, 3) = 2^3 - 12 \cdot 2 + 3^2 - 6 \cdot 3 = -25$ is a local minimum.

For the critical point (–2, 3):

$f_{xx}(-2, 3) = -12 < 0$

$f_{xy}(-2, 3) = 0$

$f_{yy}(-2, 3) = 2$

$f_{xx}(-2, 3) \cdot f_{yy}(-2, 3) - [f_{xy}(-2, 3)]^2 = -12 \cdot 2 - 0 = -24 < 0$

Thus, f has a saddle point at (–2, 3). (8-3)

16. Step 1. Maximize $f(x, y) = xy$

Subject to: $g(x, y) = 2x + 3y - 24 = 0$

Step 2. $F(x, y, \lambda) = f(x, y) + \lambda g(x, y) = xy + \lambda(2x + 3y - 24)$

Step 3. $F_x = y + 2\lambda = 0$ (1)

$F_y = x + 3\lambda = 0$ (2)

$F_\lambda = 2x + 3y - 24 = 0$ (3)

From (1) and (2), we obtain:

$y = -2\lambda$ and $x = -3\lambda$

Substituting these into (3), we have:

$-6\lambda - 6\lambda - 24 = 0$

$\lambda = -2$

Thus, the critical point is (6, 4, –2).

Step 4. Since (6, 4, –2) is the only critical point for F, we conclude that $\max f(x, y) = f(6, 4) = 6 \cdot 4 = 24.$ (8-4)

17. Step 1. Minimize $f(x, y, z) = x^2 + y^2 + z^2$
Subject to: $2x + y + 2z = 9$ or $g(x, y, z) = 2x + y + 2z - 9 = 0$

Step 2. $F(x, y, z, \lambda) = x^2 + y^2 + z^2 + \lambda(2x + y + 2z - 9)$

Step 3.
$F_x = 2x + 2\lambda = 0$ (1)
$F_y = 2y + \lambda = 0$ (2)
$F_z = 2z + 2\lambda = 0$ (3)
$F_\lambda = 2x + y + 2z - 9 = 0$ (4)

From equations (1), (2), and (3), we have:
$x = -\lambda,\ y = -\frac{\lambda}{2}$, and $z = -\lambda$

Substituting these into (4), we obtain:

$$-2\lambda - \frac{\lambda}{2} - 2\lambda - 9 = 0$$
$$\frac{9}{2}\lambda = -9$$
$$\lambda = -2$$

The critical point is: (2, 1, 2, –2)

Step 4. Since (2, 1, 2, –2) is the only critical point for F, we conclude that
$\min f(x, y, z) = f(2, 1, 2) = 2^2 + 1^2 + 2^2 = 9.$ (8-4)

18.

	x_k	y_k	$x_k y_k$	x_k^2
	10	50	500	100
	20	45	900	400
	30	50	1,500	900
	40	55	2,200	1,600
	50	65	3,250	2,500
	60	80	4,800	3,600
	70	85	5,950	4,900
	80	90	7,200	6,400
	90	90	8,100	8,100
	100	110	11,000	10,000
Totals	550	720	45,400	38,500

Thus, $\sum_{k=1}^{10} x_k = 550,\ \sum_{k=1}^{10} y_k = 720,\ \sum_{k=1}^{10} x_k y_k = 45{,}400,\ \sum_{k=1}^{10} x_k^2 = 38{,}500.$

Substituting these values into the formulas for a and b, we have:

$$a = \frac{10(45,400) - (550)(720)}{10(38,500) - (550)^2} = \frac{58,000}{82,500} = \frac{116}{165}$$

$$b = \frac{720 - \left(\frac{116}{165}\right)550}{10} = \frac{100}{3}$$

Therefore, the least squares line is:

$$y = \frac{116}{165}x + \frac{100}{3} \approx 0.703x + 33.33 \qquad (8\text{-}5)$$

19. $$\frac{1}{(b-a)(d-c)}\iint_R f(x, y)\, dA = \frac{1}{[8-(-8)](27-0)}\int_{-8}^{8}\int_0^{27} x^{2/3}y^{1/3}\,dy\,dx$$

$$= \frac{1}{16\cdot 27}\int_{-8}^{8}\left[\frac{3}{4}x^{2/3}y^{4/3}\right]_0^{27} dx = \frac{1}{16\cdot 27}\int_{-8}^{8}\frac{3^5}{4}x^{2/3}\,dx = \frac{9}{64}\int_{-8}^{8} x^{2/3}dx$$

$$= \frac{9}{64}\cdot\frac{3}{5}x^{5/3}\Big|_{-8}^{8} = \frac{9}{64}\cdot\frac{3}{5}[2^5 - (-2)^5] = \frac{9}{64}\cdot\frac{3}{5}\cdot 2^6 = \frac{27}{5} \qquad (8\text{-}6)$$

20. $$V = \iint_R (3x^2 + 3y^2)\, dA = \int_0^1\int_{-1}^1 (3x^2 + 3y^2)\, dy\, dx = \int_0^1\left[\int_{-1}^1 (3x^2 + 3y^2)dy\right]dx$$

$$= \int_0^1 \left[3x^2y + y^3\right]_{-1}^1 dx = \int_0^1 [3x^2 + 1 - (-3x^2 - 1)]\, dx = \int_0^1 (6x^2 + 2)dx = (2x^3 + 2x)\Big|_0^1 = 4 \text{ cubic units}$$

(8-6)

21. $f(x, y) = x + y;\ -10 \le x \le 10,\ -10 \le y \le 10$

Prediction: average value = $f(0, 0) = 0$.

Verification:

$$\text{average value} = \frac{1}{[10-(-10)][10-(-10)]}\int_{-10}^{10}\int_{-10}^{10}(x + y)\, dy\, dx$$

$$= \frac{1}{400}\int_{-10}^{10}\left[xy + \frac{1}{2}y^2\right]_{-10}^{10} dx = \frac{1}{400}\int_{-10}^{10} 20x\, dx = \left[\frac{1}{400}(10x^2)\right]_{-10}^{10} = 0 \qquad (8\text{-}6)$$

22. $$f(x, y) = \frac{e^x}{y+10}$$

(A) $S = \{x, y) | -a \le x \le a, -a \le y \le a\}$

The average value of f over S is given by:

$$\frac{1}{[a-(-a)][a-(-a)]}\int_{-a}^{a}\int_{-a}^{a}\frac{e^x}{y+10}dx\ dy = \frac{1}{4a^2}\int_{-a}^{a}\left[\frac{e^x}{y+10}\right]_{-a}^{a} dy$$

$$= \frac{1}{4a^2}\int_{-a}^{a}\left(\frac{e^a}{y+10} - \frac{e^{-a}}{y+10}\right)dy = \frac{e^a - e^{-a}}{4a^2}\int_{-a}^{a}\frac{1}{y+10}dy$$

$$= \left[\frac{e^a - e^{-a}}{4a^2}\ln|y+10|\right]_{-a}^{a} = \frac{e^a - e^{-a}}{4a^2}[\ln(10+a) - \ln(10-a)] = \frac{e^a - e^{-a}}{4a^2}\ln\left(\frac{10+a}{10-a}\right)$$

Now, $\frac{e^a - e^{-a}}{4a^2}\ln\left(\frac{10+a}{10-a}\right) = 5$ is equivalent to $(e^a - e^{-a})\ln\left(\frac{10+a}{10-a}\right) - 20a^2 = 0.$

Using a graphing utility, the graph of

$$f(x) = (e^x - e^{-x})\ln\left(\frac{10+x}{10-x}\right) - 20x^2$$

is shown at the right and $f(x) = 0$ at $x \approx \pm 6.28$.

The dimensions of the square are:
12.56 × 12.56.

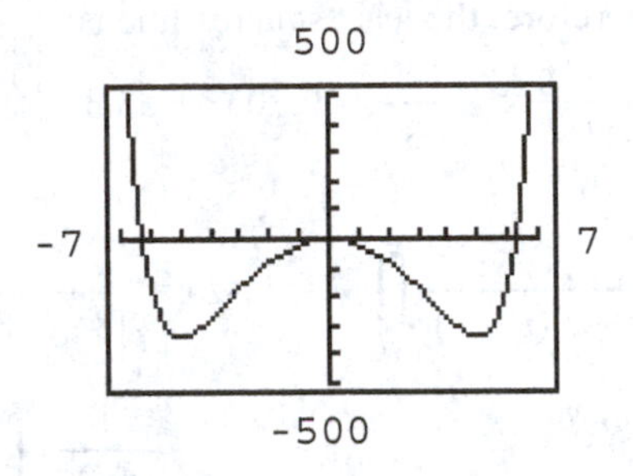

(B) To determine whether there is a square centered at (0, 0) such that

$$\frac{e^a - e^{-a}}{4a^2}\ln\left(\frac{10+a}{10-a}\right) = 0.05,$$

graph,

$$f(x) = (e^x - e^{-x})\ln\left(\frac{10+x}{10-x}\right) - 0.20x^2$$

The result is shown at the right and $f(x) = 0$ only at $x = 0$. Thus, there does not exist a square centered at (0, 0) such that the average value of $f = 0.05$.

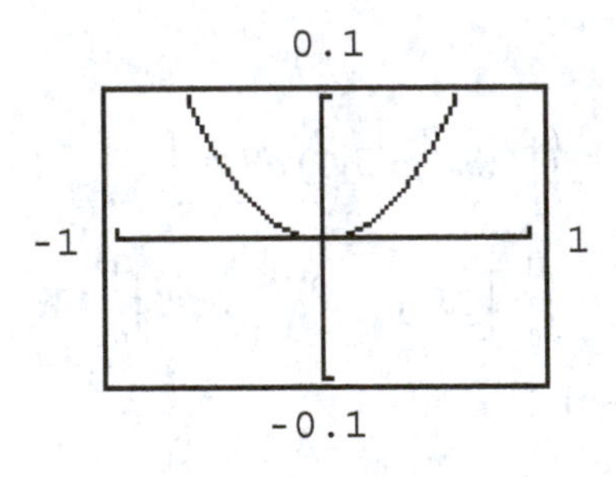

(8-6)

23. Step 1. Extremize $f(x, y) = 4x^3 - 5y^3$ subject to $g(x, y) = 3x + 2y - 7 = 0$.

Step 2. $F(x, y, \lambda) = 4x^3 - 5y^3 + \lambda(3x + 2y - 7)$

Step 3. $F_x = 12x^2 + 3\lambda = 0$ (1)

$F_y = -15y^2 + 2\lambda = 0$ (2)

$F_\lambda = 3x + 2y - 7 = 0$ (3)

From (1), $\lambda = -4x^2 \le 0$; from (2), $\lambda = \frac{15}{2}y^2 \ge 0$. This implies $x = y = \lambda = 0$ and $x = y = 0$ does not satisfy (3). The system (1), (2), (3) does not have a simultaneous solution. (8-4)

24. R is the region shown in the figure; it is a regular x-region and a regular y-region. $F(x, y) = 60x^2y \ge 0$ on R so

$$V = \iint_R 60x^2y\,dA = \int_0^1 \int_0^{1-x} 60x^2y\,dy\,dx \quad \text{(treating } R \text{ as an } x\text{-region)}$$

$$= \int_0^1 \left[30x^2y^2\right]_0^{1-x} dx = \int_0^1 30x^2(1-x)^2 dx$$

$= \int_0^1 30x^2(1 - 2x + x^2)\,dx = \int_0^1 (30x^2 - 60x^3 + 30x^4)\,dx$

$= (10x^3 - 15x^4 + 6x^5)\Big|_0^1 = 1$

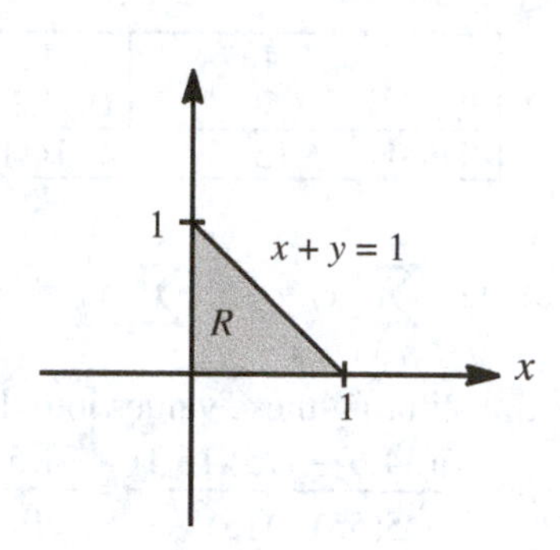

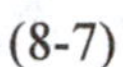
(8-7)

25. $P(x, y) = -4x^2 + 4xy - 3y^2 + 4x + 10y + 81$

(A) $P_x(x, y) = -8x + 4y + 4$
$P_x(1, 3) = -8\cdot 1 + 4\cdot 3 + 4 = 8$

At the output level (1, 3), profit will increase by \$8000 for 100 units increase in product A if the production of product B is held fixed.

(B) $P_x = -8x + 4y + 4 = 0 \quad (1)$
$P_y = 4x - 6y + 10 = 0 \quad (2)$

Solving (1) and (2) for x and y, we obtain $x = 2,\ y = 3$.
Thus, (2, 3) is a critical point.

$P_{xx} = -8$ $\qquad$ $P_{yy} = -6$ $\qquad$ $P_{xy} = 4$

$P_{xx}(2, 3) = -8 < 0$ $\qquad$ $P_{yy}(2, 3) = -6$ $\qquad$ $P_{xy}(2, 3) = 4$

$P_{xx}(2, 3)\cdot P_{yy}(2, 3) - [P_{xy}(2, 3)]^2 = (-8)(-6) - 4^2 = 32 > 0$

Thus, $P(2, 3)$ is a maximum and

$\max P(x, y) = P(2, 3) = -4\cdot 2^2 + 4\cdot 2\cdot 3 - 3\cdot 3^2 + 4\cdot 2 + 10\cdot 3 + 81 = -16 + 24 - 27 + 8 + 30 + 81$
$= 100.$

The maximum profit is \$100,000. This is obtained when 200 units of A and 300 units of B are produced per month. (8-2, 8-3)

26. Minimize $S(x, y, z) = xy + 4xz + 3yz$
Subject to: $V(x, y, z) = xyz - 96 = 0$
Put $F(x, y, z, \lambda) = S(x, y, z) + \lambda V(x, y, z) = xy + 4xz + 3yz + \lambda(xyz - 96)$. Then, we have:

$F_x = y + 4z + \lambda yz = 0 \quad (1)$

$F_y = x + 3z + \lambda xz = 0 \quad (2)$

$F_z = 4x + 3y + \lambda xy = 0 \quad (3)$

$F_\lambda = xyz - 96 = 0 \quad (4)$

Solving the system of equations, (1) – (4), simultaneously, yields $x = 6,\ y = 8,\ z = 2$, and $\lambda = -1$. Thus, the critical point is (6, 8, 2, –1) and $S(6, 8, 2) = 6\cdot 8 + 4\cdot 6\cdot 2 + 3\cdot 8\cdot 2 = 144$ is the minimum value of S subject to the constraint $V = xyz - 96 = 0$.

The dimensions of the box that will require the minimum amount of material are:
length $x = 6$ inches; width $y = 8$ inches; height $z = 2$ inches. (8-3)

27.

x_k	y_k	$x_k y_k$	x_k^2
1	2.0	2.0	1
2	2.5	5.0	4
3	3.1	9.3	9

	4	4.2	16.8	16
	5	4.3	21.5	25
Totals	15	16.1	54.6	55

Thus, $\sum_{k=1}^{5} x_k = 15$, $\sum_{k=1}^{5} y_k = 16.1$, $\sum_{k=1}^{5} x_k y_k = 54.6$, $\sum_{k=1}^{5} x_k^2 = 55$.

Substituting these values into the formulas for a and b, we have:

$$a = \frac{5(54.6) - (15)(16.1)}{5(55) - (15)^2} = \frac{31.5}{50} \approx 0.63$$

$$b = \frac{16.1 - (0.63)(15)}{5} = 1.33$$

Therefore, the least squares line is: $y = 0.63x + 1.33$

When $x = 6$, $y = 0.63(6) + 1.33 = 5.11$, and the profit for the sixth year is estimated to be \$5.11 million. (8-4)

28. $N(x, y) = 10x^{0.8}y^{0.2}$

(A) $N_x(x, y) = 8x^{-0.2}y^{0.2}$

$N_x(40, 50) = 8(40)^{-0.2}(50)^{0.2} \approx 8.37$

$N_y(x, y) = 2x^{0.8}y^{-0.8}$

$N_y(40, 50) = 2(40)^{0.8}(50)^{-0.8} \approx 1.67$

Thus, at the level of 40 units of labor and 50 units of capital, the marginal productivity of labor is approximately 8.36 and the marginal productivity of capital is approximately 1.67. Management should encourage increased use of labor.

(B) Step 1. Maximize the production function $N(x, y) = 10x^{0.8}y^{0.2}$
Subject to the constraint: $C(x, y) = 100x + 50y = 10{,}000$
i.e., $g(x, y) = 100x + 50y - 10{,}000 = 0$

Step 2. $F(x, y, \lambda) = 10x^{0.8}y^{0.2} + \lambda(100x + 50y - 10{,}000)$

Step 3.

$$F_x = 8x^{-0.2}y^{0.2} + 100\lambda = 0 \quad (1)$$

$$F_y = 2x^{0.8}y^{-0.8} + 50\lambda = 0 \quad (2)$$

$$F_\lambda = 100x + 50y - 10{,}000 = 0 \quad (3)$$

From equation (1), $\lambda = \frac{-0.08y^{0.2}}{x^{0.2}}$, and from (2), $\lambda = \frac{-0.04x^{0.8}}{y^{0.8}}$.

Thus, $\frac{0.08y^{0.2}}{x^{0.2}} = \frac{0.04x^{0.8}}{y^{0.8}}$ and $x = 2y$.

Substituting into (3) yields:

$$200y + 50y = 10{,}000$$
$$250y = 10{,}000$$
$$y = 40$$

Therefore, $x = 80$ and $\lambda \approx -0.0696$. The critical point is $(80, 40, -0.0696)$. Thus, we conclude that $\max N(x, y) = N(80, 40) = 10(80)^{0.8}(40)^{0.2} \approx 696$ units.

Production is maximized when 80 units of labor and 40 units of capital are used.

The marginal productivity of money is $-\lambda \approx 0.0696$. The increase in production resulting from an increase of \$2000 in the budget is: $0.0696(2000) \approx 139$ units.

(C) Average number of units

$$= \frac{1}{(100-50)(40-20)}\int_{50}^{100}\int_{20}^{40} 10x^{0.8}y^{0.2}\,dy\,dx = \frac{1}{(50)(20)}\int_{50}^{100}\left[\frac{10x^{0.8}y^{1.2}}{1.2}\right]_{20}^{40} dx$$

$$= \frac{1}{1000}\int_{50}^{100} \frac{10}{1.2}x^{0.8}(40^{1.2}-20^{1.2})\,dx = \frac{40^{1.2}-20^{1.2}}{120}\int_{50}^{100} x^{0.8}dx = \frac{40^{1.2}-20^{1.2}}{120}\cdot\left[\frac{x^{1.8}}{1.8}\right]_{50}^{100}$$

$$= \frac{(40^{1.2}-20^{1.2})(100^{1.8}-50^{1.8})}{216} \approx \frac{(47.24)(2837.81)}{216} \approx 621$$

Thus, the average number of units produced is approximately 621. (8-4)

29. $T(V,x) = \dfrac{33V}{x+33} = 33V(x+33)^{-1}$

$T_x(V,x) = -33V(x+33)^{-2} = \dfrac{-33V}{(x+33)^2}$

$T_x(70, 17) = \dfrac{-33(70)}{(17+33)^2} = \dfrac{-33(70)}{2500} = -0.924$ minutes per unit increase in depth when $V = 70$ cubic feet and $x = 17$ ft. (8-2)

30. $C = 100 - 24d^2 = 100 - 24(x^2 + y^2)$

$C(x, y) = 100 - 24(x^2 + y^2)$, $-2 \le x \le 2$, $-2 \le y \le 2$

Average concentration

$$= \frac{1}{4(4)}\int_{-2}^{2}\int_{-2}^{2}[100-24(x^2+y^2)]\,dy\,dx = \frac{1}{16}\int_{-2}^{2}\left[100y - 24x^2y - 8y^3\right]_{-2}^{2} dx$$

$$= \frac{1}{16}\int_{-2}^{2}[400-96x^2-128]\,dx = \frac{1}{16}\int_{-2}^{2}(272-96x^2)\,dx$$

$$= \frac{1}{16}\left[272x - 32x^3\right]_{-2}^{2} = \frac{1}{16}(544-256) - \frac{1}{16}(-544+256) = 18 + 18 = 36 \text{ parts per million.}$$

(8-6)

31. $n(P_1, P_2, d) = 0.001\dfrac{P_1P_2}{d}$

$n(100{,}000, 50{,}000, 100) = 0.001\dfrac{100{,}000\times 50{,}000}{100} = 50{,}000$ (8-1)

32.

	x_k	y_k	$x_k y_k$	x_k^2
	30	60	1,800	900
	50	75	3,750	2,500
	60	80	4,800	3,600
	70	85	5,950	4,900
	90	90	8,100	8,100
Totals	300	390	24,400	20,000

Thus, $\sum_{k=1}^{5} x_k = 300,\ \sum_{k=1}^{5} y_k = 390,\ \sum_{k=1}^{5} x_k y_k = 24{,}400,\ \sum_{k=1}^{5} x_k^2 = 20{,}000.$

Substituting these values into the formulas for a and d, we have:

$$a = \frac{5(24{,}400) - (300)(390)}{5(20{,}000) - (300)^2} = \frac{5000}{10{,}000} = 0.5$$

$$d = \frac{390 - 0.5(300)}{5} = \frac{240}{5} = 48$$

Therefore, the least squares line is: $y = 0.5x + 48$

When $x = 40$, $y = 0.5(40) + 48 = 68$. (8-5)

33. (A) We use the linear regression feature on a graphing utility with 1960 as $x = 0$.

LinReg
y=ax+b
a=.734
b=49.93333333

Thus, the least squares line is: $y = 0.734x + 49.93$.

(B) The year 2025 corresponds to $x = 65$; $y(65) \approx 97.64$ people/sq. mi.

(C)

QuadReg
y=ax²+bx+c
a=.0026428571
b=.6018571429
c=50.81428571

Quadratic regression:
$y(65) \approx 101.10$ people/sq. mi.

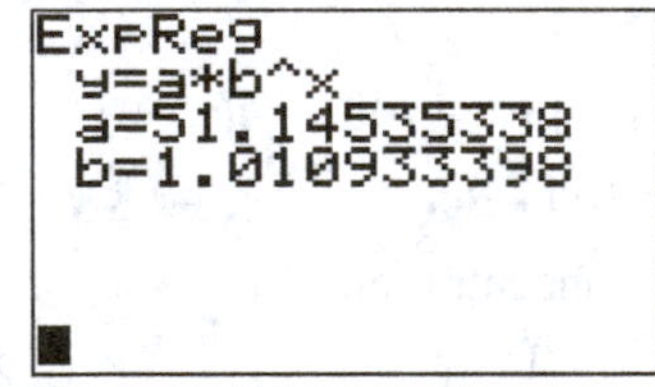

Exponential regression:
$y(65) \approx 103.70$ people/sq. mi. (8-5)

34.

LinReg
y=ax+b
a=1.069267604
b=.5223226384
r=.9793163189

(A) The least squares line is $y \approx 1.069x + 0.522$.

(B) Evaluate the result in (A) at $x = 60$: $y \approx 64.68$ yr

(C)

Evaluate at $x = 60$: $y \approx 64.78$ yr

Evaluate at $x = 60$: $y \approx 64.80$ yr (8-5)

CHAPTER 8 TRIGONOMETRIC FUNCTIONS

EXERCISE 8-1

Things to remember:

1. θ_{rad} = radian measure of θ

$$= \frac{\text{Arc length}}{\text{Radius}}$$

$$= \frac{s}{R}$$ [Note: If $R = 1$, then $\theta_{\text{rad}} = s$.]

2. DEGREE-RADIAN CONVERSION FORMULA

$$\frac{\theta_{\text{deg}}}{180^\circ} = \frac{\theta_{\text{rad}}}{\pi \text{ rad}}$$

3. On a unit circle, where the origin is at the center, $\cos\theta$ and $\sin\theta$ are measured by the abscissa and ordinate of point P, respectively, as shown in the figure.

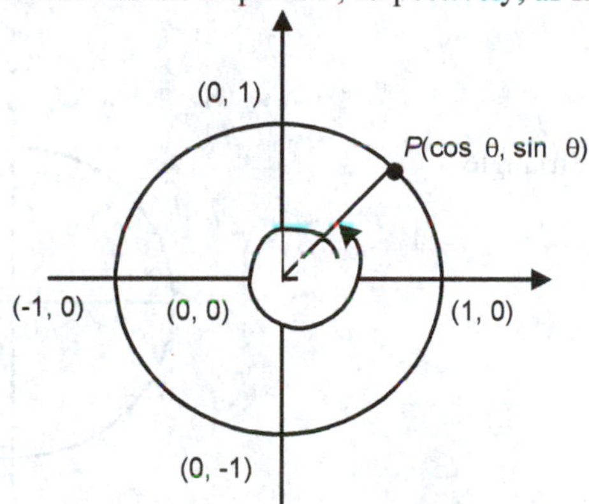

For any real number x, $\sin x = \sin(x \text{ radians})$, $\cos x = \cos(x \text{ radians})$.

4. $\tan x = \frac{\sin x}{\cos x}$, $\cos x \neq 0$

5. $\cot x = \frac{\cos x}{\sin x}$, $\sin x \neq 0$

6. $\sec x = \frac{1}{\cos x}$, $\cos x \neq 0$

7. $\csc x = \frac{1}{\sin x}$, $\sin x \neq 0$

In Problems 1 - 7, use $\theta_{rad} = \frac{\pi}{180}\theta_{deg}$ *and* $\theta_{deg} = \frac{180^\circ}{\pi}\theta_{rad}$

1. $\theta_{\text{rad}} = \frac{\pi}{180°} \cdot 60°$
$= \frac{\pi}{3}$

3. $\theta_{\text{rad}} = \frac{\pi}{180°} \cdot 135°$
$= \frac{3\pi}{4}$

5. $\theta_{\text{deg}} = \frac{180°}{\pi} \cdot \frac{-\pi}{4}$
$= -45°$

7. $\theta_{\text{deg}} = \frac{180°}{\pi} \cdot \frac{3\pi}{2}$
$= 270°$

9. $\sin 60° = \frac{\sqrt{3}}{2}$
(Figure 8)

11. $\cos 135°) = \frac{-1}{\sqrt{2}}$
(Figures 6 and 8)

13. $\sin 90° = 1$
(Figure 6)

15. $\cos(-90°) = 0$
(Figure 6)

17. $\cos\left(\frac{5\pi}{4}\right) = \frac{-1}{\sqrt{2}}$
(Figures 6 and 8)

19. $\sin\left(-\frac{\pi}{6}\right) = -\frac{1}{2}$
(Figures 6 and 8)

21. $\sin\left(\frac{3\pi}{2}\right) = -1$
(Figure 6)

23. $\cos\left(-\frac{11\pi}{6}\right) = \frac{\sqrt{3}}{2}$
(Figures 6 and 8)

25. Applying the Pythagorean Theorem to the triangle *OPQ*, we obtain:

$(\sin x)^2 + (\cos x)^2 = 1$

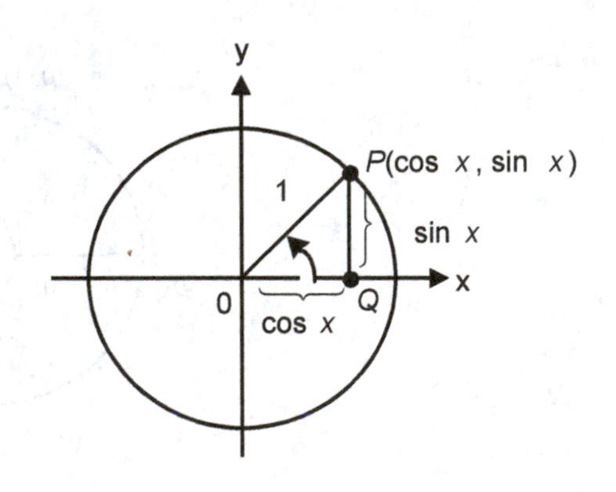

27. $\tan\left(\frac{3\pi}{4}\right) = -1$

29. $\csc\left(\frac{2\pi}{3}\right) = \frac{2}{\sqrt{3}}$

31. $\sec 90°$ not defined

33. $\cot(-150°) = \sqrt{3}$

35. $\csc\left(\frac{7\pi}{6}\right) = -2$

37. $\sec(-\pi) = -1$

39. $\tan 120° = -\sqrt{3}$

41. $\cot(-45°) = -1$

43. $\sin 10° = 0.1736$

45. $\cos(-52°) = 0.6157$

47. $\tan 1 = 1.5574$

49. $\sec(-1.56) = 92.6259$

51. $\csc 1° = 57.2987$

53. $\cot\left(\frac{\pi}{10}\right) = 3.0777$

55. $y = 2\sin \pi x$;
$0 \le x \le 2, -2 \le y \le 2$

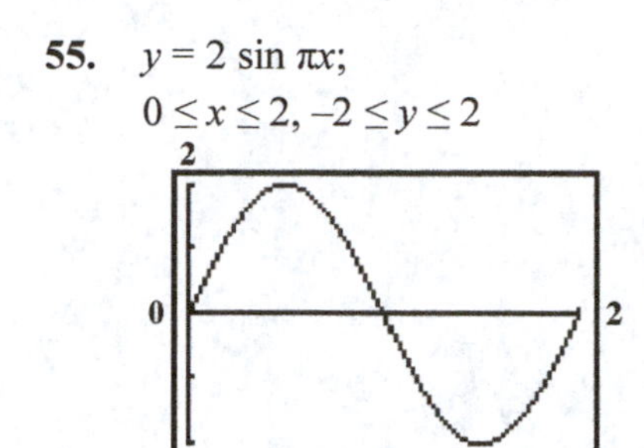

57. $y = 4 - 4\cos\frac{\pi x}{2}$;
$0 \le x \le 8, 0 \le y \le 8$

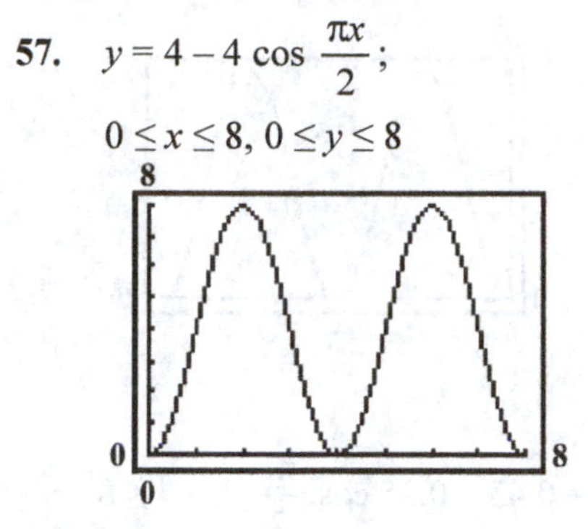

59. $\tan x = \frac{\sin x}{\cos x}$; dom $(\tan x) = \{x : x \neq \pm\frac{\pi}{2}, \pm\frac{3\pi}{2}, \ldots\} = \{x : x \neq (2n+1)\frac{\pi}{2}, n = 0, \pm 1, \pm 2, \ldots\}$

61. $\sec x = \frac{1}{\cos x}$; dom $(\sec x) = \{x : x \neq \pm\frac{\pi}{2}, \pm\frac{3\pi}{2}, \ldots\} = \{x : x \neq (2n+1)\frac{\pi}{2}, n = 0, \pm 1, \pm 2, \ldots\}$

63. $\csc x = \frac{1}{\sin x}$. Since $-1 \le \sin x \le 1$ for all x, $\frac{1}{\sin x}$ is either ≤ -1 or ≥ 1 for all x.

65. $\cot x = \frac{\cos x}{\sin x}$; $\lim_{x\to 0^+} \frac{\cos x}{\sin x} = \infty$, $\lim_{x\to \pi^-} \frac{\cos x}{\sin x} = -\infty$, the cotangent function decreases from $+\infty$ to $-\infty$ on the interval $(0, \pi)$.

67. (A) $P(t) = 5 - 5\cos\left(\frac{\pi}{26}t\right)$

$P(13) = 5 - 5\cos\left(\frac{\pi}{26}\cdot 13\right) = 5 - 5\cos\left(\frac{\pi}{2}\right) = 5 - 5(0) = 5$

$P(26) = 5 - 5\cos\left(\frac{\pi}{26}\cdot 26\right) = 5 - 5\cos(\pi) = 5 - 5(-1) = 10$

$P(39) = 5 - 5\cos\left(\frac{\pi}{26}\cdot 39\right) = 5 - 5\cos\left(\frac{3\pi}{2}\right) = 5 - 5(0) = 5$

$P(52) = 5 - 5\cos\left(\frac{\pi}{26}\cdot 52\right) = 5 - 5\cos(2\pi) = 5 - 5(1) = 0$

(B) $P(30) = 5 - 5\cos\left(\frac{\pi}{26}\cdot 30\right) \approx 5 - 5(-0.886) \approx 5 + 4.43 \approx 9.43$

$P(100) = 5 - 5\cos\left(\frac{\pi}{26}\cdot 100\right) \approx 5 - 5(0.886) \approx 5 - 4.43 \approx 0.57$

Interpretation: 30 weeks after January 1, the profit on a week's sales of bathing suits is \$943; 100 weeks after January 1, the profit on a week's sales of bathing suits is \$57.

(C)

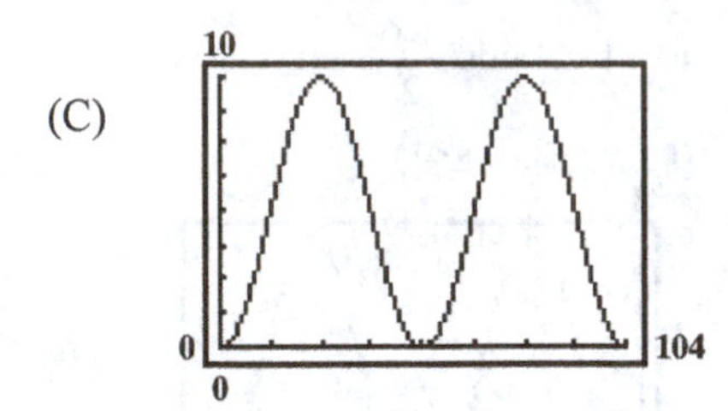

69. $V(t) = 0.45 - 0.35 \cos \frac{\pi t}{2}, 0 \le t \le 8$

(A) $V(0) = 0.45 - 0.35 \cos(0) = 0.45 - 0.35 = 0.10$

$V(1) = 0.45 - 0.35 \cos\left(\frac{\pi}{2}\right) = 0.45 - 0 = 0.45$

$V(2) = 0.45 - 0.35 \cos(\pi) = 0.45 + 0.35 = 0.80$

$V(3) = 0.45 - 0.35 \cos\left(\frac{3\pi}{2}\right) = 0.45 - 0 = 0.45$

$V(7) = 0.45 - 0.35 \cos\left(\frac{7\pi}{2}\right) = 0.45 - 0 = 0.45$

(B) $V(3.5) = 0.45 - 0.35 \cos\left(\frac{3.5\pi}{2}\right) \approx 0.45 - 0.2475 \approx 0.20$

$V(5.7) = 0.45 - 0.35 \cos\left(\frac{5.7\pi}{2}\right) \approx 0.45 + 0.3119 \approx 0.76$

Interpretation: The volume of air in the lungs of a normal seated adult 3.5 seconds after exhaling is approximately 0.20 liters; the volume of air is approximately 0.76 liters 5.7 seconds after exhaling.

(C)

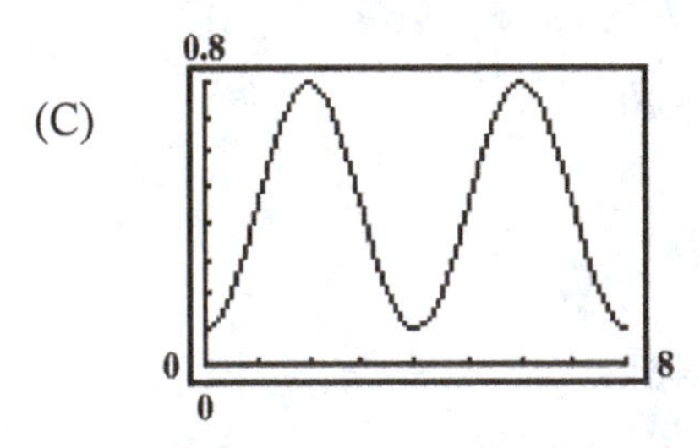

71. $d = -2.1 - 4 \sin 4\theta$

(A) $\theta = 30°$

$d = -2.1 - 4 \sin 4(30°)$

$= -2.1 - 4 \sin 120°$

$= -2.1 - 4 \cdot \frac{\sqrt{3}}{2} \approx -5.6°$

(B) $\theta = 10°$

$d = -2.1 - 4 \sin 4(10°)$

$= -2.1 - 4 \sin 40°$

$-2.1 - 4(0.6428) \approx -4.7°$

EXERCISE 8-2

Things to remember:

DERIVATIVE FORMULAS FOR SINE AND COSINE

1. Basic Form

$$\frac{d}{dx}\sin x = \cos x \qquad \frac{d}{dx}\cos x = -\sin x$$

2. Generalized Form
   For $u = u(x)$:

$$\frac{d}{dx}\sin u = \cos u\frac{du}{dx} \qquad \frac{d}{dx}\cos u = -\sin u\frac{du}{dx}$$

---

1. $y = \cos x$ is decreasing on $(0, \pi)$.

3. $y = \sin x$ is increasing on $\left(\frac{-\pi}{2}, \frac{\pi}{2}\right)$.

5. $y = \sin x$ is concave down on $(0, \pi)$.

7. $y = \cos x$ is concave down on $\left(\frac{-\pi}{2}, \frac{\pi}{2}\right)$

9. $\frac{d}{dx}(5\cos x) = 5\frac{d}{dx}\cos x = -5\sin x$ (using 1)

11. $\frac{d}{dx}\cos 5x = -\sin 5x\frac{d}{dx}5x = -5\sin 5x$ (using 2)

13. $\frac{d}{dx}\sin(x^2+1) = \cos(x^2+1)\frac{d}{dx}(x^2+1) = \cos(x^2+1)(2x) = 2x\cos(x^2+1)$

15. $\frac{d}{dw}\sin(w+\pi) = \cos(w+\pi)\frac{d}{dw}(w+\pi) = \cos(w+\pi)(1) = \cos(w+\pi)$

17. $\frac{d}{dt}t\sin t = t\frac{d}{dt}\sin t + \sin t\frac{d}{dt}t = t\cos t + \sin t$

19. $\frac{d}{dx}\sin x\cos x = \sin x\frac{d}{dx}\cos x + \cos x\frac{d}{dx}\sin x = \sin x(-\sin x) + \cos x(\cos x) = (\cos x)^2 - (\sin x)^2$

21. $\frac{d}{dx}(\sin x)^5 = 5(\sin x)^4\frac{d}{dx}\sin x = 5(\sin x)^4\cos x$

23. $\frac{d}{dx}\sqrt{\sin x} = \frac{d}{dx}(\sin x)^{1/2} = \frac{1}{2}(\sin x)^{-1/2}\frac{d}{dx}\sin x = \frac{1}{2}(\sin x)^{-1/2}\cos x = \frac{\cos x}{2\sqrt{\sin x}}$

25. $\frac{d}{dx}\cos\sqrt{x} = -\sin\sqrt{x}\frac{d}{dx}\sqrt{x}$ (using 2)

$$= -\frac{1}{2}x^{-1/2}\sin\sqrt{x} = -\frac{\sin\sqrt{x}}{2\sqrt{x}}$$

27. $f(x) = \sin x, \quad f'(x) = \cos x$

The slope of the graph of f at $x = \frac{\pi}{6}$ is: $f'\left(\frac{\pi}{6}\right) = \cos\frac{\pi}{6} = \frac{\sqrt{3}}{2} \approx 0.866$

29. f is increasing on $[-\pi, 0]$ $(f'(x) > 0)$; f is decreasing on $[0, \pi]$ $(f'(x) < 0)$; f has a local maximum at $x = 0$; the graph of f is concave upward on $\left(-\pi, -\frac{\pi}{2}\right)$ and on $\left(\frac{\pi}{2}, \pi\right)$ (f' is increasing on these intervals); the graph of f is concave downward on $\left(-\frac{\pi}{2}, \frac{\pi}{2}\right)$ (f' is decreasing on this interval); $f(x) = \cos x$, $f'(x) = -\sin x$.

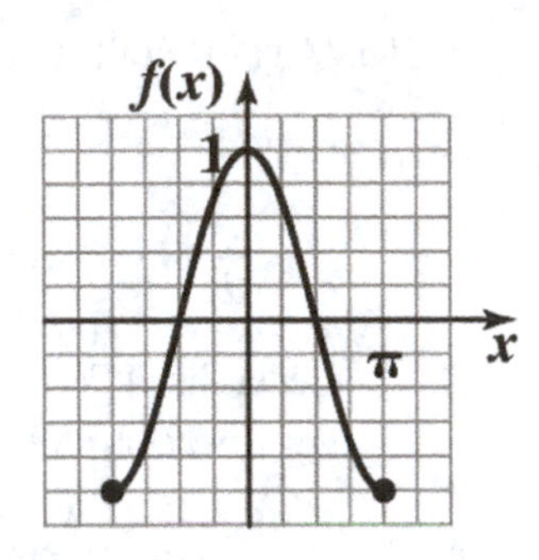

31. $\frac{d}{dx}\csc(\pi x) = -\csc(\pi x)\cot(\pi x)\frac{d}{dx}(\pi x) = -\csc(\pi x)\cot(\pi x)(\pi) = -\pi\csc(\pi x)\cot(\pi x)$

33. $\frac{d}{dx}\cot\left(\frac{\pi x}{2}\right) = -\csc^2\left(\frac{\pi x}{2}\right)\frac{d}{dx}\left(\frac{\pi x}{2}\right) = -\csc^2\left(\frac{\pi x}{2}\right)\left(\frac{\pi}{2}\right) = -\frac{\pi}{2}\csc^2\left(\frac{\pi x}{2}\right)$

35. $\frac{d}{dx}\cos(xe^x) = -\sin(xe^x)\frac{d}{dx}(xe^x) = -\sin(xe^x)(xe^x + e^x) = -(x+1)e^x\sin(xe^x)$

37. $\frac{d}{dx}\tan(x^2) = \sec^2(x^2)\frac{d}{dx}x^2 = \sec^2(x^2)(2x) = 2x\sec^2(x^2)$

39. $f(x) = e^x \sin x$

$f'(x) = e^x(\cos x) + (\sin x)e^x = e^x(\sin x + \cos x)$
$f''(x) = e^x(\cos x - \sin x) + (\sin x + \cos x)e^x = 2e^x \cos x$

41. $y = x \sin \pi x$;

$0 \le x \le 9, -9 \le y \le 9$

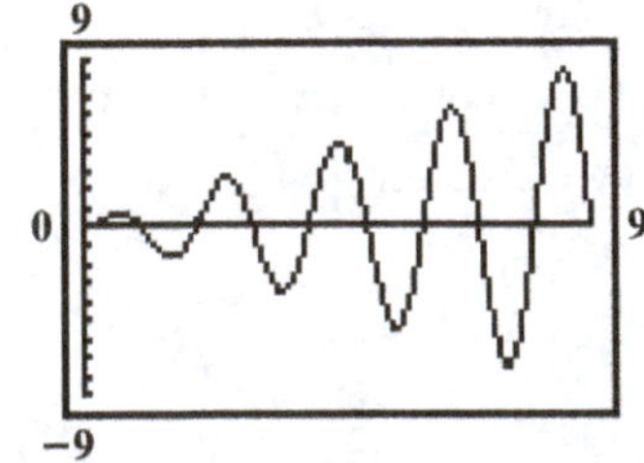

43. $y = \frac{\cos \pi x}{x}$;

$0 \le x \le 8, \ -2 \le y \le 3$

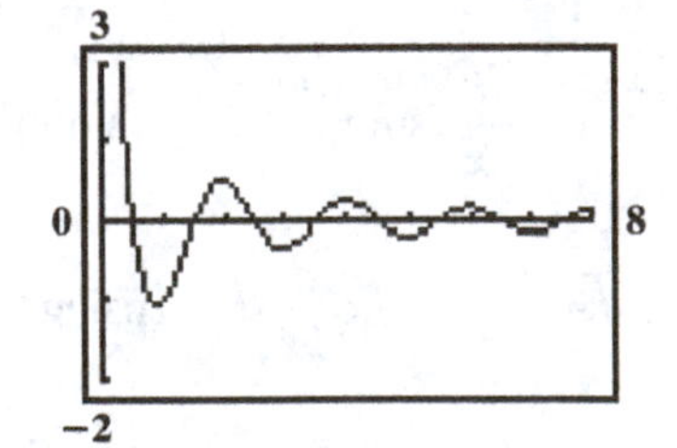

45. $y = e^{-0.3x} \sin \pi x$,
$0 \le x \le 10, -1 \le y \le 1$

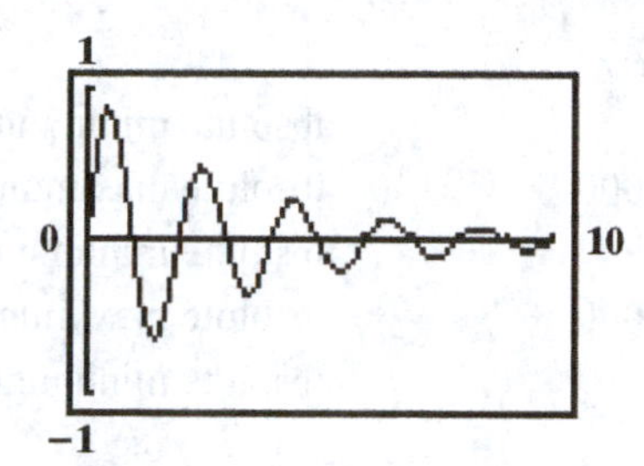

47. $P(t) = 5 - 5\cos\left(\frac{\pi t}{26}\right), 0 \le t \le 104$

(A) $P'(t) = -5\left[-\sin\left(\frac{\pi t}{26}\right)\left(\frac{\pi}{26}\right)\right] = \frac{5\pi}{26}\sin\left(\frac{\pi t}{26}\right), 0 \le t \le 104$

(B) $P'(8) = \frac{5\pi}{26}\sin\left(\frac{8\pi}{26}\right) \approx 0.50$ (hundred) or \$50 per week

$P'(26) = \frac{5\pi}{26}\sin\left(\frac{26\pi}{26}\right) = 0$ or \$0 per week

$P'(50) = \frac{5\pi}{26}\sin\left(\frac{50\pi}{26}\right) \approx -0.14$ (hundred) or –\$14 per week

(C) $P'(t) = \frac{5\pi}{26}\sin\left(\frac{\pi t}{26}\right) = 0, 0 < t < 104$

$\sin\left(\frac{\pi t}{26}\right) = 0.$

Therefore, the critical values are:

$\frac{\pi t}{26} = \pi$, or $t = 26$; $\frac{\pi t}{26} = 2\pi$ or $t = 52$; $\frac{\pi t}{26} = 3\pi$ or $t = 78$.

Now,

$P''(t) = \frac{5\pi^2}{676}\cos\left(\frac{\pi t}{26}\right)$

$P''(26) = \frac{5\pi^2}{676}\cos(\pi) = -\frac{5\pi^2}{676} < 0$

$P''(52) = \frac{5\pi^2}{676}\cos(2\pi) = \frac{5\pi^2}{676} > 0$

$P''(78) = \frac{5\pi^2}{676}\cos(3\pi) = -\frac{5\pi^2}{676} < 0$

Thus,

t	$P(t)$	
26	\$1000	local maximum
52	\$0	local minimum
78	\$1000	local maximum

(D)

t	$P(t)$	
0	\$0	absolute minimum
26	\$1000	absolute maximum
52	\$0	absolute minimum
78	\$1000	absolute maximum
104	\$0	absolute minimum

(E) The results in part (C) are illustrated by the graph of f shown at the right.

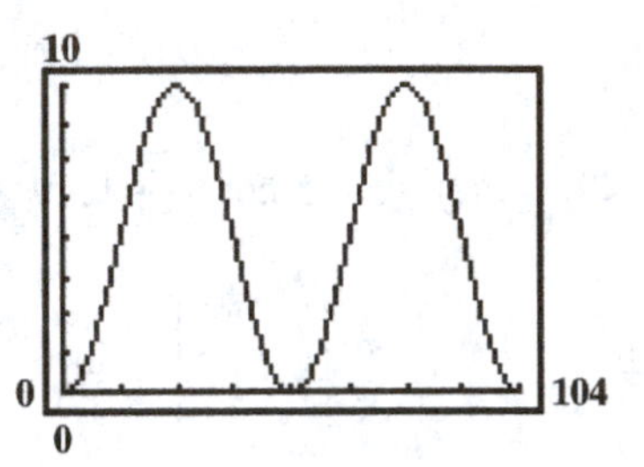

49. $V(t) = 0.45 - 0.35\cos\left(\frac{\pi t}{2}\right), 0 \le t \le 8$

(A) $V'(t) = -0.35\left[-\sin\left(\frac{\pi t}{2}\right)\right]\left(\frac{\pi}{2}\right) = \frac{0.35\pi}{2}\sin\left(\frac{\pi t}{2}\right), 0 \le t \le 8$

(B) $V'(3) = \frac{0.35\pi}{2}\sin\left(\frac{3\pi}{2}\right) = -\frac{0.35\pi}{2} \approx -0.55$ liters per second

$V'(4) = \frac{0.35\pi}{2}\sin\left(\frac{4\pi}{2}\right) = 0.00$ liters per second

$V'(5) = \frac{0.35\pi}{2}\sin\left(\frac{5\pi}{2}\right) = \frac{0.35\pi}{2} \approx 0.55$ liters per second

(C) $V'(t) = \frac{0.35\pi}{2}\sin\left(\frac{\pi t}{2}\right) = 0, 0 < t < 8$

$\sin\left(\frac{\pi t}{2}\right) = 0$

Therefore, the critical values are:

$\frac{\pi t}{2} = \pi$ or $t = 2$; $\frac{\pi t}{2} = 2\pi$ or $t = 4$; $\frac{\pi t}{2} = 3\pi$ or $t = 6$.

Now,

$V''(t) = \frac{0.35\pi}{2}\cos\left(\frac{\pi t}{2}\right)\left(\frac{\pi}{2}\right) = \frac{0.35\pi^2}{4}\cos\left(\frac{\pi t}{2}\right)$

$V''(2) = \frac{0.35\pi^2}{4}\cos(\pi) = -\frac{0.35\pi^2}{4} < 0$

$V''(4) = \frac{0.35\pi^2}{4}\cos(2\pi) = \frac{0.35\pi^2}{4} > 0$

$V''(6) = \frac{0.35\pi^2}{4}\cos(3\pi) = -\frac{0.35\pi^2}{4} < 0$

Thus,

t	$V(t)$	
2	0.80	local maximum
4	0.10	local minimum
6	0.80	local maximum

(D)

t	$V(t)$	
0	0.10	absolute minimum
2	0.80	absolute maximum
4	0.10	absolute minimum
6	0.80	absolute maximum
8	0.10	absolute minimum

Thus, 0.10 liters is the absolute minimum and 0.80 liters is the absolute maximum of V for $0 \le t \le 8$.

(E) The results in part (C) are illustrated by the graph of f shown at the right.

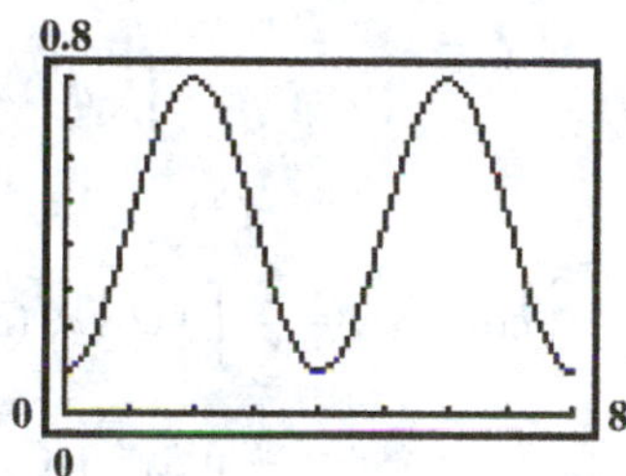

EXERCISE 8-3

Things to remember:

INDEFINITE INTEGRAL FORMULAS FOR SINE AND COSINE

1. Basic Formulas

$$\int \sin x\, dx = -\cos x + C \quad \text{and} \quad \int \cos x\, dx = \sin x + C$$

2. Generalized Formulas
For $u = u(x)$

$$\int \sin u\, du = -\cos u + C \quad \text{and} \quad \int \cos u\, du = \sin u + C$$

1. $\sin x$ on $[0, 2\pi]$: $\sin x \ge 0$ on $[0, \pi]$.

3. $\cos x$ on $[0, 2\pi]$; $\cos x > \frac{1}{2}$ on $[0, \pi/3) \cup (5\pi/3, 2\pi]$.

5. $|\sin x|$ on $[0, 2\pi]$; $|\sin x| = \frac{\sqrt{2}}{2}$, $x = \frac{\pi}{4}, \frac{3\pi}{4}, \frac{5\pi}{4}, \frac{7\pi}{4}$.

7. $\tan x$ on $(-\pi/2, \pi/2)$: $\tan x \le 1$ on $(-\pi/2, \pi/4]$

9. $\int \sin t\, dt = -\cos t + C$ (using 1)

11. $\int \cos 3x\, dx = \frac{1}{3}\int \cos 3x(3\, dx)$ [Let $u = 3x$. Then $du = 3\, dx$.]

$= \frac{1}{3}\sin 3x + C$ (using 2)

13. $\int (\sin x)^{12}\cos x\, dx = \int u^{12} du$ [Let $u = \sin x$. Then $du = \cos x\, dx$.]

$= \frac{u^{13}}{13} + C = \frac{(\sin x)^{13}}{13} + C$

15. $\int \sqrt[3]{\cos x}\sin x\, dx = -\int \sqrt[3]{\cos x}(-\sin x)dx$ [Let $u = \cos x$. Then $du = -\sin x\, dx$.]

$= -\int u^{1/3} du = -\frac{u^{4/3}}{\frac{4}{3}} + C = -\frac{3}{4}(\cos x)^{4/3} + C$

17. $\int x^2 \cos x^3 dx = \frac{1}{3}\int \cos x^3(3x^2)dx$ [Let $u = x^3$. Then $du = 3x^2 dx$.]

$= \frac{1}{3}\sin x^3 + C$

19. $\int_0^{\pi/2} \cos x\, dx = \sin x\Big]_0^{\pi/2} = \sin\frac{\pi}{2} - \sin 0 = 1 - 0 = 1$

21. $\int_{\pi/2}^{\pi} \sin x\, dx = -\cos x\Big]_{\pi/2}^{\pi} = -\left(\cos\pi - \cos\frac{\pi}{2}\right) = -(-1-0) = 1$

23. The shaded area $= \int_{\pi/6}^{\pi/3} \cos dx = \sin x\Big]_{\pi/6}^{\pi/3} = \sin\frac{\pi}{3} - \sin\frac{\pi}{6} = \frac{\sqrt{3}}{2} - \frac{1}{2} \approx 0.866 - 0.5 = 0.366$

25. $\int_0^2 \sin x\, dx = -\cos x\Big]_0^2 = -(\cos 2 - \cos 0) \approx -(-0.4161 - 1) \approx 1.4161$

27. $\int_1^2 \cos x\, dx = \sin x\Big]_1^2 = \sin 2 - \sin 1 \approx 0.9093 - 0.8415 \approx 0.0678$

29. $\int e^{\sin x}\cos x\, dx = \int e^u du = e^u + C = e^{\sin x} + C$

[Let $u = \sin x$. Then $du = \cos x\, dx$.]

31. $\int \frac{\cos x}{\sin x} dx = \int \frac{du}{u} = \ln|u| + C = \ln|\sin x| + C$

[Let $u = \sin x$. Then $du = \cos x\, dx$.]

33. $\int \tan x\, dx = \int \frac{\sin x}{\cos x} dx = -\int \frac{du}{u} = -\ln|u| + C = -\ln|\cos x| + C$

[Let $u = \cos x$. Then $du = -\sin x\, dx$.]

35. (A)

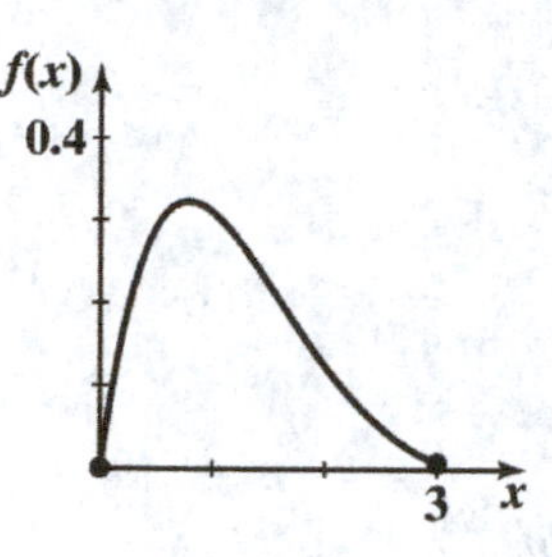

(B) $\Delta x = \frac{3-0}{6} = \frac{1}{2}$

$$L_6 = f(0)\frac{1}{2} + f\left(\frac{1}{2}\right)\frac{1}{2} + f(1)\frac{1}{2} + f\left(\frac{3}{2}\right)\frac{1}{2} + f(2)\frac{1}{2} + f\left(\frac{5}{2}\right)\frac{1}{2}$$

$$\approx [0 + 0.291 + 0.310 + 0.223 + 0.123 + 0.049]\frac{1}{2} = 0.498$$

37. $P(t) = 5 - 5\cos\left(\frac{\pi t}{26}\right), 0 \le t \le 104$

(A) Total profit during the two-year period:

$$T = \int_0^{104}\left[5 - 5\cos\left(\frac{\pi t}{26}\right)\right]dt = \int_0^{104} 5\,dt - 5\int_0^{104}\cos\left(\frac{\pi t}{26}\right)dt$$

$$= 5t\Big]_0^{104} - 5\left(\frac{26}{\pi}\right)\int_0^{104}\cos\left(\frac{\pi t}{26}\right)\left(\frac{\pi}{26}\right)dt = 520 - \left[\frac{130}{\pi}\sin\frac{\pi t}{26}\right]_0^{104} = 520$$

Thus, $T = \$520$ hundred or \$52,000.

(B) Total profit earned from $t = 13$ to $t = 26$:

$$T = \int_{13}^{26}\left[5 - 5\cos\left(\frac{\pi t}{26}\right)\right]dt = \int_{13}^{26} 5\,dt - 5\int_{13}^{26}\cos\left(\frac{\pi t}{26}\right)dt$$

$$= 5t\Big]_{13}^{26} - \frac{5(26)}{\pi}\int_{13}^{26}\cos\left(\frac{\pi t}{26}\right)\left(\frac{\pi}{26}\right)dt = 65 - \left[\frac{130}{\pi}\sin\frac{\pi t}{26}\right]_{13}^{26} = 65 + \frac{130}{\pi} \approx 106.38$$

Thus, $T = \$106.38$ hundred or \$10,638.

(C)

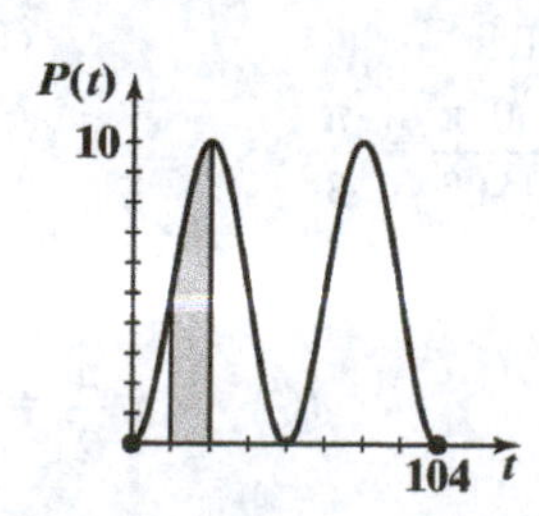

39. $P(n) = 1 + \cos\left(\frac{\pi n}{26}\right), 0 \le n \le 104$

(A) Total amount of pollutants during the two-year period:

$$T = \int_0^{104}\left[1+\cos\left(\frac{\pi n}{26}\right)\right]dn = \int_0^{104} dn + \int_0^{104}\cos\left(\frac{\pi n}{26}\right)dn$$

$$= n\Big]_0^{104} + \frac{26}{\pi}\int_0^{104}\cos\left(\frac{\pi n}{26}\right)\left(\frac{\pi}{26}\right)dn = 104 + \frac{26}{\pi}\sin\left(\frac{\pi n}{26}\right)\Big|_0^{104} = 104 \text{ tons}$$

(B) Total amount of pollutants from $t = 13$ to $t = 52$:

$$T = \int_{13}^{52}\left[1+\cos\left(\frac{\pi n}{26}\right)\right]dn = \int_{13}^{52} dn + \int_{13}^{52}\cos\left(\frac{\pi n}{26}\right)dn$$

$$= n\Big]_{13}^{52} - \frac{5(26)}{\pi}\int_{13}^{52}\cos\left(\frac{\pi n}{26}\right)\left(\frac{\pi}{26}\right)dn = 39 - \left[\frac{26}{\pi}\sin\frac{\pi n}{26}\right]_{13}^{52} = 39 + \frac{26}{\pi} \approx 30.72 \text{ tons}$$

(C)

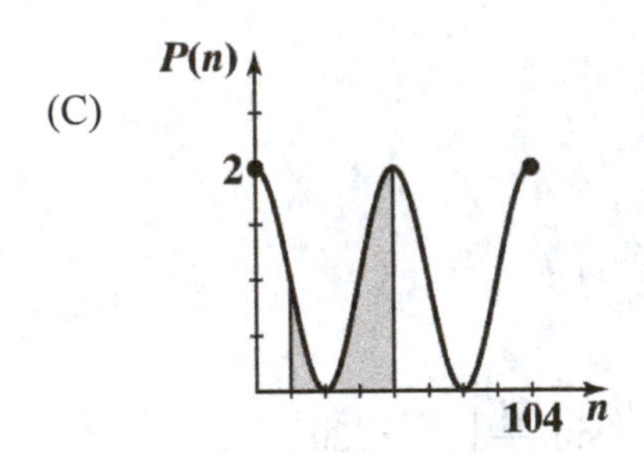

CHAPTER 8 REVIEW

1. (A) $\frac{\theta°}{180°} = \frac{\theta_{rad}}{\pi}$

$\theta_{rad} = \frac{\theta°\pi}{180°}$ [Note: θ_{rad} = radian measure of θ.]

$\theta° = 30°$

$\theta_{rad} = \frac{30°\pi}{180°} = \frac{\pi}{6}$

(B) $\theta° = 45°$

$\theta_{rad} = \frac{45°\pi}{180°} = \frac{\pi}{4}$

(C) $\theta° = 60°$

$\theta_{rad} = \frac{60°\pi}{180°} = \frac{\pi}{3}$

(D) $\theta° = 90°$

$\theta_{rad} = \frac{90°\pi}{180°} = \frac{\pi}{2}$

(8-1)

2. (A) $\cos\pi = -1$ (B) $\sin 0 = 0$ (C) $\sin\frac{\pi}{2} = 1$ (8-1)

3. $\frac{d}{dm}\cos m = -\sin m$ (8-2)

4. $\frac{d}{du}\sin u = \cos u$ (8-2)

5. $\frac{d}{dx}\sin(x^2-2x+1)=\cos(x^2-2x+1)\frac{d}{dx}(x^2-2x+1)=(2x-2)\cos(x^2-2x+1)$ (8-2)

6. $\int \sin 3t\,dt=\frac{1}{3}\int \sin(3t)3dt=\frac{1}{3}\int \sin u\,du=-\frac{1}{3}\cos u+C=-\frac{1}{3}\cos 3t+C$

[Let $u=3t$. Then $du=3\,dt$.] (8-3)

7. (A) $\theta° = \frac{180°\theta_{\text{rad}}}{\pi}$

$\theta_{\text{rad}}=\frac{\pi}{6}$

$\theta° = \frac{180°\left(\frac{\pi}{6}\right)}{\pi}=30°$

(B) $\theta_{\text{rad}}=\frac{\pi}{4}$

$\theta° = \frac{180°\left(\frac{\pi}{4}\right)}{\pi}=45°$

(C) $\theta_{\text{rad}}=\frac{\pi}{3}$

$\theta° = \frac{180°\left(\frac{\pi}{3}\right)}{\pi}=60°$

(D) $\theta_{\text{rad}}=\frac{\pi}{2}$

$\theta° = \frac{180°\left(\frac{\pi}{2}\right)}{\pi}=90°$ (8-1)

8. (A) $\sin\frac{\pi}{6}=\frac{1}{2}$ (B) $\cos\frac{\pi}{4}=\frac{\sqrt{2}}{2}$ (C) $\sin\frac{\pi}{3}=\frac{\sqrt{3}}{2}$ (8-1)

9. (A) $\cos 33.7\approx -0.6543$ (B) $\sin(-118.4)\approx 0.8308$ (8-1)

10. $\frac{d}{dx}(x^2-1)\sin x=(x^2-1)\frac{d}{dx}\sin x+\sin x\frac{d}{dx}(x^2-1)=(x^2-1)\cos x+2x\sin x$ (8-2)

11. $\frac{d}{dx}(\sin x)^6=6(\sin x)^5\frac{d}{dx}\sin x=6(\sin x)^5\cos x$ (8-2)

12. $\frac{d}{dx}\sqrt[3]{\sin x}=\frac{d}{dx}(\sin x)^{1/3}=\frac{1}{3}(\sin x)^{-2/3}\frac{d}{dx}\sin x=\frac{\cos x}{3(\sin x)^{2/3}}$ (8-2)

13. $\int t\cos(t^2-1)dt=\frac{1}{2}\int \cos(t^2-1)2t\,dt=\frac{1}{2}\int \cos u\,du=\frac{1}{2}\sin u+C=\frac{1}{2}\sin(t^2-1)+C$ (8-3)

[Let $u=t^2-1$. Then $du=2t\,dt$.]

14. $\int_0^{\pi}\sin u\,du=-\cos u\Big|_0^{\pi}=-[\cos\pi-\cos 0]=2$ (8-3)

15. $\int_0^{\pi/3} \cos x\, dx = \sin x \Big|_0^{\pi/3} = \sin\frac{\pi}{3} - \sin 0 = \frac{\sqrt{3}}{2}$ (8-3)

16. $\int_1^{2.5} \cos x\, dx = \sin x \Big|_1^{2.5} = \sin(2.5) - \sin(1) \approx 0.5985 - 0.8415 = -0.2430$ (8-3)

17. $y = \cos x,\ \ y' = -\sin x$

$y'\Big|_{x=\pi/4} = -\sin\left(\frac{\pi}{4}\right) = -\frac{\sqrt{2}}{2}$ (8-2)

18. $A = \int_{\pi/4}^{3\pi/4} \sin x\, dx = -\cos x\Big]_{\pi/4}^{3\pi/4} = \left(-\cos\frac{3\pi}{4}\right) - \left(-\cos\frac{\pi}{4}\right) = \sqrt{2}$ (8-3)

19. (A)

f(x)
1
5
x

(B) $\Delta x = \frac{5-1}{4} = 1$

$R_4 = f(2)1 + f(3)1 + f(4)1 + f(5)1 \approx 0.455 + 0.047 + (-0.189) + (-0.192) = 0.121$ (5-4, 8-3)

20. $\theta_{\text{rad}} = \frac{\theta°\pi}{180°}$

Set $\theta° = 15°$. Then $\theta_{\text{rad}} = \frac{15°\pi}{180°} = \frac{\pi}{12}$ (8-1)

21. (A) $\sin\left(\frac{3\pi}{2}\right) = -1$ (B) $\cos\left(\frac{5\pi}{6}\right) = -\frac{\sqrt{3}}{2}$ (C) $\sin\left(-\frac{\pi}{6}\right) = -\frac{1}{2}$ (8-1)

22. $\frac{d}{du}\tan u = \frac{d}{du}\frac{\sin u}{\cos u} = \frac{\cos u \frac{d}{du}\sin u - \sin u \frac{d}{du}\cos u}{[\cos u]^2} = \frac{\cos u(\cos u) - \sin u(-\sin u)}{\cos^2 u}$

$= \frac{[\cos u]^2 + [\sin u]^2}{[\cos u]^2} = \frac{1}{[\cos u]^2} = [\sec u]^2$ (8-2)

23. $\frac{d}{dx}e^{\cos x^2} = e^{\cos x^2}\frac{d}{dx}\cos x^2 = e^{\cos x^2}(-\sin x^2)\frac{d}{dx}x^2 = e^{\cos x^2}(-\sin x^2)2x = -2x\sin x^2 e^{\cos x^2}$ (8-2)

24. $\int e^{\sin x} \cos x \, dx = \int e^u \, du = e^u + C = e^{\sin x} + C$

[Let $u = \sin x$. Then $du = \cos x \, dx$.] (8-3)

25. $\int \tan x \, dx = \int \frac{\sin x}{\cos x} dx = -\int \frac{1}{u} du = -\ln u + C = -\ln \cos x + C$

[Let $u = \cos x$. Then $du = -\sin x \, dx$.] (8-3)

26. $\int_2^5 (5 + 2\cos 2x)dx = \int_2^5 5\, dx + \int_2^5 \cos(2x)2\, dx = 5x \Big|_2^5 + \sin 2x \Big|_2^5$

[Note:In the second integral, we let $u = 2x$ and $du = 2dx$.]

$= 25 - 10 + \sin 10 - \sin 4 = 15 - 0.5440 + 0.7568 = 15.2128$ (8-3)

27. $y = \frac{\sin \pi x}{0.2x}$;
$1 \le x \le 8, -4 \le y \le 4$

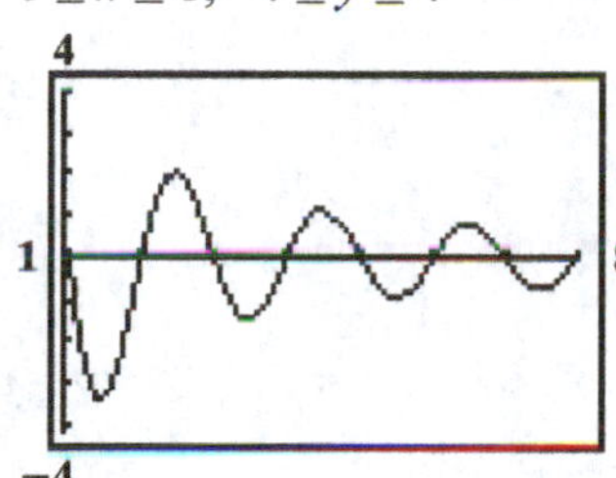

28. $y = 0.5x \cos \pi x$;

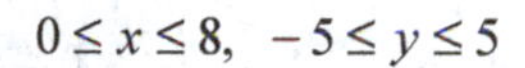
$0 \le x \le 8, \; -5 \le y \le 5$

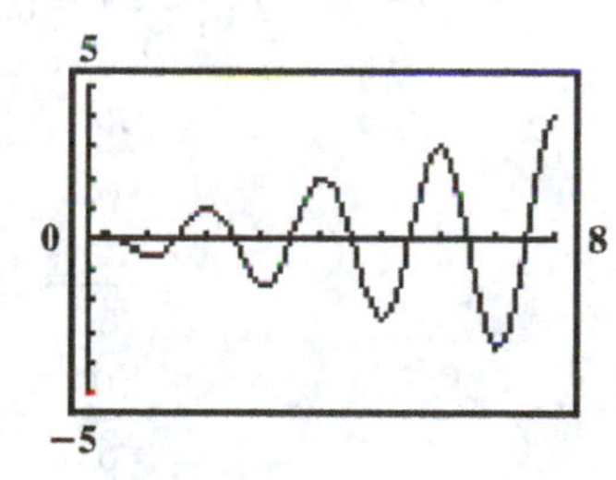

(8-2, 8-3)

29. $y = 3 - 2\cos \pi x$;
$0 \le x \le 6, 0 \le y \le 5$

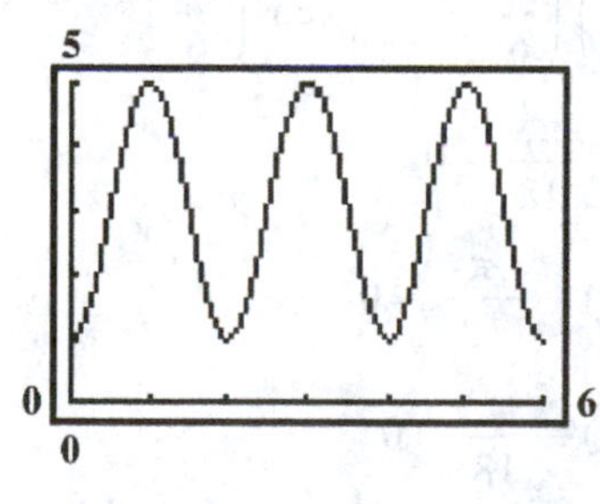

(8-2, 8-3)

30. $R(t) = 3 + 2\cos\left(\frac{\pi t}{6}\right), 0 \le t \le 24$

(A) $R(0) = 3 + 2\cos(0) = 3 + 2(1) = \$5$ thousand

$R(2) = 3 + 2\cos\left(\frac{2\pi}{6}\right) = 3 + 2\left(\frac{1}{2}\right) = \$4$ thousand

$R(3) = 3 + 2\cos\left(\frac{3\pi}{6}\right) = 3 + 2(0) = \$3$ thousand

$R(6) = 3 + 2\cos\left(\frac{6\pi}{6}\right) = 3 + 2(-1) = \$1$ thousand

(B) $R(1) = 3 + 2\cos\left(\frac{\pi}{6}\right) = 3 + 2\left(\frac{\sqrt{3}}{2}\right) \approx \$4.732$ thousand

$R(22) = 3 + 2\cos\left(\frac{22\pi}{6}\right) = 3 + 2\left(\frac{1}{2}\right) = \$4$ thousand

Interpretation: The revenue is $4,732 for a month of sweater sales 1 month after January 1; the revenue is $4,000 for a month of sweater sales 22 months after January 1. (8-1)

31. (A) $R'(t) = 2\left[-\sin\left(\frac{\pi t}{6}\right)\left(\frac{\pi}{6}\right)\right] = -\frac{\pi}{3}\sin\left(\frac{\pi t}{6}\right), 0 \le t \le 24$

(B) $R'(3) = -\frac{\pi}{3}\sin\left(\frac{3\pi}{6}\right) = -\frac{\pi}{3} \approx -\$1.047$ thousand or –$1047 per month

$R'(10) = -\frac{\pi}{3}\sin\left(\frac{10\pi}{6}\right) = -\frac{\pi}{3}\left(-\frac{\sqrt{3}}{2}\right) \approx \$0.907$ thousand or $907 per month

$R'(18) = -\frac{\pi}{3}\sin\left(\frac{18\pi}{6}\right) = -\frac{\pi}{3}(0) = \$0.00$

(C) Critical values: $R'(t) = -\frac{\pi}{3}\sin\left(\frac{\pi t}{6}\right) = 0$

$$\sin\left(\frac{\pi t}{6}\right) = 0, 0 < t < 24$$

$$\frac{\pi t}{6} = \pi \text{ or } t = 6$$

$$\frac{\pi t}{6} = 2\pi \text{ or } t = 12$$

$$\frac{\pi t}{6} = 3\pi \text{ or } t = 18$$

$$R''(t) = -\frac{\pi}{3}\cos\left(\frac{\pi t}{6}\right)\left(\frac{\pi}{6}\right) = \frac{-\pi^2}{18}\cos\left(\frac{\pi t}{6}\right)$$

$$R''(6) = \frac{-\pi^2}{18}\cos\pi = \frac{\pi^2}{18} > 0$$

$$R''(12) = \frac{-\pi^2}{18}\cos(2\pi) = \frac{-\pi^2}{18} < 0$$

$$R''(18) = \frac{-\pi^2}{18}\cos(3\pi) = \frac{\pi^2}{18} > 0$$

Thus,

t	$R(t)$	
6	$1000	local minimum
12	$5000	local maximum
18	$1000	local minimum

(D)

t	$R(t)$	
0	$5000	absolute maximum
6	$1000	absolute minimum
12	$5000	absolute maximum
18	$1000	absolute minimum
24	$5000	absolute maximum

(E) The results in part (C) are illustrated by the graph of R shown at the right.

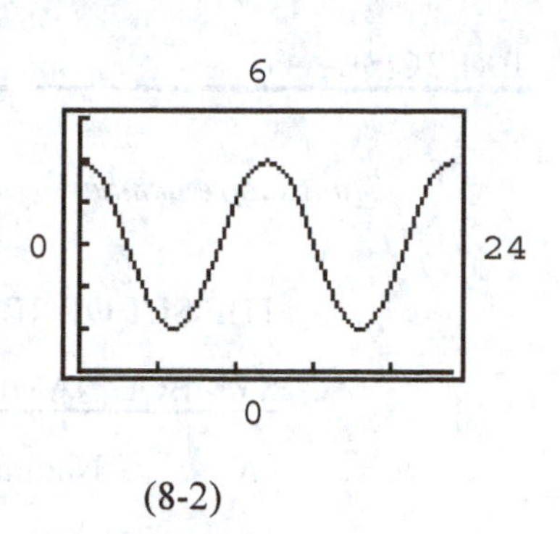

(8-2)

32. (A) Total revenue: $T = \int_0^{24}\left[3+2\cos\left(\frac{\pi t}{6}\right)\right]dt = \int_0^{24} 3\,dt + 2\int_0^{24}\cos\left(\frac{\pi t}{6}\right)dt$

$$= 3t\Big]_0^{24} + \frac{12}{\pi}\int_0^{24}\cos\left(\frac{\pi t}{6}\right)\left(\frac{\pi}{6}\right)dt = 72 + \frac{12}{\pi}\sin\left[\frac{\pi t}{6}\right]_0^{24}$$

$$= 72 + \frac{12}{\pi}[\sin 4\pi - \sin 0] = \$72 \text{ thousand or } \$72{,}000.$$

(B) Total revenue: $T = \int_5^{9}\left[3+2\cos\left(\frac{\pi t}{6}\right)\right]dt = \int_5^{9} 3\,dt + \frac{12}{\pi}\int_5^{9}\cos\left(\frac{\pi t}{6}\right)\left(\frac{\pi}{6}\right)dt$

$$= 3t\Big|_5^9 + \frac{12}{\pi}\sin\left[\frac{\pi t}{6}\right]_5^9$$

$$= 3(9-5) + \frac{12}{\pi}\left[\sin\left(\frac{9\pi}{6}\right) - \sin\left(\frac{5\pi}{6}\right)\right]$$

$$= 12 + \frac{12}{\pi}\left(-1-\frac{1}{2}\right) = 12 - \frac{18}{\pi}$$

$$= \$6.270 \text{ thousand or } \$6270.$$

(C)

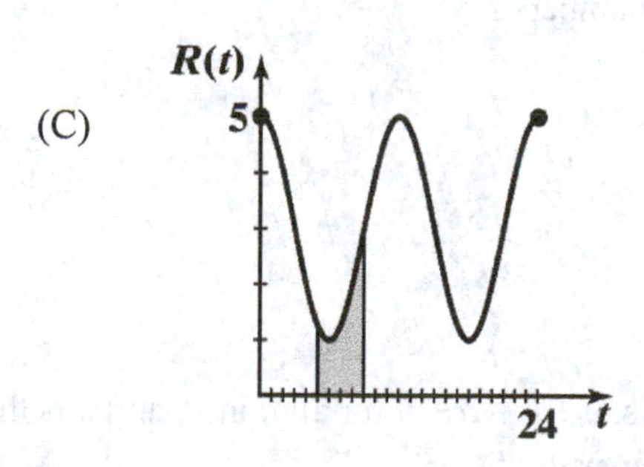

(8-2)

EXERCISE A–1

Things to remember:

1. THE SET OF REAL NUMBERS

SYMBOL	NAME	DESCRIPTION	EXAMPLES
N	Natural numbers	Counting numbers (also called positive integers)	1, 2, 3, ...
Z	Integers	Natural numbers, their negatives, and 0	… –2, –1, 0, 1, 2, …
Q	Rational numbers	Any number that can be represented as $\frac{a}{b}$, where a and b are integers and $b \neq 0$. Decimal representations are repeating or terminating.	$-4; 0; 1; 25; \frac{-3}{5}; \frac{2}{3}$; $3.67; -0.333\overline{3}; 5.2727\overline{27}$
I	Irrational numbers	Any number with a decimal representation that is nonrepeating and non–terminating.	$\sqrt{2}$; π; $\sqrt[3]{7}$; 1.414213…; 2.718281828…
R	Real numbers	Rationals and irrationals	

2. BASIC PROPERTIES OF THE SET OF REAL NUMBERS

Let a, b, and c be arbitrary elements in the set of real numbers R.

ADDITION PROPERTIES

ASSOCIATIVE: $(a + b) + c = a + (b + c)$

COMMUTATIVE: $a + b = b + a$

IDENTITY: 0 is the additive identity; that is, $0 + a = a + 0$ for all a in R, and 0 is the only element in R with this property.

INVERSE: For each a in R, $-a$ is its unique additive inverse; that is, $a + (-a) = (-a) + a = 0$, and $-a$ is the only element in R relative to a with this property.

MULTIPLICATION PROPERTIES

ASSOCIATIVE: $(ab)c = a(bc)$

COMMUTATIVE: $ab = ba$

IDENTITY: 1 is the multiplicative identity; that is, $1a = a1 = a$ for all a in R, and 1 is the only element in R with this property.

INVERSE: For each a in R, $a \neq 0$, $\frac{1}{a}$ is its unique multiplicative inverse; that is, $a\left(\frac{1}{a}\right) = \left(\frac{1}{a}\right)a = 1$, and $\frac{1}{a}$ is the only element in R relative to a with this property.

DISTRIBUTIVE PROPERTIES

$$a(b + c) = ab + ac$$

$$(a + b)c = ac + bc$$

3. SUBTRACTION AND DIVISION

For all real numbers a and b.

SUBTRACTION: $a - b = a + (-b)$

$$7 - (-5) = 7 + [-(-5)] = 7 + 5 = 12$$

DIVISION: $a \div b = a\left(\frac{1}{b}\right),\ b \neq 0$

$$9 \div 4 = 9\left(\frac{1}{4}\right) = \frac{9}{4}$$

NOTE: 0 can never be used as a divisor!

4. PROPERTIES OF NEGATIVES

For all real numbers a and b.

a. $-(-a) = a$

b. $(-a)b = -(ab) = a(-b) = -ab$

c. $(-a)(-b) = ab$

d. $(-1)a = -a$

e. $\frac{-a}{b} = -\frac{a}{b} = \frac{a}{-b},\ b \neq 0$

f. $\frac{-a}{-b} = -\frac{-a}{b} = -\frac{a}{-b} = \frac{a}{b},\ b \neq 0$

5. ZERO PROPERTIES

For all real numbers a and b.

a. $a \cdot 0 = 0$

b. $ab = 0$ if and only if $a = 0$ or $b = 0$ (or both)

6. FRACTION PROPERTIES

For all real numbers a, b, c, d, and k (division by 0 excluded).

a. $\frac{a}{b} = \frac{c}{d}$ if and only if $ad = bc$

b. $\frac{ka}{kb} = \frac{a}{b}$

c. $\frac{a}{b} \cdot \frac{c}{d} = \frac{ac}{bd}$

d. $\frac{a}{b} \div \frac{c}{d} = \frac{a}{b} \cdot \frac{d}{c}$

e. $\frac{a}{b} + \frac{c}{b} = \frac{a+c}{b}$

f. $\frac{a}{b} - \frac{c}{b} = \frac{a-c}{b}$

g. $\frac{a}{b} + \frac{c}{d} = \frac{ad+bc}{bd}$

---

1. $uv = vu$

3. $3 + (7 + y) = (3 + 7) + y$

5. $1(u + v) = u + v$

7. T; Associative property of multiplication

9. T; Distributive property

11. F; $-2(-a)(2x - y) = 2a(2x - y)$

13. T; Commutative property of addition

15. T; Property of negatives

17. T; Multiplicative inverse property

19. T; Property of negatives

21. F; $\frac{a}{b} + \frac{c}{d} = \frac{ad+bc}{bd}$

23. T; Distributive property

25. T; Zero property

27. No. For example: $2\left(\frac{1}{2}\right) = 1$. In general $a\left(\frac{1}{a}\right) = 1$ whenever $a \neq 0$.

29. (A) False. For example, –3 is an integer but not a natural number.

(B) True

(C) True. For example, for any natural number n, $n = \frac{n}{1}$.

31. $\sqrt{2}$, $\sqrt{3}$, … ; in general, the square root of any rational number that is not a perfect square; π, e.

33. (A) $8 \in N, Z, Q, R$

(B) $\sqrt{2} \in R$

(C) $-1.414 = -\frac{1414}{1000} \in Q, R$

(D) $\frac{-5}{2} \in Q, R$

35. (A) F; $a(b-c) = ab - ac$; for example $2(3-1) = 2 \cdot 2 = 4 \neq 2 \cdot 3 - 1 = 5$

(B) F; for example, $(3-7) - 4 = -4 - 4 = -8 \neq 3 - (7-4) = 3 - 3 = 0$.

(C) T; this is the associative property of multiplication.

(D) F; for example, $(12 \div 4) \div 2 = 3 \div 2 = \frac{3}{2} \neq 12 \div (4 \div 2) = 12 \div 2 = 6$.

37.
$$\begin{aligned} C &= 0.090909\ldots \\ 100C &= 9.090909\ldots \\ 100C - C &= (9.090909\ldots) - (0.090909\ldots) \\ 99C &= 9 \\ C &= \frac{9}{99} = \frac{1}{11} \end{aligned}$$

39. (A) $\frac{13}{6} \approx 2.166\,666\,666$ (B) $\sqrt{21} \approx 4.582\,575\,695$

(C) $\frac{7}{16} = 0.4375$ (D) $\frac{29}{111} \approx 0.261\,261\,261$

41. (A) $\frac{43}{13} \approx 3$ (B) $\frac{37}{19} \approx 2$

43. (A) $\frac{7}{8} + \frac{11}{12} \approx 2$ (B) $\frac{55}{9} - \frac{7}{55} \approx 6$

45. Tax: $182.39(0.09) \approx 16.4151$; \$16.42

47. Increase $4.37 - 4.25 = 0.12$; $\frac{0.12}{4.25} \approx 0.02824$; 2.8%

EXERCISE A–2

Things to remember:

1. NATURAL NUMBER EXPONENT

   For n a natural number and b any real number,

   $$b^n = b \cdot b \cdot \ldots \cdot b, \; n \text{ factors of } b.$$

   For example, $2^3 = 2 \cdot 2 \cdot 2\ (= 8)$,

   $3^5 = 3 \cdot 3 \cdot 3 \cdot 3 \cdot 3\ (= 243)$.

   In the expression b^n, n is called the EXPONENT, and b is called the BASE.

2. FIRST PROPERTY OF EXPONENTS

   For any natural numbers m and n, and any real number b,

   $$b^m \cdot b^n = b^{m+n}.$$

   For example, $3^3 \cdot 3^4 = 3^{3+4} = 3^7$.

3. POLYNOMIALS

   a. A POLYNOMIAL IN ONE VARIABLE x is constructed by adding or subtracting constants and terms of the form ax^n, where a is a real number and n is a natural number.

   b. A POLYNOMIAL IN TWO VARIABLES x AND y is constructed by adding or subtracting constants and terms of the form $ax^m y^n$, where a is a real number and m and n are natural numbers.

   c. Polynomials in more than two variables are defined similarly.

   d. A polynomial with only one term is called a MONOMIAL.
   A polynomial with two terms is called a BINOMIAL.
   A polynomial with three terms is called a TRINOMIAL.

4. DEGREE OF A POLYNOMIAL

   a. A term of the form ax^n, $a \neq 0$, has degree n. A term of the form $ax^m y^n$, $a \neq 0$, has degree $m + n$. A nonzero constant has degree 0.

   b. The DEGREE OF A POLYNOMIAL is the degree of the nonzero term with the highest degree. For example, $3x^4 + \sqrt{2}\,x^3 - 2x + 7$ has degree 4; $2x^3y^2 - 3x^2y + 7x^4 - 5y^3 + 6$ has degree 5; the polynomial 4 has degree 0.

   c. The constant 0 is a polynomial but it is not assigned a degree.

5. Two terms in a polynomial are called LIKE TERMS if they have exactly the same variable factors raised to the same powers. For example, in

   $$7x^5y^2 - 3x^3y + 2x + 4x^3y - 1,$$

   $-3x^3y$ and $4x^3y$ are like terms.

6. To multiply two polynomials, multiply each term of one by each term of the other, and then combine like terms.

7. SPECIAL PRODUCTS

   a. $(a - b)(a + b) = a^2 - b^2$

   b. $(a + b)^2 = a^2 + 2ab + b^2$

   c. $(a - b)^2 = a^2 - 2ab + b^2$

8. ORDER OF OPERATIONS

   Multiplication and division precede addition and subtraction, and taking powers precedes multiplication and division.

---

1. The term of highest degree in $x^3 + 2x^2 - x + 3$ is x^3 and the degree of this term is 3.

3. $(2x^2 - x + 2) + (x^3 + 2x^2 - x + 3) = x^3 + 2x^2 + 2x^2 - x - x + 2 + 3 = x^3 + 4x^2 - 2x + 5$

5. $(x^3 + 2x^2 - x + 3) - (2x^2 - x + 2) = x^3 + 2x^2 - x + 3 - 2x^2 + x - 2 = x^3 + 1$

7. Using a vertical arrangement:

$$\begin{array}{r} x^3 + 2x^2 - x + 3 \\ \underline{2x^2 - x + 2} \\ 2x^5 + 4x^4 - 2x^3 + 6x^2 \\ - x^4 - 2x^3 + x^2 - 3x \\ \underline{2x^3 + 4x^2 - 2x + 6} \\ 2x^5 + 3x^4 - 2x^3 + 11x^2 - 5x + 6 \end{array}$$

9. $2(u - 1) - (3u + 2) - 2(2u - 3) = 2u - 2 - 3u - 2 - 4u + 6 = -5u + 2$

11. $4a - 2a[5 - 3(a + 2)] = 4a - 2a[5 - 3a - 6] = 4a - 2a[-3a - 1] = 4a + 6a^2 + 2a = 6a^2 + 6a$

13. $(a + b)(a - b) = a^2 - b^2$ (Special product 7a)

15. $(3x - 5)(2x + 1) = 6x^2 + 3x - 10x - 5 = 6x^2 - 7x - 5$

17. $(2x - 3y)(x + 2y) = 2x^2 + 4xy - 3xy - 6y^2 = 2x^2 + xy - 6y^2$

19. $(3y + 2)(3y - 2) = (3y)^2 - 2^2 = 9y^2 - 4$ (Special product 7a)

21. $-(2x - 3)^2 = -[(2x)^2 - 2(2x)(3) + 3^2] = -[4x^2 - 12x + 9] = -4x^2 + 12x - 9$ (Special product 7c)

23. $(4m + 3n)(4m - 3n) = 16m^2 - 9n^2$ (Special product 7a)

25. $(3u + 4v)^2 = 9u^2 + 24uv + 16v^2$ (Special product 7b)

27. $(a - b)(a^2 + ab + b^2) = a(a^2 + ab + b^2) - b(a^2 + ab + b^2) = a^3 + a^2b + ab^2 - a^2b - ab^2 - b^3$
$= a^3 - b^3$

29. $[(x - y) + 3z][(x - y) - 3z)] = (x - y)^2 - 9z^2 = x^2 - 2xy + y^2 - 9z^2$ (Special product 7a)

31. $m - \{m - [m - (m - 1)]\} = m - \{m - [m - m + 1]\} = m - \{m - 1\} = m - m + 1 = 1$

33. $(x^2 - 2xy + y^2)(x^2 + 2xy + y^2) = (x - y)^2(x + y)^2 = [(x - y)(x + y)]^2 = [x^2 - y^2]^2 = x^4 - 2x^2y^2 + y^4$

35. $(5a - 2b)^2 - (2b + 5a)^2 = 25a^2 - 20ab + 4b^2 - [4b^2 + 20ab + 25a^2] = -40ab$

37. $(m - 2)^2 - (m - 2)(m + 2) = m^2 - 4m + 4 - [m^2 - 4] = m^2 - 4m + 4 - m^2 + 4 = -4m + 8$

39. $(x - 2y)(2x + y) - (x + 2y)(2x - y) = 2x^2 - 4xy + xy - 2y^2 - [2x^2 + 4xy - xy - 2y^2]$
$= 2x^2 - 3xy - 2y^2 - 2x^2 - 3xy + 2y^2 = -6xy$

41. $(u + v)^3 = (u + v)(u + v)^2 = (u + v)(u^2 + 2uv + v^2) = u^3 + 3u^2v + 3uv^2 + v^3$

43. $(x - 2y)^3 = (x - 2y)(x - 2y)^2 = (x - 2y)(x^2 - 4xy + 4y^2) = x(x^2 - 4xy + 4y^2) - 2y(x^2 - 4xy + 4y^2)$
$= x^3 - 4x^2y + 4xy^2 - 2x^2y + 8xy^2 - 8y^3$
$= x^3 - 6x^2y + 12xy^2 - 8y^3$

45. $[(2x^2 - 4xy + y^2) + (3xy - y^2)] - [(x^2 - 2xy - y^2) + (-x^2 + 3xy - 2y^2)]$
$= [2x^2 - xy] - [xy - 3y^2] = 2x^2 - 2xy + 3y^2$

47. $[(2x-1)^2 - x(3x+1)]^2 = [4x^2 - 4x + 1 - 3x^2 - x]^2 = [x^2 - 5x + 1]^2 = (x^2 - 5x + 1)(x^2 - 5x + 1)$
$= x^4 - 10x^3 + 27x^2 - 10x + 1$

49. $2\{(x-3)(x^2 - 2x + 1) - x[3 - x(x-2)]\} = 2\{x^3 - 5x^2 + 7x - 3 - x[3 - x^2 + 2x]\}$
$= 2\{x^3 - 5x^2 + 7x - 3 + x^3 - 2x^2 - 3x\} = 2\{2x^3 - 7x^2 + 4x - 3\} = 4x^3 - 14x^2 + 8x - 6$

51. $m + n$

53. Given two polynomials, one with degree m and the other of degree n, their product will have degree $m + n$ regardless of the relationship between m and n.

55. Since $(a+b)^2 = a^2 + 2ab + b^2$, $(a+b)^2 = a^2 + b^2$ only when $2ab = 0$; that is, only when $a = 0$ or $b = 0$.

57. Let x = amount invested at 9%.
Then $10{,}000 - x$ = amount invested at 12%.
The total annual income I is:

$I = 0.09x + 0.12(10{,}000 - x)$
$= 1{,}200 - 0.03x$

59. Let x = number of tickets at \$20.
Then $3x$ = number of tickets at \$30 and $4{,}000 - x - 3x = 4{,}000 - 4x$ = number of tickets at \$50.
The total receipts R are:

$R = 20x + 30(3x) + 50(4{,}000 - 4x)$
$= 20x + 90x + 200{,}000 - 200x = 200{,}000 - 90x$

61. Let x = number of kilograms of food A.
Then $10 - x$ = number of kilograms of food B.
The total number of kilograms F of fat in the final food mix is:

$F = 0.02x + 0.06(10 - x) = 0.6 - 0.04x$

EXERCISE A-3

Things to remember:

1. The discussion is limited to polynomials with integer coefficients.

2. FACTORED FORMS

   A polynomial is in FACTORED FORM if it is written as the product of two or more polynomials. A polynomial with integer coefficients is FACTORED COMPLETELY if each factor cannot be expressed as the product of two or more polynomials with integer coefficients, other than itself and 1.

3. METHODS

   a. Factor out all factors common to all terms, if they are present.

   b. Try grouping terms.

   c. ac–Test for polynomials of the form

   $$ax^2 + bx + c \quad \text{or} \quad ax^2 + bxy + cy^2$$

   If the product ac has two integer factors p and q whose sum is the coefficient b of the middle term, i.e., if integers p and q exist so that

   $$pq = ac \quad \text{and} \quad p + q = b$$

   then the polynomials have first–degree factors with integer coefficients. If no such integers exist then the polynomials will not have first–degree factors with integer coefficients; the polynomials are *not factorable.*

4. SPECIAL FACTORING FORMULAS

a.	$u^2 + 2uv + v^2 = (u + v)^2$	Perfect square
b.	$u^2 - 2uv + v^2 = (u - v)^2$	Perfect square
c.	$u^2 - v^2 = (u - v)(u + v)$	Difference of squares
d.	$u^3 - v^3 = (u - v)(u^2 + uv + v^2)$	Difference of cubes
e.	$u^3 + v^3 = (u + v)(u^2 - uv + v^2)$	Sum of cubes

---

1. $3m^2$ is a common factor: $6m^4 - 9m^3 - 3m^2 = 3m^2(2m^2 - 3m - 1)$

3. $2uv$ is a common factor: $8u^3v - 6u^2v^2 + 4uv^3 = 2uv(4u^2 - 3uv + 2v^2)$

5. $(2m - 3)$ is a common factor: $7m(2m - 3) + 5(2m - 3) = (7m + 5)(2m - 3)$

7. $4ab(2c + d) - (2c + d) = (4ab - 1)(2c + d)$

9. $2x^2 - x + 4x - 2 = (2x^2 - x) + (4x - 2) = x(2x - 1) + 2(2x - 1) = (2x - 1)(x + 2)$

11. $3y^2 - 3y + 2y - 2 = (3y^2 - 3y) + (2y - 2) = 3y(y - 1) + 2(y - 1) = (y - 1)(3y + 2)$

13. $2x^2 + 8x - x - 4 = (2x^2 + 8x) - (x + 4) = 2x(x + 4) - (x + 4) = (x + 4)(2x - 1)$

15. $wy - wz + xy - xz = (wy - wz) + (xy - xz) = w(y - z) + x(y - z) = (y - z)(w + x)$

Or $wy - wz + xy - xz = (wy + xy) - (wz + xz) = y(w + x) - z(w + x) = (w + x)(y - z)$

17. $am - 3bm + 2na - 6bn = m(a - 3b) + 2n(a - 3b) = (a - 3b)(m + 2n)$

19. $3y^2 - y - 2$

$a = 3, b = -1, c = -2$

Step 1. Use the ac-test to test for factorability

$ac = (3)(-2) = -6$

pq
$(1)(-6)$
$(-1)(6)$
$\boxed{(2)(-3)}$
$(-2)(3)$

Note that $2 + (-3) = -1 = b$. Thus, $3y^2 - y - 2$ has first-degree factors with integer coefficients.

Step 2. Split the middle term using $b = p + q$ and factor by grouping.

$-1 = -3 + 2$

$3y^2 - y - 2 = 3y^2 - 3y + 2y - 2 = (3y^2 - 3y) + (2y - 2) = 3y(y - 1) + 2(y - 1)$

$= (y - 1)(3y + 2)$

21. $u^2 - 2uv - 15v^2$
$a = 1, b = -2, c = -15$

Step 1. Use the *ac*–test
$ac = 1(-15) = -15$

pq
(1)(–15)
(–1)(15)
[(3)(–5)]
(–3)(5)

Note that $3 + (-5) = -2 = b$. Thus $u^2 - 2uv - 15v^2$ has first- degree factors with integer coefficients.

Step 2. Factor by grouping
$-2 = 3 + (-5)$
$u^2 + 3uv - 5uv - 15v^2 = (u^2 + 3uv) - (5uv + 15v^2) = u(u + 3v) - 5v(u + 3v)$
$= (u + 3v)(u - 5v)$

23. $m^2 - 6m - 3$
$a = 1, b = -6, c = -3$

Step 1. Use the *ac*–test
$ac = (1)(-3) = -3$

pq
(1)(–3)
(–1)(3)

None of the factors add up to $-6 = b$. Thus, this polynomial is *not factorable*.

25. $w^2x^2 - y^2 = (wx - y)(wx + y)$ (difference of squares)

27. $9m^2 - 6mn + n^2 = (3m - n)^2$ (perfect square)

29. $y^2 + 16$
$a = 1, b = 0, c = 16$

Step 1. Use the *ac*–test
$ac = (1)(16)$

pq
(1)(16)
(–1)(–16)
(2)(8)
(–2)(–8)
(4)(4)
(–4)(–4)

None of the factors add up to $0 = b$. Thus this polynomial is *not factorable*.

31. $4z^2 - 28z + 48 = 4(z^2 - 7z + 12) = 4(z - 3)(z - 4)$

33. $2x^4 - 24x^3 + 40x^2 = 2x^2(x^2 - 12x + 20) = 2x^2(x - 2)(x - 10)$

35. $4xy^2 - 12xy + 9x = x(4y^2 - 12y + 9) = x(2y - 3)^2$

37. $6m^2 - mn - 12n^2 = (2m - 3n)(3m + 4n)$

39. $4u^3v - uv^3 = uv(4u^2 - v^2) = uv[(2u)^2 - v^2] = uv(2u - v)(2u + v)$

41. $2x^3 - 2x^2 + 8x = 2x(x^2 - x + 4)$ [Note: $x^2 - x + 4$ is *not factorable*.]

43. $8x^3 - 27y^3 = (2x)^3 - (3y)^3 = (2x - 3y)[(2x)^2 + (2x)(3y) + (3y)^2]$
$= (2x - 3y)[4x^2 + 6xy + 9y^2]$ (difference of cubes)

45. $x^4y + 8xy = xy[x^3 + 8] = xy(x + 2)(x^2 - 2x + 4)$

47. $(x + 2)^2 - 9y^2 = [(x + 2) - 3y][(x + 2) + 3y] = (x + 2 - 3y)(x + 2 + 3y)$

49. $5u^2 + 4uv - 2v^2$ is *not factorable.*

51. $6(x - y)^2 + 23(x - y) - 4 = [6(x - y) - 1][(x - y) + 4] = (6x - 6y - 1)(x - y + 4)$

53. $y^4 - 3y^2 - 4 = (y^2)^2 - 3y^2 - 4 = (y^2 - 4)(y^2 + 1) = (y - 2)(y + 2)(y^2 + 1)$

55. $15y(x - y)^3 + 12x(x - y)^2 = 3(x - y)^2 [5y(x - y) + 4x] = 3(x - y)^2 [5xy - 5y^2 + 4x]$

57. True: $u^n - v^n = (u - v)(u^{n-1} + u^{n-2}v + \dots + uv^{n-2} + v^{n-1})$

59. False; For example, $u^2 + v^2$ cannot be factored.

EXERCISE A-4

Things to remember:

1. FUNDAMENTAL PROPERTY OF FRACTIONS

If a, b, and k are real numbers with b, $k \neq 0$, then

$$\frac{ka}{kb} = \frac{a}{b}.$$

A fraction is in LOWEST TERMS if the numerator and denominator have no common factors other than 1 or –1.

2. MULTIPLICATION AND DIVISION

For a, b, c, and d real numbers:

a. $\frac{a}{b} \cdot \frac{c}{d} = \frac{ac}{bd}$, $b, d \neq 0$

b. $\frac{a}{b} \div \frac{c}{d} = \frac{\frac{a}{b}}{\frac{c}{d}} = \frac{a}{b} \cdot \frac{d}{c}$, $b, c, d \neq 0$

The same procedures are used to multiply or divide two rational expressions.

3. ADDITION AND SUBTRACTION

For a, b, and c real numbers:

a. $\frac{a}{b} + \frac{c}{b} = \frac{a+c}{b}$, $b \neq 0$

b. $\frac{a}{b} - \frac{c}{b} = \frac{a-c}{b}$, $b \neq 0$

The same procedures are used to add or subtract two rational expressions (with the same denominator).

4. THE LEAST COMMON DENOMINATOR (LCD)

The LCD of two or more rational expressions is found as follows:

a. Factor each denominator completely, including integer factors.

b. Identify each different factor from all the denominators.

c. Form a product using each different factor to the highest power that occurs in any one denominator. This product is the LCD.

The least common denominator is used to add or subtract rational expressions having different denominators.

---

1. $\frac{5\cdot 9\cdot 13}{3\cdot 5\cdot 7} = \frac{\cancel{5}\cdot \cancel{9}^{3}\cdot 13}{\cancel{3}\cdot \cancel{5}\cdot 7} = \frac{3\cdot 13}{7} = \frac{39}{7}$

3. $\frac{\cancel{12}\cdot 11\cdot \cancel{10}\cdot 9}{\cancel{4}\cdot \cancel{3}\cdot \cancel{2}\cdot 1} = 11\cdot 5\cdot 9 = 495$

5.

$$\frac{d^5}{3a} \div \left(\frac{d^2}{6a^2}\cdot\frac{a}{4d^3}\right) = \frac{d^5}{3a} \div \left(\frac{\cancel{a}\,\cancel{d^2}}{24\,\cancel{a^2}_{a}\,\cancel{d^3}_{d}}\right) = \frac{d^5}{3a} \div \frac{1}{24ad} = \frac{d^5}{\cancel{3}\,\cancel{a}}\cdot\frac{\cancel{24}^{8}\,\cancel{a}d}{1} = 8d^6$$

7. $\frac{x^2}{12} + \frac{x}{18} - \frac{1}{30} = \frac{15x^2}{180} + \frac{10x}{180} - \frac{6}{180}$

$= \frac{15x^2 + 10x - 6}{180}$

We find the LCD of 12, 18, 30:
$12 = 2^2\cdot 3$, $18 = 2\cdot 3^2$, $30 = 2\cdot 3\cdot 5$
Thus, LCD $= 2^2\cdot 3^2\cdot 5 = 180$.

9. $\frac{4m-3}{18m^3} + \frac{3}{4m} - \frac{2m-1}{6m^2}$

$= \frac{2(4m-3)}{36m^3} + \frac{3(9m^2)}{36m^3} - \frac{6m(2m-1)}{36m^3}$

$= \frac{8m - 6 + 27m^2 - 6m(2m-1)}{36m^3}$

$= \frac{8m - 6 + 27m^2 - 12m^2 + 6m}{36m^3} = \frac{15m^2 + 14m - 6}{36m^3}$

Find the LCD of $18m^3$, $4m$, $6m^2$:
$18m^3 = 2\cdot 3^2 m^3$, $4m = 2^2 m$,
$6m^2 = 2\cdot 3m^2$
Thus, LCD $= 36m^3$.

11. $\frac{x^2-9}{x^2-3x} \div (x^2 - x - 12) = \frac{\cancel{(x-3)}\,\cancel{(x+3)}}{x\,\cancel{(x-3)}}\cdot\frac{1}{(x-4)\,\cancel{(x+3)}} = \frac{1}{x(x-4)}$

13. $\frac{2}{x} - \frac{1}{x-3} = \frac{2(x-3)}{x(x-3)} - \frac{x}{x(x-3)}$

$= \frac{2x - 6 - x}{x(x-3)} = \frac{x-6}{x(x-3)}$

LCD $= x(x-3)$

15. $\dfrac{2}{(x+1)^2}-\dfrac{5}{x^2-x-2}=\dfrac{2}{(x+1)^2}-\dfrac{5}{(x+1)(x-2)}$ LCD = $(x+1)^2(x-2)$

$$=\frac{2(x-2)}{(x+1)^2(x-2)}-\frac{5(x+1)}{(x+1)^2(x-2)}=\frac{2x-4-5x-5}{(x+1)^2(x-2)}=\frac{-3x-9}{(x+1)^2(x-2)}$$

17. $\dfrac{x+1}{x-1}-1=\dfrac{x+1}{x-1}-\dfrac{x-1}{x-1}=\dfrac{x+1-(x-1)}{x-1}=\dfrac{2}{x-1}$

19. $\dfrac{3}{a-1}-\dfrac{2}{1-a}=\dfrac{3}{a-1}-\dfrac{-2}{-(1-a)}=\dfrac{3}{a-1}+\dfrac{2}{a-1}=\dfrac{5}{a-1}$

21. $\dfrac{2x}{x^2-16}-\dfrac{x-4}{x^2+4x}=\dfrac{2x}{(x-4)(x+4)}-\dfrac{x-4}{x(x+4)}$ LCD = $x(x-4)(x+4)$

$$=\frac{2x(x)-(x-4)(x-4)}{x(x-4)(x+4)}=\frac{2x^2-(x^2-8x+16)}{x(x-4)(x+4)}=\frac{x^2+8x-16}{x(x-4)(x+4)}$$

23. $\dfrac{x^2}{x^2+2x+1}+\dfrac{x-1}{3x+3}-\dfrac{1}{6}=\dfrac{x^2}{(x+1)^2}+\dfrac{x-1}{3(x+1)}-\dfrac{1}{6}$ LCD = $6(x+1)^2$

$$=\frac{6x^2}{6(x+1)^2}+\frac{2(x+1)(x-1)}{6(x+1)^2}-\frac{(x+1)^2}{6(x+1)^2}$$

$$=\frac{6x^2+2(x^2-1)-(x^2+2x+1)}{6(x+1)^2}=\frac{7x^2-2x-3}{6(x+1)^2}$$

25. $\dfrac{1-\frac{x}{y}}{2-\frac{y}{x}}=\dfrac{\frac{y-x}{y}}{\frac{2x-y}{x}}=\dfrac{y-x}{y}\cdot\dfrac{x}{2x-y}=\dfrac{x(y-x)}{y(2x-y)}$

27. $\dfrac{c+2}{5c-5}-\dfrac{c-2}{3c-3}+\dfrac{c}{1-c}=\dfrac{c+2}{5(c-1)}-\dfrac{c-2}{3(c-1)}-\dfrac{c}{c-1}$ LCD = $15(c-1)$

$$=\frac{3(c+2)}{15(c-1)}-\frac{5(c-2)}{15(c-1)}-\frac{15c}{15(c-1)}=\frac{3c+6-5c+10-15c}{15(c-1)}=\frac{-17c+16}{15(c-1)}$$

29. $\dfrac{1+\frac{3}{x}}{x-\frac{9}{x}}=\dfrac{\frac{x+3}{x}}{\frac{x^2-9}{x}}=\dfrac{x+3}{x}\cdot\dfrac{x}{x^2-9}=\dfrac{\cancel{x+3}}{\cancel{x}}\cdot\dfrac{\cancel{x}}{\cancel{(x+3)}(x-3)}=\dfrac{1}{x-3}$

31. $\dfrac{\frac{1}{2(x+h)}-\frac{1}{2x}}{h}=\left(\dfrac{1}{2(x+h)}-\dfrac{1}{2x}\right)\div\dfrac{h}{1}=\dfrac{x-x-h}{2x(x+h)}\cdot\dfrac{1}{h}=\dfrac{-\cancel{h}}{2x(x+h)\cancel{h}}=\dfrac{-1}{2x(x+h)}$

33. $\dfrac{\frac{x}{y}-2+\frac{y}{x}}{\frac{x}{y}-\frac{y}{x}}=\dfrac{\frac{x^2-2xy+y^2}{xy}}{\frac{x^2-y^2}{xy}}=\dfrac{\cancel{(x-y)}^2}{\cancel{xy}}\cdot\dfrac{\cancel{xy}}{\cancel{(x-y)}(x+y)}=\dfrac{x-y}{x+y}$

35. (A) $\dfrac{x^2+4x+3}{x+3} = x+4$: Incorrect

(B) $\dfrac{x^2+4x+3}{x+3} = \dfrac{\cancel{(x+3)}(x+1)}{\cancel{x+3}} = x+1 \quad (x \neq -3)$

37. (A) $\dfrac{(x+h)^2-x^2}{h} = 2x+1$: Incorrect

(B) $\dfrac{(x+h)^2-x^2}{h} = \dfrac{x^2+2xh+h^2-x^2}{h} = \dfrac{2xh+h^2}{h} = \dfrac{\cancel{h}(2x+h)}{\cancel{h}} = 2x+h \quad (h \neq 0)$

39. (A) $\dfrac{x^2-3x}{x^2-2x-3} + x - 3 = 1$: Incorrect

(B) $\dfrac{x^2-3x}{x^2-2x-3} + x - 3 = \dfrac{x\cancel{(x-3)}}{\cancel{(x-3)}(x+1)} + x - 3 = \dfrac{x}{x+1} + x - 3 = \dfrac{x+(x-3)(x+1)}{x+1} = \dfrac{x^2-x-3}{x+1}$

41. (A) $\dfrac{2x^2}{x^2-4} - \dfrac{x}{x-2} = \dfrac{x}{x+2}$: Correct

$$\frac{2x^2}{x^2-4} - \frac{x}{x-2} = \frac{2x^2}{(x-2)(x+2)} - \frac{x}{x-2} = \frac{2x^2-x(x+2)}{(x-2)(x+2)}$$

$$= \frac{x^2-2x}{(x-2)(x+2)} = \frac{x\cancel{(x-2)}}{\cancel{(x-2)}(x+2)} = \frac{x}{x+2}$$

43. $$\frac{\dfrac{1}{3(x+h)^2} - \dfrac{1}{3x^2}}{h} = \left[\frac{1}{3(x+h)^2} - \frac{1}{3x^2}\right] \div \frac{h}{1} = \frac{x^2-(x+h)^2}{3x^2(x+h)^2} \cdot \frac{1}{h} = \frac{x^2-(x^2+2xh+h^2)}{3x^2(x+h)^2 h}$$

$$= \frac{-2xh-h^2}{3x^2(x+h)^2 h} = \frac{-\cancel{h}(2x+h)}{3x^2(x+h)^2 \cancel{h}} = -\frac{(2x+h)}{3x^2(x+h)^2} = \frac{-2x-h}{3x^2(x+h)^2}$$

45. $$x - \frac{2}{1-\dfrac{1}{x}} = x - \frac{2}{\dfrac{x-1}{x}} = x - \frac{2x}{x-1} \qquad (\text{LCD} = x-1)$$

$$= \frac{x(x-1)}{x-1} - \frac{2x}{x-1} = \frac{x^2-x-2x}{x-1} = \frac{x(x-3)}{x-1}$$

EXERCISE A-5

Things to remember:

1. DEFINITION OF a^n, where n is an integer and a is a real number:

   a. For n a positive integer,
   $$a^n = a \cdot a \cdot \cdots \cdot a, \ n \text{ factors of } a.$$

   b. For $n = 0$,
   $$a^0 = 1, a \neq 0, 0^0 \text{ is not defined.}$$

   c. For n a negative integer,
   $$a^n = \frac{1}{a^{-n}}, a \neq 0.$$

   [Note: If n is negative, then $-n$ is positive.]

2. PROPERTIES OF EXPONENTS

   GIVEN: n and m are integers and a and b are real numbers.

   a. $a^m a^n = a^{m+n}$ $\qquad a^8 a^{-3} = a^{8+(-3)} = a^5$

   b. $(a^n)^m = a^{mn}$ $\qquad (a^{-2})^3 = a^{3(-2)} = a^{-6}$

   c. $(ab)^m = a^m b^m$ $\qquad (ab)^{-2} = a^{-2}b^{-2}$

   d. $\left(\frac{a}{b}\right)^m = \frac{a^m}{b^m}, b \neq 0$ $\qquad \left(\frac{a}{b}\right)^5 = \frac{a^5}{b^5}$

   e. $\frac{a^m}{a^n} = a^{m-n} = \frac{1}{a^{n-m}}, a \neq 0$ $\qquad \frac{a^{-3}}{a^7} = \frac{1}{a^{7-(-3)}} = \frac{1}{a^{10}}$

3. SCIENTIFIC NOTATION

   Let r be any finite decimal. Then r can be expressed as the product of a number between 1 and 10 and an integer power of 10; that is, r can be written $r = a \times 10^n$, $1 \leq a < 10$, a in decimal form, n an integer. A number expressed in this form is said to be in SCIENTIFIC NOTATION.

   Examples:

   $7 = 7 \times 10^0$ $\qquad 0.5 = 5 \times 10^{-1}$

   $67 = 6.7 \times 10$ $\qquad 0.45 = 4.5 \times 10^{-1}$

   $580 = 5.8 \times 10^2$ $\qquad 0.0032 = 3.2 \times 10^{-3}$

   $43{,}000 = 4.3 \times 10^4$ $\qquad 0.000\,045 = 4.5 \times 10^{-5}$

---

1. $2x^{-9} = \frac{2}{x^9}$

3. $\frac{3}{2w^{-7}} = \frac{3w^7}{2}$

5. $2x^{-8}x^5 = 2x^{-8+5} = 2x^{-3} = \frac{2}{x^3}$

7. $\frac{w^{-8}}{w^{-3}} = \frac{1}{w^{-3+8}} = \frac{1}{w^5}$

9. $(2a^{-3})^2 = 2^2(a^{-3})^2 = 4a^{-6} = \frac{4}{a^6}$

11. $(a^{-3})^2 = a^{-6} = \frac{1}{a^6}$

13. $(2x^4)^{-3} = 2^{-3}(x^4)^{-3} = \frac{1}{8} \cdot x^{-12} = \frac{1}{8x^{12}}$

15. $82{,}300{,}000{,}000 = 8.23 \times 10^{10}$

17. $0.783 = 7.83 \times 10^{-1}$

19. $0.000\,034 = 3.4 \times 10^{-5}$

21. $4 \times 10^4 = 40{,}000$

23. $7 \times 10^{-3} = 0.007$

25. $6.171 \times 10^7 = 61{,}710{,}000$

27. $8.08 \times 10^{-4} = 0.000\,808$

29. $(22 + 31)^0 = (53)^0 = 1$

31. $\frac{10^{-3} \times 10^4}{10^{-11} \times 10^{-2}} = \frac{10^{-3+4}}{10^{-11-2}} = \frac{10^1}{10^{-13}} = 10^{1+13} = 10^{14}$

33. $(5x^2y^{-3})^{-2} = 5^{-2}x^{-4}y^6 = \frac{y^6}{5^2x^4} = \frac{y^6}{25x^4}$

35. $\left(\frac{-5}{2x^3}\right)^{-2} = \frac{(-5)^{-2}}{(2x^3)^{-2}} = \frac{\frac{1}{(-5)^2}}{\frac{1}{(2x^3)^2}} = \frac{\frac{1}{25}}{\frac{1}{4x^6}} = \frac{4x^6}{25}$

37. $\frac{8x^{-3}y^{-1}}{6x^2y^{-4}} = \frac{4y^{-1+4}}{3x^{2+3}} = \frac{4y^3}{3x^5}$

39. $\frac{7x^5 - x^2}{4x^5} = \frac{7x^5}{4x^5} - \frac{x^2}{4x^5} = \frac{7}{4} - \frac{1}{4x^3} = \frac{7}{4} - \frac{1}{4}x^{-3}$

41. $\frac{5x^4 - 3x^2 + 8}{2x^2} = \frac{5x^4}{2x^2} - \frac{3x^2}{2x^2} + \frac{8}{2x^2} = \frac{5}{2}x^2 - \frac{3}{2} + 4x^{-2}$

43. $\frac{3x^2(x-1)^2 - 2x^3(x-1)}{(x-1)^4} = \frac{x^2(x-1)[3(x-1) - 2x]}{(x-1)^4} = \frac{x^2(x-3)}{(x-1)^3}$

45. $2x^{-2}(x-1) - 2x^{-3}(x-1)^2 = \frac{2(x-1)}{x^2} - \frac{2(x-1)^2}{x^3} = \frac{2x(x-1) - 2(x-1)^2}{x^3}$

$= \frac{2(x-1)[x - (x-1)]}{x^3} = \frac{2(x-1)}{x^3}$

47. $\frac{9{,}600{,}000{,}000}{(1{,}600{,}000)(0.00000025)} = \frac{9.6 \times 10^9}{(1.6 \times 10^6)(2.5 \times 10^{-7})} = \frac{9.6 \times 10^9}{1.6(2.5) \times 10^{6-7}}$

$= \frac{9.6 \times 10^9}{4.0 \times 10^{-1}} = 2.4 \times 10^{9+1} = 2.4 \times 10^{10} = 24{,}000{,}000{,}000$

49. $\frac{(1{,}250{,}000)(0.00038)}{0.0152} = \frac{(1.25 \times 10^6)(3.8 \times 10^{-4})}{1.52 \times 10^{-2}} = \frac{1.25(3.8) \times 10^{6-4}}{1.52 \times 10^{-2}} = 3.125 \times 10^4 = 31{,}250$

51. On a calculator: $2^{3^2} = 64$. A calculator interprets 2^{3^2} as $(2^3)^2 = 8^2 = 64$.

53. $a^m a^0 = a^{m+0} = a^m$. Therefore, $a^m a^0 = a^m$ which implies $a^0 = 1$.

55. $$\frac{u+v}{u^{-1}+v^{-1}} = \frac{u+v}{\frac{1}{u}+\frac{1}{v}} = \frac{u+v}{\frac{v+u}{uv}} = (u+v)\cdot\frac{uv}{v+u} = uv$$

57. $$\frac{b^{-2}-c^{-2}}{b^{-3}-c^{-3}} = \frac{\frac{1}{b^2}-\frac{1}{c^2}}{\frac{1}{b^3}-\frac{1}{c^3}} = \frac{\frac{c^2-b^2}{b^2c^2}}{\frac{c^3-b^3}{b^3c^3}} = \frac{(c-b)(c+b)}{b^2c^2}\cdot\frac{b^3c^3}{(c-b)(c^2+cb+b^2)} = \frac{bc(c+b)}{c^2+cb+b^2}$$

59. (A) Per capita debt: $\frac{1.606600\times10^{13}}{3.13\times10^8} \approx 0.51329\times10^5 = 51{,}329$ or \$51,329

(B) Per capita interest: $\frac{3.60\times10^{11}}{3.13\times10^8} \approx 1.15016\times10^3 = 1{,}150$ or \$1,150

(C) Percentage interest paid on debt: $\frac{3.60\times10^{11}}{1.606\times10^{13}} \approx 2.24159\times10^{-2} \approx 0.0224$ or 2.24%

61. (A) $9 \text{ ppm} = \frac{9}{1{,}000{,}000} = \frac{9}{10^6} = 9\times10^{-6}$ (B) 0.000 009 (C) 0.0009%

63. $$\frac{404}{100{,}000}\times309{,}000{,}000 = \frac{4.04\times10^2}{10^5}\times3.09\times10^8 = 12.4836\times10^5 \approx 1{,}248{,}000$$

To the nearest thousand, there were 1,248,000 violent crimes committed in 2010.

EXERCISE A-6

Things to remember:

1. *n*th ROOT

   Let b be a real number. For any natural number n,

   r is an nth ROOT of b if $r^n = b$

   If n is odd, then b has exactly one real nth root.
   If n is even, and $b < 0$, then b has NO real nth roots.
   If n is even, and $b > 0$, then b has two real nth roots; if r is an nth root, then $-r$ is also an nth root.
   0 is an nth root of 0 for all n

2. NOTATION

Let b be a real number and let $n > 1$ be a natural number. If n is odd, then the nth root of b is denoted

$$b^{1/n} \text{ or } \sqrt[n]{b}$$

If n is even and $b > 0$, then the PRINCIPAL nth ROOT OF b is the positive nth root; the principal nth root is denoted

$$b^{1/n} \text{ or } \sqrt[n]{b}$$

In the $\sqrt[n]{b}$ notation, the symbol $\sqrt{\ }$ is called a RADICAL, n is the INDEX of the radical and b is called the RADICAND.

3. RATIONAL EXPONENTS

If m and n are natural numbers without common prime factors, b is a real number, and b is nonnegative when b is even, then

$$b^{m/n} = \begin{cases} \left(b^{1/n}\right)^m = \left(\sqrt[n]{b}\right)^m \\ \left(b^m\right)^{1/n} = \sqrt[n]{b^m} \end{cases} \quad \text{and} \quad b^{-m/n} = \frac{1}{b^{m/n}},\ b \neq 0$$

The two definitions of $b^{m/n}$ are equivalent under the indicated restrictions on m, n, and b.

4. PROPERTIES OF RADICALS

If m and n are natural numbers greater than or equal to 2 and x and y are positive real numbers, then

a. $\sqrt[n]{x^n} = x$ $\qquad\qquad \sqrt[3]{x^3} = x$

b. $\sqrt[n]{xy} = \sqrt[n]{x}\sqrt[n]{y}$ $\qquad\qquad \sqrt[5]{xy} = \sqrt[5]{x}\sqrt[5]{y}$

c. $\sqrt[n]{\dfrac{x}{y}} = \dfrac{\sqrt[n]{x}}{\sqrt[n]{y}}$ $\qquad\qquad \sqrt[4]{\dfrac{x}{y}} = \dfrac{\sqrt[4]{x}}{\sqrt[4]{y}}$

---

1. $6x^{3/5} = 6\sqrt[5]{x^3}$

3. $(32x^2y^3)^{3/5} = \sqrt[5]{(32x^2y^3)^3}$

5. $(x^2 + y^2)^{1/2} = \sqrt{x^2 + y^2}$

[Note: $\sqrt{x^2 + y^2} \neq x + y$.]

7. $5\sqrt[4]{x^3} = 5x^{3/4}$

9. $\sqrt[5]{(2x^2y)^3} = (2x^2y)^{3/5}$

11. $\sqrt[3]{x} + \sqrt[3]{y} = x^{1/3} + y^{1/3}$

13. $25^{1/2} = (5^2)^{1/2} = 5$

15. $16^{3/2} = (4^2)^{3/2} = 4^3 = 64$

17. $-49^{1/2} = -\sqrt{49} = -7$

19. $-64^{2/3} = -(\sqrt[3]{64})^2 = -16$

21. $\left(\dfrac{4}{25}\right)^{3/2} = \left(\left(\dfrac{2}{5}\right)^2\right)^{3/2} \left(\dfrac{2}{5}\right)^3 = \dfrac{2^3}{5^3} = \dfrac{8}{125}$

23. $9^{-3/2} = (3^2)^{-3/2} = 3^{-3} = \dfrac{1}{3^3} = \dfrac{1}{27}$

25. $x^{4/5}x^{-2/5} = x^{4/5-2/5} = x^{2/5}$

27. $\dfrac{m^{2/3}}{m^{-1/3}} = m^{2/3-(-1/3)} = m^1 = m$

29. $(8x^3y^{-6})^{1/3} = (2^3x^3y^{-6})^{1/3} = 2^{3/3}x^{3/3}y^{-6/3} = 2xy^{-2} = \frac{2x}{y^2}$

31. $\left(\frac{4x^{-2}}{y^4}\right)^{-1/2} = \left(\frac{2^2x^{-2}}{y^4}\right)^{-1/2} = \frac{2^{2(-1/2)}x^{-2(-1/2)}}{y^{4(-1/2)}} = \frac{2^{-1}x^1}{y^{-2}} = \frac{xy^2}{2}$

33. $\frac{(8x)^{-1/3}}{12x^{1/4}} = \frac{\frac{1}{(8x)^{1/3}}}{12x^{1/4}} = \frac{\frac{1}{2x^{1/3}}}{12x^{1/4}} = \frac{1}{24x^{1/4+1/3}} = \frac{1}{24x^{7/12}}$

35. $\sqrt[5]{(2x+3)^5} = [(2x+3)^5]^{1/5} = 2x+3$

37. $\sqrt{6x}\sqrt{15x^3}\sqrt{30x^7} = \sqrt{6(15)(30)x^{11}} = \sqrt{3(30)^2x^{11}} = 30x^5\sqrt{3x}$

39. $\frac{\sqrt{6x}\sqrt{10}}{\sqrt{15x}} = \sqrt{\frac{60x}{15x}} = \sqrt{4} = 2$

41. $3x^{3/4}(4x^{1/4} - 2x^8) = 12x^{3/4+1/4} - 6x^{3/4+8} = 12x - 6x^{3/4+32/4} = 12x - 6x^{35/4}$

43. $(3u^{1/2} - v^{1/2})(u^{1/2} - 4v^{1/2}) = 3u - 12u^{1/2}v^{1/2} - u^{1/2}v^{1/2} + 4v = 3u - 13u^{1/2}v^{1/2} + 4v$

45. $(6m^{1/2} + n^{-1/2})(6m - n^{-1/2}) = 36m^{3/2} + 6mn^{-1/2} - 6m^{1/2}n^{-1/2} - n^{-1} = 36m^{3/2} + \frac{6m}{n^{1/2}} - \frac{6m^{1/2}}{n^{1/2}} - \frac{1}{n}$

47. $(3x^{1/2} - y^{1/2})^2 = (3x^{1/2})^2 - 6x^{1/2}y^{1/2} + (y^{1/2})^2 = 9x - 6x^{1/2}y^{1/2} + y$

49. $\frac{\sqrt[3]{x^2}+2}{2\sqrt[3]{x}} = \frac{x^{2/3}+2}{2x^{1/3}} = \frac{x^{2/3}}{2x^{1/3}} + \frac{2}{2x^{1/3}} = \frac{1}{2}x^{1/3} + \frac{1}{x^{1/3}} = \frac{1}{2}x^{1/3} + x^{-1/3}$

51. $\frac{2\sqrt[4]{x^3}+\sqrt[3]{x}}{3x} = \frac{2x^{3/4}+x^{1/3}}{3x} = \frac{2x^{3/4}}{3x} + \frac{x^{1/3}}{3x} = \frac{2}{3}x^{3/4-1} + \frac{1}{3}x^{1/3-1} = \frac{2}{3}x^{-1/4} + \frac{1}{3}x^{-2/3}$

53. $\frac{2\sqrt[3]{x}-\sqrt{x}}{4\sqrt{x}} = \frac{2x^{1/3}-x^{1/2}}{4x^{1/2}} = \frac{2x^{1/3}}{4x^{1/2}} - \frac{x^{1/2}}{4x^{1/2}} = \frac{1}{2}x^{1/3-1/2} - \frac{1}{4} = \frac{1}{2}x^{-1/6} - \frac{1}{4}$

55. $\frac{12mn^2}{\sqrt{3mn}} = \frac{12mn^2}{\sqrt{3mn}} \cdot \frac{\sqrt{3mn}}{\sqrt{3mn}} = \frac{12mn^2\sqrt{3mn}}{3mn} = 4n\sqrt{3mn}$

57. $\frac{2(x+3)}{\sqrt{x-2}} = \frac{2(x+3)}{\sqrt{x-2}} \cdot \frac{\sqrt{x-2}}{\sqrt{x-2}} = \frac{2(x+3)\sqrt{x-2}}{x-2}$

59. $\frac{7(x-y)^2}{\sqrt{x}-\sqrt{y}} = \frac{7(x-y)^2}{\sqrt{x}-\sqrt{y}} \cdot \frac{\sqrt{x}+\sqrt{y}}{\sqrt{x}+\sqrt{y}} = \frac{7(x-y)^2(\sqrt{x}+\sqrt{y})}{x-y} = 7(x-y)(\sqrt{x}+\sqrt{y})$

61. $\frac{\sqrt{5xy}}{5x^2y^2} = \frac{\sqrt{5xy}}{5x^2y^2} \cdot \frac{\sqrt{5xy}}{\sqrt{5xy}} = \frac{5xy}{5x^2y^2\sqrt{5xy}} = \frac{1}{xy\sqrt{5xy}}$

63. $\frac{\sqrt{x+h}-\sqrt{x}}{h} = \frac{\sqrt{x+h}-\sqrt{x}}{h} \cdot \frac{\sqrt{x+h}+\sqrt{x}}{\sqrt{x+h}+\sqrt{x}} = \frac{x+h-x}{h(\sqrt{x+h}+\sqrt{x})} = \frac{h}{h(\sqrt{x+h}+\sqrt{x})} = \frac{1}{\sqrt{x+h}+\sqrt{x}}$

65. $\frac{\sqrt{t}-\sqrt{x}}{t^2-x^2} = \frac{\sqrt{t}-\sqrt{x}}{(t-x)(t+x)} \cdot \frac{\sqrt{t}+\sqrt{x}}{\sqrt{t}+\sqrt{x}} = \frac{t-x}{(t-x)(t+x)(\sqrt{t}+\sqrt{x})} = \frac{1}{(t+x)(\sqrt{t}+\sqrt{x})}$

67. $(x+y)^{1/2} \stackrel{?}{=} x^{1/2} + y^{1/2}$

Let $x = y = 1$. Then

$(1+1)^{1/2} = 2^{1/2} = \sqrt{2} \approx 1.414$

$1^{1/2} + 1^{1/2} = \sqrt{1} + \sqrt{1} = 1 + 1 = 2$; $\sqrt{2} \neq 2$

71. $\sqrt{x^2} = x$ for all real numbers x: False

$\sqrt{(-2)^2} = \sqrt{4} = 2 \neq -2$

73. $\sqrt[3]{x^3} = |x|$ for all real numbers x: False

$\sqrt[3]{(-1)^3} = \sqrt[3]{-1} = -1 \neq |-1| = 1$

75. False: $(-8)^{1/3} = -2$ since $(-2)^3 = -8$

77. True: $r^{1/2} = \sqrt{r}$ and $-r^{1/2} = -\sqrt{r}$ are each square roots of r.

79. True: $(\sqrt{10})^4 = (10^{1/2})^4 = 10^2 = 100$

$(-\sqrt{10})^4 = (-1)^4(\sqrt{10})^4 = (1)(100) = 100$

81. False: $5\sqrt{7} - 6\sqrt{5} \approx -0.1877$; $\sqrt{a}$ is never negative.

83. $-\frac{1}{2}(x-2)(x+3)^{-3/2} + (x+3)^{-1/2} = \frac{-(x-2)}{2(x+3)^{3/2}} + \frac{1}{(x+3)^{1/2}} = \frac{-x+2+2(x+3)}{2(x+3)^{3/2}} = \frac{x+8}{2(x+3)^{3/2}}$

85.

$$\frac{(x-1)^{1/2} - x\left(\frac{1}{2}\right)(x-1)^{-1/2}}{x-1} = \frac{(x-1)^{1/2} - \frac{x}{2(x-1)^{1/2}}}{x-1} = \frac{\frac{2(x-1)^{1/2}(x-1)^{1/2}}{2(x-1)^{1/2}} - \frac{x}{2(x-1)^{1/2}}}{x-1}$$

$$= \frac{\frac{2(x-1)-x}{2(x-1)^{1/2}}}{x-1} = \frac{x-2}{2(x-1)^{3/2}}$$

87.

$$\frac{(x+2)^{2/3} - x\left(\frac{2}{3}\right)(x+2)^{-1/3}}{(x+2)^{4/3}} = \frac{(x+2)^{2/3} - \frac{2x}{3(x+2)^{1/3}}}{(x+2)^{4/3}} = \frac{\frac{3(x+2)^{1/3}(x+2)^{2/3} - 2x}{3(x+2)^{1/3}}}{(x+2)^{4/3}}$$

$$= \frac{3(x+2)-2x}{3(x+2)^{5/3}} = \frac{x+6}{3(x+2)^{5/3}}$$

89. $22^{3/2} = 22^{1.5} \approx 103.2$ or $22^{3/2} = \sqrt{(22)^3} = \sqrt{10{,}648} \approx 103.2$

91. $827^{-3/8} = \frac{1}{827^{3/8}} = \frac{1}{827^{0.375}} \approx \frac{1}{12.42} \approx 0.0805$

93. $37.09^{7/3} \approx 37.09^{2.3333} \approx 4{,}588$

95. (A) $\sqrt{3} + \sqrt{5} \approx 1.732 + 2.236 = 3.968$

(B) $\sqrt{2+\sqrt{3}} + \sqrt{2-\sqrt{3}} \approx 2.449$

(C) $1 + \sqrt{3} \approx 2.732$

(D) $\sqrt[3]{10+6\sqrt{3}} \approx 2.732$

(E) $\sqrt{8+\sqrt{60}} \approx 3.968$

(F) $\sqrt{6} \approx 2.449$

(A) and (E) have the same value:

$$\left(\sqrt{3}+\sqrt{5}\right)^2 = 3 + 2\sqrt{3}\,\sqrt{5} + 5 = 8 + 2\sqrt{15}$$

$$\left[\sqrt{8+\sqrt{60}}\right]^2 = 8 + \sqrt{4\cdot 15} = 8 + 2\sqrt{15}$$

(B) and (F) have the same value.

$$\left(\sqrt{2+\sqrt{3}}+\sqrt{2-\sqrt{3}}\right)^2 = 2 + \sqrt{3} + 2\sqrt{2+\sqrt{3}}\,\sqrt{2-\sqrt{3}} + 2 - \sqrt{3} = 4 + 2\sqrt{4-3} = 4 + 2 = 6$$

$(\sqrt{6})^2 = 6.$

(C) and (D) have the same value.

$$\left(1+\sqrt{3}\right)^3 = \left(1+\sqrt{3}\right)^2\left(1+\sqrt{3}\right) = (1 + 2\sqrt{3} + 3)(1 + \sqrt{3}) = (4 + 2\sqrt{3})(1 + \sqrt{3})$$
$$= 4 + 6\sqrt{3} + 6 = 10 + 6\sqrt{3}$$

$$\left(\sqrt[3]{10+6\sqrt{3}}\right)^3 = 10 + 6\sqrt{3}$$

EXERCISE A-7

Things to remember:

1. A QUADRATIC EQUATION in one variable is any equation that can be written in the form

   $ax^2 + bx + c = 0,\ a \neq 0$ STANDARD FORM

   where x is a variable and a, b, and c are constants.

2. Quadratic equations of the form $ax^2 + c = 0$ can be solved by the SQUARE ROOT METHOD. The solutions are:

   $$x = \pm\sqrt{\frac{-c}{a}} \quad \text{provided} \quad \frac{-c}{a} \geq 0;$$

   otherwise, the equation has no real solutions.

3. If the left side of the quadratic equation when written in standard form can be FACTORED,

   $$ax^2 + bx + c = (px + q)(rx + s),$$

   then the solutions are

   $$x = \frac{-q}{p} \quad \text{or} \quad x = \frac{-s}{r}.$$

4. The solutions of the quadratic equation written in standard form are given by the QUADRATIC FORMULA:

$$x = \frac{-b \pm \sqrt{b^2 - 4ac}}{2a}$$

The quantity $b^2 - 4ac$ under the radical is called the DISCRIMINANT and the equation:

(i) Has two real solutions if $b^2 - 4ac > 0$.

(ii) Has one real solution if $b^2 - 4ac = 0$.

(iii) Has no real solution if $b^2 - 4ac < 0$.

5. FACTORABILITY THEOREM

The second-degree polynomial, $ax^2 + bx + c$, with integer coefficients, can be expressed as the product of two first-degree polynomials with integer coefficients if and only if $\sqrt{b^2 - 4ac}$ is an integer.

6. FACTOR THEOREM

If r_1 and r_2 are solutions of $ax^2 + bx + c = 0$, then

$ax^2 + bx + c = a(x - r_1)(x - r_2)$.

---

1. $2x^2 - 22 = 0$

$x^2 - 11 = 0$

$x^2 = 11$

$x = \pm\sqrt{11}$

3. $(3x - 1)^2 = 25$

$3x - 1 = \pm\sqrt{25} = \pm 5$

$3x = 1 \pm 5 = -4$ or 6

$x = -\frac{4}{3}$ or 2

5. $2u^2 - 8u - 24 = 0$

$u^2 - 4u - 12 = 0$

$(u - 6)(u + 2) = 0$

$u - 6 = 0$ or $u + 2 = 0$

$u = 6$ or $u = -2$

7. $x^2 = 2x$

$x^2 - 2x = 0$

$x(x - 2) = 0$

$x = 0$ or $x - 2 = 0$

$x = 2$

9. $x^2 - 6x - 3 = 0$

$$x = \frac{-b \pm \sqrt{b^2 - 4ac}}{2a}, \quad a = 1,\ b = -6,\ c = -3$$

$$= \frac{-(-6) \pm \sqrt{(-6)^2 - 4(1)(-3)}}{2(1)} = \frac{6 \pm \sqrt{48}}{2} = \frac{6 \pm 4\sqrt{3}}{2} = 3 \pm 2\sqrt{3}$$

11. $3u^2 + 12u + 6 = 0$

Since 3 is a factor of each coefficient, divide both sides by 3.

$u^2 + 4u + 2 = 0$

$$u = \frac{-b \pm \sqrt{b^2 - 4ac}}{2a}, \quad a = 1,\ b = 4,\ c = 2$$

$$= \frac{-4 \pm \sqrt{4^2 - 4(1)(2)}}{2(1)} = \frac{-4 \pm \sqrt{8}}{2} = \frac{-4 \pm 2\sqrt{2}}{2} = -2 \pm \sqrt{2}$$

13.

$$\frac{2x^2}{3} = 5x$$

$2x^2 = 15x$

$2x^2 - 15x = 0$

$x(2x - 15) = 0$

$x = 0$ or $2x - 15 = 0$

$x = \frac{15}{2}$

15. $4u^2 - 9 = 0$

$4u^2 = 9$ (solve by square root method)

$u^2 = \frac{9}{4}$

$u = \pm\sqrt{\frac{9}{4}} = \pm\frac{3}{2}$

17. $8x^2 + 20x = 12$

$8x^2 + 20x - 12 = 0$

$2x^2 + 5x - 3 = 0$

$(x + 3)(2x - 1) = 0$

$x + 3 = 0$ or $2x - 1 = 0$

$x = -3$ or $2x = 1$

$x = \frac{1}{2}$

19. $x^2 = 1 - x$

$x^2 + x - 1 = 0$

$$x = \frac{-b \pm \sqrt{b^2 - 4ac}}{2a}, \quad a = 1,\ b = 1,\ c = -1$$

$$= \frac{-1 \pm \sqrt{(1)^2 - 4(1)(-1)}}{2(1)} = \frac{-1 \pm \sqrt{5}}{2}$$

21. $2x^2 = 6x - 3$

$2x^2 - 6x + 3 = 0$

$$x = \frac{-b \pm \sqrt{b^2 - 4ac}}{2a}, \quad a = 2,\ b = -6,\ c = 3$$

$$= \frac{-(-6) \pm \sqrt{(-6)^2 - 4(2)(3)}}{2(2)} = \frac{6 \pm \sqrt{12}}{4} = \frac{6 \pm 2\sqrt{3}}{4} = \frac{3 \pm \sqrt{3}}{2}$$

23. $y^2 - 4y = -8$

$y^2 - 4y + 8 = 0$

$$y = \frac{-b \pm \sqrt{b^2 - 4ac}}{2a}, \quad a = 1, \quad b = -4, \quad c = 8$$

$$= \frac{-(-4) \pm \sqrt{(-4)^2 - 4(1)(8)}}{2(1)} = \frac{4 \pm \sqrt{-16}}{2}$$

Since $\sqrt{-16}$ is not a real number, there are no real solutions.

25. $(2x + 3)^2 = 11$

$2x + 3 = \pm\sqrt{11}$

$2x = -3 \pm \sqrt{11}$

$x = -\frac{3}{2} \pm \frac{1}{2}\sqrt{11}$

27. $\frac{3}{p} = p$

$p^2 = 3$

$p = \pm\sqrt{3}$

29. $2 - \frac{2}{m^2} = \frac{3}{m}$

$2m^2 - 2 = 3m$

$2m^2 - 3m - 2 = 0$

$(2m + 1)(m - 2) = 0$

$m = -\frac{1}{2}, 2$

31. $x^2 + 40x - 84$

Step 1. Test for factorability

$\sqrt{b^2 - 4ac} = \sqrt{(40)^2 - 4(1)(-84)} = \sqrt{1936} = 44$

Since the result is an integer, the polynomial has first-degree factors with integer coefficients.

Step 2. Use the factor theorem

$x^2 + 40x - 84 = 0$

$$x = \frac{-40 \pm 44}{2} = 2, -42 \text{ (by the quadratic formula)}$$

Thus, $x^2 + 40x - 84 = (x - 2)(x - [-42]) = (x - 2)(x + 42)$

33. $x^2 - 32x + 144$

Step 1. Test for factorability

$\sqrt{b^2 - 4ac} = \sqrt{(-32)^2 - 4(1)(144)} = \sqrt{448} \approx 21.166$

Since this is not an integer, the polynomial is not factorable.

35. $2x^2 + 15x - 108$

Step 1. Test for factorability

$\sqrt{b^2 - 4ac} = \sqrt{(15)^2 - 4(2)(-108)} = \sqrt{1089} = 33$

Thus, the polynomial has first-degree factors with integer coefficients.

Step 2. Use the factor theorem

$2x^2 + 15x - 108 = 0$

$$x = \frac{-15 \pm 33}{4} = \frac{9}{2}, -12$$

Thus, $2x^2 + 15x - 108 = 2\left(x - \frac{9}{2}\right)(x - [-12]) = (2x - 9)(x + 12)$

37. $4x^2 + 241x - 434$

Step 1. Test for factorability

$$\sqrt{b^2 - 4ac} = \sqrt{(241)^2 - 4(4)(-434)} = \sqrt{65025} = 255$$

Thus, the polynomial has first-degree factors with integer coefficients.

Step 2. Use the factor theorem

$4x^2 + 241x - 434 = 0$

$$x = \frac{-241 \pm 255}{8} = \frac{14}{8}, -\frac{496}{8} \text{ or } \frac{7}{4}, -62$$

Thus, $4x^2 + 241x - 434 = 4\left(x - \frac{7}{4}\right)(x + 62) = (4x - 7)(x + 62)$

39.

$$A = P(1 + r)^2$$
$$(1 + r)^2 = \frac{A}{P}$$
$$1 + r = \sqrt{\frac{A}{P}}$$
$$r = \sqrt{\frac{A}{P}} - 1$$

41. $x^2 + 4x + c = 0$

The discriminant is: $16 - 4c$

(A) If $16 - 4c > 0$, i.e., if $c < 4$, then the equation has two distinct real roots.

(B) If $16 - 4c = 0$, i.e., if $c = 4$, then the equation has one real double root.

(C) If $16 - 4c < 0$, i.e., if $c > 4$, then there are no real roots.

43. $x^3 + 8 = (x + 2)(x^2 - 2x + 4) = 0; \quad x = -2$

45. $5x^4 - 500 = 0, \quad x^4 - 100 = 0, \quad (x^2 - 10)(x^2 + 10) = 0, \quad x = \pm\sqrt{10}$

47. $x^4 - 8x^2 + 15 = 0, \quad (x^2 - 5)(x^2 - 3) = 0, \quad x = \pm\sqrt{5}, \ x = \pm\sqrt{3}$

49. Setting the supply equation equal to the demand equation, we have

$$\frac{x}{450} + \frac{1}{2} = \frac{6{,}300}{x}$$
$$\frac{1}{450}x^2 + \frac{1}{2}x = 6{,}300$$
$$x^2 + 225x - 2{,}835{,}000 = 0$$

$$x = \frac{-225 \pm \sqrt{(225)^2 - 4(1)(-2{,}835{,}000)}}{2} \quad \text{(quadratic formula)}$$

$$= \frac{-225 \pm \sqrt{11{,}390{,}625}}{2} = \frac{-225 \pm 3375}{2} = 1{,}575 \text{ units}$$

Note, we discard the negative root since a negative number of units cannot be produced or sold. Substituting $x = 1{,}575$ into either equation (we use the demand equation), we get

$$p = \frac{6{,}300}{1{,}575} = 4$$

Supply equals demand at $4 per unit.

51. $A = P(1+r)^2 = P(1+2r+r^2) = Pr^2 + 2Pr + P$
Let $A = 625$ and $P = 484$. Then,

$484r^2 + 968r + 484 = 625$

$484r^2 + 968r - 141 = 0$

Using the quadratic formula,

$$r = \frac{-968 \pm \sqrt{(968)^2 - 4(484)(-141)}}{968} = \frac{-968 \pm \sqrt{1{,}210{,}000}}{968}$$

$$= \frac{-968 \pm 1100}{968} \approx 0.1364 \text{ or } -2.136$$

Since $r > 0$, we have $r = 0.1364$ or 13.64%.

53. $v^2 = 64h$

For $h = 1$, $v^2 = 64(1) = 64$. Therefore, $v = 8$ ft/sec.

For $h = 0.5$, $v^2 = 64(0.5) = 32$.

Therefore, $v = \sqrt{32} = 4\sqrt{2} \approx 5.66$ ft/sec.

APPENDIX B SPECIAL TOPICS

EXERCISE B-1

Things to remember:

1. SEQUENCES
A SEQUENCE is a function whose domain is a set of successive integers. If the domain of a given sequence is a finite set, then the sequence is called a FINITE SEQUENCE; otherwise, the sequence is an INFINITE SEQUENCE. In general, unless stated to the contrary or the context specifies otherwise, the domain of a sequence will be understood to be the set N of natural numbers.

2. NOTATION FOR SEQUENCES

Rather than function notation $f(n)$, n in the domain of a given sequence f, subscript notation a_n is normally used to denote the value in the range corresponding to n, and the sequence itself is denoted $\{a_n\}$ rather than f or $f(n)$. The elements in the range, a_n, are called the TERMS of the sequence; a_1 is the first term, a_2 is the second term, and a_n is the nth term or general term.

3. SERIES
Given a sequence $\{a_n\}$. The sum of the terms of the sequence,
$a_1 + a_2 + a_3 + \cdots$ is called a SERIES. If the sequence is finite,
the corresponding series is a FINITE SERIES; if the sequence is infinite, then the corresponding series is an INFINITE SERIES.
Only finite series are considered in this section.

4. NOTATION FOR SERIES
Series are represented using SUMMATION NOTATION.
If $\{a_k\}$, $k = 1, 2, \ldots, n$ is a finite sequence, then the series

$$a_1 + a_2 + a_3 + \cdots + a_n$$

is denoted

$$\sum_{k=1}^{n} a_k.$$

The symbol $\sum$ is called the SUMMATION SIGN and k is called the SUMMING INDEX.

5. ARITHMETIC MEAN
If $\{a_k\}$, $k = 1, 2, \ldots, n$, is a finite sequence, then the ARITHMETIC MEAN $\bar{a}$ of the sequence is defined as

$$\bar{a} = \frac{1}{n}\sum_{k=1}^{n} a_k.$$

1. $a_n = 2n + 3;$ $a_1 = 2 \cdot 1 + 3 = 5$
$a_2 = 2 \cdot 2 + 3 = 7$
$a_3 = 2 \cdot 3 + 3 = 9$
$a_4 = 2 \cdot 4 + 3 = 11$

3. $a_n = \frac{n+2}{n+1};$ $a_1 = \frac{1+2}{1+1} = \frac{3}{2}$
$a_2 = \frac{2+2}{2+1} = \frac{4}{3}$
$a_3 = \frac{3+2}{3+1} = \frac{5}{4}$
$a_4 = \frac{4+2}{4+1} = \frac{6}{5}$

5. $a_n = (-3)^{n+1};$ $a_1 = (-3)^{1+1} = (-3)^2 = 9$
$a_2 = (-3)^{2+1} = (-3)^3 = -27$
$a_3 = (-3)^{3+1} = (-3)^4 = 81$
$a_4 = (-3)^{4+1} = (-3)^5 = -243$

7. $a_n = 2n + 3; a_{10} = 2 \cdot 10 + 3 = 23$

9. $a_n = \frac{n+2}{n+1}; a_{99} = \frac{99+2}{99+1} = \frac{101}{100}$

11. $\sum_{k=1}^{6} k = 1 + 2 + 3 + 4 + 5 + 6 = 21$

13. $\sum_{k=4}^{7} (2k-3) = (2 \cdot 4 - 3) + (2 \cdot 5 - 3) + (2 \cdot 6 - 3) + (2 \cdot 7 - 3) = 5 + 7 + 9 + 11 = 32$

15. $\sum_{k=0}^{3} \frac{1}{10^k} = \frac{1}{10^0} + \frac{1}{10^1} + \frac{1}{10^2} + \frac{1}{10^3} = 1 + \frac{1}{10} + \frac{1}{100} + \frac{1}{1000} = \frac{1111}{1000} = 1.111$

17. $a_1 = 5, a_2 = 4, a_3 = 2, a_4 = 1, a_5 = 6$. Here $n = 5$ and the arithmetic mean is given by:
$$\bar{a} = \frac{1}{5} \sum_{k=1}^{5} a_k = \frac{1}{5}(5 + 4 + 2 + 1 + 6) = \frac{18}{5} = 3.6$$

19. $a_1 = 96, a_2 = 65, a_3 = 82, a_4 = 74, a_5 = 91, a_6 = 88, a_7 = 87, a_8 = 91, a_9 = 77$, and $a_{10} = 74$. Here $n = 10$ and the arithmetic mean is given by:
$$\bar{a} = \frac{1}{10} \sum_{k=1}^{10} a_k = \frac{1}{10}(96 + 65 + 82 + 74 + 91 + 88 + 87 + 91 + 77 + 74) = \frac{825}{10} = 82.5$$

21. $a_n = \frac{(-1)^{n+1}}{2^n};$ $a_1 = \frac{(-1)^2}{2^1} = \frac{1}{2}$
$a_2 = \frac{(-1)^3}{2^2} = -\frac{1}{4}$
$a_3 = \frac{(-1)^4}{2^3} = \frac{1}{8}$
$a_4 = \frac{(-1)^5}{2^4} = -\frac{1}{16}$
$a_5 = \frac{(-1)^6}{2^5} = \frac{1}{32}$

23. $a_n = n[1 + (-1)^n]$;

$$a_1 = 1[1 + (-1)^1] = 0$$
$$a_2 = 2[1 + (-1)^2] = 4$$
$$a_3 = 3[1 + (-1)^3] = 0$$
$$a_4 = 4[1 + (-1)^4] = 8$$
$$a_5 = 5[1 + (-1)^5] = 0$$

25. $a_n = \left(-\frac{3}{2}\right)^{n-1}$;

$$a_1 = \left(-\frac{3}{2}\right)^0 = 1$$
$$a_2 = \left(-\frac{3}{2}\right)^1 = -\frac{3}{2}$$
$$a_3 = \left(-\frac{3}{2}\right)^2 = \frac{9}{4}$$
$$a_4 = \left(-\frac{3}{2}\right)^3 = -\frac{27}{8}$$
$$a_5 = \left(-\frac{3}{2}\right)^4 = \frac{81}{16}$$

27. Given –2, –1, 0, 1, … The sequence is the set of successive integers beginning with –2. Thus, $a_n = n - 3$, $n = 1, 2, 3, \ldots$.

29. Given 4, 8, 12, 16, … The sequence is the set of positive integer multiples of 4. Thus, $a_n = 4n$, $n = 1, 2, 3, \ldots$.

31. Given $\frac{1}{2}, \frac{3}{4}, \frac{5}{6}, \frac{7}{8}, \ldots$ The sequence is the set of fractions whose numerators are the odd positive integers and whose denominators are the even positive integers. Thus,

$a_n = \frac{2n-1}{2n}$, $n = 1, 2, 3, \ldots$.

33. Given 1, –2, 3, –4, … The sequence consists of the positive integers with alternating signs. Thus,

$a_n = (-1)^{n+1} n$, $n = 1, 2, 3, \ldots$.

35. Given 1, –3, 5, –7, … The sequence consists of the odd positive integers with alternating signs. Thus,

$a_n = (-1)^{n+1}(2n - 1)$, $n = 1, 2, 3, \ldots$.

37. Given $1, \frac{2}{5}, \frac{4}{25}, \frac{8}{125}, \ldots$ The sequence consists of the nonnegative integral powers of $\frac{2}{5}$. Thus,

$a_n = \left(\frac{2}{5}\right)^{n-1}$, $n = 1, 2, 3, \ldots$.

39. Given $x, x^2, x^3, x^4, \ldots$ The sequence is the set of positive integral powers of x. Thus, $a_n = x^n$, $n = 1, 2, 3, \ldots$.

41. Given $x, -x^3, x^5, -x^7, \ldots$ The sequence is the set of positive odd integral powers of x with alternating signs. Thus,

$a_n = (-1)^{n+1} x^{2n-1}$, $n = 1, 2, 3, \ldots$.

43. $$\sum_{k=1}^{5} (-1)^{k+1}(2k-1)^2 = (-1)^2(2\cdot 1-1)^2 + (-1)^3(2\cdot 2-1)^2 + (-1)^4(2\cdot 3-1)^2 + (-1)^5(2\cdot 4-1)^2 + (-1)^6(2\cdot 5-1)^2 = 1 - 9 + 25 - 49 + 81$$

45. $$\sum_{k=2}^{5} \frac{2^k}{2k+3} = \frac{2^2}{2\cdot 2+3} + \frac{2^3}{2\cdot 3+3} + \frac{2^4}{2\cdot 4+3} + \frac{2^5}{2\cdot 5+3} = \frac{4}{7} + \frac{8}{9} + \frac{16}{11} + \frac{32}{13}$$

47. $$\sum_{k=1}^{5} x^{k-1} = x^0 + x^1 + x^2 + x^3 + x^4 = 1 + x + x^2 + x^3 + x^4$$

49. $$\sum_{k=0}^{4} \frac{(-1)^k x^{2k+1}}{2k+1} = \frac{(-1)^0 x}{2\cdot 0+1} + \frac{(-1)^1 x^3}{2\cdot 1+1} + \frac{(-1)^2 x^5}{2\cdot 2+1} + \frac{(-1)^3 x^7}{2\cdot 3+1} + \frac{(-1)^4 x^9}{2\cdot 4+1} = x - \frac{x^3}{3} + \frac{x^5}{5} - \frac{x^7}{7} + \frac{x^9}{9}$$

51. (A) $2 + 3 + 4 + 5 + 6 = \sum_{k=1}^{5} (k+1)$ (B) $2 + 3 + 4 + 5 + 6 = \sum_{j=0}^{4} (j+2)$

53. (A) $1 - \frac{1}{2} + \frac{1}{3} - \frac{1}{4} = \sum_{k=1}^{4} \frac{(-1)^{k+1}}{k}$ (B) $1 - \frac{1}{2} + \frac{1}{3} - \frac{1}{4} = \sum_{j=0}^{3} \frac{(-1)^j}{j+1}$

55. $$2 + \frac{3}{2} + \frac{4}{3} + \ldots + \frac{n+1}{n} = \sum_{k=1}^{n} \frac{k+1}{k}$$

57. $$\frac{1}{2} - \frac{1}{4} + \frac{1}{8} - \ldots + \frac{(-1)^{n+1}}{2^n} = \sum_{k=1}^{n} \frac{(-1)^{k+1}}{2^k}$$

59. False: $$1 + \frac{1}{2} + \frac{1}{3} + \frac{1}{4} + \frac{1}{5} + \frac{1}{6} + \ldots + \frac{1}{64}$$
$$= 1 + \frac{1}{2} + \left(\frac{1}{3} + \frac{1}{4}\right) + \left(\frac{1}{5} + \frac{1}{6} + \frac{1}{7} + \frac{1}{8}\right) + \left(\frac{1}{9} + \cdots + \frac{1}{16}\right) + \left(\frac{1}{17} + \cdots + \frac{1}{32}\right) + \left(\frac{1}{33} + \cdots + \frac{1}{64}\right)$$
$$> 1 + \frac{1}{2} + \frac{1}{2} + \frac{1}{2} + \frac{1}{2} + \frac{1}{2} + \frac{1}{2} = 4$$

61. True: $$\frac{1}{2} - \frac{1}{4} + \frac{1}{8} - \frac{1}{16} + \frac{1}{32} - \ldots$$
$$= \left(\frac{1}{2} - \frac{1}{4}\right) + \left(\frac{1}{8} - \frac{1}{16}\right) + \left(\frac{1}{32} - \frac{1}{64}\right) + \text{(positive terms)}$$
$$= \frac{1}{4} + \frac{1}{16} + \frac{1}{64} + \ldots > \frac{1}{4}$$

63. $a_1 = 2$ and $a_n = 3a_{n-1} + 2$
for $n \geq 2$.
$a_1 = 2$
$a_2 = 3 \cdot a_1 + 2 = 3 \cdot 2 + 2 = 8$
$a_3 = 3 \cdot a_2 + 2 = 3 \cdot 8 + 2 = 26$
$a_4 = 3 \cdot a_3 + 2 = 3 \cdot 26 + 2 = 80$
$a_5 = 3 \cdot a_4 + 2 = 3 \cdot 80 + 2 = 242$

65. $a_1 = 1$ and $a_n = 2a_{n-1}$
for $n \geq 2$.
$a_1 = 1$
$a_2 = 2 \cdot a_1 = 2 \cdot 1 = 2$
$a_3 = 2 \cdot a_2 = 2 \cdot 2 = 4$
$a_4 = 2 \cdot a_3 = 2 \cdot 4 = 8$
$a_5 = 2 \cdot a_4 = 2 \cdot 8 = 16$

67. If $a_1 = \frac{A}{2}$, $a_n = \frac{1}{2}\left(a_{n-1} + \frac{A}{a_{n-1}}\right)$, $n \geq 2$, let $A = 2$. Then:

$$a_1 = \frac{2}{2} = 1$$

$$a_2 = \frac{1}{2}\left(a_1 + \frac{A}{a_1}\right) = \frac{1}{2}(1 + 2) = \frac{3}{2}$$

$$a_3 = \frac{1}{2}\left(a_2 + \frac{A}{a_2}\right) = \frac{1}{2}\left(\frac{3}{2} + \frac{2}{3/2}\right) = \frac{1}{2}\left(\frac{3}{2} + \frac{4}{3}\right) = \frac{17}{12}$$

$$a_4 = \frac{1}{2}\left(a_3 + \frac{A}{a_3}\right) = \frac{1}{2}\left(\frac{17}{12} + \frac{2}{17/12}\right) = \frac{1}{2}\left(\frac{17}{12} + \frac{24}{17}\right) = \frac{577}{408} \approx 1.414216$$

and $\sqrt{2} \approx 1.414214$

69. $a_1 = 1,\ a_2 = 1,\ a_n = a_{n-1} + a_{n-2},\ n \geq 3$
$a_3 = a_2 + a_1 = 2,\ a_4 = a_3 + a_2 = 3,\ a_5 = a_4 + a_3 = 5,$
$a_6 = a_5 + a_4 = 8,\ a_7 = a_6 + a_5 = 13,\ a_8 = a_7 + a_6 = 21$
$a_9 = a_8 + a_7 = 34,\ a_{10} = a_9 + a_8 = 55$

EXERCISE B-2

Things to remember:

1. A sequence of numbers $a_1, a_2, a_3, \ldots, a_n, \ldots$, is called an ARITHMETIC SEQUENCE if there is constant d, called the COMMON DIFFERENCE, such that
$$a_n - a_{n-1} = d,$$
that is,
$$a_n = a_{n-1} + d$$
for all $n > 1$.

2. A sequence of numbers $a_1, a_2, a_3, \ldots, a_n, \ldots$, is called a GEOMETRIC SEQUENCE if there exists a nonzero constant r, called the COMMON RATIO, such that
$$\frac{a_n}{a_{n-1}} = r,$$
that is,
$$a_n = ra_{n-1}$$
for all $n > 1$.

3. nTH TERM OF AN ARITHMETIC SEQUENCE

If $\{a_n\}$ is an arithmetic sequence with common difference d, then

$$a_n = a_1 + (n-1)d$$

for all $n > 1$.

4. nTH TERM OF A GEOMETRIC SEQUENCE

If $\{a_n\}$ is a geometric sequence with common ratio r, then

$$a_n = a_1 r^{n-1}$$

for all $n > 1$.

5. SUM FORMULAS FOR FINITE ARITHMETIC SERIES

The sum S_n of the first n terms of an arithmetic series $a_1 + a_2 + a_3 + \ldots + a_n$ with common difference d, is given by

(a) $S_n = \frac{n}{2}[2a_1 + (n-1)d]$ (First Form)

or by

(b) $S_n = \frac{n}{2}(a_1 + a_n)$. (Second Form)

6. SUM FORMULAS FOR FINITE GEOMETRIC SERIES

The sum S_n of the first n terms of a geometric series $a_1 + a_2 + a_3 + a_n$ with common ratio r, is given by:

$$S_n = \frac{a_1(r^n - 1)}{r-1}, \; r \neq 1, \quad \text{(First Form)}$$

or by

$$S_n = \frac{ra_n - a_1}{r-1}, \; r \neq 1. \quad \text{(Second Form)}$$

7. SUM OF AN INFINITE GEOMETRIC SERIES

If $a_1 + a_2 + a_3 + \ldots + a_n + \ldots$, is an infinite geometric series with common ratio r having the property $-1 < r < 1$, then the sum S_∞ is defined to be:

$$S_\infty = \frac{a_1}{1-r}.$$

1. (A) $-11, -16, -21, \ldots$
This is an arithmetic sequence with common difference $d = -5$;
$a_4 = -26,\ a_5 = -31.$

(B) $2, -4, 8, \ldots$
This is a geometric sequence with common ratio $r = -2$;
$a_4 = -16,\ a_5 = 32.$

(C) $1, 4, 9, \ldots$
This is neither an arithmetic sequence $(4 - 1 \neq 9 - 4)$ nor a geometric sequence $\left(\frac{4}{1} \neq \frac{9}{4}\right)$.

(D) $\frac{1}{2}, \frac{1}{6}, \frac{1}{18}, \ldots$
This is a geometric sequence with common ratio $r = \frac{1}{3}$;
$a_4 = \frac{1}{54}, a_5 = \frac{1}{162}$.

3. $\sum_{k=1}^{101} (-1)^{k+1} = 1 - 1 + 1 - 1 + \ldots + 1$
This is a geometric series with $a_1 = 1$ and common ratio $r = -1$.
$$S_{101} = \frac{1[(-1)^{101} - 1]}{-1-1} = \frac{-2}{-2} = 1$$

5. This series is neither arithmetic nor geometric.

7. $5 + 4.9 + 4.8 + \ldots + 0.1$ is an arithmetic series with $a_1 = 5$, $a_{50} = 0.1$ and common difference $d = -0.1$:
$$S_{50} = \frac{50}{2}[5 + 0.1] = 25(5.1) = 127.5$$

9. $a_2 = a_1 + d = 7 + 4 = 11$
$a_3 = a_2 + d = 11 + 4 = 15$ (using 1)

11. $a_{21} = a_1 + (21 - 1)d = 2 + 20 \cdot 4 = 82$ (using 3)
$$S_{31} = \frac{31}{2}[2a_1 + (31-1)d] = \frac{31}{2}[2 \cdot 2 + 30 \cdot 4] = \frac{31}{2} \cdot 124 = 1922 \text{ [using 5(a)]}$$

13. Using 5(b), $S_{20} = \frac{20}{2}(a_1 + a_{20}) = 10(18 + 75) = 930$

15. $a_2 = a_1 r = 3(-2) = -6$
$a_3 = a_1 r^2 = 3(-2)^2 = 3 \cdot 4 = 12$
$a_4 = a_1 r^3 = 3(-2)^3 = 3(-8) = -24$ (using 4)

17. Using 6, $S_7 = \frac{-3 \cdot 729 - 1}{-3 - 1} = \frac{-2188}{-4} = 547.$

19. Using 4, $a_{10} = 100(1.08)^9 = 199.90.$

21. Using 4, $200 = 100\,r^8$. Thus, $r^8 = 2$ and $r = \pm\sqrt[8]{2} \approx \pm 1.09$.

23. Using 6, $S_{10} = \dfrac{500[(0.6)^{10} - 1]}{0.6 - 1} \approx 1242$,

$$S_\infty = \frac{500}{1 - 0.6} = 1250.$$

25. $S_{41} = \sum_{k=1}^{41}(3k+3)$. The sequence of terms is an arithmetic sequence. Therefore,

$$S_{41} = \frac{41}{2}(a_1 + a_{41}) = \frac{41}{2}(6 + 126) = \frac{41}{2}(132) = 41(66) = 2{,}706$$

27. $S_8 = \sum_{k=1}^{8} (-2)^{k-1}$. The sequence of terms is a geometric sequence with common ratio $r = -2$ and

$a_1 = (-2)^0 = 1$.

$$S_8 = \frac{1[(-2)^8 - 1]}{-2 - 1} = \frac{256 - 1}{-3} = -85$$

29. Let $a_1 = 13$, $d = 2$. Then, using 3, we can find n:

$$67 = 13 + (n-1)2 \quad \text{or} \quad 2(n-1) = 54$$
$$n - 1 = 27$$
$$n = 28$$

Therefore, using 5(b), $S_{28} = \dfrac{28}{2}[13 + 67] = 14 \cdot 80 = 1120$.

31. (A) $2 + 4 + 8 + \cdots$. Since $r = \dfrac{4}{2} = \dfrac{8}{4} = \ldots = 2$ and $|2| = 2 > 1$, the sum does not exist.

(B) $2, -\dfrac{1}{2}, \dfrac{1}{8}, \ldots$. In this case, $r = \dfrac{-1/2}{2} = \dfrac{1/8}{-1/2} = \ldots = -\dfrac{1}{4}$.

Since $|r| < 1$, $S_\infty = \dfrac{2}{1 - (-1/4)} = \dfrac{2}{5/4} = \dfrac{8}{5} = 1.6$.

33. $f(1) = -1$, $f(2) = 1$, $f(3) = 3, \ldots$. This is an arithmetic sequence $a_1 = -1$, $d = 2$. Thus, using 5(a),

$$f(1) + f(2) + f(3) + \cdots + f(50) = \frac{50}{2}[2(-1) + 49 \cdot 2] = 25 \cdot 96 = 2400.$$

35. $f(1) = \dfrac{1}{2}$, $f(2) = \left(\dfrac{1}{2}\right)^2 = \dfrac{1}{4}$, $f(3) = \left(\dfrac{1}{2}\right)^3 = \dfrac{1}{8}, \ldots$. This is a geometric sequence with $a_1 = \dfrac{1}{2}$ and $r = \dfrac{1}{2}$. Thus, using 6:

$$f(1) + f(2) + \cdots + f(10) = S_{10} = \frac{\frac{1}{2}\left[\left(\frac{1}{2}\right)^{10} - 1\right]}{\frac{1}{2} - 1} \approx 0.999$$

37. Consider the arithmetic sequence with $a_1 = 1$, $d = 2$. This is the sequence of odd positive integers. Now, using 5(a), the sum of the first n odd positive integers is:

$$S_n = \frac{n}{2}[2\cdot 1 + (n-1)2] = \frac{n}{2}(2+2n-2) = \frac{n}{2}\cdot 2n = n^2$$

39. $S_n = a_1 + a_1r + \ldots + a_1r^{n\text{-}1}$. If $r = 1$, then $S_n = na_1$.

41. No: $\frac{n}{2}(1 + 1.1) = 100$ implies $(2.1)n = 200$ and this equation does not have an integer solution.

43. Yes: Solve the equation $6 = \frac{10}{1-r}$ for r. This yields $r = -\frac{2}{3}$.

The infinite geometric series: $10 - 10\left(\frac{2}{3}\right) + 10\left(\frac{2}{3}\right)^2 - 10\left(\frac{2}{3}\right)^3 + \ldots$

has sum $S_\infty = 6$.

45. Consider the time line:

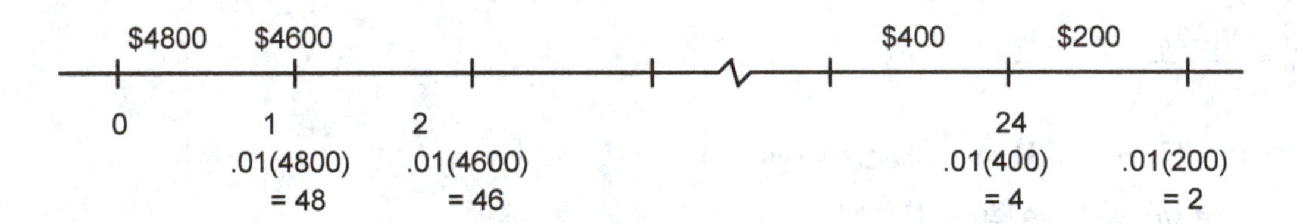

The total cost of the loan is $2 + 4 + 6 + \cdots + 46 + 48$. The terms form an arithmetic sequence with $n = 24$, $a_1 = 2$, and $a_{24} = 48$. Thus, using 5(b):

$$S_{24} = \frac{24}{2}(2+48) = 24\cdot 25 = \$600$$

47. This is a geometric sequenceprogression with $a_1 = 3{,}500{,}000$ and $r = 0.7$.
Thus, using 7:

$$S_\infty = \frac{3{,}500{,}000}{1-0.7} \approx \$11{,}666{,}666.67$$

49.

$1000 $1050 $1102.50 $1157.63

1: (1.05)1000 = 1050

2: (1.05)(1050) = $1000(1.05)^2$ = 1102.50

3: 1.05(1102.50) = $1000(1.05)^3$ = 1157.63

In general, after n years, the amount A_n in the account is:

$$A_n = 1000(1.05)^n$$

Thus, $A_{10} = 1000(1.05)^{10} \approx \$1628.89$

and $A_{20} = 1000(1.05)^{20} \approx \$2653.30$

EXERCISE B-3

Things to remember:

1. If n is a positive integer, then n FACTORIAL, denoted $n!$, is the product of the integers from 1 to n; that is,

$$n! = n \cdot (n-1) \cdot \ldots \cdot 3 \cdot 2 \cdot 1 = n(n-1)!$$

Also, $1! = 1$ and $0! = 1$.

2. If n and r are nonnegative integers and $r \le n$, then:

$$C_{n,r} = \frac{n!}{r!(n-r)!}$$

3. BINOMIAL THEOREM

For all natural numbers n:

$$(a+b)^n = {}_nC_0a^n + {}_nC_1a^{n-1}b + {}_nC_2a^{n-2}b^2 + \cdots + {}_nC_{n-1}ab^{n-1} + {}_nC_nb^n$$

1. $6! = 6\cdot5\cdot4\cdot3\cdot2\cdot1 = 720$

3. $\frac{10!}{9!} = \frac{10\cdot9!}{9!} = 10$

5. $\frac{12!}{9!} = \frac{12\cdot11\cdot10\cdot9!}{9!} = 1320$

7. $\frac{5!}{2!3!} = \frac{5\cdot4\cdot3!}{2\cdot1\cdot3!} = 10$

9. $\frac{6!}{5!(6-5)!} = \frac{6\cdot5!}{5!1!} = 6$

11. $\frac{20!}{3!17!} = \frac{20\cdot19\cdot18\cdot17!}{3!17!}$

$= \frac{20\cdot19\cdot18}{3\cdot2\cdot1} = 1140$

13. ${}_5C_3 = \frac{5!}{3!(5-3)!} = \frac{5!}{3!2!} = 10$ (see Problem 7)

15. ${}_6C_5 = \frac{6!}{5!(6-5)!} = 6$ (see Problem 9)

17. ${}_5C_0 = \frac{5!}{0!(5-0)!} = \frac{5!}{1\cdot5!} = 1$

19. ${}_{18}C_{15} = \frac{18!}{15!(18-15)!} = \frac{18\cdot17\cdot16\cdot15!}{15!3!} = \frac{18\cdot17\cdot16}{3\cdot2\cdot1} = 816$

21. Using 3,

$(a+b)^4 = {}_4C_0a^4 + {}_4C_1a^3b + {}_4C_2a^2b^2 + {}_4C_3ab^3 + {}_4C_4b^4 = a^4 + 4a^3b + 6a^2b^2 + 4ab^4 + b^4$

23. Using 3,

$(x-1)^6 = [x+(-1)]^6$

$= {}_6C_0 x^6 + {}_6C_1 x^5(-1) + {}_6C_2 x^4(-1)^2 + {}_6C_3 x^3(-1)^3 + {}_6C_4 x^2(-1)^4 + {}_6C_5 x(-1)^5 + {}_6C_6(-1)^6$

$= x^6 - 6x^5 + 15x^4 - 20x^3 + 15x^2 - 6x + 1$

25. $(2a-b)^5 = [2a+(-b)]^5$

$= {}_5C_0(2a)^5 + {}_5C_1(2a)^4(-b) + {}_5C_2(2a)^3(-b)^2 + {}_5C_3(2a)^2(-b)^3 + {}_5C_4(2a)(-b)^4 + {}_5C_5(-b)^5$

$= 32a^5 - 80a^4b + 80a^3b^2 - 40a^2b^3 + 10ab^4 - b^5$

27. The fifth term in the expansion of $(x-1)^{18}$ is:

$${}_{18}C_4\, x^{14}(-1)^4 = \frac{18\cdot17\cdot16\cdot15}{4\cdot3\cdot2\cdot1}x^{14} = 3060x^{14}$$

29. The seventh term in the expansion of $(p+q)^{15}$ is:

$${}_{15}C_6\, p^9q^6 = \frac{15\cdot14\cdot13\cdot12\cdot11\cdot10}{6\cdot5\cdot4\cdot3\cdot2\cdot1}p^9q^6 = 5005p^9q^6$$

31. The eleventh term in the expansion of $(2x+y)^{12}$ is:

$${}_{12}C_{10}\,(2x)^2y^{10} = \frac{12\cdot11}{2\cdot1}4x^2y^{10} = 264x^2y^{10}$$

33. $${}_nC_0 = \frac{n!}{0!(n-0)!} = \frac{n!}{1\cdot n!} = 1, \quad {}_nC_n = \frac{n!}{n!(n-n)!} = \frac{n!}{n!0!} = 1$$

35. The next two rows are:

1 5 10 10 5 1 and 1 6 15 20 15 6 1,

respectively. These are the coefficients in the binomial expansions of $(a+b)^5$ and $(a+b)^6$.

37. The nth row of Pascal's triangle gives the coefficients of $(a+b)^k$.
If we let $a = 1$ and $b = -1$, we get

$$0 = (1-1)^n = {}_nC_0\,1^n + {}_nC_1\,(1)^{n-1}(-1) + {}_nC_2\,1^{n-2}(-1)^2 + \ldots + {}_nC_n\,(-1)^n$$

$$= {}_nC_0 - {}_nC_1 + {}_nC_2 - {}_nC_3 + \ldots + (-1)^n\,{}_nC_n.$$

39. $${}_nC_{r-1} + {}_nC_r = \frac{n!}{(r-1)!(n-[r-1])!} + \frac{n!}{r!(n-r)!} = \frac{n!}{(r-1)!(n-r+1)!} + \frac{n!}{r!(n-r)!}$$

$$= \frac{r\cdot n! + (n-r+1)n!}{r!(n-r+1)!} = \frac{(n+1)n!}{r!(n-r+1)!} = \frac{(n+1)!}{r!(n+1-r)!} = {}_{n+1}C_r$$